THE POPULATION STRUCTURE OF VEGETATION

HANDBOOK
OF
VEGETATION SCIENCE
PART III

Editor in chief
Helmut Lieth

THE POPULATION STRUCTURE OF VEGETATION

edited by

J. White

1985 SPRINGER-SCIENCE+BUSINESS MEDIA, B.V.

Library of Congress Cataloging in Publication Data

Main entry under title:

The Population structure of vegetation.

(Handbook of vegetation science ; 3)
Includes index.
1. Plant populations. 2. Vegetation dynamics.
I. White, J. II. Series.
QK911.H3 pt. 3 [QK910] 581 s [582'.05248] 85-222

ISBN 978-94-010-8927-2 ISBN 978-94-009-5500-4 (eBook)
DOI 10.1007/978-94-009-5500-4

Т. Работнов

This book is dedicated to the distinguished Soviet geobotanist
Professor T. A. RABOTNOV

His researches on the demography of plant coenopopulations have been its inspiration.

CONTENTS

Series Editor's preface IX
Foreword XI
List of Contributors XIII
The population structure of vegetation
by J. White 1
On the significance of demographic processes in phytosociology
by O. Wilmanns 15
The census of plants in vegetation
by J. White 33
Vegetation changes in herbaceous communities
by R. Bornkamm 89
Research on permanent quadrats in the USSR
by M.V. Markov 111
Dynamics of plant coenotic populations
by T.A. Rabotnov 121
Population biology of steppe plants
by L.I. Vorontzova and L.B. Zaugolnova 143
Structure and dynamics of coenopopulations of some temperate grasses
by L.A. Zhukova and I.M. Ermakova 179
Coenopopulation structure of *Agrostis* species
by E.I. Kurchenko 207
The population biology of ephemeroids
by N.I. Shorina and O.V. Smirnova 225
Demography of coenopopulations of forest herbs
by K. Falińska 241
Population structure of psammophyte vegetation
by E. Symonides 265
Vegetation dynamics of prairie glacial marshes
by A.G. van der Valk 293
Analysis and prediction of population and community change: a grassland case study
by D.R. Peart and T.C. Foin 313

Demography of short-lived forbs in chalk grassland in relation to vegetation structure
by H.J. During, A.J. Schenkeveld, H.J. Verkaar and J.H. Willems 341
Life cycle ecology of annual plant species of cedar glades of southeastern United States
by J.M. Baskin and C.C. Baskin 371
Population structure and processes of tundra plants and vegetation
by T.V. Callaghan and U. Emanuelsson 399
Demographic, genetic and ecophysiological variation in *Plantago major* and *P.lanceolata* in relation to habitat type
by P.J.M. van der Aart 441
Population biology and habitat relations of some hemiparasitic Scrophulariaceae
by S.J. ter Borg 463
Demography of *Bromus tectorum* in *Artemisia* communities
by J.A. Young and R.A. Evans 489
Towards a functional description of vegetation
by J.P. Grime 503
Life-history characteristics of temperate woodland plants in Japan
by S. Kawano 515
Buried viable seeds in temperate forests
by N. Nakagoshi 551
Patterns of flowering phenology in communities, populations, individuals, and single flowers
by R.B. Primack 571
Plant populations and vegetation in relation to habitat, disturbance and competition: problem of generalisation
by P.J. Grubb 595
Population dynamics of Australian plant communities, with special reference to the invasion of neophytes
by O.B. Williams 623
Patterns in vegetation succession: an ecomorphological study
by P. Hogeweg, B. Hesper, C. van Schaik and W.G. Beeftink 637
Index 667

SERIES EDITOR'S PREFACE

The Handbook of Vegetation Science is growing. After the first volumes under my editorship have appeared the interest of the scientific community has been revived and many new volume editors have started their work. The present volume was jointly designed by Drs. J. White and W. Beeftink. Due to unforseen developments Dr. White signs now as the sole editor.

The development of this volume within the series had a special history as Dr. White points out in his preface. Adding to this I need only to state that I found it essential to include the topic of this volume into a Handbook of Vegetation Science. It was included therefore in my first revised list of topics to be included in the Handbook when I took over from Dr. Tüxen.

It is a great pleasure for me to see this volume appear. Having read through the many contributions to this volume I can certainly congratulate Drs. White and Beeftink for their success in generating so much interest in this volume among their colleagues. The cooperation on this volume is for me the first sign that the new concept of the Handbook has been understood by the generation of scientists which I have to address. The influence this volume will have on the field of plant population studies only time can tell. It appears to me, however, that this volume will become a standard resource for some future.

Dr. White asked me to have this volume dedicated to Dr. Rabotnov. This honor given to Dr. Rabotnov is equally strong supported by me. I learned of Dr. Rabotnov's work while I worked as an assistant to Dr. Walter in Hohenheim. Dr. Walter had always close contact to contemporary Russian colleagues and his scientific correspondence with Dr. Rabotnov was intensive, indeed.

Finally I thank everybody who has contributed to this volume, especially Dr. White for carrying the editorial work through to final success and wish this volume a full acceptance in our scientific community.

Osnabrück, October 1984 H. LIETH

FOREWORD

During a visit to Reinhold Tüxen in July 1975 on my way to the Botanical Congress in Leningrad, I expressed my surprise to him that the outline for a comprehensive series of Handbooks on Vegetation Science which he had initiated had no provision for plant population biology. He at once suggested that I undertake to remedy the omission and begin by sending him a manuscript for *Excerpta Botanica* listing all the studies on plant populations undertaken in natural vegetation. I did neither during his lifetime, but this Handbook makes good the commitment he sought of me; I am happy that his successor as editor-in-chief of this series, Helmut Lieth, has given me the opportunity to do so. I have incorporated into it the second part of Tüxen's request, insofar as I have tried to collate a substantial number of studies concerning the census of plants in vegetation, at least until 1970; I have not attempted to emulate T.A. Rabotnov's remarkable catalogue of coenopopulation studies from the Soviet Union in *Excerpta Botanica* for 1980 and 1981. I feel that Tüxen would have been happy to see in this Handbook another strand being woven into the woof and warp of his own rich tapestry of vegetation science. I wish here first to acknowledge Tüxen's influence as a teacher; this was largely channelled to me, but with similar fervour and stimulus, by John Moore.

A few days after that meeting with Tüxen, I was cordially greeted in Moscow by T.A. Rabotnov, whose publications I had come to know in 1969. More than any other geobotanist at the time his researches combined the approaches to vegetation analysis with which this present Handbook is concerned – the demography of plants in natural vegetation. Although his work was already widely emulated in the Soviet Union by 1970, it has only more recently received widespread international recognition. It is a great pleasure to be able to dedicate this Handbook to him in his eightieth year, as a token in recognition of his considerable contribution to its subject matter. His important role as a conductor of ecological ideas from abroad to his Soviet colleagues (*Botaniska Zhurnal* 60(9), 1975) is echoed in his many contributions in the English-language to those who might otherwise be unaware of Soviet researches; he has, for example, contributed to 4 of the 7 published volumes of this Handbook series.

When I decided to attempt the compilation of a series of studies on the population structure of vegetation, I received ready encouragement from several colleagues. Conceived in 1975 with some urging from R. Tüxen and nurtured by practical examples from the Soviet Union, the development of the project has been rather slow, however. Participants were gradually enlisted by a commitment more definite than words of support. Many have been able eventually to meet it by written essays: I am extremely grateful to them and particularly to those whose manuscripts were among the first to arrive for their patience in seeing them published.

At a crucial stage in the planning in 1979 Dr. W.G. Beeftink accepted an active role in editorial matters. He was responsible for negotiations with our publisher to accept the project in the Handbook of Vegetation Science series, and with some subsequent matters which then arose on the composition of the book. As manuscripts began to arrive in 1982 the process of editing them to achieve the (somewhat idealistic) goals we had set ourselves began. I wish to record my appreciation of his assistance, especially during an intensive two-week period at the Delta Instituut, Yerseke and in Middelburg during August 1983, when several aspects of editorial policy were determined. This was possible in a most congenial and stimulating atmosphere; the contributions of Jeanne Beeftink and of Ad and Loekie Huiskes to this happy period are also gratefully acknowledged. His active field research programme and administrative responsibilities curtailed his planned involvement in this Handbook, much to my regret. He had been able to undertake the primary editing of several manuscripts and has given valuable advice on some others. I am very grateful for his help.

J. WHITE

LIST OF CONTRIBUTORS

P.J.M. van der Aart, Department of Plant Ecology, University of Utrecht, Lange Nieuwstraat 106, 3512PN Utrecht, The Netherlands.

C.C. and J.M. Baskin, Thomas Hunt Morgan School of Biological Sciences, University of Kentucky, Lexington, Kentucky 40506, U.S.A.

W.G. Beeftink, Delta Instituut voor Hydrobiologisch Onderzoek, 4401EA Yerseke, The Netherlands.

S.J. ter Borg, Department of Vegetation Science, Plant Ecology and Weed Science, Agricultural University, Bornsesteeg 69, 6708 PD Wageningen, The Netherlands.

R. Bornkamm, Institut für Ökologie-Botanik, Technische Universität Berlin, Rothenburgstr. 12, 1000 Berlin 41, Federal Republic of Germany.

T.V. Callaghan, The Institute of Terrestrial Ecology, Merlewood Research Station, Grange-over-Sands, Cumbria LA11 6JU, U.K.

H.J. During, Department of Plant Ecology, University of Utrecht, Lange Nieuwstraat 106, 3512 PN Utrecht, The Netherlands.

U. Emanuelsson, Department of Plant Ecology, University of Lund, Ostra Vallgatan 14, S-223 61 Lund, Sweden.

I.M. Ermakova, Biological Problems Laboratory, Lenin State Pedagogical Institute, Kibaltchicha d.6 kor. 5, 129243 Moscow, USSR.

R.A. Evans, Renewable Resource Center, USDA/SEA, 920 Valley Road, Reno, Nevada 89512, USA.

T.C. Foin, Division of Environmental Studies, University of California, Davis, California 95616, USA.

K. Falińska, Stacja Geobotaniczna UW, 17-230 Białowieża, Poland.

J.P. Grime, Unit of Comparative Plant Ecology (NERC), Department of Botany, The University of Sheffield, Sheffield S10 2TN, UK.

P.J. Grubb, Botany School, University of Cambridge, Downing Street, Cambridge CB2 3EA, UK.

B. Hesper, Bioinformatica, Padualaan 8, 3584 CH Utrecht, The Netherlands.

P. Hogeweg, Bioinformatica, Padualaan 8, 3584 CH Utrecht, The Netherlands.

S. Kawano, Department of Botany, Kyoto University, Kyoto 606, Japan.

E.I. Kurchenko, Biological Problems Laboratory, Lenin State Pedagogical Institute, Kibaltchicha d.6 kor. 5, 129243 Moscow, USSR.
M.V. Markov, Department of Botany, Kazan State University, Ulitsa Lenina 18, 42008 Kazan, USSR.
N. Nakagoshi, Botanical Institute, Hiroshima University, Higashi-senda, Nakaku, Hiroshima 730, Japan.
D.R. Peart, Department of Biological Sciences, Dartmouth College, Hanover, New Hampshire 03755, USA.
R. Primack, Biological Science Center, Boston University, 2 Cummington Street, Boston, Mass. 02215, USA.
T.A. Rabotnov, Department of Geobotany, Moscow University, 117234 Moscow, USSR.
C. van Schaik, Bioinformatica, Padualaan 8, 3548CH Utrecht, The Netherlands.
A.J. Schenkeveld, Department of Plant Ecology, University of Utrecht, Lange Nieuwstraat 106, 3512 PN Utrecht, The Netherlands.
N.I. Shorina, Department of Botany, Lenin State Pedagogical Institute, Kibaltchicha d.6 kor. 5, 129243 Moscow, USSR.
O.V. Smirnova, Biological Problems Laboratory, Lenin State Pedagogical Institute, Kibaltchicha d.6 kor. 5, 129243 Moscow, USSR.
E. Symonides, Instytut Botaniki UW, Al. Ujazdowskie 4, 00-478 Warsaw, Poland.
A. van der Valk, Department of Botany, Iowa State University, Bessey Hall, Ames, Iowa 50011, USA.
H.J. Verkaar, Department of Plant Ecology, University of Utrecht, Lange Nieuwstraat 106, 3512 PN Utrecht, The Netherlands.
L.I. Vorontzova, Biological Problems Laboratory, Lenin State Pedagogical Institute, Kibaltchicha d.6 kor. 5, 129243 Moscow, USSR.
J. White, Department of Botany, University College Dublin, Dublin 4, Ireland.
J.H. Willems, Department of Plant Ecology, University of Utrecht, Lange Nieuwstraat 106, 3512 PN Utrecht, The Netherlands.
O.B. Williams, Division of Water and Land Resources, CSIRO, PO Box 1666, Canberra City, ACT 2601, Australia.
O. Wilmanns, Lehrstuhl für Geobotanik, Institut für Biologie II, Albert-Ludwigs-Universität, Schänzlestrasse 1, D-7800 Freiburg, Federal Republic of Germany.
J.A. Young, Renewable Resource Center, USDA/SEA, 920 Valley Road, Reno, Nevada 89512, USA.
L.B. Zaugolnova, Biological Problems Laboratory, Lenin State Pedagogical Institute, Kibaltchicha d.6 kor. 5, 129234 Moscow, USSR.
L.H. Zhukova, Department of Botany, Lenin State Pedagogical Institute, Kibaltchicha d.6 kor. 5, 129243 Moscow, USSR.

THE POPULATION STRUCTURE OF VEGETATION

JAMES WHITE

INTRODUCTION

The primary purpose of this Handbook is to urge the complementarity of demographic and sociological approaches to vegetation science. This was suggested by A.S. Watt (1964) some twenty years ago, but his plea fell on deaf ears. Why is such advocacy necessary? Eighty years ago Frederic Clements in the first American textbook of ecology (*Research Methods in Plant Ecology*, 1905) wrote of the need for exact methods of vegetation analysis: he extolled the value of the quadrat which he had lately introduced with Roscoe Pound. He used it to list the plant species in a small, representative sample of vegetation and to provide a numerical estimate of their abundance, either absolutely or relatively. The 'list quadrat', so-called, allowed 'the determination of the greatest variable in vegetation, namely number'. But this was to be augmented, he urged, by the 'chart quadrat', to indicate the areal extent of each species and to 'furnish a valuable check upon mere number'. Ideally, in Clements' view, list and chart quadrats should be permanent, that is, located in a fixed position and revisited regularly, to follow processes of migration, succession and competition in careful detail. These simple techniques, prescriptive in Clements' opinion for accurate vegetation analysis, have underlain two of the major strands in vegetation science which are brought together in this Handbook, plant sociology and plant demography. Both are concerned with sampling vegetation to abstract coherent generalizations about the dynamic behaviour and interrelationships of species: the diversity of life-history phenomena, the manner of organization of different species into assemblages which we refer to as communities. But how rarely (e.g., Watt 1955) have these strands interacted with each other during the development of geobotany!

Species lists in quadrats of defined size coupled with some quasi-quantitative estimates of their abundance have been given the special names of *relevés* or *Aufnahmen* by phytosociologists, which serve almost to distinguish them as a special type of quadrat sampling. The tabulation and rearrangement of such relevés and comparison of the resultant abstracted tables constitute a large part of phytosociology. While there is

White, J. (ed.), The Population Structure of Vegetation.
© 1985, Dr W. Junk Publishers, Dordrecht. ISBN 978-94-010-8927-2

now an international system of vegetation units as the product of several decades of intensive research, the demographic attributes of the constituent plant species have played a minor role in this process. Only some rather general life-history features (annual vs. perennial, herb vs. shrub vs. tree) are usually considered in the synsystematic arrangement of vegetation types. Disproportionately less attention has been devoted by geobotanists to the life cycle and population biology of species than to their assembly into communities, notwithstanding that a knowledge of the demographic properties of species is essential for a full understanding of community composition and dynamics.

On the other hand, records obtained from quadrats by precise enumeration or charting of individuals have been used to make generalisations about the population biology and life history of separate plant species. There has been little evidence to date of attempts by plant demographers to integrate such data into the community matrix from which they have been derived. It is perhaps understandable why this has been so, since, as will be outlined in the third chapter of the Handbook, the census of plants in vegetation has been conducted fitfully until recent times. Only in the past two decades have sufficient numbers of studies been published to enable us to appreciate the range and diversity of demographic phenomena in plants. It seems to me that the time has come to indicate that the separation between investigations of the demography and sociology of plants in natural vegetation is artificial, unnecessary and undesirable. A change in the detail or scale of quantitative recording of even the same quadrat has long been associated with quite different sorts of subsequent analyses, either sociological or demographic. Relative generality and relative precision in sampling have characterised (albeit somewhat crudely, I acknowledge) a polarity between plant sociology and plant demography. In principle, it seems to me, there is no barrier whatsoever to the conduct of both levels of vegetation analysis concomitantly.

THE NATURE OF VEGETATION: HISTORICAL PERSPECTIVE

The origins of this dichotomy lie in a long-standing debate about the nature of vegetation; this in turn has deeper philosophical roots. I shall outline both briefly, since they seem to have an abiding influence to our own time (Harper 1977a, 1982, McIntosh 1980, Richardson 1980). In the same textbook (*Research Methods in Plant Ecology*) which was so inspirational in its practical techniques for early plant ecologists, Clements also introduced his philosophical notion of vegetation as an organism. The essence of this view was that populations of individual species in nature are assembled into a well-defined and closely integrated ecological community. This found its mature expression in his remarkable *Plant Succession* (1916) and forthwith drew the celebrated response of H.A. Gleason (1917):

> Vegetation, in its broader aspects, is composed of a number of plant individuals. The development and maintenance of vegetation is

therefore merely the resultant of the development and maintenance of the component individuals, and is favoured, modified, retarded, or inhibited by all causes which influence the component plants. According to this view, the phenomena of vegetation depend completely upon the phenomena of the individual. It is in sharp contrast with the view of Clements that the unit of vegetation is an organism, which exhibits a series of functions distinct from those of the individual and within which the individual plants play a part as subsidiary to the whole as that of a single tracheid within a tree . . . . Certain common phenomena of the plant individual, namely, migration of germules, germination and growth, when performed *en masse* by numerous individuals, may produce visible effects which are worthy of special study and which demand a special terminology. Among these effects may be mentioned the structure of the plant association and of the subsidiary assemblages of plants within it, its space relation to neighbouring associations, its development, and its ultimate disappearance.

It would be unmindful to neglect Clements' own appreciation of the significance of the population biology of species for interpreting community patterns and processes; as the third chapter of this Handbook attempts to illustrate, he fostered plant censusing actively for several years. But undoubtedly his name among professional ecologists came to be associated with a holistic philosophy of vegetation which devoted more attention to the ensemble characteristics of communities than to those of populations of separate constituent species. This holistic view has remained popular to the present time, as the extensive use of ecosystem approaches to natural communities attests.

Gleason's antithesis with its emphasis on individuals or populations as the constitutive elements and determinative forces of vegetation leads naturally and more easily to demographic concepts. Curiously, Gleason sustained his argument (1917, 1926, 1939) by attacking Clements' theoretical writings, but neglected the relevance and importance of Clements' own census researches, which were far more extensive and detailed than any he ever conducted himself. In England Clements' great admirer A.G. Tansley, while not wholeheartedly espousing an organismic analogy of vegetation, was more holistic than reductionist in his philosophy of nature: the use of the ecosystem concept owes much to him (Tansley 1935). The debate seems never to have been as sharp in England as in North America, but one can also identify in the philosophy and practice of E.J. Salisbury the same antipathy expressed by Gleason. Indeed, unlike Gleason who actually wrote relatively little on plant ecology (McIntosh 1975), Salisbury stands as one of the outstanding figures of plant population biology, years 'ahead of his time' in many respects (Jackson 1981): his knowledge of the demographic properties of plants, expressed in numerous publications (e.g., Salisbury 1929, 1942, 1952, 1961, 1970, to cite only a handful), was probably unrivalled for many years. There is little published evidence during the period 1930–1970 that they were emulated or even appreciated by his contemporary plant ecologists.

The protracted debate on the nature of vegetation and the most suitable means of studying it is usually characterised as between Clements and Gleason; perhaps it might also be personalised in other ways such as Tansley–Salisbury, though I know of no published polemic between them. There is little evidence of the debate being conducted in continental Europe in similar terms, where there seems to have been an overwhelming acceptance of a community-based system of vegetation analysis, even allowing for some regional variants in the methodology of sampling and synthesis (Whittaker 1962). Braun-Blanquet (1928, 1951, 1964) followed Clements' holistic philosophy and this has been transmitted to modern times in almost metaphysical terms by Tüxen in his remarkable 'seven laws of coexistence' (Tüxen 1977). The major debate among continental ecologists centred on the relative significance of descriptive phytosociology and of physiological experimentation for vegetation science. Demographic concepts which were focussed on populations of single species played little part in the discussion; even today, as Wilmanns suggests in her chapter here, plant demography is underdeveloped among geobotanists in continental Europe.

The origins of the controversy have, however, a greater antiquity; but these nonetheless seem to maintain a contemporary relevance. The belief that the integrated plant community has an ontological reality as a unit beyond the individual plant can be traced to the nineteenth century plant geographers von Humboldt, Grisebach and Drude. Tobey (1981) has argued in an informative historical analysis that this 'idealistic' philosophy of vegetation, as he refers to it, was derived from Kant in philosophical terms. He counterposes this with a view of nature characterised as 'realistic naturalism'. Its principal exponent for natural vegetation was the plant geographer Alphonse de Candolle: the individual plant and its relation to habitat was the focus of his discourse, conducted in quite mechanistic terms. Only the individual had ontological reality. Darwin was, of course, the most influential exponent of realistic naturalism in biology: he too applied it to vegetation, quite explicitly in Chapter 3 of *On the Origin of Species*. His view of nature was, as Tobey (1981) has shown, transmitted to Warming, widely acknowledged as a founder of scientific plant ecology, through the influence of Nägeli at Münich in the 1870s. Warming became an ardent advocate of the Darwinian interpretation of natural phenomena: the 1896 German edition of his famous textbook of plant ecology was rigorously reductionist in a Darwinian (or de Candollean) sense. Warming did not believe that plant communities had supra-organismic mechanisms of their own which enabled them to interact with each other: the interactions between individuals in populations were the causes of community change. This edition influenced Cowles and his colleagues at Chicago, whence came the major source of the alternative paradigm to Clements' in the development of plant ecology in North America (Tobey 1981). The English language edition of Warming's textbook (1909) incorporated some of the new terminology of plant ecology which had been invented by Clements; this, in Tobey's view, 'imparted to it a faint flavor of teleology'.

By the turn of the century, therefore, the two distinct approaches to geobotany which characterise in various degrees a philosophical dichotomy among vegetation scientists to the present day, were well established. For convenience, and without serious misrepresentation I believe, they can be referred to as sociological and demographic. In many respects Clements, curiously, embodied both elements of idealism and reductionism in his theory of vegetation: a supra-organismic community was coupled to population dynamics of individual species (e.g., Clements 1910), but he never effectively showed their common ground, their dialectical interaction or complementarity for understanding the organisation of complex systems. This was unfortunate for the development of geobotany, because Clements was (and has remained I believe) unrivalled in the subject intellectually: 'by far the greatest creator of the modern science of vegetation' (Tansley 1947). Ultimately he opted for a non-Darwinian approach to vegetation and in his later years became a confused Lamarckian as an unhappy result of his transplant experiments (e.g., Hagen 1984).

THE PROSPECT FOR SYNTHESIS

The reader of this Handbook will find it quite unabashed in its 'realistic naturalism', to use Tobey's (1981) expression. The contributors take as a point of departure (implicitly, I think, if not explicitly) the view that the demographic attributes of species are important in understanding the manner in which plant communities are constituted or change; some have expressed this opinion for several years (e.g., Foin and Jain 1977, Grubb 1977, Grime 1979, van der Valk 1981). This is an attitude shared by many contemporary plant ecologists not represented here, who see the merits of a Darwinian (or Gleasonian/Salisburyan), population-centred explanation of community phenomena (e.g., Drury and Nisbet 1973, Raup 1975, Noble and Slatyer 1980, Shmida and Ellner, 1984, to cite but a few examples in the past decade or so). McIntosh (1975) has argued that few plant ecologists had needed Gleason's earlier advice and that the tide of professional opinion did not begin to turn in his favour until the late 1940s. In my view this is not surprising, since Gleason had given little or no practical example apart from an early, rather desultory effort (Gleason in Hart and Gleason 1907), apparently his only published work on the demography (census) of plants in vegetation.

But I admit that it is not clear to me how far this approach can take us: can a knowledge of the demographic properties of species be used to predict the dynamic behaviour of communities? This may be possible in certain cases where particular species, for whatever reason, determine community properties to an overwhelming extent. This is a venerable idea among some plant ecologists, especially in the Soviet Union where the concept of 'edificator species' has been extensively discussed (e.g., Rabotnov 1975). It has been independently and profitably used by many zoologists studying the structure of intertidal ecosystems (e.g., Paine 1980), who use the phrase 'keystone species'.

Goodman (1975) believes, following Frank (1968), that the population characteristics of one or a few critical species may determine the stability of the community as a whole. There is, undoubtedly, increasing evidence for this belief in some plant communities (Watt 1947, Foin and Jain 1977, Turkington et al. 1977, Turkington and Harper, 1979) but it is far from being convincingly documented. In this Handbook Symonides, van der Valk and Peart provide some instructive examples, using a diversity of conceptual approaches.

I believe that, on balance, the rhetorical question I posed above is unlikely to be answered generally with a simple 'yes', and may indeed be a forlorn hope despite expressions of optimism (e.g., Harper 1982); Snaydon (1984) seems to share my doubts. The reasons for my reservations are concerned with what may be termed 'competitive indeterminancy' and 'demographic stochasticity' respectively. It has become increasingly evident to plant population biologists working with experimental monocultures and mixtures of two or few species that the performance of species in mixtures is not easily predictable from their performance in pure stands. This insight, first published by Montgomery (1912), may almost be elevated to a general principle and has even been eponymously named the 'Montgomery Effect' by de Wit (1960). Harper (1977b) reviewed some of the extensive evidence now known to substantiate it. The effects of interference on life-history parameters of all sorts are often density-dependent; the relative abundance and fitness of species in mixtures are invariably frequency-dependent (e.g., Antonovics and Levin 1980). The extent of both density- and frequency-dependent effects in field conditions remains very poorly understood: the prospect of describing community dynamics in terms of the dynamics of constituent species studied *in isolation* seems to me remote.

This is probably the most important reason for investigating the population behaviour of species in their natural environments. Several of the contributors to this Handbook illustrate how this is done in a variety of vegetation types. There has indeed been a perceptible shift in interest of plant population biologists from artificial systems (flower pots) to natural communities in recent years. The insights gained from techniques of agronomic experimentation may still be usefully employed in field conditions, but the overwhelming and dominant influence of agricultural science on the formative years of plant population biology (e.g., Harper 1977b) has declined rapidly.

A major trend in ecology and evolutionary biology in modern times has been the replacement of deterministic models by stochastic ones: the behaviour of populations subject to changes in environmental conditions are, at best, predictable in probabilistic terms. This may simply reflect the inherent genetic diversity of most populations, although demographic stochasticity is not necessarily eliminated even in populations which are thought to be uniform genetically (Lerner and Dempster 1962). Much of the early evidence for demographic stochasticity came from zoologists (Mertz 1973, Mertz et al. 1976) but there are now well-investigated examples for plant populations: these include *Anthoxanthum odoratum* (Antonovics 1972), *Bromus tectorum* (Mack and Pyke 1983), *Cakile edentula* (Keddy 1982), *Chamaelirium luteum* (Meagher and Antonovics

1982), *Carduus nutans* (Lee and Hamrick 1983), *Fragaria* spp. (Angevine 1983), *Plantago lanceolata* (Antonovics and Primack 1982), *Poa annua* (Law 1979), *Stellaria media* (van der Vegte 1978).

A further problem is posed by the discordance between the physiological and ecological tolerances of individuals, a distinction which owes much to continental plant ecologists (Leith and Ellenberg 1958, Ellenberg 1953, 1974). The behaviour of species grown in artificial monocultures in an array of environments is not necessarily a good guide to their ability to survive in the presence of other species. This matter is considered in more detail by Grubb in this Handbook. The ecological amplitude of individuals seems invariably to be more restricted than their physiological amplitude, with diverse patterns of displacement along environmental gradients that are not easy to predict a priori (e.g., Austin and Austin 1980). It may be remarked in passing, however, that there is some scepticism about the general validity of this principle (Ernst 1978). Nonetheless it raises further important doubt about the usefulness of studying the population dynamics of natural plant populations in isolation from their phytosociological environment: some demographic attributes of a population in artificial monoculture may have no counterpart to those normally expressed in nature. This is less likely to be true, of course, for those species which normally grow in monoculture in nature: perhaps not surprisingly much of our foundation knowledge of plant demography has been derived from weedy species or their crop derivatives, or from species which commonly grow in high density patchy monocultures (many winter annuals and monocarpic perennials) (Harper 1977b).

Plants respond to changes in their natural environments within the context of the vegetation of which they are a part (Pigott 1982). No longer can plant demographers accept (if they ever did) any simplistic notion that suites of demographic properties of a species are narrowly circumscribed or absolute. Modern studies have affirmed and extended the more general genecological patterns of life-history diversity within species that have been known for many years (e.g., Böcher 1949, Clausen 1951). We may expect that insofar as there is genetic diversity between populations of a species (and this is now extensively recorded for plants as Briggs and Walters (1984) have conveniently summarised) so far will there be variation in their life-history features. Rather few plant demographers have addressed this problem explicitly in natural vegetation (some of those listed above; see also Gray et al. 1979 and references therein). It is by no means clear that the population biology of a species well-studied in one type of vegetation is similar to its behaviour in another: this is especially likely to be true of species which have a physiological and sociological amplitude which enables them to grow in several habitats and with diverse combinations of species. Wilmanns raises this problem in her paper here with respect to Sarukhán's celebrated study of *Ranunculus* species which grow in several types of grassland vegetation, but it could as fittingly be asked of any of dozens of comparable studies. The suggestion by Watt (1964) that 'it would be interesting to compare the social status of species over the whole range of their distribution to see whether the status of the species varied from one community to another', simply rephrased to accompany 'social status' with 'demographic behaviour', provides a continuing challenge to geobotanists. I think that the social status of many species is now understood in terms of Watt's suggestion by phytosociologists, especially in northwest Europe, but the extent to which variations in 'social status' are consequent on variations in demographic properties is virtually unknown. Ter Borg gives an example in our Handbook of the wide scale of investigation which must be undertaken to attempt to address this problem, while van der Aart illustrates the variety of demographic, genetic and physiological techniques which must be applied to determine some conclusive answers. It seems appropriate to note that plant sociologists have repeatedly drawn attention to the genetic variety evident within some species found in diverse plant communities (Guinochet 1973, Landolt 1977) and may even have noted the extent to which certain demographic patterns vary between ecotypes (e.g., Jaeger 1978).

Accordingly, it seems to me that the conduct of plant demography in isolation from the sociological context of the community is most unlikely to provide a rounded view of the nature of vegetation. I share the view expressed by Levins and Lewontin (1980) that the community is a dialectical whole: 'the whole is a contingent structure in reciprocal interaction with its own parts and with the greater whole of which it is a part; whole and part do not completely determine each other'. A pure mechanistic materialism which views the parts of a whole as separate and distinct and believes that the whole can be reconstructed as the sum of the individual parts, and no more, is scarcely appropriate to what we now know of the complexity of natural vegetation and of the dynamic behaviour of individual species. A holistic materialism (Allen 1983) which views the whole as a sum of the parts plus their interactions, interactions which *cannot* be predicted from a knowledge only of the parts themselves, necessitates a study of the parts *in situ*. The complex whole has, in this view, so-called emergent properties. It is well appreciated by plant sociologists that levels of integration of communities vary widely. Noy-Meir (1980) has, at the one extreme, questioned whether there is any integration whatsoever in some desert communities in which populations are strongly and (it seems) independently regulated by abiotic forces. There is increasing speculation, and a little evidence, that the coexistence of species in some communities results from chronic disturbance; it is believed that this enables a greater diversity of species to inhabit a site than might otherwise be possible if they were able to interact under more benign conditions (Grubb 1977, Pickett 1980, Braakhekke 1980, Caswell 1982, Del Moral 1983). On the other hand the lesson to be learned from the Park Grass Experiment at Rothamsted in England, continuously monitored since 1856, has been that the assembly of groups of species may be remarkably stable (even in small plots) under relatively invariant treatments (Thurston et al. 1976, Silvertown 1980); and this despite the turnover of individual ramets or genets. Detectable genetic change has taken place in this particular experiment (Snaydon 1970) and this is likely to be a general consequence of the interactions among diverse genotypes in complex communities (e.g., Turkington and Aarssen 1984). Internal *qualitative* changes within a system which are the results of the *quantitative* interactions of opposing forces within it characterise the dialectical materialism which Allen (1983) identifies as the philosophical basis of Darwin's view of nature. This is congruent with the concept of the community as a dialectical whole espoused by Levins and Lewontin (1980). A unified description of vegetation which involves both the demographic dynamics of constituent species and their ensemble relationships is, I believe, a desirable goal of geobotany which can now be realistically achieved by a synthesis of plant demography and plant sociology.

COMPOSITION OF THE HANDBOOK

At this stage of knowledge the main aim of our Handbook is to urge that the demographic patterns of plant populations be described with some reference to the community of species in which they have been studied, and conversely, to explore the degree to which a knowledge of demographic patterns enables us to understand, maybe even to predict, community processes. The papers illustrate the many ways in which population phenomena of various sorts are being investigated in a diversity of natural vegetation types. Since the application of demographic concepts to vegetation analysis has languished for so long and has become popular only in the past ten years, most of the contributions should be regarded as case-studies rather than as consolidated accounts of concepts, practices and prescriptive techniques which are the usual content of Handbooks. The authors have been selected deliberately to illustrate the various ways in which the population structure of vegetation is being investigated: the diversity is somewhat striking. Certain topics may be conspicuous to some readers by their absence. The compilation of multi-authored books such as this has certain problems, not least the failure to obtain commissioned contributions in time: I am conscious of several omissions, I acknowledge. But one type of omission was deliberate: there are no abstract theoretical or too-mathematical papers, since I wished to stay as close as possible to the reality of natural history in this particular volume.

In an opening chapter Wilmanns welcomes the prospect of fertile interchange between plant demographers and sociologists, in the spirit of Reinhold Tüxen who encouraged his students to explore all means of arriving at a comprehensive understanding of vegetation. The next chapter outlining the history of plant census in herbaceous vegetation shows that only sporadic or isolated attempts have been made to do so until modern times. (This is not to say that plant population biology as a whole was undeveloped. Many aspects of plant reproductive biology and seed biology have a venerable history. Plant census is not entirely synonymous with plant demography, though it is undoubtedly an important constitutive element). The paper by Bornkamm is representative of the best type of quantitative analysis of vegetation in permanent quadrats which comes from the phytosociological tradition of continental Europe: this exemplifies the compromise between precision and generality which so often seems unavoidable in the analysis of vegetation if one is to investigate the dynamics of all the species simultaneously in a single community.

A series of papers follows on the demography of various species in natural vegetation, to illustrate part of the variety of techniques and ideas which are increasingly evident among geobotanists. The level of integration of demography and sociology varies in all these papers, but I believe there is a common thread throughout: the attempt to understand population biology not in isolation from, but in relation to the community or ecosystem in which the dynamics of particular species have unfolded. It is a particular pleasure for me to be able to include several papers from colleagues in the Soviet Union whose researches have been much neglected because of linguistic barriers. These are introduced by T.A. Rabotnov, the doyen of Soviet plant demographers. I think it can be said without contradiction that nowhere else has the study of plant demography in natural or seminatural vegetation been so extensively conducted as in the Soviet Union. Closely related conceptually to the Soviet coenopopulation tradition is the work of Falińska, some of whose researches are summarised here. The link between plant sociology and demography is made explicit in the special term which characteristics their writing about natural populations, *coenopopulations*: these are populations of a species sharing a particular habitat in common with other species, usually of similar ecological preferences.

I believe the word coenopopulation deserves a wider currency, to be used of populations of plants in natural vegetation: often the genetic identity of the individuals is obscure and the word embraces both ramets and genets, that is, all individuals which are denumerable as ecological entities. There are two cognate words. 'Population' is most often used to denote groups of organisms known to be genetically distinct. It is entirely appropriate for groups of plants where this is evident: many annual herbs and trees present little difficulty in this respect, and it is usually clearcut in experimental circumstances where plants are grown from seeds. 'Metapopulation' (White 1979) is a neologism for the assemblage of parts which constitute a genetic individual: the serial sequence of metamers and their modular organisation are almost universal features of the construction of plants (White 1984). I pass over here the more difficult problem of circumscribing either populations or coenopopulations in natural vegetation. This issue has been reviewed by Crawford (1984). The reader will find from the essays in this Handbook a widespread pragmatism among plant ecologists on the problem: a (coeno) population is a group of organisms of a well-defined taxon, often of unknown genetic identity at an individual level, found in a particular place for a period of time convenient to the observer. More precise delimitations based on shared genetic identity or gene flow present considerable difficulties: genetic 'neighbourhood' sizes are so far known for very few plant species (Levin 1981, Crawford 1984).

The confines of a Clementsian square-frame quadrat often determine its limits unambiguously, with brute simplicity. Whether or not the constitution and dynamics of the population of a particular species within that quadrat are typical of anything other than that population are poorly known. The replication by Sarukhán in his analysis of *Ranunclus* species, while not exceptional, remains rare. Even here, the demonstration of the remarkable consistency of population dynamics within a small field (Sarukhán and Harper 1973) provides no evidence that they are similar for populations of the same (widespread) species growing in different plant communities.

A smaller group of papers by Grime, Kawano, Nakagoshi, Primack and Grubb takes a more community-centered perspective on our twin-theme: to what extent can communities be characterised by the ensemble properties of constituent species. Two final papers hint at the scope of relevant contemporary research which may be included under the rubric of the 'population structure of vegetation': Williams reminds us of the importance of demographic studies in man-modified landscapes, being created today at an accelerating pace.

The recent pages of the ecological journals readily testify that one could make a companion volume with a completely different set of contributors. Our present volume will have served its purpose if it demonstrates that a synthesis of plant demography and sociology is a realistic enterprise for the future development of geobotany. In some cases the reader will see that only tentative steps in this direction have been possible, but some essays are, I believe, fine examples of its application. As compiler and editor of this particular group of essays I share, at least, the conviction of Marc Bloch (1931):

> Dans le développment d'une discipline, il est des moments où un synthèse, fût-elle en apparence prématurée, rend plus de services que beaucoup de travaux d'analyse, où, en d'autres termes, il import surtout de bien énoncerl les questions, plutôt, pour l'instant, que de chercher à les résoudre.

REFERENCES

Allen, G.E. (1983) The several faces of Darwin: materialism in nineteenth and twentieth century evolutionary theory. pp. 81–102 in D.S. Bendall (Ed.) Evolution from Molecules to Man. Cambridge: Cambridge University Press.

Angevine, M.W. (1983) Variations in the demography of natural populations of wild strawberries *Fragaria vesca* and *F. virginiana*. J. Ecol. 71: 959–974.

Antonovics, J. (1972) Population dynamics of the grass *Anthoxanthum odoratum* on a zinc mine. J. Ecol. 60: 351–366.

Antonovics, J. and Levin, D.A. (1980) The ecological and genetic consequences of density-dependent regulation in plants. Ann. Rev. Ecol. Syst. 11: 411–452.

Antonovics, J. and Primack, R.B. (1982) Experimental ecological genetics in *Plantago. VI.* The demography of seedling transplants of *P. lanceolata*. J. Ecol. 70: 55–75.

Austin, M.P. and Austin, B.O. (1980) Behaviour of experimental plant communities along a nutrient gradient. J. Ecol. 68: 891–918.

Bloch, M. (1931) Les Caractères Originaux de l'Histoire rural Française. Oslo.

Böcher, T.W. (1949) Racial divergences in *Prunella vulgaris* in relation to habitat and climate. New Phytol. 48: 285–324.
Braakhekke, W.G. (1980) On coexistence: a causal approach to diversity and stability in grassland vegetation. Wageningen: Centre for Agricultural Publishing and Documentation. 164 pp.
Braun-Blanquet, J. (1928) Pflanzensoziologie. Grundzüge der Vegetationskunde. Berlin: Springer (2nd ed. 1951, 3rd ed. 1964).
Briggs, D. and Walters, S.M. (1984) Plant Variation and Evolution. 2nd edition. Cambridge: Cambridge University Press. 412 pp.
Caswell, H. (1982) Life history theory and the equilibrium status of populations. Amer. Nat. 120: 317–339.
Clausen, J. (1951) Stages in the Evolution of Plant Species. Ithaca, New York: Cornell University Press. 206 pp.
Clements, F.E. (1905) Research Methods in Plant Ecology. Lincoln, Nebraska: University Publishing Co. 334 pp.
Clements, F.E. (1910) The life history of lodgepole pine burn forests. U.S. Dept. Agric. For. Serv. Bull. 79. 56 pp.
Clements, F.E. (1916) Plant Succession. An Analysis of the Development of Vegetation. Washington: Carnegie Institution of Washington. 512pp.
Crawford, T.J. (1984) What is a population? pp. 135–173 in B. Shorrocks (Ed.) Evolutionary Ecology. Oxford: Blackwell Scientific Publications.
Del Moral, R. (1983) Competition as a control mechanism in subalpine meadows. Amer. J. Bot. 70: 232–245.
Drury, W.H. and Nisbet, I.C.T. (1973) Succession. J. Arnold Arb. 54: 331–368.
Ellenberg, H. (1953) Physiologisches und ökologisches Verhalten derselben Pflanzenarten. Ber. Deutsch. Bot. Ges. 65: 351–362.
Ellenberg, H. (1974) Vegetation Mitteleuropas mit den Alpen. 2nd ed. Stuttgart: Eugen Ulmer. 943 pp.
Ernst, W. (1978) Discrepancy between ecological and physiological optima of plant species. A reinterpretation. Oecol. Plant 13: 175–188.
Foin, T.C. and Jain, S.K. (1977) Ecosystem analysis and population biology: lessons for the development of community ecology. Bioscience 27: 532–538.
Frank, P.W. (1968) Life histories and community stability. Ecology 49: 355–357.
Gleason, H.A. (1917) The structure and development of the plant association. Bull. Torrey Bot. Club 44: 463–481.
Gleason, H.A. (1926) The individualistic concept of the plant association. Bull. Torrey Bot. Club 53: 7–26.
Gleason, H.A. (1939) The individualist concept of the plant association. Amer. Midl. Nat. 21: 92–108.
Goodman, D. (1975) The theory of diversity-stability relationships in ecology. Q. Rev. Biol. 50: 237–266.
Gray, A.J., Parsell, R.J. and Scott, R. (1979) The genetic structure of plant populations in relation to the development of salt marshes. pp. 43–64 in R.L. Jefferies and A.J. Davy (Ed.) Ecological Processes in Coastal Environments. Oxford: Blackwell Scientific Publications.
Grime, J.P. (1979) Plant Strategies and Vegetation Processes. Chichester: John Wiley. 222 pp.
Grubb, P.J. (1977) The maintenance of species-richness in plant communities: the importance of the regeneration niche. Biol. Rev. 52: 107–145.
Guinochet, M. (1973) Phytosociologie. Paris: Masson et Cie. 227 pp.
Hagen, J.B. (1984) Experimentalists and naturalists in twentieth-century botany: experimental taxonomy, 1920–1950. J. Hist. Biol. 17: 249–270.
Harper, J.L. (1977a) The contributions of terrestrial plant studies to the development of the theory of ecology. pp. 139–157 in C.E. Goulden (Ed.) Changing Scenes in the Life Sciences, 1776–1976. Philadelphia: Academy of Natural Sciences.
Harper, J.L. (1977b) Population Biology of Plants. London: Academic Press. 892 pp.
Harper, J.L. (1982) After description. pp. 11–25 in E.I. Newman (Ed.) The Plant Community as a Working Mechanism. Oxford: Blackwell Scientific Publications.
Hart, C.A. and Gleason, H.A. (1907) On the biology of the sand areas of Illinois. Bull. Ill. State Lab. Nat. Hist. 7(7): 131–272.

Jackson, J.B.C. (1981) Interspecific competition and species' distributions: the ghosts of theories and data past. Amer. Zool. 21: 889–901.
Jaeger, P. (1978) Action favorisante du fauchage sur la diversification de certains ecotypes dans les Arrhénatheraies de la Plaine d'Alsace. Coll. Phytosoc. 5: 199–203.
Keddy, P.A. (1982) Population ecology on an environmental gradient: *Cakile edentula* on a sand dune. Oecologia 52: 348–355.
Landolt, E. (1977) The importance of closely related taxa for the delimitation of phytosociological units. Vegetatio 34: 179–189.
Law, R. (1979) Ecological determinants in the evolution of life histories. pp. 81–103 in R.M. Anderson, B.D. Turner and L.R. Taylor (Ed.) Population Dynamics. Oxford: Blackwell Scientific Publications.
Lee, J.M. and Hamrick, J.L. (1983) Demography of two natural populations of musk thistle (*Carduus nutans*). J. Ecol. 71: 923–936.
Lerner, I.M. and Dempster, E.R. (1962) Indeterminism in interspecific competition. Proc. Natl. Acad. Sci. U.S.A. 48: 821–826.
Levin, D.A. (1981) Dispersal versus gene flow in plants. Ann. Missouri Bot. Gdn. 68: 233–253.
Levins, R. and Lewontin, R. (1980) Dialectics and reductionism in ecology. Synthese 43: 47–78.
Lieth, H. and Ellenberg, H. (1958) Konkurrenz und Zuwanderung von Wiesenpflanzen. Ein Beitrag zum Problem der Entwicklung neu angelegten Grünlandes. Z. f. Acker-u. Pflanzenbau. 106: 205–223.
Mack, R.N. and Pyke, D.A. (1983) The demography of *Bromus tectorum*: variation in time and space. J. Ecol. 71: 69–93.
McIntosh, R.P. (1975) H.A. Gleason – "individualistic ecologist" 1882–1975: his contributions to ecological theory. Bull. Torrey Bot. Club 102: 253–273.
McIntosh, R.P. (1980) The background and some current problems of theoretical ecology. Synthese 43: 195–255.
Meagher, T.R. and Antonovics, J.T. (1982) Life history variation in dioecious plant populations: a case study of *Chamaelirium luteum*. pp. 139–154 in J. Dingle and J. Hegmann (Ed.) Evolution and Genetics of Life Histories. New York: Springer.
Mertz, D.B. (1973) The *Tribolium* model and the mathematics of population growth. Ann. Rev. Ecol. Syst. 3: 51–78.
Mertz, D.B., Cawthon, D.A. and Park, T. (1976) An experimental analysis of competitive indeterminacy in *Tribolium*. Proc. Natl. Acad. Sci. U.S.A. 73: 1368–1372.
Montgomery, E.G. (1912) Competition in cereals. Nebraska Agric. Expt. Stn. Bull. 127: 3–22.
Noble, I.R. and Slatyer, R.O. (1980) The use of vital attributes to predict successional changes in plant communities subject to recurrent disturbances. Vegetatio 43: 5–21.
Noy-Meir, I. (1980) Structure and function of desert ecosystems. Israel. J. Bot. 28: 1–19.
Paine, R. (1980) Food webs: linkage, interaction, strength and community infrastructure. J. Anim. Ecol. 49: 667–685.
Pickett, S.T.A. (1980) Non-equilibrium coexistence of plants. Bull. Torrey Bot. Club 107: 238–248.
Piggott, C.D. (1982) The experimental study of vegetation. New Phytol. 90: 389–404.
Rabotnov, T.A. (1975) On phytocoenotypes. Phytocoenologia 2: 66–72.
Raup, H.M. (1975) Species versatility in shore habitats. J. Arnold Arb. 56: 126–163.
Richardson, J. L. (1980) The organismic community: resilience of an embattled ecological concept. BioScience 30: 465–470.
Salisbury, E.J. (1929) The biological equipment of species in relation to competition. J. Ecol. 17: 197–222.
Salisbury, E.J. (1942) The Reproductive Capacity of Plants. London: Bell. 244pp.
Salisbury, E.J. (1952) Downs and Dunes: their Plant Life and its Environment. London: Bell. 328 pp.
Salisbury, E.J. (1961) Weeds and Aliens. London: Collins.
Salisbury, E.J. (1970) The pioneer vegetation of exposed mud and its biological features. Phil. Trans. Roy. Soc. Lond. B259: 207–255.
Sarukhán, J. and Harper, J.L. (1973) Studies on plant demography: *Ranunculus repens* L., *R. bulbosus* L. and *R. acris* L. I. Population flux and survivorship. J. Ecol. 61: 675–716.

Shmida, A. and Ellner, S.P. (1984) Coexistence of plant species with similar niches. Vegetatio 58: 29–55.

Silvertown, J. (1980) The dynamics of a grassland ecosystem: botanical equilibrium in the Park Grass Experiment. J. Appl. Ecol. 17: 491–504.

Snaydon, R.W. (1970) Rapid population differentiation in a mosaic environment. 1. The response of *Anthoxanthum odoratum* populations to soils. Evolution 24: 257–269.

Snaydon, R.W. (1984) Plant demography in an agricultural context. pp. 389–407 in R. Dirzo and J. Sarukhán (Ed.) Perspectives on Plant Population Ecology. Sunderland, Mass: Sinauer Assoc.

Tansley, A.G. (1935) The use and abuse of vegetational concepts and terms. Ecology 16: 284–307.

Tansley, A.G. (1947) Frederic Edward Clements 1874–1945. J. Ecol. 34: 194–196.

Thurston, J.M., Williams, E.D. and Johnston, A.E. (1976) Modern developments in an experiment on permanent grassland started in 1856: effects of fertilisers and lime on botanical composition and crop and soil analyses. Ann. Agron. 27: 1043–1082.

Tobey, R.C. (1981) Saving the Prairies. The Life Cycle of the Founding School of American Plant Ecology, 1895–1955. Berkeley: University of California Press. 315 pp.

Turkington, R. and Aarssen, L.W. (1984) Local-scale differentiation as a result of competitive interactions. pp 107–127 in R. Dirzo and J. Sarukhán (Ed.) Perspectives on Plant Population Ecology. Sunderland, Mass: Sinauer Assoc.

Turkington, R.A., Cavers, P.B. and Aarssen, L.W. (1977) Neighbour relationships in grass-legume communities. I. Interspecific contacts in four grassland communities near London, Ontario. Can. J. Bot. 55: 2701–2711.

Turkington, R.A. and Harper, J.L. (1979) The growth, distribution and neighbour relationships of *Trifolium repens* in a permanent pasture. I. Ordination, pattern and contact. J. Ecol. 67: 201–218.

Tüxen, R. (1977) The significance of phytosociology for the protection of environment. pp. 13–20 in A. Miyawaki and R. Tüxen (Ed.) Vegetation Science and Environment Protection. Tokyo: Maruzen & Co.

van der Valk, A.G. (1981) Succession in wetlands: a Gleasonian approach. Ecology 62: 688–696.

van der Vegte, F.W. (1978) Population differentiation and germination ecology in *Stellaria media* (L.) Vill. Oecologia 37: 231–245.

Warming, E. (1909) Oecology of Plants. An Introduction to the Study of Plant Communities. Oxford: Clarendon Press, 422 pp.

Watt, A.S. (1947) Pattern and process in the plant community. J. Ecol. 35: 1–22.

Watt, A.S. (1955) Bracken versus heather, a study in plant sociology. J. Ecol. 43: 490–506.

Watt, A.S. (1964) The community and the individual. J. Ecol. 52 (Suppl.): 203–211.

White, J. (1979) The plant as a metapopulation. Ann. Rev. Ecol. Syst. 10: 109–145.

White, J. (1984) Plant metamerism. pp. 15–47 in R. Dirzo and J. Sarukhán (Ed.) Perspectives on Plant Population Ecology. Sunderland, Mass: Sinauer Assoc.

Whittaker, R.H. (1962) Classification of natural communities. Bot. Rev. 28: 1–239.

Wit, C.T. de (1960) On competition. Versl. Landbouwk. Onderz. 66: 1–82.

ON THE SIGNIFICANCE OF DEMOGRAPHIC PROCESSES IN PHYTOSOCIOLOGY

O. WILMANNS

ABSTRACT

Though demography is essentially a quantitative and phytosociology a qualitative discipline, they may fruitfully be combined: the knowledge of whose life cycles of characteristic species and the correlation of various demographic types with certain habitats and communities yield a better understanding of the endogenous connections in the biocoenose and of its evolution. On the other hand phytosociology offers a clear reference system to judge the scope of validity of demographic statements; it is of heuristic value for demographic comparisons.

These general opinions are illustrated by numerous examples: origin of populations (dispersal, seed-bank), development of populations (life cycles, clone formation, flowering period) and age structure and extinction.

INTRODUCTION

The results of the research of phytosociologists and of demographers are building stones for their own systems of theories. The aim of the phytosociologist (vegetation ecologist) is to describe and to interpret all facets of the life of plant communities, that is, to recognise the endogenous and exogenous factors which determine the spatial and temporal order of the communities, and not least their significance for animals and man. Tüxen (1965) for example attempted to encompass the characteristic features of the biocoenose in seven 'laws of coexistence of plants and animals'. The aim of the demographer on the other hand is to construct laws (if possible, quantitative ones) of population dynamics, by comparative analyses of many species in different environments. Often such building stones may be put into the construction of theories of related disciplines and there fill previously empty gaps. Phytosociology, in spite of its typically more holistic approach, can profitably incorporate the results of demography, a more reductionist enterprise, into its own theoretical system.

The phytosociologist willingly takes up all results which render a better understanding of the role of each member of the community. A full

White, J. (ed.), The Population Structure of Vegetation.
 ISBN 978-94-010-8927-2

understanding of the whole life cycles of dominating (key) species and of character species is especially useful, as Ernst (1979) showed for *Allium ursinum*. Of high interest is the possible coincidence of distinct demographic types with distinct habitat types, and their accumulation in distinct community types. Additionally, demographic features provide a bridge between vegetation science and evolutionary biology. Such questions as the ecological forces of selection, the possible coexistence among species and their niche structure (see for example Wilmanns 1983), and the equivalence of species, structures and functions in different regions and different communities can all be precisely phrased in demographic terms.

But the flow of information also runs in the opposite direction, for the development of populations depends both on the physical environment and on the interactions of members of the biocoenose. If we wish really to understand a demographic fact, to see its teleonomic significance and its meaning in the ecosystem, we shall only succeed if the context in which it is realized is taken into consideration (see Grubb et al. 1982). Phytosociology with its floristically defined syntaxa offers a clear reference system and a background for further studies. A demographic statement is valid at first only for a specific environment, usually within the framework of the stand studied in a syntaxon of low rank. The scope of its generalisation can be amplified only if it is checked in other stands both (a) of the same syntaxon and (b) of syntaxa of graded similarity. The phytosociological system provides 'pre-performed work' (Tüxen 1970) as a basis for inductive generalisation, by which a specific fact may become a broader scientific statement.

I shall try to demonstrate the fertility of such mutual interrelations by some examples selected from a phytosociologist's point of view. The interrelated disciplines of demography, population dynamics and evolutionary biology will not be sharply separated in discussing their links to phytosociology. That this enterprise is not entirely novel is shown by Schwabe-Braun and Tüxen (1980) in their treatment of the class Lemnetea minoris. But, in general, the intellectual cooperation between demographers and phytosociologists has not been very close until recently. One may ask for the reasons. An essential cause, though probably not the only one, lies in their different approaches. The demographer must work according to the method of census and usually needs an amenable system. Consequently a particularly high number of examples (e.g., Harper and White 1974) belong to horticulture and forestry, where at least quasi-experimental conditions are usual: monocultures (often even of cultivars), uniformity of soil, climate and management, facilities for the exact counting or weighing the products are typical. Corresponding situations studied in nature are often communities with few or even one clearly dominant species. As a rule such systems are much more complicated than horticultural ones. Phytosociology, on the other hand, is essentially qualitative. Tüxen's seven 'laws', for example, are of a qualitative nature. It is true that communities which are relatively simple and easy to survey offer methodological advantages, as is demonstrated by the Lemnetea study referred to earlier. Quantitative determinations such as the numerals of Tüxen and Ellenberg (1937) by which an exact description of synthetic characteristics of communities was intended, were clearly thought of as an aid, not as an end. The ecological code numbers of Ellenberg (1979, 1982) give scaled numerals for essentially qualitative features, which can be described in more detail in words (e.g., Oberdorfer 1983). But this does not mean that many qualitative statements in phytosociological literature are inapplicable in demography: semi-quantitative estimations of species cover or abundance, qualitative data of vitality, or silvicultural-sociological tree classes as expressions of age structure, sociability estimates as hints of clonal size, all are examples which sometimes offer preliminary insights on demographic parameters.

Until recently the emphasis in phytosociological studies was more on the spatial distribution of communities than on their temporal sequence. The former can be investigated with precision in a short time and was promoted through the practical application of phytosociology by means of vegetation mapping (see Tüxen 1963). Demography, however, describes developmental processes, so its close connection to phytosociology will be particularly expected in studies of successional stages. But vegetation dynamics has been developed in recent years mostly in connection with problems of nature conservation: abandonment of meadows, intensification of agriculture, all sorts of rapid destruction of vegetation (e.g., Westhoff 1979, Böttcher 1974, Beeftink 1975, Schmidt 1981, Fischer 1982). Consequently, census of individuals and other detailed estimations of cover are becoming increasingly available.

ORIGIN OF POPULATIONS

The establishment and rise of populations play a phytosociological role in the following syndynamical processes:

(a) A stand of a community develops on a substrate previously without plants, the first step in a primary succession.

(b) The succession starts on a previously occupied substrate, from a former, completely destroyed, and not regenerating vegetation; this is the first step in a secondary succession.

Mixed types are often found in the industrial countries. The pioneer plants of these types are called 'primary pioneers' (Tüxen 1979). Plants with a ruderal 'strategy-type' (Grime 1979) dominate. Examples of phytosociological studies on these communities which contain demographic elements are those of Fischer (1982), van Noordwijk-Puijk et al. (1979) and Schmidt (1981).

(c) Later steps in both primary and secondary successions come about if other plants invade the existing vegetation, followed by a direct or indirect change of species combination. These 'invading pioneers' (Tüxen 1979) are expected to have a different type of demography from that of the first group, for they must be strong 'competitors' (Grime 1979). In any case, demographic parameters are of interest: number of seeds, size of seeds and their dispersal type. Dispersal and immigration show many connections to vegetation history, biocoenology and applied phytosociology.

(d) Another case is the regeneration of more or less destroyed vegetation, which has, however, left viable seeds and dormant buds in the soil. The regeneration may begin at once (as on fields) or after long intervals (as in clearings, after a century of forest). In these cases the temporal gap must be bridged by a reservoir of viable seeds or buds.

Dispersal and immigration of species in various communities by seeds or fruits

Dispersal by generative and by vegetative diaspores (= disseminules) is considered separately, for the distinction is ecologically and genetically important. Although both types of diaspores may have low viability and a short life-span, impressively long life-spans (up to 1700 years) and transport distances (several kilometers by zoochory, probably several hundreds of kilometers by migrant birds: Müller-Schneider 1977) are performed only by well-proected diaspores with a low respiratory burden, that is, typically by seeds or fruits. Comparative demographical studies of plants which are able to reproduce in both ways are rare but they would be informative, as the paper of Petersen (1981) on *Polygonum viviparum* suggests.

A detailed study by Hard (1972a) may serve as a typical example of the methodological connection between syndynamics and demography. He investigated forest stands of pine (*Pinus sylvestris*), which had developed spontaneously on fallow fields running perpendicular to a forest with older pines. Near the old forest the pines on these tongue-like strips (mostly 100 m long) reached nearly double the height of those at the end of the tongue, so that one might suppose them to have advanced gradually from the forest edge, year by year. Determinations of height diameter, density, and age demonstrated, however, that more than 90% of the pine trees had begun their growth within a span of 3 to 4 years, nearly simultaneously. Height growth had been increased where the young trees had reached their highest density, in the immediate neighbourhood of seed dispersing parental trees in the old stand. Growth of seedlings at a greater distance later became almost impossible when the formerly open field was conquered simultaneously by scrub of *Sarothamnus scoparius* and *Rubus fruticosus*, by grasses and other species: their competition was too strong for the pioneer *Pinus sylvestris*. This example shows, additionally, the limited scope of anemochory.

Obviously we have so far only few generalizations which combine the general patterns of seed demography with general synecology. Salisbury (1942) indicated one correlation: plants of later succession stages tend to have bigger seeds than plants of early stages. This is easily understood by the necessity to match the stronger interference of already established members of the vegetation. However his conclusions seem not to be relevant for the drier climate of California (Baker 1972).

A regular relationship between the number of seeds and the synecology of a species remains to be examined. As shown by van Andel and Vera (1977) and van Baalen (1982), studies of the energy allocation of a plant during its life cycle, its syntaxonomical (and therefore synecological) position and its mode of dispersal are valuable in this respect.

The mode of dispersal is often decisive for founding a new population. There are several syntaxa with one prevailing mode of dispersal, as shown by dissemination spectra (Müller-Schneider 1977). Rhamno–Prunetea communities of forest margins and hedges are predominantly endozoochorous: many birds not only use this biotope as a place for feeding, mating and breeding, but also help to constitute it by leaving the undigested kernels of stone-fruits, berries and analogous fruit-types (Tüxen 1977). This poses questions for demographic studies (besides those of pure census) which are of interest for the student of population dynamics, of phytosociology and evolutionary biology combined. Is there competition among plants for the dispersing animals, and consequently niche segregation of shrub species with respect to their fruit-ripening time? Some observations point to this being so (Müller-Schneider 1977). Where do we find generative rejuvenation in the stand itself? It is supposedly a rare event, because the shrub growth-form with its basitonous branching hinders the development of seedlings. Does the passage through the bird's gut influence seed viability, and therefore rejuvenation of the vegetation? Myrmecochory remains demographically poorly investigated, though it is estimated to be a mode of dispersal of 30% of the central European forest herbs (Wilmanns 1984).

The dispersal mode of founding populations of rock and wall crevice communities is different: the Parietarietea have a high proportion of myrmecochorous plants, as one may see in Segal's (1969) study; the Asplenietea are essentially anemochorous. May one conclude that their synevolution (the origin and development of syntaxa in geological times, Wilmanns 1984) has been different? The dispersal of species of the Isoeto-Nanojuncetea by swamp and water birds was appreciated a century ago (Müller-Schneider 1977), but still only in bare outlines in quantitative demographic terms.

Two examples which have recently been thoroughly studied – though without mentioning the word demography – are worth treating in some detail, because they show that in spite of having the same dissemination type the integration of species into the vegetation mosaic may be completely different. The studies concern the jay (*Garrulus glandarius*) (Bossema 1979) and the nutcracker (*Nucifraga caryocatactes*) (Mattes 1982). In both cases the birds gather diaspores of a few distinct woody plant species and hide them as a food stock. Only that proportion which is not refound and eaten by birds or other animals will be able to germinate. The jay has the following plant demographic and phytosociological effects: Distances of at least 4 kilometers from the parent tree can be spanned by the bird's flights. The area where the jay hides the acorns in the soil must have a woodland character. The Mantel grassland strip or dwarf shrub/grass facies are favoured. Every acorn is pushed separately into the soil 0.5–2 m from one another. Owing to the rich food reserves of the acorn, the seedling is able to grow through a grassy covering of vegetation, so this invasion pioneer may give an impulse to an otherwise stagnant grassland succession.

For a long time the nutcracker has been a competitor of man for the nutrient-rich 'nuts' of *Pinus cembra* as food. The nutcracker causes afforestation not only inside the forest, but even outside and above the actual timber line, in the dangerous zone of avalanches and erosion. Some of the results of Mattes (1982) demonstrate a long coevolution of plant and bird species and the following items are of a demographical and phytosociological interest. Hiding (and therefore 'sowing') of *Pinus cembra* (and of *Corylus avellana*, too) is possible even over distances of 5–12 km. Foundation of new populations of the tree happens when hungry nutcrackers from suboptimal habitats, especially spruce forests, collect nuts and store them in their own home range, or if they bring the nuts to the zone above the existing forest. It is estimated that a couple of nutcrackers set up at least 15,000 hiding-places annually and may refind as many as 80%. Only a small percentage of the annual nut-production remains for the reproduction of the tree species. The seeds germinate to give groups of 4 or 5 plants. Consequently, there is on the one hand competition between them but, on the other hand, their primary roots unite so that the strongest sapling is promoted. The nutcracker has a capacity to select habitat types: it does not hide seeds in higher, dwarf-shrubby places (as in *Rhododendron ferrugineum*) nor in mires or in closed grassy stands of *Nardus stricta* or *Deschampsia caespitosa*, places in which pine seedlings would be lost.

Bulbils and turions

These types of diaspores may be treated under the same functional aspects as the seeds. The formation of bulbils instead of flowers is called 'false vivipary' (in contrast to the 'true vivipary', which is limited nearly totally to the mangrove: see van der Pijl 1982). It is genetically clone formation and therefore like the formation of stolons and ramets; ecologically however it is perhaps more analogous to the formation of seeds. In European vegetation it has evidently a positive selection value for species in two vegetation types: in subarctic spring vegetation of the Montio-Cardaminetea (several *Saxifraga* species, *Deschampsia alpina*), and in dry grassland, species of which pass partly into vineyards (*Allium carinatum, A. oleraceum, A. pulchellum, A. vineale, Gagea villosa, Poa bulbosa*). In other communities we find only scattered representatives: *Dentaria bulbifera* and *Ranunculus ficaria* in nutrient-rich forests, *Lilium bulbiferum* and *Poa alpina* in alpine grassland. In all cases the biological significance of these diaspores can be seen in the store of reseve substances for a quick start in a short vegetation period. The demographical research on *Allium vineale* by Håkansson (1963) demonstrates the use of this approach.

Turions as buds for overwintering and for dispersal of water plants are very frequent. Their demography should be analysed together with ramets: both may easily float away and found new populations.

The seed-bank as a part of the phytocoenose

During past years phytosociologists have focussed their attention on the seed bank in the soil but rarely on the shorter-living bud bank. The main reason for this has been the retrogression or even disappearance of numerous species and communities, the observation of their occasional reappearance, and the hope they might spontaneously re-establish themselves on biotopes created by the hand of man. Another reason was an economic one: the capacity of several weeds to hold their ground, in general considered unfavourably, has recently been more positively appreciated. A census of seeds is not only of theoretical but also of practical interest.

If roads are constructed through a former field, red ribbons of *Papaver rhoeas* sometimes develop at the margins, though the species has become rare in the fields themselves. This is a conspicuous argument for the wealth of seeds in the soil. The longevity of seeds of weeds is well known (e.g., Harper and White 1974). It may be mentioned that those two field weeds in Germany which do not have a seed-bank are nearly extinct: *Agrostemma githago* and *Bromus secalinus* var. *grossus* (Sukopp et al. 1978). Nature conservancy now tries to save the old field weeds by paying farmers for not using herbicides on the marginal strips of some cornfields. This experiment can be successful, but only in places where an old seed-bank had been preserved (Schumacher 1980). So we ask the demographer: what is the rate of seed mortality under various circumstances and management (e.g., Roberts and Dawkins 1967)? How many years or generations does it take to build up a new seed-bank?

In viticulture a green plant cover is desirable for soil improvement and to retard erosion; the species should not, however, compete with the vine plants for water. Stalder, Potter and Barben (1976) have developed a procedure in Switzerland by which they promoted harmless annuals which regenerated easily from the seed bank (*Stellaria media*, *Veronica persica* and others) as 'soil-coverers', whilst suppressing *Convolvulus arvensis*, a very problematical weed, by herbicides.

In some central European agricultural regions with a former combination of coppice clearing and burning in alternation with field-cropping, the clearings even today shine red with *Digitalis purpurea* in the second and third year, and from the fourth year are golden with *Sarothamnus scoparius*. Plants of clearings (Epilobietalia angustifolii) exist plentifully in the seed bank of woods (Brown and Oosterhuis 1981). *Digitalis purpurea* was examined demographically and experimentally by van Baalen (1982), who demonstrated a resistance of the seeds to fungi and phytophagous animals in the soil: he explains this by their content of effective chemical substances. Does this substantiate the idea of Tüxen (1985) that a high proportion of the Epilobietalia character species are poisonous and/or medicinal herbs?

Brown and Oosterhuis (1981) found species typical of the class Isoeto-Nanojuncetea in their samples of forest soil. Krause (1979) observed that after soil disturbance species of the Littorellion and the Artemisietea vulgaris, which had vanished for decades, unexpectedly reappeared. These and other examples suggest the idea, which must be demographically verified, that plants typical of small, ephemeral habitats in the original landscape of the temperature zone tend to build up a seed bank.

Phytosociological relevance of life cycles

Species with a very short life cycle occur in a formation-determining abundance only in short-term habitats. These habitats may be periodically or episodically free from competitors, but yet colonizable. This is shown by an analysis of the classes Thero-Salicornietea, Saginetea maritimae, Cakiletea, Bidentetea tripartitae, Polygono-Poetea annuae, Stellarietea mediae and Isoeto-Nanojuncetea (see Wilmanns 1984). In the latter two classes species whose life cycle exceeds one year can survive because the soil disturbance is usually not a radical one.

Another correlation between life cycle and sociological position is seen with trees. Following Firbas (1949), the ability to flower begins with the pioneer species such as *Betula pendula* and *Corylus avellana*, as a rule at 10 years, with *Populus tremula* at 20–25 years and with *Fraxinus excelsior* at 25 years. The species in the forest communities of the regional natural vegetation begin flowering much later: *Abies alba* 30 yr., *Fagus sylvatica* 30–40 yr., *Picea abies* 30–40 yr., *Quercus robur* 30–60 yr. These are numbers for individuals, growing in isolation; in a closed stand they flower for the first time even later: *Abies alba* at 60–70 yr., *Quercus robur* at 80 yr.

An individual may reach its reproductive maturity earlier than the average within its species; this may be, but is not necessarily, an unequivocal advantage of selection. Rather, one has to consider a possible correlation between this quick development and the vitality on a long-term basis: *Abies alba* (which is together with *Fagus sylvatica* the strongest competitor among the trees of Central Europe, Ellenberg 1982) is able to grow a few decades in the undergrowth in the shade and in the shelter of the canopy. It is well-known to foresters that those individuals with a delayed youth development and correspondingly narrow annual rings reach a higher total age and with better health.

Demographically insufficiently known is the phytosociological process known in German as *Versaumung*: *Saum* species, originally living at natural margins of woods, penetrate into closed meadows and pastures no longer cut and grazed (Wilmanns et al. 1977, Schwabe-Braun 1980, Bürger 1984). These are species of the same life-form, but of different ecological constitution and therefore syntaxonomical position. The change of environment is caused only by the lack of management. As these Saum species enter the turf over the course of some years, other species disappear in consequence. Mesobrometa are invaded by forbs of the Trifolio--Geranietea such as *Anthericum ramosum*, *Aster amellus*, *Geranium sanguineum*, *Origanum vulgare*; character species of the Festuco-Brometea and other species which are weak in competition (such as *Anthyllis vulneraria*, *Briza media*, *Hieracium pilosella*, *Orchis ustulata*) decrease. The increase in species number demonstrates that the input of species is higher than the output. The reason is that the competitive strength of Saum species is very high, unless they are damaged by agricultural practices such as mowing in early Summer, as most flower in July or even later.

Vegetative growth of populations (formation of clones)

The type of growth which leads to an increasing number of shoots is extremely frequent in cormophytes. This may give rise to communities which are dominated by them and consist of only one or a few species: *Pteridium* facies, subtropical Gleichenietea, herds of *Brachypodium pinnatum* and *Prunus spinosa*-shrub in fallow limestone grassland, *Rubus fruticosus*-slopes and so on. Exact analyses such as Watt (1976) gave for *Pteridium aquilinum*, would be informative in view of the significance of these species in understanding the structure and successional relationships of communities. Species with vegetative reproduction have focal points in certain communities and habitat types.

The separation of whole daughter-plants of freely floating water plants (*Lemna, Salvinia, Azolla, Eichhornia, Pistia*) is ecologically understandable because competition for space is weakened by wind-drift, and water and nutrients usually are not limited. (Conversely, *Lemna gibba* produces seeds under amphibious conditions in the subtropics; Schwabe-Braun and Tüxen 1980).

The formation of overground runners or soil-adjacent horizontal shoots is especially frequent in (and even characteristic of) the Lolio-Potentillion (former Agropyro-Rumicion crispi p.p.), the vegetation of flooded meadows with a high 'variety in time' (van Leeuwen 1965) and in pastures where there are places disturbed by animals' hooves (Sýkora 1982). Similar conditions exist in amphibious riparian habitats of the Littorelletea, where *Juncus bulbosus, Ranunculus reptans* and *Elatine hexandra* are representatives of this growth-form, in the spring biotopes are the Montio-Cardaminetea and sometimes in the nutrient-rich habitats of the Glechometalia (Tüxen and Brun-Hool 1975).

Subterranean runners, often at the same time with a storage function, dominate in three syntaxonomical classes of a specialized character: in the Ammophiletea on the coastal dunes with moving sand, in the Spartinetea on moving mud and in the Agropyretea intermedio-repentis, an anthropogenously disturbed, fire-resistant, semi-ruderal, semi-dry grassland (Fischer 1982). Supply of sand by storms enhances the vitality of *Ammophila arenaria* (Lux 1964) and *Carex arenaria* (Noble et al. 1979).

In the absolutely natural communities of alpine talus (Thlaspietea rotundifolii), the characteristic plants have clones which are easily regenerated when damaged; they creep through the scree and use the scanty soil in the hollows (see Wilmanns 1984).

A few demographical studies have compared genetically related species with different growth-forms. Those species with abundant vegetative propagation have less intensive generative propagation, and vice versa: *Agropyron caninum* versus *Agropyron repens* (Tripathi and Harper 1973), *Ranunculus acris* and *Ranunculus bulbosus* versus *Ranunculus repens* (Sarukhán and Harper 1973). It would be interesting to know if this regularity is valid even for one and the same ramet-producing species in different habitats, that is, in different communities. Does it depend on the quality of the environment? The thorough studies of Sarukhán were carried out in a single vegetation type. Consequently the synecological (biotic and abiotic) conditions cannot have been optimal for each of the three *Ranunculus* species, since normally they do not live together in the same community. Would the differences have been greater or smaller, if they had been tested in their respective synecological optima: *Ranunculus acris* (a character species of the class Molinio-Arrhenatheretea) in a community, or indeed several communities, of this economic grassland; *R. bulbosus* (a character species of the Mesobromion) in a dry grassland of this alliance or in a dry subassociation of the Arrhenatheretum; and *R. repens* (a character species of the Agrostietea stoloniferae) in the Lolio-Plantaginetum? Both answers are theoretically conceivable: the difference should be smaller, if in the studied stand with joint occurrence character displacement (*Kontrastbetonung*) had taken place; the difference would be greater if, for example, *R. repens* were to produce more seeds under unfavourable conditions, and even more stolons in its optimum habitat.

As an instructive example for a phytosociological and demographical contrast to *Carex arenaria* (Noble et al. 1979) we may cite *Carex curvula*, a character species of acid grassland in the alpine mountains of Europe (Grabherr et al. 1978). The dune pioneer *C. arenaria* with its long rhizomes can grow a little more than 3 m annually whereas the individual rhizomatous shoot of *C. curvula* grows barely 1 mm annually. *C. arenaria* produces 400–700 daughter shoots/m^2 in the year, *C. curvula* likewise about 500. Their weights, however, are completely different, as might be expected considering the length of time favourable for photosynthesis: nearly the whole year in Atlantic Europe and about 115 days in the alpine zone of the Central Alps. The subterranean standing crop of *C. arenaria* is about 700 g/m^2. If one may take this value roughly as productivity, this is eight times that of the Caricetum curvulae (with many cryptograms) and the 25-fold greater than *C. curvula* itself. The age structures are also totally different: while all shoots of more than 9 months age in *Carex arenaria* amount to no more than 20% of the cohort, the shoots of *C. curvula* build up 2 new leaves per annum for 10 years and each leaf has a lifespan of 2–3 years.

Clone formation in woody plants

This type of propagation has a great advantage for the demographer, who can easily and reliably define the age structure in temperate zones. The following two examples indicate the close relationship between demography and phytosociology in elucidating the dynamics of vegetation. Syntaxonomically the examples refer to the classes Rhamno-Prunetea and Vaccinio-Piceetea; sigmasociologically they refer to characteristic vegetation mosaics in the potential forest area and in the subalpine upper tree limit zone.

Shrubs such as *Prunus spinosa*, *Ligustrum vulgare*, *Cornus sanguinea* and trees such as *Populus tremula* and *Robinia pseudacacia* are invasion pioneers which build up new woody plant vegetation. They push forward from woodland edges, banks and hedges as 'cores' in fallow grassland

(Hard 1972b, Lohmeyer and Bohn 1973, Wilmanns et al. 1977, Bürger 1984). Their morphology and rate of expansion are very different. The surface of *Prunus spinosa* scrub slopes gently to the margin as it advances regularly; in Mesobrometa we found a rate of advance of ca. 0.5 m per year. The blackthorn shades strongly and displaces the herb layer quickly, corresponding to the 'phalanx-type' described by Harper (1978). Aspen on the contrary advances in a manner similar to the 'guerilla-type'. *Rubus fruticosus* 'walks' by rooting of its tips; it is (together with bracken and tree species) one of the strongest competitors in the European flora, often forming one-species scrub: the demography of this *Scheinstrauch* (false shrub) would be fascinating.

The genus *Picea* is able to produce clones by rooting branch layers, this mostly happens near the natural timber line, but occasionally, too, in lower situations. Holtmeier (1982) described 'atolls', 'ribbon forests' and 'hedges' of *Picea engelmannii*, a characteristic aspect of the landscape in the Rocky Mountains. The most thorough investigations using phytosociological, demographic and silvicultural methods, both descriptive and experimental, were made by Kuoch and Amiet (1970) with *Picea abies* in the Swiss Alps. Living branches on isolated trees contacting the soil are required for the formation of such 'conical collectives' ('Rotten'). The increase in height in the clones was only 1–3 cm per year. Analyses of year rings in the main stem of the plant showed that vegetative propagation begins only at age 50–130 years; the branch layer is 30–100 years old when it forms the first roots; this happens on shoots 5–9 years behind the tip of that branch. After 10–15 years at the earliest, sometimes only after 70 years, the parent tree ceases to transfer nutrients to the vegetative offspring. The changes of survival of the clone in the plant communities studied are much higher than those of individual treelets, because the groups are better protected against storm, driving or gliding snow and probably frost drought. So the frequent growth leewards, like a string of pearls, is understandable. Altogether, the formation of resistant conical spruce-clones with a few 'generations' of layers needs several centuries. For *Picea abies* this is "the only possible way to reconquer former forest areas by woodland and to preserve the woodland of the struggle zone" (translated from Kuoch and Amiet 1970).

These examples call attention to the different distribution of individuals of the same species in a stand, their dispersion. This is recorded in a rough manner as sociability in phytosociological relevés using the Braun–Blanquet scale and should be heeded more in demographical studies, for it may be an expression of the growth-form and/or the environmental mosaic (in a large sense) (Kershaw 1963).

Flowering period

Interest in the periodicity and spatial distribution of flower quantities and flower types seems to have awakened only recently among phytosociologists. For the demographer flowering and the production of seeds is important and is well documented. Methods of semi-quantitative symphenological recording (Balátová 1970, Dierschke 1972) reveal rhythms of flowering to be community-specific. A more detailed analysis needs both exact quantitative dates and good entomological information. Kratochwil (1983, 1984) provided a good example. He determined the quantities of flowers in a Mesobrometum which was interspersed with Saum plants over the whole vegetation period and found qualitatively a continuous sequence of flowering. But quantitatively there were four waves of flowering: in the first two, yellow flowers predominated, *Primula veris* (in April) and *Hippocrepis comosa* (in May). In the later two waves mauve flowers were common: *Coronilla varia* (in July) and *Origanum vulgare* (in August/September). At first, grassland plants of the Festuco-Brometea mostly flowered but afterwards Saum plants of the Trifolio-Geranietea were more evident. The flowering period of the grassland plants occurs before mowing times in July/August. The Saum plants, on the whole, flower and fruit later and characterise a later stage of succession without mowing. The biosociological importance of the Saum plants is great, since they provide a large supply of food for social hymenoptera (Apoidea) just at the time when they have to care for their broods.

The biocoenological effects of the early mowing of nutrient-rich grassland (Arrhenatheretalia) for silage in modern agriculture are little thought about, and even less studied. From ancient times these plants have been selected for a quick development to flowering and fruiting (e.g., Ellenberg 1982), but nowadays we have to expect a strong decline or even disappearance of those species which typically have a high population flux unless they get an opportunity to reproduce sexually. Studies similar to those of Sarukhán and Harper (1973) on *Ranunculus* species would be informative in this respect, but they remain very rare. In the long run, too, species with a long individual life time and less dependent on sexual replacement will also be influenced: this might be forecast by knowledge of their depletion curves. Moreover the speed of modern mowing with powerful machines causes abrupt and disastrous destruction of the food resources of flower visitors, such as butterflies. The preservation and accessibility of unmanaged biotopes such as clearings and tracks in the forests, and ruderal areas with rich vegetation become increasingly important as refuges for flower pollination. A special mosaic of plant communities complementary in time to man-modified vegetation has consequently to be ensured or created. There is a serious lack of scientific basic knowledge for these tasks of nature conservation.

CHANGES IN THE AGE STRUCTURE OF POPULATIONS IN PLANT COMMUNITIES

If the rate of mortality exceeds the rate of natality, a population not only diminishes, but in the course of time it may also change its age structure. This may result in qualitative and quantitative variations in the vegetation. If a certain age class distribution characterises a species in a particular type of vegetation, the question arises how it has been developed and how it is regulated under natural conditions. This cannot be investigated without demographic methods. The problem, however, is so vast and complicated that no more than some vague outline of an answer is yet possible. One may generally conclude that interferences by man, at least as performed during the past century, can result in a rough mosaic of vegetation, often with relatively uniform age classes, as in the management of forests, or in populations established after ploughing the soil, or after burning, or following elimination of wetland habitats by drainage. On the contrary, with natural regulation one may expect a mixture of many age groups often arranged in very fine spatial and temporal mosaics, as was described in the classical study by Watt (1957).

A modern example, founded on 20 years' succession studies on permanent plots, points to the link between phytosociology and demography of research on population age structure: it concerns the heather beetle (*Lochmaea suturalis*) in the Genisto-Callunetum in the Netherlands (de Smidt 1977). Larvae and adults of this beetle feed exclusively on *Calluna vulgaris*. The beetle prefers old and closed patches of *Calluna*, with a moss layer beneath and a humid microclimate. Only under these conditions does the beetle cause the death of the heather plants. *Calluna* subsequently rejuvenates, however, by seed. The next outbreak of the beetle is usually 5 to 10 years later but at this stage the young patches of *Calluna* remain uninjured. Other plants, now grown older, provide food. The resulting mosaic is blurred by fire, which gives rise to an age-independent, large rejuvenescence of the Genisto-Callunetum. Sheep, however, tend to stabilize the age structure, because they graze the young shoots of *Calluna* (and of grasses) from the regenerating patches.

A last example will be given, one which is a challenge for the phytosociologically interested demographer and the demographically interested phytosociologist because of its implications for theory and practice: the 'dieback model' of Mueller-Dombois, constructed for rain forests on Hawaii and recently extended on other Pacific islands (Mueller-Dombois 1984). For some years death of the tree species *Metrosideros polymorpha* on the Hawaiian islands has been noted. It happens to smaller or bigger groups in the stands. In an individual stand, trees of the same age die, although the ages of dying groups of trees differ between stands, implying a simultaneous rejuvenation of the population. This so-called cohort senescence – thus the hypothesis – may be triggered by different exogenous or endogenous stress factors. Details of the interesting combination with factors in the ecosystem are given in the original paper.

Besides the disappearance of whole stands of many communities by abrupt destruction of their biotopes, the decline and (local) extinction of populations are often acute (e.g., Sukopp et al. 1978, Westhoff 1979). The effect of different factors (and therefore, too, of different human activities) on the age structure of populations of endangered species may be very different. (As to demographic studies, see Synge 1981.) An instructive example is provided by Schwabe-Braun (1980) for *Arnica montana*, a threatened species listed in the German Red Data Book. It lives on acid soils in extensively grazed, unfertilized pastures of the Nardetalia and occasionally still imparts to them a bright yellow colour. This species is threatened by destruction of its biotope, as indeed is the whole of this semi-natural vegetation. The threats include afforestation with *Picea abies*; intensification of grazing together with intensified fertilizing which tends to change the vegetation into a Cynosurion community; the cessation of the agricultural use of land for economic reasons (*Sozialbrache*); whether or not cropping of flowers for folk medicine is a relevant factor in its decline is still an open question. The first threat (afforestation) leads to a rapid reduction of the population, independent of its age structure. The second threat involves the sensitivity of the rosettes to frequent treading; presumably this, too, is age-independent. The third threat causes *Arnica* to increase at first, because it is totally relieved from treading by cattle; but the competitive interrelationships change as some grasses are favoured and gradually increase: the rosettes of *Arnica* are shaded too strongly. The development of grass litter prevents the establishment of young *Arnica* plants and the population grows older without replacement. It is important for the conservation and preservation of this species to know under which conditions the process of decline is reversible.

CONCLUSION

This sketch cannot treat all aspects of the relationships between demography and phytosociology, but its aim is to attempt to stimulate an increasing intellectual and practical co-operation between scientists engaged in both disciplines. It is dedicated to the memory of Reinhold Tüxen, who all his life sought actively to combine the different branches of ecological research with phytosociology.

ACKNOWLEDGEMENT

Dr. James White (Dublin) made a linguistic revision of the manuscript. He, Dr. W.G. Beeftink (Yerseke) and my former students, now colleagues, Dr. A. Fischer (Gießen) and Dr. A. Schwabe-Kratochwil (Freiburg i.Br.) stimulated and encouraged me by constructive discussion of the topics: I am grateful to them.

REFERENCES

Van Andel and Vera, F. (1977) Reproductive allocation in *Senecio sylvaticus* and *Chamaenerion augustifolium* in relation to mineral nutrition. J. Ecol. 65: 747–758.

Van Baalen, J. (1982) Germination ecology and seed population dynamics of *Digitalis purpurea*. Oecologia 53: 61–67.

Baker, H.G. (1972) Seed weight in relation to environmental conditions in California. Ecology 53: 997–1010.

Balátová-Tuláčková, Emilie (1970) Beitrag zur Methodik phänologischer Beobachtungen. In: Tüxen, R. (Ed.): Gesellschaftsmorphologie (Strukturforschung) (Ber. Symp. IVV Rinteln 1966): 108–121. Junk, Den Haag.

Beeftink, W.G. (1975) Vegetationskundliche Dauerquadratforschung auf periodisch überschwemmten und eingedeichten Salzböden im Südwesten der Niederlande. In: W. Schmidt (Red.): Sukzessionsforschung (Ber. Symp. IVV Rinteln 1973, R. Tüxen (Ed.)): 567–578. Cramer, Vaduz.

Böttcher, H. (1974) Bibliographie zum Problem der Sukzessions-Forschung mit Hilfe von Dauerquadraten und der Vegetationskartierung. Exc. Bot., Sectio B. 14: 33–56.

Bossema, I. (1979) Jays and Oaks: an eco-ethological study of a symbiosis. Behaviour 70: 1–117.

Brown, A.H.F. and Oosterhuis, Lenoor (1981) The role of buried seed in coppicewoods. Biol. Cons. 21: 19–38.

Bürger, R. (1984) Successional limestone grassland communities of the Kaiserstuhl with regard to their conservation management. In: Géhu, J.-M. (ed.) La végétation des pelouses calcaires. Coll. Phytosoc. 9: 405–419.

Dierschke, H. (1972) Zur Aufnahme und Darstellung phänologischer Erscheinungen in Pflanzengesellschaften. In: v.d. Maarel, E. and Tüxen, R. (Red.): Grundfragen und Methoden in der Pflanzensoziologie (Ber. Symp. IVV Rinteln 1970, R. Tüxen (Ed.)): 291–312. Junk, Den Haag.

Ellenberg, H. (1979) Zeigerwerte der Gefäßpflanzen Mitteleuropas. 2. Aufl. 122 S. Goltze, Göttingen.

Ellenberg, H. (1982) Vegetation Mitteleuropas mit den Alpen in ökologischer Sicht. 3. Aufl. 989 S. Ulmer, Stuttgart.

Ernst, W.H.O. (1979) Population biology of *Allium ursinum* in northern Germany. J. Ecol. 67: 347–362.

Firbas, F. (1949/52) Spät- und nacheiszeitliche Waldgeschichte Mitteleuropas nördlich der Alpen. 2 Bd. Fischer, Jena.

Fischer, A. Mosaik und Syndynamik der Pflanzengesellschaften von Lößböschungen im Kaiserstuhl (Südbaden). Phytocoenologia 10: 73–256.

Grabherr, G., Mähr, E. and Reisigl, H. (1978) Nettoprimärproduktion und Reproduktion in einem Krummseggenrasen (Caricetum curvulae) der Ötztaler Alpen, Tirol. Oecol. Plant. 13: 227–251.

Grime, J.P. (1979) Plant Strategies and Vegetation Processes. 222pp. Wiley, Chichester.

Grubb, P.J., Kelly, D. and Mitchley, J. (1982) The control of relative abundance in communities of herbaceous plants. In: Newman, E.I. (Ed.): The Plant Community as a Working Mechanism: 79–97. Blackwell, Oxford.

Håkansson, S. (1963) *Allium vineale* as a weed with special reference to the conditions in south-eastern Sweden. Växtodling. Plant Husbandry 19: 208pp. Almquist & Wiksell, Uppsala.

Hard, G. (1972a) Wald gegen Driesch. Das Vorrücken des Waldes auf Flächen junger "Sozialbrache". Ber. z. dt. Landeskd. 46: 49–80.

Hard, G. (1972b) Vegetationsdynamik und Verwaldungsprozesse auf den Brachflächen Mitteleuropas. Erde 106: 243–276.

Harper, J.L. (1977) Population Biology of Plants. 892pp. Academic Press, London.

Harper, J.L. (1978) The demography of plants with clonal growth. Verh. Kon. Nederl. Akad. Wetensch. Afd. Natuurk. 70: 27–48.

Harper, J.L. and White, J. (1974) The demography of plants. Ann. Rev. Ecol. Syst. 5: 419–463.

Holtmeier, F.K. (1982) "Ribbon-Forest" und "Hecken". Streifenartige Verbreitungsmuster an der oberen Waldgrenze in den Rocky Mountains. Erdkunde 36: 142–153.

Kershaw, J. (1963) Pattern in vegetation and its causality. Ecology 44: 377–388.

Kratochwil, A. (1983) Zur Phänologie von Pflanzen und blütenbesuchenden Insekten (Hymenoptera, Lepidoptera, Diptera, Coleoptera) eines versaumten Halbtrockenrasens im Kaiserstuhl-ein Beitrag zur Erhaltung brachliegender Wiesen als Lizenz-Biotope gefährdeter Tierarten. Beih. Veröff. Naturschutz Landschaftspflege Baden-Württ. 34: 57–108. Karlsruhe.

Kratochwil, A. (1984) Pflanzengesellschaften und Blütenbesucher-Gemeinschaften--biozönologische Untersuchungen in einem nicht mehr bewirtschafteten Halbtrockenrasen (Mesobrometum) im Kaiserstuhl (Südwestdeutschland). Phytocoenolgia 11: 455–669.

Krause, W. (1979) Das Wiederauftauchen aussterbender Pflanzen und Pflanzengesellschaften auf anthropogenen Standorten. In: Wilmanns, O. and Tüxen, R. (Red.): Werden und Vergehen von Pflanzengesellschaften (Ber. Intern. Symp. IVV Rinteln 1978, Tüxen, R. (Ed.)): 315–326. Cramer, Vaduz.

Kuoch, R. and Amiet, R. (1970) Die Verjüngung im Bereich der oberen Waldgrenze der Alpen. Mitt. Schweiz. Anstalt f.d. Forstl. Versuchswesen 46: 159–342.

Van Leeuwen, C.G. (1965) A Relation theoretical approach to pattern and process in vegetation. Wentia 15: 25–46.

Lohmeyer, W. and Bohn, U. (1973) Wildsträucher-Sproßkolonien (Polycormone) und ihre Bedeutung für die Vegetationsentwicklung auf brachgefallenem Grünland. Natur u. Landschaft 48: 75–79.

Lux, H. (1964) Die biologischen Grundlagen der Strandhaferpflanzung und Silbergrasansaat im Dünenbau. Angew. Pflanzensoziologie (Stolzenau/W.) 20: 5–53.

Mattes, H. (1982). Die Lebensgemeinschaft von Tannenhäher und Arve. Ber. Eidg. Anst. forstl. Versuchswesen 241: 3–74.

Mueller-Dombois, D. (1984) Zum Baumgruppensterben in pazifischen Inselwäldern. Phytocoenologia 12: 1–8.

Müller-Schneider, P. (1977) Verbreitungsbiologie (Diasporologie) der Blütenpflanzen. Veröff. Geobot. Inst. ETH Rübel Zürich 61: 226 S. 2. Aufl. Zürich.

Noble, J.C., Bell, A.D. and Harper, J.L. (1979) The population biology of plants with clonal growth. 1. The morphology and structural demography of *Carex arenaria*. J. Ecol. 67: 983–1008.

Van Noordwijk-Puijk, K., Beeftink, W.G. and Hogeweg, P. (1979) Vegetation development on salt-marsh flats after disappearance of the tidal factor. Vegetatio 39: 1–13.

Oberdorfer, E. (1983) Pflanzensoziologische Exkursionsflora. 5. Aufl. 1051 S. Ulmer, Stuttgart.

Petersen, P.M. (1981) Variation of the population structure of *Polygonum viviparum* L. in relation to certain environmental conditions. Meddr. Grønland, Biosci. 4: 19pp.

Van der Pijl, L. (1982) Principles of Dispersal in Higher Plants. 215pp. Springer, Berlin, Heidelberg, New York.

Roberts, H.A. and Dawkins, P.A. (1967) Effect of cultivation on the numbers of viable weed seeds in soil. Weed Res. 7: 290–301.

Salisbury (1942) The Reproductive Capacity of Plants. 244pp. Bell, London.

Sarukhán, J. and Harper, J.L. (1973) Studies on plant demography: *Ranunculus repens* L., *R. bulbosus* L. and *R. acris* L. Population flux and survivorship, J. Ecol. 61: 657–716.

Schmidt, W. (1981) Ungestörte und gelenkte Sukzession auf Brachäckern. Scripta Geobot. 15: 199 S. Goltze, Göttingen.

Schumacher, W. (1980) Schutz und Erhaltung gefährdeter Ackerwildkräuter durch Integration von landwirtschaftlicher Nutzung und Naturschutz. Natur u. Landschaft 55: 447–453.

Schwabe-Braun, Angelika (1980) Eine pflanzensoziologische Modelluntersuchung als Grundlage für Naturschutz und Planung. Weidfeld-Vegetation im Schwarzwald. Urbs et Regio 18: 212 S.

Schwabe-Braun, Angelika and Tüxen, R. (1980) Prodromus der europäischen Pflanzengesellschaften: Tüxen, R. (ed.), Lief 4; Lemnetea minoris. 141 S. Cramer, Vaduz.

Segal, S. (1969) Ecological Notes on Wall Vegetation. 325pp. Junk, The Hague.

De Smidt, J.T. (1977) Interaction of *Calluna vulgaris* and the heather beetle (*Lochmaea suturalis*). In: Tüxen, R. (Ed.): Vegetation und Fauna (Ber. Intern. Sympos. IVV Rinteln 1976), 179–186. Cramer, Vaduz.

Stalder, L., Potter, C.A. and Barben, E. (1977) Neue Erfahrungen mit integrierten Maßnahmen zur Bekämpfung der Acker- und Zaunwinde (*Convolvulus arvensis, C. sepium*) im Weinbau. Proc. EWRS Symp. Methods Weed Control and their Integration, Uppsala: 221–228.

Sukopp, H., Trautmann, W. and Korneck, D. (1978) Auswertung der Roten Liste gefährdeter Farn- und Blütenpflanzen in der Bundesrepublik Deutschland für den Arten- und Biotopschutz. Schriftenr. Veget. kd. 12: 138 S.

Sýkora, K.V. (1982) Syntaxonomy and synecology of the Lolio-Potentillion anserinae Tüxen 1947 in the Netherlands. Acta bot. Neerl. 31: 65–95.

Synge, H. (Ed.) (1981) The Biological Aspects of Rare Plant Conservation. 558pp. Wiley, Chichester.

Tripathi, R.S. and Harper, J.L. (1973) The comparative biology of *Agropyron repens* (L.) Beauv. and *A. caninum* (L.) Beauv. J. Ecol. 61: 353–368.
Tüxen, R. (Ed.) (1963) Bericht über das Internationale Symposion für Vegetationskartierung 1959 in Stolzenau/Weser. 500 S. Cramer, Weinheim.
Tüxen, R. (1965) Wesenszüge der Biozönose. In: Tüxen, R. (Ed.): Biosoziologie (Ber. Intern. Symp. Stolzenau/W. 1960), 10–13. Junk, Den Haag.
Tüxen, R. (1970) Pflanzensoziologie als synthetische Wissenschaft. Meded. Botan. Tuinen en Belmonte Arboretum Landbouwhogeschool Wageningen XII (1968) 1970: 141–159.
Tüxen, R. (1977) Diskussionsbemerkung zu Müller-Schneider. In: Tüxen, R. (Ed.): Vegetation und Fauna (Ber. Symp. IVV Rinteln 1976): 127–129. Vaduz.
Tüxen, R. (1979) Die Pflanzengesellschaften Nordwestdeutschlands. Lief. 2: Bidentetea tripartitae. 2. Aufl. 212 S. Cramer, Vaduz.
Tüxen, R. (1985) Unser Buchenwald im Jahreslauf. Im Druck.
Tüxen, R. and Brun-Hool, J. (1975) *Impatiens noli-tangere* – Verlichtungsgesellschaften. Mitt. Flor.-soz. Arbeitsgem. N.F. 18: 133–155.
Tüxen, R. and Ellenberg, H. (1937) Der systematische und der ökologische Gruppenwert. Mitt. Flor.-soz. Arbeitsgem. Nieders. 3: 171–183.
Watt, A.S. (1957) Pattern and process in the plant community. J. Ecol. 35: 1–22.
Watt, A.S. (1976) The ecological status of bracken. Bot. J. Linnean Soc. 73: 217–239.
Westhoff, V. (1979) Bedrohung und Erhaltung seltener Pflanzengesellschaften in den Niederlanden. In: Wilmanns, O. and Tüxen, R. (Red.): Werden und Vergehen von Pflanzengesellschaften (Ber. Symp. IVV Rinteln 1978): 285–313. Vaduz.
Wilmanns, O. (1983) Lianen in mitteleuropäischen Pflanzengesellschaften und ihre Einnischung. Tuexenia 3: 343–358.
Wilmanns, O. (1984) Ökologische Pflanzensoziologie. 3. Aufl. 372 S. Quelle and Meyer, Heidelberg.
Wilmanns, O., Wimmenauer, W., Fuchs, G., Rasbach, H. and K. (1977) Der Kaiserstuhl – Gesteine und Pflanzenwelt. 2. Aufl. 261 S. Landesanstalt f. Umweltschutz Baden-Württemberg, Karlsruhe.

THE CENSUS OF PLANTS IN VEGETATION*

JAMES WHITE

Plants stand still and wait to be counted.
Population Biology of Plants
John L. Harper (1977: 515)

ABSTRACT

The history of plant census in natural or semi-natural vegatation excluding forests is recounted. A few examples of single, non-repetitive censuses of grassland plants are known from the nineteenth century. Following the introduction of the permanent quadrat by F.E. Clements, the practice of repeated censuses of individually located plants became widespread by the 1920s. Intensive research was conducted in desert of semi-desert vegetation, mostly by rangeland scientists: some of these studies lasted for several decades. The various contributions of plant ecologists (other than range scientists) are recorded, ceasing with studies initiated before 1970, since when plant census has become a major element of vegetation science.

INTRODUCTION

The simultaneous recording at a particular time of demographic data on all the individuals living in a particular place constitutes a census. Censuses are most useful if they are made at regular intervals, since they reveal trends of population development over a period. A census is an essential prerequisite of population analysis, no less for plants than for human beings and animals. It lies at the heart of any enterprise to describe the population structure of vegetation.

In this essay I shall review the history of plant census in natural vegetation, insofar as I can bring together the many fragments of which it consists. Whereas plant enumeration has a long tradition (as I shall

*This paper is dedicated to John L. Harper, whose teaching has inspired much of the contemporary revival of plant census in vegatation.

White, J. (ed.), The Population Structure of Vegetation.
 ISBN 978-94-010-8927-2

demonstrate), only in the past decade has plant demography emerged as a coherent body of theory and practice, contributing significantly to vegetation science. I shall not investigate here the sociological reasons for either its former neglect or present popularity, but some hints will become apparent from the narrative. I would like to disavow at the outset that I regard this essay as merely historical. The importance of censuses grows with their repetition and I shall show that many early plant census records either continue to be kept or have been kept for very long periods. Plant demographers have access to a few extensive series of censuses (up to 70 years) which have hitherto been inadequately explored or appreciated by all but a handful of biologists. The importance of these data increases with their updating: I hope that by documenting them I shall contribute to their continued maintenance as irreplaceable information for a science of plant demography.

I have imposed two restrictions on myself in this essay. I have excluded forest trees. A few references to plant ecologists who have counted trees escape this restriction, for reasons evident later, but a review of tree census would greatly enlarge my account: there are many demographic data for trees but few have been systematically collated beyond preliminary statements (e.g., Harper and White 1974, Silvertown 1980). The census of trees may often be virtually synonymous with some aspects of forest management on which there is a very large and well known literature (e.g., Assmann 1970, Smith 1962). My second restriction is to limit my survey to studies initiated in the period before 1970, since by that date it appears to me that plant demography had become well established as a major element in contemporary ecology (Harper and White 1974). Thereafter the number of published papers on plant census seems to have grown exponentially, although I have not formally quantified its growth, as Schoener (1974) did for studies on resource partitioning among animals.

Some characteristic features of plant census

There are three types of census which I shall consider: (1) a single census never repeated and (2) repeated censuses of the same area, either (a) with or (b) without precise location and identification of individuals. Only type 2 provides data suitable for formal demographic analyses; type 2a conducted regularly for long periods provides information on survivorship, life expectancy, age structure, immigration and emigration rates, sex ratio (for dioecious species) and so on, and permits the estimation of future population trends under similar conditions. All three types of census may be found reported in the botanical literature, but I shall pay most attention to type 2a, that is, the repeated census of populations whose members are identified and can be precisely relocated over time. The relative paucity until recent times of published demographic data for plants compared to those for animals belies the ease of this protocol for plants compared to many animals.

There are, however, a number of elements in the formal definition of census given in the introduction which ought to be considered in a little more detail as they apply to plants: the nature of the demographic data that can be conveniently collected, the definition of 'individuals' and the usual meaning of 'a particular place'. There are some distinctions perhaps worth making about, on the one hand, plant census and on the other, animal and human censuses. Usually a census of human beings is exhaustive, in the sense that it seeks to record not merely numbers but also divers other demographic data on *all* the individuals in a large territory; excluding small countries, this is typically on a geographical scale of 10^{10}–10^{14} m^2. Though not without its difficulties (e.g., Petersen 1975), this is straightforward in principle. Animal censuses vary widely in their accuracy, comprehensiveness and sophistication. They depend largely on the problems of sampling in aerial, aquatic or terrestrial environments and on the size and mobility of the organisms: an immense literature testifies to the various means by which the technical difficulties are solved (e.g., Caughley 1977, Southwood 1978). But most plant censuses have certain features in common which demarcate them technically, if not necessarily conceptually, from those of animals and human beings.

From the evidence I shall review in the pages to follow, I can confidently assert that most plant censuses have been conducted on a small scale: since the introduction of the quadrat by Pound and Clements about 1900 this has tended to be on the almost uniformly small scale of 1 m^2. Except for forest trees (which I am not considering in this paper) there are relatively few examples of plant census on a scale much larger than 10 m^2: the most notable concern the enumeration of the saguaro cactus, discussed later. It seems to be generally true that a census of plants is regarded as a *sample* of the typical composition of the vegetation of a more extensive, contiguous area. (I acknowledge F.N. Egerton for making this point specifically to me.) Nonetheless, the phrase 'plant census' may be properly used, I believe, as long as the sampling conforms to the other features of the formal definition: the 'particular place' for plant census is, however, usually small (1–10 m^2), seldom exceeds 10^2 m^2 and rarely, even for trees, exceeds 10^4 m^2.

For both annual plants and trees there is, usually, little difficulty in defining another element of a plant census, the limits of genetically distinct individuals. The demarcation of denumerable individuals is, however, a major difficulty in conducting a census of many herbaceous perennial plants: the most basic requirement of a census is compromised unless some convention is adopted about what to count. Since perennial herbs comprise a dominant part of the landscape in those regions where vegetation science developed, this problem may have inhibited the earlier growth of plant demography. But as the nature of plant construction becomes better understood by plant ecologists (e.g., White 1979, 1984), we may expect that more sophisticated census techniques will be devised to cope with the problems of individuality, coloniality, and size plasticity in plant morphology.

Another feature of a plant census as traditionally conducted, if one may judge from the published examples, is that it typically involves enumeration alone, to the exclusion of other demographic data. Age structures of populations, for example, have been much more rarely recorded in censuses of plants than of animals. The increased sophistication of plant census in the past ten years is tending to blur this distinction, but there are very few examples in the period to 1970 where plant census is not simply synonymous with plant enumeration. Some reasons for this are readily apparent, I believe. Cohorts of annual plants may be followed through time from birth to death in a matter of months: repeated enumeration in this case provides copious demographic data directly, and indeed of a quality not easily obtained for many animal populations. The chief difficulty in plant census arises with the widespread occurrence of coloniality in plants, particularly in many perennial herbs and shrubs, where demographic data comparable to those commonly recorded in animals are often impossible to obtain. Since the main focus of plant census has been on such plants, as I shall show later, the *absence* of information on age structure has been a typical distinction between it and animal census. There are, of course, comparable problems with colonial animals, but these have not, until recently, received the attention devoted to insects, birds, fish, reptiles and mammals by animal demographers. Botanists have attempted to devise solutions to this traditional demographic difficulty of ageing perennial herbs: notable in this respect has been the development of the concept of age-states by Soviet botanists (Gatsuk et al. 1980, Rabotnov 1978b). Ages may be directly determined for temperate trees, but measures of size (such as stem diameter at breast height) have been much more frequently recorded by forest scientists. There are more substantial difficulties in estimating ages of trees in tropical regions (Bormann and Berlyn 1981). Excellent census data of various sorts are, of course, available for many tree populations, but I shall not review them here.

Plant census, then, has been primarily concerned with enumeration, often with no clear distinction between genetically different individuals. But insofar as enumeration of known individuals (however defined) in a population has been repeated, direct and unambiguous demographic information can be derived on longevity, life expectancy, (calendar) age or age-state structure, reproductive rates, patterns of recruitment and mortality, the rate and scale of migration, changes in sex ratio (of dioecious species) and so on. A comprehensive knowledge of all these parameters constitutes the basis for understanding the population structure of vegetation.

There is no obvious reason why plants have not played a larger part in the empirical foundations of demographic theory in biology, the problems of individuality or plasticity in morphological expression notwithstanding. This becomes even more apparent when one discovers the considerable wealth of census data compiled in the first quarter of this century by plant scientists. Much of it seems to have remained unpublished for many years, or published outside the mainstream of ecology and never known or soon forgotten by plant biologists. I shall try to bring to light some of it in the following pages.

PLANT CENSUS TO ABOUT 1900

Just as the record of human census is lost in tradition, we may assume that it will be difficult to recapture fully the history of plant census, except from printed sources. The modern science of human demography was inaugurated about 1662 by John Graunt in London. The first census was conducted in Sweden in 1751 and by the early nineteenth century human demography was well established in northern Europe and North America (e.g., Petersen 1975, Hutchinson 1978). Animal demography was well developed by the 1940s (e.g., Allee et al. 1949, Hutchinson 1978) and could draw on a long and even ancient tradition (Egerton 1968). From the scraps of information I have gleaned from the literature, I think it may be stated with some confidence at this stage that no formal science of plant census was initiated before the seminal work of Pound and Clements which they first reported in 1898: I shall discuss this in the next section.

An important source for the numerology of plants is Darwin's *On the Origin of Species* (1859), as Harper (1967) vividly illustrated, since he was among the first to count plants in natural vegetation, sometimes in defined and delimited areas small enough to make accurate and repeated records through a season. It becomes clearer from his *Natural Selection* (Stauffer 1975), which unlike *Origin* contains references, that Darwin could draw on earlier exemplars. The agricultural journals of the day contained several examples of plant census in crop experiments, which I shall not review here; we know that Darwin read these journals (Stauffer 1975). In *Natural Selection* he cites some of the observations of George Sinclair (1824) who sought to determine the botanical composition of natural grassland to enable him to devise suitable species mixtures and sowing densities for establishing new pastures. Table 1 reproduces his results, perhaps the earliest census of plants in semi-natural vegetation.

Table 1. The number of distinct plants and species found combined in a space of one foot square of the turf of natural and artificial pastures (Sinclair 1824)

Pasture Type	Species number	Number of distinct rooted plants per 1 foot^2 ($\equiv 0.09\,m^2$)	
		Grasses	Other Herbs
Richest natural pasture	20	940	60
Rich ancient pasture	–	1032	58
Ancient pasture	12	880	30
Ancient pasture	8	510	124
Water meadow	–	1702	96
Artificial pasture, 2 yr. old, formed of ryegrass and white clover	2	452	18
6 year old monocultures			
Poa augustifolia	1	192	
Alopecurus pratensis	1	80	
Lolium perenne	1	75	

Contemporaneous with Darwin's interest in this subject (e.g., Darwin 1855), the celebrated Park Grass Experiment was established at Rothamsted in 1856 by Lawes and Gilbert. It still continues, with annual records of the botanical composition of small plots under a diversity of fertilizer regimens. But plants were never counted in this experiment (Thurston et al. 1976). In a review of its early years Lawes et al. (1882) also called attention to Sinclair's work and mentioned (p. 1208) that of Oemler and Fuchs in Schleswig about 1870, who counted the number of plants growing in one foot square ($\equiv 0.09\,m^2$) of meadowland, listed by species from most to least numerous: 431 plants altogether. Further work of this kind is rare and sporadic and not part of any intellectual tradition obvious to me at present. Botanists rarely counted plants in the nineteenth century, it seems. Darwin is a striking, but not entirely singular exception. Hoffmann was counting plants in a competition experiment under semi-natural conditions at Giessen during the 1860s (Hoffmann 1865; see also a brief anonymous and untitled report in *The Gardener's Chronicle* for 14 May 1870). This appears to be among the earliest designed experiments on plant competition; but the history of such experiments is poorly known and has not been reviewed since the account given by Clements et al. (1929).

I have collated in Table 2 the examples I know so far of plant census in grasslands in Europe during the last century. I have extended the time limit a little to incorporate the closely related research of two English grassland scientists, since subsequent research of this type seems to have died out by about 1920, and results published in plant units (rather than relative frequency) seems to have ceased by 1912 (Stapledon 1912).

It is apparent from such data that the typical range of plant density in established grasslands is within the range 10^3–$10^4\,m^{-2}$. Ellenberg

Table 2. Census of plant in grasslands. Densities are expressed as 10^3 plants m^{-2} equivalent

Density		
6.8–11.2	four old pastures	Sinclair 1824
11.9	water meadow	Sinclair 1824
5.1	2 yr old pasture	Sinclair 1824
0.8–2.1	three 6 yr old monocultures of grasses	Sinclair 1824
8.0	fertile meadow	Schnizlein and Frickhinger 1848 in Ellenberg 1952
4.4	(not given)	Oemler and Fuchs (undated) in Lawes et al. 1882
5.4–20.2	six plots in alpine pastures at 730–1031 m, with 17–22 spp. per sample plot	Stebler and Schröter 1887
2.0–6.3	ten old pastures	Armstrong 1907
0.7–1.3	four new pastures	Armstrong 1907
3.7–6.6	four old pastures	Stapledon 1912
2.9, 3.6	two sown pastures, 17 and 20 yr old	Stapledon 1912

(1952: 17) cites a range (7–12) × $10^3 m^{-2}$ for grasslands generally, without specific evidence. But what were these grassland scientists counting? The indeterminacy of plant size and the difficulty of demarcating a denumerable individual have typically been regarded as stumbling blocks to plant census by plant ecologists (Greig-Smith 1964). Jenkin (1919), on the other hand, seemed to have had no problem in making a choice suitable to his purpose: 'for most plants, the plant unit shall consist of any portion of a plant possessing an independent root, which can be separated from others with at least three rootlets still attached to it'; for *Trifolium repens* it was 'any portion of runner up to three inches in length and bearing functioning rootlets'. However these were not necessarily the criteria adopted by other workers. A tiller was usually the denumerable unit in grasses it seems, without regard to whether or not it had roots. On the other hand, as indicated in Table 1, Sinclair counted distinct rooted plants. Consequently, the data in Table 2 are not necessarily comparable and provide no more than order-of-magnitude estimates of shoot density in grasslands. Indeed to the present day there are no reliable estimates of the number of genotypes of various species that can be accommodated on a square metre of natural or semi-natural grassland! One species alone (*Trifolium repens*) can be represented by no fewer than 50 distinct genotypes per m^2 (Trathan, quoted in Harper 1983) in an old grassland with over 30 species (Turkington and Harper 1979).

Jenkin's name is omitted from Table 2: although he conducted many censuses of grasslands in Wales, sometimes in collaboration with Stapledon (the doyen of English grassland scientists), his results were condensed for publication to what he referred to as 'percentage frequency' measures. This practice was also adopted by Stapledon (Stapledon and Jenkin 1916) and by subsequent workers: this has led us, I believe, to neglect their contribution to plant census hitherto. Other measures of abundance became more popular among agricultural scientists (Davies 1933), such as percentage cover or biomass (which had, of course, a long tradition at least back to Lawes et al. 1882), or frequency based on simple presence/absence data rather than on numbers of each species. It is however worth noting for the record the interest expressed by the English agricultural scientist Engledow (1926) in plant census, even though his attention was directed to a crop plant.

All these grassland censuses I have quoted were of a simple, non-repetitive type and could never in themselves have provided an adequate basis for demographic analyses of plant populations. It is interesting to record that, as a lecturer at the University College of North Wales, Bangor, Jenkin was the first to count plants at the College Farm, Aber, a location which was to become celebrated, half a century later, for plant census by a distinguished school associated with John Harper. But in the intervening period, the focus of plant census was elsewhere: repetitive censuses of plants were first conducted in the deserts and prairies of North America, not in the pastures of Europe.

PLANT CENSUS IN PERMANENT QUADRATS: THE INNOVATION OF POUND AND CLEMENTS

'In determining the abundance of species, appearances are extremely deceptive . . . closer analysis of the floral covering proved that the conclusions formed from looking at the prairie formations and from long field experience, without actual enumeration of individual plants, were largely erroneous' (Pound and Clements 1898). 'Actual field experience has shown that species which appear most prominent in the constitution of the prairies, even to the careful observer, are not necessarily the most abundant . . . To ensure accurate or even approximately accurate results, it is necessary to resort to some method of actual count' (Pound and Clements 1900).

The invention of the quadrat

The tradition of enumerating plants in plots ('quadrats', they came to be called) is due mainly to Pound and Clements in their botanical inventory of Nebraska in the 1890s. Whereas earlier botanists (such as Darwin) had counted plants in small areas or turfs, Pound and Clements advocated and practised the method to the extent that within a decade it had become a standard technique in the emerging sciences of plant and rangeland ecology. Their priority and influence has been widely acknowledged (e.g., Schroeter 1910, Tansley 1923, Lüdi 1930, du Rietz 1930) although Clements himself (1916: 425) noted earlier attempts by Sinclair, Darwin and others. Pound did not continue his researches on vegetation, devoted less and less time to them after 1898 and by 1907 had abandoned them completely in favour of a legal career: he later became a famous jurisprudent and Dean of Harvard Law School. Clements on the other hand soon became the most influential ecologist of his time, 'by far the greatest creator of the modern science of vegetation' (Tansley 1947), with a prodigious output of research. The credit for the promulgation of the permanent quadrat and plant census approach to vegetation may be confidently ascribed to Clements, while noting Pound's early contribution. Clements is nowadays regarded (and correctly in my opinion) as having a too-idealistic notion of vegetation development, but he has received considerably less than due credit as the creator of the science of plant demography. His innovation of repeated census of mapped individuals belies his reputation nowadays as a metaphysical theorist of vegetation. To redress the omission I shall give a somewhat detailed, though given his influence still rather cursory, account of his particular contributions to plant census.

Already by 1897 Pound and Clements had made 'a large number of enumerations of the individual plants . . . in plots 5 meters square' in Nebraska, although 'the method involves no little labour' (Pound and Clements 1898). These were all simple censuses and did not involve mapping. Tobey (1981) in an important historical and sociological analysis has meticulously traced the intellectual history which led to Pound and Clements' development of both census and quadrat techniques. He argues, convincingly I believe, that they reformulated plant geography on a new conceptual base: their novel methodology prised plant ecology from its geographical antecedents and thrust it as a new science into the twentieth century. Tobey (1981: 48) dates their 'leap into numerical quantification of ecology' to 1896, when both were members of C.E. Bessey's botanical group at the University of Nebraska and were engaged on writing *The Phytogeography of Nebraska.* The major influence appears to have been Pound's reading of Drude's *Deutschlands Pflanzengeographie* (1896) which he reviewed for *American Naturalist* in June 1896. Their 'profound epistemological shift' was, Tobey argues, a continuation of the intellectual movement initiated by Darwin to a statistical theory of natural selection, though neither was consciously aware of themselves as bringing to ecology the Darwinian approach. They were, it seems, unaware of Darwin's (1859) quadrat-like method, though Clements (1916) acknowledged it later.

Within a decade the quadrat approach to vegetation analysis was in full bloom and was consolidated with typical pedagogical thoroughness by Clements in the first American textbook of ecology, his widely influential *Research Methods in Plant Ecology* (1905, it is now back in print.) This was one of the finest first fruits of the new science of plant ecology, a fundamental break with nineteenth century naturalism: sense impressions of vegetation could not be trusted without accurate counting and other quantitative techniques. The new quadrat method provided the foundation for a dynamic approach to understanding vegetation change. And despite the more recondite features of Clements' subsequent theory of vegetation succession (notably its analogy with the ontogeny of an organism), it was founded on a dynamic population concept, quite Darwinian in its thrust: 'the most striking feature of succession lies in the movement of populations, the waves of invasion, which rise and fall through the habitat from initiation to climax' (Clements 1916: 3).

Research Methods in Plant Ecology was published at a propitious moment, a year after W.G. Smith and A.G. Tansley had initiated the committee that eventually (in 1913) became The British Ecological Society. Blackman and Tansley (1905) hailed its publication in a twenty-seven page review, Tansley condensed it into a pamplet for fieldworkers (Smith 1905) and later incorporated it into his *Practical Plant Ecology* (Tansley 1923). In a chapter which opens with a statement on 'the need of exact methods', Clements (1905) described the various types of quadrat record which might be employed. The simplest was the *list quadrat*, a list of the numbers of each species at one particular time. Certain rules of thumb were proposed for that traditional difficulty of plant census: the delimitation of plant individuals. He had a low opinion (p. 167) of 'mere number', without some ancillary indication of plant size. The *chart quadrat*, typically using a 1 metre-square plot, was used to map individual plants and the outline of clusters of shoots if discrete individuals were indistinguishable. Clements saw its purpose for comparing the developmental stages of various communities within the same vegetation formation. If either list or chart quadrats were recorded more than once they became *permanent quadrats*, but in practice only chart quadrats could meet the usual requirement of following identifiable individuals. Permanent quadrats were the key to understanding 'invasion or succession . . . the gradual effects of competition . . . minute changes as they are ocurring'. So simple a technique we now take for granted, but it was invented by Clements! Finally, he advocated the use of *denuded quadrats*, a permanent quadrat from which plants have been removed after being mapped. His strong experimental bias to ecological investigation saw in these the means to study the details of invasion and competition in vegetation.

The intellectual debt of present day plant demographers to Clements is scarcely acknowledged, perhaps because his advocacy for census was channelled mainly through rangeland scientists in the western United States rather than through the academic schools of plant ecology at Chicago, Cambridge or Montpellier. Tobey (1981) has admirably shown how influential he was among US grassland ecologists, chiefly through his student John Weaver. Clements taught at the University of Nebraska from 1898–1907, then at the University of Minnesota until 1917 where Weaver received his Ph.D. under him in 1916. Weaver then taught at Nebraska for the next 40 years where he promulgated Clements' ideas and became the most productive research worker on prairie grasslands (Tobey 1981). Weaver (1918) reviewed the progress to that date on the use of quadrats which indicated its widespread acceptance. Clements himself, we may assume, continued to use them from the 1890s, but it is not until he joined the Carnegie Institution of Washington in 1917 that we can follow his activities easily in the published record. We know, however, that during the summers of 1907 and 1908 he made frequent use of quadrats to count tree seedlings for his study of the life history of lodgepole pine forests (Clements 1910), the earliest detailed plant demographic study known to me. While this paper has neither the precision nor polish of modern accounts of plant population biology, it is remarkable for its time, with details of age structures of trees, ages of bushes and herbs, cone and seed production, seedling numbers and survivorship and effects of interspecific competition on seedlings. It remained unique for many years. Clements himself never published a comparable study again.

Clements and rangeland science

Clements joined the Carnegie Institution of Washington in 1917 and stayed in its employment until he retired in 1941. His activities are recorded annually in the Yearbooks of the Institution, the accounts of plant ecology under his direction swelling progressively from 4 pages in 1918 to 25 pages by 1921 and 37 pages by 1926. During his early years in the Institution we may glimpse his activities in permanent quadrat research, often in collaboration with rangeland ecologists. Clements found a ready audience among these scientists, not least because many of them were trained by Weaver in the Clementsian tradition and well prepared for employment in experiment stations and federal research agencies (Tobey 1981). Tobey has documented the collaborative research networks and coauthorships of the Clements–Weaver 'invisible college', which was extensive throughout the western United States.

The problem they faced is simply stated. By the 1880s ranges in the western states had been overstocked; the era of free and abundant grass and of great buffalo herds was past. Droughts and blizzards exacerbated the difficulties for grazing cattle. A botanical evaluation of the rangelands was initiated by the Department of Agriculture and conducted by F.V. Coville about 1898 and within the next ten years the science of range management was progressively established by extensive surveys of range conditions from Texas to Oregon (Chapline et al. 1944, Talbot and Cronemiller 1961). Tracts of land were set aside to study ways and means of improving worn-out ranges; a wide variety of experimental observations on exclosures was begun from about 1907 onwards (Chapline et al. 1944). Clements' *Research Methods in Plant Ecology* (1905) was the leading available pedagogic manual for analysing the vegetation.

In his first year at the Carnegie Institution Clements 'made a close scrutiny of grazing conditions' in Arizona, New Mexico, Colorado, Utah, Wyoming, Nebraska, North Dakota, Montana, Oregon and California; 'the grazing industry is in a critical condition throughout the entire West and immediate action in accordance with the scientific results is imperative' (Clements 1918). He mentions studies already initiated on ranges at Santa Rita (Arizona), Jornada (New Mexico), Mandan (North Dakota) and at the Utah Grazing Experiment Station. (All of these, as we shall see later, have provided significant data on plant census.) A year later he reported (Clements 1919) that considerable numbers of permanent quadrats were being installed and charted in several states, often in co-operation with range scientists, a theme repeated in successive years (Clements 1920, 1921, 1922, 1923, 1924a, b, 1925). His collaborators within the Carnegie Institution were his wife Edith and J.G.V. Loftfield. He appears to have supervised two series of permanent quadrats, one for the empirical study of succession in natural or semi-natural vegetation and another for grazing research; these are reported on separately for a number of years. He recognized (Clements 1919) that 'the unique value of each permanent quadrat increases with each year's change and the record of it, and it is proposed to summarize the results at intervals of 5 to 10 years'. He stated that in 1919 the grass quadrats at Santa Rita near Tucson, Arizona were charted for the first time; this was a co-operative project he had begun with the U.S. Forest Service, the U.S. Biological Survey and the University of Arizona in 1918 (Clements 1927). Though it has not hitherto been appreciated, some of these are the quadrats on which Canfield (1957) based his analysis of survivorship of 48,000 plants, a paper now celebrated among plant demographers since its 'rediscovery' by Williams (1970a) and reanalysis by Sarukhán and Harper (1973).

By 1920 the use of permanent quadrats had been extended by Clements to the majority of the plant associations of the west; about 200 in all were being recorded (Clements 1921). He had initiated transplant and seeded quadrats with Weaver in 1918 and these were expanded in 1919 to sites in Nebraska, Kansas and Colorado, and were recorded annually. Their diversity had increased greatly by 1925 to include studies on competition with Weaver and Hanson (Clements 1924a, b, 1925, 1926). Clements (1921) emphasized the need to identify each individual plant on the charts. An innovation had been introduced at Santa Rita in 1919 to which he drew attention: the use of the pantograph by Robert Hill to chart the quadrats (Hill 1920). Clements was now using this and making records on tracing sheets to allow superimposition of annual records (Clements 1922). A considerable number of new quadrats installed in 1922 were being 'adapted to life-history studies by means of charting at intervals of 2 weeks or a month during the growing season and at longer intervals between' (Clements 1923). By 1923 the work had been 'expanded until 300 quadrats are being charted at the end of each growing season or oftener, and almost as many more are charted at longer intervals. Many of these have been charted for 6 years at regular intervals. They have been tabulated during the year and the data are at present being compiled' (Clements 1924a). This refers to the work on natural vegetation alone, I think, exclusive of the grazing research quadrats at Santa Rita and elsewhere. Loftfield (1924) reported on his continuing work in Northern Arizona, where he had been investigating the effects of rodents on vegetation: 'the life histories of *Stipa commata, Agropyron smithii, Sporobolus cryptandrus, Bouteloua gracilis, B.eriopoda* and others, have been worked out in considerable detail'; the results had been summarized and would soon be published, we are informed. A more complete report of these particular investigations (Taylor and Loftfield 1924) did not include census data from the quadrats; Clements was acknowledged for his advice and assistance in organising the experiment and for 'making provisions for charting the vegetation'.

Where are these reports? Where are the summaries promised at 5 to 10 year intervals? They constitute some of the most comprehensive and sustained research on plant census ever undertaken, but as far as I know, have never been published. Clements had been with the Carnegie Institution in Tuscon, Arizona from 1917 to 1924, although he travelled widely and had a separate Alpine Laboratory in Colorado. (His wife later gave a highly personalized narrative of their travels together (Clements 1960) but it lacks a useful account of his research on quadrats.) In 1925 he founded a Coastal Laboratory near Santa Barbara, California and moved there. The diversity of his research by this time was amazing and permanent quadrat research was much less conspicuous than formerly: there is almost nothing about it in the Yearbooks of the Institution after 1925, beyond a few retrospective comments on the Santa Rita quadrats (Clements 1927, 1930). Again publication was promised: 'the first progress report on the exclosures and related problems is being prepared for eventual publication' (Clements 1930), referring to the first 12 years of the Santa Rita experiment. But that seems to be the last we hear of them until Canfield (1957) resuscitated some of them. In 1925 also Loftfield resigned according to the Yearbook and in 1926 the U.S. Department of Agriculture took over the mapping of his quadrats in N. Arizona (Clements 1926). The detailed history of all this research remains to be investigated, but it seems at first sight that Clements' research programme on permanent quadrat charting, in which he was evidently assisted greatly by Loftfield, ceased in 1925. The whereabouts of his mapping records and research summaries are not known to me. If they still exist, they would provide a fascinating record of Clements as a pioneer plant demographer and a treasury of plant census.

CENSUS OF DESERT PLANTS

The Carnegie Institution of Washington was probably the most important source of funding for plant ecology in the first thirty years of this century (McIntosh 1983). For a time it employed the best group of plant ecologists (in a broad sense) in the world, including Cannon, Clements, McDougal and Shreve. It published many of the early classics of the new science, such as Clements' massive *Plant Succession* (1916), and maintained a diversity of botanical research laboratories. Almost as soon as it was established in 1902 a group of botanical advisors (chaired by Coville) recommended the establishment of a desert research laboratory (Cittadino 1980). This was constructed the following year on Tumamoc Hill, near Tucson, Arizona and still exists (McGinnies 1981). It soon became the focus for outstanding botanical research on deserts, as McGinnies (1981) has shown. And this included plant census. Clements appears to have been only casually associated with it although he lived in Tucson and wrote his annual reports on ecology from 1917 to 1924 from Tucson. McDougal provided laboratory facilities for him (J.E. Bowers, personal communication) and he did experimental work on water relations of plants there; but he conducted extensive researches elsewhere and is not considered to have been regularly associated with the Desert Laboratory (W.G. McGinnies, personal communication).

The first census was conducted on a large scale under the supervision of V.M. Spalding after he joined the Desert Botanical Laboratory in 1905. The most notable species in the vicinity was the giant saguaro cactus, *Carnegiea gigantea* and this was mapped plant by plant over an area of about 700 ha on a scale of 1 : 6000. Thousands of plants are shown on the published map, the earliest example known to me of a detailed plant census on this scale (Spalding 1909). The more generalized distributions of other species were also reported, with indications of their relative density over the area, but precise data were only published for *Cercidium torreyanum* (now *C. floridum*).

In 1964 and again in 1970 J.R. Hastings and R.M. Turner remapped (on aerial photographs) and remeasured all saguaros on four plots of c. 11 ha. each, facing the four cardinal points on the slopes of Tumamoc Hill. These overlap part of the extensive area mapped by Spalding. From direct observations of individual growth rates over time, a height-age relationship may be calculated for saguaro (Hastings and Alcorn 1961) and using such a conversion factor, Turner has been able to estimate the age structure of saguaros on these four plots. In addition, he has remapped (in

1964) all the surviving individuals of *Cercidium floridum* originally surveyed by Spalding, and found a marked decline in population since 1908. (I am indebted to R.M. Turner for information on the modern census of plants at Tumamoc Hill.)

In 1906 Spalding established and mapped nineteen 10 × 10 m^2 quadrats near the laboratory to enumerate the perennial plants (Spalding 1909), but he did not repeat the census because he retired soon afterwards, incapacitated by illness. The plots received further attention from Forrest Shreve who joined the Laboratory in 1908 and remained with it until the Carnegie Institution withdrew its support in 1940. The census approach begun by Spalding was extended by Shreve. At long intervals (1910, 1928 and 1936) he remapped five of the original 100 m^2 plots and laid out two new areas, known as A and B. Area A was 557 m^2, a slightly irregular rectangle, and established in 1910; plants were counted but never mapped (Shreve 1917). Area B consisted of 8 contiguous 10 × 10 m^2 plots and was established in 1928; plants were mapped and counted. The history of all these permanent plots was summarized to 1936 by Shreve and Hinckley (1937).

From the outset Shreve was interested in plant demography, as one of his first papers indicates: this was a study of seedling survivorship of paloverde (*Parkinsonia microphylla*, now *Cercidium microphyllum*) (Shreve 1911). By 1917 he had sufficient data from his plot (Area A) on Tumamoc Hill to present a table on the germination and survivorship of annual cohorts of this species. Although they constitute the first survivorship curves for a plant species known to me (Table 3) they have been hitherto neglected by plant demographers. Of 1370 germinated seedlings in the preceding eight years only 19 survived in 1917 (Shreve 1929). Census details of the perennials on some of Spalding's plots (nos. 11, 12, 15, 16 and 17) were given by Shreve, following his remeasurements in 1910 and 1928 (Shreve 1929): deaths and additions are noted. By this time he was coming to appreciate the longevity and dynamics of the desert shrubs and cacti; this was given its final expression after his census of 1936 (Shreve and Hinckley 1937). Over the years the five Spalding areas he mapped had

Table 3. Survivorship of cohorts of *Parkinsonia microphylla* which germinated on plot A (557 m^2) at Tumamoc Hill near Tucson, Arizona from 1910 to 1917. The numbers shown for 1909 include survivors from previous years. From Shreve 1917.

Year	1909	1910	1911	1912	1913	1914	1915	1916	1917
1909	303	189	19	16	7	6	6	6	6
1910		542	62	35	16	4	3	3	2
1911			122	49	6	1	0	0	0
1912				151	24	3	3	3	2
1913					34	5	2	2	2
1914						7	2	2	2
1915							0	0	0
1916								29	5
1917									38

increased their total plant populations from 530 to 1401 individuals, an increase which was due primarily to the absence of grazing since 1907 around the Desert Laboratory. The average length of life of the six large perennial species on the plots he calculated to be about 90 years, based on his census data. These estimates are no more than inferences, I should add, since Shreve gives no evidence of applying formal demographic theory to his data. Nonetheless they are remarkable for their time. The continuing records from these sites should provide further information, but no details have been published since 1937. The census history of the mapped plots to the present time is given in Table 4 (Martin and Turner 1977; R.M. Turner, personal communication): they are the longest observed permanent plots on which plants have been censused. The numbers of saguaros on plot 15 from 1906 to 1977 have been published by Martin and Turner (1977) and a more comprehensive report is in preparation (D.E. Goldberg and R.M. Turner, personal communication).

The saguaro cactus, mapped so thoroughly around Tumamoc Hill since 1906, has perhaps attracted more attention than any other single species from plant census takers. There is another remarkable series of records from the Saguaro National Monument near Tucson, Arizona which has remained poorly known to plant demographers. During four months from November 1941 to March 1942 12,898 saguaros were censused and mapped on an area of 640 acres (≡ 259 ha) by two plant pathologists, L.S. Gill and P.C. Lightle. This was part of an investigation of the occurrence of bacterial necrosis in this huge population, following extensive deaths of saguaros in 1939 and 1940 (Steenbergh and Lowe 1977). Each plant was examined annually until 1945 and thereafter annual records have been kept until the present time (except in 1949 and 1953) on plants in only 6 of the original 64 ten-acre plots. Mortality by height class of plants and by

Table 4. Dates of mapping and census on permanent plots at the Desert Botanical Laboratory, Tumamoc Hill near Tucson, Arizona (post-1936 dates courtesy of R.M. Turner, personal communication).

Plot No.	1906	1910	1928/9	1936	1948	1957	1959	1960	1967/8/9	1974/5	1978
4*	+								+		+
7	+								+		+
9	+								+		+
10	+								+		+
11	+	+	+	+	+			+	+	+	+
12	+		+	+	+			+	+	+	+
14	+								+		+
15	+	+	+	+	+		+		+	+	+
16	+		+	+	+	+	+		+	+	+
17	+		+	+	+	(destroyed)					
B			+	+	+	+			+		+

*Of Spalding's original 19 plots nos. 1, 2, 3 and 5 were destroyed and not recensused; nos. 6, 8, 13, 18 and 19 have not been relocated.

causal agent was carefully documented. The original detailed reports from 1941 to 1951 have been reprinted as a lengthy appendix to the study on the demography of saguaro by Steenbergh and Lowe (1983). An interim account was given by Alcorn and May (1962) on the subpopulation on the 6 ten acre plots. Between 1942 and 1961 29% of the 1475 plants died, most of them over 4 m. high: the depletion pattern was linear on a linear plot, from which the authors projected the complete destruction of the population in a further thirty-five years, if the same rate of decline continued. This is one of the few examples of population projection, albeit somewhat crude, for a plant which is based on long-term census data. Examples of the decline in this particular population are shown photographically by Hastings and Turner (1965: Plate 61).

Steenbergh and Lowe (1983) report further demographic details of some of these plots to 1975, and the records continue to be updated (R.M. Turner, personal communication). It is now believed that bacterial necrosis is not a primary cause of saguaro deaths, but is, in adult individuals, a decomposition process which results from damage to living tissues, primarily by freezing temperatures (Steenbergh and Lowe 1983). Nonetheless, the pathological symptoms stimulated the undertaking of the very detailed census originally. The various threats to the saguaros have prompted several studies of their life history, some of which have been published (e.g., Niering et al. 1963, Turner et al. 1969, Steenbergh and Lowe 1977, 1983) and yet others remain unpublished. J.R. Hastings and R.M. Turner established several permanent plots about 1960 to chart the growth and mortality of populations as part of a long term study: some 2,000 individuals are observed at irregular intervals. An almost eighty-year old record of demographic research on saguaro has provided us with one of the most comprehensively studied examples of the population biology of any plant species, but curiously it has remained almost neglected by plant demographers.

Desert plants are relatively easy to count even from photographs and there is a long tradition of doing so by this means, as Hastings and Turner (1965) have shown in their splendid photographic analysis. Steenbergh and Lowe (1983) give some examples for saguaro. Martin and Turner (1977) have provided several further examples: changes in abundance of *Haplopappus tenuisectus* (burroweed) on the Santa Rita Range from 1922 to 1975 are sufficiently recorded photographically to provide a useful account of its long-term dynamics, which appear to be governed by cool-season rainfall; comparable photographic data are shown also by them for *Opuntia fulgida* from 1905 to 1975 and for *Prosopis juliflora* from 1903 to 1975. The use of photographs for monitoring changes in perennial vegetation became widespread in rangeland ecology from an early date and insofar as individual plants can be recognized they constitute important evidence of long term trends in plant numbers. Photography was recommended by Clements as a normal part of recording permanent quadrats.

PLANT CENSUS ON NORTH AMERICAN RANGELANDS

In a lengthy report prepared for the United States Senate in 1936 on the quality of grazing lands in the western states, the authors of the chapter entitled 'The White Man's Toll' noted that their conclusions were based on the observations of 'more than 100 Forest Service officials, skilled in judging range conditions . . . through periodic reexamination over many years of 6300 permanently marked sample plots' (United States Senate 1936). In the extensive prairies and semi-desert grasslands in the western half of the United States the census of plant populations on permanent quadrats was practised on a very large scale during the period 1910–1950. The tradition which fostered it was associated with the schools of ecology and range science centered in the western Universities, notably Nebraska (Ares and Campbell 1974: 23). 'Nebraska graduate students were well prepared by their training in the Clementsian microparadigm for employment in experiment stations and in federal research agencies, such as the Forest Service . . . and the Department of Agriculture' (Tobey 1981). The 'Clementsian microparadigm' to which Tobey refers was the approach to the study of plant-habitat interrelations which Clements had advocated in numerous publications, such as *Plant Succession* (1916). The techniques had been clearly set out in *Research Methods in Plant Ecology* (1905) and subsequently elaborated. They included, as I have shown earlier, a strong emphasis on the use of permanent quadrats and quantitative measures of plant abundance, such as census. It is a small wonder then that over the years a huge fund of information on plant census in permanent quadrats was accumulated. But much of it remains to this day unpublished and is only gradually being rediscovered by contemporary plant demographers. In this section I shall call attention to some of it, insofar as I have gleaned it from the published record: though meagre compared with what evidently remains unpublished, this was the most comprehensive (but neglected) source of information on plant population biology until recent times.

The general tenor of the report prepared for the U.S. Senate in 1936 was that there had been serious depletion in range quality in the previous few decades: ranges had undergone about a 50% reduction in grazing capacity for domestic livestock compared to their virgin condition a half century or more earlier. The severity of the depletion increased from northern to southern states. This had been caused by excessive stocking of animals and almost unrestricted grazing. The unregulated use of ranges had been a growing problem since about 1880. Following an investigation by F.V. Coville a programme of range management and research was initiated about 1900 and the first decade of the century saw the genesis of a new science of range management (Chapline et al. 1944, Talbot and Cronemiller 1961). Experimental reserves on which grazing was restricted were established. Among the first of these was the Santa Rita Range Reserve in southern Arizona, which was fenced out of the public domain in 1903 and set up by Presidential proclamation in 1910. In 1907 a series of studies was begun by the Forest Service in co-operation with the Department of Agriculture to determine the grazing capacity of land within the national forests of the western states. Over 500 reseeding experiments with cultivated forage plants were begun in 1907, for example (Coville in Sampson 1913a).

An intrinsic and quite explicit part of these studies was to seek an understanding of the growth habits and life cycles of important range plants. Coville selected one of Clements' students at Nebraska, A.W. Sampson, to carry out such an investigation on the sheep ranges of the Wallowa National Forest in northeastern Oregon (Coville in Sampson 1908). In the same summer (1907) as Clements was studying the life history of lodgepole pine, Sampson began charting permanent quadrats at Wallowa, paying special attention to seedling numbers on about 300 metre-square quadrats (Sampson 1914). 'Special attention was directed to the time and conditions under which the seeds germinate in the spring, when the flower stalks are sent up and when the seeds are matured and disseminated . . . it was considered of high importance to secure accurate and reliable information on the rate at which plants are invading overgrazed areas and becoming established upon them' (Sampson 1908). By following the fate of seedlings for several years Sampson determined the survivorship and age of first reproduction for several species (Sampson 1914). Officially entitled 'Expert in Plant Ecology' in 1907, and later 'Plant Ecologist', he must have been one of the earliest botanists to have such a professsional appellation, a sign of the impact of the developing science. He believed that the growth requirements of range plants could 'best be determined by a study of individual species throughout their life-cycle'. But his papers contain few quantitative demographic data, unfortunately, and leave one with an impression that he became more interested in morphological generalities than in precise details of life-cycle parameters (e.g., Sampson 1913b, 1917, 1919, 1924, 1926). One may at least observe that he did not report the details: maybe they were accumulated and in the absence of any formal theory of plant demography were drawn upon to make those more general statements on grazing tolerance which were required by the cattlemen to whom they were primarily directed. Sampson later became one of the leaders of the new science of range management (e.g., Sampson 1952) but was also, clearly, a pioneer plant demographer.

An expanding organization of research led to the establishment of numerous experimental areas throughout the western States in the following decade (Chapline et al. 1944). As I have indicated earlier, Clements was a participant in this research until about 1924.

Santa Rita Experimental Range

This range was fenced and cattle removed in 1903. Grazing was permitted again as grazing studies began about 1912 and several exclosures were established on the grazed areas between 1916 and 1920. Photographs were taken at regular intervals to monitor changes in the vegetation (Martin and Reynolds 1973, Turner et al. 1980). A large series of metre-square quadrats was set up on some of the 235 study areas located throughout the range and they probably still exist (W.H. Kruse, Santa Rita, personal communication 1983). Charting was initiated by R.R. Hill in 1918, in association with Clements (Clements 1920). (Hill (1917) had earlier shown his interest in plant census by reporting the causes of mortality of trees.) Eighty quadrats were mapped each year according to Hill (1920) who also stated that as many more were to be established. On these plots, Hill introduced the pantograph to plant ecology. Data were recorded annually until 1935 when the last charting was done. It is not clear who carried out the recording: Canfield (1957) mentions Hill and M.J. Culley; it was not done by Canfield himself, since he did not move to Tucson until 1935 from the Jornada range in New Mexico (Ares and Campbell 1974). Some were examined in later years by researchers working on separate, independent studies, but there is no record of any formal publications. The basic field data are still available, along with 80 years of other data on the exclosures.

Hill appears to have published nothing further on the quadrats. A report on their floristic composition appeared (Canfield 1948) before Canfield's now celebrated paper containing records on the life-span of 48,000 individual plants (Canfield 1957). This paper remained in obscurity to plant demographers until it was rediscovered by Williams (1970a) and used by Sarukhán and Harper (1973), thus forging a remarkable link in a research tradition of plant census back to Clements in 1918. Canfield's paper is less than satisfactory for a full understanding of the extent of quadrat research at Santa Rita, because he did not indicate precisely the scope of his database: 'meter-square quadrats were established at numerous sites on rangelands subject to yearlong grazing'. The data he reported cover the 17-year 'chartograph' record, from which he was able to determine the life spans of individual plants and survivorship of cohorts. But his account is clearly only preliminary and the original data would doubtless provide much new information to a demographer. Canfield (1948) had previously indicated that the composition of mesquite grassland under heavy grazing was observed on 127 metre-square quadrats charted annually from 1915 to 1935, and that annual chartings were done for the same period on foothill type ranges, without stating the numbers of quadrats involved; presumably Hill's (1920) intention to establish further quadrats had been fulfilled.

The Santa Rita Range has been the focus of several studies on the demography of shrubs. Brown (1950) reported changes in shrub numbers from censuses of mapped plants in three one-hectare blocks (variously protected from grazing) in 1931, 1940 and 1949. Shrub numbers increased from 1305 to 2443 under open grazing, from 651 to 1343 under protection from cattle grazing and from 616 to 930 under complete protection from cattle and rodent grazing, during the 18-year period. Two of the commonest shrubs involved were mesquite (*Prosopis juliflora*) and burroweed (*Haplopappus tenuisectus*), both of which were the subject of more detailed investigations. Parker and Martin (1952) reported the results of observations on the numbers of mesquite plants from 1932 to 1949. The influence of increasing abundance of mesquite on perennial grasses was recorded in large numbers of permanent quadrats, but only ground cover data were collected, it seems. Glendening (1952) also recorded census data of mesquite and other desert shrubs and cacti over a 17-year period, but used only two censuses, 1932 and 1949. Further details of the demography of mesquite, including the survivorship of seedlings in permanent quadrats subjected to various treatments, are reported by Glendening and Paulsen (1955). Tschirley and Martin (1961) reported the dynamics of burroweed on permanent plots using photographic records, from 1922 to 1958, and these have been updated to 1975 by Martin and Turner (1977). Survivorship of grasses associated with burroweed was monitored for four years (Tschirley and Martin 1961).

More recently Cable (1979) has recorded the depletion of 250 marked plants of *Trichachne california* (one of the grasses dealt with by Canfield in 1957) at Santa Rita from 1961 to 1975: after 15 years 31 of 200 grazed plants were still alive, including one of the largest and one of the smallest plants originally tagged; 17 of 50 plants protected from grazing were still alive.

Reading these studies on the demography of plant species from the Santa Rita Experimental Range, one is continually reminded that in southern Arizona at least Clements' advocacy of a numerate vegetation science had found a receptive audience. Many of these range scientists were plant demographers in all but name: they seem to have taken it for granted that a proper understanding of plant life history was the secure foundation for scientific range management. Their approach to the analysis of vegetation provides a remarkable constrast to that of more academic plant ecologists at about the same time, as I shall show later.

Jornada Experimental Range

The Jornada Experimental Range near Las Cruces in New Mexico was established in 1912 and was operated until 1915 by the Bureau of Plant Industry, when its administration passed to the Forest Service. Its history has been vividly recounted by Ares and Campbell (1974), and contains among its many illustrations one of Campbell himself and R.H. Canfield in an 'air-conditioned' Model-T Ford car 'en route to field, charting quadrats' in 1928. Charting quadrats began at Jornada under the direction of W.R. Chapline in 1915 and continued for many years as an important part of its research activity. About 90 quadrats have been mapped and censused for several decades and a slightly greater number have been observed for periods of 5–30 years. With the exception of a few years between 1954 and 1967, the records on the long-term quadrats had been maintained annually since 1915 and were last made in 1968 (Wright 1972), a remarkable history of plant census, unrivalled so far to my knowledge for its frequency and duration.

The Jornada range is an extensive area (7.3×10^4 ha) of semi-desert rangeland; the history of vegetation changes on it in modern times has been comprehensively summarised by Buffington and Herbel (1965). More detailed changes on mapped belt transects over the period 1935 to 1980 have been described by Hennessey et al. (1983). The open nature of the vegetation facilitated the accurate mapping and census of various bunch (tussock) grasses, of which by far the most important species for animal grazing were *Bouteloua eriopoda* (black grama) and *Hilaria mutica* (tobosa grass). The life history of black grama was the focus of the earliest investigations on numerous permanent quadrats, subject to various degrees of protection from grazing. The first thirteen years of census data (to 1927) on 38 metre-square quadrats were reported by Nelson (1934), a valuable account of the morphology and population dynamics of black grama. Though the paper lacks formal demographic concepts, it shows clearly the sophistication achieved in plant life-history studies by 1930 among range scientists: the relative importance of revegetation by seed and by vegetative propagation (stolons) were carefully investigated for example, and have scarcely, in my opinion, been emulated to the present day. This paper is a neglected minor classic of plant demography. It was, in fact, completed (anonymously) by W.R. Chapline, since Nelson had left Jornada in 1924 (Ares and Campbell 1974).

About this time Campbell (1931) published information (based on permanent quadrat censuses) on *Scleropogon brevifolius* and *Hilaria mutica.* We also learn from him that Hill's pantograph technique was introduced to the Jornada range in 1925. Further census studies on black grama and tobosa grass followed: for example, Canfield (1939) reported 11 years' census data of plants under various clipping treatments. Long-term trends (1916–1953) in the performance of black grama on quadrats were summarized by Paulsen and Ares (1961).

The recent resuscitation of the permanent quadrat census records is due to C.H. Herbel, who began a comprehensive research programme on the Jornada range in 1956 (Ares and Campbell 1974). The first fruit of this was the synthesis of the records by Dittberner (1971) under Herbel's direction: for the first time the data from 53 years of charted quadrats were treated by modern demographic techniques. The results of this thesis mostly remain unpublished, unfortunately, since they are based on uniquely detailed and long-term census data. About 80 quadrat maps were photographed, scanned electronically and the co-ordinates of each plant recorded on magnetic tape for computer analysis (Dittberner 1971, Wright 1972). Life-tables were constructed for 17 species of grasses and other herbs. Maximum lifespans of grasses ranged from 11 years for *Aristida divaricata* to about 25 years for *Bouteloua eriopoda*; the longest lifespan of a forb (non-grass herb) was 15 years, for *Croton corymbulosus.* (I am grateful to P.L. Dittberner for permission to cite his thesis.) Some data on mean lifespan (e_x at birth) were reported by Herbel et al. (1970), but most of these were revised by Dittberner (1971), which should be regarded as the definitive analysis. Grasses had mean lifespans ranging from 1.7 years for *Sporobolus flexuosus* to 2.2 years for *Bouteloua eriopoda* (Wright 1972). Survivorship curves were typically Deevey Type III (Deevey 1947), and mean lifespans varied from 8–29% of maximum lifespans (Wright 1972). The need for long-term census records to calculate these demographic parameters is manifest: they exist for very few plant species.

Further information on the life history of seven perennial grasses using census data from 35 of the long term permanent quadrats has been reported by Wright and van Dyne (1976). Age-specific survival rates increased, on average for all species, from 0.39 at age 1 to a maximum of 0.68 at age 7 and declined subsequently to 0.48 at age 12; the risk of death was related to the availability of water, the period between 3 and 5 year old being particularly vulnerable. These conclusions were based on observations of 12,437 plants. The influence of the shrub *Prosopis juliflora* (mesquite, a vigorous invader of semi-desert grassland; Parker and Martin 1952), on two important range grasses was investigated by Wright and van Dyne (1981), again drawing on the long-term chart records at Jornada. From the census data they computed the stable age structure for *Bouteloua eriopoda*: it takes about 15 years to achieve for all but the older age classes. They could find no stable age structure for *Sporobolus flexuosus*, because of its shorter life-span and oscillatory pattern of establishment. The influence of mesquite invasion on the structure of black grama populations was simulated by altering the age-specific survival rates, since it is known that mesquite affects water availability, a key factor in grass survival. So far, however, 'only a portion of the vast amount of data available on perennial grasses on the Jornada Experimental Range has been used' (R.G. Wright, personal communication).

Other U.S. range research using plant census on permanent quadrats

The extensive research on permanent quadrats at Santa Rita and Jornada was by no means untypical of range science practice elsewhere in the western United States. These were the two principal range research areas of the Forest Service, which administered both of them from 1915 onwards. (This did not necessarily mean that they had forests; such a nice distinction had little bureaucratic significance!)

Some of the 6,300 permanent plots referred to in the U.S. Senate report of 1936 (and these were from the Forest Service alone) probably came from these two ranges, but I have not seen any breakdown of their origins. R.S. Campbell (in Ares and Campbell 1974: 48) refers to some 20,000 field plots, which I presume is the total number on rangelands whether or not under Forest Service control. The precise scale of charting and plant census on such field plots remains to be evaluated, but there are some indications that it is very large indeed.

Among the other range research stations from which plant census data have been recorded are:

1. Utah (later Great Basin) Experiment Station, established in 1912,
2. Northern Great Plains Field Station, Mandan, North Dakota, established in 1915,
3. Sheep Experiment Station, Dubois, Idaho, established in 1917,
4. Desert Experimental Range, Utah, established in 1933.

The valuable history written by Chapline et al. (1944) mentions several others. I have not investigated the literature published from these places in much detail and here only place some preliminary notes on record.

A.W. Sampson, referred to earlier, was in charge of the Great Basin

Experiment Station until 1922 and was presumably responsible for some of the census studies initiated there: his first major textbook (Sampson 1923) gives an example of a chart record made at the station and illustrates a pantograph in use on a chart plot. Already by 1923 he hinted that a 'distinct disadvantage in the extensive use of chart plots is the great amount of tedious work involved in the mapping and crystallization of the data recorded. For this reason the regular chart plot is less popular than formerly, for certain other short-cut methods have been developed in grazing studies which show well the changes in vegetation'. This signalled the beginning of the end of detailed plant census on American rangelands, at a time when, as we shall see, it was virtually unknown to plant ecologists! But it was not until the Second World War that the shortage of cheap labour caused its virtual demise; meanwhile many valuable long-term census records continued to be kept.

Mapping of permanent quadrats began in 1915 at Mandan, where 'liberal use was made of square metre list quadrats' (Sarvis 1941). J.T. Sarvis, who was in charge of agronomic investigations at Mandan from 1915 to 1941, reported some of the early observations and later results to 1935 (Sarvis 1920, 1923, 1941), though it is clear from his papers that only a small fraction of the census data which he collected was published. The new Hill pantograph was tested at Mandan in 1918 (Sarvis 1923), probably introduced by Clements. We may recall that in Clements' first report to the Carnegie Institution (Clements 1918) he stated that he had made a close scruitiny of grazing conditions throughout the western states: undoubtedly his extensive peripatetic researches fostered a widespread application of his theories and techniques among range scientists. He had many co-operative projects with Government research scientists (e.g., Clements 1927).

The example of Sarvis at Mandan was the inspiration for similar work at the Dominion Range Experiment Station in Alberta, Canada: the plants in large numbers of permanent quadrats were charted and censused from 1928; the results to 1939 were reported by Clarke et al. (1943).

I have not found any published records of early permanent quadrat research at the Sheep Experiment Station at Dubois, Idaho. Blaisdell (1958) reported extensive data on numerous grass species for 23 years, from 1932 to 1954. West et al. (1979) have used further long-term pantograph charts, mapped from 1930 to 1956 and again by themselves in 1973 to determine longevities of five perennial grasses and three shrubs: mean lifespans of grasses varied from 2.3 to 3.7 years, but the maximum longevities varied from 19 to 43 years. Generally speaking, grazing increased grass longevity.

Some long-term census data from the Desert Experimental Range have been summarized by Norton (1978) and West (1979). 128 plots each 9.3 m^2 in area were charted at irregular intervals from 1935 to 1970, and despite substantial gaps in the records, cohorts of seedlings that were established in 1935–37 were followed until 1968–70: relatively few individuals had died since the second year after establishment. West (1979) also cites unpublished theses which have analysed more of these data.

The bibliography prepared by Renner et al. (1938) gives a convenient insight into the use of quadrats on western rangelands, but a comprehensive account seems never to have been compiled. Other examples of plant censusing I have come across from this period are those of Ellison and Woolfolk (1937), Williams and Post (1945) and Ellison (1949), but I have no doubt that more exist and that my review is incomplete on this topic.

The early investigators at such research stations occasionally introduced technical innovations for charting the quadrats (e.g., Pearse 1935, Pearse et al. 1935). The use of tracing overlays was re-introduced by Ellison (1942) to follow the fate of individual plants more easily; their earlier use in 1921 by Clements (1922) was not mentioned. The technique continues to be valuable (Sarukhán and Harper 1973).

By the 1930s the assessment of the practical use of charting quadrats, foreshadowed by Sampson (1923), was in full swing. The decade began with the evaluation by Hanson and Love (1930), based on their four years' study of hundreds of quadrats in Colorado: their conclusions emphasised the need for a careful choice of methodology for well-specified research purposes, with counting being suitable only in certain circumstances. But a more radical critique followed: arguing that the charted quadrat was too time-consuming to be widely used for surveys, Stewart and Hutchings (1936) introduced the point-observation-plot (square-foot density) method. Plant abundance was estimated visually for each species in terms of the number of square feet of cover per $100\,\text{ft}^2$ ($\equiv 9.3\,\text{m}^2$) circular plot. This parameter was termed 'square-foot density', or simply 'density'. The method rapidly became popular and the use of the word 'density' from then on typically did *not* involve plant enumeration among range scientists (e.g., Stewart et al. 1940, Lang 1945, Moore and Reid 1951, Johnson 1953). Stewart and Hutchings (1936) were among the first to introduce randomization and replication of sample plots for range survey and the growing awareness of the desirability for methods of statistical adequacy during the 1930s seems to have accelerated the demise of tedious plant census on relatively few sample sites. Other sampling methods, such as the line intercept method, were introduced about this time (e.g., Sampson 1952).

Weaver's research on prairie grasslands

Quite the most productive grassland ecologist during the period 1920–1950 was J.E. Weaver at the University of Nebraska (Tobey 1981). He was Clements' most distinguished student and prolonged the influence of his teacher on the training of range scientists in the western United States until about 1950 (Tobey 1981). He was an early enthusiast of plant census in permanent quadrats (Weaver 1918). (Among his many illustrations of the value of census records was one which showed the inverse relationship between declining plant density and increasing size of survivors (Weaver 1918).) Weaver continued to use permanent quadrats for nearly 40 years, but a full account of his very extensive researches on them is not possible here. He himself summarized his work on prairie grasslands

(Weaver 1954a, 1968, Weaver and Albertson 1956), but only a detailed examination of his many papers reveals the scope of his permanent quadrat studies (e.g. Weaver 1924, Weaver and Hanson 1941). His conclusions about the effects of drought on plants in one paper alone were based on 160 quadrats charted for three years (Weaver and Albertson 1936). Sometimes he censused quadrats without charting them, it appears: annual census data for seven years on 43 quadrats in pastures once dominated by *Andropogon* spp. and *Poa pratensis* were used to determine successional changes in relative abundances of species (Weaver and Bruner 1945). A further 10 years' census data on these quadrats were reported later (Weaver 1954b), with the observation that 'succession in grassland can be followed only by exact methods of counting, measuring, mapping, and comparing the vegetation year after year'. This philosophy, first expounded a half-century earlier by his mentor, had long since become commonplace practice in U.S. rangelands through their influence. But Weaver was preoccupied with the Clementsian paradigm of succession (Tobey 1981) and never, so far as I know, attempted any formal demographic analyses of his vast census data.

Range reference areas

I have excluded from this account of US range research publications which do not include census data. Other measures of plant performance on long-term permanent quadrats have been reported frequently. The measures are reviewed in most range science textbooks and include biomass, height, basal area and projective cover; height-volume relationships were particularly popular among southwestern range scientists (Sampson 1952). Where basal areas of tussock grasses in open semi-desert habitats are mapped, the charts may often be used to extract census data; indeed, mapped quadrats typically contain both types of information. By consulting the original charts, census data may be obtained from sources from which perhaps only basal area information has been published: Smeins et al. (1976), for example, report basal area changes of herbaceous species on 36 square-foot ($\equiv 0.09\,m^2$) quadrats for 25 years.

I have no doubt that much more remains to be known about the extent of plant demographic research conducted for decades in several range research stations throughout the western United States and that my search for data already published is quite incomplete. Fortunately the Society for Range Management has embarked on a programme to record all rangeland research areas, both in the United States and abroad (Laycock 1975). It is particularly concerned to list all exclosure plots and permanent quadrats (especially the older ones), given their importance as long-term reference and control areas. It recognizes the problems involved in access to data, 'often buried in old literature or in unpublished theses' (Laycock 1975: 29). The preservation of all exclosures on rangelands is a primary aim of the Society, reflecting a concern of the American Association for the Advancement of Science (1963) for the recognition and retention of such areas for research on natural environments. Among the States for which inventories have been compiled to date are Wyoming (Williams 1963), Utah (Laycock 1969), Nebraska (see Laycock 1975) and Arizona (Turner et al. 1980).

Not all of these exclosures necessarily contain permanent quadrats and among those which do, it is difficult to ascertain whether or not census data have been recorded. Given the open-nature of the vegetation on many ranges, especially in arid areas, it is extremely likely that permanent quadrats were charted at regular intervals by some means (including photographs) from which demographic information can be extracted (e.g., Laycock 1975: Figs 9 and 10). The compilation on the range reserve areas in Arizona (Turner et al. 1980) is instructive in revealing the pattern of establishment of these areas: 3 before 1910, including the Tumamoc Hill area discussed earlier; 23 between 1910 and 1919, most of them within the Santa Rita Range, and also including sites studied by Clements and collaborators; 24 between 1920 and 1929; 77 between 1930 and 1939; 29 between 1940 and 1949, and about 170 between 1950 and 1979. Whether or not charted permanent quadrats are present is not indicated, but I suspect there is a rich store of demographic information on plants still to be revealed from U.S. rangelands.

PLANT CENSUS BY VEGETATION SCIENTISTS

In this section I shall attempt to give a coherent account of census by botanists, working mostly in an academic tradition of plant ecology. The distinction between 'pure' and 'applied' vegetation scientists is, I readily acknowledge, no more than an imprecise, even artificial, one. Its boundary was breached most conspicuously by Clements, who combined the most intellectualised with the most practical approaches to vegetation analysis. But it is clear from all the classical textbooks of vegetation science that a 'self-conscious' intellectual discipline loosely known as 'plant ecology' had emerged by 1920; its history has been partly recounted by McIntosh (1976). Most (though not all) of the scientists were associated with University departments of botany. Much of the research I shall refer to is well within this academic mainstream, and some of it will doubtless be familiar to my readers. But since no comprehensive resumé of it is known to me I believe it is worthwhile to attempt a definitive account. I shall deal with the research chronologically and geographically. For convenience I shall also include here accounts of some investigations which do not quite meet the criterion of 'academic' plant ecology.

Plant census in the period to 1930

North America

The most striking aspect of this period is the paucity of census information: Clements' writings seem (among ecologists) to have fallen on ears deaf to this aspect of his research, but more receptive to his polemical opinions on theoretical ecology. One of Clements' earliest students at Nebraska, H.L. Shantz, did follow his teacher's precept. Following his doctoral thesis in 1905 in which he reported the numbers of plants in many metre-square quadrats (Shantz 1906), he undertook large-scale descriptions of vegetation in several areas of the western United States and made regular use of quadrats to make maps of 'typical portions of plant associations'; these usually indicated plant numbers. His studies continued for several years, but his census records appear not to have been repeated at any given site (Shantz 1911, Shantz and Piemeisel 1924, 1940). His colleague Piemeisel later continued census studies of shrub-grasslands in Idaho on permanent quadrats for many years (Piemeisel 1951, Hironaka and Tisdale 1963). H.A. Gleason scarcely ever attempted to census plants, it seems (Hart and Gleason 1907), nor did he refer to Clements' own demographic researches: he might have sustained his arguments against Clements' view of the nature of vegetation more cogently had he done so.

A notable academic exponent of census on quadrats in the early years of the emerging science of plant ecology was W.S. Cooper. His interest in plant census owed nothing to Clements, it seems: he was trained in plant ecology by Cowles at Chicago where the Clementsian philosophy of vegetation was ill-received (Tobey 1981). Although he joined the University of Minnesota in 1915 as a teaching assistant to Clements, war service interrupted his teaching and Clements had left Minnesota by the time he returned (Lawrence 1979a). His doctoral thesis under Cowles' supervision was conducted on Isle Royale, Lake Superior in 1909–1910: among his observations, he charted the position and kind of every tree, down to the smallest seedling, on sixteen 100 m^2 quadrats. The ages of about 900 trees were determined. His results are among the earliest on plant demography by a plant ecologist (Cooper 1913). On a field trip to Alaska in 1914 he found a site at Glacier Bay on which to study the pattern and rate of vegetation development from bare ground, following glacial recession. He returned in 1916 and established nine permanent metre-square quadrats along the shore of Glacier Bay which he hoped would provide 'accurate unimpeachable data as to the movements and activities of the plant population of small areas typical of the whole' (Cooper 1923). He summarized his census data for 1916, 1921, 1929 and 1935 (Cooper 1939) on all but one of these quadrats, destroyed by erosion of the site. His celebrated results are well-known and widely quoted. Observations on these eight plots have continued to the present day by D.B. Lawrence, who has remapped and rephotographed them four times between 1941 and 1972. He believes, I think correctly, that they are the oldest permanent plots on terrain of known age following glacial recession (Lawrence 1979b). They have been recharted in June 1982 by I.A. Worley and M.G. Noble, to whom Lawrence introduced them (D.B. Lawrence, personal communication), so that a 66 year history of precise floristic change is now available. A further quadrat was added to the series by Lawrence in 1941 and has been recharted at intervals (Lawrence 1979b and personal communication). No details of the floristic changes have been published since Cooper's last paper on the plots in 1939; Lawrence (1979b) has given a synopsis of the vegetation development on plot 1.

Cooper also experimented with photography to chart quadrats quickly and accurately (Cooper 1924). This was the forerunner of many photographic innovations for this purpose: e.g., Booth 1943, Claveran A 1966, Pierce and Eddleman 1970, 1973. Other innovations for mapping also continue to be introduced: Cullen et al. 1978, Mack and Pyke 1979.

Northern Europe

I have remarked earlier that Clements' *Research Methods in Plant Ecology* (1905) was published about the time that plant ecology was becoming established as a scientific activity in Britain and Ireland. A.G. Tansley was an enthusiast for permanent quadrats: he was an admirer of Clements' approach to vegetation science. He even attempted to chart permanent quadrats himself, but his patrician temperament (see Godwin 1977) seems to have found this an inconducive tedium. He charted and censused four small 10 inch square ($\equiv$ 0.06 m^2) quadrats in 1914 and 1920 on chalk grasslands but these areas were, 'unfortunately not the same because the wooden marking pegs disappeared during the six-year interval' (Tansley and Adamson 1925). (These (impermanent) quadrat diagrams are still, however, reproduced as examples of permanent quadrat charts!) The quality of this work was simply not comparable to the standard by then current among range scientists in the U.S. Tansley also recorded the invasion of woody species into two exclosures (820 m^2 and 390 m^2 in size) on chalk grassland by charting individuals in 1909, 1914 and 1920; this seems to be his only other attempt at mapping and census (Tansley 1922).

Of course it is well-known that Tansley fostered research on permanent plots and encouraged two notable English ecologists in particular to undertake it. He helped H. Godwin to fence a small area of Wicken Fen near Cambridge; here Godwin mapped the bushes > 1 m tall and > 50 cm diameter at irregular intervals, but it was never comprehensively censused and a discussion of the results falls outside the scope of this paper. The results to 1972 were reported by Godwin et al. (1974). Of more importance for the history of plant census in vegetation was Tansley's influence on A.S. Watt, who was conducting research on seedling survivorship of beech (*Fagus silvatica*) on permanent quadrats in 1920–1922 (Watt 1923). In 1935 Watt began a series of studies on grasslands in East Anglia which partly used census techniques: I shall deal with them later.

The two major reviews by leading plant ecologists on succession research which appeared in 1930 (Lüdi 1930, du Rietz 1930) give a revealing insight into the extent of census on permanent quadrats by 'academic' botanists to that time: Clements, Weaver, Cooper and Tansley exhaust the list of names. There is no evidence whatsoever that they were aware of the voluminous U.S. range research, not a little of which had been already published by then. The dichotomy between the activities of botanists and range scientists is starkly revealed by these reviews, which continued for many years to be the definitive sources for the early history of plant ecology in continental Europe. Even in the United States this was reflected in the textbooks. Oosting (1956) stated that 'the use of permanent quadrats has been advocated by many plant ecologists but few have followed their own excellent advice' which indicates to me that he did not have range scientists in mind as plant ecologists. To this day there is little evidence that the scope of published (not to say unpublished) research on plant census in permanent quadrats has been appreciated by academic plant ecologists (e.g., Daubenmire 1968, Mueller-Dombois and Ellenberg 1974, Barbour et al. 1980). There was of course an established tradition of permanent quadrat research in continental Europe by 1930, but most of this involved the recording of changes over time in floristic composition at the level of species identity rather than in numbers of individuals (Braun-Blanquet 1964). Braun-Blanquet himself laid out permanent quadrats in open areas in the Alps in 1921 and recorded numbers of individuals at a few irregular intervals until 1947 (Braun-Blanquet 1964), but this seems to have been untypical. The use of permanent quadrats continues to be widespread in continental Europe among phytosociologists, but only very little has been published about exact observations, and census data remain, to my knowledge, extremely rare (Böttcher 1974, 1975).

In Leningrad in 1926 a French woman Yvonne Bogdanowskaya-Guihéneuf published a paper which recorded seeds, seedlings and juvenile plants on 4 small quadrats (20 × 20 cm^2) during 1923–1924. She concluded that the study of the population dynamics of seedlings was as important for understanding the dynamics of grassland vegetation as it was for forests. Despite its simple message, this paper had an important influence in northern Europe, directly and indirectly, on the future development of plant demography, as I shall record in a later section. Y. Guihéneuf was born in Nantes in 1886, emigrated to Russia in 1902 and died in 1968. Her French name is usually retransliterated from a Russian transliteration of the original, so her paper is usually cited as Bogdanowskaya-Gienef (1926). Most of her research, begun in 1915 under the direction of the geobotanist A.P. Shennikov, was on swamp vegetation, but some of her later work was republished under the guidance of T.A. Rabotnov (Bogdanowskaya-Gienef 1954: the original publication in 1941 was destroyed in the siege of Leningrad). Her 1926 paper does not refer to non-Russian researches, so there is no evidence that she was influenced by Clements or others in western Europe or the United States at the time she undertook her investigations.

Australia

One of the most significant long-term census studies in vegetation was begun in 1925 at Koonamore, north of Adelaide, Australia by T.G.B. Osborn of the University of Adelaide. An area of 390 ha was enclosed by rabbit- and sheep-proof fencing and a series of quadrats of varying sizes was laid out in 1926–1927: five 1 ha quadrats, four 100 m^2 quadrats and several 1 m^2 quadrats. Depending on quadrat size, trees, tall shrubs, low shrubs and herbs were censused, charted and photographed at regular intervals. The vegetation was arid chenopod shrub steppe, with a scattered tree layer. The area chosen had been heavily overgrazed and the primary object of the project was to study the growth and regeneration of plants protected from grazing by cattle and, as far as possible, from rabbits. The experiment was inspired by U.S. rangeland research (Osborn et al. 1932). Much of the field work was conducted by T.B. Paltridge and records were kept regularly by him and by J.G. Wood until 1936 (Osborn et al. 1935, Wood 1936). The first paper on the important fodder grass *Stipa nitida* reported census data on 124 metre-square quadrats at three-monthly intervals from May 1928 to March 1931; these quadrats were not permanent, but were on well-defined sites (Osborn et al. 1931). This paper comes close to the type and quality of work by then familiar for U.S. range grasses. Subsequently only shrubs on the larger quadrats appear to have been censused. Some further quadrats and permanent transects were later established (Hall et al. 1964).

It is not clear why Osborn began these studies. He may have been influenced by Tansley (see Osborn in Tansley and Chipp 1926), by Tansley's erstwhile colleague R.S. Adamson with whom he wrote two vegetation papers and by W.A. Cannon, a founder-member of the Carnegie Desert Botanical Laboratory in Arizona and a colleague of Shreve and Clements. Cannon had visited south Australia and in his subsequent report particularly acknowledged Osborn's assistance (Cannon 1921); Cannon may have brought the Desert Botanical Laboratory tradition to Australia. T.G.B. Osborn (1887–1973) was an Englishman and Professor of Botany at Adelaide and at Sydney before succeeding Tansley as Sherardian Professor at Oxford (1937–1955). Whatever the reasons for Osborn's interest in plant census, he was probably thoroughly familiar with the opinions on the importance of this type of research among the leaders of academic plant ecology at the time. Later at Oxford he never spoke about this work to his students, among whom were J.L. Harper and G.R. Sagar (personal communications). He had little influence on the development of plant ecology in England. His departure from Australia led to a decline in the early intensity of research activity at Koonamore. Nonetheless the larger quadrats were regularly photographed and charted by C.M. Eardley and students at the University of Adelaide and now provide one of the few well-documented examples of long-term plant census. Various aspects of these data were reported by Hall et al. (1964) and, with the contemporary interest in plant demography, have ben diligently re-examined in a series of recent papers (Crisp and Lange 1976, Noble 1977, Crisp 1978, Fatchen 1978, Noble and Crisp 1979, Osmond et al. 1980, Silander 1983). Silander's study of *Cassia nemophila* is particularly interesting since it is based on accurate census data of recruitment and mortality of over 3000 individuals from 1925 to 1978. The Koonamore Reserve is now named the T.G.B. Osborn Vegetation Reserve.

Plant census in the period 1930–1960

By 1930 as we have seen, only a handful of University-based plant ecologists had contributed to the census of plants in vegetation, and their efforts were quite uneven in quality and quantity. During the following thirty years a few significant long-term studies were initiated, in northern

Europe, the Soviet Union, England and Australia. I shall deal with these in turn. They appear to have been conducted to a large extent quite independently of one another, with little or no interaction between the participants. (This can be judged from the absence of cross-citations, although I have verified it also from some personal communications.) Only in the past twenty years have these several diverse strands of plant demography been drawn slowly together into a comprehensive framework; but even to-day this task is still incomplete.

Scandinavia and the Soviet Union

By the early 1930s a well-developed Finnish School of plant ecology was flourishing (Whittaker 1962). The dynamics of seedling establishment in vegetation received detailed attention from the group of botanists associated with K. Linkola at Helsinki. Two publications in particular were influential, those of Linkola (1935) and of his student Perttula (1941). It is evident from the citations in Perttula's paper that there was a developing Finnish tradition of research on plant population biology. This continued into modern times in the work of Oinonen (see Harper and White 1974). Linkola (1935) reported the results of his observations of seedling survival on eighteen $0.1\,m^2$ quadrats in 1931–32; he explicitly stated (p. 6) that his model was the study published by Bogdanowskaya-Guihéneuf in 1926. He also attempted to age the various herbaceous species by morphological criteria: his paper contains some of the earliest published examples of age-structures for herbaceous plants (see Harper 1977 for examples of some of them). From 1928 to 1934 Perttula (1941) conducted a comprehensive study of the reproductive output, seed dispersal patterns and seedling establishment of numerous species. His valuable monograph remains too little appreciated among plant demographers.

The influence of these two papers remained rather local but their example helped to inaugurate a major school of plant demography, that founded by T.A. Rabotnov (b. 1904), whose name is now virtually synonymous with the study of plant populations in natural vegetation (coenopopulations). His most celebrated work on the population dynamics of grassland plants was begun about 1940 and first published at length in 1950 in his doctoral dissertation (Rabotnov 1950). This reported his investigations on subalpine meadows of the northern Caucasus during the early 1940s. Most of Rabotnov's research on coenopopulations was conducted in the State Meadow Institute (now Williams All-Union Research Institute of Fodders), where he worked for over 40 years until his appointment to the Chair of Geobotany at Moscow University in 1967. His work has, therefore, been more in the tradition of applied plant science, akin perhaps to that conducted on U.S. rangelands. But it is evident that the demarcation between pure and applied vegetation science, which was a feature of the development of Anglo-American plant ecology had little or no meaning for Rabotnov or other Soviet geobotanists.

Rabotnov (1950) has acknowledged the formative influence of Linkola on the development of his own approach to vegetation analysis. The hallmarks of this were, as is now well known, the long-term census of marked individuals and the use of morphological criteria to estimate the developmental stages (age-states) of perennial herbs. The publication of his thesis (1950) laid the foundation of a new approach to vegetation analysis among plant ecologists, particular in the Soviet Union. This was in no small measure due to his influence as a teacher and his voluminous publications: his outstanding knowledge of world literature has been especially appreciated among Soviet geobotanists. His 1950 paper is revealing in this respect: of the numerous references to foreign authors, Linkola and other Finnish botanists account for 12, Weaver for 10, Clements 4, Irmisch (on morphology) 5; there is an evident familiarity with much U.S. rangeland literature. I know of no other paper on vegetation analysis at that time which reveals such a catholicity of quotation, all brought to bear on 'the life-cycle of perennial herbaceous plants in meadow coenoses'.

Rabotnov (1980, 1981) has recently provided a valuable guide to the literature of coenopopulation research in the Soviet Union. Although publications on trees are excluded, it contains nearly 900 entries. Less than 6% of these are dated 1950 or earlier, which gives some indication of the quickening pace of this research in the past 30 years (and, I believe, of Rabotnov's extensive influence intellectually). How many of these publications are actually concerned with the census of plants in vegetation is not easy for me to calculate (since I have read no more than 10% of them) but there is a considerable number, probably hundreds (see also Gatsuk et al. 1980). Much of this research remained unknown to English-speaking plant ecologists, despite Rabotnov's publications in English (e.g., Rabotnov 1961, 1969a, b) until the early 1970s (Harper and White 1971, 1974). (It was, however, cited in some detail by Walter (1968), so it was not entirely neglected outside the Soviet Union.) It is now well-recognized that Rabotnov's long-term census research in natural vegetation is among the most distinguished in the history of plant demography: his studies on *Ranunculus* species (Rabotnov 1958, 1978a, b, Rabotnov and Saurina 1971) for a period of ten years are well-known examples, but many more deserve wider recognition outside the Soviet Union than they have received to date.

Both Linkola's and Perttula's studies were also influential on a young Swedish botanist C.O. Tamm, searching for a subject for research in the early 1940s. Discouraged by his Professor's poor opinion on the scientific value of plant census on permanent quadrats he concentrated his research on plant nutrition and forest ecology. Nonetheless he began to chart the fate of marked plants on permanent quadrats in meadows and forests about 100 km NE of Stockholm from 1943 onwards, once a year (C.O. Tamm, personal communication). The publication of his results (Tamm 1948, 1956) passed almost unnoticed in the ecological literature until, after his lecture at the Botanical Congress in Edinburgh (Tamm 1964), they were reanalysed and brought vividly to the attention of plant ecologists by Harper (1967). Two further papers subsequently appeared (Tamm 1972a, b) and the results of his 'spare time' research are now justly celebrated. Observations on the permanent quadrats still continue annually. They are the only long-term plant census plots in herbaceous vegetation in Europe on which records have been kept for over 40 years, to the best of my knowledge.

England

After a period of research (since about 1915) on problems of woodland regeneration A.S. Watt turned his attention to grasslands in the mid-1930s. During 1935–36 he set up five rabbit-proof exclosures ($6 \times 6\,m^2$) on grass-heath on acid soils in the Breckland region of East Anglia; only one of these remained undamaged by 1949. Small ($0.16\,m^2$) permanent quadrats, one inside and one outside each exclosure were charted annually. Watt (1960) used frequency measures (presence or absence of a plant shoot or leaf in $1.25 \times 1.25\,cm^2$ grid squares within the quadrat) rather than census of individual plants, except for *Agrostis* spp. and *Aira praecox*, in which shoots were counted. He believed that the assessment of abundance by a count of the number of shoots was less accurate than by frequency estimates (Watt 1971a). Further records from the pair of undamaged quadrats were subsequently reported (Watt 1971a), but it appears that these plots have since been abandoned.

Two further exclosures ($6 \times 6\,m^2$) were established by Watt in 1936 on ungrazed chalk grassland at Lakenheath Warren, also in the Breckland. The changes in floristic composition of one of these have been reported (Watt 1957, 1974), but no long-term charting or censusing of permanent quadrats was undertaken on it; some frequency data were recorded on small ($10 \times 50\,cm^2$) quadrats for about six years and thereafter sporadically (Watt 1957). On the second exclosure, however, accurate chart records were maintained by Watt annually for all but two years from 1936 to 1973 (Watt 1962, 1981) on two small quadrats ($0.16\,m^2$) one inside and one outside the exclosure. These records continue to be kept since 1974 by Davy and Jefferies (1981) and are now the only surviving examples of the grassland quadrats established by Watt. In fact they are to my knowledge the only two quadrats in England established before 1960 on which some long-term plant census data have been maintained to the present day. As in the case of the long-term quadrats observed by Watt on acid grass-heath, the abundances of species on these two quadrats were mostly determined by frequency of occurrence of a plant part in each $1.25 \times 1.25\,cm^2$ grid square within the $160 \times 10\,cm^2$ area. Three species were dominant at different periods, *Festuca ovina*, *Hieracium pilosella* and *Thymus drucei*. Given the tiny size of the quadrats, however, it is questionable whether the population dynamics recorded on them by frequency estimates of plant parts reflect (especially in the case of *Thymus*) more than the waxing and waning of the morphological expression of a few long-lived genets. For some (mostly annual) species the fluctuations in abundance *were* based on census records, made once a year in early July. The survivorship of seedling cohorts of a few species was recorded at regular (often monthly) intervals during the years 1961–1969. Although they are primarily plant frequency quadrats, and almost outside the scope of this essay, their continued maintenance in view of the rarity of long-term census records in England is commendable.

Throughout the 1930s, Watt made very detailed studies of the morphology and growth dynamics of *Pteridium aquilinum* (bracken fern) on a series of permanent plots in Breckland. By comparing the length of rhizome per unit area with the average length of rhizome per individual and then finding the number of fronds per unit length of rhizome, Watt (1940) computed the number of individual plants from the number of fronds per unit area. He estimated that there were the equivalent of $(2.8–4.7) \times 10^3$ independent plants per hectare. His work focussed on differences in frond and rhizome morphology across transects from advancing margins to long established hinterlands of populations (Watt 1943, 1945, 1947). Census observations continued on various permanent plots into the 1960s (Watt 1967). The long series of papers produced until 1971 (Watt 1971b) is remarkable for its combination of morphology and demography, forerunner of some recent developments in dynamic morphology among plant ecologists (e.g., Harper and Bell 1979, Harper 1981).

Buttercups, as is now well-known, have been a popular subject for demographic research in England. John Harper began to count them while still at school (personal communication) and one of his first papers on plant population biology reported census data on three species across a ridge and furrow system in a seasonally-flooded grassland (Harper and Sagar 1953). A quite extensive census of *Ranunculus bulbosus* was conducted by Barling (1955) during the 1940s on a wide range of grassland types in southern Britain. Though not quite clear from his text, at least 130 swards of permanent grasslands and a further 5 areas of temporary grassland were surveyed; at each site fully established plants (not seedlings) were recorded on at least 100 square-foot ($\equiv 0.09\,m^2$) quadrats. None of his quadrats was permanent, since the object of his investigation was simply to make quantitative statements about the sizes of natural plant populations in relation to various habitat conditions. Some swards had exceptional concentrations of *R.bulbosus*, sometimes exceeding the equivalent of 2×10^5 per acre ($\equiv 0.4\,ha$). Contiguous populations of up to 14×10^6 plants (over what area is unclear) were 'in no way exceptional' in the Cotswold area. He also independently detected the pattern of *R.bulbosus* and *R.acris* on ridge and furrow systems reported by Harper and Sagar (1953).

A detailed demographic analysis of *Plantago* species was conducted by G.R. Sagar during 1957–59. It was notable in several respects. Its inspiration was due almost exclusively to animal demographers, especially to David Lack (see Sagar 1970; 972). Sagar's teacher, John Harper, was also strongly influenced by the thinking of zoologists on natural populations, notably by Elton, Lack and Varley, all Oxford colleagues. (This influence is clearly seen in the major theoretical paper of Harper's Oxford group at the time: Harper et al. 1961.) Only a handful of botanists were quoted by Sagar (1959) as exemplars of plant population biology; almost none of those whose names are now familiar as pioneer plant demographers in the previous half-century are cited. This probably reflected the special interests at that time of Harper and his students in experimental plant competition and in the biology of weeds (Harper 1960, 1961). There was then great interest and support among scientists concerned with weed control for the type of work his group were doing, which accounts for the publication of several of their papers in the Proceedings of the British Weed Control Conferences (J.L. Harper, personal communication). But the absence of the names of plant demographers from Sagar's thesis is not unexpected on more general grounds either: the plant ecology literature of this period in England gives little evidence that the well-established tradition of plant census in natural vegetation had been transmitted to English botanists, even at Oxford where Osborn had succeeded Tansley. Tansley himself had provided almost no examples of a personal interest in it, though he had advocated the use of quadrats and census in general terms. Watt, as we have seen, had mixed feelings about census, at least in grassland. Clements was an almost discredited figure in Anglo-American plant ecology, as Gleason's star shone brilliantly after 1947 (McIntosh 1975, 1980). Rabotnov's papers were unknown.

Sagar's thesis was, therefore, all the more significant by being the first, well-considered attempt by a plant ecologist in England to apply demographic theory (derived from zoologists) to plant populations. To this was added an experimental approach derived from the example of English agricultural botanists. But perhaps most significantly, Sagar's research was a harbinger of an attempt by John Harper to direct the attention of plant ecologists more keenly towards plant population biology, as a means of understanding the evolutionary mechanisms which give rise to the composition and diversity of vegetation (Harper 1964, 1967). By the time the demographic results of Sagar's thesis were published in some detail (Sagar 1970) a renewed interest in plant demography had become widespread.

Australia

In 1937 R. Roe, influenced by the work of Osborn and his colleagues at Koonamore, set out a grazing experiment in south western Queensland in which he charted and censused the chenopod weed *Bassia birchii*. He continued these observations at regular intervals (up to 5 times annually) but the life-table data he compiled remain unpublished (Williams 1981). (I am grateful to O.B. Williams for bringing the original data to my attention.) Roe expanded his researches to census a variety of grass species, by setting up groups of permanent transects and quadrats in *Astrebla* grasslands in Queensland. Charting and censusing of *Astrebla*, *Dichantium* and *Eragrostis* spp. began on five 1 m^2 permanent quadrats in 1941 and continued almost annually until 1958. 45 band transects, each with 100 $20 \times 20\,cm^2$ quadrats and 45 metre-square quadrats were censused from 1944 to 1970 at irregular intervals by Roe and thereafter from 1973 to 1981 by Williams (Williams and Mackey 1982). Results from some of the band transects and quadrats have been summarized by Williams and Roe

(1975). Another series of 25 $2\,m^2$ quadrats was charted at irregular intervals from 1949 to 1973 (Williams and Roe 1975). These various investigations were stimulated originally by a long series of drought years in western Queensland in the mid 1930s. In many respects these were the Australian counterparts of those U.S. rangeland studies which were also stimulated by the problems of drought (e.g., Weaver and Albertson 1936). The studies of Roe were preceded by some investigations of J.G. Davies, brother of the notable English grassland scientist William Davies (e.g., Davies 1933, referred to earlier): he is credited with the introduction of the Hill pantograph to Australian pasture research (O.B. Williams, personal communication).

During 1946–47 O.B. Williams worked with Roe on his grazing experiments in *Astrebla* grasslands. When he moved to Deniliquin in New South Wales he began similar research on *Danthonia* grasslands. This was on a large (indeed Clementsian) scale and involved the pantograph charting of 324 metre-square quadrats on grazed and ungrazed areas. Plant numbers were recorded at regular intervals until 1968 and the results have been reported in a series of papers (Williams 1966, 1968, 1970a, b, Williams and Roe 1975, Austin et al. 1981). Williams was the first Australian scientist to apply formal demographic methods to plant census data and has been chiefly instrumental in highlighting the hitherto neglected research on plant census by Roe.

Neither the design of these experiments nor the techniques of recording quadrats, nor indeed the ecological thinking which inspired them owed anything to European exemplars. They were all directly derived from U.S. rangeland studies. As such, they sit uneasily in this section of my narrative, except insofar as Osborn's research at Koonamore may have been their direct, local inspiration. But their publication in recent years also owes a little to the renewed interest in plant demography among plant ecologists, especially to Harper's (1967) call for such information (O.B. Williams, personal communication 1980).

Other census studies 1930–1960

Few other long-term studies on plant census in permanent quadrats appear to have been initiated in the period 1930–1960. A notable example, however, is the 14-year census of the annual *Linanthus parryae* conducted on a long 800 m transect by Epling, Lewis and Ball from 1944. The numbers of blue- and white-flowered plants were sampled within $10 \times 10\,ft^2$ ($= 9.3\,m^2$) quadrats along the length of the transect, 260 quadrats in all. The density of plants varied greatly from year to year, but the frequencies of blue and white flowers remained remarkably stable within local patches (Epling et al. 1960). Lewis (1962) later censused the persistence of sown populations of *Clarkia* spp. in several metre-square quadrats for nine years (1953–61) in an attempt to understand the evolutionary consequences of large fluctuations in population size.

Runge has reported his long-term observations on several permanent quadrats in a variety of vegetation types in NW Germany since 1955. The numbers of plants on a 3 m^2 permanent quadrat in grassland have been recorded for five years, 1957–1961 (Runge 1963); the fluctuations in density of some herbaceous plants on a permanent quadrat in heathland have been recorded from 1955 to 1969 (Runge 1971); investigations on 68 permanent quadrats for varying periods of time (5 to 17 years), which include some census information of herbaceous plants, have been briefly reviewed (Runge 1975).

Among plant ecologists plant census studies of any type were, it seems, relatively infrequent during this period, apart from those I have already mentioned above. A few further examples which I have seen are by Wager (1938), who counted plants by age-states in quadrats in Arctic fjaeldmark in 1935–36; by Whitford (1949), who made a simple, non-repeated census of woodland herbs in 1947; and by Keever (1950), who made a short-term study of plant survivorship in abandoned fields in 1948–49. Some examples from the ecological literature may not be too obvious, since plant census may be undertaken as part of a wider research activity. In assessing the primary productivity of the desert shrub *Larrea tridentata*, for example, Chew and Chew (1965) censused plants. In a notable study on the biological control of *Senecio jacobaea* Cameron (1935) reported several census data; similar work was later undertaken by Dempster and Lakhani (1979). The literature on biological control of weeds may indeed have extensive plant census data, but I have not systematically searched for it: Tisdale (1976), for example, recorded the fluctuations of *Hypericum perforatum* over a 15 year period.

Occasional papers on plant census by grassland scientists can also be noted from this period, but the popularity of other measures of plant performance was widespread. Stone and Fryer (1935) reported census results on a number of pasture mixtures from 1931 to 1934: they were particularly interested in the relative survival of species in different mixtures. They attempted to count distinct plants with independent root systems, but could do so only approximately after the first year. The highest density achieved in their mixtures was 699 plants per m^2, low by the standards cited earlier in this paper (Table 2) but perhaps a more realistic approximation of the number of genets per m^2 than those counts of shoots. This study seems to be one of the earliest of its kind on the survival of plants in grassland swards; later studies, it seems, estimated population changes by random sampling of tillers (e.g., Brougham et al. 1960, Charles 1961, 1964) rather than by detailed census of plants in situ.

Counts of seedlings and mature plants on Californian grasslands dominated by annual species have been reported by Biswell and Graham (1956) and by Heady (1958).

A unique and remarkable census of *Asclepias syriceca* (milkweed) was undertaken in the southern part of Ontario and Quebec, Canada during the summers of 1943 and 1944. During the war, in the search for native sources of rubber, milkweed was regarded as a potentially useful species and an experimental pilot plant at Ottawa produced 'milkweed rubber'. The fibres associated with its seeds were used as substitutes for kapok. In order to determine the abundance of the plant, Groh and Dore (1945) surveyed the extent of wild populations by travelling 6,132 miles (c. 10,000 km) by car, train and bus. Being a tall and conspicuous plant, wherever it could be seen its density was estimated: the 'speed of some trains detracted from, but seldom wholly prevented reasonably satisfactory work'! But this unique method of plant census the numbers of stalks per mile were calculated and scaled up (using field samples) to the density per township: stalks per mile varied from 160 to 3×10^4; stalks per township (of unequal size) varied from 6.2×10^4 to 1.2×10^8. The data were used to organize a leaf and pod collection campaign in 1944. While it is commonplace for foresters to estimate the number of trees over large areas, this is the only example I know of an attempt to make an assessment of the numbers of an herbaceous plant over an area of several thousand hectares.

Although its title fails to indicate its demographic content, the study by Stewart et al. (1940) is noteworthy for the presentation of age structures for five desert shrubs, based on several hundred age determinations in a variety of vegetation types. As one might expect from the scientists who introduced the 'square-foot density' method (Stewart and Hutchings 1936), their 'density' estimates for shrubs and grasses are not based on census. Their paper, however, indicates once again that range scientists were actively recording plant demographic data at a time when it was unfashionable among plant ecologists.

Despite Pelton's (1953) call to plant ecologists for a study of the life histories of plants there is little evidence of such an interest among them in the published literature of the period 1950–1960. Pelton enunciated clearly, and with obvious appreciation of the work of foresters and range scientists, the various stages in the life-cycle of a plant which merited particular attention. Rangeland studies on plant census continued both in the United States and in Australia; Rabotnov's grassland studies continued and his ideas gained adherents in the Soviet Union; elsewhere isolated studies were conducted and occasionally published.

Plant census in vegetation, 1960–1970

The decade from 1960 to 1970 witnessed the emergence of what, to echo a phrase of Allee et al. (1949), may be called a 'self-conscious' science of plant demography among vegetation scientists. The earlier part of the decade cannot be demarcated intellectually from the preceding period, but by its close there was an unmistakable change. This was the decade which saw among plant ecologists the flourishing expansion of plant population studies in artificial environments (in flower pots, typically). It was a decade in which plant competition studies were conducted with increasing sophistication (Harper 1977 reviewed the various developments), but also one in which the investigation of plant populations in natural environments grew steadily in popularity among a new generation of students. It was ushered in by the widely influential papers of de Wit (1960) and Harper (1961); by its close the researches of Sharitz and of Sarukhán were well-advanced (Sharitz and McCormick 1972, Sarukhán and Harper 1973).

I have ventured in this section beyond 1970 only to note publication of studies *initiated* in the 1960s. The proliferation of new investigations on plant demography, not least on the census of plants in vegetation, has expanded greatly since 1970. The International Society for Plant Population Biologists (founded at the Botanical Congress in Leningrad in 1975) has recently issued a list of scientists who conduct research on plant population biology (considered in a rather broad sense): it consists of some 450 names (Sharitz et al. 1984). The evaluation of what has been accomplished on plant census in vegetation in recent years is a task which I shall not embark on here.

During the 1960s some well-established plant census records continued to be maintained, as will have already been evident from some of the citations in the previous sections of this paper. A notable increase of plant census research took place in the Soviet Union. Apart from the continuing work of Rabotnov and his students, the botanist A.A. Uranov (1901–1975) turned his attention to coenopopulation studies about 1950. (He had previously been more interested in statistical plant ecology.) Until his death he was the leader of an active research school at Lenin State Pedagogical Institute in Moscow from which hundreds of studies on coenopopulation dynamics have emerged since about 1960 (Rabotnov 1980, 1981). Several of these have appeared in collective works (Uranov 1967, 1968, Smirnova et al. 1976, Uranov et al. 1977) and a hint of their scope has been provided by Gatsuk et al. (1980). Uranov had some 30 graduate students, of whom about 10 still conduct coenopopulation research in the same Institute (L.B. Zaugolnova, personal communication 1981). Uranov's close collaboration from 1961 with the plant morphologist I.G. Serebryakov (1914–1969) is reflected in the quality of the age-state discriminations which have been a hallmark of this particular research school. The careful study of age-state ontogenesis and census of plant coenopopulation persists to the present time in the Soviet Union with undiminished vigour, to judge from Rabotnov's (1980) bibliography.

Studies initiated during 1960–1966

Among the census studies initiated in the early 1960s only two have continued to the present time, to my knowledge. Following the devastating North Sea storm floods in 1953 the Dutch Government began large-scale civil engineering works to protect the southern delta region of the Rhine-Meuse estuary. Since 1961 a series of barrages has been constructed in parts of the delta, progressively isolating large areas from tidal influence. The consequent changes in salt marsh vegetation have been monitored regularly on nearly 500 permanent quadrats (Beeftink 1975, 1978). Although vegetation changes have been recorded mainly by cover/abundance estimates of the Braun-Blanquet type, many census records have been combined with them to produce a rather detailed record of the dynamics of individual species. These studies provide a good example of the nice balance between precision and generality which is often necessary to understand vegetation dynamics (Harper 1982). It is expected that these observations will be maintained for at least another ten years (Beeftink, personal communication); some results to 1983 are reported by Hogeweg et al. (1985).

The population dynamics of rare or attractive plants have frequently been commented upon by botanists in an anecdotal manner: fluctuations in abundance of orchids, for example, are well known and rather precise census data have been recorded (Curtis, 1946, Curtis and Greene 1953, Summerhayes 1968). Annual records of the numbers of rare species may be found in the publications of conservation organizations or local natural history societies (see Bradshaw and Doody 1978a for some details of English examples). An accurate, mapped census of orchids was initiated by Wells on permanent quadrats in chalk grassland in 1963. Some preliminary observations on *Spiranthes spiralis* (Wells 1967) have now been augmented by a long series of annual census data to 1979 (Wells 1981), especially notable for the survivorship details of successive annual cohorts. Similar studies on two further orchids, *Aceras anthropophorum* and *Herminium monorchis*, have been conducted since 1966 (Wells 1981). A census of *Pulsatilla vulgaris* on a sheep- and rabbit-proof exclosure (540 m^2) for six years (1963–1968) showed that while the absence of grazing led to an increase in the number of plants, the enhanced vigour of other vegetation led to a large decrease in flowering plants (Wells 1968).

Detailed mapping of *Anthoxanthum odoratum* along six 9.1 m transects, radiating from a fixed point on a derelict zinc mine spoil tip provided Antonovics (1972) with census data over the period 1964–1970. Of the total 546 individual plants recorded, only 16 lived for more than 5 years. Antonovics demonstrated that population maintenance was a very dynamic process: recruitment of new plants was mainly from seed rather than by vegetative propagation.

The population dynamics of *Senecio jacobaea* have been investigated using census data from 1966 to 1974 by Dempster and Lakhani (1979), an example of the interest of animal ecologists in such records of food plants. There may be many comparable examples in the literature of animal ecology: I have pointed out previously (White 1979) that several zoologists have counted leaves on trees, a task which rather few botanists seem ever to have undertaken. I would not be surprised to learn of more plant census data collected by zoologists. Among other short-term plant census data of this time are those of Beatley (1967), Matthews and Conrad (1968) and McCarthy and Scifres (1969). A single census of *Liatris aspera* was reported by Kerster (1968).

The emergence of plant demography as a science

In his Presidential address to The British Ecological Society on 5 January 1967, John Harper seems to have caught brilliantly the temper of the time among plant ecologists of a new generation in northwest Europe and North America. His call for a revivification of the Darwinian approach to plant ecology, for a better understanding of the demography of plant species in natural environments, was widely influential (Harper 1967). It also represented a reorientation of his own research interests from empirical studies of plant competition (e.g., Harper 1961) to a greater emphasis on the role of natural selection in plant communities. Within a further three years (the limit of my present narrative) several studies on plant demography, all of them using census techniques, were initiated: the publication of the results in recent years has demonstrated beyond question the emergence of a fully-fledged science of plant demography. Almost all of these papers refer to Harper (1967) (if not always to Darwin!) in such a way that it is clear that his paper is a significant element of the 'foundation literature' (Tobey 1981 explains this concept in detail) of plant demography. Now (1985) in retrospect, we can see this paper as remarkable in another respect: Tamm, Sagar and Antonovics (and Darwin) provided the key examples of the Darwinian approach to natural vegetation. Only later did a wider appreciation of the (preexisting) relevant literature come about (Harper and White 1971, 1974) and plant demographers have continued to prise it piece by piece from forgotten sources or unfamiliar literature. This gradual collation of source material is not unique to plant census: Levin and Kerster (1974) drew on a very diverse, often agronomic, literature to seek quantitative information on gene flow in plants; Willson and Burley (1983) have brought an older classical botanical literature on conifer embryology to bear on the very modern discussion of mate choice in plants. *Annual Review of Ecology and Systematics* is an annual reminder of the process of discovery, recension and synthesis of earlier publications which characterizes the search for pattern and coherence in the great diversity of natural history phenomena. It is a hallmark of the emerging and developing science of plant population biology as a whole that it seeks to explore and integrate all aspects and all examples of plant life-history studies within an evolutionary framework; its most coherent theory is Darwinian as Harper (1967) reminded us (though he was not the first nor the only one – see Clements 1909). Plant ecologists have undoubtedly given widespread assent to this general research programme in recent years. Those more particularly interested in plant demography have contributed significantly to 'the Darwinian approach to plant ecology' since the late 1960s. This present essay is an attempt in the same spirit, to add a little more to the intellectual edifice, by placing our recent efforts in some perspective which I trust will be a prolegomenon to a more definitive historical account as our knowledge increases.

Studies initiated during 1967–1969

By the time Harper's lecture was published in 1967, R.R. Sharitz had begun a very detailed census of two winter annual species, *Minuartia uniflora* and *Sedum smallii* on granite outcrops in Georgia, U.S.A. Though the field study was conducted for only a year (June 1967–June 1968), Sharitz and McCormicks' (1972) treatment of the census results represented a quite new departure in plant demographic analysis, in their rigorous application of life-table techniques. (Previous attempts by Hett and Loucks 1968 or by Harcourt 1970 are outside the scope of my essay but were, nonetheless, by no means as comprehensive as Sharitz and McCormicks'.) Their study remains an exemplary model of the formal treatment of plant census data.

Several plant census studies were initiated in 1968. The most comprehensive series was that of E. Symonides at Warsaw, who conducted a very detailed analysis of the annual fluctuations in numbers of several species on charted quadrats; results for 1968–1975 were reported in a series of papers (Symonides 1979, 1983). Silvertown (1982) has further analysed Symonides' data on the annual *Spergula vernalis* as an example of the application of key-factor analysis to plant populations, the first such published example. Only with long-term data of the outstanding quality of Symonides is such an analysis possible, and it represents a significant example of the high standard of demographic information that is becoming increasingly available from plant census data.

Ernst (1979) has reported census results of *Allium ursinum* for 1968–1977. Van Andel (1975) recorded census data on permanent quadrats for *Chamaenerion angustifolium* from 1968 to 1973. Also in 1968, Ericson began long-term studies on the rising shore lines of the northern Baltic coast of Sweden; these include census of several plant species on about 200 permanent quadrats, and observations still continue (Ericson 1981 and personal communication 1982). After the decision in 1965 to build a reservoir in Upper Teesdale in northern England, M.E. Bradshaw began a detailed study of the population biology of the rare species which are a notable feature of the vegetation there. Following the example of Tamm, demographic data were gathered from permanent quadrats on eight perennial herbaceous species from 1968 onwards, several of which were reported by Bradshaw and Doody (1978a, b). Census data from 1968 to 1980 were published by Bradshaw (1981). Thomas recorded census details on *Hieracium floribundum* populations for seven years beginning in 1968, although only data for 1970–71 were reported in detail (Thomas and Dale 1975).

Another important group of plant census investigations was begun in 1969. The most lengthy was on *Narcissus pseudonarcissus*: detailed demographic information for 1969–1978 has been provided by Barkham (1980). Census studies on *Ranunculus* species by Sarukhán during 1969–1971 were notable for the high frequency (at 2–4 week intervals) of pantograph mapping of individuals in permanent quadrats: this enabled sophisticated analyses of their demography to be undertaken which have since proven to be as influential as Sharitz and McCormicks' analysis on the development of plant population biology (Sarukhán and Harper 1973, Sarukhán and Gadgil 1974). Probably not since Clements' (1923) report on two-weekly quadrat charting had such a detailed census of perennial plants been undertaken. The census records of Sarukhán's plots of *Ranunculus repens* were continued for a further year (Soane and Watkinson 1979). Fluctuations in the density of several populations of *Anthyllis vulneraria* were recorded from 1969 to 1976 by Sterk (Sterk 1975, Sterk et al. 1982). Individuals of *Plantago major* and *P. rugelii* were mapped at frequent intervals on permanent quadrats during the growing seasons of 1969–1971 by Hawthorn and Cavers (1976).

Among other census investigations which were conducted in the 1960s the following may also be mentioned. Wallace and Romney (1972) undertook a very large census in 1968 of all perennial plants in 25 circular plots, each about 730 m^2, on the Nevado Test Site in the Mojave Desert: 19,000 plants representing 28 different species were censused and mapped as part of a study to investigate the spatial relationships between species. Both significantly positive and significantly negative associations occur among various combinations of species, reflecting some of the later conclusions of Turkington and Harper (1979) on grassland swards. I do not know if Wallace and Romneys' census has been repeated on the same site. Such a study on what is sometimes called 'pattern analysis' by plant ecologists is not unique in its compilation of census data. Although I have not searched the relevant literature systematically, I can cite two further papers of this type with census data: Barbour (1969) recorded the numbers of *Larrea tridentata* shrubs on hundreds of quadrats throughout the range of the species in the United States; Malik et al. (1976) enumerated *Atriplex vesicaria* bushes in Australia. Undoubtedly further examples of this genre exist.

Census of cryptograms

A feature of the period under consideration in this section has been the general expansion of plant census in vegetation. This has not been confined to flowering plants. Several census studies on marine algae have been reported, many of them indeed dating from the 1950s, as Kain (1971) has shown for *Laminaria hyperborea* alone. North (1971) quotes publications which have reported half-life values for *Macrocystis* populations in several places: they vary from two months in shallow waters along exposed coasts to several years on protected coasts. A detailed census study of *Macrocystis pyrifera* by Rosenthal et al. (1973) from 1967 to 1973 provided copious life history data from which Chapman (1979) has constructed a cohort life table. Dayton (1973) reported census data obtained in 1968 for the annual alga *Postelia palmaeformis*. During 1968–1970 Collins (1976) made a study of the population dynamics of the moss *Polytrichum alpestre*, the first of a series of census investigations on mosses; his analyses of population flux were modelled closely on the example of Sarukhán and Harper (1973). In the years since 1970 several further investigations on the demography of cryptogamic plants (mostly mosses and marine algae) have been undertaken. Their results are not qualitatively different from those now familiar to us from flowering plants and serve to demonstrate how techniques of plant demography based on accurate census data have universal application in the plant kingdom.

CONCLUSION

As a botanist trained originally in the academic tradition of Braun-Blanquet and Tüxen, but later refashioned somewhat by the Darwinian philosophy of John Harper, I should not perhaps express surprise at the biases of the textbooks of plant ecology. But I have come in the past few years to realize the extent of their bias through their virtual neglect of research on plant demography. And this despite the considerable volume of published work that has been available for a balanced view of the whole scope of vegetation ecology. It has now become clear to me that the failure to incorporate census studies earlier into the academic mainstream has hindered the development of vegetation science. Several ecologists from Tansley onwards have advocated a numerate, analytical science of plant ecology, but somehow never grasped that it was flourishing among rangeland ecologists of their own generation. In only one instance was there a significant language barrier to the transmission of information (from the Soviet Union to North America and most of Western Europe) so this cannot be held as a responsible cause. A sociological explanation may be forthcoming in future as we come to know more of the intellectual contacts between vegetation scientists. In this respect Tobey's (1981) study is a valuable source of information on the competing ideologies of plant ecologists associated either with Nebraska or with Chicago, despite Egerton's (1983) reservations about his analysis. Egerton (1983) partly identifies the problem, I believe: 'the history of terrestrial plant ecology seems more difficult to understand and appreciate than that of the other subdivisions of ecology . . . the main difficulty appears to be a parochialism that arose from the feeling among plant ecologists that different methods were appropriate for different kinds of vegetation and that the descriptive phase of their work need not await the time when a consensus on proper methods was reached'.

But I think a deeper issue is involved: the extent to which botanists influential in the development of vegetation science had a Darwinian view of nature. Botanists have been traditionally lukewarm to Darwin's theory of evolution by natural selection, as I have mentioned elsewhere (White 1984). Only a few plant ecologists in the past have seen in Chapters 3 and 4 of *On the Origin of Species* guidelines for a research programme on the life histories of plants, skilfully sketched by Darwin, but needing years of careful investigation. The first, I think, was Clements who had a marvellous appreciation of the value of experimental approaches to natural vegetation and of the necessity for a numerate ecological science. Despite his success in promoting his ideas on plant census research his students and followers seem, in the main, to have regarded the techniques as a means to a severely practical end, the increase of the grazing capacity of rangelands. There is little evidence (e.g., Canfield 1957) that the possession of uniquely detailed demographic information on plants was used by them to answer 'Darwinian' questions. Perhaps, too, the failure of Clements to transmit his vision of vegetation science through academic channels (he was not associated with University teaching after 1917), did not allow his ideas on plant demography to escape beyond the relative intellectual parochialism of the prairie and semi-desert grasslands. His grander theoretical writings were, on the other hand, widely debated. Tansley's (1947) encomium of him as 'by far the greatest creator of the modern science of vegetation' recognized his creation of 'a permanent structure of science, without which the amassing of detailed knowledge . . . can have no coherent meaning'. Perhaps, with fellow-feeling, Tansley revealed as much about himself in these words (see Godwin 1977): he had more sympathy with Clements' holistic philosophy of vegetation than with his practical expertise.

A half-century or more elapsed until in the 1960s the Darwinian approach to vegetation was resuscitated by John Harper, now more skilfully and with more conviction I believe than by any previous vegetation scientist. We live today in the midst of the development of a new style of plant ecology. I cannot claim to be a disinterested observer of its achievements, but it is incontrovertible that the demography of plants has now become a well-established feature of vegetation science.

ACKNOWLEDGEMENTS

I am grateful to those scientists, mentioned in the text, who have contributed information to this paper by their personal communications. I am especially grateful to T.A. Rabotnov for tireless correspondence over the years since 1970.

I acknowledge the comments of F.N. Egerton, J.L. Harper, J. Major, S.C. Martin, R.P. McIntosh, E.W. Tisdale, R.C. Tobey, R.M. Turner and O.B. Williams on an earlier version of the paper.

REFERENCES

Alcorn, S.M. and May, C. (1962) Attrition of a saguaro forest. Plant Dis. Rep. 46(3): 156–158.

Allee, W.C., Park, O., Emerson, A.E., Park, T. and Schmidt, K.P. (1949) Principles of Animal Ecology. Philadelphia: W.B. Saunders Co. 837 pp.

American Association for the Advancement of Science. (1963) Natural Areas as Research Facilities. Washington: AAAS. 193 pp.

Andel, J. van. (1975) A study on the population dynamics of the perennial plant species *Chamaenerion angustifolium* (L.) Scop. Oecologia 19: 329–337.

Antonovics, J. (1972) Population dynamics of the grass *Anthoxanthum odoratum* on a zinc mine. J. Ecol. 60: 351–365.

Ares, F.N. and Campbell, R.S. (1974) The Jornada Experimental Range: an epoch in the era of southwestern range management. Range Monograph 1. 74 pp. Denver, Colorado: Society for Range Management.

Armstrong, S.F. (1907) The botanical and chemical composition of the herbage of pastures and meadows. J. Agric. Sci. 2: 283–304.

Assmann, E. (1970) The Principles of Forest Yield Study. Oxford: Pergamon Press. 506pp.

Austin, M.P., Williams, O.B. and Belbin, L. (1981) Grassland dynamics under sheep grazing in an Australian Mediterranean type climate. Vegetatio 47: 201–211.

Barbour, M.G. (1969) Age and space distribution of the desert shrub *Larrea divaricata.* Ecology 50: 679–685.

Barbour, M.G., Burk, J.H. and Pitts, W.D. (1980) Terrestrial Plant Ecology. Menlo Park, California: The Benjamin/Cummings Publ. Co. 604pp.

Barkham, J.P. (1980) Population dynamics of the wild daffodil (*Narcissus pseudonarcissus*). I. Clonal growth, seed production, mortality and the effects of density. II. Changes in number of shoots and flowers, and the effect of bulb depth on growth and reproduction. J. Ecol. 68: 670–633; 635–664.

Barling, D.M. (1955) Some population studies in *Ranunculus bulbosus* L. J. Ecol. 43: 207–218.

Beatley, J.C. (1967) Survival of winter annuals in the northern Mojave Desert. Ecology 48: 745–750.

Beeftink, W.G. (1975) Vegetationskundliche Dauerquadratforschung auf periodisch überschwemmten und eingedeichten Salzböden in Südwesten der Niederlande. pp. 567–578 in W. Schmidt (Ed.) Sukzessionsforschung. Vaduz: J. Cramer.

Beeftink, W.G. (1978) Aspects of population dynamics in *Halimione portulacoides* communities. Vegetatio 36: 31–43.

Biswell, H.H. and Graham, C.A. (1956) Plant counts and seed production on California annual-type ranges. J. Range Mgmt. 9: 116–118.

Blackman, F.F. and Tansley, A.G. (1905) Ecology in its physiological and phytogeographical aspects. New Phytol. 4: 199–203; 232–253.

Blaisdell, J.P. (1958) Seasonal development and yield of native plants on the upper Snake River plains and their relation to certain climatic factors. U.S. Dept. Agric. Tech. Bull. 1190, 68 pp.

Bogdanovskaya-Gienef, I.D. (1926) (Contribution to seed renewal in meadow communities). Zap. Lening. Sel. Khoz. Inst. 3: 216–253. (in Russian).

Bogdanovskaya-Gienef, I.D. (1954) (Seed renewal of plants in the meadow coenoses of the forest zone). Uch. Zap. Lening. Univ. Biol. 34: 3–47. (in Russian).

Booth, W.E. (1943) Tripod method of making chart quadrats. Ecology 24: 262.

Bormann, F.H. and Berlyn, G. (Ed.) (1981) Age and growth rates of tropical trees: new directions for research. Yale Sch. For. Envir. Stud. Bull. 94. 137pp.

Böttcher, H. (1974) Bibliographie zum Problem der Sukzessionsforschung mit Hilfe von Dauerquadraten und der Vegetationskartierung. Excerp. Bot. B14: 35–56.

Böttcher, H. (1975) Stand der Dauerquadratforschung in Mitteleuropa. pp. 31–36 in W. Schmidt (Ed.) Sukzessionsforschung. Vaduz: J. Cramer.

Bradshaw, M.E. (1981) Monitoring grassland plants in Upper Teesdale, England. pp. 241–251 in H. Synge (Ed.). The Biological Aspects of Rare Plant Conservation. Chichester: John Wiley.

Bradshaw, M.E. and Doody, J.P. (1978a) Plant population studies and their relevance to nature conservation. Biol. Conserv. 14: 223–242.

Bradshaw, M.E. and Doody, J.P. (1978b) Population dynamics and biology. pp. 48–63 in A.R. Chapman (Ed.) Upper Teesdale. London: Collins.

Braun-Blanquet, J. (1964) Pflanzensoziologie. 3rd ed. Wien: Springer. 865pp.

Brougham, R.W., Glenday, A.C. and Fejer, S.O. (1960) The effects of frequency and intensity of grazing on the genotypic structure of a ryegrass population. N.Z. J. Agric. Res. 3: 442–453.

Brown, A.L. (1950) Shrub invasion of southern Arizona desert grassland. J. Range Mgmt. 3: 172–177.

Buffington, L.C. and Herbel, C.H. (1965) Vegetational changes on a semidesert grassland range from 1858 to 1963. Ecol. Mon. 35: 139–164.

Cable, D.R. (1979) Ecology of Arizona cottontop. U.S. For. Serv. Res. Pap. RM-209. 21pp.

Cameron, E. (1935) A study of the natural control of ragwort (*Senecio jacobaea* L.). J. Ecol. 23: 265–322.

Campbell, R.S. (1931) Plant succession and grazing capacity on clay soils in southern New Mexico. J. Agric. Res. 43: 1027–1051.

Canfield, R.H. (1939) The effect of intensity and frequency of clipping on density and yield of black grama and tobosa grass. U.S. Dept. Agric. Tech. Bull. 681. 32pp.

Canfield, R.H. (1948) Perennial grass composition as an indicator of condition of southwestern mixed grass ranges. Ecology 29: 190–204.

Canfield, R.H. (1957) Reproduction and life span of some perennial grasses of southern Arizona. J. Range Mgmt. 10: 199–203.
Cannon, W.A. (1921) Plant habits and habitats in the arid portions of South Australia. Carnegie Inst. Wash. Publ. 308. 139pp.
Caughley, G. (1977) Analysis of Vertebrate Populations. London: Wiley. 234pp.
Chapline, W.R., Campbell, R.S., Price, R. and Stewart G. (1944) The history of western range research. Agric. Hist. 18: 127–143.
Chapman, A.R.O. (1979) Biology of Seaweeds. Baltimore: University Park Press. 134pp.
Charles, A.H. (1961) Differential survival of cultivars of *Lolium*, *Dactylis* and *Phleum*. J. Brit. Grassl. Soc. 16: 69–75.
Charles, A.H. (1964) Differential survival of plant types in swards. J. Brit. Grassl. Soc. 19: 198–204.
Chew, R.M. and Chew, A.E. (1965) The primary productivity of a desert shrub (*Larrea tridentata*) community. Ecol. Mon. 35: 355–375.
Cittadino, E. (1980) Ecology and the professionalization of botany in America, 1890–1905. pp. 171–198 in W. Coleman and C. Limoges (Eds.) Studies in History of Biology, Vol. 4. Baltimore: The Johns Hopkins University Press.
Clarke, S.E., Tisdale, E.W. and Skoglund, N.A. (1943) The effects of climate and grazing practices on short-grass prairie vegetation in southern Alberta and southwestern Saskatchawan. Domin. Dept. Agric. Tech. Bull. 46. 53pp.
ClaveranA, R. (1966) Two modifications to the vegetation photographic charting method. J. Range Mgmt. 19: 371–373.
Clements, E.S. (1960) Adventures in Ecology: Half a Million Miles from Mud to Macadam. New York: Pageant Press. 244pp.
Clements, F.E. (1905) Research Methods in Ecology. Lincoln, Nebraska: University Publishing Co. 334pp. (Reprint 1977, New York: Arno Press.)
Clements, F.E. (1909) Darwin's influence upon plant geography and ecology. Amer. Nat. 43: 143–151.
Clements, F.E. (1910) The life history of lodgepole pine burn forests. U.S. Dept. Agric. For. Serv. Bull. 79. 56pp.
Clements, F.E. (1916) Plant Succession. An Analysis of the Development of Vegetation. Washington: Carnegie Institution of Washington. 512pp.
Clements, F.E. (1918) Ecology. Carn. Inst. Wash. Yrbk. 16: 303–306.
Clements, F.E. (1919) Ecology. Carn. Inst. Wash. Yrbk. 17: 287–297.
Clements, F.E. (1920) Ecology. Carn. Inst. Wash. Yrbk. 18: 330–343.
Clements, F.E. (1921) Ecology. Carn. Inst. Wash. Yrbk. 19: 341–366.
Clements, F.E. (1922) Ecology. Carn. Inst. Wash. Yrbk. 20: 389–411.
Clements, F.E. (1923) Ecology. Carn. Inst. Wash. Yrbk. 21: 337–358.
Clements, F.E. (1924a) Ecology. Carn. Inst. Wash. Yrbk. 22: 302–322.
Clements, F.E. (1924b) Ecology. Carn. Inst. Wash. Yrbk. 23: 246–269.
Clements, F.E. (1925) Ecology. Carn. Inst. Wash. Yrbk. 24: 309–343.
Clements, F.E. (1926) Ecology. Carn. Inst. Wash. Yrbk. 25: 335–371.
Clements, F.E. (1927) Ecology. Carn. Inst. Wash. Yrbk. 26: 305–337.
Clements, F.E. (1930) Ecology. Carn. Inst. Wash. Yrbk. 29: 228–241.
Clements, F.E., Weaver, J.E. and Hanson, H.C. (1929) Plant Competition: an Analysis of Community Functions. Washington: Carnegie Institution of Washington. 304pp.
Collins, N.J. (1976) Strategies of growth and population dynamics of tundra plants. 2. Growth and population dynamics of the moss *Polytrichum alpestre* in the maritime Antarctic. Oikos 27: 389–401.
Cooper, W.S. (1912) The climax forest of Isle Royale, Lake Superior, and its development. I. Bot. Gaz. 55: 1–44.
Cooper, W.S. (1923) The recent ecological history of Glacier Bay, Alaska: III. Permanent quadrats at Glacier Bay: an initial report upon a long-period study. Ecology 4: 355–365.
Cooper, W.S. (1924) An apparatus for photographic recording of quadrats. J. Ecol. 12: 317–321.
Cooper, W.S. (1939) A fourth expedition to Glacier Bay, Alaska. Ecology 20: 130–155.
Crisp, M.D. (1978) Demography and survival under grazing of three Australian semi-desert shrubs. Oikos 30: 520–528.

Crisp, M.D. and Lange, R.T. (1976) Age structure, distribution and survival under grazing of the arid-zone shrub *Acacia burkittii*. Oikos 27: 86–92.
Cullen, J.M., Weiss, P.W. and Wearne, G.R. (1978) A plotter for plant population studies. New Phytol. 81: 443–448.
Curtis, J.T. (1946) Use of mowing in management of white ladyslipper. J. Wildlife Mgmt. 10: 303–308.
Curtis, J.T. and Greene, H.C. (1953) Population changes in some native orchids of southern Wisconsin, especially in the University of Wisconsin Arboretum. Orchid J. 2: 152–155.
Darwin, C. (1855) Vitality of seeds. Gdn. Chron. Agric. Gaz. 1855, no. 46: 758. Reprinted in P.H. Barrett (ed.) 1977. The Collected Papers of Charles Darwin. Chicago: University of Chicago Press. 277+326pp.
Darwin, C. (1859) On the Origin of Species by Means of Natural Selection. London: John Murray. 490pp.
Daubenmire, R. (1968) Plant Communities. A Textbook of Plant Synecology. New York: Harper & Row. 300pp.
Davies, W. (1933) The history of pasture analysis. Agric. Prog. 10: 225–231.
Davy, A.J. and Jefferies, R.L. (1981) Approaches to the monitoring of rare plant populations. pp. 219–232 in H. Synge (Ed.) The Biological Aspects of Rare Plant Conservation. Chichester: John Wiley.
Dayton, P.K. (1973) Dispersion, dispersal, and persistence of the annual intertidal alga, *Postelia palmaeformis* Ruprecht. Ecology 54: 433–438.
Deevey, E.S. (1947) Life tables for natural populations of animals. Q. Rev. Biol. 22: 283–314.
Dempster, J.P. and Lakhani, K.H. (1979) A population model for cinnabar moth and its food plant, ragwort. J. Anim. Ecol. 48: 143–163.
Dittberner, P.L. (1971) A Demographic Study of Some Semidesert Grassland Plants. M.S. Thesis, New Mexico State University, Las Cruces, 77pp.
Du Rietz, G.E. (1930) Vegetationsforschung auf soziationsanalytischer Grundlage. Handb. biol. Arbeitsmeth. 11(5,2): 293–480.
Egerton, F.N. (1968) Ancient sources for animal demography. Isis 59: 175–189.
Egerton, F.N. (1983) The history of ecology: achievements and opportunities, part one. J. Hist. Biol. 16: 259–310.
Ellenberg, H. (1952) Wiesen und Weiden und ihre standörtliche Bewertung. Stuttgart: Ulmer. 143pp.
Ellison, L. (1942) Overlays as a visual aid in analysis of permanent quadrat records. Ecology 23: 482–484.
Ellison, L. (1949) Establishment of vegetation on depleted subalpine range as influenced by microenvironment. Ecol. Mon. 19: 95–121.
Ellison, L. and Woolfolk, E.J. (1937) Effects of drought on vegetation near Miles City, Montana. Ecology 18: 329–336.
Engledow, F.L. (1926) A census of an acre of corn (by sampling). J. Agric. Sci. 16: 166–195.
Epling, C., Lewis, H. and Ball, F.M. (1960) The breeding group and seed storage: a study in population dynamics. Evolution 14: 238–255.
Ericson, L. (1981) Aspects of the shore vegetation of the Gulf of Bothnia. Wahlenbergia 7: 45–60.
Ernst, W.H.O. (1979) Population biology of *Allium ursinum* in northern Germany. J. Ecol. 67: 347–362.
Fatchen, T.J. (1978) Changes in grazed *Atriplex vesicaria* and *Kochia astrotricha* (Chenopodiaceae) populations, 1929–1974. Trans. R. Soc. S. Aust. 102: 39–42.
Gatsuk, L.E., Smirnova, O.V., Vorontzova, L.I., Zaugolnova, L.B. and Zhukova, L.A. (1980) Age states of plants of various growth forms: a review. J. Ecol. 68: 675–696.
Glendening, G.E. (1952) Some quantitative data on the increase of mesquite and cactus on a desert grassland range in southern Arizona. Ecology 33: 319–328.
Glendening, G.E. and Paulsen, H.A. (1955) Reproduction and establishment of velvet mesquite as related to invasion of semidesert grasslands. U.S. Dept. Agric. Tech. Bull. 1127. 50pp.
Godwin, H. (1977) Sir Arthur Tansley: the man and the subject. J. Ecol. 65: 1–26.
Godwin, H., Clowes, D.R. and Huntley, B. (1974) Studies in the ecology of Wicken Fen. V. Development of fen carr. J. Ecol. 62: 197–214.
Greig-Smith, P. (1964) Quantitative Plant Ecology 2nd ed. London: Butterworths. 256pp.

Groh, H. and Dore, W.G. (1945) A milkweed survey in Ontario and adjacent Quebec. Sci. Agri. 25: 463–481.
Hall, E.A.A., Specht, R.L. and Eardley, C.M. (1964) Regeneration of the vegetation on Koonamore Vegetation Reserve, 1926–1962. Aust. J. Bot. 12: 205–264.
Hanson, H.C. and Love, L.D. (1930) Comparison of methods of quadratting. Ecology 11: 734–748.
Harcourt, D.G. (1970) Crop life tables as a pest management tool. Can. Entomol. 102: 950–955.
Harper, J.L. (1960) Factors controlling plant numbers. pp. 119–132 in J.L. Harper (Ed.) The Biology of Weeds. Oxford: Blackwell Scientific Publications.
Harper, J.L. (1961) Approaches to the study of plant competition. Symp. Soc. Expt. Biol. 15: 1–39.
Harper, J.L. (1964) The individual in the population. J. Ecol. 52 (Suppl): 149–158.
Harper, J.L. (1967) A Darwinian approach to plant ecology. J. Ecol. 55: 247–270.
Harper, J.L. (1977) Population Biology of Plants. London: Academic Press. 892pp.
Harper, J.L. (1981) The concept of population in modular organisms. pp. 53–77 in R.M. May (Ed.) Theoretical Ecology. Principles and Application (2nd ed.). Oxford: Blackwell Scientific Publications.
Harper, J.L. (1982) After description. pp. 11–25 in E.I. Newman (Ed.) The Plant Community as a Working Mechanism. Oxford: Blackwell Scientific Publications.
Harper, J.L. (1983) A Darwinian plant ecology. pp. 323–345 in D.S. Bendall (Ed.) Evolution from Molecules to Men. Cambridge: Cambridge University Press.
Harper, J.L. and Bell, A.D. (1979) The population dynamics of growth form in organisms with modular construction. pp. 29–52 in R.M. Anderson, B.D. Turner and L.R. Taylor (Ed.) Population Dynamics. Oxford: Blackwell Scientific Publications.
Harper, J.L., Clatworthy, J.N., McNaughton, I.H. and Sagar, G.R. (1961) The evolution and ecology of closely related species living in the same area. Evol. 15: 209–227.
Harper, J.L. and Sagar, G.R. (1953) Some aspects of the ecology of buttercups in permanent grassland. Proc. Brit. Weed Control Conf. 1: 256–263.
Harper, J.C. and White, J. (1971) The dynamics of plant populations. pp. 41–63 in P.J. den Boer and G.R. Gradwell (Ed.). Dynamics of Populations. Wageningen: Centre for Agricultural Publishing and Documentation.
Harper, J.L. and White, J. (1974) The demography of plants. Ann. Rev. Ecol. Syst. 5: 419–463.
Hart, C.A. and Gleason, H.A. (1907) On the biology of the sand areas of Illinois. Bull. Ill. State Lab. Nat. Hist. 7(7): 131–272.
Hastings, J.R. and Alcorn, S.M. (1961) Physical determinations of growth and age in the giant cactus. J. Ariz. Acad. Sci. 2: 32–39.
Hasting, J.R. and Turner, R.M. (1965) The Changing Mile. An Ecological Study of Vegetation Change with Time in the Lower Mile of an Arid and Semiarid Region. Tucson: University of Arizona Press. 317pp.
Hawthorn, W.R. and Cavers, P.B. (1976) Population dynamics of the perennial herbs *Plantago major* L. and *P.rugellii* Decne. J. Ecol. 64: 511–527.
Heady, H.F. (1958) Vegetational changes in the California annual type. Ecology 39: 402–416.
Herbel, C.H., Dittberner, P.L. and Bickle, T.S. (1970) A quantitative ecology of the Jornada Experimental Range. pp. I 133–I 178 in R.G. Wright and G.M. van Dyne (Eds.) Simulation and analysis of dynamics of a semi-desert grassland. Colorado State Univ. Range Sci. Dept. Report No. 6.
Hennessy, J.T., Gibbens, R.P., Tromble, J.M. and Cardenas, M. (1983) Vegetation changes from 1935 to 1980 in mesquite dunelands and former grasslands of southern New Mexico. J. Range Mgmt. 36: 370–374.
Hett, J.M. and Loucks, O.L. (1968) Application of life-table analyses to tree seedlings in Quetico Provincial Park, Ontario. For. Chron. 4: 29–32.
Hill, R.R. (1917) Effects of grazing upon western yellow-pine reproduction in the national forests of Arizona and New Mexico. U.S. Dept. Agric. Bull. 580. 27pp.
Hill, R.R. (1920) Charting quadrats with a pantograph. Ecology 1: 270–273.
Hironaka, M. and Tisdale, E.W. (1963) Secondary succession in annual vegetation in Idaho. Ecology 44: 810–812.

Hoffmann, H. (1865) Untersuchungen zur Klima- und Bodenkunde mit Rüchsicht auf die Vegetation. Bot. Zeit. 23 (Beilage): 1–124.

Hogeweg, P., Hesper, B., van Schaik, C. and Beeftink, W.G, (1985) Patterns in vegetation succession, an ecomorphological study. pp. 637–666 in J. White (Ed.) The Population Structure of Vegetation. Dordrecht: Dr. W. Junk.

Hutchinson, G.E. (1978) An Introduction to Population Ecology. New Haven: Yale University Press. 260pp.

Jenkin, T.J. (1919) Pasture Studies: Some Results. Bangor: Jarvis and Foster. 58pp.

Johnson, W.M. (1953) Effect of grazing intensity upon vegetation and cattle gains on ponderosa pine-bunchgrass ranges of the Front Range of Colorado. U.S. Dept. Agric. Circ. 929. 36pp.

Kain, J.M. (1971) Synopsis of biological data on *Laminaria hyperborea*. FAO Fish. Synop. 87. 74pp.

Keever, C. (1950) Causes of succession on old fields of the Piedmont, North Carolina. Ecol. Mon. 20: 231–250.

Kerster, H.W. (1968) Population age structure in the prairie forb, *Liatris aspera*. BioSci. 18: 430–432.

Lang, R.L. (1945) Density changes in native vegetation in relation to precipitation. Wyom. Agric. Expt. Stn. Bull. 272. 31pp.

Lawes, J.B., Gilbert, J.H. and Masters, M.T. (1882) Agricultural, botanical and chemical results of experiments on the mixed herbage of permanent meadow, conducted for more than twenty years in succession on the same land. Phil. Trans. R. Soc. Lond. 173: 1181–1413.

Lawrence, D.B. (1979a) William Skinner Cooper (1884–1978). Bull. Ecol. Soc. Amer. 60: 18–19.

Lawrence, D.B. (1979b) Primary versus secondary succession at Glacier Bay National Monument, Southeastern Alaska. pp. 213–224 in R.M. Linn (Ed.) Proceedings of the First Conference on Scientific Research in the National Parks, New Orleans, Louisiana, 1976. Washington, D.C.: US Department of the Interior, National Park Service.

Laycock, W.A. (1969) Exclosures and natural areas on rangelands in Utah. U.S. For. Serv. Res. Pap. INT-62. 44pp.

Laycock, W.A. (1975) Range reference areas. Soc. Range Mgmt., Range Sci. Ser. 3. 66pp.

Levin, D.A. and Kerster, H.W. (1974) Gene flow in seed plants. Evol. Biol. 7: 139–220.

Lewis, H. (1962) Catastrophic selection as a factor in speciation. Evol. 16: 257–271.

Linkola, K. (1935) Über die Dauer und Jahresklassenverhältnisse des Jugendstadiums bei einigen Wiesenstauden. Acta For. Fenn. 42: 1–56.

Loftfield, J.V.G. (1924) Grazing ranges of Northern Arizona. Carn. Inst. Wash. Yrbk. 23: 264–265.

Lüdi, W. (1930) Die Methoden der Sukzessionsforschung in der Pflanzensoziologie. Handb. biol. Arbeitsmeth. 11(5,3): 527–728.

Mack, R.N. and Pyke, D.A. (1979) Mapping individual plants with a field-portable digitizer. Ecology 60: 459–461.

Malik, A.R., Anderson, D.J. and Myerscough, P.J. (1976) Studies on structure in plant communities. VII. Field and experimental analyses of *Atiplex vesicaria* populations from the Riverine plain of New South Wales. Aust. J. Bot. 24: 265–280.

Martin, S.C. and Reynolds, H.G. (1973) The Santa Rita range: your facility for research on semidesert ecosystems. J. Ariz. Acad. Sci. 8: 56–67.

Martin, S.C. and Turner, R.M. (1977) Vegetation changes in the Sonoran Desert region, Arizona and Sonora. J. Ariz. Acad. Sci. 12: 59–69.

Matthews, V.B. and Conrad, P.W. (1968) An ecological life history of tall bluebell (*Mertensia arizonica* var. *Leonardi*) in Utah. Ecology 49: 1113–1123.

McCarthy, M.K. and Scifres, C.J. (1969) Life cycle studies with musk thistle. Nebr. Agric. Expt. Stn. Res. Bull. 230. 15pp.

McGinnies, W.G. (1981) Discovering the Desert: Legacy of the Carnegie Desert Botanical Laboratory. Tuscon: University of Arizona Press. 276pp.

McIntosh, R.P. (1975) H.A. Gleason – "individualistic ecologist" – 1882–1975: his contributions to ecological theory. Bull. Torrey Bot. Club 102: 253–273.

McIntosh, R.P. (1976) Ecology since 1900. pp. 353–372 in B.J. Taylor and T.J. White (Eds.) Issues and Ideas in America. Norman, Oklahoma: University of Oklahoma Press.

McIntosh, R.P. (1980) The background and some current problems of theoretical ecology. Synthese 43: 195–255.

McIntosh, R.P. (1983) Pioneer support for ecology. Bioscience 33(2): 107–112.

Moore, A.W. and Reid, E.H. (1951) The Dallas pocket gopher and its influence on forage production of Oregon mountain meadows. U.S. Dept. Agric. Circ. 884. 36pp.

Mueller-Dombois, D. and Ellenberg, H. (1974) Aims and Methods of Vegetation Ecology. New York: John Wiley 547pp.

Nelson, E.W. (1934) The influence of precipitation and grazing upon black grama grass range. U.S. Dept. Agric. Tech. Bull. 409. 32pp.

Niering, W.A., Whittaker, R.H. and Lowe, C.H. (1963) The saguaro: a population in relation to environment. Science 142: 15–23.

Noble, I.R. (1977) Long-term biomass dynamics in an arid chenopod shrub community at Koonamore, South Australia. Aust. J. Bot. 25: 639–653.

Noble, I.R. and Crisp, M.D. (1979) Germination and growth models of short-lived grass and forb populations based on long term photo-point data at Koonamore, South Australia. Israel J. Bot. 28: 195–210.

North, W.J. (1971) Introduction and background. pp. 1–97 in W.J. North (Ed.) The Biology of Giant Kelp Beds (*Macrocystis*) in California. Lehre: J. Cramer.

Norton, B.E. (1978) The impact of sheep grazing on long-term successional trends in salt desert shrub vegetation of southwestern Utah. Proc. 1st Intern. Rangeland Cong., Denver, Colorado, pp. 610–613.

Oosting, H.J. (1956) The Study of Plant Communities. 2nd ed. San Francisco: W.H. Freeman. 440pp.

Osborn, T.G.B., Wood, J.G. and Paltridge, T.B. (1931) On the autecology of *Stipa nitida*, a study of a fodder grass in arid Australia. Proc. Linn. Soc. N.S.W. 56: 299–324.

Osborn, T.G.B., Wood, J.G. and Paltridge, T.B. (1932) On the growth and reaction to grazing of the perennial saltbush, *Atriplex vesicarium*. An ecological study of the biotic factor. Proc. Linn. Soc. N.S.W. 57: 377–402.

Osborn, T.G.B., Wood, J.G. and Paltridge, T.B. (1935) On the climate and vegetation of the Koonamore Vegetation Reserve to 1931. Proc. Linn. Soc. N.S.W. 60: 392–427.

Osmond, C.B., Bjorkman, O. and Anderson, D.J. (1980) Physiological Processes in Plant Ecology: Towards a Synthesis with Atriplex. Berlin: Springer. 468pp.

Parker, K.W. and Martin, S.C. (1952) The mesquite problem on southern Arizona ranges. U.S. Dept. Agric. Circ. 908. 70pp.

Paulsen, H.A. and Ares, F.N. (1961) Trends in carrying capacity and vegetation on an arid southwestern range. J. Range Mgmt. 14: 78–83.

Pearse, K. (1935) An area-list method of measuring range plant populations. Ecology 16: 573–579.

Pearse, K., Pechanec, J.F. and Pickford, G.D. (1935) An improved pantograph for mapping vegetation. Ecology 16: 529–530.

Pelton, J. (1953) Ecological life cycle of seed plants. Ecology 34: 619–628.

Perttula, U. (1941) Untersuchungen über die generative und vegetative Vermehrung der Blütenpflanzen in der Wald-, Hainswiesen und Hainfelsenvegetation. Ann. Acad. Sci: Fenn., Ser A, 58: 1–387.

Petersen, W. (1975) Population. 3rd ed. New York: Macmillan. 784pp.

Piemeisel, R.L. (1951) Causes affecting change and rate of change in a vegetation of annuals in Idaho. Ecology 32: 53–72.

Pierce, W.R. and Eddleman, L.E. (1970) A field stereophotographic technique for range vegetation analysis. J. Range Mgmt. 23: 218–220.

Pierce, W.R. and Eddleman, L.E. (1973) A test of stereophotographic sampling in grasslands. J. Range Mgmt. 26: 148–150.

Pound, R. and Clements, F.E. (1898) A method of determining the abundance of secondary species. Minn. Bot. Stud. 2(1): 19–24.

Pound, R. and Clements, F.E. (1900) The Phytogeography of Nebraska. 2nd edition. Lincoln, Nebraska: Published by the Seminar.

Rabotnov, T.A. (1950) (The life cycle of perennial herbaceous plants in meadow coenoses). Trudy Bot. Inst. A.N. SSSR., Ser. 3, 6: 7–204. (in Russian).

Rabotnov, T.A. (1958) (The life cycle of *Ranunculus acer* L. and *R.auricomus* L.). Biull. Mosk. Obshch. Ispy. Prir. Otd. Biol. 63(6): 77–86. (in Russian).

Rabotnov, T.A. (1961) Some problems of increasing the proportion of leguminous species in permanent meadows. Proc. 8th Int. Grassl. Cong. pp. 260–264.
Rabotnov, T.A. (1969a) On coenopopulations of perennial herbaceous plants in natural coenoses. Vegetatio 19: 87–95.
Rabotnov, T.A. (1969b) Plant regeneration from seed in meadows of the USSR. Herbage Abs. 39: 269–277.
Robotnov, T.A. (1978a) (Structure and method of study of coenotic populations of perennial herbaceous plants). Ekol. 2: 5–13. (in Russian; English translation in Soviet J. Ecol. 9: 99–105).
Rabotnov, T.A. (1978b) On coenopopulations of plants reproducing by seeds. pp. 1–26 in A.H.J. Freysen and J.W. Woldendorp (Eds.) Structure and Functioning of Plant Populations. Amsterdam: North-Holland Publ. Co.
Rabotnov, T.A. (1980) Bibliography of papers on the problems of coenopopulations published in the U.S.S.R. Part I. Part II. Excerp. Bot. B20: 71–96; 171–206.
Rabotnov, T.A. (1981) Bibliography of papers on the problems of coenopopulations published in the U.S.S.R. Part III. Excerp. Bot. B21: 91–119.
Rabotnov, T.A. and Saurina, N.I. (1971) (Number of individuals and age spectrum of some populations of *Ranunculus acris* L. and *R.auricomus* L.) Bot. Zhur. 56: 476–484. (in Russian).
Renner, F.G., Crafts, E.C., Hartman, T.C. and Ellison, L.C. (1938) A selected bibliography on management of western ranges, livestock, and wildlife. U.S. Dept. Agric. Misc. Publ. 281. 486pp.
Rosenthal, R.J., Clarke, W.D. and Dayton, P.K. (1974) Ecology and natural history of a stand of giant kelp, *Macrocystis pyrifera*, off Del Mar, California. Fish. Bull. 72: 670–684.
Runge, F. (1963) Die Artmächtigkeitsschwankungen in einem nordwestdeutschen Enzian-Zwenkenrasen. Vegetatio 11: 237–240.
Runge, F. (1971) Jährliche Schwankungen der Individuenzahl in einer nordwestdeutschen Trockenenheide. III. Vegetatio 23: 71–76.
Runge, F. (1975) 18-Jährige Erfahrungen mit Dauerquadraten. pp. 39–51 in W. Schmidt (Ed.) Sukzessionsforschung. Vaduz: J. Cramer.
Sagar, G.R. (1959) The Biology of Some Sympatric Species of Grassland. D. Phil. Thesis, University of Oxford.
Sagar, G.R. (1970) Factors controlling the size of plant populations. Proc. 10th Brit. Weed Control Conf. 1970, pp. 959–979.
Sampson, A.W. (1908) The revegetation of overgrazed range areas. Preliminary report. U.S. Dept. Agric. For. Serv. Circ. 158. 21pp.
Sampson, A.W. (1913a) The reseeding of depleted grazing lands to cultivated forage plants. U.S. Dept. Agric. Bull. 4. 34pp.
Sampson, A.W. (1913b) Range improvement by deferred and rotational grazing. U.S. Dept. Agric. Bull. 34. 14pp.
Sampson, A.W. (1914) Natural revegetation of range lands based upon growth requirements and life history of the vegetation. J. Agric. Res. 3: 93–147.
Sampson, A.W. (1917) Important range plants: their life history and forage value. U.S. Dept. Agric. Bull. 545. 63pp.
Sampson, A.W. (1919) Plant succession in relation to range management. U.S. Dept. Agric. Bull. 790. 76pp.
Sampson, A.W. (1923) Range and Pasture Management. New York: John Wiley. 421pp.
Sampson, A.W. (1924) Native American Forage Plants. New York: John Wiley. 435pp.
Sampson, A.W. (1926) Grazing periods and forage production on the national forests. U.S. Dept. Agric. Bull. 1405. 54pp.
Sampson, A.W. (1952) Range Management Principles and Practices. New York: John Wiley. 570pp.
Sarukhán, J. and Gadgil, M. (1974) Studies on plant demography: *Ranunculus repens* L., *R.bulbosus* L. and *R.acris* L. III. A mathematical model incorporating multiple modes of reproduction. J. Ecol. 62: 921–936.
Sarukhán, J. and Harper, J.L. (1973) Studies on plant demography: *Ranunculus repens* L., *R.bulbosus* L. and *R.acris* L. I. Population flux and survivorship. J. Ecol. 61: 675–716.
Sarvis, J.T. (1920) Composition and density of the native vegetation in the vicinity of the Northern Great Plains Field Station. J. Agric. Res. 19: 63–72.

Sarvis, J.T. (1923) Effects of different systems and intensities of grazing upon the native vegetation at the Northern Great Plains Field Station. U.S. Dept. Agric. Bull. 1170. 45pp.
Sarvis, J.T. (1941) Grazing investigations on the northern Great Plains. North Dakota Agric. Expt. Stn. Bull. 308. 110pp.
Schoener, T.W. (1974) Resource partitioning in ecological communities. Science 185: 27–39.
Schroeter, C. (1910) Ueber pflanzengeographische Karten. Actes III Cong. Int. Bot., Bruxelles. 1: 97–154.
Shantz, H.L. (1906) A study of the vegetation of the Mesa Region east of Pike's Peak: the Bouteloua formation. I. Structure of the formation. II. Development of the formation. Bot. Gaz. 42: 16–47, 179–207.
Shantz, H.L. (1911) Natural vegetation as an indicator of the capabilities of land for crop production in the Great Plains area. U.S. Dept. Agric. Bur. Plant Ind. Bull. 201. 100pp.
Shantz, H.L. and Piemeisel, R.L. (1924) Indicator significance of the natural vegetation of the southwestern desert region. J. Agric. Res. 28: 721–801.
Shantz, H.L. and Piemeisel, R.L. (1940) Types of vegetation in Escalante Valley, Utah, as indicators of soil conditions. U.S. Dept. Agric. Tech. Bull. 713. 46pp.
Sharitz, R.R., Fransham, D. and Patterson, K.K. (Eds.) (1984) International Society of Plant Population Biologists: Membership Directory 1984. Aiken, South Carolina: Savannah River Ecology Laboratory. 54pp.
Sharitz, R.R. and McCormick, J.F. (1972) Population dynamics of two competing annual plant species. Ecology 54: 723–740.
Shreve, F. (1911) Establishment behavior of the palo verde. Plant World 14: 289–299.
Shreve, F. (1917) The establishment of desert perennials. J. Ecol. 5: 210–216.
Shreve, F. (1929) Changes in desert vegetation. Ecology 10: 364–373.
Shreve, F. and Hinckley, A.L. (1937) Thirty years of change in desert vegetation. Ecology 18: 463–478.
Silander, J.A. (1983) Demographic variation in the Australian desert cassia under grazing pressure. Oecologia 60: 227–233.
Silvertown, J. (1980) The evolutionary ecology of mast seeding in trees. Biol. J. Linn. Soc. London. 14: 235–250.
Silvertown, J.W. (1982) Introduction to Plant Population Ecology. London: Longman. 209pp.
Sinclair, G. (1824) Hortus Gramineus Woburnensis. London: James Ridgway. 438pp.
Smeins, F.E., Taylor, T.W. and Merrill, L.B. (1976) Vegetation of a 25-year exclosure on the Edwards Plateau, Texas. J. Range Mgmt. 29: 24–29.
Smirnova, O.V., Zaugolnova, L.B. and Ermakova, I.M. (Eds.) (1976) (Plant Coenopopulations: Basic Concept and Structure). Moscow: Nauka. 216pp. (in Russian).
Smith, D.M. (1962) The Practice of Silviculture. 7th ed. New York: John Wiley. 578pp.
Smith, W.G. (1905) The Central Committee for the Survey and Study of British Vegetation. New Phytol. 4: 254–257.
Soane, I.D. and Watkinson, A.R. (1979) Clonal variation in populations of *Ranunculus repens*. New Phytol. 82: 557–573.
Southwood, T.R.E. (1978) Ecological Methods with Particular Reference to the Study of Insect Populations (2nd ed.). London: Chapman & Hall.
Spalding, V.M. (1909) Distribution and Movements of Desert Plants. Washington: Carnegie Institution of Washington. 144pp.
Stapledon, R.G. (1912) Pasture problems: drought resistance. J. Agric. Sci. 5: 129–151.
Stapledon, R.G. and Jenkin, T.J. (1916) Pasture problems: indigenous plants in relation to habitat and sown species. J. Agric. Sci. 8: 26–64.
Stauffer, R.C. (1975) Charles Darwin's Natural Selection. Cambridge: Cambridge University Press. 694pp.
Stebler, F.G. and Schröter, C. (1887) Einfluss des Beweidens auf die Zusammensetzung des Rasens. Landwirts. Jb. Schweiz 1: 178–190.
Steenberg, W.F. and Lowe, C.H. (1977) Ecology of the saguaro. II. Reproduction, germination, establishment, growth and survival of the young plant. U.S. Dept. Interior, Natl. Park. Serv., Sci. Mon. 8. 242pp.
Steenbergh, W.F. and Lowe, C.H. (1983) Ecology of the saguaro. III. Growth and demography. U.S. Dept. Interior, Natl. Park. Serv., Sci. Mon. 17. 228pp.
Sterk, A.A. (1975) Demographic studies of *Anthyllis vulneraria* L. in the Netherlands. Acta Bot. Neerl. 24: 315–337.

Sterk, A.A., van Duijkeren, A., Hogervorst, J. and Verbeek, E.D.M. (1982) Demographic studies of *Anthyllis vulneraria* in The Netherlands. II. Population density fluctuations and adaptations to arid conditions, seed populations, seedling mortality, and influence of the biocoenosis on demographic features. Acta Bot. Neerl. 31: 11–40.

Stewart, G., Cottam, W.P. and Hutchings, S.H. (1940) Influence of unrestricted grazing on northern salt desert plant associations in western Utah. J. Agric. Res. 60: 289–316.

Stewart, G. and Hutchings, S.S. (1936) The point-observation-plot (square-foot density) method of vegetation survey. J. Amer. Soc. Agron. 28: 714–722.

Stone, W.C. and Fryer, J.R. (1935) A botanical study of pasture mixtures. Sci. Agric. 15: 777–805.

Summerhays, V.S. (1968) Wild Orchids of Britain. 2nd Ed. London: Collins. 366pp.

Symonides, E. (1979) The structure and population dynamics of psammophytes on inland dunes. I. Populations of initial stages. II Loose-sod populations. III Populations of compact psammophyte communities. IV Population phenomena as a phytocoenose-forming factor (a summing-up discussion). Ekol. Polska 21: 3–37; 191–234; 235–257; 259–281.

Symonides, E. (1983) Population size regulation as a result of intra-population interactions. I. Effect of density on the survival and development of *Erophila verna* (L.) C.A.M. Ekol. Pol. 31: 839–881.

Talbot, M.W. and Cronemiller, F.P. (1961) Some of the beginnings of range management. J. Range Mgmt. 14: 95–102.

Tamm, C.O. (1948) Observations on reproduction and survival of some perennial herbs. Bot. Not. 1948: 305–321.

Tamm, C.O. (1956) Further observations on the survival and flowering of some perennial herbs. I. Oikos 7: 273–292.

Tamm, C.O. (1964) Survival of individual perennial herbs, with some remarks on the role of plant persistance in vegetation dynamics. 10th Intern. Bot. Cong. Abst. 629.

Tamm, C.O. (1972a) Survival and flowering of some perennial herbs. II. The behaviour of some orchids on permanent plots. Oikos 23: 23–28.

Tamm, C.O. (1972b) Survival and flowering of some perennial herbs. III. The behaviour of *Primula veris* on permanent plots. Oikos 23: 159–166.

Tansley, A.G. (1922) Studies of the vegetation of the English chalk. II. Early stages of redevelopment of woody vegetation on chalk grassland. J. Ecol. 10: 168–177.

Tansley, A.G. (1923) Practical Plant Ecology: A Guide for Beginners in Field Study of Plant Communities. London: G. Allen and Unwin. 228pp.

Tansley, A.G. (1947) Frederic Edward Clements 1874–1945. J. Ecol. 34: 194–196.

Tansley, A.G. and Adamson, R.S. (1925) Studies of the vegetation of the English chalk. III. The chalk grasslands of the Hampshire–Sussex border. J. Ecol. 13: 173–223.

Tansley, A.G. and Chipp, T.F. (1926) Aims and Methods in the Study of Vegetation. London: The British Empire Vegetation Committee. 383pp.

Taylor, W.P. and Loftfield, J.V.G. (1924) Damage to range grasses by the zuni prairie dog. U.S. Dept. Agric. Bull. 1227. 15pp.

Thomas, A.G. and Dale, H.M. (1975) The role of seed reproduction in the dynamics of established populations of *Hieracium floribundum* and a comparison with that of vegetative reproduction. Can. J. Bot. 53: 3022–3031.

Thurston, J.M., Williams, E.D. and Johnstone, A.E. (1976) Modern developments in an experiment on permanent grassland started in 1856: effects of fertilisers and lime on botanical composition and crop and soil analyses. Ann. Agron. 27: 1043–1082.

Tisdale, E.W. (1976) Vegetational responses following biological control of *Hypericum perforatum* in Idaho. Northwest Sci. 50: 61–75.

Tobey, R.C. (1981) Saving the Prairies. The Life Cycle of the Founding School of American Plant Ecology, 1895–1955. Berkeley: University of California Press. 315pp.

Tschirley, F.H. and Martin, S.C. (1961) Burroweed on southern range lands. Arizona Agric. Expt. Stn. Tech. Bull. 146. 34pp.

Turkington, R. and Harper, J.L. (1979) The growth, distribution and neighbour relations of *Trifolium repens* in a permanent pasture. I. Ordination, pattern and contact. J. Ecol. 67: 201–218.

Turner, R.M., Alcorn, S.M. and Olin, G. (1969) Mortality of transplanted saguaro seedlings. Ecology 50: 835–844.

Turner, R.M., Applegate, L.H., Bergthold, P.M., Gallizioli, S. and Martin, S.C. (1980) Arizona range reference areas. U.S. For. Serv. Gen. Tech. Rep. RM-79. 34pp.
United States Senate (1936) The western range: a great but neglected natural resource. U.S. Senate Doc. 199. 620pp.
Uranov, A.A. (Ed.) (1967) (Ontogenesis and Age-State Composition of Flowering Plant Populations). Moscow: Nauka. 155pp. (in Russian).
Uranov, A.A. (Ed.) (1968) (Problems of Morphogenesis of Flowering Plants and Composition of their Populations). Moscow: Nauka. 228pp. (in Russian).
Uranov, A.A., Zaugolnova, L.B. and Smirnova, O.V. (Eds.) (1977) (Plant Coenopopulations: Development and Interrelations). Moscow: Nauka. 134pp. (in Rusian).
Wager, H.G. (1938) Growth and survival of plants in the Arctic. J. Ecol. 26: 390–410.
Wallace, A. and Romney, E.M. (1972) Radioecology and ecophysiology of desert plants at the Nevada Test Site. U.S. Atomic Energy Commission Report TID-25954. 439pp.
Walter, H. (1968) Die Vegetation der Erde in öko-physiologischer Betrachtung. Band II. Die gemässigten und arktischen Zonen. Stuttgart: Gustav Fischer. 1001pp.
Watt, A.S. (1923) On the ecology of British beechweeds with special reference to their regeneration. Part I. J. Ecol. 11: 1–48.
Watt, A.S. (1940) Contributions to the ecology of bracken (*Pteridium aquilinum*). I. The rhizome. New Phytol. 39: 401–422.
Watt, A.S. (1943) Contributions to the ecology of bracken (*Pteridium aquilinum*). II. The frond and the plant. New Phytol. 42: 103–126.
Watt, A.S. (1945) Contributions to the ecology of bracken (*Pteridium aquilinum*). III. Frond types and the make up of the population. New Phytol. 44: 156–178.
Watt, A.S. (1947) Contributions to the ecology of bracken (*Pteridium aquilinum*). IV. The structure of the community. New Phytol. 46: 97–121.
Watt, A.S. (1957) The effect of excluding rabbits from grassland B (Mesobrometum) in Breckland. J. Ecol. 45: 861–878.
Watt, A.S. (1960) Population changes in acidophilous grass-heath in Breckland, 1936–1957. J. Ecol. 48: 605–629.
Watt, A.S. (1962) The effect of excluding rabbits from grassland A (Xerobrometum) in Breckland (1936–1960). J. Ecol. 50: 181–198.
Watt, A.S. (1967) The differentiation and fate of the bracken (*Pteridium aquilinum*) frond and their relation to the age-structure of the shoot and frond population. New Phytol. 66: 75–84.
Watt, A.S. (1971a) Factors controlling the floristic composition of some plant communities in Breckland. pp. 137–152 in E. Duffey and A.S. Watt (Eds.) The Scientific Management of Animal and Plant Communities for Conservation. Oxford: Blackwell Scientific Publications.
Watt, A.S. (1971b) Contributions to the ecology of bracken (*Pteridium aquilinum*). VIII. The marginal and the hinterland plant: a study in senescence. New Phytol. 70: 967–986.
Watt, A.S. (1974) Senescence and rejuvenation in ungrazed chalk grassland (grassland B) in Breckland: the significance of litter and of moles. J. Appl. Ecol. 11: 1157–1171.
Watt, A.S. (1981) Further observations on the effects of excluding rabbits from grassland A in East Anglian Breckland: the pattern of change and factors affecting it (1936–73). J. Ecol. 69: 509–536.
Weaver, J.E. (1918) The quadrat method in teaching ecology. Plant World 21: 267–283.
Weaver, J.E. (1924) Plant production as a measure of environment. A study in crop ecology. J. Ecol. 12: 205–237.
Weaver, J.E. (1954a) North American Prairie. Lincoln, Nebraska: Johnsen Publ. Co. 348pp.
Weaver, J.E. (1954b) A seventeen-year study of plant succession in prairie. Amer. J. Bot. 41: 31–38.
Weaver, J.E. (1968) Prairie Plants and Their Environment: A Fifty-Year Study in the Midwest. Lincoln, Nebraska: University of Nebraska Press. 276pp.
Weaver, J.E. and Albertson, F.W. (1936) Effects of the great drought on the prairies of Iowa, Nebraska and Kansas. Ecology 17: 567–639.
Weaver, J.E. and Albertson, F.W. (1956) Grasslands of the Great Plains: Their Nature and Use. Lincoln, Nebraska: Johnsen Publ. Co. 395pp.
Weaver, J.E. and Bruner, W.E. (1945) A seven-year quantitative study of succession in grassland. Ecol. Mon. 15: 297–319.
Weaver, J.E. and Hanson, W.W. (1941) Native midwestern pastures. Their origin, composition and degeneration. Nebraska Conserv. Bull. 22. 93pp.

Wells, T.C.E. (1967) Changes in a population of *Spiranthes spiralis* (L.) Chevall. at Knocking Hoe National Nature Reserve, Bedfordshire, 1962–65. J. Ecol. 55: 83–89.
Wells, T.C.E. (1968) Land-use changes affecting *Pulsatilla vulgaris* in England. Biol. Conser. 1: 37–43.
Wells, T.C.E. (1981) Population ecology of terrestrial orchids. pp. 281–295 in H. Synge (Ed.) The Biological Aspects of Rare Plant Conservation. Chichester: John Wiley.
West, N.E. (1979) Survival patterns of major perennials in salt desert shrub communities of southwestern Utah. J. Range Mgmt. 32: 442–445.
West, N.E., Rea, K.H. and Harniss, R.O. (1979) Plant demographic studies in sagebrush-grass communities of southeastern Idaho. Ecology 60: 376–388.
White, J. (1979) The plant as a metapopulation. Ann. Rev. Ecol. Syst. 10: 109–145.
White, J. (1984) Plant metamerism. pp. 15–47 in R. Dirzo and J. Sarukhán (Eds.) Perspectives on Plant Population Ecology. Sunderland, Mass.: Sinauer Assoc.
Whitford, P.B. (1949) Distribution of woodland plants in relation to succession and clonal growth. Ecology 30: 199–208.
Whittaker, R.H. (1962) Classification of natural communities. Bot. Rev. 28: 1–239.
Williams, C. (1963) Range exclosures within the state of Wyoming. Wyom. Range Mgmt. 170: 1–19.
Williams, O.B. (1966) Population changes of the perennial tussock-grass *Danthonia caespitosa* Gaudich in a semi-arid environment between the years 1949 and 1965. Proc. Ecol. Soc. Aust. 1: 111–117.
Williams, O.B. (1968) Studies in the ecology of the Riverine Plain. IV. Basal area and density changes of *Danthonia caespitosa* Gaudich. in a natural pasture grazed by sheep. Aust. J. Bot. 16: 565–578.
Williams, O.B. (1970a) Population dynamics of two perennial grasses in Australian semi-arid grassland. J. Ecol. 58: 869–875.
Williams, O.B. (1970b) Longevity and survival of some dietary constituents in a natural semi-arid grassland grazed by sheep. Proc. 11th Intern. Grassl. Cong., pp. 85–88.
Williams, O.B. (1981) Monitoring changes in populations of desert plants. pp. 233–240 in H. Synge (Ed.) The Biological Aspects of Rare Plant Conservation. Chichester: John Wiley.
Williams, O.B. and Mackey, B. (1983) Easy-care, no-hassle conservation – Mitchell grass (*Astrebla*). A case study from the 'South Oestrus' exclosure, 'Gilruth Plains', Cunnamulla, Queensland. pp. 141–145 in J. Messer and G. Moseley (Eds.) What Future for Australia's Arid Lands? Proceedings of the National Arid Lands Conference, Broken Hill, New South Wales, 1982. Australian Conservation Foundation.
Williams, O.B. and Roe, R. (1975) Management of arid grasslands for sheep: plant demography of six grasses in relation to climate and grazing. Proc. Ecol. Soc. Austr. 9: 142–156.
Williams, R.M. and Post, A.H. (1945) Dry land pasture experiment at the Central Montana Branch Station, Moccasin, Montana. Mont. Agr. Expt. Stn. Bull. 431. 31pp.
Willson, M.F. and Burley, N. (1983) Mate Choice in Plants. Tactics, Mechanisms and Consequences. Princeton, New Jersey: Princeton University Press. 251pp.
Wit, C.T. de (1960) On Competition. Versl. Landbouwk. Onderz. 66: 1–82.
Wood, J.G. (1936) Regeneration of the vegetation of Koonamore Vegetation Reserve, 1926–1936. Trans. R. Soc. S. Aust. 60: 96–111.
Wright, R.G. (1972) Computer processing of chart quadrat maps and their use in plant demographic studies. J. Range Mgmt. 25: 476–478.
Wright, R.G. and van Dyne, G.M. (1976) Environmental factors influencing semidesert grassland perennial grass demography. Southwest. Nat. 21: 259–274.
Wright, R.G. and van Dyne, G.M. (1981) Population age structure and its relationship to the maintenance of a semidesert grassland undergoing invasion by mesquite. Southwest. Nat. 26: 13–22.

VEGETATION CHANGES IN HERBACEOUS COMMUNITIES

REINHARD BORNKAMM

ABSTRACT

The stages of successional change of vegetation are described from a number of experimental plots observed over several years. Changes in succession which may be described phytosociologically as a sequence of vegetation types are founded on the demographic properties of individual species.

INTRODUCTION

In vegetation science Warming (1896) was the first to stress the general nature of succession as a feature of vegetation. Warming thus founded the European tradition of the science of succession. (The results of Lawes and Gilbert in the Rothamsted experiment, see Williams 1978, seem not to have been a constituent part of knowledge in botanical science at that time.) In the young discipline of phytosociology Braun-Blanquet (1928) devoted 10% of his book to the chapter on community development or syndynamics; it is interesting to note that this was three times more than he needed for the chapter on synsystematics. Under the influence of Clements (1916) and other American workers and the publications of Lüdi (1924) and Furrer (1922), Braun-Blanquet's notions of seral development were based on spatial gradients of vegetation (zonations) that had been interpreted as temporal gradients (successions), where the time scale was arbitrary. But at the same time Braun-Blanquet and many others started research on permanent plots, as witnessed by the second and third editions of the same book (Braun-Blanquet 1951, 1964). The increasing experimental evidence concentrated the interest of plant ecologists on more detailed patterns within the succession in order to gain an understanding of the real process and its precise mechanisms. The principles of this type of research have been pointed out several times (e.g. Ellenberg 1956, Westhoff 1969, Feoli et al. 1975, Schmidt 1981). The methodical aspects were given first by Christiansen (1930) and by Lüdi (1932) in a lengthy monograph. Permanent plots now exist in all parts of the world.

White, J. (ed.), The Population Structure of Vegetation.
 ISBN 978-94-010-8927-2

The present paper deals with herbaceous communities in the lowlands of Central Europe. Several reviewers have summarized the work done so far in this area (Böttcher 1974, 1975, Schmidt 1974). In his compilation Schmidt (1974) mentioned 801 plots: the largest portion (512) was situated in salt-marsh communities, mostly in the Netherlands; 110 plots were established in grassland, 60 in weed communities, 58 in forest and similar communities, 61 plots in all other types of vegetation. If we disregard the special case of salt marshes we can observe that research has been done nearly exclusively in secondary succession, indeed mostly in man-made vegetation, and that most of the plots belong to a partial sere of a forest sere (in the sense of Braun-Blanquet 1964).

"Vegetation succession is the successive occurrence of phytocoena at a given site" according to van der Maarel and Werger (1980), following a formulation of F. McCormick. Phytosociological succession research starts with the plant community as a whole. It describes not only the sequence of stages (characterized by changes in the dominant species and dominant life forms) (Knapp 1974), but also the sequence of plant associations or of vegetation units below the association level. The analysis of the succession process leads to the detection of continuous small scale changes that have been described as micro-successions ("Kleinsukzessionen", Bornkamm 1962). The driving forces of these events are in fact the populations of species which grow in the plot. In the present paper this analysis will be carried out for a number of examples. It will be shown that phytosociological succession research, beginning at the community level and advancing through analytical steps, finally meets population ecology and autecology.

THE SEQUENCE OF STAGES

It was Warming (1896) who early designated the sequence of stages "annuals/biennals – perennial forbs and grasses – woody vegetation" as the overall sequence of seres in a forest biome. This sequence is exemplified by a series of life forms that characterize the various species at different sites. Such a sequence is probably generally correct for all succession investigations done so far in our region, but the duration of stages (Major 1974a) differs widely as a consequence of the variation of external factors, including the activity of man. Some communities with frequent disturbance (e.g., certain agricultural and ruderal communities) never reach the perennial stage. In others the perennials germinate even in the first year and reach dominance in the second. Similar differences exist for the change from herbaceous to woody vegetation.

For a number of examples the sequence of stages is recorded in Table 1. The experiments compared here were of different design. Mahn (1966) observed one plot of 1 m^2 size in a xerothermic grassland on a porphyry hill near Halle (GDR) for 5 years after burning. Schmidt (1981) set up a multifaceted experiment (total size 3450 m^2) in the New Botanical Garden of the University of Göttingen. The habitat is characterized by a deep

Table 1. Duration (years) of stages in a number of experimental succession plots

Site	Management	Annual	Perennial	Woody	Years recorded	Author
Dry grassland on porphyry near Halle (DDR)	Not used	1–2 (3 spring)	4–5 (3 autumn)	–	5	Mahn (1966)
Old field near Göttingen	Heat sterilized I	1	2–8	9	9	Schmidt (1981)
	Heat Sterilized II	1–3	4–9	–	9	
	Herbicide sterilized	1–3	3–9	–	9	
	Ploughed	1–3	4–9	–	9	
	Mown or milled	1–2	3–9	–	9	
Horticultural soil in Cologne	–	1	2–10	–	10	Bornkamm (1981b)
Dry grassland on limestone near Göttingen	–	1	2–29	–	29	Bornkhamm (1975 and unpublished)
Former garden in Berlin with different imported soils:						
sand	–	1–2	1–6	7–9	9	Bornkamm and Hennig (1982)
silty sand	–	1–2	3–9	–	9	
silty clay loam	–	1–2	3–9	–	9	
clay loam	–	1	2–9	–	9	
loamy sand	–	1	2–9	–	9	

loamy soil in the valley of a small stream in a (shell) limestone area. Some blocks were sterilized before starting the experiment, and different kinds of management were used (see Table 1). The third experiment (Bornkamm 1981b) consisted in ten plots of 1 m^2 size each on a nutrient-rich horticultural soil in the garden of the Botanical Institute of the University of Cologne. The succession was started by complete removal of the existing vegetation. After 2 years a stand of high forbs developed, dominated by *Urtica dioica*, which remained dominant until the end of the experiment.

The next experiment (Bornkamm 1975 and unpublished) was established in a dry calcarous grassland on rendzina soil on a south-facing slope near Göttingen. The vegetation before the war was grazed by cattle; it was ploughed once or twice after the last war and not used later on. Three very small plots were used: in plot B (2 m^2) the existing vegetation was removed completely; in plot A (2 m^2) the vegetation was removed except for the two dominating species *Bromus erectus* and *Brachypodium pinnatum*. In the first year of succession the stands were characterized by a great number of annual weeds of arable fields. In the second year perennial grasses became dominant, starting a long period of grassland vegetation. The smallest plot C (0.56 m^2) was a 40 cm deep soil core bearing a portion of a mesic meadow dominated by *Arrhenatherum elatius*. Plot C was transplated from its original site at the valley bottom to the dry slope near plots A and B. (Plot C is not included in Table 1 but will be dealt with later.)

The last experiment (Bornkamm and Hennig 1982, Bornkamm 1984) was established in the experimental area of the Institute of Ecology of the Technical University of Berlin. Here 10 m^2-blocks were available each containing 50 cm deep layers of one of five different imported soils; each block was separated into ten plots of 1 m^2 size. Since two blocks were used for each soil type the whole experiment initially comprised 100 plots (for the lay-out of these plots see Fig. 4). Here after a short period with stands of different species of annuals (corresponding to the different soil types) a vegetation of perennial forbs developed, dominated by *Solidago canadensis*. It was only in the sand plots that after less than one decade a shrubby vegetation was established which was dominated by *Sarothamnus scoparius*.

The sequential transition from bare soil to woody (or at least shrubby) vegetation has been established by direct visual evidence on permanent plots in only a few cases. Many investigations start with a stabilized vegetation type and the stage does not change during the period of observation. The transition from the perennial to the woody stage is, consequently, usually concluded from a comparison of plots of different known age. The change from the annual to the perennial stage is relatively well represented by direct data. This transition was already recorded in one of the earliest observations of this type: the conversion of a former garden into a road during the years 1908–1910 (Hildebrand 1911; see Table 2).

The differentiation of the 'annual' life-form, which in fact comprises a

Table 2. Number of species of various life-forms in a former garden changed into a road construction area (according to Hildebrand 1911)

Spec	1908	1909	1910*
Therophyta	9	19	9
Hemicryptophyta	6	23	12
Geophyta	1	1	1
Phanerophyta	–	1	1

*Incomplete because road work was finished.

number of varying potential life-forms (Sissingh 1952) enables us to look closer at this first change. Schmidt (1981) presented data from old fields where such a differentiation was made. It can be seen (Table 3A) that summer annuals (TA) decrease very rapidly whereas wintergreen annuals overwintering in a juvenile phase (TH) remain longer in the succession, taking advantage of their earlier growth. Annuals overwintering in their fully developed (summer) habit until they are stopped and occasionally killed by frost (TE), and bi-annuals (TB) do not play an important role. Later in succession the autumn germinating species decrease: Borstel (1974) investigated the offspring of seedlings in soil cores taken either in spring or in autumn from old fields bearing different succession phases; he observed a decrease of seedlings in the autumn samples as compared with the spring samples. This is not necessarily so under other conditions, as indicated by Table 3B which contains data from a dry grassland only 2 km away from Schmidt's plots (Bornkamm 1961, 1974).

The variation of annual life cycles is more important for the development of vegetation if open space is available during different seasons. In the experimental plots on different soils in Berlin (Bornkamm and Hennig 1982; see also Table 1) succession started in spring 1968. From year to year a number of plots were harvested and were allowed to start a new succession. Between the first series (that started growth in April) and the second series (starting growth in August, after harvesting) a distinct floristic difference was recognizable. As far as annuals were concerned the harvested plots showed increased cover values of *Apera spica-venti*, *Bromus hordeaceus*, *Bromus sterilis*, *Vicia sativa* ssp. *angustifolia* and *Vicia tetrasperma*, all of them winter annuals. A great number of species showed a decrease: most of them were summer annuals (*Anagallis arvensis*, *Euphorbia helioscopia*, *Galinsoga parviflora*, *Myosotis stricta*, *Sisymbrium officinale*, *Atriplex patula*, *Chenopodium album*, *Digitaria ischaemum* and *Gnaphalium uliginosum*) or Therophyta epeteia (*Poa annua*, *Senecio vulgaris* and *Capsella bursa-pastoris*). There was one biennial plant which preferred the harvested plots (*Oenothera biennis*) and three others preferred the non-harvested plots (*Conyza canadensis*, *Daucus carota*, *Verbascum lychnitis*). Differences remained over a number of years in the perennial vegetation also.

Table 3. Number of species of various life-forms growing in experimental plots near Göttingen: (A) in old fields in a stream valley (Schmidt 1981), (B) on a dry slope (Bornkamm 1974). TA = Therophyta aestivalia, TH = Therophyta hivernalia, TE = Therophyta epeteia, TB = Therophyta bisannualia, H = Hemicryptophyta, C = Chamaephyta, G = Geophyta, P = Phanerophyta.

	A.										B.									
Years	1	2	3	4	5	6	7	8	9	10	1	2	3	4	5	6	7	8	9	10
TA	9	9	7	5	3	2	0	0	2	1	6	5	4	3	2	2	1	–	2	1
TH	5	5	5	5	4	5	3	2	5	1	4	5	3	–	1	–	–	1	1	1
TE	3	3	2	2	–	–	–	–	–	–	–	1	–	–	–	–	–	–	–	–
TB	1	1	1	1	1	–	–	–	–	–	–	–	–	–	1	1	–	–	1	1
H	10	14	14	15	15	15	15	14	15	15	17	14	17	13	16	18	18	19	22	26
C	–	1	1	1	1	1	1	1	1	1	–	–	–	1	–	–	–	1	–	–
G	3	3	4	4	4	4	4	4	4	4	1	1	1	1	1	1	1	1	1	1
P	5	6	7	9	9	9	9	9	9	9	–	3	1	1	1	1	1	1	–	1

THE SEQUENCE OF PLANT COMMUNITIES

In spite of the fact that the study of succession was one of the first research topics of early phytosociology, as it was for other branches of vegetation science, it should not be overlooked that it raises considerable theoretical problems. Let us recall that the phytosociological method (see Westhoff and van der Maarel 1973) is based on relevés which must contain three basic elements: a chosen area, a complete list of species, and an estimation of the quantity of each species. At least the first two are absolutely necessary. In syntaxonomy vegetation types are defined as abstract units in a hierarchical system chiefly by floristic similarity (Braun-Blanquet 1964). It is assumed that floristic similarities and differences are caused by environmental factors, including human management. The many observations of the correlations between floristic composition and the quantity of a given environmental factor justify this assumption. The relation between vegetation unit and habitat is one of the crucial problems in phytosociology: not surprisingly, vegetation mapping is one of its widespread applications.

During succession in a given habitat, on the other hand, different stages of one stand follow each other in time. The stages can be described as vegetation types because of their floristic similarities with phytosociological units. But they also bear floristic similarities to each other, caused by their own history, by their own successional connections. It is well known that a number of vegetation systems are based on these syndynamic connections. This holds true even for some branches of phytosociology (Aichinger 1951, 1974), but this is an exception. Both views shall be combined in analysing a number of succession experiments listed in Table 1.

As an example we can consider first the Berlin experiment outlined earlier (details in Bornkamm & Hennig 1982). The sociological status of the species according to Oberdorfer (1979) shows that on sand (Fig. 1a) the vegetation started with a dry type of Sisymbrion community (first and

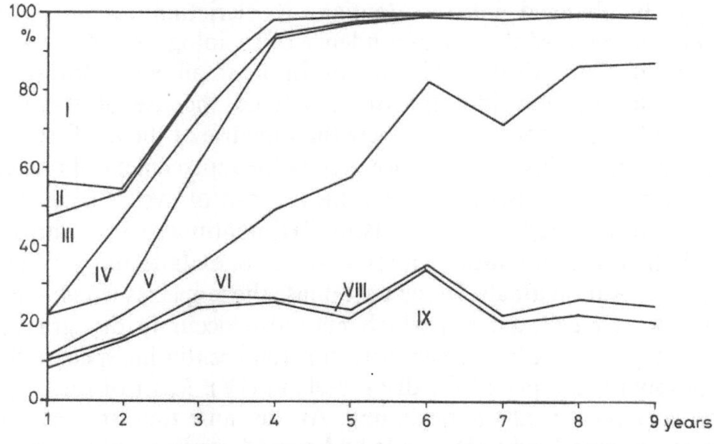

Fig. 1a. Change of sociological groups during the first years of a succession on sand in Berlin.

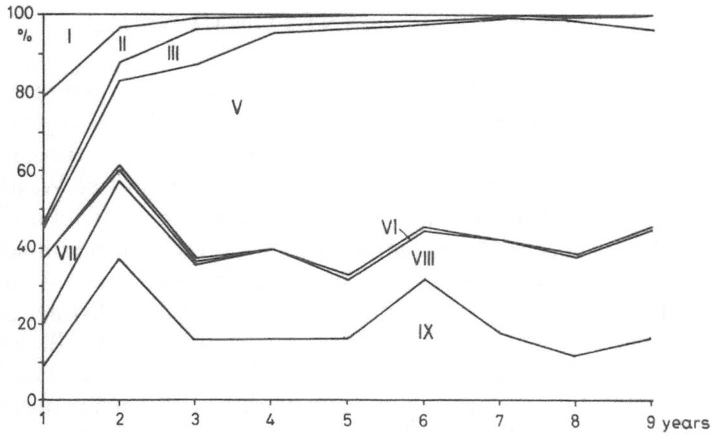

Fig. 1b. Change of sociological groups during the first years of succession on loam in Berlin. I, Sisymbrion and Sisymbrietalia; II, Polygono-Chenopodion and Polygono-Chenopodietalia; III, Chenopodietea class; IV, differentiating species of drier subunits of Chenopodietea communities; V, Arction and Artemisietea class; VI, Sarothamnion and Querco-Fagetea; VII, Plantaginetea class; VIII, Agropyretea class; IX, others.

second year, transition in the third year) changing into an Arction community by the fourth and fifth year. In the sixth year shrubs became dominant, notably *Sarothamnus scoparius*, an indicator of the Sarothamnion alliance. On loam (Fig. 1b) the vegetation started with a Polygono-Chenopodion community, changing in the second year into an Arction community that turned out to be dominant for a number of years. In the first years species of trodden habitats (Plantaginetea class) were present in considerable amounts, whereas species of ruderal grassland communities (Agropyretea class) invaded and remained constant throughout the experiment.

In the Göttingen experiment on limestone outlined earlier (details in Bornkamm 1974, 1975) a Caucalido-Adonidetum (alliance Caucalidion) very quickly changed into a Gentiano-Koelerietum (alliance Mesobromion). In this case the correspondence of sociological indicators with changes of life forms (Table 1) is evident. In the small plot C transplanted from a mesic meadow into the dry grassland the use of sociological indicators allows a more detailed view than the use of the life forms only. The mechanism of this change is not merely the replacement of one group of characteristic species by another, but a series of events (Fig. 2). The dominance of the mesic meadow plants (III) continues to the third year. Some of them die out rather quickly, other ones fade out or keep on growing as tiny individuals. In the next phase those species which typically grow in moist meadows, but which may also occur in dry grasslands, increase: they can here be used as indicators (differentiating species) for the more mesophilous types of the dry grassland (IV). Most of these species were already present at the beginning. At the same time species characteristic of dry grassland (VI) invade and spread considerably. Groups IV and VI match each other after 7 years, followed by a steady dominance of

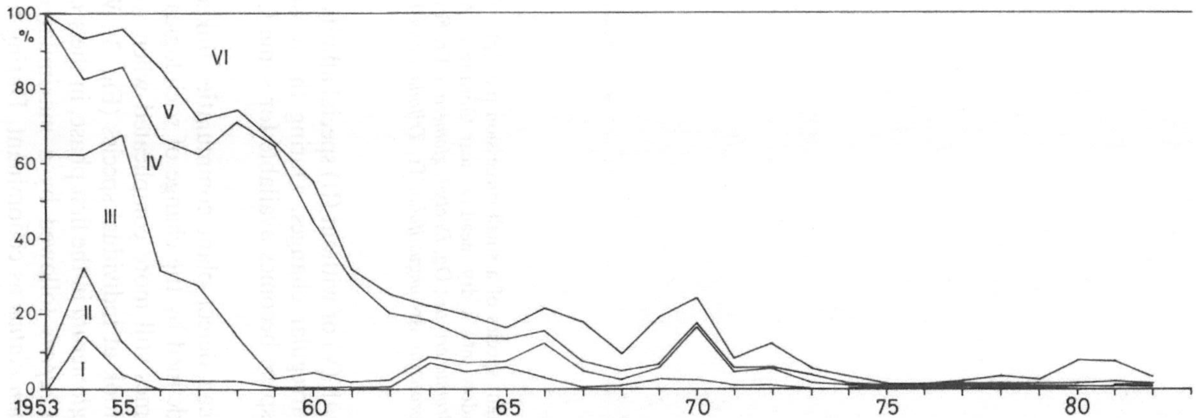

Fig. 2. Change of sociological groups during a succession on a south-facing slope in a limestone area near Göttingen. I, Caucalido-Adonidetum and Secalinetea class; II, species not included in groups I or III–VI; III, Dauco-Arrhenatheretum and Molinio-Arrhenatheretea class; IV, like III, but at the same time with differentiating species of the mesophilous subunits of the Gentiano-Koelerietum; V, like VI but at the same time with differentiating species of the drier subunits of the Dauco-Arrhenatheretum; VI, Gentiano-Koelerietum and Festuco-Brometea class.

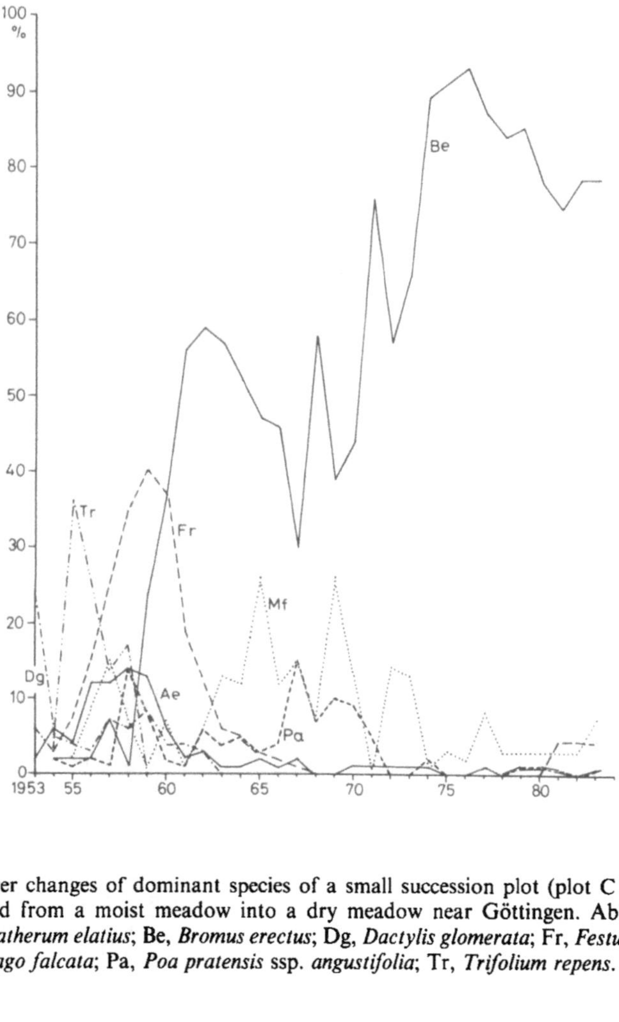

Fig. 3. Cover changes of dominant species of a small succession plot (plot C = $0.56\,m^2$) transplanted from a moist meadow into a dry meadow near Göttingen. Abbreviations: Ae, *Arrhenatherum elatius*; Be, *Bromus erectus*; Dg, *Dactylis glomerata*; Fr, *Festuca rupicola*; Mf, *Medicago falcata*; Pa, *Poa pratensis* ssp. *angustifolia*; Tr, *Trifolium repens*.

group VI. Other species with (V) or without (II) special affinity to the two types of grassland show irregular changes. During the most dramatic changes (first years) free space becomes available for some Caucalidion species (I).

In these examples we can detect plant communities linked together through different phases defined by the change of sociological indicator species. The picture becomes still more complicated when we move to the population level and look at individual species (Fig. 3). We can see the maximum of *Dactylis glomerata* in the first phase, in the second phase a maximum of *Trifolium repens* followed by a maximum of *Festuca rupicola* with *Arrhenatherum elatius* as codominant. During ten years of increasing dominance of *Bromus erectus* two other species (*Medicago falcata* and *Poa pratensis* ssp. *augustifolia*) pass through lower maxima. The position of *Bromus* as the sole exclusive dominant is established after 21 years.

A very similar sequence of plant populations with increasing dominance has been found during succession of salt marshes (Beeftink et al. 1971, Hogeweg et al. 1985). After frequent changes in the first years, a rather stable phase of *Poa pratensis* grassland is reached.

These series of population changes raise the question of the mechanisms in action during these processes. In the present paper attention will be paid to the change from the short-lived species to the perennials, and to changes between the herbaceous perennials. The change from herbaceous perennials to woody plants will not be discussed.

THE TRANSITION FROM SHORT-LIVED TO PERENNIAL PLANTS

It is very well established that the first vegetation cover by annuals in a succession is regulated by the immigration of seeds or by the soil seed bank, and by the special characteristics of germination and growth (Knapp 1967, Raynal and Bazzaz 1973, Harper 1977, Bazzaz 1979). The initial steps of the succession in the first year are primarily brought about by the life cycles of the short-lived plants. This agrees with the earlier findings of Keever (1950).

In our Berlin experiment annuals and biennials germinated in the first year. The annuals mostly showed maximum cover and frequency in the first year, except for *Vicia sativa* ssp. *angustifolia* and *Bromus sterilis* which exhibited maxima after c. 6 years. Both these species produce very large seeds that permit germination within established perennial vegetation. In the case of *Bromus* longevity of seed caused by light-induced dormancy (Pollard 1982) may also play a role. In the first year the most important perennials like *Solidago canadensis* and *Artemisia vulgaris* were present in a minority (c. 30%) of the plots only (Table 4). In most plots they invaded in the course of the following years. For this reason the dominance of short-lived plants, especially biennials, lasted more than one year in a number of plots. Within the experiment the most important annuals were *Senecio vulgaris* (on sand and loam) and *Sonchus asper* (more on loam); the most important biennials were *Conyza canadensis* and *Oenothera biennis*; the most important perennials were *Artemisia vulgaris* and *Solidago canadensis*. The behaviour of these six species is clearly reflected in the figures in Table 4. There is an obvious parallelism between *Senecio* and *Sonchus*: in the first year both species exhibited high frequency and biomass as well as considerable mean cover and number of individuals. From the first to the second year mean cover and biomass decreased by nine-tenths of the former values; the species were still rather frequent but showed smaller individual size. The shoot : root ratio remained high. In the harvested plots these species do not occur any more in the third year.

Conyza canadensis is usually supposed to be a biennial plant but is well able to flower in the first year. Its occurrence was more extended than that of the first two species. Mean cover, frequency and biomass did not

Table 4. Development of the six most important species in the first 4 years of the Berlin experiment

		All plots:		Harvested plots only:							
		Mean cover (%)	Mean frequency (%)	Mean biomass (dry weight) ($g \cdot m^{-2}$)		Mean number[1] of individuals (m^{-2})		Mean dry weight per individual (g)		Shoot : root ratio[2]	
	Year			R	P	R	P	R	P	R	P
Senecio vulgaris											
	1	12.3	64	1.5	36.6	66	69	0.02	0.53	4.1	9.0
	2	1.7	27	–	3.9	–	16	–	0.25	–	8.2
	3	0.1	5	–	–	–	–	–	–	–	–
	4	0.0	2	–	–	–	–	–	–	–	–
Sonchus asper											
	1	10.6	61		26.8		19		1.37		4.8
	2	0.9	24		3.8		4		0.85		6.1
	3	0.2	7		–		–		–		–
	4	0.0	1		–		–		–		–
Conyza canadensis											
	1	7.0	54		13.7		18		0.78		5.7
	2	7.6	56		14.5		302		0.05		8.7
	3	2.8	31		0.6		35		0.02		6.8
	4	1.2	5		–		–		–		–
Oenothera biennis											
	1	3.0	26	0.4	–	0.7	–	0.58	–	9.3	–
	2	4.7	42	–	5.1	–	0.2	–	23.8	–	10.0
	3	5.3	46	0.2	12.6	0.6	0.6	0.38	21.0	7.3	6.4
	4	6.5	44	0.3	24.4	5	14	0.07	1.78	2.9	5.7

Artemisia vulgaris											
	1	1.5	29	0.1	17.3	0.1	0.5	1.01	37.0	2.2	4.0
	2	4.4	47	0.04	3.3	0.4	0.4	0.11	9.1	3.9	3.4
	3	11.8	65	0.01	140.0	0.2	4	0.03	31.8	2.0	2.6
	4	11.3	68	0.4	3.9	0.8	0.2	0.45	19.3	2.3	5.5
Solidago canadensis											
	1	2.4	33	0.5	3.0	5	0.7	0.11	4.6	4.2	3.1
	2	9.9	72	0.1	98.2	0.6	15	0.17	6.6	5.4	3.8
	3	26.2	84	0.3	326.8	4	36	0.07	9.1	1.8	3.6
	4	42.2	86	0.04	743.0	0.2	55	0.22	13.6	2.1	2.3

R = non-flowering rosettes or seedlings; P = vegetative or flowering plants.
[1]Number of shoots in *Solidago*; in all other cases plants are separate genetic individuals.
[2]Ratio of above ground to below ground parts.

alter from the first to the second year. The number of individuals and shoot : root ratio reached a maximum in the second year. This indicates a greater number of flowering plants in the second year, because the shoot : root ratio is much higher in reproductive specimens than in non-flowering rosettes (6.1 vs. 3.3 in a sample where the ratios have been determined separately). The individual size decreased rapidly from the first to the second year. Evidently germination and growth become more and more difficult for the short-lived plants.

Oenothera biennis as a true biennial holds an intermediate position between annuals and perennials. It increased its cover, frequency, number of individuals and biomass up to the fourth year. But the size of the plants and their shoot : root ratio decreased. In the second year it was most frequently the highest plant among the vegetation of its plot. It showed maximum cover values around the years 4–6 and decreased after that (Bornkamm and Hennig 1982). *Artemisia vulgaris* and *Solidago canadensis* differed in the more irregular distribution of *Artemisia* and its shorter duration in the succession. They both started with a frequency of about 30%, but a cover of only 2%, and both parameters increased rapidly with time. With *Solidago* the biomass, number of shoots size and increased simultaneously. This stands in direct contrast to the behaviour of the annuals.

Bornkamm and Schrade (1981) carried out experiments with *Galinsoga parviflora* as an annual, *Oenothera biennis* as a biennial and *Solidago canadensis* as a perennial in pure and mixed cultivations through one vegetation period. The annual plant dominated with respect to cover and biomass until summer, when it flowered. In autumn the biennial plant exhibited the greatest amount of biomass with large rosettes, whereas the perennial plant was rather inconspicuous. The shoot : root ratio was high in *Oenothera*, intermediate in *Galinsoga* and declined from 2 to < 1 in *Solidago* over 12 weeks. This means that the perennial plant stored most of the photosynthetic products in the underground parts, thus preparing for its success in the next growing season. This is known for a number of perennial plants like *Tussilago farfara* and *Agropyron repens* (Hahn et al. 1979), *Chamaenerion angustifolium* (van Andel et al. 1979, van Andel and Nelissen 1979), and *Solidago canadensis* itself (Schäfer and Werner 1979). A number of *Oenothera* plants flowered in the first year and behaved like annuals. This observation points to the fact that our knowledge of the multiplicity of short life-cycles is very incomplete.

TRANSITIONS BETWEEN HERBACEOUS PERENNIALS

The invasion of perennials is regulated in the same way as annuals by the supply of seeds. In addition invasion by rhizomes or other types of clonal growth play a role. Once established, the perennials form a rather irregularly dispersed pattern (microfacies, Knapp 1967, 1974) changing by numerous small steps (microsuccession, Bornkhamm 1962) to a more complex community structure. For the first phases it is very important which species invade first or are present from the beginning. Perennial

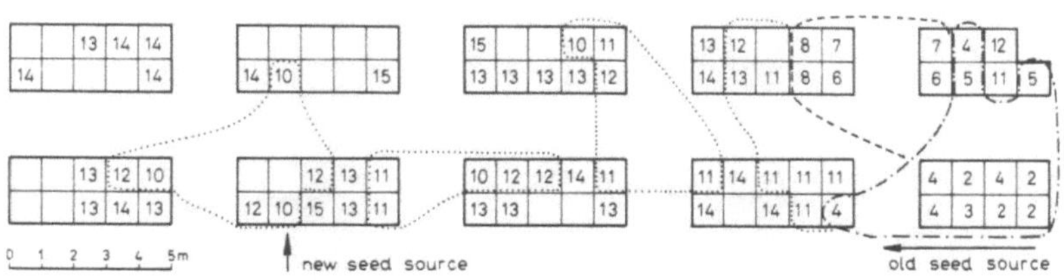

Fig. 4. Spread of *Arrhenatherum elatius* in the experimental plots in Berlin. Numbers indicate the year of the first appearance in each plot (year 1 = 1968). The lines enclose plots colonized by *Arrhenatherum* within periods of 5, 8 and 12 years. Plots without number are devoid of *Arrhenatherum*.

plant communities are 'founder controlled' (Yodzis 1978) to a great extent. Experimental data for this assertion have been given by Arens (1962) and others. With the vigorous spread of a small number of dominant plants, types of herbaceous vegetation can rather quickly reach a considerable structural stability (Bornkamm 1981a), but internal changes are nevertheless occurring. Both processes, invasion and repartition of species already existing in the plot, shall be discussed in more detail.

Figure 4 shows the invasion of *Arrhenatherum elatius* in the Berlin experiment. Its fruits were probably present in the seed bank at the beginning of the experiment in some plots at the southern end (right hand side of Fig. 4). It spread slowly during the next few years. In the vicinity of the northern part of the plots a new seed source became available in 1975 by setting up a separate experiment containing *Arrhenatherum*. After the very dry year in 1976, the grass invaded several new plots: two distinct waves in the moist years 1977 (year 10) and 1980 (year 13) can be observed in several plots.

The change and replacement of species already existing in the plots is accomplished by clonal growth rather than by seed reproduction. In the transition from the rough pattern of microfacies to a more finely patterned community structure, Knapp (1967) formally discerned the following processes: overgrowing (e.g., high forb over small grass), undergrowing (the reverse), penetration (e.g., mixing up of rhizome systems), lateral crowding, outliving. These all are means of interference of the shoots; considerably less is known about the interference of roots. Early successional plants show a large overlap of their root systems, whereas later successional species exhibit a stronger niche separation (Parrish and Bazzaz 1976). For example, the interaction of roots of the perennial grass *Poa pratensis* led to a restricted root growth and especially to lower lateral growth of the annual grass *Bromus tectorum* (Bookman and Mack 1982). An important factor for the year-to-year fluctuation of both roots and shoots is supposed to be the distribution of available soil space in spring. Early root growth seems to be an advantage in this process (Bornkamm 1961). Important too is the fact that roots (or rhizomes) as truly perennial organs are accumulated to a higher extent than the predominantly summergreen shoots (Bornkamm 1981b). It has been shown (Bornkamm

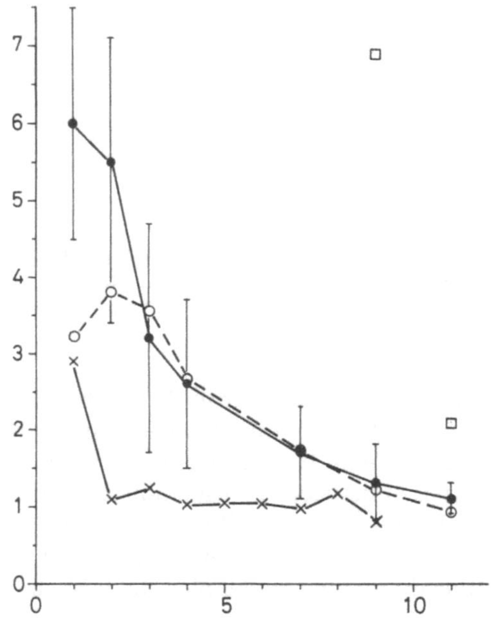

Fig. 5. Changes in shoot : root ratios during succession. Berlin experiment: ●, whole herbaceous plots; □, whole shrubby plots; ○, *Solidago canadensis* alone. Cologne experiment: ×, whole plots. Plot size was 1 m^2. Ratios are shown on the ordinate and time in years on the abscissa.

1981b) that this accumulation can be responsible for the accumulation of carbon and nitrogen in a succession (Major 1974b, c). In our Berlin experiment (Table 4) *Solidago canadensis* increased its biomass during the first 4 years. This increase kept on until the ninth year (Bornkamm 1984). At the same time it decreased its shoot : root ratio to < 1 (Fig. 5). The shoot : root ratio of the whole vegetation on the plots decreased from 6 to 1 in the Berlin experiment, and from 3 to less than 1 in the Cologne experiment, where *Urtica dioica* was the dominant species. In two of the Berlin plots, where shrubby vegetation had developed the ratio increased again (Fig. 5). Thus the perennial herbaceous communities pass through a phase of very low shoot : root ratios before the ratio is raised again by the woody plants.

Comparing the two stages discussed so far the conclusion can be drawn that the annual stage is regulated primarily by the seed bank (Raynal and Bazzaz 1973) and by the germination and growth characteristics (Bazzaz 1979) of the given species, whereas plant interactions become more and more important with the establishment of the perennial vegetation. This agrees with the opinions expressed by Keever (1950) and by Parrish and Bazzaz (1976). The different kinds of interactions (Malcolm 1966) may lead to niche separation and avoidance of interference. If niche separation becomes progressively more distinct during succession a more and more

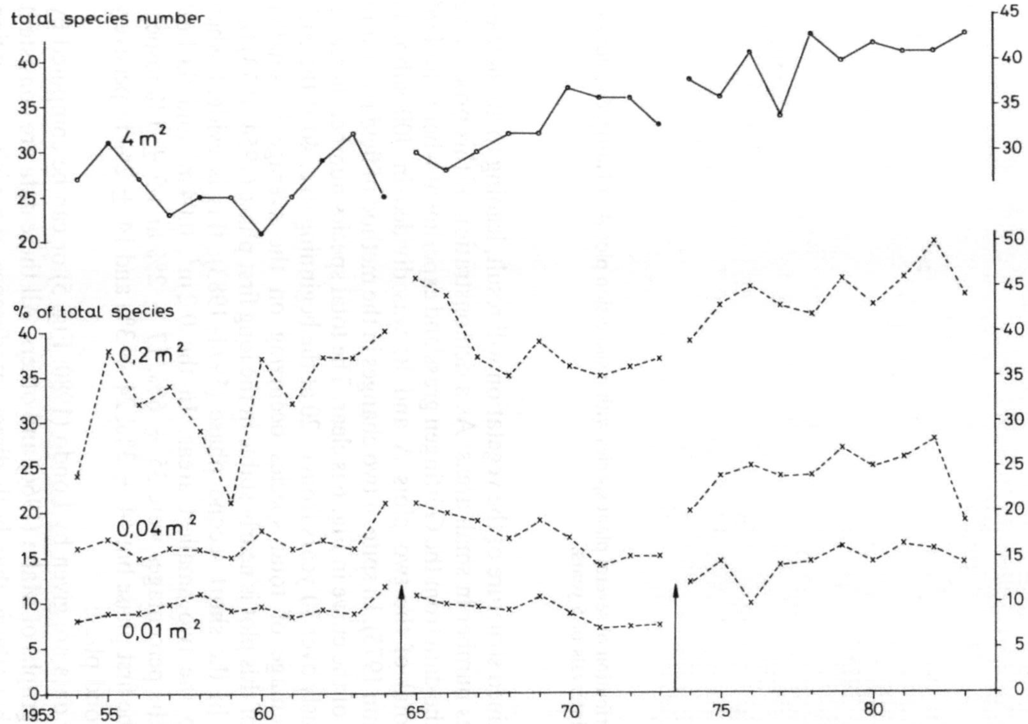

Fig. 6. Species number in a succession plot ($4\,m^2$) in a dry meadow site near Göttingen, and percentage of these species (on average) occurring in smaller proportions ($0.2\,m^2$, $0.04\,m^2$ and $0.01\,m^2$) of this plot. Arrows indicate time when changes in the method of field recording were made.

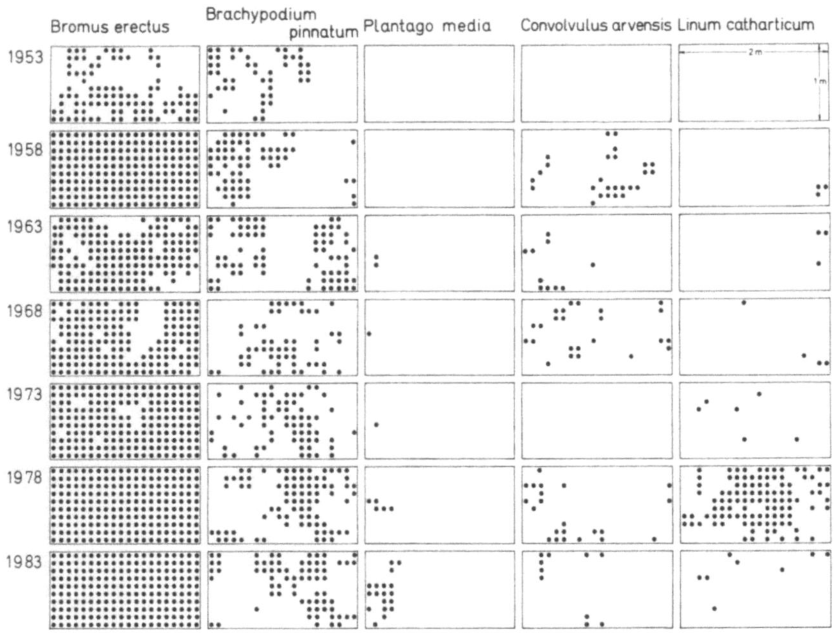

Fig. 7. Distribution of several plant species within succession plot A at Göttingen (plot size $= 2\,m^2$) at intervals of 5 years.

finely-grained structure of the vegetation will result, leading to an increasing species number in small areas. As a demonstration of this process we can take the data from the Göttingen grassland experiment where one $4\,m^2$ plot (= total of the two plots A and B) was divided in 400 subplots (Bornkamm 1975). In spite of two changes in the method of field recording the shape of the curve in Fig. 6 is clear. The total species number increases continuously over 30 years from c. 20 at the beginning to c. 40 at the end. The percentage of total species occurring on the average in smaller portions of this plot is nearly stable in the long first phase (1954–1973) but increases in the short second phase (1974–1983); this is evident when related to the two smallest areas. In the $0.2\,m^2$, $0.04\,m^2$ and $0.01\,m^2$ subplots the percentages were $35 \pm 6\%$, $17 \pm 2\%$ and $9 \pm 1\%$ respectively in the first phase but $44 \pm 3\%$, $24 \pm 3\%$ and $14 \pm 2\%$ respectively in the second phase.

Similar data are given by Londo (1980: Fig. 5) or can be computed by using the graphs of Mahn (1966) and others. All these data are consistent with the hypothesis that both direct interference and niche separation share in creating the conditions for coexistence of a great number of species within a small area.

This complex pattern is achieved by the different reactions of all the plants within the community. As an example, the distribution of a number of species of contrasting behaviour in plot A of the Göttingen experiment is plotted in Fig. 7 at 5 year intervals. As mentioned earlier, the vegetation was removed on this plot in 1953, except for the two dominating grasses *Bromus erectus* and *Brachypodium pinnatum*. *Bromus* is a 'stress-tolerant competitor' according to Grime (1979) and after 5 years it covers the whole area and remains dominant. *Brachypodium* shows considerable changes of cover within a short time. This variability is even more distinct in *Convolvulus arvensis*: in dry years (like 1973 and 1983) it is scarcely visible above the surface but becomes widespread in moist years (1958, 1968, 1978). *Linum catharticum* (an annual) shows low frequency in most years but an outburst of specimens occurred in the moist year 1978. *Plantago media* is included in Fig. 7 as an example of a settlement of constant locality and slow spread. Unfortunately the below-ground dynamics of these species are not known, but with regard to the above-ground dynamics, it is obvious that both environmental factors and life-cycles of species play a role in the distribution and abundance of each species. Bornkamm (1974) found that cover changes in a grassland succession (Göttingen plots A + B) were correlated only to a small extent with climatological factors (like precipitation and temperature) but to a large extent with the duration of succession.

REFERENCES

Aichinger, E. (1951) Soziationen, Assoziationen und Waldentwicklungstypen. Angew. Pflanzensoz. (Wien) 1: 21–68.

Aichinger, E., (1974) Vegetationsentwicklungetypen. Handb. Veget. Sci. 8: 179–183.

Andel, J. van, Ernst, W. and Nelissen, H. (1979) Mineral stoffkreislauf in Populationen von Kahlschlagarten in Beziehung zur Sukzession. Verh. Ges. f. Oekologie 7: 361–368.

Andel, J. van and Nelissen, H.J.M. (1979) Nutritional status of soil and plant species in several clearings in coniferous woods compared to that in two related habitats. Vegetatio 39: 115–121.

Arens, R. (1962) Auswirkungen der Saatstärke auf das Konkurrenzverhalten der Arten und die erste Bestandsbildung bei Weideaussaaten. Zeitschr. Acker- u.Pflanzenbau 115: 357–374.

Bazzaz, F.A. (1979) The physiological ecology of plant succession. Ann. Rev. Ecol. Syst. 10: 351–371.

Beeftink, W.G., Daane, M.C. and Munck, W. (1971) Tien jaar botanisch oecologische Verkenningen langs het Veersche Meer. Natuur en Landschap 25: 50–63.

Bookman, P.A. and Mack, R.N. (1982) Root interaction between *Bromus tectorum* and *Poa pratensis*: A three-dimensional analysis. Ecology 63: 640–646.

Bornkamm, R. (1961) Zur Konkurrenzzkraft von *Bromus erectus*. Ein sechsjähriger Dauerversuch. Bot. Jahrb. Syst. 80: 466–479.

Bornkamm, R. (1962) Öber die Rolle der Durchdringungsgeschwindigkeit bei Klein-Sukzessionen. Veröff. Geobot. Inst. ETH (Rübel) Zürich 37: 16–26.

Bornkamm, R. (1974) Zur Konkurrenzkraft von *Bromus erectus* II. Ein zwanzigjähriger Dauerversuch. Bot. Jahrb. Syst. 94: 391–412.

Bronkamm, R. (1975) Zwanzig Jahre Vegetationsentwicklung in einem mittel europäischen Halbtrockenrasen. In: W. Schmidt (Ed.). Sukzessionsforschung. Vaduz: Cramer. pp. 535–537.

Bornkamm, R. (1981a) Rates of change in vegetation during secondary succession. Vegetatio 47: 213–220.

Bornkamm, R. (1981b) Zusammensetzung, Biomasse und Inhaltsstoffe der Vegetation während zehnjähriger Sukzession auf Gartenboden in Köln. Decheniana (Bonn) 134: 34–48.

Bornkamm, R. (1984) Experimentell-ökologische Untersuchungen zur Sukzession von ruderalen Pflanzengesellschaften auf unterschiedlichen Böden. II. Flora 175: 45–74.
Bornkamm, R. and Hennig, U. (1982) Experimentell-ökologische Untersuchungen zur Sukzession von ruderalen Pflanzengesellschaften auf unterschiedlichen Böden. I. Flora 172: 267–316.
Bornkamm, R. and Schrade, W. (1981) Öber die Interferenz von Arten unterschiedlichen Lebensform bei Beginn einer ruderalen Sukzession. Verh. Ges. f. Ökol. 9: 197–201.
Borstel, H.-O. von (1974) Untersuchungen zur Vegetationsentwicklung auf ökologisch verschiedenen Grünland- und Ackerbrachen hessischer Mittelgebirge. (Westerwald, Rhön, Vogelsberg). Diss. Gießen. 159 pp.
Böttcher, H. (1974) Bibliographie zum Problem der Sukzessions-Forschung mit Hilfe von Dauerquadraten und der Vegetationskartierung. Excerpta Bot. B 14: 35–56.
Böttcher, H. (1975) Stand der Dauerquadrat-Forschung in Mitteleuropa. In: W. Schmidt (Ed.): Sukzessionsforschung. Vaduz: Cramer. pp. 31–38.
Braun-Blanquet, J. (1928) Pflanzensoziologie. 1st ed. Berlin: Springer 330 pp.; 2nd ed. 1951. Wien: Springer. 631 pp.; 3rd ed. 1964. Wien–New York: Springer. 865 pp.
Christiansen, W. (1930) Zur Methodik der Untersuchung von Dauerquadraten. Repert. spec. nov. regni veget. Beih. 61: 178–180.
Clements, F.E. (1916) Plant succession. Carnegie Inst. Washington 242. 512 pp.
Ellenberg, H. (1956) Aufgaben und Methoden der Vegetationsgliederung. Stuttgart: Ulmer. 136 pp.
Feoli, E., Lausi, D. and Pignatti, S. (1975) Grundsätze einer kausalen Erforschung der Vegetationsdynamik. In: W. Schmidt (Ed.). Sukzessionsforschung. Vaduz: Cramer. pp. 1–12.
Furrer, E. (1922) Begriff und System der Pflanzensukzession. Vierteljahrschr. d. naturf. Ges. Zürich 67: 132–156.
Grime, J.P. (1979) Plant Strategies and Vegetation Processes. Chichester: J. Wiley 222pp.
Hahn, W., Wolf, A. and Schmidt, W. (1979) Untersuchungen zum Stickstoff-Umsatz von *Tussilago farfara*- und *Agropyron repens*-Beständen. Verh. Ges. f. Oekologie 7: 369–380.
Harper, J. (1977) Population Biology of Plants. London: Academic Press. 892pp.
Hildebrand, F. (1911) Beobachtungen über das Vorkommen von Pflanzenarten auf einem nicht mehr in Kultur befindlichen Gelände. Mitt. Bad. Landesver. f. Naturk. 6: 97–104.
Hogeweg, P., Hesper, B., van Schaik, C.P. and Beeftink, W.G. (1985) Patterns in vegetation succession: an ecomorphological study. In: J. White (Ed.) The Population Structure of Vegetation, pp. 637–666. Dordrecht: Dr. W. Junk.
Keever, C. (1950) Causes of succession in old fields of the Piedmont, North Carolina. Ecol. Monogr. 20: 229–250.
Knapp, R. (1967) Experimentelle Soziologie. 2nd ed. Stuttgart: Ulmer. 266 pp.
Knapp, R. (1974) Some principles of classification and of terminology in successions. Handb. Veget. Sci. 8: 169–177.
Londo, G. (1980) A method of studying vegetation dynamics in a two-dimensional microgradient. Phytocoenosis (Warszawa-Białowieża) 7: 165–175.
Lüdi, W. (1924) Die Untersuchung und Gliederung der Sukzessionsvorgänge in unserer Vegetation. Verh. naturf. Ges. Basel 35: 277–302.
Lüdi, W. (1932) Die Methoden der Sukzessionsforschung in der Pflanzensoziologie. Handb. biol. Arbeitsmeth. XI/5: 527–728.
Maarel, E. van der and Werger, M.J.A. (1980) On the treatment of succession data. Phytocoenosis (Warszawa–Białowieża) 7: 257–278.
Mahn, E.-G. (1966) Beobachtungen über die Vegetations- und Bodenentwicklung eines durch Brand gestörten Silikattrockenrasenstandort. Arch. Natursch. u. Landschaftsforsch. 6: 61–90.
Major, J. (1974a) Kinds and rates of changes in vegetation and chronofunctions. Handb. Veget. Sci. 8: 7–18.
Major, J. (1974b) Biomass accumulation in successions. Handb. Veget. Sci. 8: 195–203.
Major, J. (1974c) Nitrogen accumulation in successions. Handb. Veget. Sci. 8: 205–213.
Malcolm, W.M. (1966) Biological interactions. Botan. Rev. 32: 243–254.
Oberdorfer, E. (1979) Pflanzensoziologische Exkursionsflora. 3rd ed. Stuttgart: Ulmer. 997 pp.

Parrish, J.A.D. and Bazzaz, F.A. (1976) Underground niche separation in successional plants. Ecology 57: 1281–1288.
Pollard, F. (1982) Light induced dormancy in *Bromus sterilis*. J. appl. Ecol. 19: 563–568.
Raynal, D.J. and Bazzaz, F.A. (1973) Establishment of early successional plant populations on forest and prairie soil. Ecol. 54: 1335–1341.
Schäfer, J. and Werner, W. (1979) Stickstoffhaushalt von *Solidago canadensis*-Beständen. Verh. Ges. f. Oekologie 7: 381–392.
Schmidt, W. (1974) Bericht über die Arbeitsgruppe für Sukzessionsforschung auf Dauerflächen der internationalen Vereinigung für Vegetationskunde. Vegetatio 29: 69–73.
Schmidt, W. (1981) Ungestörte und gelenkte Sukzession auf Brachäckern. Scripta Geobotanica (Göttingen) 15: 1–199.
Sissingh, G. (1952) Ethologische synoecologie van enkele onkruid-associaties in Nederland. Meded. Landbouwhogesch. Wageningen 52: 167–206.
Warming, E. (1896) Lehrbuch der ökologischen Pflanzengeographie. Berlin: Bornträger. 412 pp.
Westhoff, V. (1969) Langjährige Beobachtungen an Aussüßungs-Dauerprobe-flächen beweideter und unbeweideter Vegetation an der ehemaligen Zuiderzee. In: R. Tüxen (Ed.). Experimentelle Pflanzensoziologie. Den Haag: Junk. pp. 246–253.
Westhoff, V. and Maarel, E. van der (1973) The Braun-Blanquet approach. Handb. Veget. Sci. 5: 617–726.
Williams, E.D. (1978) Botanical composition of the Park Grass plots at Rothamsted 1956–1976. Harpenden: Rothamsted Exp. Sta. 61 pp.
Yodzis, P. (1978) Competition for the space and the structure of ecological communities. Lecture Notes in Biomath. 25. Berlin–Heidelberg–New York. Springer. 191 pp.

RESEARCH ON PERMANENT QUADRATS IN THE USSR

MICHAEL V. MARKOV

ABSTRACT

Some of the early researches on permanent quadrats in the USSR are reviewed. The long tradition established by L.G. Ramenskii about 1910 has continued to the present day, and has received new impetus since the publication of the classical studies of T.A. Rabotnov in 1950.

PHYTOCOENOLOGY: THE CONTRIBUTION OF L.G. RAMENSKI

Phytocoenology, generally defined as the science of plant communities, has a long history in the USSR. L.G. Ramenskii (1925) introduced the term to replace the name 'phytosociology' which he considered to be undesirably anthropocentric. According to Ramenskii (1934) phytocoenology consists of two parts: (a) the description of the species composition, types and structures of vegetation and (b) the study of interactions between components of the community, and patterns of development and succession (dynamic phytocoenology). Causal experimental phytocoenology is linked to dynamic phytocoenology, since the processes of interactions between populations as components of a plant community can be properly investigated only by a combination of descriptive and experimental methods.

The plant cover of any region is not rigid and stable; each plant community changes from year to year and from season to season. This phenomenon is especially pronounced in steppe and grassland communities. The seasonal changes of steppe communities were described in detail many years ago by Alekhin (1909). For the northern steppes he attempted to distinguish 10–12 phases or aspects in vegetation development, each exhibiting a different character through the physiognomical dominance of some species group. The nonsynchronous development of species in steppe communities is of great importance, as a way to more rational utilization of the resources; Markov (1962) named this phenomenon 'layering in time'. Alekhin's ideas were derived by observing

White, J. (ed.), The Population Structure of Vegetation.
 ISBN 978-94-010-8927-2

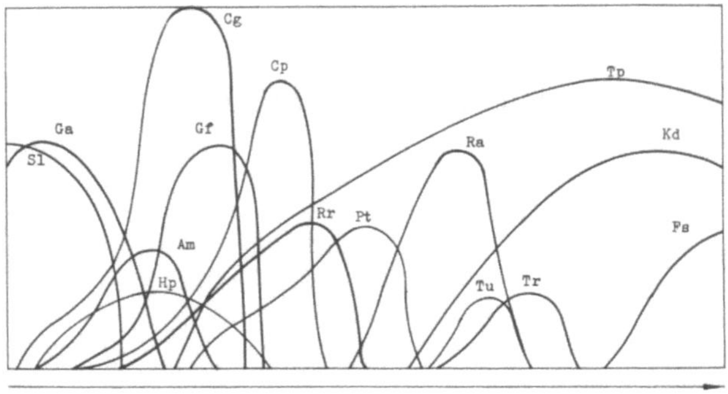

Fig. 1. The projective cover (relative scale) of species along a gradient (indicated by the arrow) from a swamp meadow to a plateau of mesoxerophytic grassland along the Oskol River.

Fs, *Festuca sulcata*; Kd, *Koeleria delavignei*; Tp, *Trifolium pratense*; Tr, *Trifolium repens*; Tv, *Taraxacum vulgare*; Ra, *Ranunculus acer*; Pt, *Poa trivialis*; Rr, *Ranunculus repens*; Cp, *Caltha palustris*; Gf, *Glyceria fluitans*; Cg, *Carex gracilis*; Hp, *Heleocharis palustris*; Am, *Alisma michaletii*; Ga, *Glyceria aquatica*; Sl, *Scirpus lacustris* (Ramenskii 1925).

the same plots from time to time but he did not use any quantitative methods.

Many interesting and important phytocoenological ideas were put forward by Ramenskii. He was the first Soviet researcher to use long-term observations on permanent quadrats by recording the behaviour of different species: he was the first to understand the necessity to apply objective methods to study the complex relationship between vegetation and habitat. From 1913 onwards Ramenskii conducted his investigations of meadow, fen and steppe vegetation for many years. He employed a method of 'projective abundance' or 'projective cover' estimation which he used to determine the status of each species in a community. The extensive data collected during 10 years of observation (70,000 estimations) allowed him to draw a number of interesting phytocoenological conclusions. These confirmed his earlier ideas which he had enunciated in 1910 in a report to XIIth Congress of Russian naturalists and doctors. Ramenskii (1910) wrote that each plant species has its own performance curve on the co-ordinates of a synecological diagram in consequence of its ecological individuality, and that the set of curves for different species in a community represents gradual spatial changes of vegetation according to changes in environmental factors. The distribution of each plant species in, for example, a meadow community along an environmental gradient was considered to be unrelated to distributions of others: the peaks of curves do not coincide, as the example in Fig. 1 demonstrates. A natural outcome of his postulate of the individuality of plant species was Ramenskii's recognition of a continuity (continuum) of plant cover. By recording the proportion of species on permanent quadrats he believed that processes of succession in meadow communities undermined statements about stable vegetation units: 'There is no stable groups of plants. Everything

Table 1. Succession (as cover changes) in a meadow community on a permanent plot 4 m^2 near Pavlovsk (Ramenskii 1925).

Species	1914	1915	1916	1917	1918	1919	1920	1921	1922	1923
Galium rubioides	60.2	35.6	32.5	10.3	4.1	2.6	2.9	2.3	1.3	1.6
Alopecurus pratensis	4.9	16.8	3.5	4.5	3.5	24.5	2.2	0.3	9.7	0.1
Agropyrum repens	2.2	7.7	15.8	7.6	13.9	29.2	6.8	9.2	24.1	22.4
Carex precox	3.1	5.0	13.4	7.4	16.8	5.3	10.7	3.6	10.6	14.3
Agrostis alba	0.1	0.1	13.6	14.5	13.3	4.2	3.1	7.9	4.8	0
Lysimachia nummularia	1.9	5.5	2.5	0.6	2.0	2.0	0.9	1.2	7.3	29.9
Filipendula ulmaria	3.4	1.9	9.0	6.2	1.9	2.0	3.8	3.4	5.3	9.8
Sanguisorba officinalis	0.8	0.4	2.3	2.4	1.6	3.1	2.1	4.2	3.8	5.2

flows, breaking all the conditional borders. The laws of formation of species groups but not the groups themselves are stable and these former must be investigated' (Ramenskii 1925: 20). His observations on permanent plots ($4\,m^2$) were carried out in various communities and showed clearly that a mere list of species however complete was insufficient to determine the full composition and structure of a herbaceous community; it was necessary to observe a community over a period of several years under different weather conditions. Quantitative relations between species may be significantly altered from year to year, as the example in Table 1 indicates even though the species composition may remain relatively constant. Ramenskii believed that a community resists the penetration of alien species.

Ramenskii (1938) later introduced the term 'phytocoenotypes' for the groups of plant species showing certain life-history patterns. He differentiated three phytocoenotypes: 'violents' – species of high competitive ability, related to their high level of resource utilization (high absorption capacity of the roots, heavily shading foliage etc.); 'patients' – species of high tolerance to adverse conditions; 'explerents' – species with the ability to invade rapidly into disturbed vacant habitats where competition was reduced. It is not difficult to see a correlation between the coenotypes of Ramenskii and the 'strategies' characterized by Grime (1979).

RESEARCH ON PERMANENT QUADRATS

Some investigators interested in seed reproduction in natural meadows began to use permanent quadrats from an early date. The periodic registrations on 1 square foot ($\equiv 0.09\,m^2$) quadrats in grassland communities in 1917–1918 showed that many seedlings appeared during the spring (Shennikov and Baratynskaya 1924). These seedlings represented only 13 species (although the total number of species in the communities investigated was as high as 86) and were found mainly in microsites where the natural cover had been disturbed by cattle. Most seedlings were recorded during the first dates of observation but then their number decreased abruptly as rapid canopy formation and overshading became obstacles for seedlings to develop. Seedlings of species that were in small numbers in a community (*Geum rivale*, *Ranunculus acer*) demonstrated a high tolerance to shade. However, the majority of species, even those widely distributed in the vegetation, were not represented by seedlings. The conclusion was drawn that seed renewal in two grassland communities was of low significance and only took place when there was disturbance.

For two years (1923–1924) Bogdanovskaya-Gienef (1926, 1954) examined the seed renewal process in three types of meadow communities (Mixto-herbetum humidum, Nardeto-succisetum and Deschampsietum). As she stated herself, she used the Clementsian method of permanent quadrats, with a frame of $4\ dm^2$ being divided into 16 quadrats (each $25\ cm^2$) by strings. All seedlings and young individuals were mapped, and

Table 2. The number of seedlings and juvenile plants on four permanent quadrats in Mixto-herbetum humidum (Bogdanovskaya-Gienef 1926, 1954).

Quadrat No.	1923	1924
1	105	104
2	66	61
3	166	183
4	152	175

stage of development and size (leaf number and width of largest leaf) of each seedling were recorded. These very detailed observations were repeated every two weeks but the initial number of ten quadrats in each community had to be reduced to four each. Bogdanovskaya-Gienef showed there were significant differences in the seed renewal processes in the various meadow communities. The number of seedlings and juvenile plants differed, as well as their viability and survivorship. The number of seedlings emerging from year to year in a given community was maintained fairly constant in spite of different meteorological conditions and different rates of propagule production from year to year (Table 2). The mortality rate of seedlings and juvenile individuals in the meadows was high (29–90%). The majority of young plants died as a result of alterations in environmental factors by adult plants.

Seed renewal processes in the three meadow communities were compared in relation to the degree of space occupied aboveground and belowground by adult plants. The total volume of dominant plants was negatively correlated with the maximum number of seedlings emerged and number of juvenile plants, but positively correlated with mortality among young plants (Table 3). The development of seedlings under grassland canopy was extremely slow: phytocoenotic conditions 'cause the prolongation of the juvenile phase in perennial herbs for several years' (Bogdanovskaya-Gienef 1926: 252).

Table 3. The relation between total volume of dominant species and the number and mortality of young plants in three meadow communities (Bogdanovskaya-Gienef 1926).

	Mixto-herbetum humidum	Nardeto-succisetum	Deschampsietum
Total volume (cm^3) of dominant species per 16 sq. dm	1085	1465	1485
Total number of seedlings and juvenile plants per 16 sq. dm	523	118	112
Mortality of young plants emerging in 1923 at the end of summer of 1924 (per cent)	67	80	90

The difficulties of long-term observations on the mapping and drawing of numbers of individuals led to a restriction in the number of species chosen for observation rather than to a rejection of the method. One may observe such a trend in the fact that some authors choose to investigate natural communities with a small number of species, e.g., communities of halophytes. Zakaryan (1930, 1934) studied the seedling emergence, growth, intra- and interspecific relationships of species of solonchak halophytes (*Salsola crassa, Suaeda splendens, S. altissima, Bassia hyssopifolia, Petrosimonia brachiata, Atriplex incisa*) on 47 permanent quadrats (1 dm^2) in Mugan steppe. The growth dynamics of all these species is presented in general outline in Table 4. Species differed in the time taken for transition to dormancy or semi-dormancy, in their time of seedling emergence and in their rhythm of development. Many quadrats supported individuals of a single species only. By weekly mapping of plants on permanent quadrats it became possible for Zakaryan to show that the probability of death of individuals increased with density: in overcrowded stands of *S. splendens* 91.5% of individuals died within 5 months; from 867 individuals of *B. hyssopifolia* only 41 remained alive in autumn and only 5 were fertile. He also found that severe elimination occurs during periods of rapid growth and that the rate of elimination slowed sharply as individuals came into a dormant state: 50% of individuals of *S. crassa* died during times of rapid growth and only 8% during dormancy. The light requirements of seedlings also changed with age. During the first period of growth seedlings were suppressed by the intense light but later the strong insolation became a necessary factor for their good development. On some quadrats which were placed on open areas, a high number of seedlings perished during the first phase of growth, and only few survived. On the contrary, shaded seedlings did not perish until the second period of growth, when severe mortality suddenly began. Surviving individuals of different species showed varied reproductive ability. On two quadrats 238 of 256 seedlings of *P. brachiata* and *B. hyssopifolia* survived: these were small and approximately of equal height (5 cm). A large group of the latter species remained sterile while all the individuals of the former developed fruits.

A.A. Shakov (1947) studied the population dynamics of two-, three- and four-species communities of halophytes on the bottom of dried lakes in Northern Kazakhstan. Data were obtained from detailed long-term quantitative investigation of *Aster tripolium, Chenopodium chenopodioides, Suaeda maritima* and *Atriplex litoralis* during the season and through several generations. For example, the population dynamics in a two-species community of halophytes was studied for three years: besides the yearly recorded decrease in number of individuals in both species an obvious trend could be seen in the gradual replacement of *Salicornia* by *Suaeda* (Table 5). In a few of his investigations Shakhov removed some chosen species to create single- or two-species stands. He found, for example, that (a) mortality of *C. chenopodioides* was higher in two-species than in monospecific stands (Table 6) and (b) *S. maritima* caused a greater suppression of *C. chenopodioides* than did *A. tripolium*. Investigations such as these in which straightforward observations on permanent quadrats are

Table 4. Different growth phases and their duration in some halophytes (Zakaryan 1930, 1934).

Growth phases	Germination and a first period of growth	Semi-dormancy (with hardly visible changes in height of seedlings)	Second period of growth	Dormancy	Intensive growth and reproduction
Duration (calendar months)	II–III	III–IV	IV–VII	VII–VIII–IX	VIII–IX–XI

Table 5. The number dynamics of *Salicornia* and *Suaeda* during three years on a permanent plot 16 sq. dm (Shakhov 1947).

1938			1939			1940		
Dates	*Salicornia*	*Suaeda*	Dates	*Salicornia*	*Suaeda*	Dates	*Salicornia*	*Suaeda*
						28.V	453	758
			14.VII	906	1275	18.VI	221	514
			24.VII	915	1229	8.VII	207	532
14.VIII	809	948	14.VIII	870	1193	28.VII	122	480
24.VIII	774	783	24.VIII	817	1063	18.VIII	96	455
14.IX	716	655	14.IX	729	883	8.IX	65	393
24.IX	680	515	24.IX	700	812	28.IX	46	367

Table 6. The number dynamics of halophytes in artificially created mono- and two-specific communities (Shakhov 1947).

No. of permanent plot	Species	Dates of observation		Mortality (%)
		10.VI	9.IX	
1	*Chenopodium chenopodioides*	342	36	89.5
	Aster tripolium	15	12	20.0
2	*Suaeda maritima*	205	93	54.7
	Aster tripolium	12	7	49.2
3	*Chenopodium chenopodioides*	470	77	83.7
4	*Suaeda maritima*	138	30	79.3
	Chenopodium chenopodioides	233	17	92.7

combined with experimental manipulations are important in understanding the dynamics of vegetation. The number of species chosen for such investigations is, however, usually small.

T.A. Rabotnov's paper of 1950 is now recognized as a classical one on permanent quadrat research. His painstaking observations on permanent quadrats of the population biology of some perennials in subalpine meadows of Northern Caucasus initiated a new direction in Soviet phytocoenology. The results of his extensive work are now well known and widely appreciated. Among the most important conclusions of his research is the understanding that variation in population composition is typical for plants that grow in meadow communities. This variation is closely connected with their ability to establish different morphological states depending on their age and growth conditions. In the process of population structure formation the ability of seeds to germinate gradually and sustain their viability for a long time, the ability of young plants to persist for a long time in the juvenile or semi-mature state, and the ability of mature plants to endure severe growth conditions (without mortality) for a long time are especially important. The analysis of species behaviour must be based upon the estimation of numbers of individuals. Rabotnov considers the long-term observation of marked individuals on permanent quadrats to be of especial importance in the study of plant communities (Rabotnov 1951).

REFERENCES

(English translations of Russian-language titles are given in brackets)

Alekhin, V.V. (1909) (Vegetation and succession on a plot "Streletskaya steppe" near Kursk.) Trudy Sanct-Peterburg-skogo Obshchestva estestvoispytatelei 60(1): 11–20.

Bogdanovskaya-Gienef, I.D. (1926) (Contribution to seed renewal in meadow communities.) Zap. Leningrad. Sel. khoz. Inst. 3: 216–253.

Bogdanovskaya-Gienef, I.D. (1954) (Seed renewal of plants in the meadow coenoses of the forest zone.) Uchen. Zap. Leningrad. Univers. Biol. 34: 3–47.

Grime, J.P. (1979) Plant Strategies and Vegetation Processes. John Wiley and Sons Publ., New York.
Markov, M.V. (1962) (Common geobotany.) Moscow. 450 pp.
Rabotnov, T.A. (1950) (The life cycle of perennial herbaceous plants in meadow coenoses.) Trudy. Bot. Inst. A.N.SSSR, Ser. 3. 6: 7–204.
Rabotnov, T.A. (1951) (Contribution to the methods of observation of herbaceous plants on permanent plots.) Bot. Zhur. 36: 643–645.
Ramenskii, L.G. (1910) (On comparative method in ecological investigation of plant communities.) Proc. of XII Congress of Russian naturalists and doctors. Otdel II, 9: 389–390.
Ramenskii, L.G. (1925) (The main peculiarities in vegetation cover and their investigation.) Separated copy, Voronej., 1–37.
Ramenskii, L.G. (1934) (On ecology) Sov. Bot. 3: 68–71.
Ramenskii, L.G. (1938) (Introduction to the geobotanical study of complex vegetations.) Moscow: Selkhozgiz. 620 pp.
Shakhov, A.A. (1947) (Struggle for existence in the halophytic coenoses.) Sov. Bot. 4: 185–198.
Shennikov, A.P. and Baratynskaya, E.P. (1924) (Some results of studies on the structure and variability of meadow communities.) Zhur. Russk. Bot. Obshch. 9: 75–82.
Zakaryan, S.F. (1930) (Struggle for life between seedlings of halophytes.) Trudy Azerbaijan. Otdel. Zakavkaz. Filiala A.N. SSSR I: 25–39.
Zakaryan, S.F. (1934) (Struggle for life between seedlings of halophytes.) Trudy Azerbaijan. Otdel. Zakavkaz. Filiala A.N. SSSR II: 89–102.

DYNAMICS OF PLANT COENOTIC POPULATIONS

T.A. RABOTNOV

ABSTRACT

A coneopopulation is the collective term for the sum of the individuals of a species in a plant community. It is typically composed of varying numbers of individuals in particular age-states, which are defined on morphological and ontogenetical criteria. Since the coenopopulation structure of a species reflects the conditions of the biocoenosis in which it lives, the age-state spectrum of a species varies in space and in time. Some of the varieties of coenopopulation composition and dynamics are discussed in this paper.

THE CONCEPT OF A COENOTIC POPULATION

The concept of a population as the sum of the individuals of a species, differing in their age and vitality state, in a plant community was established for the first time in studies of subalpine meadow plants in the North Caucasus (Rabotnov 1945, 1950a, 1950b). Such populations were considered as structural elements of phytocoenoses. It was, therefore, suggested to name them 'coenotic populations' or *coenopopulations* (Petrovskii 1961, Korchagin 1964).* It was recognized that any species in a phytocoenosis is represented by a coenopopulation special to that phytocoenosis and characterized by the number of individuals belonging to different age-states (Rabotnov 1945, 1950a). For this purpose a classification of four main periods of the life (age-states) of plants reproducing by seeds was used (traditional abbreviations are indicated in brackets): (1) Latent period (period of primary dormancy when individuals exist as viable dormant seeds) (se); (2) Virginile period, from germination of seeds to formation of generative shoots; (3) Generative period of reproducing by seeds; (4) Senile period, when plants lose their ability to reproduce by seeds because of senescence (Rabotnov 1945, 1950a).

*Development of the concept of coenopopulation was influenced by studies of plant reproduction by seeds in meadow phytocoenoses by Bogdanovskaya-Gienef (1926, 1941), Linkola (1930, 1936) and Perttula (1941).

White, J. (ed.), The Population Structure of Vegetation.
 ISBN 978-94-010-8927-2

Within the virginile period the following groups are distinguished: seedlings (**pl**), juveniles (**j**), transitional forms from juveniles to matures (immatures (**im**) according to Uranov 1960), and mature virginile (**v**) individuals. The generative period can, if it is prolonged, be divided into three subperiods (Rabotnov 1949, Trulevich 1960): a subperiod of increasing generative and vegetative vigour of individuals ($\mathbf{g_1}$); a subperiod of maximal generative and vegetative vigour of individuals (subperiod of life culmination) ($\mathbf{g_2}$); a subperiod of senescence ($\mathbf{g_3}$). Individuals of polycarpic species, having attained the ability to flower, often develop generative organs not every year but at intervals of one or several years. Thus the group of generative individuals can consist of individuals both with and without generative shoots at any given time. Furthermore adult individuals of several species in unfavourable conditions can turn into a dormant state persisting in unfavourable periods in the form of dormant underground organs. Senile individuals (**s**) are, by definition, absent in coenopopulations of monocarpic species. They are usually absent also in coenopopulations of species with short duration of life, and in coenopopulations of trees. The existence of individuals in the postgenerative state can be determined both by their senescence and the influence of unfavourable environmental conditions. The group of such 'quasisenile' individuals is apparently widespread, particularly among herbaceous species of broad-leaved forests (Smirnova 1983).

The structure of a particular coenopopulation of a species is determined by counting the number of individuals per unit area belonging to each age-state group. This entails a careful and detailed analysis of the morphology and morphological changes of individual genets as they grow older. Coenopopulation age-state spectrum is investigated usually either by observing all individuals of a species on permanent plots over a period of several years, or by a single inventory of all individuals of a species within plots situated in different environments and their classification into age-state groups. In all cases a morphological study of ontogenesis is an essential preliminary step.

According to the biomorphological properties of species and the aim of investigation different age-state groups of individuals are distinguished in coenopopulations. It is often expedient to unite into one group seedlings and juveniles; it is sometimes impossible to demarcate a boundary between juveniles and immatures. It is usually necessary to unite in the one group of adult vegetative individuals both mature virginile and those 'generative' individuals which in the particular season of investigation did not develop generative shoots: it is impossible to distinguish them sufficiently accurately on their aboveground organs.

In Tables 1, 2 and 3 data on the coenopopulations of four meadow species are presented. These are selected species, typical for different types of meadows: *Anemone fasciculata* is a species distributed widely on subalpine meadows of the North Caucasus; *Potentilla erecta* is a species typical of upland (not flooded) meadows of the forest zone; *Ranunculus acris* and *R. auricomus* are species typical both of upland and of some types of flooded valley meadows. Both numbers of individuals and

Table 1. Changes of age-state spectrum of coenopopulations of *Anemone fasciculata* on different types of subalpine meadows during 1945–1947 (Rabotnov 1950a)

Types of meadows	Years	Investigated area m^2	Total number of individuals	Including %: Generative	Mature vegetative	Immatures	Juveniles	% Generatives of total mature individuals	Seeds/m^2
I. *Bromus variegatus* +	1945	50	1250	14.9	31.8	8.8	44.5	26.8	42
Carex humilis +	1946	50	1256	12.9	34.2	9.0	43.9	23.0	
"dry" herbs	1947	50	1180	5.3	42.8	10.0	41.3	9.0	
II. *Bromus variegatus* +	1945	34	5890	4.7	25.4	12.9	57.0	11.0	
"moist"	1946	34	5000	2.9	29.2	10.9	57.0	6.7	
herbs	1947	34	3660	2.4	29.9	11.9	55.8	5.3	
III. "Moist"	1945	30	1280	30.3	18.7	1.3	54.7	66.2	288
herbs +	1946	30	1400	27.4	10.5	2.4	59.7	68.0	
Brachypodium pinnatum	1947	30	750	47.4	21.8	4.6	26.3	64.0	
IV. *Carex humilis* +	1945	50	820	18.2	27.5	6.1	48.2	35.0	128
Festuca varia	1946	50	850	26.1	17.8	6.1	50.1	52.0	
	1947	50	598	28.6	31.0	6.4	34.0	43.4	
V. *Bromus variegatus* +	1945	50	486	5.0	33.0	10.3	51.7	10.2	
herbs-*Rhytidium*	1946	50	516	3.9	29.0	12.1	55.0	8.6	
rugosum + *Pleurozium schreberi*	1947	50	552	6.2	24.3	15.5	54.0	13.4	

Table 2. Age-state spectrum of different coenopopulations of *Potentilla erecta* (mean number per m²) (Varlygina 1982)

Types of phytocoenoses / Age-state groups	Moscow region			Western Estonia		
	Potentilla erecta – – Sphagnum girgensohnii	*Alchemilla monticola + Succisa pratensis*	Grasses + herbs meadow community	*Molinia coerulea + Carex tomentosa + Sesleria coerulea*	*Scorzonera humilis + Leontodon hispidus + Brachypodium pinnatum*	*Schoenus ferrugineus + Molinia + Sesleria coerulea*
Seedlings	11.7	170.7	197.0	11.7	11.0	157.3
Juveniles + immatures	87.0	79.3	70.3	26.0	25.0	184.0
Mature virginile	7.3	70.3	93.7	24.3	89.5	257.3
Generative temporarily without generative shoots	14.0	25.3	52.3	16.0	91.0	193.3
Generative	77.3	93.7	111.0	61.0	60.5	174.0
Total	197.3	439.3	524.3	139.0	267.5	965.9
Viable seeds in 0–10 cm soil	1960	1240	1140			

Table 3. Changes of age-state spectrum of coenopopulations *Ranunculus acris* and *R. auricomus* on a flood-plain meadow during 1949–1958 on a plot of $10\,m^2$ (Rabotnov 1958)

Years	Total number of individuals	Including								% Generatives of total number of mature individuals
		Generative		Mature vegetative		Immatures		Juveniles		
		Number	%	Number	%	Number	%	Number	%	
				Ranunculus acris						
1949	312	10	3.2	104	33.3	31	10.0	167	53.5	7.0
1950	555	25	4.5	137	24.7	145	26.1	248	44.7	8.0
1951	479	66	13.8	176	36.8	209	43.6	28	5.8	14.6
1952	395	53	13.4	239	60.5	87	22.1	16	4.0	14.0
1953	208	27	13.0	155	74.4	22	10.6	4	2.0	13.2
1954	327	37	11.3	127	38.8	10	3.1	153	46.8	21.2
1955	133	6	4.5	117	88.0	1	0.8	9	6.7	5.0
1956	101	21	21.0	73	72.0	6	6.0	1	1.0	21.0
1957	152	5	3.3	78	51.3	3	2.0	66	43.4	6.0
1958	130	9	6.9	67	51.5	7	5.4	47	36.2	10.8
				Ranunculus auricomus						
1949	25	1	4.0	8	32.0	4	16.0	12	48.0	11.0
1950	78	5	6.4	9	11.5	11	14.0	53	68.1	35.7
1951	91	16	17.6	11	12.0	37	40.7	27	29.7	59.2
1952	94	20	21.3	30	31.0	23	24.5	21	22.3	40.0
1953	90	22	24.4	37	41.2	18	20.0	13	14.4	37.0
1954	174	50	28.7	18	10.3	23	13.2	83	47.8	73.0
1955	188	49	26.0	28	14.9	19	10.1	92	49.0	64.0
1956	179	57	31.8	34	19.0	64	35.8	24	13.4	62.0
1957	286	59	20.6	56	19.5	47	16.4	124	43.5	51.0
1958	269	86	32.0	55	20.5	20	7.4	108	40.1	61.0

age-state spectrum in the coenopopulations of these species vary from one type of phytocoenosis to another, and from year to year. Table 3 shows the different directions of change of coenopopulations of the two species of buttercups on the same plot during a ten-year period.

Coenopopulations are subject to changes: they arise, reach maximal development and then can degrade to the point of disappearance. In the first paper devoted to the elaboration of the coenopopulation concept three types were distinguished according to the stages of their development (Rabotnov 1945):

(1) invasive, in which the age-state groups characteristic of early ontogenesis are predominant;
(2) normal, in which all age-state groups are represented to varying degrees;
(3) regressive, in which age-state groups characteristic of later ontogenesis are predominant.

Within these groups many variants have been distinguished and more comprehensive classifications have been proposed (Rabotnov 1950b; Rysin and Rysina 1965; Rysin and Kazantseva 1975; Uranov and Smirnova 1969).

The coenopopulation represents the existence of a species in a particular phytocoenosis and reflects its coexistence with other plant species and with other organisms. The composition and dynamics of coenopopulations reflect 'the totality of adaptations ensuring the possibility for species to coexist with other organisms and to occupy definite positions in biocoenoses" (Rabotnov 1975).

The sum of successive coenopopulations *in the one place* is sometimes referred to as the 'coenopopulation flow' (Uranov and Smirnova 1969) or the 'great cycle of coenopopulations' (Rabotnov 1969). This includes the subcycles of successive (temporal) coenopopulations and comprises the periods of formation, persistence and decline, corresponding to invasive, normal and regressive age-state spectra sequentially.

The scheme outlined above of the age-state spectrum of coenopopulations (sometimes with minor alterations) is widely used in the USSR. The bibliography of papers on the problem of coenopopulations (Rabotnov 1980, 1981) gives an idea of the large volume of work expended in the studies on coenopopulations in the USSR. An especially large contribution has been made by A.A. Uranov and his collaborators (e.g., Uranov and Serebryakova 1976).

THE IMPORTANCE OF LIFE HISTORY PATTERNS FOR DETERMINING COENOPOPULATION COMPOSITION

As a basis for distinguishing the types of life history patterns of species the three phytocoenotypes proposed by Ramenskii (1935, 1938) are useful. These are violents, patients and explerents characterized in the following way. Violents (Latin: violentum = tend to violence) develop vigorously, occupy territory and retain it, and suppress their rivals by the energy of their vital activity and by their thoroughness in the utilization of the environmental resources (high absorption capacity of the roots, heavily shading foliage, etc). Patients (latin: patiens = endurance) succeed in the struggle for existence not by virtue of vigorous vital activity and growth but by their tolerance for extreme environmental conditions at all times or temporarily. Explerents (latin: expleo = to fill up) have a low competitive capacity but are able to invade vacant territory, quickly filling the gaps between strong plants, although being as easily displaced by them. It is a pity that the remarkable papers by Ramenskii and some other Soviet phytocoenologists (including V.N. Sukachev and A.P. Shennikov) published in Russian have remained unknown to plant ecologists outside the USSR. Grime (1979) distinguished, as did Ramenskii, three main types of plant 'strategies', but he named them not quite correctly. The 'violents' of Ramenskii he named 'competitors', but any species regardless of its 'strategy' competes in a plant community with other plants and is thus a competitor. Instead of 'explerents' Grime used the term 'ruderals', but in botany for a long time this term has been used for so-called 'plants of waste places'.

Among the main types of life-history pattern distinguished by Ramenskii and Grime there are transitions. This question was discussed by Ramenskii (1938) but it has been considered especially thoroughly by Grime (1978). Furthermore, violents in unfavourable environmental conditions tend to become patients. The ability to be patient ('patiency'), that is, for individuals to persist for a long time in juvenile, immature or mature virginile age-states, is inherent in virginile individuals of many typical violents. The transition of individuals into the secondary dormant state can be considered as a manifestation of their ability to be patient. A temporary existence in the patient state is typical of many explerents. Consequently, patiency is a widespread phenomenon. The ability of plant species to be violents, patients and explerents (constantly or temporarily) are important coenotic properties which change according to the environment and age-state of individuals. This is of importance for the study of coenopopulations.

One can understand explerents more widely than did Ramenskii and distinguish among them seasonal, demutational and fluctuational types which differ both in the composition and in the dynamics of their coenopopulations.

Spring ephemeroids of broadleaved forests (for example *Corydalis solida*) are seasonal explerents. They are characterised by an ability to respond to a short term decrease of competition (especially for light) early in spring. This is responsible for pronounced seasonal alterations in the composition of their coenopopulations: in early spring coenopopulations consist chiefly of individuals in an active state (virginile, generative) and probably of some dormant individuals (seeds, tubers), while later in the year there are only individuals in seasonal or long term dormant states. According to observations in the forest of *Quercus robur* in the reserve "Les on the Vorskla", coenopopulations of *Corydalis solida* consist chiefly of virginile individuals. The following is a typical age-state spectrum:

seedlings 88, juveniles 99, generatives 61 – in total 248/m^2 (Goryshina 1969). Maximal development of aboveground organs occurs at the end of April and beginning of May; by the middle of May a visible decrease of aboveground/organs of *Corydalis* was observed and by 1–7 June they had died completely (Neshataev et al. 1974).

Chamaenerion angustifolium (fireweed) is a typical demutational* explerent. It dominates phytocoenoses of the first stage of secondary succession, after forest destruction by fire or felling. In this period its coenopopulations arise as a result of mass germination of seeds carried by wind from other places. In the first 2–3 years after fire favourable conditions (abundance of nutrients especially nitrogen, and an absence of competition with other species) are created for establishment of seedlings and their rapid transformation into mature individuals.

By rapid vegetative propagation a dense sward (200–300 shoots/m^2), chiefly of generative shoots, is formed within three or four years. In spite of its enormous seed productivity (up to 10,000 seeds per shoot) reproduction by seeds under the parental cover is absent. Vegetative reproduction slowly declines in an increasingly unfavorable environment and the stands of fireweed slowly die out. Only some dormant underground organs are preserved for a long time (Danilov 1938).

Species of *Betula*, which form groves on the sites of coniferous forests after fire or felling, can also be considered as demutational explerents. In a fully developed state they form coenopopulations exclusively of generative individuals, fruiting abundantly every year. The reserve of viable seeds is annually renewed. This is of no importance however for the stability of birch coenopopulations since there is no regeneration by seed under mature trees. *Prunus pensylvanica* is a special type of demutational explerent, since its coenopopulations in the natural forest consist for long periods of dormant seeds only (a cryptic period of coenopopulation existence). After destruction of the forest these dormant seeds germinate and the coenopopulation passes into an active state consisting of generative, abundantly fruiting individuals. However coenopopulations of *Prunus pensylvanica* exist in this state for a relatively short time, and during this period a considerable reserve of dormant seeds is accumulated in the soil (Marks 1974).

Some pyrophytes are also demutational explerents. They are characterised by fast formation of all-component coenopopulations, as a result of germination of dormant seeds after fire followed by quick degradation of coenopopulations to the cryptic state. They are exemplified by many species of *Adenostoma fasciculatum* chapparal (Haines 1971).

Thus, among demutational explerents there are both species with autonomous provision of their coenopopulations by seeds and species which depend for initial formation of their populations on an immigrant supply of seeds. The latter in their turn produce a great quantity of seeds for establishment of their coenopopulations in new places.

*The term demutation was proposed by Vysotzkii (1915) for changes of disturbed vegetation in the direction of its initial state; it is widely used in Russian phytocoenological publications.

Ranunculus repens is a typical fluctuational explerent. In years of mass death of the principal meadow plants (as a result of temporary water-logging) it spreads widely and becomes a dominant plant. Its dominance lasts for a short period and as soon as the conditions become favourable again for the other species its importance in the plant community suddenly decreases. However during the period of its dominance a considerable reserve of its viable seeds is accumulated in the soil. In undisturbed meadow communities coenopopulations of *Ranunculus repens* consist of dormant seeds and of few adult, chiefly vegetative, individuals. For example in the central part of the flood valley of the Oka river near Dedinovo (Moscow region), meadows with dominance of *Alopecurus pratensis* and *Phalaris arundinacea* are found. As a result of long term spring flooding in 1951–1953 mass death of both these species occurred, and *Ranunculus repens* became dominant. Its dominance lasted for three years (1952–1954). Already in the second half of the growth season in 1954 favourable conditions for regrowth of grasses caused a rapid decline in the participation of *Ranunculus* in the sward. Grasses again became dominants: in 1956 participation of grasses had increased 3.6 times in comparison with 1953, whereas buttercup had decreased 100-fold. During the period of buttercup dominance many of its seeds were incorporated in the soil. Just after cutting of herbage in 1954 there were about 5000 seeds/m^2 on the surface of the meadow (Rabotnov 1958, 1974).

As indicated above, the content of dormant seeds in coenopopulations is of considerable importance for the stability of both demutational (*Prunus pensylvanica*) and fluctuational (*Ranunculus repens*) explerents: when phytocoenoses are disturbed conditions arise which are favourable for germination of dormant seeds and development of seedlings into adult individuals. Dormant seeds are of importance for many species of the forest understory. They germinate en masse on felling sites and form coenopopulations of numerous abundantly fruiting generative individuals. This is typical for species of the herbaceous stage of secondary succession after forest destruction (*Deschampsia caespitosa*, *Agrostis tenuis* and others). During the short period of their abundant participation in phytocoenoses a considerable reserve of their seeds can be accumulated in the soil. Then after formation of the tree layer the coenopopulations of such species gradually degrade to the intitial state: they consist of dormant seeds and some quasi-senile adult individuals. There are two phases in the dynamics of such coenopopulations: a patient phase under forest cover and an explerent phase after disturbance of the forest. Such species can be considered as demutational explerents. Viable seeds are of importance not only for recovery of vegetation after total disturbance but also after small local ones (such as the activity of rodents). A considerable quantity of stunted virginile (chiefly immature) individuals is typical of coenopopulations of many violents and patients (but not of explerents), including trees of climax forests (*Picea*, *Quercus*, etc.) and herbaceous species of steppes and meadows. Immature individuals are able to turn into mature ones as soon as appropriate ecological circumstances permit, usually as a result of dying of neighbouring adult individuals. This type of life history pattern (which is also distinguished by Grime 1979) occurs in phytocoenoses where gaps of different sizes often develop as a result of a local disturbance (for example as a result of windfall in forests).

Individuals in the state of secondary dormancy are of importance only for some species, chiefly those growing in sites of unstable water supply (i.e., changing greatly from year to year or from one period to another). In the years of deficient environmental resources (eg. water) mass transition of mature individuals from an active into dormant state occurs. However as soon as environmental conditions become more favourable these dormant individuals again turn into active ones. Such populations are characterised by alternation of two phases: dominance of mature individuals in an active state and prevalence of mature individuals in a dormant state (with different content of dormant seeds). Such a life history pattern often occurs among species growing in depressions of semiarid regions, for example in the Baraba of West Siberia (Kurkin 1976). Among species of this type are *Alopecurus arundinaceus*, *Hordeum brevisubulatum*, *Puccinellia dolicholepis*, *Eleocharis palustris*, *E. uniglumis*, *Scirpus maritimus*, *Butomus umbellatus*, etc. (Kurkin 1976, Rabotnov 1978a).

THE MAIN TYPES OF COENOPOPULATION DYNAMICS

In any coenopopulation death and transition from one age-state into another occur among every age group of individuals. This process can be traced by longterm observations of marked individuals on permanent plots. Such observations reveal the heterogeneity of their age-state groups. For example, in the meadow in the flood valley of the Oka river in 1950 all individuals of *Ranunculus acris* on a $10\,m^2$ plot were marked: 178 juveniles, 122 immatures, 122 adult vegetative and 25 generative specimens. The observations made during the next ten-year period (1951–1960) showed that on the basis of the subsequent behaviour of individuals during this period juveniles could be classified into 24 groups (Table 4), immatures into 31 groups (Table 5), adult vegetative into 57 categories (Rabotnov 1978c) and generative individuals into 17 groups (Rabotnov 1978c). Similar data were obtained from longterm observations of *Anemone fasciculata* in a subalpine meadow (Table 6). Longterm observations of marked individuals can also give an idea of the different behaviour in the past of individuals of the same age-state group in a contemporary coenopopulation. For example, 22 adult individuals of *Ranunculus acris* which were in a vegetative state in 1960 could be classified into 19 groups based on their behaviour in the previous 10 years (Table 7).

In coenopopulations of some species considerable changes occur from year to year in the distribution of individuals of different age-state groups, including changes which depend on the alteration of their vitality state with age (Watt 1947). Coenopopulations vary according with seasonal, fluctuational and successional changes of phytocoenoses.

Table 4. Changes during 1950–1960 in the age-states of individuals of *Ranunculus acris*, which were in a juvenile state in 1950 (Rabotnov 1978b)

Years											Number of individuals
1950	1951	1952	1953	1954	1955	1956	1957	1958	1959	1960	
j	0										80
j	j										9
j	j	im	0								1
j	j	v	v	v	v	v	v	g	g	g	1
j	im	0									30
j	im	im	0								17
j	im	im	im	0							1
j	im	im	im	im	0						1
j	im	v	0								15
j	im	im	v	0							2
j	im	im	v	v	0						1
j	im	v	v	0							2
j	im	v	v	v	0						2
j	im	v	v	v	v	0					2
j	im	v	v	v	v	v	v	v	v	v	1
j	im	v	v	g	g	g	g	g	g	g	1
j	im	im	im	v	v	v	v	v	v	v	1
j	v	0									1
j	v	v	0								5
j	v	v	v	0							1
j	v	v	v	g	0						1
j	v	v	v	v	v	0					1
j	v	v	v	v	v	g	v	v	v	v	1
j	v	v	v	g	v	v	v	v	v	g	1

j, juvenile; im, immature; v, vegetative; g, generative; 0, dead.

Table 5. Changes during 1950–1960 of individuals of *Ranunculus acris*, which were in the immature state in 1950 (Rabotnov 1978b).

Years											Number of individuals
1950	1951	1952	1953	1954	1955	1956	1957	1958	1959	1960	
im	0										27
im	im	0									10
im	im	im	0								10
im	im	im	im	0							2
im	im	im	im	im	0						3
im	im	v	0								12
im	im	v	v	0							1
im	im	v	v	v	0						7
im	im	im	v	v	v	v	v	0			1
im	im	im	im	v	0						1
im	v	0									1
im	v	im	0								1
im	v	v	0								22
im	v	g	0								4
im	v	v	v	0							1
im	v	v	g	0							2
im	v	v	v	v	0						4
im	v	v	v	v	v	v	v	0			2
im	v	v	v	v	v	v	v	v	v	v	1
im	v	v	g	g	v	g	v	v	g	v	1
im	v	v	g	v	g	g	v	v	v	0	1
im	v	g	v	v	v	v	v	v	g	0	1
im	v	v	v	g	v	v	v	im	0		1
im	v	v	v	g	v	g	v	v	v	v	1
im	v	v	v	g	v	v	v	v	v	v	2
im	v	v	v	g	v	g	v	g	v	g	1
im	v	v	v	v	v	v	im	v	v	v	1
im	v	v	v	v	v	v	v	g	g	v	1
im	v	v	v	v	v	v	v	v	v	g	1
im	g	v	0								1
im	g	v	g	g	g	0					1

im, immature; v, vegetative; g, generative; 0, dead.

Table 6. Changes of age-states of mature individuals of *Anemone fasciculata* in a subalpine meadow during the period 1943–1947 (Rabotnov 1950b).

Years of observations					Number of individuals
1943	1944	1945	1946	1947	
g	g	g	g	g	1
g	g	g	g	0	1
g	g	v	g	0	1
g	g	g	v	0	1
g	g	g	v	v	4
g	v	g	v	0	3
g	v	g	v	v	4
g	g	v	v	v	2
g	g	v	v	0	2
g	v	v	v	v	11
g	v	v	0	0	2
g	v	v	v	0	10
g	v	v	v	j	1
v	v	v	v	v	26
v	v	v	v	0	12
v	v	v	0	0	17
v	v	0	0	0	3
v	0	0	0	0	3
v	g	g	0	0	1
v	g	g	v	0	1
v	v	g	g	v	1
v	v	g	v	v	2
v	v	g	v	0	1
v	v	v	v	j	4
v	v	v	j	j	1
v	v	v	j	0	5
v	v	j	j	j	4
v	v	j	j	0	2
v	v	j	0	0	4
v	j	j	j	0	2
v	j	0	0	0	1

g, generative state; v, vegetative state; j, secondary immature state; 0, dead.

Seasonal changes

Seasonal changes consist both of variations of the numbers of individuals (by birth or death) and transitions of individuals from one age-state to another. An especially significant variation occurs in the number of viable seeds. There is a great quantity of them just after seeding of plants and their number gradually or suddenly decreases, sometimes to the point of total disappearance as a result of germination, destruction or loss of viability.

Based on seasonal variations three types of coenopopulations can be distinguished: coenopopulations of annual plants, coenopopulations of perennial plants with a seasonally dormant state (e.g., spring ephemeroids), and coenopopulations of perennial plants without a seasonal period of dormancy, charactised by a relative stability in number of

Table 7. The behaviour during 1950–1960 of a group of individuals of *Ranunculus acris*, all of which were in the mature vegetative age-state in 1960 (Rabotnov 1978b).

Years											Number of individuals
1960	1959	1958	1957	1956	1955	1954	1953	1952	1951	1950	
v	v	v	v	v	v	v	v	v	im	j	1
v	v	v	v	v	v	v	v	im	im	j	1
v	v	v	v	g	v	v	v	v	v	j	1
v	v	v	v	v	v	v	v	v	v	im	1
v	g	v	v	g	v	g	g	v	v	im	1
v	v	v	v	g	v	g	v	v	v	im	1
v	v	v	v	v	v	g	v	v	v	im	2
v	v	v	d	v	v	v	v	v	v	im	1
v	g	g	v	v	v	v	v	v	v	im	1
v	v	v	v	v	v	v	v	v	v	v	2
v	v	v	v	v	v	g	v	g	g	v	1
v	v	v	v	g	v	g	g	g	g	v	1
v	v	v	g	g	v	g	v	v	v	v	1
v	v	v	v	v	v	v	v	v	g	v	1
v	v	v	v	v	v	v	v	g	v	v	1
v	v	v	v	v	v	v	v	v	g	g	1
v	v	v	v	g	v	v	v	v	g	g	1
v	v	v	v	v	v	v	v	g	g	g	2
v	v	v	v	v	v	g	g	g	g	g	1

j, juvenile state; im, immature state; v, vegetative state; g, generative state; d, dormant state.

mature individuals but by changes in the numbers of viable seeds, seedlings and juveniles.

Fluctuations

Fluctuations (alterations from year to year) are especially pronounced in herbaceous phytocoenoses. There are three main types: oscillations, cyclic fluctuations and digressive-demutational fluctuations.

Oscillations occur in connection with the changes of environmental conditions which do not result in a disturbance of phytocoenoses, for example the alternation of moist and dry years. This is responsible for changes in the abundance of species to the extent of a change of dominants. Consequently the composition of coenopopulations changes from year to year in either direction from the average. Usually this is followed by changes in the vitality state of mature individuals: by their more luxuriant development (with abundant formation of generative shoots) in favourable years and by decrease of vitality (including transition from generative to vegetative state) in less favourable years for certain species. At the same time the composition of the virginile group (especially seedlings and juveniles) changes. For example Shennikov (1930) described the alternation of dominants from year to year on some types of meadows in the flood valley of the Volga river near Ulyanovsk. In moist years the dominant species was *Alopecurus pratensis*, in the dry ones *Elytrigia repens* and *Bromus intermis* dominated. In the moist years *Alopecurus* developed many generative tillers while populations of *Bromus* and *Elytrigia* consisted chiefly of vegetative ones. On the contrary in the dry years, populations of *Bromus* and *Elytrigia* were characterized by abundance of generative tillers, but populations of *Alopecurus* by vegetative ones.

Cyclic changes of coenopopulations are typical of *Trifolium pratense* and *T. hybridum*. The cycles consist of two periods: (1) the gradual formation of a group of actively growing individuals by germination of seeds buried in the soil – the population consists mainly of virginile individuals together with a few specimens of weakly developed generative ones; and (2) the mass transition of virginile individuals into luxuriantly developed generative ones ('clover year'), followed by their mass death in the same or in the following year. The cycle is then repeated (Rabotnov 1961). The recurrence of cycles is caused by competition between clovers and other components of phytocoenoses, chiefly grasses. It is probable that the relationship of clovers with the clover nematode (*Heterodera trifolii*) is of significance, since this species of nematode consumes clover roots and may cause the death of plants (Krall and Ryisnere 1966).

The germination of seeds of some annuals occurs only when there is a fair probability of the seedlings developing to reproductive maturity. This adaptive property has evolved in many desert annuals. For several years their coenopopulations may consist only of dormant seeds, which germinate when the conditions are favourable for development of seedlings into mature individuals (Went 1948, 1949, Walter 1973). In such years coenopopulations consist of actively growing individuals and the dormant seed reserve is replenished. Analogous changes were observed in coenopopulations of some annual legumes (*Vicia hirstua*, *V. tetrasperma*, *V. villosa*) in steppe and steppe meadows. For long periods their coenopopulations consist only of dormant seeds, which germinate in years of high precipitation or as a result of flooding, when transition of populations from cryptic into an active state and replenishment of seed reserves occur (Smirnov 1958, Kurkin 1966).

Digressive-demutational fluctuations, following temporary disturbances of phytocoenoses, are caused by significant deviations of meteorological or hydrological conditions from the usual ones or by mass reproduction of some animals. During the period of disturbance coenopopulations of violents and patients undergo mass death of individuals and a decrease in their vitality state and reproduction. In contrast, in coenopopulations of explerents the number, vitality state and reproductive capacity of growing individuals increase considerably. During the period of demutation phytocoenoses return to a state near to that of the initial one, as changes in coenopopulations occur. An example of digressive-demutational fluctuation has been described above for the periodical changes between *Alopecurus pratensis* and *Ranunculus repens* in the flood valley of the Oka river.

One may suggest that periodic disturbances of phytocoenoses may result in a specific life cycle and special type of fluctuation of the coenopopulations of some species. An illustrative example is the monocarpic species *Chaerophyllum prescottii*. In its life cycle there are three stages: (1) Growth of tap-rooted virginile individuals from seeds; they exist like this for several years before the tap root dies and the caudex is transformed into a spherical dormant organ; (2) persistence in a dormant state; (3) mass transformation of dormant organs into luxuriantly developed and abundantly seeding generative individuals. Usually this last stage is connected with a disturbance of phytocoenoses which occurred in the past, evidently in years of mass reproduction of burrowing rodents. The first two phases can be rather prolonged, the third phase is short. Dormant seeds are absent. Observations on the dry (but occasionally flooded) meadows in the flood valley of the Oka river have confirmed this life history pattern of *Chaerophyllum prescottii*. Viable seeds in soil were absent here. Oversowing of its seeds on the surface of the meadow for two years was unsuccessful in establishing plants. There were only negligible numbers of virginile individuals and only in some years a single weakly-developed generative individual appeared. There were, however, about one hundred dormant individuals per m^2 which were transformed into luxuriantly developed generative ones by disturbance of the meadow as a result of ploughing. It is very probable that the life history pattern of *Chaerophyllum prescottii* was coevolved with rodents (probably voles) in the steppe regions. Voles in the years of their mass reproduction greatly disturbed steppe soil and vegetation as a result of burrowing activity, creating a 'fallow regime', as

Table 8. Changes of age-state spectra (%) of *Filifolium sibiricum* and *Festuca lenensis* as influenced by grazing of increasing intensity (Gorshkova and Grineva 1977).

Age-state groups	Stages of increasing grazing intensity							
	I	II	III	IV	I	II	III	IV
	Filifolium sibiricum				*Festuca lenensis*			
Seedlings	5.1	0.4						
Juveniles	4.0	3.0	1.1	1.5	4.3			
Immatures	5.5	4.5	6.5	11.9	10.7	3.9	4.3	
Mature vegetative	75.1	82.8	91.9	86.6	75.3	53.4	46.8	65.4
Generative	10.3	9.3	0.5	–	9.7	42.7	48.9	34.6
Total number on 1 m^2	72.7	55.8	51.9	42.9	46.5	51.5	47.0	40.5
Cover (%)	30	25	10	3	8	9	6	4

it was described by Lavrenko and Yunatov (1952) in the Mongolian steppes. Dormant spherical organs of *Chaerophyllum* are not damaged, or only weakly damaged, by the burrowing activity of voles, which in fact creates very favourable conditions for their transformation into generative, abundantly-seeded individuals, and for establishment of seedlings from these seeds (Rabotnov 1964).

Successional changes

Successional changes of coenopopulations (particularly those caused by different forms of human activity) are widespread: they are a normal expression of the dynamics of coenopopulations. Therefore there are no grounds to consider such states of coenopopulations as populations of the secondary type (Uranov and Smirnova 1969). They are stages of degradation or, in constrast, of further development of coenopopulations occurring during successions.

Unfortunately there is very little quantitative information on the successional changes of coenopopulations. Some data are given in Table 8 on the changes of coenopopulations of *Filifolium sibiricum* and *Festuca lenensis* under increasing intensity of grazing in one type of Trans-Baikalian steppe pasture. In coenopopulation of *Filifolium sibiricum* increased grazing results in a decrease of cover, total number of individuals, and quantities of generative individuals and seedlings. The average size of individuals is also decreased greatly, as is evident from the greater decrease of cover than of total number of individuals. *Festuca lenensis* is more tolerant to an increase of grazing intensity than *Filifolium sibiricum*: the age-state spectrum of its coenopopulations changes less than that of *Filifolium.*

In an experiment with different times of cutting of subalpine meadow herbage during the period 1943–1947, the age-states of the mature individuals of *Anemone fasciculata* were changed differently, according to time of cutting (Table 9). The earlier the time of cutting the more oppressive

Table 9. Age-state (%) of mature individuals of *Anemone fasciculata* in 1947 as influenced by cutting at different times during the period 1943–1947 (Rabotnov 1974).

Time of cutting	Individuals which were in 1943:							
	In generative state				In vegetative state			
	Age states in 1947							
	Generative	Vegetative	Secondary immature	Dead	Generative	Vegetative	Secondary immature	Dead
Early	0	22.5	12.5	65.0	0	14.0	30.0	56.0
Normal	2.0	49.0	2.0	47.0	0	31.5	10.0	58.5
Late	36.0	32.0	0	32.0	4.0	46.0	4.0	46.0
Very late	41.0	23.0	0	36.0	10.5	47.5	3.0	39.0

was its influence on the vitality state of *Anemone*. In 1947 at the early cutting there were no individuals in the generative state and most individuals developed leaves of immature type instead of mature ones (they were accordingly in the secondarily immature age-state). There was a difference in the reaction to time of cutting between individuals which were generative or vegetative in 1943: the vegetative individuals responded to earlier cutting more adversely than generative ones. This is an indication of the higher vitality state of generative individuals than of vegetative ones. As a result of cutting at different times the age-state spectrum of the group of mature individuals was greatly changed. The high percentage of dead individuals in all variants of the experiment was related to the burrowing activity of water voles, since 1943–1947 was a period of its mass reproduction.

Different changes occur as a result of the 'catastrophic successions' (Alexandrova 1964) which are especially widespread in forests after fires or felling. Cuttings of climax forests involve destruction of adult trees. Consequently in coenopopulations of climax tree species only virginile suppressed individuals are preserved, since viable seeds are usually absent. The new populations are formed either as a result of the development of new individuals from seeds transported from outside (for example *Picea abies*) or by sprouting from the remnant stool as in *Quercus*. In climax forests in the boreal and temperate zones gaps appearing on the sites where single adult trees die off are filled by rapid growth of virginile individuals which are usually present in the understory in considerable quantities.

After destruction of tree layers the growth conditions for forest shrubs, dwarfshrubs and herbaceous plants change. Some species respond to this alteration positively, others negatively. As soon as the initial tree layer is restored the composition of their coenopopulations usually returns to the initial state (Korchagin 1954). Increasing importance of species after cutting of forests can be attained by different pathways, including an increase of the vigor of adult individuals without an increase in their number. A positive response can be exemplified by *Corylus avellana* which begins to grow luxuriantly when trees are removed. However, reproduction of hazel by seeds is absent as establishment of its seedlings is possible only in the conditions of forest environment where the upper soil horizon is only poorly penetrated by roots of herbaceous plants. Observations near Moscow indicated that a stable coenopopulation of *Corylus avellana* with adult individuals only is formed in open habitats with meadow-like vegetation between the shrubs. As soon as the tree layer is restored they can be transformed into coenopopulations of normal type. Among forest herbaceous plants and dwarf shrubs the positive responses to destruction of the tree layer result in increase of both the quantity of individuals and their vigour. Increase of the number of individuals in the active state of species which reproduce exclusively or chiefly by seeds depends on germination of dormant seeds and sometimes on transformation of dormant underground organs into active states (*Deschampsia caespitosa*, *Carex* spp.) Moreover, the vitality state of many individuals, formerly suppressed in the forest, rises and they acquire the ability to flower and fruit; from their seeds new growing individuals develop (e.g. *Calamagrostis arundinacea*). As a result coenopopulations of normal type are formed.

A special type of coenopopulation dynamics is exemplified by a demutational explerent, *Prunus pensylvanica*. The following stages can be distinguished: (1) formation of a many-component coenopopulation as a result of mass germination of dormant seeds after destruction of the forest; (2) development of the most complete coenopopulation, together with accumulation of a new reserve of dormant viable seed; (3) degradation of the population after restoration of the forest, and its transformation into a cryptic state consisting only of dormant seeds. Some species of explerents of this group of demutational explerents, for example *Rubus ideus*, are characterised by germination of dormant seeds both after complete destruction of forests and in locally disturbed areas (sites of uprooted trees, etc.). After destruction of forests favourable conditions arise for the formation of populations of such demutational explerents as *Chamaenerion angustifolium* and *Betula* spp., from seeds transported from other sites.

Some herbaceous forest species (*Oxalis acetosella*, *Trientalis europea*, *Maianthemum bifolia* and others) respond negatively to the destruction of the tree layer chiefly as a result of their inability to compete with the species spreading in felling areas whose roots form a rather dense turf. Both the number and the vitality of individuals in coenopopulations of such species gradually or quickly decrease and they 'disappear' from phytocoenoses, though in some cases they survive as cryptic populations for long periods as dormant seeds and/or dormant underground organs.

The important parameter for characterisation of coenopopulations is the number of individuals in an active state. There are populations both with great and negligible numbers of individuals; the latter are not clearly understood yet. Coenopopulations of nondominant species are likely to be both normal and regressive. Depending on the fluctuations of phytocoenoses in some periods they can be transformed from normal to regressive or vice versa from regressive to normal.

The dynamics of coenopopulations are, so far, imperfectly understood. To date attention has been focused on the study of coenopopulations of species with different life forms, but in future, more consideration should be given to the study of coenopopulations of species of different coenotic importance.

REFERENCES

(English translations of Russian-language titles are given in brackets)

Aleksandrova, V.D. (1964) (Dynamics of the plant cover.) Polevaya geobotanica 3: 300–447.

Bogdanovskaya-Gienef, I.D. (1926) (Contribution to seed renewal in meadow communities.) Zap. Leningrad. Sel.-khoz. Insit. 3: 216–253.

Bogdanovskaya-Gienef, I.D. (1941) (Seed renewal of plants in the meadow coenoses of the forest zone.) Uch. Zap. Leningrad. Univ., Biol. 20: 93–133.

Danilov, M.D. (1938) (Vegetative and seed reproduction of fireweed (*Epilobium angustifolium* L.). Priroda, 3: 99–106.

Gorshkova, A.A. and Grineva, N.F. (1977) (Changes of ecology and structure of steppe communities under influence of grazing.) (Ecology and pasture digression of Trans-Baikalian steppe communities.) Novosibirsk, pp. 157–182.

Goryshina, T.K. (1969) (Early spring ephemeroides of forest-steppe oak forests.) Leningr. Univ. 232 pp.

Grime, J.P. (1979) Plant Strategies and Vegetation Processes. Chichester: Wiley. 222 pp.

Haines, T.L. (1971) Succession after fire in the chaparral of southern California. Ecol. Monogr. 41: 27–52.

Korchagin, A.A. (1954) (Effect of fires on forest vegetation and its restoration after fire in European North.) Trudy Bot. Inst. Acad. Nauk SSSR. Ser. 3, 9: 75–149.

Korchagin, A.A. (1964) (Intraspecific (population) composition of plant communities and the methods of its investigation.) Polevaya Geobotanica 3: 63–131.

Krall, E. and Ryisnere, U. (1966) (Investigation on hostparasite relationships of *Heterodera trifoli* on legumes.) Izvest. Acad. Nauk Eston. SSR XV Scr. biol. 1: 83–89.

Kurkin, K.A. (1966) (Factors of the closeness of meadow biogeocoenoses.) Trudy Moskov. Obshch. Ispyt. Prirody 27: 98–116.

Kurkin, K.A. (1976) (System investigation of dynamics of meadows.) Nauka. 284 pp.

Lavrenko, E.M. and Yunatov, A.A. (1952) (Fallow regime in steppes as result of influence of Brandt vole on steppe sward and soil.) Botan. Jhurn. 37, 2: 128–138.

Linkola, K. (1930) Über das Vorkommen von Samenkeimlingen bei Pollakanthen in der natürlichen Pflanzengesellschaften. – Ann. Soc. Zool. Bot. Fenn. Vanamo 11: 150–172.

Linkola, K. (1936) Über die Dauer und Jahresklassen – verhältnisse des Jugendesstadiums bei einigen Wiesenstauden. Acta Forest. Fenn. 42 (2): 1–56.

Marks, P.L. (1974) The role of pin cherry (*Prunus pensylvanica* L.) in the maintenance of stability in northern hardwood ecosystems. Ecol. Monogr. 44: 73–88.

Neshataev, J.N., Dobretsova, L.A. and Samilyak, S.U. (1974) (Biological productivity and its seasonal dynamics in different layers of oak forest.) (Biological productivity and its factors in forest-steppe oak forest.) Leningr. Univ., pp. 119–152.

Perttula, U. (1941) Untersuchungen über die Generative und Vegative Vermehrung der Blütenpflanzen in der Wald-Hain-Wiesen und Hainfelsenvegetation. Annal. Acad. Sci. Fenn, Ser. A, 58 (1): 1–388.

Petrovskii, V.V. (1961) (Synusiae as the forms of coexistence of plants.) Botan. Jhurnal 46: 1615–1626.

Rabotnov, T.A. (1945) (Biological observations on the subalpine meadows of the Northern Caucasus.) Botan. Jhurnal 30: 167–177.

Rabotnov, T.A. (1949) (The principal problems and methods of study of life cycle of perennial herbaceous plants and composition of their populations.) Nauchn. Meth. Zap. Glavn. Upravlenya po Zapovednikam 12: 41–48.

Rabotnov, T.A. (1950a) (The life cycle of perennial herbaceous plants in meadow coenoses.) Trudy Bot. Inst. Acad. Nauk SSSR, Ser. 3, 6: 7–204.

Rabotnov, T.A. (1950b) (The problems of study of population composition for purposes of phytocoenology.) Problemy Botaniki 1: 465–483.

Rabotnov, T.A. (1958) (The life cycle of *Ranunculus acer* L. and *R. auricomus* L.) Bjul. Moskav. Obschch. Ispyt. Prirody 63, 6: 77–86.

Rabotnov, T.A. (1961) Some problems in increasing the proportion of leguminous species in permanent meadows. Proc. 8th Intern. Grassland Congress, Reading: pp. 260–264.

Rabotnov, T.A. (1964) (On the biology of monocarpous perennial meadow plants.) Bjul. Moskov. Obschch. Ispyt. Prirody 69, 6: 47–55.

Rabotnov, T.A. (1969) (Some problems of study of coenotic populations.) Bjul. Moskov. Obschch. Ispyt. Prirody 74, 1: 141–149.

Rabotnov, T.A. (1974) (Meadow science.) Moskow. 384 pp.

Rabotnov, T.A. (1975) (Study of the coenotic populations for the purposes of elucidation of strategy of life of plant species.) Bjul. Moskov. Obschch. Ispyt. Prirody 80 (2): 5–17.

Rabotnov, T.A. (1978a) (Phytocoenology.) Moscow: 384 pp. 2nd Ed. 1983, 292 pp.

Rabotnov, T.A. (1978b) (Structure and method of study of coenotic populations of perennial herbaceous plants.) Ekologia 2: 5–13; English translation: Sov. J. Ecology, 9: 99–105.

Rabotnov, T.A. (1978c) On coenopopulations of plants reproducing by seeds. pp. 1–26 in A.H.J. Freysen and J.W. Woldendorp (Eds.) Structure and Functioning of Plant Populations. Amsterdam: North-Holland.

Rabotnov, T.A. (1980, 1981) Bibliography of papers on the problems of coenopopulations published in the USSR. Part I, Part II, Part III. Excerpta Botanica B 20: 71–96, 171–206; 21: 91–119.
Ramenskii, L.G. (1935) (On the principal approaches, main concepts and terms of economic typology of lands.) Sovet. botanika 4: 25–42.
Ramenskii, L.G. (1938) (Introduction to the complex soil-geobotanical investigation of lands.) Moscow. 620 pp.
Rysin, L.P. and Kazantseva, T.N. (1975) (Method of coenopopulational analysis in the geobotanical investigations.) Botan. Jhurnal 60: 199–209.
Rysin, L.P. and Rysina, G.P. 1965. (An attempt of populational analysis of forest communities.) Bjul. Mosk. Obshch. Ispyt. Prirody 71(1): 84–94.
Shennikov, A.P. (1930) (Volgian meadows of the Middle-volgian region.) Leningrad 386 pp.
Smirnov, P.A. (1958) (Flora of the Prioksko-terrasnogo Reseve.) Moscow. 247 pp.
Smirnova, O.V. (1983) (Behaviour of species and functional organization of herbaceous cover of broadleaved forests.) Leningr. Univ. 21 pp.
Trulevich, N.V. (1960) (The structure of bush and population composition of *Artemisia tianschanica* Krash. in stages of pasture digression.) Uch. Zap. Fakult. Estestvozn. Moskov. Pedag. Inst. V.I. Lenina 4: 107–129.
Uranov, A.A. (1960) (The vitality of species in a plant community.) Bjul. Mosk. Obshch. Ispyt. Prirody 65: 77–92.
Uranov, A.A. and Serebryakova, T.I. (Eds.) (1976) (Coenopopulations of plants.) Moscow, Nauka 216 pp.
Uranov, A.A. and Smirnova, O.V. (1969) (The classification and the principal features of population development of the perennial plants.) Bjul. Moskov. Obshch. Ispyt. Prirody 74 (1): 119–133.
Varlygina, T.I. (1982) (Comparative characteristic of coenopopulations of *Potentilla erecta* (L.) Raeusch.) in Western Estonia and Moscow region.) Bjul. Mosk. Obshch. Ispyt. Prirody 875: 66–72.
Vysotzkii, G.N. (1915) (Ergenya – cultural phytological essay.) Trudy byuro prikl. Bot. 8(10–11): 1113–1443.
Walter, H. (1973) Die Vegetation der Erde in ökophysiologisher Bedeutung. Bd 1. Die tropischen und subtropischen Zonen. 3 Auf. G. Fischer 743 pp.
Watt, A.S. (1947) Pattern and process in the plant community. J. Ecol. 35: 1–22.
Went, F.W. (1948, 1949) Ecology of desert plants I, II. Ecology 29: 242–353; 30: 1–13, 26–38.

POPULATION BIOLOGY OF STEPPE PLANTS

L.I. VORONTZOVA and L.B. ZAUGOLNOVA

ABSTRACT

Five types of life history are distinguished among herbaceous species growing in steppe grasslands in Kazakhstan: they include tussock plants, tap-rooted plants of various kinds and long-rhizome herbs. They vary in the structure and dynamics of their coenopopulations and in the extent to which they are maintained in coenoses by seed or by vegetative growth. An understanding of the morphology and ontogenetic development of plants in natural coenoses is an indispensible prerequisite for describing their coenopopulation characteristics. These in turn reveal how plant communities are organised as dynamic systems.

INTRODUCTION

The existence of multispecies communities is possible because the species of the community are ecologically and biologically unique (Ramenskii 1925, Golubev 1959, 1970). The demographic characteristics of species ensure the segregation of their ecological attributes in time and space (Harper 1977, Grubb 1977), so that individual plants use the environmental resources in a complementary fashion (Ramenskii 1925). Although the totality of life history characters of populations is unique in each species, a considerable similarity and overlap of several features among different species cannot be ruled out. Such differences and similarities allow us to differentiate several types of life history attributes (see review by Harper 1977). These are commonly based on the reproductive biology of species and on the division of resources between reproductive and vegetative organs. Ramenskii (1938) seems to have been the first to initiate a functional typology of plant behaviour and to specify three phytocoenotypes known as 'patients', 'violents' and 'explerents'. The contemporary concepts of types of plant 'strategy' (Grime 1978), correspond on the whole to Ramenskii's ideas.

Studies of coenopopulations of perennial herbs in different coenoses (Smirnova et al. 1976, Uranov et al. 1977) and the review of literature by

White, J. (ed.), The Population Structure of Vegetation.
 ISBN 978-94-010-8927-2

Harper (1977) seem to suggest that a behavioural typology of perennial herbs is possible. The main characters used for differentiation of various types must be relevant to and decisive for the continued existence of a species in a coenosis and help to identify and determine its role in the community. In our opinion, such characters are: 1. The life-form and the course of plant ontogenesis; 2. The manner of species self-maintenance in a coenosis, that is, the relationship between the type of vegetative propagation and the frequency of seed reproduction; 3. Duration of ontogenesis; 4. Intensity of plant mortality during ontogeny.

In this paper we confine our attention to steppe plants and according to the characteristics listed above, we differentiate the following types of life history in these plants: (I) tussock plants, (II) tap-rooted perennial plants with seed self-maintenance, (III) tap-rooted perennial plants with 'mixed' self-maintenance, (IV) tap-rooted plants with a short duration of ontogeny, (V) long-rhizome herbs.

STEPPE GRASSLANDS

The steppe environment

Our material was collected in Naurzum Preserve (North Kazakhstan), located in the subzone of dry steppes. The dry steppes ('tussock-grasslands' of Lavrenko 1954) stretch in latitude from the Black Sea to the Baikal Region. They are confined to southern chernozems and chestnut soils with a very deep underground water table. The dry steppe region is characterized by a continental climate with hot dry summer and cold winter. The average annual rainfall is 260–300 mm. Although maximum precipitation falls in summer, air humidity and soil moisture content decrease significantly in this period because of higher temperature and evaporation. The dry period falls between June and September, but in some years it is less pronounced. The arid climate is the basic factor determining plant characteristics in this zone. In spite of the considerable geographical extent and ecological heterogeneity of dry steppes they have a number of common characteristics: predominance of tussock grasses of the genera *Stipa*, *Festuca*, *Koeleria*, *Agropyron* (Lavrenko 1940); low floristic diversity of communities; discontinuous overground plant cover; atmospheric water supply; lack of forests.

In Naurzum Preserve the investigations were conducted on sandy soils. These psammophyte variants of zonal steppes are reported in different regions of North Kazakhstan (Karamysheva and Rachkovskaya 1973).

Vegetation

Stipa pennata, *Festuca beckeri* and in places *F. valesiaca* (subsp. *sulcata*) provide the predominant cover; *Koeleria glauca*, *K. macrantha*, *Phleum phleoides* occur continuously but in low abundance. Of the various perennial herbs, *Potentilla cinerea*, *Centaurea sibirica*, *Artemisia campestris*,

Onosma simplicissimum occur most continuously. Annual plants are limited to *Bassia sedoides*, *Coryspermum hyssopifolium*, *Polygonum arenarium*. The number of species reaches 40–44 per 100 m^2. The total projective cover varies from 50% to 80%. The plants were studied in one particular phytocoenosis; all individuals of each species in that phytocoenosis were collectively regarded as a 'coenopopulation'.

AGE-STATE ANALYSIS OF COENOPOPULATIONS

Numbers and age structure are the most essential coenopopulation parameters. The investigation of age structure in perennial herbs is rather difficult since the calendar age can be determined in only few cases. In our studies we used the concept of age-states developed by T.A. Rabotov and other Soviet botanists. Age-states are the successive stages of ontogeny of a genetic individual (see Gatsuk et al. 1980 for a review). The individual plant is characterized by morphological integrity and it may be of seed or vegetative origin. A vegetatively derived individual is given the name 'particule'. On the basis of morphological characters the following age states and corresponding groups of individuals are differentiated (with their standardised abbreviation): seeds (**se**); seedlings (**pl**); juvenile (**j**); immature (**im**); virginile (**v**); reproductive, young (**g_1**), mature (**g_2**), and old (**g_3**); subsenile (**ss**) and senile (**s**). Investigations of age-state structure must be based on a clear understanding of plant morphology, to enable one to recognize the individual unit and the changes during its ontogeny.

The following four systems of 'calculation units' are used for enumering of plants in coenopopulation studies (with the age-state categories in brackets), illustrated in Fig. 1.

I Individuals of seed origin during the whole ontogeny;

II Individuals derived from seeds (**j**–**g_1**), compact clones (**g_2**, **g_3**, **ss**) or single particules (**s**). Since a single particule may be considered as a residue of a formerly existing clone, this second system of calculation units corresponds to the enumeration of genets in Harper's (1977) sense;

III Individuals of seed origin (**j**–**g_1**), clones (**g_2**), particules within the clones (**g_3**, **ss**, **s**) or single particules (**s**).

IV 'Partial tufts' or shoots within one or different individuals. A partial tuft (Serebryakov 1962) is a cluster of shoots with its own roots, connected with the same tuft by long rhizomes. Particules and partial tufts and shoots seem to correspond to ramets (Harper 1977).

The different types of calculation units reflect the various degrees of shoot concentration in plants of different life-forms. The autonomy of shoots and the morphological integrity of the plant as a whole change during ontogeny. The shoots may all be concentrated in one place, or they may be dispersed in several clusters linked to one another. The centre of shoot concentration is regarded as the source of a plant's phytocoenotic influence. As a consequence two types of calculation unit are distinguished by us, a morphological unit and a phytocoenotic unit. The former corre-

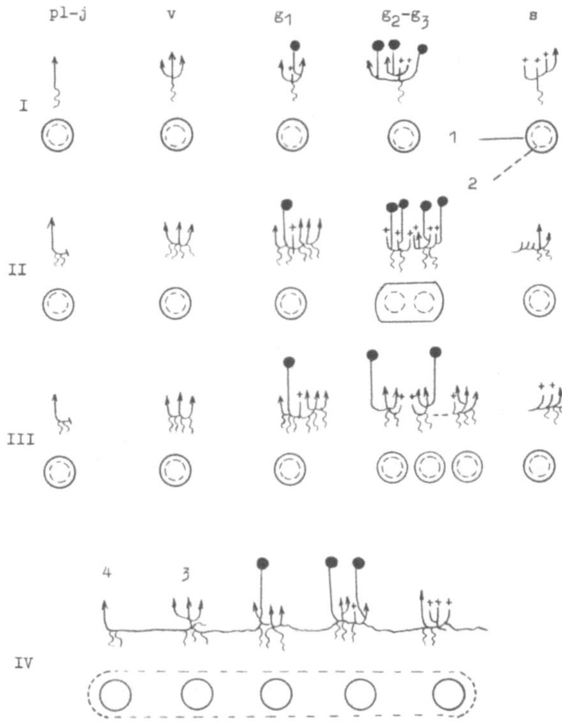

Fig. 1. Calculation systems used in studies of plant coenopopulations. Two types of units are distinguished, phytocoenotic (1) and morphological (2) (see text). I: Both types are coincident during the entire ontogeny and are of seed origin (genets). II and III: The units are distinct at later stages of ontogeny; morphological units may remain closely united to give a single phytocoenotic unit (II) or may fragment to give separate particules, each of which becomes an independent phytocoenotic unit (III). IV: A morphological unit, while retaining physical integrity may behave as multiple phytocoenotic units, each consisting of a partial tuft (3) or shoot (4).

sponds to the individual as a morphological and physical entity, the latter to the centre of shoot and root concentration. The correlation between the two units is different for various life forms and is determined by careful mapping and excavation of plants in natural phytocoenoses (Fig. 1).

System I is used for tap-rooted perennials, systems II and III for tussock grasses and species with short-rhizomes and system IV for species with long-rhizomes and those with root suckers.

The total plant number (in terms of calculation units) and the percentage of calculation units of different age-state groups (age spectra according to Uranov 1967) are determined usually on small plots (0.25–1 m^2) placed by a random-regular method (Vasilevich 1969) within a uniform area of a phytocoenosis. In our studies the number of such plots ranged from 20 to 160 for each species. Most observations were carried out at permanent transects (0.5 × 10 m^2) where the positions of plants were mapped annually, once a season in June.

The seed and fruit numbers of all flowering individuals were estimated on permanent $1 \times 10\,m^2$ transects. The number of seedlings was determined not less than twice a season on either permanent or nonpermanent plots, forming a $0.2 \times 10\,m^2$ transect. Seedlings are usually not recorded in the estimations of age-state spectra because of their ephemeral existence: a great number of seedlings die off in the first months of life, and those that survive pass on to the next age-state.

TYPE I – TUSSOCK PLANTS

The life history features discussed in this section are typical of:

(a) firm and loose tussock grasses, for example, *Stipa lessingiana, S. sareptana, S. pennata, S. rubens, S. capillata, Festuca valesiaca,* subsp. *sulcata, F. beckeri, Cleistogenes squarrosa, Koeleria macrantha, K. glauca, Phleum phleoides,* and

(b) herbs with short-rhizomes, for example, *Potentialla cinerea, Centaurea sibirica, Arenaria procera.*

The species of *Stipa* and *Festuca* are usually the dominants of steppe communities. The rest of the grasses are, as a rule, in a subordinate position although they may become dominants in some geographical and ecological steppe types: for example, there are steppes with predominance of *Cleistogenes squarrosa* in the Transbaikal Region and in Mongolia. In the arid steppes herbs with short-rhizomes are, as a rule, only moderately abundant, but are permanently present in the community.

Self-maintenance of coenopopulations of these species is by seed. At a certain stage of ontogeny the seedling transforms itself to a clone by vegetative propagation. An adult individual of seed origin is a more or less compact shoot system with regeneration buds at the base of shoots, and each shoot has its own root system. The death of shoots in the centre of the individual fragments it into particules. However, the particules are not rejuvenated after separation, as a rule, and ageing processes result in their death. This vegetative growth cannot provide prolonged existence of the species in a given habitat in the absence of regeneration by seeds, although the longevity of seed-derived individuals and of clones in these species may be considerable, from 30 to 80 years. More detailed life-history characteristics are given separately for the two groups of species.

Firm-tussock grasses

Seeds and seedlings

Steppe tussock grasses are characterised by seed regeneration whose efficiency depends on (a) potential and actual seed productivity, (b) seed germination capacity, (c) seed bank in the soil and (d) survival of seedlings and juvenile plants. 'Potential seed productivity' is defined as the number of ovules formed per plant; 'potential seed yield' is the number of ovules per $1\,m^2$. 'Actual seed productivity' is defined as the number of healthy

Table 1. Potential and actual seed yields in some firm-tussock grasses in the Naurzum Preserve.

Species	Year of observation	Number of flowering plants per 10 m²	Number of reproductive shoots per 10 m²	% Caryopses			Seed yield per 1 m²	
				Healthy	Damaged	Puny	Potential	Actual
Stipa pennata	1973	22	26	19.8	54.0	26.2	24.4	4.7
	1974	50	129	64.8	29.5	5.7	92.0	54.9
	1975	27	64		No data		66.1	No data
	1976	9	15	36.9	0	63.1	13.0	4.9
	1977	34	93	60.7	32.2	7.1	83.3	50.6
*Festuca beckeri**	1973	2	2	23.4	17.4	59.2	18.4	4.3
	1974	2	2	9.3	55.0	35.7	12.9	1.2
	1975	2	4	17.0	28.6	54.4	24.1	4.1
*Koeleria macrantha**	1974	3	6	4.7	85.3	10.0	55.7	2.6
	1975	3	4	8.3	83.6	8.1	66.4	4.8

*In these species reproductive shoots were absent in 1976 and 1977.

seeds per plant; 'actual seed yield' is the number of healthy seeds per 1 m^2.

The limits of potential seed productivity are genetically fixed, and specific for each species and habitat. In sandy steppes the value is 10–30 ovules per plant in *Stipa pennata*, 60–120 in *Festuca beckeri*, 180–200 in *Koeleria macrantha*. The number of healthy (intact) seeds formed depends on pollination conditions and the degree of ovary damage caused by phytophagous insects. Firm-tussock grasses are all wind-pollinated. For many species of *Stipa* and *Festuca* different rates of cleistogamy have been reported: it is unevenly pronounced in different species and varies within one species in different years (Ponomarev and Zvorygina 1949, Solntseva 1968). Cleistogamy increases under poor flowering conditions (excessive rain or drought for example). Many species of *Stipa* are characterised by rapid ripening of seeds in pollinated flowers and by non-simultaneous ripening in different parts of the inflorescence, which prolongs flowering and seed ripening (Solntseva 1968). These seed development characteristics may be considered as adaptations to the rather changeable spring conditions in the subzone of dry steppes.

The development of phytophagous insects and parasitic fungi in the reproductive parts of steppe grases is a typical phenomenon. The reproductive shoots of *Stipa pennata* are damaged by rust fungi; the ovaries are eaten by gallfly larvae, tripsids and aphids. Tripsids are significant pests in *Festuca beckeri* and *Koeleria* species. Inflorescence predation and poor pollination may result in extremely poor actual seed yield, from 4 to 60% of the potential level; this varies considerably from year to year (Table 1).

The seed yield depends not only on the level of potential seed productivity but also on the number of flowering plants in the coenopopulation and the number of reproductive shoots on each flowering individual; the fluctuations of these parameters will be mentioned later. In *Stipa pennata* coenopopulations a rather high actual seed yield is developed, in spite of the relatively low potential seed productivity (Table 1). In this species the percentage of healthy seeds was highest compared to other grass species. It seems to be one of the reasons ensuring the predominant role of *Stipa pennata*. The subordinated position of *Koeleria macrantha* in the community investigated is primarily related to the fact that an insignificant number of healthy seeds is formed on the plant (Table 1), although the species showed high levels of potential seed productivity. The potential seed productivity of steppe grasses is, as a rule, stable, while the actual productivity changes considerably according to the phytocoenotic conditions and regime of economic activcities (Osychnyuk and Shupranov 1978).

Caryopsis dispersion of most steppe grasses (*Festuca*, *Koeleria*, *Agropyron*) is by barochory (Levina 1956). In *Stipa* the seeds are wind dispersed, but even with strong wind they are not usually carried further than 3–4 (rarely 11) m (Ponomarev and Zvorygina 1949). The fate of the fallen fruits is different: some caryopses are retained in the litter and a great number of them are eaten by rodents or damaged by phytophagous insects. Less than 3% of *Stipa* caryopses falling on the ground are

Table 2. Seed bank of some grasses in the soil of Central Kazakhstan steppes in vegetation of *Stipa lessingiana* + *Festuca valesiaca* + *Artemisia gracilescens* (Borissova and Popova 1972).

Species	Year	Seed stock (no. m^{-2}) in soil layers				
		0–2 cm	2–4 cm	4–6 cm	6–8 cm	8–10 cm
Stipa lessingiana	1960	140	80	0	0	0
	1961	50	19	14	0	0
Festuca valesiaca	1960	300	100	80	80	0
subsp. *sulcata*	1961	1133	128	50	43	11

transformed into viable seedlings (Ponomarev and Zvorygina 1949). Evidently at this stage of ontogeny the death rate is the highest. *Stipa* caryopses become submerged in the soil to a depth of 2–4 cm owing to its 'screwing-in' mechanism. In *Festuca* and *Koeleria* the caryopses germinate directly on the soil suface: this factor increases the risk of death in seedlings of these species in the first stages of development and determines the subsequent tussock position. The seeds of many species of *Stipa* are characterised by inherent dormancy of different duration: some may germinate after nine or even 16–17 years dormancy in laboratory conditions (Bespalova and Borissova 1979). The germination capacity and the onset of germination in caryopses change according to the time of ripening, depending on the meteorological conditions in the year of fruit ripening (Bespalova and Borissova 1979). Because of the long retention of viability a seed bank can be developed in the soil. In grasses growing in communities on loamy soils the seed bank is rather large and can vary significantly from year to year (Table 2). Of the total seed number found in soil, the live caryopses of *Stipa lessingiana* and *Festuca valesiaca* represented 7% and 58% respectively (Borisova and Popova 1971). According to our observations, there is practically no seed reserve of most species in sandy steppe soil. The seeds of the current or previous year are the basic source of regeneration. In some years seedlings appear only on account of the seed reserves in the loamy soil of the steppes.

Steppe plant seedlings appear rather sporadically (not annually) according to moisture conditions. The seeds of most grass species germinate both in spring and autumn, but *Stipa* only in spring. Seedling number reaches 6–8 per 1 m^2 in sandy steppes. If the seedlings appear in spring most of them (50–84%) die in the course of the summer. Thus even in the dominant grasses the number of juvenile plants is rather low (2–6 per 1 m^2).

Age-state ontogeny

Ontogenetic development and morphological changes of individuals reflect the plant's life form. Ontogeny is rather similar in different species of firm-tussock grasses (Fig. 2, see also Gatsuk et àl. 1980). The uniaxial juvenile plant passes into the virginile state, as a result of shoot tillering

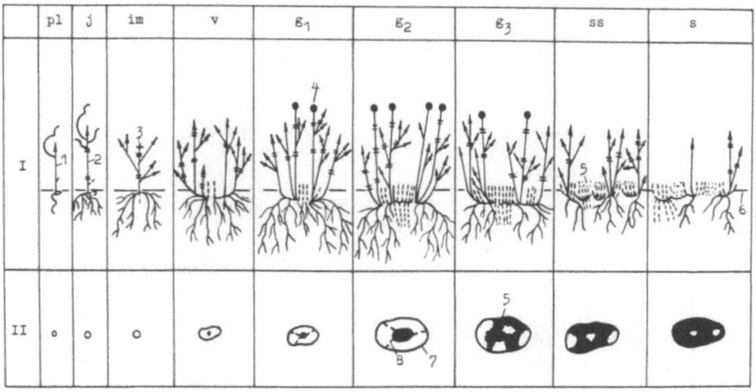

Fig. 2. Age-states in firm-tussock grasses (for example *Festuca beckeri* from Mikhailova (1977)). I, schematic diagram of tussock ontogeny; II, horizontal tussock projection. 1, vegetative first year shoot with leaves; 2, vegetative perennial shoot; 3, dead shoot, 4, reproductive shoot; 5, dead part of a tussock; 6, soil level; 7, boundary of an individual of seed origin or of the clone (genet); 8, boundary of the particules, the individuals of vegetative origin (ramets).

and forms a tussock. Then it begins to flower (**g_1**). Shoot tillering becomes more intensive, the tussock increases, but simultaneously dead shoots accumulate (**g_2**). The tussock divides into particules which together compose a clone. The ageing processes in shoot and root systems are intensified, the frequency of shoot formation is reduced, the number of live shoots declines (**g_3**, **ss**). Tussock ageing results in a reduction of live shoot number to 2–3, and eventually only one particule remains from the clone (**s**).

The difference in ontogeny between the species of dense-tussock grasses is due primarily to the different durations of each age-state: these can be established approximately by using the methods of calendar age determination (Persikova 1959, Bedanokova et al. 1975). Life-span varies under different conditions (Table 3), increasing usually in unfavourable conditions. In some species (*Koeleria glauca*, *K. macrantha*) the longevity of clones in Naurzum Preserve may reach 100 years. In firm tussock grasses the old age period (**g_3** + **ss** + **s**) is the longest. In many natural communities the age-state structure of coenopopulations is determined by this peculiarity. There are cases, however, when the old age period is greatly reduced, and the senile state is less evident, as reported for *Stipa lessingiana* and *Festuca valesiaca* in Central Kazakhstan steppes (Borissova and Popova 1971). The pre-reproductive period is usually extended in grasses of natural communities. It is significantly reduced in culture, where *Stipa* species for example flower in the second year of life (Ivashin and Chuprina 1976).

Firm-tussock grasses are generally characterised by a considerable life-span which can be considered as an adaptive character. In subarid and arid conditions this guarantees a stable species position even with relatively low actual seed productivity. In many species the tussock constitutes a clone in the mature state and with ageing the particules becomes more

Table 3. Duration (years) of the age-state and ontogeny is some steppe grasses.

Species	Habitat	Age-States							Duration of full ontogeny
		j	im	v	g_1	g_2	g_3 + ss	s	
Stipa pennata	Dry steppes (Kustanai Region)	3	3	8	8	9	14 + 20	10	75
	Northern steppes* (Kursk Region	1–4		2–3		20–25		10	45
Festuca valesiaca subsp. *sulcata*	Semidesert (Caspian Lowland)	1		2	5	10	6	5	30
	Northern steppes** (Kursk Region)	0.1–0.6		1–2	2–4	18–19	5–8	5	40
Festuca beckeri	Dry steppes (Kustanai Region)	2–3	3	5	2	6	11 + 8	4	40

*Data of Golubeva (1964).
**Data of Persikova (1959).

and more isolated. However, little rejuvenation occurs, the particules age and die off. Such vegetative propagation cannot ensure prolonged existence of the coenopopulation in the absence of reproduction of seeds.

Survival of age-states

The individuals of different age-states are characterised by uneven rates of annual survival which increase in the first part of ontogeny and reach 100% in the mature state. This mature period is associated with active shoot formation in the tussock and growth of the root system which may reach depths of 100–150 cm and diameters of 60–80 (up to 100) cm in *Stipa pennata* and *Festuca valesiaca*. The plants become weaker when they reach the senile state, a considerable number of shoots and roots die off and the survival rate is consequently reduced.

To characterise the death of coenopopulations of different steppe grass species depletion curves (Harper 1977) were drawn. They correspond to the negative exponential (Fig. 3A), indicating a relatively constant death rate; even an abrupt fluctuation of conditions (such as a drought in 1975) did not affect the depletion rate of plants. The half-life period was 20.5 years in *Stipa pennata* and 13.7 years in *Festuca beckeri*. In steppe firm-tussock grasses the decay of the coenopopulation is, as a rule, rather slow; this is due both to the stability of individuals and to the considerable duration of age-state ontogeny. Using the data on the death rate of plants of different age-states in *Stipa pennata* coenopopulations (**se** = 90–97%; **pl** = 84, **j** = 66, **im** = 40, **v**–$\mathbf{g_2}$ = 0, $\mathbf{g_3}$ = 26, **ss** = 76, **s** = 73, on the basis of genet number) we attempted to plot a survival curve for one generation of seedlings (**pl** = 100) (Fig. 3B). Biological rather than chronological time (that is, using age-states) is plotted on the horizontal axis. As might be expected, the curve shape approximates type III of Deevey (Harper 1977). This however is not necessarily contradictory to the exponential type of depletion (Fig. 3A) curves. Survival seems to correspond frequently to the exponential *within* different age-stages during ontogeny, whereas the total survivorship curve may be represented as a combination of a few such curves, with different inclination angles, each of them corresponding to a certain age-stage. Sarukhán (1978) reached a similar conclusion.

The number of plants in coenopopulations of tussock grass species does not change greatly in time (Fig. 3C). The stability is related to scanty seed reproduction, adult plant stability and considerable life span of particules and whole clones.

Reproduction

In coenopopulations of tussock grasses inflorescences are not formed by all individuals having reproductive capacity. Intermittent flowering, reported for many plant species (Rabotnov 1978) is rather typical of the dense-tussock steppe grasses. The intervals are particularly long in old plants: some individuals form 1 or 2 generative shoots only once in 7 or

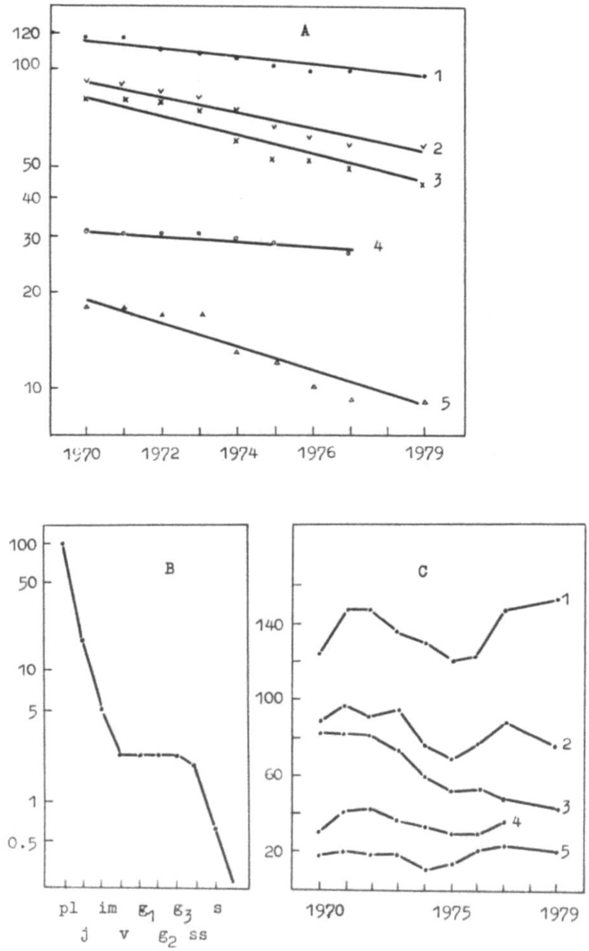

Fig. 3. Survival and dynamics of grass coenopopulations.
A. Depletion curves for coenopopulations: 1, *Stipa pennata*; 2, *Festuca beckeri*; 3, *Koeleria glauca*; 4, *Stipa capillata*; 5, *Koeleria macrantha*. The numbers of individuals on a $10 \times 0.5\,m^2$ transect from the population of 1970 are shown on the ordinate.
B. Calculated survival curve of *Stipa pennata* individuals during ontogeny; age-states are indicated on the abscissa and the number of surviving individuals on the ordinate.
C. Dynamics of plant numbers (system II calculation units) for 10 years on a $10 \times 0.5\,m^2$ transect: 1, *Stipa pennata*; 2, *Festuca beckeri*; 3, *Koeleria glauca*; 4, *Stipa capillata*; 5, *Koeleria macrantha*. The number of calculation units is shown on the ordinate.

8 years. The rise and fall of reproductive activity in different grass species may not coincide; for example, *Stipa pennata* and *Festuca beckeri* have different periods of inflorescence initiation and development. The particules of some grass species (*Festuca beckerii*, *Koeleria macrantha*, *K. glauca*, *Phleum phleoides*) retain the ability to flower almost until they die and can be observed in particules with 1–2 vegetative shoots. While the efficiency of such flowering is very low, there is still the possibility of seed formation. Furthermore, the presence of temporarily sterile individuals

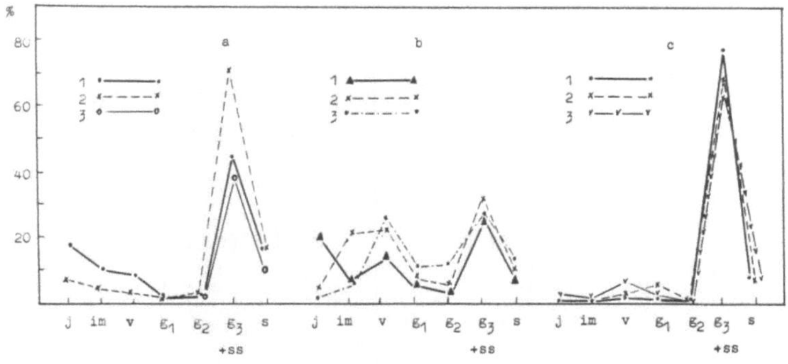

Fig. 4. Age-state spectra of coenopopulations of some grasses in sandy steppes on a preserve area (data of 1972). (a) *Stipa pennata.* The proportion of each age-state group in a coenopopulation (%) is shown on the ordinate: 1, % of total genet number; 2, % of the total number of individuals of seed and of vegetative origin (genets + ramets); 3, % of clones within genets. (b) *Stipa capillata* (1), *Festuca beckeri* (2), *Cleistogenes squarrosa* (3); (c) *Koeleria glauca* (1), *Koeleria macrantha* (2), *Phleum phleoides* (3). The proportions of each age-state group as a % of total genet number is shown for (b) and (c).

in a coenopopulation pre-empts space from occupation by other individuals and results in the maintenance of genetic diversity in the coenopopulation.

Age-state spectra

The age-state spectra of coenopopulations in many firm-tussock grasses growing in a particular community are rather similar (Fig. 4): the absolute maximum in the spectrum occurs in **g_3 + ss** specimens, and the minimum in those of **g_2**. Some years are characterised by local maxima in the group of young individuals. If system III calculation units are used (that is, including particules within clones) the predominance of old plants proves to be still more notable (Fig. 4a).

A review of age-state spectra (Fig. 5) of some firm-tussock grass species in different communities suggests that in some habitats a species is characterised by a particular relationship between plants of different age-states in the stable part of the population (**g_1 + g_2 + g_3 + ss + s**). *Festuca valesiaca* is, as a rule, characterised by total predominance of young individuals (**j + v + g_1**) over the old ones (**g_3 + ss + s**) or their more or less equal relationship. In *Stipa lessingiana* two types of spectra are distinguished – with maxima on **g_2** in the stable part (Fig. 5BI) or on old individuals (Fig. 5BII). It should be noted that the maximum on old individuals of *Festuca valesiaca* subsp. *sulcata* and *Stipa lessingiana* is typical of non-grazed localities while the maximum on **g_1 – g_2** is usually observed on continuously grazed plots. In *Stipa pennata*, the age-state spectrum in most of the habitats investigated is characterised by a predominance of old plants (Fig. 5C); the northern steppe (Kursk Region) is characterised by the predominance of reproductive plants (**g_3**) among the old ones and higher participatation of mature plants (**g_2**); in the

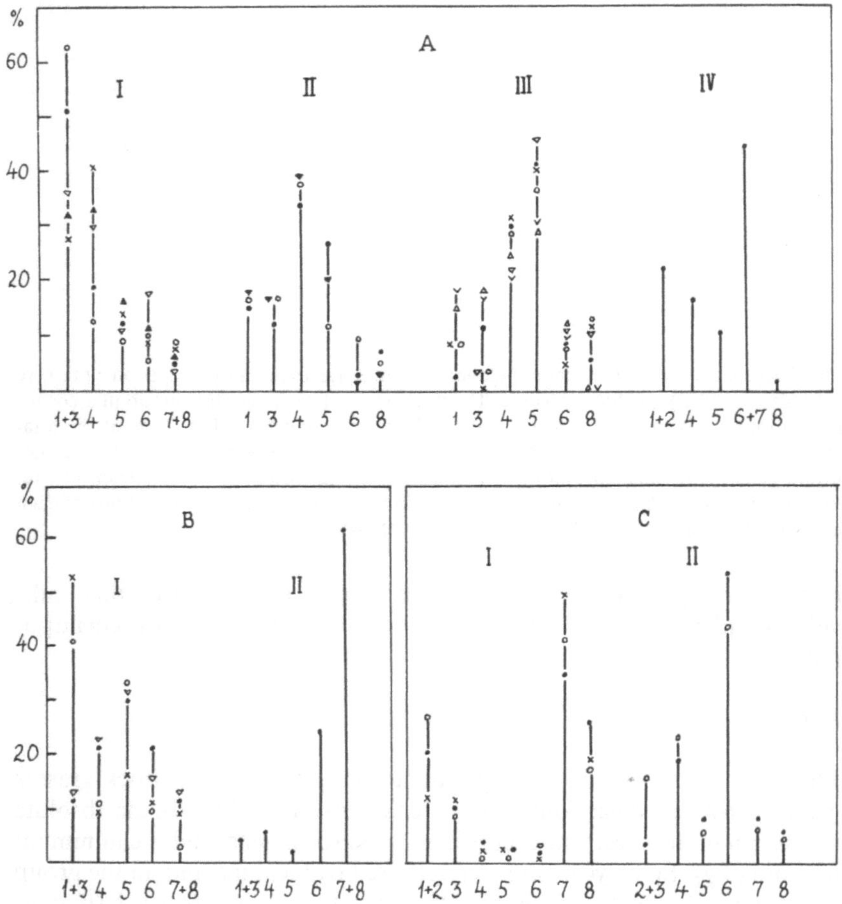

Fig. 5. The age-state spectra of coenopopulations of some firm tussock grasses. The proportions of various age-state groups in coenopopulations are shown on the ordinate. 1, **j**; 2, **im**; 3, **v**; 4, $\mathbf{g_1}$; 5, $\mathbf{g_2}$; 6, $\mathbf{g_3}$; 7, **ss**; 8, **s**.
(A) *Festuca valesiaca* subsp. *sulcata* growing in (I) steppes of Central Kazakhstan (Borissova and Popova 1972), (II) semidesert of North Caspian area under low and moderate grazing (Vorontzova and Zhukova 1976), (III) arid steppes of Inland Tian-Shan (Kozhevnikova and Trulevich 1977) and microsinks in the semidesert of North Caspian area (Vorontzova and Zhukova 1976), (IV) arid steppes of Naurzum Preserve without grazing.
(B) *Stipa lessingiana* growing in (I) arid steppes of Central Kazakhstan (Borissova and Popova 1972), (II) arid steppes of Naurzum Preserve without grazing.
(C) *Stipa pennata* growing in (I) sandy steppes of Naurzum Preserve, (II) northern steppes of the European part of the USSR. Each symbol on the vertical lines corresponds to a particular coenopopulation.

sandy steppes (Kustanai Region) subsenile plants (**ss**) are predominant (Fig. 5CI).

The following indices were used to characterise the dynamic processes in grass coenopopulations (Table 4). (On permanent grass plots system III calculation units were used.)

1. Death rate of plants, determined as $f.100/N_1$, where f is the number of plants dying in the period between two successive observations and N_1 is the plant number at the first of the two observations.
2. The rate of coenopopulation growth $(\ln N_2 - \ln N_1)/(t_2 - t_1)$, where N_2 is the number of plants during the second of two successive observations.
3. The rate of ontogenetic change $q.100/N_1$, where q is the number of plants that changed their age-state between t_1 and t_2.
4. Coenopopulation age-state index Δ (Uranov 1975), which estimates the coenopopulation age-state level on the basis of age-state group relationship: $\Delta = \Sigma\, k_i m_i / \Sigma\, k_i$, where k_i is the number in each age-state group and m_i is the age 'value' of one individual of the corresponding group. The values of m_i are calculated on the basis of the logistic curve and expressed as follows: **j** = 0.018; **im** = 0.047; **v** = 0.119; $\mathbf{g_1}$ = 0.270, $\mathbf{g_2}$ = 0.500, $\mathbf{g_3}$ = 0.731; **ss** = 0.880; **s** = 0.953. This index Δ changes from 0 to 1 and increases with coenopopulation ageing; Δ for two successive time intervals is shown as a fraction in Table 4. The theoretical basis of this calculation is discussed by Uranov (1975).
5. The number of juvenile individuals of the first year of life.
6. The number of particules arising from previously existing plants, that is, the number of individuals that appeared vegetatively.

The dynamic processes in coenopopulations of dense-tussock grasses can be characterised as follows (Table 4):

1. The death rate of seed individuals and particules varied from year to year. The maximum indices do not always coincide with the most unfavourable years: for example, after the drought in 1975–1976 the death rate of *Festuca beckeri* was no higher than in the previous years.
2. The death rate increased during the middle of the observation period and fell again by the end of it.
3. The differences between such indices as the death rate of the initial coenopopulation (or depletion rate) and the death rate of the whole coenopopulation are related to the fact that in the second index the death of all newly formed individuals is taken into account, which causes fluctuations of this index from year to year.
4. The rate of coenopopulation growth was low; coenopopulation numbers increased mostly on account of particule formation. If seed recruitment to the coenopopulation is considered alone, the growth rate becomes negative.
5. The ontogenetic development was uneven. Considerable changes in age-state composition occurred in the years after drought (1976–1977), when the plants were subjected to intensive ageing, resulting from the death of a great number of shoots in tussocks.
6. The recruitment of seed progeny was uneven.
7. The coenopopulations of *Festuca beckeri* revealed a somewhat higher death rate than those of *Stipa pennata*.
8. The rate of coenopopulation growth (assessed by number of genets)

Table 4. Parameters of grass coenopopulation dynamics on a permanent transect (0.5 × 10 m²) in Naurzum Preserve. Seedlings are excluded.

Species	Parameters	Observation periods							
		1970–1971	1971–1972	1972–1973	1973–1974	1974–1975	1975–1976	1976–1977	1970–1977
Stipa pennata	1. Death rate (%)	2.4	4.7	6.0	9.5	9.1	11.5	3.2	30.5
	2. Coenopopulation growth rate	+0.17	0	−0.08	−0.4	−0.07	+0.02	+0.18	+0.17
	3. Rate of ontogenetic change	2.0	3.3	8.0	8.3	7.4	9.7	9.4	7.6
	4. Age-state index (Δ)	0.70/0.58	0.58/0.57	0.57/0.60	0.60/0.63	0.63/0.66	0.66/0.75	0.75/0.76	0.70/0.76
	5. Number of juvenile individuals in the first year of life	25	6	0	0	2	1	1	35
	6. Number of separated particules	0	1	0	7	1	15	26	50
Festuca beckeri	1. Death rate (%)	1.1	6.2	5.5	12.6	10.4	8.6	7.8	33.6
	2. Coenopopulation growth rate	+0.07	−0.06	−0.05	−0.12	−0.09	+0.09	+0.15	0
	3. Rate of ontogenetic change	6.7	8.2	15.4	11.5	16.9	15.7	24.7	14.1
	4. Age-state index (Δ)	0.59/0.57	0.57/0.58	0.58/0.60	0.60/0.63	0.63/0.66	0.66/0.73	0.73/0.82	0.59/0.82
	5. Number of juvenile individuals in the first year of life	8	0	0	1	1	0	0	10
	6. Number of separated particules	0	1	0	0	13	24	0	38

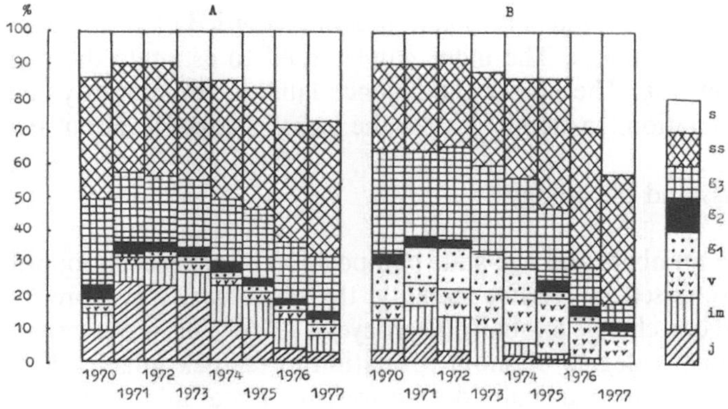

Fig. 6. Changes of some age-state spectra of coenopopulations of *Stipa pennata* (A) and *Festuca beckeri* (B). The proportions of age-state groups (% of the total number of individuals) are shown on the ordinate.

was higher in *Stipa pennata* than in *Festuca beckeri*; this reflects the more plentiful regeneration in *Stipa*.

9. The ontogenetic changes are more variable and evident in *Festuca beckeri* than in *Stipa pennata*.

The basic processes occurring in the course of time in the coenopopulation (the appearance of new individuals, their ontogenetic development, and the death rate which is specific for different age-state groups) determine the dynamics of coenopopulation age spectra. Although the main characteristics of the spectra during eight years of observations were retained in the grasses, some changes in age-state group relationships did occur (Fig. 6). The share of **j** and **im** became reduced to zero in *Festuca beckeri*. The relationships in the young part of the coenopopulation underwent wave-like changes caused by death and ontogenetic development of individuals. The share of **ss** individuals and **s** age-states increased, particularly after the drought of 1975. The consequences of drought were manifested in a reduced number of vegetative and reproductive shoots in 1975, by increased formation of particules and by particule ageing. Thus, the changes in age-state spectrum are related to the population characteristics of species and the effect of the environmental variations through time. In 1979 many seedlings (22) and juvenile individuals (17) of *Stipa pennata* were recorded on the transect, suggesting that grass coenopopulations are replenished periodically by small portions of new seed individuals, roughly after 9–10 years.

In coenopopulations of the grasses investigated the permanent individuals (that is, those retained during the whole observation period) make up not less than 50 percent of all recorded individuals: the flow of deaths and regenerations is stabilised by the permanent part of the coenopopulation. We tried to estimate these relationships quantitatively using a parameter which we call 'balance lability', calculated as $1 - (N'/N)$, where N' is the number of plants permanently retained during the whole observation period and N is total number of recorded plants (using system III calculation units). The index can be used to estimate the degree of population flux. The higher the balance lability, the more dynamic the coenopopulation. In steppe grasses the index is equal to about 0.53.

Phytomass and phenology

While the numbers and age-state composition of grass coenopopulations are characterised by relative stability, the phytomass is a more dynamic index and can change annually (Gordeyeva 1976). This is due primarily to changes in the degree of shoot formation in tussock grasses. From our observations some particules do not form shoots at all in dry years, that is, they are in a state of secondary dormancy. This characteristic is particularly abruptly manifested in grasses under very adverse conditions for growth, where the period of dormancy may last up to 3 years (Steshenko 1976, Borissova et al. 1976).

One of the essential aspects of the population biology of a species is its development rhythm throughout the year, or phenology. Most species of steppe grasses form a rather integral group with regard to the rhythm of seasonal development. The leaves start growing in April–May, reach maximum development in mid-June, while at the end of June–July the leaf tips begin to dry. In years of drought leaves start to dry earlier and more rapidly; even in less extreme years up to 80% of total leaf area may dry out by the end of the summer. The grasses can survive the warmest and driest period of the summer in a state of temporary dormancy. New leaves appear in autumn and remain green under snow. The reproductive shoots of *Stipa pennata* are initiated in autumn, those of other species are established in spring prior to flowering (*Festuca*, *Koeleria* and others). The species of *Stipa* usually flower in the middle and at the end of May, those of *Festuca* and *Koeleria* at the beginning of June. In some species (*Stipa capillata*, *Cleistogenes squarrosa*) flowering is delayed until mid-summer.

Physiological characteristics

The considerable similarity of the population biology of steppe grasses is based on their common ecological characteristics. The narrow-leaved steppe grasses were classified by Kolpikov (1960) in an ecological group which he named 'stipaxerophytes'. Their ecological and physiological characteristics explain to a considerable extent the stability of steppe grasses in arid conditions. These species are characterised by a high threshold of protoplasm coagulation (55–63°C); low water content (down to 38% of plant fresh weight, Grigoryev and Razumova 1974) 79% of which is bound water (Zhuravleva 1977), retained after water extraction by a 30% glucose solution; high protoplasm viscosity and relatively high (maximum up to 35–40 atm) osmotic pressure (Sveshnikova and Bobrovskaya 1973); the latter determines the considerable root suction in steppe grasses. The stipaxerophytes have relatively low transpiration rates (usually less than 1·5 $g.g^{-1} hr^{-1}$) particularly in the drought period

(Gorshkova and Kopytova 1977), enabling them to resist dehydration. Because of their low content of free water the stipaxerophytes have a relatively low photosynthetic rate (Spivak 1977). The rate and regulation mechanisms of water balance determine the distribution of grasses in nature and their role in coenoses (Sveshnikova 1979). *Stipa sareptana* dominates in the most xeromorphic conditions of desert steppes. Its water balance is regulated by variations of water regime indices (transpiration rate, water content in leaves, osmotic pressure), by the rhythm of stomatal functioning and by rolling and unrolling of leaf blades. *Stipa rubens* and *S. pennata* are characterised by higher transpiration rates, lower variability of water regime indices, and by absence of leaf rolling: these species are typical of more northern steppes. In the south of the steppe zone they occur either on light-textured soils or under high moisture conditions. *Festuca valesiaca* subsp. *sulcata* has a very plastic water balance that permits it to survive in a wide range of conditions (Sveshnikova 1979).

Grazing

The steppe tussock grasses make up the basis of climax communities and are subject to intensive grazing. While in the pre-agricultural period the grazing of wild animals in natural conditions was necessary for equilibrium maintenance (Semenova–Tyan–Shanskaya 1978), with modern agricultural management the development of many steppe plants is influenced by intensive grazing. The effect of grazing, however, is versatile and not unequivocal. With steppe grasses grazing affects the vitality and numbers of plants in coenopopulations and their age-state spectrum (Zaugolnova 1977). The response of species to grazing is to a great extent specific and depends on grazing intensity. Thus in some species (*Festuca valesiaca* subsp. *sulcata*, *F. beckeri*) the first stages of grazing can be associated with improved vitality of individuals and clones, an increase in the number of vegetative and reproductive shoots and their height. The basic reason for such behaviour lies evidently in the attenuation of the dominant *Stipa* species that are strong competitors. With increased grazing the vitality of the *Festuca* spp. is worsened. In *Stipa caucasia* and *Agropyron pectiniforme* even slight grazing is associated with a decrease in tussock diameter, as well as shoot number and height (Kozhevnikova and Trulevich 1971).

Intensive grazing affects not only plant vigour and number but also leads to rearrangement of the age-state spectrum (Kozhevnikova and Trulevich 1971): the number of old plants is reduced because of their high death rate; there is also a 'false rejuvenation' of coenopopulations, related to the accumulation of young plants, the impossibility of complete ontogeny and the death of plants that have not reached the natural limits of life.

Conclusion

The above review makes it possible to distinguish the most common coenopopulation characteristics of steppe tussock grasses.

1. The formation of compact tussocks with strong root systems leads to high survival of adult individuals;
2. The considerable length of tussock life allows old individuals to accumulate in a coenopopulation;
3. The numbers and age-state composition of coenopopulations are stable;
4. Both the numbers of reproductive shoots and the vegetative mass are dynamic.

These characteristics determine the dominating and edificator role of tussock grasses. Even grass species with low abundance in steppes are extremely stable components, retaining their positions in the coenosis for a long time.

Short-rhizome herbs

In sandy steppes this group is represented by *Potentilla cinerea, Centaurea sibirica, Arenaria procera, Hieracium echioides, H. virosum.* Our observations of *Potentilla cinerea* and *Centaurea sibirica* may be used to exemplify the group. While each species has its own specific characteristics of coenopopulation growth and dynamics, they are all similar in many points.

The potential seed yield in these species is rather different (Table 5); the actual yield is similar. The latter index varies slightly over time, whereas the potential yield may vary rather strongly. The fluctuations in potential yield are related primarily to the number of reproductive shoots formed on a plant. The reproductive organs are initiated in these species in the previous summer, and the drought occurring in this period may have a negative effect on the number of reproductive shoots (as for example in 1976, see Table 5). These two species flower at the end of May, and are insect pollinated. It is possible that with a small number of reproductive shoots and flowers pollination may be more efficient. Many seeds of *Potentilla* and *Centaurea* are damaged by weevil larvae. If flowering is poor, the number of phytophagous insects may drop abruptly and the seed damage is reduced in these years.

Table 5. Seed yield and its components (per 1 m^2) in short-rhizome plants on Naurzum Preserve.

Species	Year	Number of reproductive shoots	Number of flowers (inflorescences)	Potential yield	Actual yield
Potentilla	1973	27.2	61.8	1876.8	64.7
cinerea	1974	14.6	34.6	1051.2	49.8
	1976	3.4	9.3	275.4	58.7
Centaurea	1974	6.9	161.4(6.9)	161.4	56.3
sibirica	1977	11.0	249.0(11)	249.0	75.8

The seeds of *Potentilla cinerea* usually germinate in close proximity to maternal plants, giving rise to groups of 20–40 seedlings. The fruits of

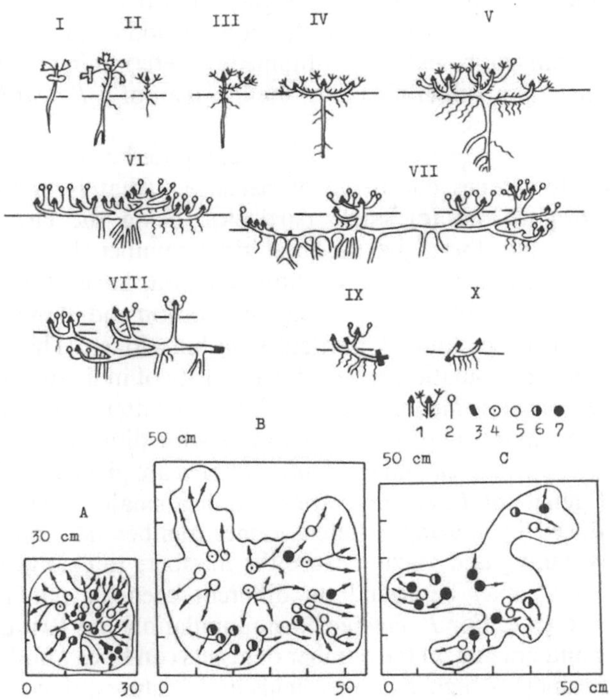

Fig. 7. Age-states and clone structure in *Potentilla cinerea.* The individuals of seed origin are shown as: I, seedling; II, juvenile; III, immature; IV, virginile; V, young; and VI, mature reproductive. Individuals of vegetative origin are: VII, mature; VIII, old strong; IX, old weak; X, senile. 1, vegetative shoot; 2, reproductive shoot; 3, dead shoot. Individuals of vegetative origin are: 4, mature; 5, old strong; 6, old weak; 7, senile. A, B, C, the stages of clone development.

Centaurea sibirica may be carried away by beetles picking the fleshy tendrils; the seedlings of this species are, as a rule, randomly distributed. The number of seedlings of *Potentilla cinerea* is rather high in some years; *Centaurea sibirica* seedlings are always sporadic. However the survival of small *Potentilla* seedlings is rather low, and by the end of summer fewer than 1–2 per 5 m^2 persist. Seedling density of *Centaurea sibirica* by the end of summer is the same.

The coenopopulations are replenished extremely slowly: during the whole observation period (1970–1979) on a transect (0.5×10 m^2) only one juvenile individual of *P. cinerea* became established but not a single one of *C. sibirica.* These species are characterised by clone formation in mature and old age-states (Fig. 7) and the clones spread rather slowly. Some particules do not undergo rejuvenation and age gradually. When they die the vacant site is taken by the same or by other species. The dimensions of clones and the number of particules in each are different in these species: clone diameter in *P. cinerea* is 20–70 cm, in *C. sibirica* 10–20 cm; the average particule number per clones is 8.1–13.2 and 3.2–3.5 respectively. The differences affect the position of the species in a coenosis:

as a rule, the cover of *P. cinerea* is higher than that of *C. sibirica*. The clones of these species exist for a rather long period of time. An estimate of their life-span can be indirectly obtained by determining the particules' 'provisory age' (Gatsuk et al. 1980): the life-span of a *Potentilla* clone is not less than 50 years.

Coenopopulations of these species are composed of (a) individuals of seed origin, (b) clones consisting of particules (that is, individuals of vegetative origin) and (c) single particules, being the last fragments of clones. Together these give the total genet number. In short-rhizome plants different patches of a coenopopulation have different genet numbers, caused by differences in seed regeneration and in the death rates of isolated particules and of particules within clones. The number of ramets in a coenopopulation reflects the intensity of individual disintegration and the average ramet number in a clone; the latter varies from species to species. The numbers of genets and ramets for different coenopopulations of *Potentilla cinerea* and *Centaurea sibirica* are given in Table 6. The number of genets of *P. cinerea* varies within a small range and is only slightly reduced by grazing, whereas ramet number falls rather significantly. Haymaking and grazing cause the numbers to change within the same ranges as may be found on different coenosis patches on the Naurzum Preserve. For *P. cinerea* coenopopulations no serious effect of haymaking and grazing on the number of genets can be established but the number of ramets is higher on hay-fields and pastures, that is, there are more particules in the clone (on average) than on a preserve plot (without mowing or grazing). The coenopopulations of these species are characterised by the prevalence of old ($\mathbf{g_3}$ + **ss** + **s**) particules (Fig. 8); the role of old particules is even higher in spectra calculated on the basis of ramet number. Seed regeneration is not abundant but it ensures the permanent presence of young individuals in the coenopopulation. This spectrum is rather stable and can last for a long period of time. The number of juvenile individuals can change both in time and space.

It is practically impossible to investigate the dynamics of individuals in short-rhizome species since their boundaries cannot be established without excavation. Thus in observations on permanent plots only the number of shoots was estimated. The numbers of vegetative shoots in *P. cinerea* and *C. sibirica* varied slightly during 8 years of observations: 334–380 in *P. cinerea* and 75–93 in *C. sibirica* on a 0.5×10 m^2 transect, indicating coenopopulation stability. These plants are characterised by well-branched, deep (160–200 cm), adventitious and fleshy roots with water reserves in them. The compound leaves of *P. cinerea* are rather hairy and leaflets can fold along the midrib when air and soil dryness increase. These features of plant texture contribute to an increase in stability of the coenopopulations in arid conditions. *P. cinerea* and *C. sibirica* are typical of psammophytic and petrophytic variants of steppes, occurring in pine forests on sandy soils, as well as in the steppe and forest-steppe zone of the European part of the USSR, Kazakhstan and Siberia. They are scarcely eaten by cattle and withstand grazing. The vitality of individuals improves in the first stages of pasture deterioration and with heavy overgrazing,

Table 6. The composition of coenopopulations of some short-rhizome plants in different coenoses, based on 100 plots, each 0.5 m^2

Species	Coenosis	Year	Individuals of seed origin	Clones	Particule number in clones	Number of single particules	Genets	Ramets (or particules)
Potentilla cinerea	Preserve	1971	221	34	348	–	255	348
	Preserve	1972	165	141	1374	–	306	1374
	Hayfield	1972	131	95	812	–	226	812
	Pasture	1972	118	85	571	–	203	571
Centaurea sibirica	Preserve	1971	54	101	357	173	328	530
	Preserve	1972	32	71	242	34	137	276
	Hayfield	1972	51	144	593	66	261	659
	Pasture	1972	80	197	789	37	314	826

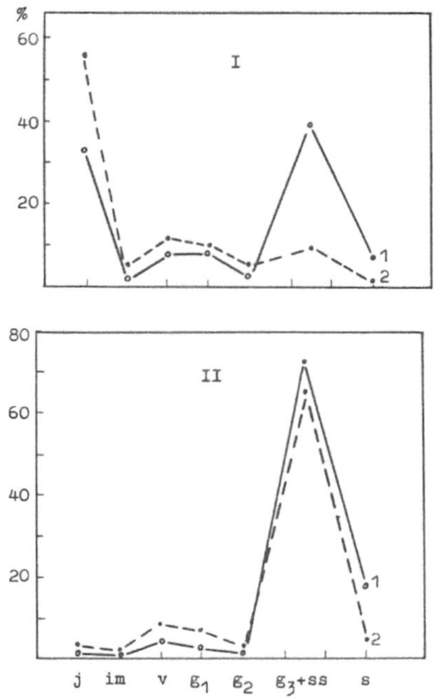

Fig. 8. Age-state spectra of short-rhizome plants: *Potentilla cinerea* (I) and *Centaurea sibirica* (II). The proportions of each age-state in a coenopopulation are shown on the ordinate; these are calculated either on the basis of total number of individuals (1) or on the basis of genet number (2). Data are shown for 1972.

when many species are eliminated from the grass stand, *P. cinerea* becomes the dominant of the coenosis. Their projective cover is not high (2–5%, maximum 12–15%), and because of their small height (5–7 cm) their occupation of space is 2–5 times lower than dense-tussock grasses. The foregoing population characteristics make them permanent components of sandy steppes.

TYPE II – TAP-ROOTED PERENNIAL PLANTS WITH SEED REGENERATION

This group is represented by species which regenerate in the climax community only by seed: *Artemisia campestris* (f.*sericea* and f.*glabra*; data reported here refer only to f.*sericea*), *Euphorbia seguieriana, Onosma simplicissimum*, *Gypsophila paniculata*. Root suckers can potentially be formed in *Euphorbia seguieriana* but is not observed in the climax community of sandy steppes. Vegetative propagation is also absent in these plants in the old age-state. The potential seed productivity varies in mature individuals of these species within a wide range: it is highest in *Artemisia campestris* (4,200–12,700 ovules per individual) and the lowest in *Euphorbia seguieriana* (24–95). Numerous seedlings are formed in favourable

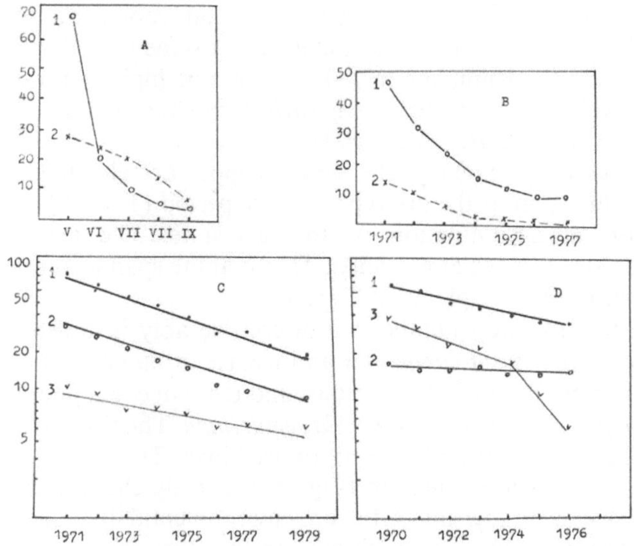

Fig. 9. Survival of tap-rooted plants in coenopopulations.
(A) Survival of seedlings during the vegetation period (May–September) on Naurzum Preserve in 1971: 1, *Artemisia campestris*; 2, *Euphorbia seguieriana*. The number of seedlings on a 10 × 0.2 m^2 transect is shown on the ordinate.
(B) Survival of juvenile individuals recorded in 1971, from 1971 to 1977. The number of individuals on a 10 × 0.5 m^2 transect is shown on the ordinate. 1, *Artemisia campestris*; 2, *Euphorbia seguieriana*.
(C) Depletion curves for all individuals recorded in 1971 on a 10 × 0.5 m^2 transect. 1, *Artemisia campestris*; 2, *Euphorbia seguieriana*; 3, *Onosma simplicissimum*.
(D) Depletion curves of: 1, *Scorzonera ensifolia*; 2, *Seseli ledebouri*; 3, *Helichrysum arenarium*.

years but most of them die off in the course of summer (Fig. 9A); the survival of seedlings is related to seed size and this is higher in *E. seguieriana*. Juvenile plants may be numerous in the population but the death rate is so high (Fig. 9B) that their participation in the coenopopulation decreases abruptly in the first few years after their appearance. Thus seed regeneration in these species is sporadic. The share of juvenile individuals recorded in the coenopopulation is higher (on average 40–50%) than in the groups of species discussed previously.

The underground part of these plants become branched with ontogenetic development and the root system reaches a depth of 160–200 cm in the mature state. In the first stages of ontogeny the stability of individuals in coenoses increases somewhat and their death rate decreases correspondingly. As plants age their shoots die and the tap root deteriorates, the depth of root penetration is reduced and plant stability declines. Plant survival changes correspondingly during ontogeny. For example in *Artemisia campestris* the % survival of each age-state changes in the following way:**j** – 27, **im** – 44, **v** and $\mathbf{g_1}$ – 72; $\mathbf{g_2}$ – 100; $\mathbf{g_3}$ and **ss** – 65; **s** – 70. One of the essential differences between this type of population behaviour and that of bunch or tussock plants is the lower survival at the beginning and the end of the ontogeny of the tap-rooted plants. In spite of the higher number of juvenile individuals at some periods in a coenopopulation the total number (per 0.5 m^2) is not high in these species: 0.1–7.3 individuals in *Artemisia campestris*, 0.5–0.8 in *Onosma simplicissimum* and 0.2–3.1 in *Euphorbia seguieriana*.

The species investigated in the sandy steppes are characterised by a rather long life span of the pre-reproductive period (Table 7), one of the reasons for an accumulation of this age-state. The ageing period (**g**$_3$ + **ss** + **s**) is somewhat extended. The total life span is, on the whole, shorter than in tussock plants of Type I.

The coenopopulation number varies considerably in time. In coenopopulation of *Artemisia campestris* an increase in number was recorded only once in nine years of observations, and twice in coenopopulations of *Onosma simplicissimum* and *Euphorbia seguieriana*. The rises are related to the sporadic formation and rooting of seedlings. The depletion rate of individuals in a coenopopulation (Fig. 9C) scarcely changes in different years and seems to depend little on environmental fluctuations. The depletion curves are in all cases close to the exponential relationship and the species differ slightly in death rate. The population half-life is shortest in *A. campestris* (4.0 years), 4.7 years in *E. seguieriana* and 5.0 years in *O. simplicissimum*.

The comparison with tussock plants reveals another essential characteristic of the population behaviour of tap-rooted plants: their number is much more dynamic, suggesting that they are relatively labile species: balance lability is 0.81–0.87 in *A. campestris* and *E. seguieriana* and 0.69 in *O. simplicissimum*.

Although the depletion rates (of plants present from the first observation) do not change in time (Fig. 9) the death rate sometimes increases greatly (Table 8). This was caused mainly by death of young plants that appeared in 1971. The growth rate of a coenopopulation varies from year to year, but negative indices prevail over positive ones: the coenopopulation number decreases significantly during the nine-year period.

The age-state spectra of coenopopulations of the tap rooted species are characterised by the following features (Fig. 10): the absolute maximum occurs as a rule among the individuals of the pre-reproductive period, or of **g**$_1$; the position of the maximum is mobile and depends on the rate of initial development of individuals, their death rate and on the sporadic appearance of seedlings; the height of the maximum in the spectrum depends on the intensity of species seed reproduction and on the time that has passed since mass seedling appearance; local maxima may appear among the old individuals, caused by a slightly extended ageing period; the longer the plant ontogeny the more important and stable is the maximum on **g**$_3$ (Table 7, Fig. 10).

The characteristics of ontogenetic development account for another essential aspect of population behaviour of this group of species, namely the dynamism of the age-state spectra in time, as may be illustrated by *Artemisia campestris*. During the whole observation period (Fig. 11) the

Table 7. The duration of age-states and ontogeny (years) in some tap-rooted species on permanent plots in the Naurzum Preserve

Species	Duration of age states								Total duration of the ontogeny	
	j	im	v	g_1	g_2	g_3	ss	s	Ranges	Average
Artemisia campestris	3–7	1–5	2–6	1–5	2–4	2–8	1–3	2–4	14–42	30
Euphorbia seguieriana	1	5–8	–	4–7	2–5	2-6	1–2	1–2	15–29	20
Onosma simplicissimum	4	4	5	10	6	3	not ascertained			about 40

Table 8. Some indices of dynamics of *Artemisia campestris* coenopopulations on a 0.5 × 10 m² transect in the Naurzum Preserve. Seedlings are omitted.

Parameters	1970–1971	1971–1972	1972–1973	1973–1974	1974–1975	1975–1976	1976–1977	1970–1977
Death rate (%)	1.2	21.1	16.9	19.7	23.5	23.1	0	67.3
Growth rate of coenopopulations	+0.12	−0.41	−0.34	−0.92	−0.69	−0.51	0	−1.1
Degree of ontogenetic change (%)	9.4	3.3	16.9	27.9	13.7	23.1	30.0	68.4
Age-state index	$\frac{0.20}{0.19}$	$\frac{0.19}{0.22}$	$\frac{0.22}{0.25}$	$\frac{0.25}{0.31}$	$\frac{0.31}{0.35}$	$\frac{0.35}{0.40}$	$\frac{0.40}{0.45}$	$\frac{0.20}{0.45}$

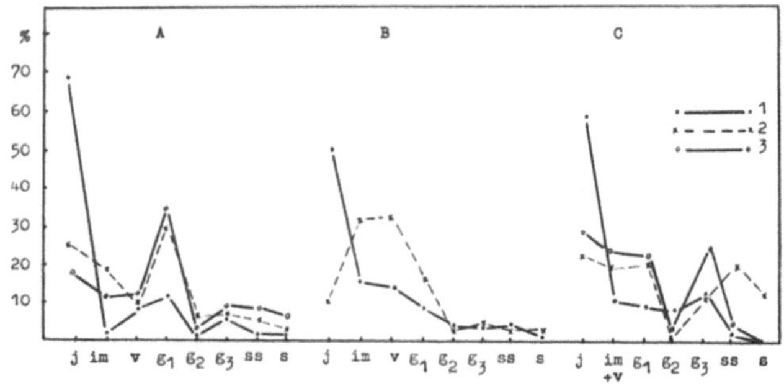

Fig. 10. Age-state spectra of coenopopulations of tap-rooted plants. The proportion (%) of the total genet number is shown on the ordinate. A, *Artemisia campestris*; B, *Euphorbia seguieriana*; C, *Onosma simplicissimum*. 1 and 2, Preserve area in 1971 and 1972; 3, hayland in 1972.

participation of juvenile individuals decreased abruptly while old ones increased. The share of **im**, **v**, $\mathbf{g_1}$ individuals is marked by wave-like changes: the absolute maximum moves from **j** to $\mathbf{g_1}$. These changes are a result of the death of individuals and their age development. We did not succeed in establishing any relationship between fluctuations in meteorological conditions and the degree of ontogenetic changes: such changes seem to depend largely on population properties of the species. On the whole, the rate of ontogenetic change is much higher in this species than in dense-tussock grasses (compare Table 4), and is related to the shorter life-span of *Artemisia campestris* individuals.

The abundance of tap-rooted species increases with coenosis disturbance. *Artemisia campetris* f. *sericea* becomes dominant in a pine forest on disturbed sandy soils; *A. campestris* f. *glabra* reaches high numbers in the first stages of old field disturbance (up to 3–8 specimens per 1 m^2), but in

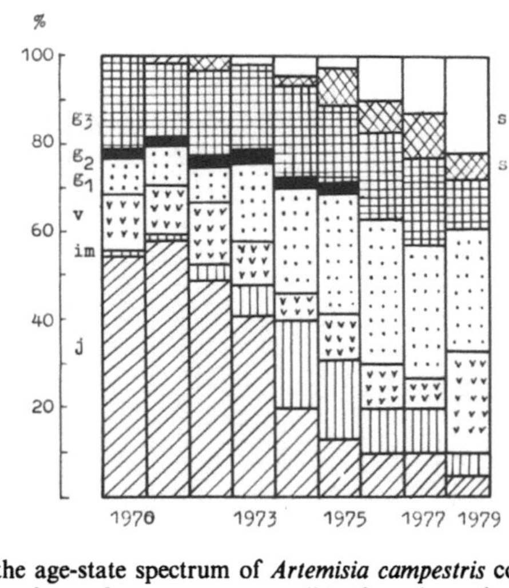

Fig. 11. Changes of the age-state spectrum of *Artemisia campestris* coenopopulation over nine years. The proportions of age-state groups (% of total number) are shown on the ordinate.

the natural community the density decreases abruptly (0.8–0.1). The numbers of *Euphorbia sequieriana* are increased by grazing 2.5–7 times, compared to the maximum number on a preserved area. The attenuation of interspecific plant interference results in their improved vegetative and reproductive growth, increased seed productivity and a correspondingly higher rate of coenopopulation growth.

The following are the most typical features of the population behaviour of these species: (1) the absence of vegetative propagation; (2) relatively high elimination of individuals; (3) a shorter ontogeny than that of tussock plants and reduced duration of the senile state; (4) greater dynamism of numbers and of age-state spectra.

TYPE III – TAP-ROOTED PERENNIALS WITH MIXED REGENERATION

This group is represented by the following species in steppe communities: *Scorzonera ensifolia, Seseli ledebouri, Helichrysum arenarium.* The potential seed productivity of species is extremely different: in *S. ensifolia* it is 20–40 ovules per specimen and in *S. ledebouri* 1000–3000 (rarely up to 6500) ovules per specimen. The actual seed productivity is very low in *S. ensifolia* (0.2–2.3% of the potential) and considerably higher in *S. ledebouri* (30–40%, though in some years it is 0.2–1.2%). The number of seedlings however is very low even in *S. ledebouri*: in favourable years, 2–4 per 10 m^2. Seedlings of *S. ensifolia* and *H. arenarium* are sparse.

Vegetative propagation was also recorded in these species. Buds on roots give rise to shoots which then form their own root system and can exist independently. According to their age-state, these plants may then be equated with juvenile or with immature plants. In some species the connection between the root-suckers and the maternal plant may last for a long time (*S. ledebouri*), while in others it is disturbed rapidly as a result of root rot or breakage. Individuals of vegetative origin may retain traces of their former connection with the maternal plant for a long time and in coenopopulation analysis such plants can be estimated separately. Plants originating from root suckers undergo age changes and therefore different age groups can be distinguished among them, similar to those derived from seed. The ratio between the individuals of seed and vegetative origin in a population varies in different plant species: the percentage vegetative progeny among the total number of individuals is 28–35% in *S. ledebouri*, 88–97% in *H. arenarium* and 99–100% in *S. ensifolia*. The vigour of vegetative propagation in these species is different and this accounts for observed variations in numbers of the species: the average number of individuals per 0.5 m^2 is 0.2–19.3 in *Helichrysum arenarium*, 0.4–1.3 in *Seseli ledebouri*, 1.4–6.1 in *Scorzonera ensifolia* (Naurzum sand steppe). In coenopopulations pre-reproductive plants are as a rule predominant. Consequently the population number of these species is rather variable in time; balance lability is 0.8–0.9. Their half-life is rather short: 3.3–2.5 years in *H. arenarium* and 7.0 in *S. ensifolia*. *S. ledebouri* is different and coenopopulations of this species are more stable in time, with a balance lability of 0.56 and half-life of 20.5 years. The root suckers of this species are more viable and have a slower rate of development than the two other species.

The calendar ages of these species are difficult to determine. Observations on permanent plots and morphological analyses have shown that the life-span of individuals in sandy steppes is roughly 18–20 years in *Helichrysum arenarium*, 15–25 years in *Scorzonera ensifolia*, 20–30 years in *Seseli ledebouri*.

Root suckers are formed actively in humid years. Newly formed individuals die off in a manner similar to those derived from seed (Fig. 9D), showing an exponential relationship that does not reveal any dependence on environmental fluctuations and plant age (*H. arenarium* is an exception). A higher death rate in populations of these species does not result from active formation of vegetative propagules. For example in *Scorzonera ensifolia* (growing on a $0.5 \times 10\,m^2$ transect) the highest number of dead plants was recorded in 1977–1979, but only two out of 19 were young (age 2–3 years), the rest being perennial individuals; out of 11 plants that died off in 1972–1973 five were 1–2 years old, the rest were older. The death of vegetatively formed individuals seems to be primarily an endogenous process in this species, determined by the survival level of individuals and the duration of their ontogeny. The survival of vegetative progeny may possibly depend on attachment to the maternal plant: the half-life is considerably greater in species where the connection is retained for a longer period of time.

Plants of the species investigated have different means of survival in arid conditions: in *Scorzonera ensifolia* the leaf bases are densely covered with long hairs which coat the young growing parts like a blanket; in *Helichrysum arenarium* both leaves and stems are hair-covered; in *Seseli ledebouri* the finely dissected leaves are covered with cuticula. The thick and succulent roots of *S. ledebouri* and *Scorzonera ensifolia* have water reserves in them that enables them to tolerate drought periods. Vegetative propagation of these species is intensified in areas where the plant cover is damaged by digging animals. Sousliks and blind rat-moles often eat the sappy roots of *S. ensifolia*, thus activating vegetative propagation: the density of this species may rise abruptly in such sites. However, these tap-rooted species are sensitive to grazing and are rapidly eliminated from the community by it.

The population behaviour of this species group reveals a considerable similarity with the previous group of species: the coenopopulations are, as a rule, characterised by a rather high lability of number and age-state composition, sporadic regeneration and predominance of young plants. The main difference lies in their dependence on vegetative propagation rather than seed reproduction. Vegetative progeny are more viable than seed progeny, and the death rates of juvenile and adult plants do not differ as sharply as in species with seed regeneration.

The tap-rooted plants with mixed self-maintenance may be considered as the relatively labile fillers of communities, rapidly colonizing disturbed areas and retaining their place for varying periods of time.

TYPE IV – TAP-ROOTED PLANTS WITH A SHORT ONTOGENY

In the steppes, this type is exemplified by *Alyssum lenense*, *A. tortuosum*, *Syrenia sessiliflora*. The density of these species on undisturbed plots of a community is usually low (approximately 0.4–1.4 specimens per 1 m^2) but it can grow abruptly after disturbance, reaching an average of 20 specimens per 1 m^2. The life-span of plants is short, reaching 2–3 years in *Syrenia sessiliflora* and under 7–9 years in *Alyssum lenense*. The senile period is not pronounced in plant ontogeny. Most individuals in populations of *Syrenia sessiliflora* are monocarpic, with only some of them flowering for a second time or vegetating one more year after flowering.

These species are marked by a high death rate of individuals: 95.4% of the initial coenopopulation of *A. lenense* died in the period 1970–1979. This is caused primarily by the thin and shallow root system which only reaches a depth of 50 cm. Coenopopulations of these species decrease catastrophically after droughts. Coenopopulation numbers and age-state spectra are rather dynamic (Fig. 12) and subject to abrupt variations in time and space (Zaugolnova 1976). Young plants are predominant. Coenopopulation lability is due to the following factors: periodicity in

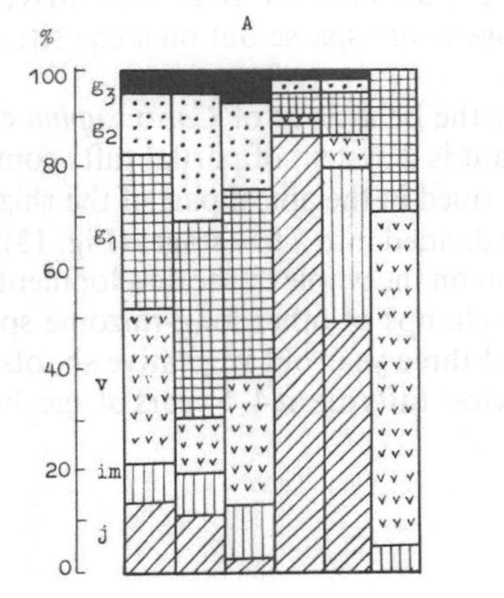

Fig. 12. Changes of the age-state spectra and the numbers of *Alyssum lenense* coenopopulations over time.
(A) Age-state spectra, showing the proportions of different age-state groups (% of the total number on a 15 × 0.5 m^2 transect); years and months of observations are shown on the abscissa.
(B) Changes in the total number of individuals on the same transect.

seedling establishment (the number of seedlings per 1 m^2 varies from 1 to 70 annually), high death rate of individuals, and rapid ontogenetic development. The balance lability of *Alyssum lenense* is about 1, that is, during the observation period (1970–1979) almost all individuals in the coenopopulation were replaced by others. The coenopopulation half-life in *A. lenense* varies from 1.2 to 2.7 years depending on the environment: death in these species is distinctly related to environmental fluctuations.

The species of this group are highly dynamic and marked by low competitive capacity. They are close in behaviour to explerent species (Ramenskii 1938): they can rapidly occupy disturbed areas but are rapidly suppressed by other species as well.

TYPE V – LONG-RHIZOME PERENNIALS

In the sandy steppes this group is represented by *Carex supina*, *Veronica spuria* and *Galium ruthenicum*. Coenopopulation maintenance is basically by vegetative propagation and the species occur permanently in steppe communities on sandy soils, although their cover is not large. The ratio between progeny derived from seed and vegetative propagules varies: *Carex supina* coenopopulations are fully vegetative, whereas in *Galium ruthenicum* seed progeny are sparse but on some sites it may prevail over vegetative progeny.

We shall consider the behaviour of *Carex supina* coenopopulations in some detail. The plant is a system of partial tufts connected by rhizomes. The partial tuft is formed in the apical part of the rhizome by tillering. Its development can be divided into a few stages (Fig. 13): young, mature and old which correspond on the whole to the development stages described by Kershaw (1973) for clumps of other long-rhizome species. The tuft consists of one, two and three year old vegetative shoots and di- or tricyclic reproductive ones. Most tufts die at 4–5 years of age, but some of them can

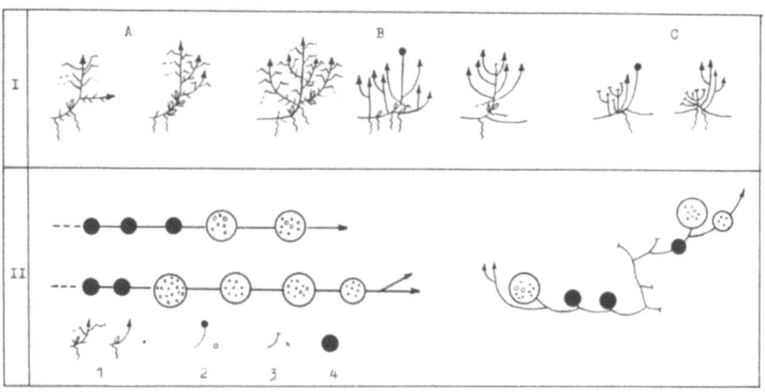

Fig. 13. The phases of development of *Carex supina* partial tufts (I) and structure of individuals. Tufts: A, young; B, mature; C, old. 1, Vegetative shoot; 2, reproductive shoot; 3, dead shoot; 4, dead tuft.

live up to 7 or 8 years. The rhizome is rarely branched and an individual has usually not more than 2–3 rhizome branches (Fig. 13, II). Although reproductive shoots with fruits can usually be seen in a coenopopulation we have never detected seedlings in natural conditions. The coenopopulation dynamics of the species are entirely related to the degree of partial tuft formation and death, both of which undergo changes in time. Sometimes these depend mainly on the weather conditions: the highest number of new tufts is formed in humid years (such as 1973), and these are years when coenopopulation rejuvenation occurs. Young tufts develop rapidly and correspondingly the number of this age-state rapidly decreases in subsequent year. Unless continual formation of young tufts occurs the population undergoes gradual ageing. Most tufts (mainly the old ones) die off in drought years (such as 1975). Since the life-span of tufts is not long the half-life of coenopopulations is short, between 1.1 and 4.0 years depending

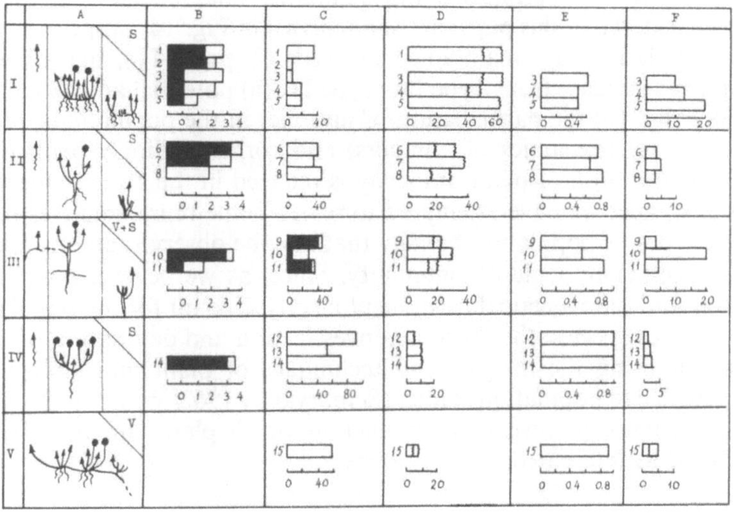

Fig. 14. Some quantitative and qualitative characteristics of the different types (I–V) of population life-history discussed in this paper.

(A) Life-form: arrows, vegetative shoots; arrows with circles, reproductive shoots; heavy black lines, dead shoots. Type of self-maintenance is shown in the upper right corner: s, seminal; v, vegetative.

(B) Seed production per plant (genet), on a logarithmic scale: open columns indicate potential seed productivity, the closed columns, actual seed productivity.

(C) Coenopopulation provision for regrowth, indicated by the number of juvenile and immature plants as a % of the total number; open columns indicate seed progeny and the closed columns indicate vegetative progeny.

(D) Duration of the ontogeny (years).

(E) Balance lability.

(F) Half-life of coenopopulation (years).

1, *Potentilla cinerea*; 2, *Centaurea sibirica*; 3, *Koeleria macrantha*; 4, *Festuca beckeri*; 5, *Stipa pennata*; 6, *Artemisia campestris*; 7, *Onosma simpliccissimum*; 8,*Euphorbia seguieriana*, 9, *Helychrysum arenarium*; 10, *Seseli ledebouri*; 11, *Scorzonera ensifolia*; 12, *Syrenia sessiliflora*; 13, *Alyssum tortuosum*; 14, *Alyssum lenense*; 15, *Carex supina*.

on environmental conditions. This factor accounts also for the observation that most partial tufts in coenopopulations were renewed during our observation period (1970–1979): the balance lability was 0.94. The coenopopulations of this species are characterised by a relatively high dynamics of number and age-state composition of partial tufts. The main causes of lability are: (a) sporadic formation of partial tufts, (b) their rather high death rate and (c) their relatively rapid ontogenetic development. Coenopopulations of *Carex supina* are extremely persistent because of their high lability: the species has the ability to utilize rapidly the favourable conditions that occur occasionally in the dry steppes.

CONCLUSIONS

The demographic analysis of steppe plants has shown that a plant community consists of coenopopulations with different types of dynamic behaviour. Some quantitative and qualitative characteristics of the five types considered in this paper are summarised in Fig. 14. In species whose coenopopulations are maintained by seeds (Types I–IV), the following parameters increase from Type I to Type IV: (a) potential and actual seed productivity, (b) the participation of juveniles in the populations, (c) the turnover rate (dynamics of numbers) and correspondingly plant death rate. The duration of plant ontogeny is reduced in the same sequence.

In this paper we have attempted to review some representatives of the diverse types of population biology that may be observed among species living together in a plant community. Since, as we believe, plant morphology and ontogenetic development largely account for the population behaviour of species, the thorough investigation and description of plant life cycles using the insights and techniques of plant morphology are indispensible to population studies (Serebryakov 1962, White 1979). These studies promise to reveal the manner in which plant communities are structured as integrated living systems.

REFERENCES

(English translations of Russian-language titles are given in brackets)

Bedanokova, O.A., Vorontzova, L.I. and Mikhailova, N.F. (1975) (Some specific biological features of *Stipa pennata* L. in the steppes of Naurzum reserve.) Bull. Mosk. obsch. isp. prir. otd. biolog. 80(2): 77–91.

Bespalova, Z.G. and Borissova, I.V. (1979) (Germination peculiarities of caryopsis of *Stipa L.* (*Poaceae*).) Botan. zhur. SSSR. 64(8): 1081–1090.

Borissova, I.V., Bespalova, Z.G. and Popova, T.A. (1976) (Peculiarities of phaenological development in plants of steppe and desert associations in northern Gobi (MPR).) Trudy Mosk. obsch. isp. prir. 42: 239–256.

Borissova, I.V. and Popova, T.A. (1971) (Age phases of bunch-formation in steppe grasses.) Botan. zhur. SSSR. 56(5): 619–626.

Borissova, I.V. and Popova, T.A. (1972) (The dynamics of quantity and age composition of bunch grass coenopopulation in desert steppes of Central Kazakhstan.) Botan. zhur. SSSR. 57(7): 799–793.

Gatsuk, L.E., Smirnova, O.V., Vorontzova, L.I., Zaugolnova, L.B. and Zhukova, L.A. (1980) Age states of plants of various growth forms: a review. J. Ecol. 68: 675–696.

Golubev, V.N. (1959) (On some characteristics of the life-form morphogenesis in the herbaceous plants of the forest-meadow zone as related to the evolution of these plants.) Botan. zhur. SSSR. 44(12): 1704–1716.

Golubev, V.N. (1970) (About elementary structural unit of plant communities and its dynamics.) "Mosaic structure of plant communities and its dynamics", 83–85. Vladimir.

Golubeva, I.V. (1964) (Age composition and number dynamics of feather-grass (*Stipa pennata L.*) in meadow steppe.) Uch. zap. Mosc. oblast. pedagog. inst. 153: 283–303.

Gordeyeva, T.K. (1976) (Phytomass.) "Main plant communities of Central Kazakhstan desert steppe", 64–68, Leningrad: Nauka.

Gorshkova, A.A. and Kopytova, L.D. (1977) (The water reserve in community and expense on the plant transpiration.) "Ecology and pasture degression of steppe communities in Transbaical region", 53–94, Novosibirsk: Nauka.

Grigorjev, U.S. and Razumova, L.M. (1974) (Comparative-ecological study of two species of *Festuca*.) Biolog. nauk. 8: 73–78.

Grime, J.P. (1978) Interpretation of small-scale patterns in the distribution of plant species in space and time. pp. 101–124 in A.H.J. Freysen and J.W. Woldendorp (Eds.) Structure and Functioning of Plant Populations. Amsterdam, Oxford, New York: North-Holland Publ. Co.

Grubb, P.J. (1977) Maintenance of species richness in plant communities: the importance of regeneration niche. Biol. Rev. 52: 107–145.

Harper, J.L. 1977. Population Biology of Plants. London: Academic Press. 892 pp.

Ivashin, D.C. and Chuprina, T.T. (1976) (Some biomorphological peculiarities of pinnate feather-grasses in Donbass, with introduction.) "Plant resource of Ukraine, their utilization and enrichment", 58–62. Kiev. (in Ukrainian).

Karamysheva, Z.V. and Rachkovskaja, E.I. (1973) (Botanical geography of the Central Kazakhstan steppe zone.) Leningrad: Nauka.

Kershaw, K.A. (1973) Quantitative and Dynamic Plant Ecology. London: Edward Arnold.

Kolpikov, D.I. (1960) (About some characteristics of the plant response on drought in phytocoenosis of arid steppe.) "Physiology of Plant Stability". Nauka.

Kozhevnikova, N.D. and Trulevich, N.V. (1971) (Dry steppes of Inner Tyan-Shan.) Frunze.

Lavrenko, E.M. (1940) (The steppes of USSR.) "The Vegetation of USSR" 2. Moscow, Leningrad.

Lavrenko, E.M. (1954) (The steppes of Euroasian region, their geography, dynamics, history.) "The question of botany", I: 157–191. Moscow, Leningrad.

Levina, R.E. (1956) (About modes of plant dispersion in steppes.) Botan. zhur. SSSR. 45(5): 619–633.

Mikhailova, N.F. (1977) The great life cycle of *Festuca beckeri* (Hack.) Trautv. in the steppes of the Naurzum Preserve. Bull. Msk. obsch. isp. prir. biolog. 82(2): 125–132.

Osychnjuk, V.V. and Shupranov, N.P. (1978) (Seed productivity and seed yield of some sod grasses in phytocoenoses of the Khomutovan steppe.) Ukrain. botan. zhur. 35(2): 122–126. (in Ukrainian).

Persikova, Z.I. (1959) (The formation and life cycle of some tussock grasses.) Biolog. nauk. 3.

Ponomarev, A.N. and Zvorygina, L.F. (1949) (About biology of feather-grasses.) Uch. zap. Molotov. univer. 5(1): 63–75.

Rabotnov, T.A. (1978) On coenopopulations of plant reproducing by seeds. pp. 1–26 in A.H.J. Freysen and J.W. Woldendorp (Eds.) Structure and Functioning of Plant Populations. Amsterdam, Oxford, New York: North-Holland Publ. Co.

Ramenskii, L.G. (1925) (The main regularities of plant cover and their study on the base of geobotanical researches in Voronezh region.) Voronezh.

Ramenskii, L.G. (1938) (Introduction to complex soil-geobotanical research of lands.) Moscow.

Sarukhán, J. (1978) Studies on the demography of tropical trees. pp. 163–184 in P.B. Tomlinson and M.H. Zimmermann (Eds.) Tropical Trees as Living Systems. London, New York, Melbourne: Cambridge University Press.

Semenova-Tyan-Shanskaya, A.M. (1978) (Protection regime of herbal communities and individual plant species.) Zhur. obsch. biol. 39(1): 3–47.

Serebryakov, I.G. (1962) (Ecological morphology of plants.) Moscow: Vysshaja shkola.
Smirnova, O.V., Zaugolnova, L.B. and Ermakova, I.M. (1976) (Plant Coenopopulations: Basic Concepts and Structure.) Moscow: Nauka. 216 pp.
Solntzeva, M.P. (1968) (The peculiarities of flowering and embryology of feather-grasses.) Uch. zap. Ulyanov. pedagog. inst. 23(3): 29–40.
Spivak, A.I. (1977) (The potential intensity of photosynthesis of xerophytes in Transbaikal steppes.) "Ecology and pasture degression of steppe communities in Transbaiksl region", 95–123, Novosibirsk: Nauka.
Stechenko, A.P. (1976) (On relation between live and dead parts of plants in the Pamir highlands.) Ecologia 6: 27–34.
Sveshnikova, V.M. (1979) (The dominants of Kazakhstan steppes.) Leningrad: Nauka.
Sveshnikova, V.M. and Bobrovskaya, N.I. (1973) (Ecology of *Stipa* species in Kazakhstan (result of water regime study).) Problems of biogeocoenology, geobotany and plant geography, 240–256.
Uranov, A.A. (1967) (Ontogeny and age-state composition of populations.) "Age-state composition of flowering plant populations in connection with ontogeny", 3–8. Moscow: Nauka.
Uranov, A.A. (1975) (Age spectrum of the phytocoenopopulation as a function of time and energetic wave processes.) Biolog. nauk. 2: 7–34.
Uranov, A.A., Zaugolnova, L.B. and Smirnova, O.V. (1977) (Plant coenopopulations: development and interrelations.) Moscow: Nauka. 134 pp.
Vasilevich, V.I. (1969) (Statistical methods in geobotany.) Leningrad: Nauka.
Vorontzova, L.I. and Zhukova, L.A. (1977) (The biomorphological peculiarities and age-state structure of coenopopulations of firm tussock grasses.) Plant coenopopulations: basic conceptions and structure, 107–129. Moscow: Nauka.
White, J. (1979) The plant as a metapopulation. Ann. Rev. Ecol. Syst. 10: 109–145.
Zaugolnova, L.B. (1976) (The heterogenity of coenopopulation in time and space (exemplified by *Alyssum lenense* Adam.).) Botan. zhur. SSSR. 61(2): 177–196.
Zaugolnova, L.B. (1977) (Analysis of plant coenopopulations as a method of studying anthropogenic effects on phytocoenoses.) Botan. zhur. SSSR. 62(12): 1767–1780.
Zhuravleva, N.A. (1977) (Interrelation between some ecological and biomorphological indices of steppe plants.) "Ecology and pasture degression of steppe communities in Transbaikal region", 124–152. Novosibirsk: Nauka.

STRUCTURE AND DYNAMICS OF COENOPOPULATIONS OF SOME TEMPERATE GRASSES

L.A. ZHUKOVA and I.M. ERMAKOVA

ABSTRACT

The coenopopulation age-state structure of five species of meadow grasses in a variety of coenoses is described. The mode of coenopopulation self-maintenance in different coenoses varied with each of the species, *Deschampsia caespitosa*, *D. flexuosa*, *Festuca pratensis*, *Agropyron repens* and *Poa annua*: seed reproduction alone, a combination of seed and vegetation reproduction, or a predominance of vegetative reproduction. The most persistent species are able to change their coenopopulation age-state spectra in response to changing ecological conditions.

INTRODUCTION

The duration of any phytocoenosis is determined by the life time of its constituent coenopopulations. This in turn depends on the potential ontogeny and persistence of plants with different life-forms and on the conditions which ensure self-maintenance of coenopopulations. Regular grazing or mowing or flooding are the basis limiting factors hindering such life forms as trees and shrubs in meadow phytocoenoses. Owing to anthropogenic influences secondary meadows dominated by perennial herbaceous mono- and polycarpic plants with a continuously renewable overground shoot system may develop on flood-plain terraces. The minimum and maximum durations of their ontogeny are 1–10 and 50–80 years respectively, that is, considerably shorter than most trees and many shrubs. Meadow phytocoenoses develop more rapidly but are shorter-lived than climax forest communities. They are also more dynamic, since they respond quickly to abrupt fluctuations of various exogenous factors. Grasses are typical of these coenoses in the temperate zone, and there have been numerous studies on their ontogeny and on their biological and ecological characteristics (Rozhevitz 1937, Smelov 1947, Larin 1950, Serebryakova 1971, Tsvelev 1976, Arber 1934). However demographic characteristics have been obtained for only a few species of grasses. The object of this study is to show the diversity of coenopopulation structure

White, J. (ed.), The Population Structure of Vegetation.
 ISBN 978-94-010-8927-2

and modes of self-maintenance for a number of grass species and to describe the basic features of their life cycles which account for the dynamic processes of phytocoenoses.

Individuals of a coenopopulation have various calendar ages, age-states, sexual and other biological characteristics. Studies now completed on over 200 species of flowering plants, including 60 meadow species, have confirmed the method of age-state definition for different life-forms (Smirnova et al. 1976, Uranov et al. 1977).

MATERIALS

Grasses of different life-form, characterised by different types of growth and coenopopulation self-maintenance were studied (Fig. 1): *Deschampsia caespitosa* P.B. (tussock grass), a compact caespitose meadow weed; *Festuca pratensis* Huds. (meadow fescue), a loose caespitose grass of high feed value; *Deschampsia flexuosa* Trin. (common hairgrass), a caespitose-stoloniferous grass of medium feed value; *Agropyron repens* L. (couch grass), a long-rhizome valuable meadow grass; *Poa anna* L. (annual meadow grass), a short-lived pasture grass of medium feed value. Meadow fescue was studied by Ermakova (1968), the other four species by Zhukova (1961, 1979, 1980).

According to the accepted subdivision of age-state ontogeny (Rabotnov 1950a, Uranov 1975, Gatsuk et al. 1980), the species were classified into 9 or 10 age-states (full details in Serebryakova 1980). Coenopopulations of the species were studied on the inundation and continental meadows in the northern, north-western and central regions of the European part of the USSR, and on the mountain meadows of the Trans-Carpathian area and the Crimea. Details of the ecology and management of the various habitats investigated are given in Tables 1–5. All the individuals collectively of a species in a particular site (phytocoenosis) belong to one type of coenopopulation, defined by its age-state spectrum. The variety of coenopopulations encountered in various habitats is indicated on the tables: their precise definition will be presented later. Over 70 coenopopulations of *D. caespitosa* were examined, together with 44 of *F.pratensis*, 21 of *D.flexuosea*, 12 of *A.repens* and 10 of *P.annua.*

In order to characterise each population 25–300 plots, each 0.25 m^2 in area, were established. For dynamic studies on the coenopopulations of tussock-grass, common hairgrass, meadow fescue and annual meadow-grass, the same coenoses were inspected for a number of years. The age-states of individuals of each species were determined on temporary or permanent plots. Absolute and relative participation of each age-state group and the coenopopulations age-state spectrum were calculated for each phytocoenosis. Averaged spectra for similar climatic conditions or similar types of pasture utilization were also calculated. Detailed methods are given in Smirnova et al. (1976).

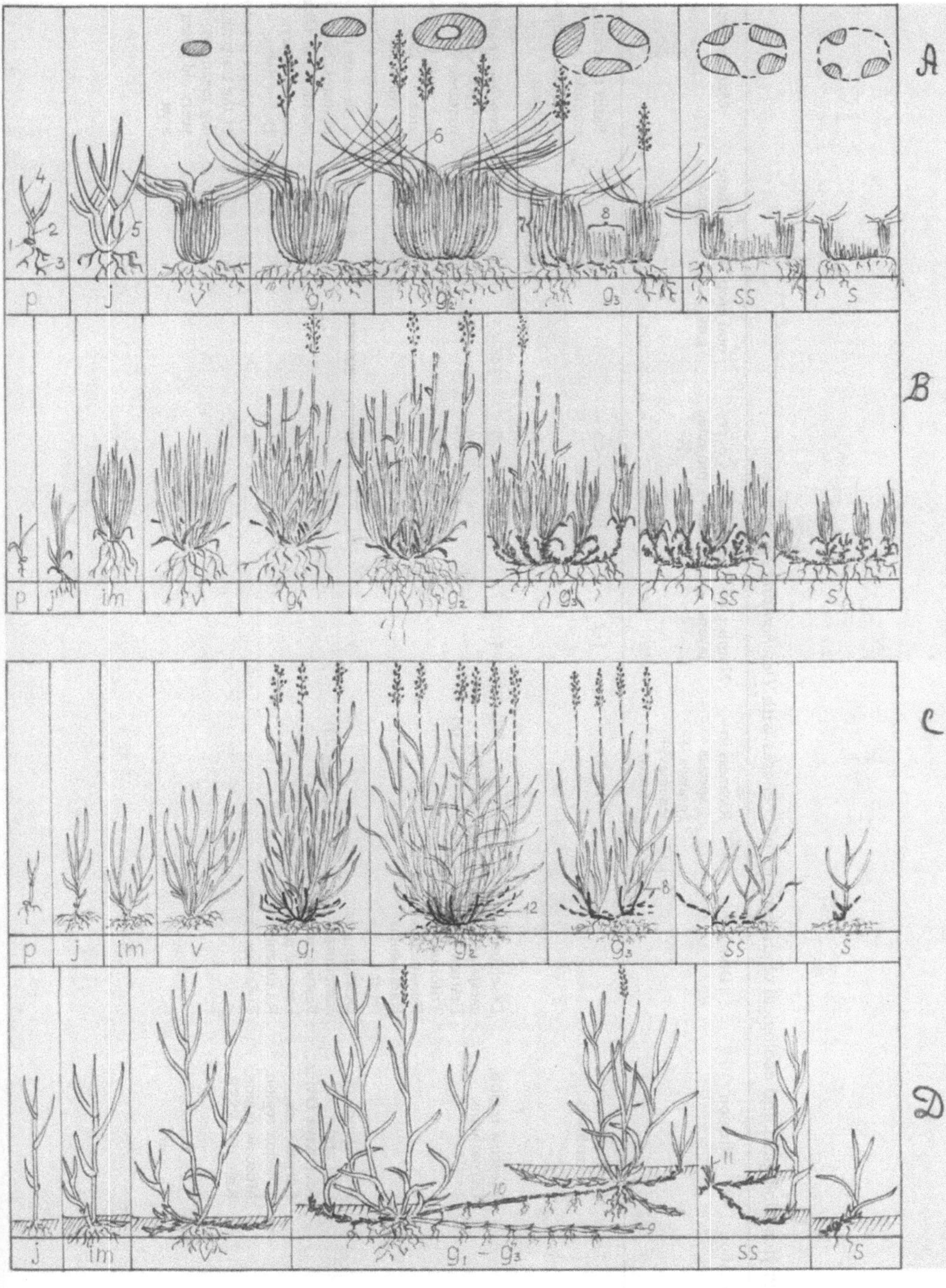

Fig. 1. Schematic diagram of age-states in the ontogeny of four grasses: (a) *Deschampsia caespitosa*, (b) *Deschampsia flexuosa*, (c) *Festuca pratensis*, (d) *Agropyron repens*. The ontogeny of *Poa annua* resembles (c). A horizontal projection of the tussocks is given in the upper part of the diagram for *D.caespitosa*. 1, caryopsis; 2, coleoptile; 3, adventitious roots; 4, assimilating leaves; 5, rosette shoot; 6, reproductive semi-rosette shoot; 7, tussock live part; 8, partial shoot, 9, offshoot; 10, communication rhizomes; 11, dead shoot; 12, decayed part.

Table 1. Phytosociological and ecological characteristics of coenoses with *Deschampsia caespitosa*.

Coenopopulation type	Region	Dominants	Richness in species (grasses + legumes + forbs)	Number of layers	Total cover %	Cover projected %	Total crop g m^{-2}	% Share in crop	Management
Invasive	Moscow region, Carpathians	–	10–70	1–2	25–30	< 5			Paths for cattle, trampled area 1–2 years old, summer cutting
Normal									
young	Moscow region, Carpathians	Deschampsia caespitosa, Festuca pratensis, Trifolium repens, Ranunculus acris, Ranunculus repens	53 (29–100) (10 + 5 + 38)	3	73	25	348–657	10–12	Improved pastures, 6–8 years old pastures, 2–5 years old felled area, 10 years old plantings of spruce
mature	Flood lands of Oka and Severnaya Dvina, Carpathians, Vologda region, Moscow region, Karelian ASSR	Deschampsia caespitosa, Festuca rubra, Festuca pratensis, Ranunculus acris, Trifolium repens	34 (8 + 3 + 23)	3	79	30	209–613	9–35	6–30 years old pastures, hay meadows of Oka and Severnaya Dvina rivers, subalpine meadows of the Carpathians, dry meadows, 5–10 years old felled area

ageing	Flood lands of Oka and Severnaya Dvina, Moscow region, Vologda region, Karelian ASSR	Deschampsia caespitosa, Festuca rubra, Agrostis stolonizans, Ranunculus acris, Trifolium repens, Alchemilla vulgaris	37 (19–50) (8 + 4 + 25)	3–4	78.2	24	250–1370	5.5–34	35–40 years old perennial pastures flood plain mainland hay meadows, 10–20 years old felled area, birch forests
old	Flood lands of Oka, Karelian ASSR, Kaluga region, Vologda region	Deschampsia caespitosa, Agrostis stolonizans, Poa trivialis, Ranunculus repens	42 (9 + 5 + 28)	3–4	83	6	120–335	2.7	Mainland and flood plain hay meadow, aspen forests, pastures, terraces

Table 2. Phytosociological and ecological characteristics of coenoses with *Deschampsia flexuosa*.

Coenopopulation type	Region	Dominants	Richness in species	Number of layers	Total cover %	Cover projected %	Management
Invasive	Arkhangelsk region Pribaltik	Deschampsia flexuosa, Epilobium angustifolium	29 (20–37)	1–2	43	8	1–2 years felled area
Normal							
young	Karelian ASSR Pribaltik Arkhangelsk region	Picea excelsa, Deschampsia flexuosa, Epilobium angustifolium	20 (15–38)	2–3	60	22	Felled area, grass heath, plantings of pine and spruce forest
mature	Pribaltik	Pinus silvestris, Deschampsia flexuosa	16 (15-20)	4	70	30	Plot of natural park
ageing	Arkhangelsk region Jaroslavl region Karelian ASSR	Deschampsia flexuosa, Vaccinium myrtillus, Nardus stricta	12 (7–22)	3–5	75	32	Felled area, plantings of spruce, island grass heath
old	Karelian ASSR Jaroslavl region Arkhangelsk region	Picea excelsa, Vaccinium vitis-idaea, Deschampsia flexuosa, Nardus stricta	17 (13–22)	3–4	60	14	Felled area, island grass heath, plantings of pine and spruce
Regressive	Moscow region	Pinus silvestris Festuca ovina, Deschampsia flexuosa	15	3	80	10	Plantings of pine

Table 3. Phytosociological and ecological characteristics of coenoses with *Festuca pratensis*.

Coenopopulation type	Region	Dominants	Richness in species (grasses + legumes + forbs)	Number of layers	Total cover %	Cover projected %	Total crop g m^{-2}	% Share in crop	Management
Invasive	Pscov region	–	–	–	–	–	118 (86–150)	2.7–59	Hay meadows (1 cutting)
Normal young	Pscov region	Dactylis glomerata Festuca pratensis Taraxum officinale	33.7 (8.4 + 3 + 22.3)	4	73	3.5	143 (97–170)	5–48	Hay meadow (3 cuttings)
mature	Moscow region Kaluga region Crimea	Festuca pratensis	42.6 (8.8 + 3.4 + 30)	4	76	37.7	67	31	Hay meadow (1c.), mowing and grazing combined
ageing	Kaluga region Arhangelsk region Rjasan region Karelian isthmus	Festuca pratensis F.rubra	40.0 (8.1 + 4.2 + 27.7)	4–5	81	11.5	(1.5–425)	1.3–29	Hay meadow (1c.), mowing and grazing combined + pasture meadow (3 cuttings)
old	Kaluga region Moscou region Rjasan region Crimea Karelium isthmus	Festuca pratensis F.rubra Zerna inermis Poa pratensis P.augustifolia	43.7 (8.1 + 4.6 + 31)	4–5	77	9.6	27 (5.6–61)	2.5–14.9	Hay meadow (1–3 cuttings) meadow pasture

Table 4. Phytosociological and ecological characteristics of coenoses with *Agropyron repens*.

Coenopopulation type	Region	Dominants	Richness in species	Numbers of layers	Total cover %	Cover projected %	Total crop $g\,m^{-2}$	% Share in crop	Habitat
Invasive									
Normal young	Kaluga region Moscow region Ural	Agropyron repens, Festuca pratensis, Deschampsia caespitosa, Trifolium repens	27 (25–30)	2–3	50–80	10–12	121–1725	6.6–32	Meadow of intensive use (Oka), fallows, flood-plain pastures
mature	Moscow region Kaluga region	Agropyron repens, Festuca rubra, Trifolium repens	18 (10–24)	2–3	70–85	20–25	361	5–27.5	Fallows, hay dry meadows
ageing	Kaluga region Pskov region Karelia ASSR	Agropyron repens, Festuca pratensis Dactylis glomerata	21 (17–30)	2–3 2–3	90–100	12–16	570–842	3.5–15	Meadow of intensive use (Oka), sown meadows (Karelian ASSR)
old	Moscow region Ural Pskov region	Agropyron repens, Festuca pratensis, Deschampsia caespitosa, Trifolium repens	23 (15–27)	2–3	70–90	8–10	616–1376	1.2–11	Fallows, flood-plain pastures, sown meadows

Table 5. Phytosociological and ecological characteristics of coenoses with *Poa annua*.

Coenopopulation type	Region	Dominants	Richness in species	Number of layers	Total cover %	Cover projected %	Total crop $g\,m^{-2}$	% Share in crop	Management
Invasive	Moscow region	Taraxacum officinale, Plantago major, Polygonum aviculare, Tripleurospermum inodorum	10–15	1–2	20–25	5			Fallow, sown meadow, dry meadow
Normal young	Moscow region	Poa annua, Agrostis tenuis, Plantago major, Taraxacum officinale	6–16	2	30–42	10–12	24–97	4.8–5.5	Fallow, sown meadow, dry meadow, roadsides
mature	Moscow region	Poa annua, Festuca pratensis, Plantago major	6–15	2–3	40–50	25	57–75	10.7–32	Fallow, dry meadow
ageing	Moscow region	Poa annua, Festuca pratensis, Poa pratensis, Taraxacum officinale	5–12	2–3	60–70	20–25	20–60	1.5–19	Fallow, dry meadow, roadsides
old	Moscow region	Poa annua, Festuca pratensis	3–5	1–2	40	10–15	74–111	0.13–2	Dry meadow

AGE-STATE SPECTRA OF COENOPOPULATIONS

Coenopopulations of any species may be considered as continously developing elements of the plant cover (Uranov 1975). The comparison of age-state spectra of the same (single) species in different meadow phytocenoses shows a wide range, caused primarily by different stages in their development. Rabotnov (1969) proposed a classification of coenopopulations into three types: invasive (I), normal (N) and regressive (R) to correspond with their long term development from emergence to maturity and to extinction. The unidirectional development of a coenopopulation ranging from the invasive to regressive state constitutes the 'large life cycle' (Rabotnov 1969) of the coenopopulation or 'big wave of population flow' (Uranov and Smirnova 1969); the development of one diaspore generation from invasion to death constitutes a 'small or reproduction wave' (Smirnova et al. 1976).

The types of coenopopulations of the various grasses we have studied are summarized in Tables 6–10 and will be discussed below in turn for each species.

(a) *Invasive coenopopulations*

Invasive coenopopulations consist of plants of the pregenerative period and are incapable of self-maintenance since they depend on continual diaspore invasion. Coenopopulations of this type are rather rare in meadow phytocoenoses and occur mainly on plots with a highly disturbed plant cover (pastures, cattle tracks) or on primary substrates such as alluvial deposits in flood-plains, spoil banks of irrigation or drainage systems, as well as in the first stages of meadow agrophytocoenoses.

Invasive coenopopulations of *D.caespitosa* constituted 6.3% of all coenopopulations of this species which we studied. They were found in cattle camps of high-mountain meadows in the Trans-Carpathian area, on cattle tracks of Oka pastures, and in the annual meadowed felling areas of forests. These invasive coenopopulations emerge either in sites of old overgrazed coenopopulations or on formerly unoccupied places such as forest felling areas. They may develop from seed (caryopsis) reserves in the soil or be dispersed from elsewhere. According to Rabotnov (1950b) the number of caryopses in soil was 20–100 m^{-2} in the central flood-plain and 220 m^{-2} in terrace-side areas. The invasive coenopopulation density is rather variable (Table 6). The first months are marked only by seedling emergence, then come the juvenile plants and after 2–4 years the age-state spectrum is represented by all groups of the pregenerative period. The phytocoenotic significance of these coenopopulations is low since their projected cover is only 1–5% (Table 1). The negative effects of various exogenous factors result in continuous elimination of young *D. caespitosa* plants and cause distinct fluctuations in the population density of all age-states groups present, particularly the seedlings. The death of the whole coenopopulation is not infrequent. Such doomed coenopopulations

Table 6. Coenopopulation characteristics of *Deschampsia caespitosa*. The values of some parameters are shown as means (and ranges).

Coenopopulation type	Age-state groups*	Duration of life (years)	Coenopopulation age level (Δ)	Population density (per 0.25 m^2)	Quantity of generative individuals %	$\frac{j + v}{g_1 + g_2 + g_3}$ (%)	Recorded occurrences N	Recorded occurrences % of total	Habitat
Invasive	p, j, v	2–5	0.05 (0.03–0.09)	72.25 (3.1–177.0)	0	–	4	6.3	Paths for cattle, trampled areas, 1–2 year-old summer cutting
Normal									
(i) young	j, v, $\underline{g_1}$, g_2, g_3, (ss, s)	3–5	0.28 (0.09–0.35)	6.1 (2.7–15.2)	52.9 (48.2–57.7)	183 (24–945)	9	14.1	Improved pastures, 6–8 year-old pastures, 2–5 year-old felled area 10 year-old spruce plantation
(ii) mature	(j), v, g_1, $\underline{g_2}$, g_3, ss, (s)	12–20	0.49 (0.27–0.63)	4.0 (2.38–8.6)	67.0 (64.8–68.6)	50 (4–96)	26	40.6	6–30 year-old pastures and hay meadows of Oka S. Dvina rivers, subalpine meadow, dry meadows, 5–10 year felled area
(iii) ageing	(j, v), g_1, g_2, $\underline{g_3}$, (ss, s)	8–15	0.59 (0.53–0.72)	3.56 (2.0–7.2)	70.3 (64.4–75.9)	27 (1–85)	17	26.6	35–40 year-old perennial pastures, flood-plain main land hay meadows, 10–20 year-old felled area, birch forests
(iv) old	(j, v, g_1), g_2, g_3, $\underline{ss}$, (s)	5–15	0.53 (0.41–0.80)	2.1 (1.6–4.9)	39.3 (32.9–43.4)	81 (12–208)	8	12.5	Mainland and flood-plain hay meadows, aspen forests, pastures, terraces

*The predominating age-state group is underlined and groups which may be absent are shown in brackets.

Table 7. Coenopopulation characteristics of *Deschampsia flexuosa*. The values of some parameters are shown as means (and ranges).

Coenopopulation type	Age-state groups*	Duration of life (years)	Coenopopulation age level (Δ)	Population density (per 0.25 m^2)	Quantity of generative individuals %	$\frac{j + im + v}{g_1 + g_2 + g_3}$ (%)	Recorded occurrences N	Recorded occurrences % of total	Habitat
Invasive	**p, j, im, v**	1–3	0.006 (0.003–0.020)	85.2 (31.7–171.7)	0	–	6	20	1–2 years felled area
Normal									
(i) young	**p, j, im, v** **g_1, g_2(g_3, ss, s)**	2–3	0.21 (0.15–0.34)	10.9 (3–25)	14.7 (9.4–25.9)	417.8 (59–832)	6	20	Felled area, grass heath, plantings of pine forest
(ii) mature	**(p, j, im), (v), g_1, g_2, g_3, (ss, s)**	1–2	0.43	4.4	42.9 (40–45.7)	59.0	4	13.3	Plot of natural park
(iii) ageing	**(p, j, im, v, g_1), g_2, g_3, ss, (s)**	7–12	0.60 (0.45–0.78)	4.36 (2–16)	34.0 (18.3–77.8)	42.0 (27–72)	6	20	Felled area, plantings of spruce, island grass heath
(iv) old	**p, j, im, v, g_1, g_2, g_3, ss, s**	3–5	0.73 (0.41–0.87)	5. (1.3–12.8)	15.2 (6.8–25.0)	20.7 (8.1–63)	7	23.4	Felled area, island grass heath, plantings of pine and spruce
Regressive		2–?	0.89	1.75	0	0	1	0.3	Plantings of pine

*The predominating age-state group is underlined and groups which may be absent are shown in brackets.

Table 8. Coenopopulation characteristics of *Festuca pratensis*. The values of some parameters are shown as means (and ranges).

Coenopopulation type	Age-state groups*	Duration of life (years)	Coenopopulation age level (Δ)	Population density (per 0.25 m^2)	Quantity of generative individuals %	$\frac{j + im + v}{g_1 + g_2 + g_3}$ (%)	Recorded occurrences N	% of total	Habitat
Invasive	**p, j, im, v**	1	0.6 (0.04–0.09)	175.7 (17–252)	0	–	16	16.6	Agrophytocoenoses Pscov region
Normal									
(i) young	**p, j, im, v, g₁, g₂, (g₃, ss, s)**	0.1–1	0.19 (0.18–0.29)	6.6 (3.1–11.1)	51.32 (20–73)	54.45 (12–158)	16	16.6	Agrophytocoenoses Pscov region
(ii) mature	**(p, j, im, v), g₁, g₂, g₃, (ss, s)**	1–5	0.45 (0.28–0.67)	11.01 (0.4–27.5)	45.41 (22–54)	51.39 (34–78)	4	5.7	Flood-plain of Ugra River, mountain meadow (Crimea), dry meadow (Karelian)
(iii) ageing	**(p, j, im, v, g,₁), (g₂), g₃, ss, (s)**	1–4	0.67 (0.28–0.98)	4.5 (1.3–9.9)	64.82 (36–85)	10.58 (0–98)	17	24.2	Flood-plain of Ugra River, Oka River, Severnaya Dvina River, Severnaya Dvina River, dry meadow (Karelian), mountain meadow (Crimea)
(iv) old	**(p, j, im, v, g₁), (g₂), g₃, ss, (s)**	1–13	0.75 (0.42–0.97)	5.5 (2–14.5)	27.78 (9–45)	22.88 (0–100)	17	24.2	Flood-plain of Ugra River, Oka River, Severnaya Dvina River, mountain meadow (Crimea), dry meadow (Karelian, Moscow region)

*The predominating age-state group is underlined and groups which may be absent are shown in brackets.

Table 9. Coenopopulation characteristics of *Agropyron repens*. The values of some parameters are shown as means (and ranges).

Coenopopulation type	Age-state groups*	Duration of life (years)	Coenopopulation age level (Δ)	Population density (per 0.25 m²)	Quantity of generative individuals %	$\frac{j + im + v}{g_1 + g_2 + g_3}$ (%)	Recorded occurrences N	Recorded occurrences % of total	Habitat
Invasive									
Normal									
young	(j, im, v), <u>g₁</u>, g₂, (g₃, ss, s)	1–2	0.31 (0.15–0.44)	90.2 (32–191)	29 (20.9–50)	234 (45–400)	5	31.3	Meadow of intensive use (Oka), fallows, flood-plain pastures
mature	(j, im), v, g₁, <u>g₂</u>, g₃, (ss, s)	0.5–3	0.47 (0.43–0.54)	72.0 (70.2–129)	53.2 (46.5–59.7)	44.3 (25–59)	3	18.7	Fallows, hay dry meadows
ageing	(j, im, v), g₁, g₂, <u>g₃</u>, (ss, s)	1–2	0.54 (0.44–0.65)	51.3 (39–60)	78.0 (53.8–100)	6.6 (0–10.1)	3	18.7	Meadows of intensive use (Oka), sown meadows (Karelian ASSR)
old	(im), v, g₁, g₂, g₃, <u>ss</u>, (s)	0.5–1	0.55 (0.46–0.68)	75.8 (25–198)	25.1 (13.6–43.5)	182.5 (8.4–363)	5	31.3	Fallows, flood-plain pastures, sown meadows

*The predominating age-state group is underlined and groups which may be absent are shown in brackets.

Table 10. Coenopopulation characteristics of *Poa annua.*

Coenopopulation type	Age-state groups*	Duration of life (years)	Coenopopulation age level (Δ) (mean)	Population density (per 0.25 m^2) (mean)	Quantity of generative individuals %	$\frac{j + im + v}{g_1 + g_2 + g_3}$ (%)	Recorded occurrences N	Recorded occurrences % of total	Habitat
Invasive	p, $\underline{j}$, im, v	0.5–1	0.025	138.2	0	–	3	18.7	Fallow, sown meadow, dry meadow
Normal									
young	j, im, v, $\underline{g_1}$, (g_2), (g_3, ss)	0.5–2	0.24	40.6	22.6	602.5	7	43.8	Fallow, sown meadow, dry meadow, roadsides
mature	(j, im, v, g_1), $\underline{g_2}$, g_3, ss	0.5–1	0.31	26.5	68.4	49.1	2	12.5	Fallow, dry meadow
ageing	v, g_1, g_2, $\underline{g_3}$, ss, s	0.5–1	0.66	14.2	76.0	1.7	3	18.7	Fallow, dry meadow, roadsides
old	g_3, $\underline{ss}$	0.5–1	0.84	18.3	8.9	0	1	6.3	Dry meadow

*The predominating age-state group is underlined and groups which may be absent are shown in brackets.

of tussock grass occur on alluvial deposits of the Oka and North Dvina rivers, as well as on spoilbanks of drainage ditches.

Invasive populations of *D.flexuosa* are common in forest felling areas and rare on sand dunes. Their population density ranges widely (Table 7) with a cover of 5–8%. However their rate of individual development is high and by the end of the first year the coenopopulation includes all groups of the pregenerative period, and in the second year some of the virginile plants proceed to flower.

The life time of *Poa annua* invasive coenopopulations is even shorter than those of *Deschampsia* (Table 10). Their population density is 57–375 individuals per $0.25\,m^2$.

Invasive coenopopulations were not found either in couch grass or in meadow fescue in natural phytocenoses. However, invasive populations of *Festuca pratensis* lasting for less than one year have been described (Ermakova and Mironova 1980) in meadow agrophytocoenoses in the sowing year; their population density was 17–252 individuals per $0.25\,m^2$.

(b) *Normal coenopopulations*

Normal coenopopulations do not depend on diaspore invasion and are capable of self-maintenance by seed or vegetative propagation, either simultaneously or in succession. This is the most frequent coenopopulation type and may be described as 'full-membered' if all age-states are present, or 'incomplete-membered' if one or several groups are absent. The incomplete-membered type can be caused either by the effects of various exogenous factors (climatic, zoogenic, anthropogenic, etc.) resulting in interruption, elimination, delay or acceleration of individual development, by the effects of endogenous factors such as the inhibition of development of young growth by adult plants, or by specific characteristics of flowering and fruit formation in different age-states. The incomplete-membered coenopopulation spectrum can either be related to a natural absence of the postgenerative period in the ontogeny, as in monocarpic plants, or to the relatively young age of a coenopopulation if no plants have reached the **ss** and **s** age-states. Spectra without young growth lack reproduction waves. Polymodal spectral reflect different waves of reproduction, possibly due to irregular, non-annual seedling establishment. Age-state spectra are shown graphically with age-states ranked from **pl** to **s** in sequence along the abscissa from left to right; the terms 'left-handed' and 'right-handed' spectra refer to the preponderance of young or of old individuals on such a graph.

Following the concepts of Zhukova (1967) and of Uranov and Smirnova (1969), we classified the normal coenopopulation into four versions according to their successive development stages.

(i) Young normal coenopopulations have the $\mathbf{g_1}$ age-state group prevalent among adult plants and show a relatively regular replenishment of the adult population. The spectrum is, as a rule, left-handed; this is due to the presence of most or all of the pregenerative age-states predominant over the postgenerative ones.

In *D.caespitosa* this version constitutes about 14% of all coenopopulations studied and was found in 2–3 year old post-felling areas and in 8–10 year spruce plantations (Table 6). These coenopopulations emerged as a result of recent species invasion to these areas and are primarily incomplete-membered, with the oldest **ss** and **s** groups being absent. The population density is rather high. Flood pastures are characterized by young, full-membered coenopopulations with a lower population density (3 or 4 individuals per 0.25 m^2) and continuous seed reproduction (Zhukova 1980). The number of young individuals averages about 2 per single generative plant here. This results primarily from coenosis unsaturation with a particular species and is due to the relatively small diameter of the grass tussock (5–10 cm) and area of its influence ('phytogenic field' sensu Uranov 1965) (8.3–115 cm^2) under the prevalent young generative plants. The openness of the vegetation ensures the persistence of new waves of reproduction and intensive growth of the existing individuals. This results in rapid development of young normal coenopopulations of *D.caespitosa*, with a life-time no longer than about 2-5 years.

The young normal coenopopulations of *D.flexuosa* were described on waste-land meadows and plots of burned forests (Table 7). The area occupied by one individual in the prevalent $\mathbf{g}_1$ age-state is 10–38 cm^2. The life-span of individuals is 1–2 years (Zhukova 1979).

The young normal coenopopulations of *Agropyron repens* are, as a rule, incomplete-membered, being devoid of seedlings and juvenile plants: this is due to the complete inhibition of seed reproduction (Table 9). Young generative and virginile groups with numerous young rhizomes are prevalent. The total number of communication rhizomes linking short clusters or 'partial tufts' (Fig. 1d) is lower than that of the offshoots. The mode of coenopopulation self-maintenance is vegetative and new offshoots are formed by the separate shoot clusters. Coenopopulation density is 5.2 shoot clusters per 0.25 m^2. The ageing of these coenopopulations is manifested in the gradual accumulation of $\mathbf{g}_3$ and **ss** age-states absent in the earlier stages.

In *Festuca pratensis* the young normal coenopopulation was identified in the second year of its occurrence in agrophytocoenoses (Table 8). The coenopopulation was characterised by the prevalence of young generative plants, the presence of virginile and sometimes immature plants, and the emergence of a few medium-aged generative plants a month later; it is quite short-lived.

The young normal coenopopulations of *Poa annua* have a rather high population density (Table 10). Initially individuals of the pregenerative period are prevalent. As the number of generative plants increases the number of seedlings and juvenile plants is drastically reduced.

(ii) Mature normal coenopopulations are characterised by the prevalence of the $\mathbf{g}_2$ age-state group, cover the greatest area in coenoses and produce the highest amount of biomass in these grasses. Their age spectra are uni- or bimodal: the second peak is among the groups of the pregenerative period whose population density is higher than the postgenerative fraction, and this indicates continual recruitment from seed or vegetative reproduction.

In *Deschampsia caespitosa* mature coenopopulations are widespread and constitute about 40% of all the coenopopulations of this species studied. Medium-aged generative plants accumulate and result in considerable duration of this age-state, for 5–15 years or longer (Table 6). The stability of mature *Deschampsia* coenopopulations reflects the competitive ability of $\mathbf{g}_2$ individuals which have large tussocks (10–15 cm diameter) and aggressive root systems (30–40 cm diameter) leading to large phytogenic fields, both under each individual (226–933 cm^2) and under the group as a whole. *Deschampsia caepitosa* is a strong edificator and produces an inhibiting effect on other species and on its own young growth.

In *D. flexuosa* mature normal coenopopulations are less frequent (about 10%). The tussock area in the $\mathbf{g}_2$ age-state is 140–270 cm^2, the phytomass is 6.9–7.3 g and seed reproduction is predominant.

In *Festuca pratensis* this coenopopulation type occurred only rarely in natural phytocoenoses; in the flood-plain of the Ugra River, where it was the dominant species, the projective coverage was about 50–60%. In the adult part of the coenopopulation all groups of generative plants were present with intensive seed reproduction. There were 3.2 group individuals per generative individual. The total population density reached a level of 27 individuals per 0.25 m^2. In meadow fescue, as in most loose-caespitose grasses, this coenopopulation version is short-lived and is replaced by the ageing one after 1–4 years.

In *Agropyron repens* the mature coenopopulation was recorded on fallow sites. It was devoid of seed reproduction and particules of vegetative origin were predominant in the $\mathbf{g}_2$ age-state. As a rule the coenopopulation is polycentric with a ratio of offshoots to communication rhizomes close to 1. Separation of non-rejuvenated shoot clusters in the generative state and of rejuvenated ones in the immature and virginile states is possible. A high intensity of vegetative reproduction was recorded.

(iii) The ageing normal coenopopulations are characterised by constant prevalence of age-state $\mathbf{g}_3$. This type may be subdivided into those with (a) a prevalence of pregenerative plants over the postgenerative ones, and (b) a prevalence of postgenerative plants over the pregenerative ones.

In *Deschampsia caespitosa* about a quarter of the coenopopulations observed were ageing normal (Table 6). From north to south there is a tendency towards a decrease in their occurrence: they make up 33%, 21% and 16% of the coenopopulation in the northern, central and Trans-Carpathian areas, respectively. Their relative frequency changes only slightly with different modes of grassland utilization. Their generalized spectrum is distinctly single-peaked, with the peak on the $\mathbf{g}_3$ age-group. Large tuft-clones are about 20–30 cm in diameter, consist of 2–5 particules and form a single phytogenic field, 360–1040 cm^2 in area. Rather large groups of dead shoots in tussock-grass tufts retain a high allelopathic activity (Zhukova and Bogdanova 1977). Consequently, in spite of the reduction in the population density (Table 6), the coenotic closure in the ageing tussock-grass coenopopulations becomes higher and seed reproduction decreases drastically. This is confirmed by the much lower participation of pregenerative plants in coenopopulations (13.5%) and a ratio of young plants per generative individual of 0.23 : 1. Nearly 50% of the incomplete-membered ageing coenopopulations studied were devoid of juvenile and/or virginile individuals. The postgenerative fraction is somewhat higher, although with intensive grazing senile plants may be absent. Full-membered ageing coenopopulations with weakly manifested reproduction waves occur on hay meadow plots which have a lower level of coenosis occupancy by tussock-grass.

The life time of the ageing tussock-grass coenopopulations is determined in the absence of reproduction waves by the duration of the old generative age-state $\mathbf{g_3}$; it may last from 3 to 15 years and probably longer. The species continues to be an edificator in the ageing tussock-grass meadows in spite of the reduction in its projective coverage by as much as 30–40%.

In *Festuca pratensis* the ageing normal coenopopulations are widely distributed in pastures and rarer in hay meadows. They are co-dominants (in 65% of cases studied) and less frequently (35%) subordinate though permanent members of the coenosis. If reproduction is annual then all age states of the pregenerative period are present in the coenopopulation, otherwise only some of them are present or pre-generative plants are altogether absent. Young, medium-aged and senile plants are few or absent and the **ss**-group is relatively abundant. Plant size, developmental potential, productivity and projective coverage are lower than in the previous coenopopulation version. The area occupied by the prevalent $\mathbf{g_3}$ individuals is 13.2–34.2 cm^2 each. The number of young individuals per one generative individual ranges from 0.01 to 4.9.

In *Deschampsia flexuosa* four ageing coenopopulations were identified in the meadowed 7–8 year-old felling areas. On the rough meadows of the White Sea Islands they are usually full-membered, less frequently incomplete-membered, by the absence of a number of groups belonging to the pregenerative period; age-state $\mathbf{g_3}$ constitutes 30–56% of the coenopopulation. They cover an area of 514 cm^2. The total population density is unstable, ranging from 2 to 16 specimens per 0.25 m^2.

In *Agropyron repens* ageing coenopopulations are rare and represented by old polycentric systems or single partial tufts in which the number of communication rhizomes is considerably higher than that of the aboveground shoots. As a rule, they occur on perennial old fallow fields.

(iv) The old normal coenopopulations are characterised by the prevalence of subsenile plants and a distinct righthanded age-state spectrum; the population density of the generative plants is lower than that of the postgenerative. Sometimes the co-prevalence of groups **ss** and $\mathbf{g_3}$ is possible in an intermediate version.

In *Deschampsia caespitosa* old normal coenopopulations occur far less frequently than mature and ageing ones (Table 6): they constitute 5.3% in pastures, 20.8% on haylands and are absent in forest felling areas. Such uneven occurrence of this type can be explained by the lower viability of the **ss** group which is fully eliminated by grazing but preserved slightly longer by hay-mowing management. Tussock-grass individuals do not manage to reach the **ss** age-state in the 10–15 year old felling areas and accumulate mainly after 20–30 years when the felling area is overgrown: thus in forest phytocoenoses old normal coenopopulations occur far more frequently (40%). The total population density of these coenopopulations is low (Table 6). This is associated with a simultaneous decrease in clone-tussock areas (to 112–136 cm^2), primarily in the area under the living part (to 19–70 cm^2). By this stage there is a decrease in coenosis saturation with tussock-grass, but at the same time there is fresh recruitment of seeds from the seed reserves in the soil. Consquently the share of the pregenerative fraction increases (to 22.6%) and two or three-peak age-state spectra develop.

The old normal coenopopulations of *D. flexuosa* occur most frequently on the island and continental rough meadows and on 12–16 year old felling areas. Their total population density is low (Table 7) and the area of one clone is 46–394 cm^2. The coenopopulation is self-maintained by vegetative reproduction; seed reproduction is inhibited.

In *Festuca pratensis* the old normal coenopopulations are widespread in all the investigated geographical localities and habitats. Sporadic groups of pregenerative individuals and $\mathbf{g}_1$ and $\mathbf{g}_2$ are either scarce or absent. According to all indicators, this coenopopulation is rather similar to the previous one (Table 8). The area under one individual is 0.7–1.6 cm^2. The old normal coenopopulations of meadow fescue are nearly equally frequent as co-dominant and as subordinate components of the communities.

In *Agropyron repens* the old normal coenopopulations were reported for old fallow fields, roadsides and artificial agrophytocoenoses. All or most of the pregenerative groups are preserved in them, with immature or virginile partial tufts occurring. The saturation of the upper soil layers with communication rhizomes increases.

(c) *Regressive coenopopulations*

Regressive coenopopulations are those that are unable to maintain themselves either through seed or vegetative reproduction and are thus dependent on immigrant diaspores. Three variants are recognised (Smirnova et al. 1976);

R_1 is devoid of viable seed and individuals of the pregenerative period; flowering (but not fruiting) individuals and groups of the postgenerative period are present.

R_2 consists of sub-senile and senile plants.

R_3 consists of senile plants.

Regressive coenopopulations are extremely rare in nature and were not identified in tufted hairgrass, meadow-fescue or couch grass. In common hairgrass only one regressive coenopopulation was reported, near the southern border of its range in a 15 year old pine tree plantation in the

Moscow region. This indicates both the low vitality of regressive coenopopulations and the fact that most coenopopulations do not survive to this age.

THE LARGE LIFE CYCLE OF A COENOPOPULATION

These selected species of grasses indicate the high diversity in coenopopulation age-state structure and population density that may be observed in nature. Each coenopopulation is in a particular stage of development at any specific moment and is marked by a 'specific age level' according to Uranov (1975). Uranov assumed that each age-state had a particular numerical value, reflecting its developmental status: the closer the individual came to completing its ontogeny, the larger its numerical value. Denoting the value of individuals in the ith age-state by m_i and the number of individuals in the same group by k_i, the 'coenopopulation age level' (Δ) was calculated as

$$\Delta = \frac{\sum k_i m_i}{\sum k_i}.$$

The higher this weighted mean, the older the population. The following values were assigned by Uranov (1975) to various age states: **pl** = 0.0067, **j** = 0.018, **im** = 0.047, **v** = 0.119, $\mathbf{g}_1$ = 0.27, $\mathbf{g}_2$ = 0.50, $\mathbf{g}_3$ = 0.731, **ss** = 0.880, **s** = 0.953.

In the course of development, the coenopopulation of any species is often subjected to unidirectional and irreversible changes in population density, age-state composition and productivity; that is, it is unstable and successive. The development may be reversible if a regressive coenopopulation is rejuvenated to the normal age-state spectrum; this may occur if ecological conditions change abruptly during succession.

Many meadow plants including grasses may be in a state of dynamic equilibrium with the habitat in one of the stages of their development: then the coenopopulation density and age-state composition undergo fluctuations only and the age-state structure remains relatively stable. According to Uranov and Smirnova (1969), those are the 'definitive coenopopulations'. The combination of reversible and irreversible changes in population density and age structure ensure cyclic coenopopulation development, accomplished by the multiple regular or (more frequently) irregular replication of reproduction waves throughout the large wave of the population flow. Only a single manifestation of the large wave is observed under extremely unfavourable conditions.

The full extent of a large coenopopulation wave can only be observed extremely rarely in natural phytocoenoses: in fact, it has not been previously described in literature for one and the same coenopopulation. Here we describe it for *Poa annua* (Fig. 2) in an establishing marginal meadow phytocoenosis in the Moscow region under conditions of intensive anthropogenic disturbance. In May the *Poa annua* coenopopulations

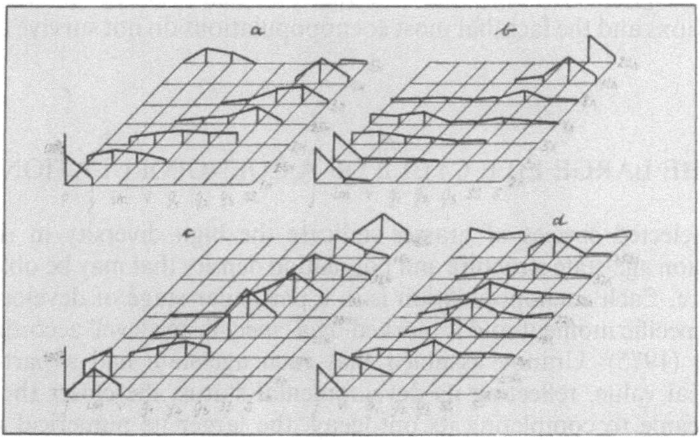

Fig. 2. Large wave of coenopopulation flux in (a) *Poa annua*, (b) *Deschampsia flexuosa*, (c) *Festuca pratensis*, (d) *D.caespitosa*. The duration of the wave varies: 5 months in *P.annua*, 4–6 years in *F.pratensis*, 20 years or more in *D.flexuosa* and 30–70 years in *D.caespitosa*. The relative percentage in each age-state group is shown on the ordinate.

were invasive with a high population density of 125–175 seedling and juvenile plants per 0.25 m^2. After two weeks some of these became immature and virginile and in mid-June about 30% of the individuals reached flowering stage (g_1 and g_2): the coenopopulation then passes into the young normal state. In early July the coenopopulations were mature-normal with a prevalent generative fraction, g_2 individuals constituting more than a half (55%). In late July and August old generative individuals were prevalent (45%) and the coenopopulations ageing; pregenerative plants were completely absent. Coenopopulation development was associated with a regular decrease in density and a gradual transfer of young individuals into the subsequent age-states. The large coenopopulation wave was essentially completed in September. No new reproduction wave was found in autumn. However a coenopopulation which included both the end of an old wave and the beginning of a new wave was described for an adjacent plot (Fig. 2a) in September 1981.

In the other species investigated only single stages of the large wave were observed. Nevertheless, essentially all stages of the large wave were traced in a spatio-temporal sequence of pasture degradation for *Deschampsia caespitosa* (Fig. 2). As already mentioned, the 1–2 year old cattle tracks and pastures are marked by the occurrence of invasive coenopopulations of this species and the 6–8, 15–30, 40 and 50 year old plots used for grazing are characterised by the young and mature, mature, ageing, old and temporarily regressive coenopopulations, respectively. The regular coenopopulation ageing was confirmed by studies of *Deschampsia* in the Oka flood-plain in 1960–1980. A mature coenopopulation on a 30-year old pasture turned into an ageing coenopopulation after seven years in the Oka meadows and an invasive coenopopulation reached a young normal state and then a mature normal state after 11 and 18–20 years, respectively.

This large wave is not highly localized, since similar stages in development of this species have been described for other parts of its range, in the north of the European part of the USSR and in Carpathian regions. It covers a period from 30 to 70 years.

A large coenopopulation wave of *D.flexuosa* was described for the meadowed felling areas (Fig. 2c). It covers a total period of 30–70 years. The old-normal coenopopulations occur in the 'definitive state' only in the forests, while the rest of the stages are successive.

In *Festuca pratensis* the large coenopopulation wave was traced in agrophytocoenoses and had a duration of 5–6 years (Fig. 2d). Therefore the absence of a number of coenopopulation age structure versions of this species in natural phytocoenoses seems to indicate the brevity of their existence and apparent instability, with the exception of the definitive ageing-normal and old-normal types. Numerous transitions of one of such coenopopulations of *Festuca pratensis* from the old normal state to the ageing state and vice versa, resulting from changes in the conditions of meadow uses and its intensity, are shown in Fig. 3a. The spectrum type

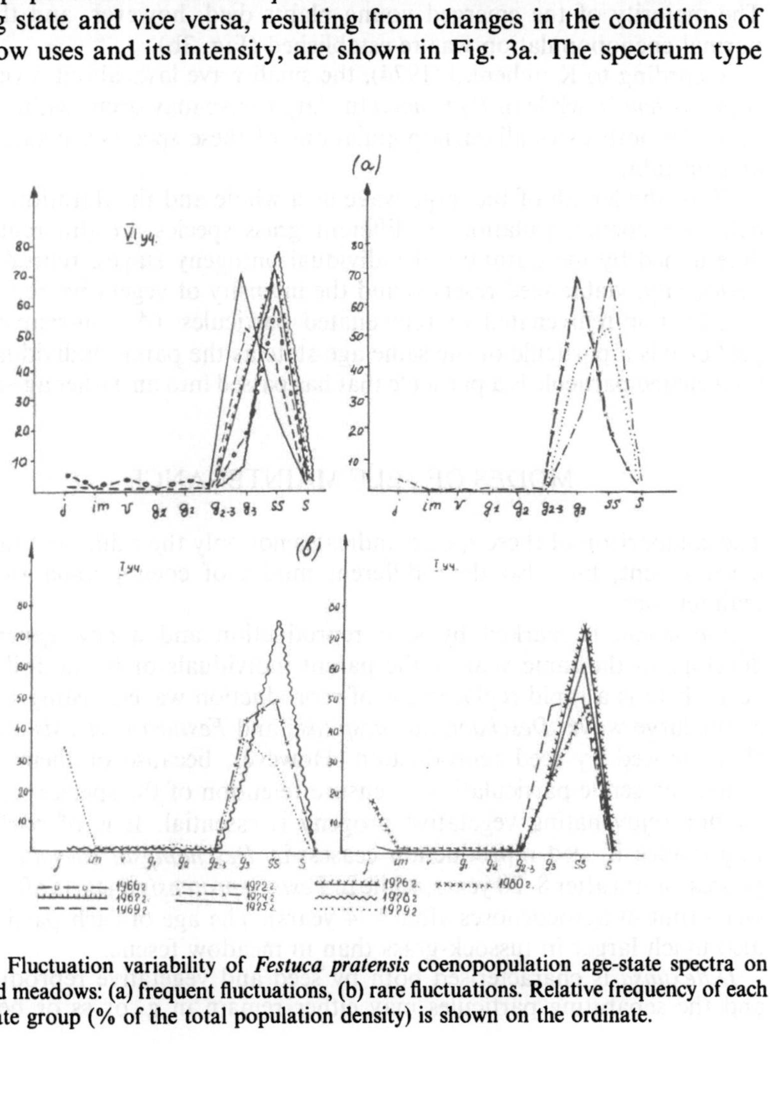

Fig. 3. Fluctuation variability of *Festuca pratensis* coenopopulation age-state spectra on flooded meadows: (a) frequent fluctuations, (b) rare fluctuations. Relative frequency of each age-state group (% of the total population density) is shown on the ordinate.

changed 5 times in 15 years. Fluctuational changes are as a rule less frequent, occurring after 5–6 and even 13 years (Fig. 3b). Such fluctuations are mainly caused by the changes in prevalence of the old generative and subsenile plants. The distribution of the roles of these two age-states depends on the death of their representatives and on the transition of individual plants from $\mathbf{g}_3$ to **ss** and vice versa, which occurs in cases of drastic environmental changes. The third and perhaps the principal cause of the changing roles of $\mathbf{g}_3$ and **ss** plants is the senile decay of the old generative and subsenile plants, resulting now and then in the prevalence of either the secondary generative particules or secondary subsenile particules. The fluctuating changes in *Festuca pratensis* spectra were less frequently caused by annual fluctuations in the young part of the coenopopulation. For example complete or nearly complete absence of seed replenishment for many years on hay meadows was followed by an extremely large reproduction wave, related to a strongly disturbed coenosis. The majority of the emerged young plants died, however, and the old normal coenopopulation was re-established (Fig. 3b).

According to Kurchenko (1974), the small wave lasts about 5 years in *Agrostis tenuis*, while in *Poa annua* the large wave may occur within 1–1.5 years. In both cases all coenopopulations of these species are successive and unstable.

Thus, the length of the large wave as a whole and the duration of the definitive coenopopulation in different grass species are different, and determined by the duration of individual ontogeny stages, reproductive periodicity, viable seed reserves and the intensity of vegetative reproduction by non-rejuvenated or rejuvenated particules. (A non-rejuvenated particule is a particule of the same age-state as the parent individual; the rejuvenated particule is a particule that has passed into an earlier age-state).

MODES OF SELF MAINTENANCE

The comparison of these species indicates not only their different rates of development, but also their different modes of coenopopulation self-maintenance.

Poa annua is marked by seed reproduction and a new generation develops in the same year as the parent individuals or in the following year: there is a rapid replacement of reproduction waves, lasting as long as the large wave. *Deschampsia caespitosa* and *Festuca pratensis* are also characterised by seed reproduction. However, because of their longer ontogeny, senile particulation to ensure retention of the species territory by non-rejuvenating vegetative progeny is essential. It is of particular importance if seed reproduction ceases: in *Deschampsia caespitosa* this process starts after 8–20 years, while in *Festuca pratensis* it starts after 7–13 years (but in agrocoenoses after 3–4 years). The age of each particule is also much larger in tussock-grass than in meadow fescue.

D.flexuosa is characterised both by seed and vegetative reproduction and the separating particules may either remain in $\mathbf{g}_3$ or **ss** or become rejuvenated to **j** or **im** age-state. Seed reproduction is prevalent in young, well-lit forest felling areas. Seed and vegetative reproduction is typical of the old felling areas overgrown with trees and bushes, but particule rejuvenescence does not occur in the initial stage. Seed reproduction is inhibited as the coenosis becomes saturated with edificator species, and then the separating individuals of vegetative origin become considerably rejuvenated (Zhukova 1979). Seed reproduction is absent in forest phytocoenoses in spite of the low species occurrence. The seedlings and drastically rejuvenated particules can only emerge in well-lit patches of the forest. Thus common hairgrass demonstrates one of the most stable coenopopulations.

In *Agropyron repens* on fallow and meadow agrophytocoenoses vegetative reproduction with rejuvenated or non-rejuvenated partial tufts is prevalent. Seed reproduction is essentially completely inhibited since no seedlings or other age-states of the young growth of seed origin were found in any of the coenopopulations described.

The concept of stable and unstable coenopopulations and of different intermediate versions (Rabotnov 1978) is based on their density dynamics and fluctuation ranges, and should be supplemented by concepts of coenopopulation age-state structure stability and lability. The stable coenopopulation must preserve the prevalence of certain groups of the adult population part. The rest of the age-state groups (primarily those belonging to the pregenerative period) are subjected to substantial fluctuation changes caused by periodicity of reproduction waves.

Coenopopulation life consists of alternating stable and unstable stages differing in duration and depending on the exogenous and endogenous factors. Among the exogenous abiotic factors affecting coenopopulation self-maintenance the most important are the weather and edaphic conditions and the mode and intensity of anthropogenic effects. The most influential exogenous biotic factors are the effects of other species of plants, animal and microorganisms, either by weakening the coenopopulation (by elimination of caryopses or vegetative organs or by diseases) or by contributing to its prosperity (mycorrhizal formation, the effects of biocoenotic microorganisms or zoocomponents, eradicating and decomposing the dead parts of grasses). Among the endogenous factors the most significant is saturation of the coenosis by the species itself. This is determined by indicators of the projective coverage, area of occupancy and phytomass, which depend on the density, vitality and age level of the coenopopulation. The higher the saturation, the weaker is rejuvenation in the coenopopulation, and vice versa. This is a regulative mechanism of self-maintenance. Simultaneously, it defines the actual role of the species in a real coenosis: potential edificators and dominants, for example, realise their possibilities most fully under high coenosis saturation.

The investigations done in different coenoses show that the dominance effect of various species depends on the development stage of a coenopopulation and not on habitat conditions. Sometimes the coenopopulation age level of dominant plants can determine phytocoenosis composition. For example, there may be decreasing diversity of species on different meadows during coenopopulation maturation, but a subsequent increase later as the coenopopulation ages.

CONCLUSIONS

1. The longevity of specific age-states determines the age level of the coenopopulation with the highest frequency in each species: for *Deschampsia caespitosa* these are the mature and ageing coenopopulations; for *D. flexuosa* and *Festuca pratensis* – the ageing and sensile coenopopulations; for *Poa annua* and *Agropyron repens* – all variants of normal coenopopulations.

2. The modes of self-maintenance of meadow grass coenopopulations depend on their ontogeny type and on their capacity for vegetative reproduction and rejuvenescence: their realisation is determined by the effect of exogenous abiotic and biotic factors and on the coenotic situation. The following coenopopulations can be distinguished on the meadows: those with seed reproduction alone, those with seed and vegetative reproduction and those with vegetative reproduction predominating.

3. The most stable species are able to change their self-maintenance modes in a strongly varying ecological situation, either by combining seed and vegetative reproduction or by alternating both types.

4. For coenopopulations of meadow grasses the basic ways of invading a territory in disturbed communities are (a) intensive seed reproduction from caryposes, stored in soil or dispersed from elsewhere, and (b) expansion of tussocks, rhizomes and stolons by intensive vegetative reproduction.

5. In meadow grasses, the principal way of holding occupied territory is by particulation of tussocks or of polycentric systems in long-rhizome and stolon-forming plants.

6. The species undergo selection in the meadows according to their labile responses to the rapidly changing environmental conditions. Rapidly developing coenopopulations are prevalent in meadow phytocoenoses.

7. The coenotic value of each species varies with the different development stages of coenopopulations. Maximum saturation of the habitat is realised by different variants of normal coenopopulations.

REFERENCES

(English translations of Russian-language titles are given in brackets)

Arber, A. (1934) The Gramineae, a Study of Cereal, Bamboo and Grass. Cambridge: Cambridge University Press. 480pp.

Ermakova, I.M. (1968) (About age-states of meadow fescue, *Festuca pratensis* Huds.) Nauchniye doklady vysshei shkoly, Biologichesckie nauki, 7: 70–77.

Ermakova, I.M. and Mironova, A.S. (1980) (Systematic approach to studying the effect of fertilizers and sowing rates on meadow fescue in pure and mixed crops.) Sistemno-ecologichesky podkhod k sovremennym problemam selskogo khozyaistva i nauki. 2: 49–51. Gorky.

Gatsuk, L.E., Smirnova, O.V., Vorontzova, L.I., Zaugolnova, L.V. and Zhukova, L.A. (1980) Age states of plants of various growth forms: a review. J. Ecol. 68: 675–696.

Kurchenko, E.I. (1974) (Ontogenesis and year-to-year changes in the structure of the coenopopulations of *Agrostis tenuis* Sibth). (Age composition of populations of flowering plants and their ontogenesis.) Moskovago gosudarstvennogo pedagogicheskogo instituta pp. 70–87.

Larin, I.V. (Ed.) (1950) (Feed plants of USSR hay-meadows and pastures.) Moscow–Leningrad: 687pp.

Rabotnov, T.A. (1950a) (Life cycle of perennial herbaceous plants in meadow coenoses.) Trudy BIN AN SSSR, ser. 3. Geobot. 6: 7–204.

Rabotnov, T.A. (1950b) (Common hairgrass.) Kormoviye rasteniya senokosov i pastbishch, pp. 305–306. Moscow–Leningrad.

Rabotnov, T.A. (1969) (Some aspects of studying coenotic populations.) Biull. Mosk. Obsch. Isp. Prir. Otd. Biol. 74 (1): 141–149.

Rabotnov, T.A. (1978) (Phytocoenology.) Moscow: Moscow Univ. Publ. 385pp.

Rozhevitz, R.U. (1937) (Grasses.) Moscow: Selkhozgis. 638pp.

Serebryakova, T.I. (1971) (Morphogeny of shoots and evolution of vital forms of grasses.) Moscow. 358pp.

Serebryakova, T.I. (Ed.) (1980) (Diagnoses and keys of meadow plant age-states.) Moscow: Lenin MGPI. 109pp.

Smelov, S.P. (1947) (Biological fundamentals of meadow culture.) Moscow: Selkhozgiz.

Smirnova, O.V.., Zaugolnova, L.B. and Ermakova, I.M. (Eds.) (1976) (Plant coenopopulations: basic concepts and structure.) Moscow: Nauka. 216pp.

Tsvelev, N.N. (1976) (USSR grasses.) Leningrad: Nauka. 488pp.

Uranov, A.A. (1965) (Phytogenic field.) (The problems of soviet botany.) 2: 251–254. Moscow–Leningrad: Nauka.

Uranov, A.A. (1975) (Age spectrum of phytocoenotic wave processes.) Nauchniye doklady vysshei shkoly, biol. nauki. 2: 7–34.

Uranov, A.A. and Smirnova, O.V. (1969) (Classification and principal characteristics of perennial plant populations.) Biull. Mosk. Obsch. Isp. Prir. Otd. Biol. 74(1): 119–132.

Uranov, A.A., Zaugolnova, L.B. and Smirnova, O.V. (Eds.) (1977) (Plant coenopopulations: development and interaction.) Moscow: Nauka. 133pp.

Zhukova, L.A. (1961) (The pecularities of age changes in tussock hairgrass.) (Morphogenesis of Plants) 2: 635–638.

Zhukova, L.A. (1967) (Changing age composition of tufted hairgrass population in Oka meadows.) Candidate thesis, Moscow. 252pp.

Zhukova, L.A. (1979) (Ontogeny and age composition of *Deschampsia flexuosa* (L.) Trin (Poaceae) coenopopulations). Bot. Zhur. 64(4): 525–540.

Zhukova, L.A. (1980) (Effects of anthropogenic factors on the age composition of tufted hairgrass coenopopulations under various geographical conditions.) Biull. Mosk. Obsch. Isp. Prir. Otd. Biol. 84(6): 87–100.

Zhukova, L.A. and Bogdanova, A.G. (1977) (Analysis of interrelations in some meadow plants using the allelopathic method). pp. 108–120 in Tsenopopulyatsii rasteny: Razvitiye i vzaimootnosheniya. Moscow: Nauka.

COENOPOPULATION STRUCTURE OF *AGROSTIS* SPECIES

E.I. KURCHENKO

ABSTRACT

The coenopopulation structure and dynamics of seven species of *Agrostis* are described. In relatively open vegetation seed renewal and a dynamic age-state spectrum are recorded, whereas in dense vegetation seed renewal is absent and population dynamics show only slight fluctuations. The consequences of these patterns for microevolutionary processes are discussed.

INTRODUCTION

Plant populations are studied at present in different ways: by phytocoenologists as structural and functional units of vegetation and by evolutionists and typologists as structural units of the evolutionary process. Much material has now been accumulated on structure, dynamics and functional characteristics of the populations of forest, steppe and meadow coenoses (Yunatov and Lavrenko 1969, Kurkin 1976, Smirnova et al. 1976, Uranov et al. 1977). Beginning with the fundamental works by Timofeyev-Resovsky (1925), Vavilov (1922, 1926), Dobzhansky (1937) and other researchers, the significance of studying natural animal and plant populations as the primary structural units of species has become obvious. The past decade has been marked by publication of monographs and collections of papers summarizing modern concepts of the microevolutionary process, mainly based on studies of various animal groups (Mayr 1970, Timofeyev-Resovsky et al. 1977) but less on studies of plant species (Zavadsky 1968, Stebbins 1974, Grant 1977).

Studies of plant microevolutionary processes are considerably complicated by the effects of coenotic factors, that is, by the presence of other species in the plant's environment. Evolutionary processes of plants are more related to the coenotic environment than that of animals since the coenosis acts not only as an arena but also as an environment for the evolutionary processes. In many meadow species, for example, the establishment of new genotypes from seeds is determined by coenotic factors

White, J. (ed.), The Population Structure of Vegetation.
 ISBN 978-94-010-8927-2

which are responsible for selecting individuals with the appropriate biological advantages for survival, such as the rate of growth or of territorial invasion. Investigations of the role of coenotic factors in evolution are still in the initial stage and publications are few (Sukachev 1927, Khokhlov 1947, Kurkin 1976).

In this paper I review studies on the evolution of coenopopulations of *Agrostis* species both in disturbed, slightly closed coenoses with weak competitive relationships among plants, and in closed plant coenoses with species subject to strong competition. Bentgrasses have a wide geographical distribution, occur in various coenoses, have broad variability of morphological characters, and show an ability for hybridization within and between species (Tsvelev 1976). The following species were studied: *Agrostis tenuis* Sibth., *A. gigantea* Roth., *A. syreistschikowii* Smirn., *A. trinii* Turcz., *A. marschalliana* Sered., *A. canina* L.; the first two are in section Agrostis, the remainder in section Agraulus of the genus.

METHODS

Material was collected in coenoses with high (80–100%) and low (40–60%) vegetation cover at the time of flowering in late June–early July in the Moscow and Kursk regions, the Trans-Baikal area, and in the Caucasus. Geobotanical descriptions of the sites were made and then transects were established. The transects were divided into plots of $1 \times 1\,m^2$, $0.5 \times 0.5\,m^2$ and $0.25 \times 0.25\,m^2$. In sites with a medium bentgrass population density (10–20 individuals per $1\,m^2$) 60 plots of $0.5 \times 0.5\,m^2$ were established, whereas with a high population density (40 or more individuals) 30–35 plots of $0.25 \times 0.25\,m^2$ were sufficient; with a low density the number of $1 \times 1\,m^2$ plots was increased to 80.

In natural populations the bentgrasses are represented by individuals of seed or vegetative origin (particules). The individuals of seed origin were distinguished according to their age-state characteristics: seedlings, juvenile, immature, virginile, young, mature reproductive, old reproductive, subsenile, senile (Fig. 1). The age-state periods were classified according to the system advocated by Uranov (1975) and his School (see also Gatsuk et al. 1980). Each age-state in *Agrostis* is characterised by a combination of the following characters:

Seedling (**pl**): a primary shoot with a coleoptile, 2–3 narrow assimilating leaves and a radicle; the seedling remains connected to the caryopsis.

Juvenile (**j**): a primary unbranched shoot with 4–5 narrow and short leaves and 2–3 adventitious roots; the connection with the caryopsis has been lost.

Immature (**im**): a primary and 3–4 branched shoots; the leaf blades are larger than those of juveniles but not as large as those of mature leaves; the number of adventitious roots is 4–5.

Virginile (**v**): shoots, leaves and roots typical of mature plants are formed; the number of shoots increases; the root system increases in width and depth by newly-formed adventitious roots; reproductive shoots are absent.

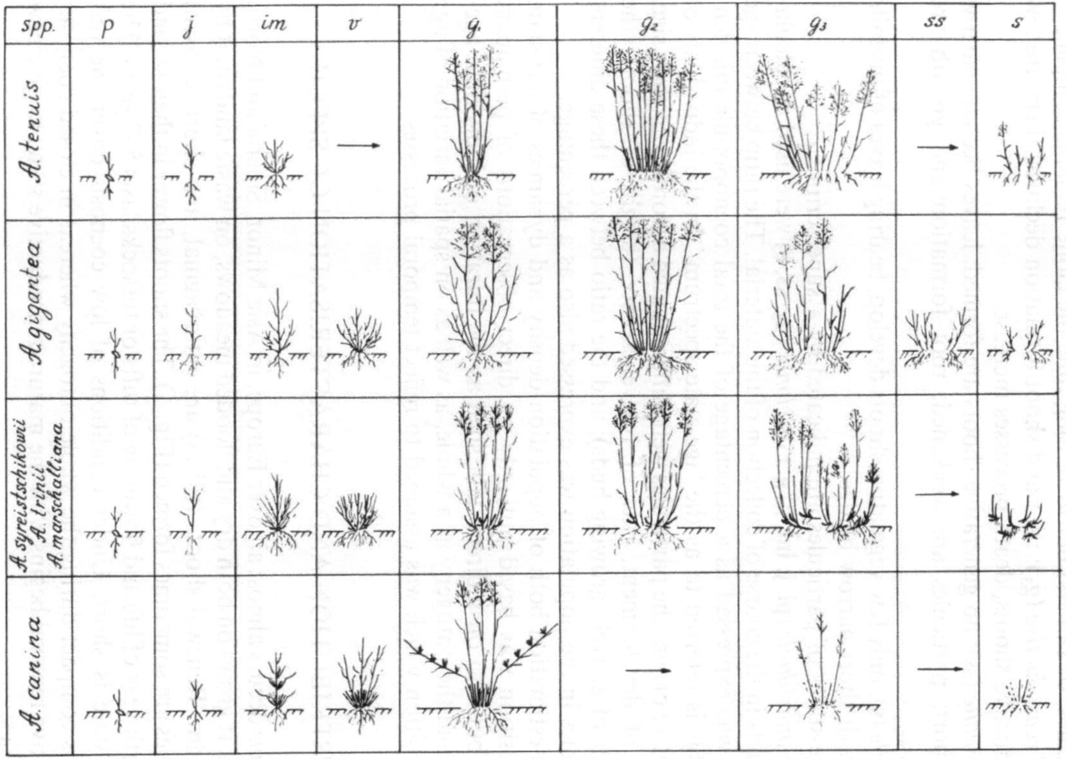

Fig. 1. Ontogenetic characteristics of *Argrostis* species. **pl**, seedling; **j**, juvenile; **im**, immature; **v**, virginile; $\mathbf{g_1}$, $\mathbf{g_2}$, $\mathbf{g_3}$, young, mature and old reproductive; **ss**, subsenile; **s**, senile.

Young reproductive (**g_1**): the first reproductive shoots appear; dying of shoots is sporadic.

Mature reproductive (**g_2**): the number of reproductive shoots and the intensity of lateral shoot and root formation are at their highest; necrotic parts are formed as some shoots die off and this results in disintegration of the primary individual (emerged from the seed) into separate pieces (particules, or ramets) but for a long time the clone remains compact.

Old reproductive (**g_3**): root and shoot formation declines; there are few reproductive shoots; death processes increase.

Subsenile (**ss**): no generative shoots are formed; leaves become narrow and short; particules are weakened; root formation and growth are reduced.

Senile (**s**): only few vegetative shoots develop, bearing leaves of juvenile type with short narrow blades.

(Age-states of particules are indicated by a superscript ′.)

The *individuals* (**pl, j, im, g, g_1**) or *clones* (**g_2, ss, s**) were used as calculation units in the course of collection of the material. The ratio between age states was expressed as a percentage of the total coenopopulation density: this is referred to as the 'age-state spectrum.' In the individuals of vegetative origin, the particules were differentiated according to their size, phase of development, rate of rejuvenescence (height of shoots, the number of actively growing buds); and the ratio between these different particules in the population was expressed also as a percentage.

The estimation both of population density and dynamics of age-state composition was based not only on direct observations of permanent plots, but also on indirect data such as the duration of individual age-states and the ontogeny as a whole, as well as on spatial variation of age composition which was assumed to reflect temporal processes.

DISTRIBUTION AND CHARACTERISATION OF SPECIES

A. tenuis occurs almost all over Europe, in Asia Minor, Siberia and North Africa. It grows on both dry and flooded meadows, on waste lands and by roadsides. Renewal shoots (tillers) are extravaginal, and short (c. 5 cm) rhizomes are sometimes formed (Fig. 1). The shoots flower in their second to fourth year of life and flowering of tufts or tussocks lasts 5–7 years. The senile state is short. Under conditions of low coenosis cover the tuft remains compact during the whole ontogeny whereas in closed coenoses vegetative renewal begins with the mature reproductive state, to give new particules.

A. gigantea, like *A. tenuis*, is characterized by a wide geographical range: it occurs all over Europe and Asia and can be found as an introduction in all non-tropical countries of both hemispheres. It grows on flood plains and mountain meadows, forest clearings, by the roads, in populated areas. Renewal shoots are extravaginal. Shoots flower in their second or third year of life and flowering of tufts lasts 3–5 years. It is characterised by a faster rate of development than *A. tenuis* (Table 1). Under conditions of

Table 1. The longevity of different age states (years, unless specified) of *Agrostis* species.

Species	*A. tenuis*	*A. syreistschikowii*	*A. gigantea*	*A. canina*
Characters				
pl, j, im	1	1–2	1	2–3 days
v	1–2	1		1.5 weeks
				–1 year
g_1		2–3	1	1–2
g_2	2–3	1–3	1	Not differentiated
g_3	2–3	1–2	1–2	0.5
The longevity of **pl**–g_3	6–8	8–10	4–5	2–3
The longevity of tufts of seed origin	7–10	10–12	6–8	3–4
The longevity of vegetative progeny	?	30–35	~10	Dozens
Periodicity of seed renewal	4–5	3–4	3–4	Annually

low coenosis cover the tuft remains compact during the whole ontogeny, but where the coenosis cover is high, tufts show vegetative renewal.

A. syreistschikowii is a typical species of meadow steppes in the European and Asian parts of the USSR; beyond the steppes to the north it occurs on water meadows. Renewal shoots are extravaginal and rhizomes are formed on sandy soils, c. 3–5 cm long. Shoots flower in their second or third year. The duration of tuft flowering is 4–7 years. Tuft decomposition and vegetative renewal are observed in the young reproductive state (g_1). In the old reproductive state (g_3), the tuft consists of individual particules and a plant can survive in a coenosis for a long period of time (Table 1). This conclusion is based on a comparison of bentgrass population density in the meadow steppe of Kursk region in 1933 and 1970, that is, 37 years later. In 1933 the population density was 642 per 1 m^2 (Alekhin and Uranov 1933). The high population density resulted from annual steppe harrowing, which provided favourable conditions for bentgrass population renewal from seeds. Beginning with 1935 some steppe plots were no longer mown. Dying plant shoots do not decay during the winter and from year to year a layer of waste (15–25 cm deep) accumulated in the grass stand, hindering seed renewal.

As a result of that, bentgrass population density on a plot, devoid of any conditions for seed renewal for 30 years, had been reduced to 4 individuals per 1 m^2 according to our estimates in 1970. On another plot not mown since 1935, *A. syreistschikowii* completely disappeared from the grass stand. It can be suggested therefore that self-maintenance of the population by vegetative renewal may last not less than about 30 years.

In the eastern Tobolsk-Omsk area, *A. syreistschchikowii* is replaced by

A. trinii, an allied species prevalent on the steppes and saline meadows of Trans-Baikal region, Yakutia and extending to the Pacific Ocean. It occurs also in more southern areas in Mongolia. The external morphology of the shoots, the characteristics of tuft development and longevity resemble those of *A. syreistschikowii.*

A. marschalliana occurs in subalpine and alpine meadows of the Caucasus. Its shoot structure and ontogeny are similar to those of *A. syreistschikowii* and *A. trinii.*

A. canina is a Euroasian boreal species prevalent in boggy areas. The shoots flower in the second year, although many of them die off in the vegetative state in the first year. Renewal shoots are intravaginal. Very damp conditions lead to shoot lodging and rooting at the nodes. Shoots can grow up to 1 m long in shallow water. Rapid growth and intensive tillering can occur in early stages of ontogeny, resulting in rapid tuft development (Fig. 1). The juvenile state lasts only a day or two (Table 1) and after little more than a week the plants pass into the virginile state. The reproductive period lasts 1–2 years. The tuft does not disintegrate into particules as it ages. With moderate moisture conditions, the longevity of tufts of seed origin is 3–4 years, but with high moisture vegetative renewal caused by the lodging shoots may last dozens of years.

The longevity of individual age-states of these bentgrasses is summarized in Table 1. The highest rate of development and, accordingly, the lowest longevity were recorded in *A. canina*; next comes *A. gigantea*; the development of *A. tenuis* and *A. syreistschikowii* is slightly longer. The more rapid the development cycle, the more frequent is the reproduction (Timofeyev-Resovsky et al. 1977). It was found in our material that seed renewal was most frequent in the juvenile *A. canina* and rarer in other species.

DYNAMICS OF COENOPOPULATIONS

The dynamics and age-state composition of coenopopulations were investigated in open coenoses with a projective cover of 40–60% and in dense coenoses with projective cover of 80–100%. Coenoses with small coverage occur usually on poor sandy soils and are characterised by low species diversity. Bentgrass is predominant in the grass stand with a density on average of 20–30 tufts per 1 m^2. The dense coenoses occur as a rule on rich or well-moistened soils and are characterised by higher species diversity. The relative contribution of bentgrass (except *A. trinii*, *A. marschalliana*) in these dense coenoses is reduced since most bentgrasses are poor competitors, but the number of separate units (particules) is doubled (on average, 40–60) while their size is far lower, compared to the tufts of seed origin. The ecology and phytosociology of the communities studied is summarized in Table 2.

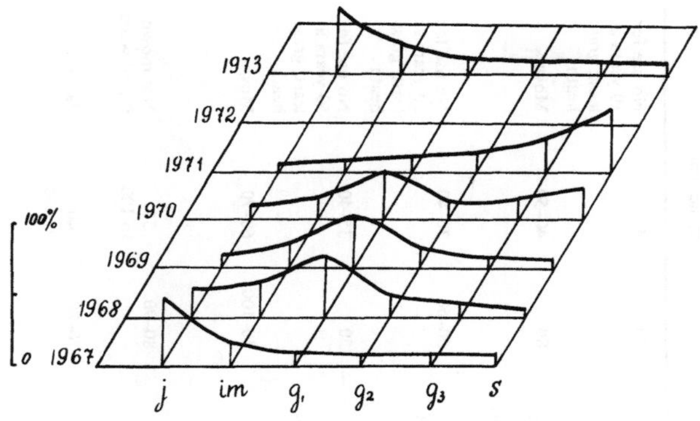

Fig. 2. Changes in age-state composition of an *Agrostis tenuis* coenopopulation from 1967 to 1973; no data were collected in 1972.

A. tenuis

Observations were carried out for seven years in the Oka-land terraced preserve, Moscow region, on a permanent transect established in a bentgrass meadow with projective cover of 50%. *A. tenuis* was predominant and made up 40% of the cover. The soils are loamy with high sand content. The population density changed slightly during the first observation years: 32.1, 30.7, 28.6, 24.1 individuals per 1 m^2 in 1967–1970, and dropped to 19.3 in 1971. The age-state spectra changed qualitatively during this period (Fig. 2). The year 1967 was marked by intensive seed renewal and group **j** made up 73.1% of the population. A decrease in the number of juvenile plants was observed in 1968 as they grew into the next age-states, giving an increase in the number of immature and young reproductive plants. An increase in the number of mature reproductive and old reproductive plants was observed in 1969. In 1971 the population passed into the state of regression (curve rises in $\mathbf{g_3}$ and **s** states). The gradual ageing of individuals could result in final degeneration of the population. However, a new phase of seed renewal was recorded in 1973. With regard to the coenopopulation as a whole it should be noted that it consists mainly of individuals of seed origin. All age-states were present annually in the population with the peak of age-state number shifting from **j** to **s** with plant development and transition into subsequent stages. The generation development lasted 7–8 years. Seed renewal occurred at the moment of ageing of most individuals.

In dense coenoses with a projective cover of 80–90% of *A. tenuis*, the coenopopulation consists of individuals of vegetative origin (a population of particules). The particules are unevenly distributed, forming separate aggregations 60–100 cm in diameter, and are evidently of clone origin. One such coenopopulations was described in the Orenburg region in a herbaceous bentgrass sward with 80% projective cover in which *A. tenuis* was prevalent (60% cover). Bentgrass population density was 68 individuals per 1 m^2. The particules differed in size, number of shoots and growth rate. Particules 1.0–2.5 cm at the base with 1–5 reproductive and

Table 2. Phytosociological and ecological characteristics of coenoses in which *Agrostis* species were studied.

Investigated species Number per m^2	Region	Soils	Ground level	Dominants cover, %	Codominants	Total cover, %	Height of grass stand, cm	Management
Agrostis tenuis 30	Moscow region, the Oka-land terraced preserve	Sandy loam	No development	A. tenuis–40	Solidago virgaurea, Agropyron repens, Calamagrostis epigeios, Anthoxantum odoratum	50	40	No use for 10 years at least; grass heath
Agrostis tenuis 68	Orenburg region, Guberlin mountain depression between hills	Meadow chernozemic	No development	A. tenuis–60	Achillea nobilis, Eryngium campestre, Cichorium intybus, Sanguisorba minor, Potentilla parviflora	80	40–50	Mown
Agrostis syreistschikowii 20	Moscow region, the Oka-land terraced preserve	Sandy loam	Lichens: Peltigera, Cetraria, Thuidium	A. syreistschikowii–40	A. tenuis, Festuca sulcata, Phleum phleoides, Koeleria grandis, Veronica incana, Dianthus borbasii	50–60	30–40	No use for 30 years at least; grass heath
Agrostis syreistschikowii 36	Moscow region v. Myachkovo	Loam	No development	A. syreistschikowii–30	A. tenuis, Berteroa incana, Stellaria graminea, Poa angustifolia, Festuca ovina, Sedum acre	50	35–40	No use for 50 years at least; grass heath
Agrostis syreistschikowii 44	Kursk region, Central Chernozem preserve, plot I	Chernozemic	No development	Bromus riparius–20 Festuca sulcata–20	A. syreistschikowii, Helictotrichon pubescens, Medicago falcata, Vicia tenuifolia	80–100	60–70	Mown
Agrostis syreistschikowii 4	Kursk region, Central chernozem preserve, plot II	Chernozemic	Herbage mulch, 15–25 cm deep	Bromus riparius–30 Poa angustifolia–20	Calamagrostis epigeios A. syreistschikowii	80–90	35 to 100	Not mown for 30 years
Agrostis syreistschikowii 46	Kursk region, Central chernozem preserve, plot III	Chernozemic	Mosses	Agrostis syreistschikowii–30, Festuca sulcata–15, Agropyron intermedium–10, Medicago falcata–10	Bromus riparius, Koeleria gracilis, Lotus corniculatus, Trifolium montanum	70–80	40 to 100	Pasture
Agrostis trinii 152	Trans-Baikal area, v. Kalenovo, plot I	Meadow salt	No development	Hordeum brevisubulatum–20, A. trinii–15, Medicago falcata–20, Trifolium montanum–10	Festuca pratensis, Bromus inermis, Potentilla multifida	90	50–60 to 100	Mown

Agrostis trinii 1022	Trans-Baikal area, v. Kalenovo, plot II	Meadow salt, irrigated	No development	A. trinii–80	Alopecurus ventricosus Hordeum brevisubulatum, Festuca pratensis, Allium schoenoprasum	100	70	Mown
Agrostis marschalliana 545	Caucasus, Beshtau, subalpine meadow	Mountain meadow chernozemic	Herbage mulch, 10 cm deep	Bromus inermis–20, Poa angustifolia–20, A. marschalliana–10	Salvia pratensis, Geranium pratense, Gallium verum, Phlomis tuberosa	90	40–70 to 100	Not mown for 10 years at least
Agrostis gigantea 13	Moscow region v. Dedinovo flood plain meadow Oka river, plot I	Soddy-loam	No development	Alopecurus pratensis–15, Phleum pratense–10, Festuca pratensis–10, Geranium pratense–15, Galium verum–10	A. gigantea Bromus inermis, Poa pratensis	95	60–70	Mown
Agrostis gigantea 35	Moscow region v. Dedinovo flood plain meadow Oka river, plot II	Soddy-loam	No development	Bromus inermis–15, A. gigantea–10, Alopecurus pratensis–10, Geranium pratense–15, Veronica longifolia–10	Poa pratensis, P. palustris, P. trivialis, Sanguisorba officinalis, Libanotis intermedia	100	70–80	Mown

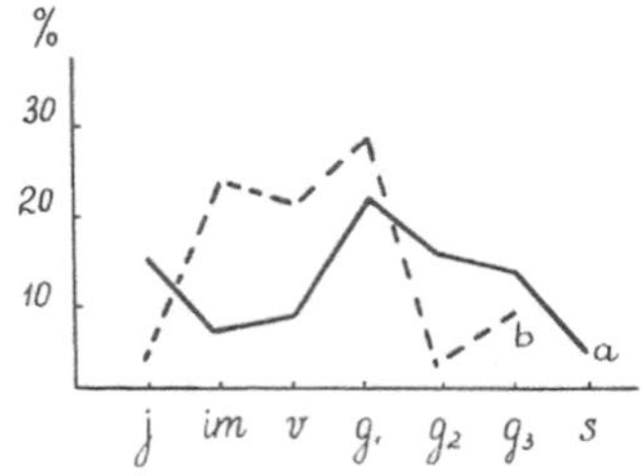

Fig. 3. Age-state composition of *Agrostis syreistschikowii* coenopopulations in (a) Oka-land terraced preserve and (b) Myachkovo, Moscow region.

3–7 vegetative shoots were characterised by a high rate of rejuvenescence. These were considered as young reproductive particules of vegetative origin ($\mathbf{g}_1'$). 1–2 renewal buds were growing on each shoot. Particules 3.0–4.5 cm at the base with 7–10 reproductive and 10–12 vegetative shoots had the highest growth rate and were referred to as mature reproductive ones ($\mathbf{g}_2'$) of vegetative origin. Each shoot had on average 1–3 actively growing buds. Particules of 1.5 cm and smaller, purely in the vegetative state, with low growth rate of shoots, weak shoots and sparse activity of buds were considered as senile ($\mathbf{s}'$).

The above particules form a sequence of development: $\mathbf{g}_1'$ passes into $\mathbf{g}_2'$; $\mathbf{g}_2'$ breaks up on ageing into smaller (1.0–1.5 cm) particules ($\mathbf{s}'$). In coenopopulations about 50% of the particules were in the reproductive state ($\mathbf{g}_1'$–35%, $\mathbf{g}_2'$–25%) and 40% in the senile. In coenoses with a large projective cover and impeded seed renewal the particule spectrum reflects such features of clone population development.

A. syreistschikowii

The material was collected in a bentgrass meadow with 50–60% projective cover in the Oka-land terraced preserve, Moscow region (Table 2). The bentgrass was predominant with 40% cover. The soils are loamy with high sand content. The population density of bentgrass was 20.5 individuals per 1 m^2. Because of poor development of a sward and weakened competitive relations, some years are marked by seed renewal of this species. Individuals of seed origin were recorded in the coenopopulation in the juvenile (**j**) age-state. **j** and $\mathbf{g}_1$ were the most numerous age-states. The longevity of certain bentgrass age-states (Table 1) can be used to interpret the spectrum shown in Fig. 3a in the following way. The observation year was marked by seed renewal (the peak in **j** group), while the previous renewal should have taken place 3–4 years before. The plants developed into young reproductive tufts (the peak in $\mathbf{g}_1$) during that time interval. Thus, the individuals in **j** and $\mathbf{g}_1$ states are members of two independent generations; $\mathbf{g}_3$ individuals belong to an earlier generation.

The spectrum type in another bentgrass meadow (Myachkovo, Moscow region) with a projective cover of 50% and *A. syreistschikowii* constituting 30% of the cover, is in general outline similar to the former and presents an earlier moment of seed invasion (Fig. 3b). Thus the coenopopulations of *A. syreistschikowii* in coenoses with a small projective cover are characterised by the presence of seed renewal covering all age groups. The full period of generation development is about 8–10 years.

The characteristics of *A. syreistschikowii* age-state spectra in coenoses with 80–100% projective cover were investigated in the Central Chernozem preserve, Kursk region; the soils are chernozemic. Material was collected on three plots of the herbs-grass association, subjected to different modes of management (Table 2). Plot I is a meadow steppe with numerous herbs and is mown. The projective cover is 80% and the height of the grass is 60–70 cm. Plot II is not mown and has not been used since 1940. The herbs are dying off, particularly the tap-root species, and rhizome grasses are predominant. The projective cover is 80–90% and the height of the basic grass stand is 35 cm, although the reproductive shoots of some species reach 100 cm. The grass stand lodges in midsummer and there is a layer of waste herbage 15–25 cm deep on the soil surface. Plot III is moderately grazed and is characterized by a reduced number of grass and herbs species, with a projective cover of 70–80%. The grass stand is 40 cm high and the soil surface is covered with moss. The contribution of bentgrass to the structure of the mown coenoses and pastures is 4–46 particules per 1 m^2. The individuals are unevenly distributed and form sporadic aggregations 30–40 cm in diameter. The non-mown plot is characterized by a low population density of particules, with only 4 individuals per 1 m^2, that can evidently be explained by the negative effect of the waste layer preventing seed renewal and hindering vegetative renewal. In all test plots, bentgrass individuals occur as particules of vegetative origin and tufts of seed origin are absent. The absence of the latter on the mown plot can be explained by the high grass stand density and low competitive ability of bentgrass seedlings. Seed renewal is particularly hindered on the non-mown plot where the seeds and roots of seedlings are separated from the soil by a layer of waste. The seedlings are also prevented from rooting on pastures; this is evidently due to the mossy cover on the soil surface and trampling by animals.

The following age-state groups of particules were distinguished according to the number of shoots, the rate of their rejuvenescence and their phase of development. Firstly, there are two groups with particules in the vegetative state with one or two buds active in each shoot: immature (**im′**) particules with 1–3 elongated 2–3 year old shoots and 1–3 shortened 1 year old shoots; virginile (**v′**) particules with 5–7 elongated and a few shortened shoots. There are also two groups of particules in the reproductive state with one or two buds active in each shoot: young reproductive ($\mathbf{g_1'}$) particules with 1–3 reproductive and 1–3 shortened vegetative shoots and mature reproductive ($\mathbf{g_2'}$) particules with 5–7 reproductive and a few shortened vegetative shoots. Apart from these, similar particules, but with signs of senility, occur on the test plots: axillary buds are weak and inactive, reproductive shoots are few, inflorescences are short, blade length and width are reduced. These particules are in their last year of life and have been referred to the senile age-state (**s′**). The particules in vegetative

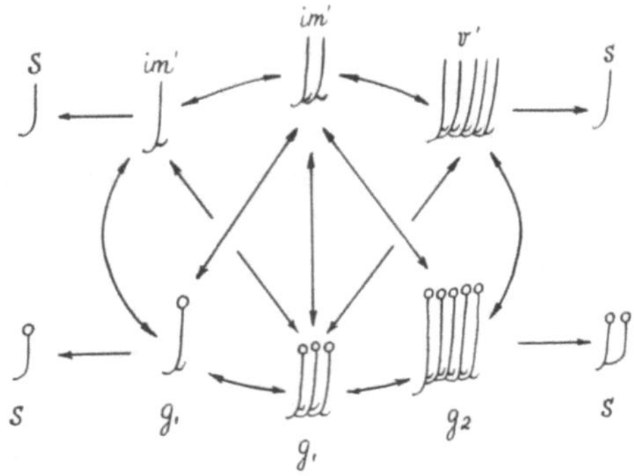

Fig. 4. Diagram of possible trends in the development of *Agrostis syreistschikowii* particules during their vegetative renewal. The upper part depicts vegetative shoots, the lower depicts reproductive shoots.

age-states (**im′**, **v′**) can pass to reproductive age-states next year and vice versa. The possible trends in particule development in the coming years are presented schematically in Fig. 4. Despite considerable variation in the number of individuals on the three plots, there is a similarity in particule age-state relationships. The **im′** and **g′** particules are most numerous (Fig. 5).

In two other species, *A. trinii* and *A. marschalliana*, taxonomically close to *A. syreistschikowii*, the relationship between particules under the conditions of high projective cover is nearly the same. Two coenopopulations of *A. trinii* were studied in the herbs-grass association in the Trans-Baikal area and a coenopopulation of *A. marschalliana* was investigated on a subalpine meadow in the Caucasus (Beshtau) (Table 2). The projective cover of the plots was 90–100% and the grass stand was 70–90 cm high. The contribution of *A. trinii* in the grass stand was abundant, 152 particules per 1 m^2. The population density was still higher (545 particules per 1 m^2) in *A. marschalliana* where the bentgrass co-dominates with other grasses in the upper grass stand layer. The populations were represented by particules of vegetative origin and individuals derived directly from

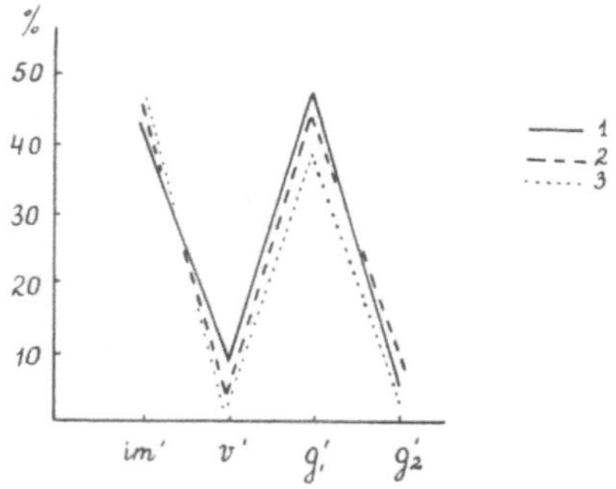

Fig. 5. Particule spectrum of *Agrostis syreistschikowii* in Central Chernozem preserve. 1, Plot I, mown; 2, Plot II, not mown; 3, Plot II, pasture.

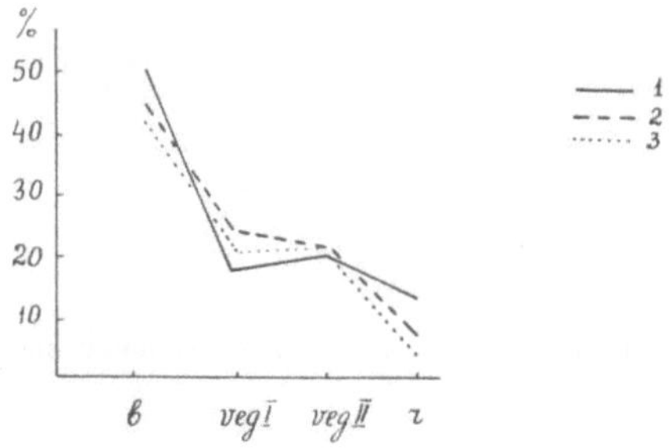

Fig. 6. Shoot spectrum of (1) *Agrostis syreistschikowii* (in Central Chernozem preserve), (2) *A. trinii* (in Trans-Baikal area) and (3) *A. marschalliana* (in subalpline Caucasus meadow). *b*, Active buds; veg I, veg II, vegetative shoots of the first and second year of life; *r*, reproductive shoot.

seed were absent. In spite of great differences in the population density of particules, the relationship between age-states in coenopopulations of *A. trinii*, *A. marschalliana*, *A. syreistschikowii* is similar. The relationship between active axillary buds, 1–2 year old vegetative shoots and reproductive shoots is also similar (Fig. 6). These characteristics can be explained by the similar external structure of shoots: metamer number, longevity of the aerial shoots and the perennial underground parts, as well as the duration and rhythm of the vegetative processes (Kurchenko 1979).

A. canina

This is the most dynamic of all species studied. Seed renewal takes place on a soil surface devoid of plants and mosses in disturbed grass stands. As a result of intensive shoot formation in early ontogeny, the tufts of *A. canina* undergo rapid development, flower in the second year of life, and die off in the third or fourth year of life. The dynamics of age composition were studied by observations on the territory of Oka-land terraced preserve (Moscow region) along the marginal land of a dry pond. In the spring of the first year of observations an exposed plot of mossy, slightly moistened soil surface was colonized by *A. canina* seeds. At the end of the vegetative period the shoots passed into the virginile state. The projective plot cover was 40%–50%. Next year most of the shoots passed into the young reproductive state and new seed individuals appeared. In the third year the young reproductive shoots passed into the senile age-state and seed-derived individuals of a third generation appeared. Thus the coenopopulation of *A. canina* was characterised by annual seed renewal and rapid plant development. The peak population density of age states shifted annually from **v** towards $\mathbf{g}_3$ (Fig. 7).

In dense grass stands, the plants undergo vegetative reproduction. The shoots are lengthened, lodge and take root at the nodes. Under such conditions, the number of reproductive shoots can only be estimated. Clone longevity is estimated at dozens of years.

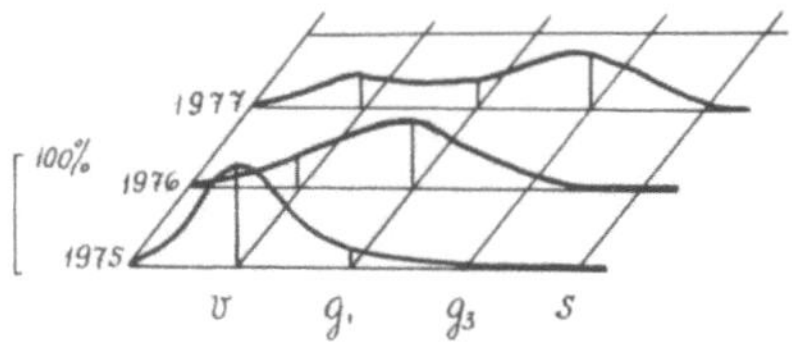

Fig. 7. Changes in age-state composition of an *Agrostis canina* coenopopulation over three years.

A. gigantea

The data on the foregoing species suggest that their population age structure and dynamics depend on the density of the grass stand. However, not all bentgrass species are related identically to the coenotic factor and *A. gigantea* is an exception to the species so far described. The age composition and population density of this species were studied on inundation meadows of the Oka river (Dedinovo v., Moscow region). Two plots were selected in the transitional and central parts of the floodplain (Table 2). The associations are of herbs-and-grass type. The projective cover on both plots was 95%–100%, with scattered bentgrass. On the first plot, the density of bentgrass individuals of seed origin is only 5 individuals per 1 m^2 and of vegetative origin 8.4; on the second plot the corresponding densities are 5.2 and 29.8 individuals respectively. In spite of high grass stand density, the coenopopulations contain not only particules but individuals of seed origin as well. The relationships between the individuals of both seed and vegetative origins were estimated separately (Fig. 8). The presence of **v** individuals in the age-state spectrum of seed individuals on the first plot indicates seed renewal and establishment of seedlings. The duration of **v** (1 year) and $\mathbf{g_1}$ (2 years) states (Table 1) can be used to explain the rise in population density of $\mathbf{g_1}$ by seed replenishment of the population two or three years earlier. On the second plot, seed renewal was recorded for the last time 3–4 years earlier (peak in $\mathbf{g_3}$ population density). The longevity of the seed-derived generation totals 5–6 years.

The relationship of particules on both plots (Fig. 8B) reveals signs of similarity, manifested in slight **v′** contribution and maximum $\mathbf{g_1'}$ population density and seems to be subject to slight annual fluctuation.

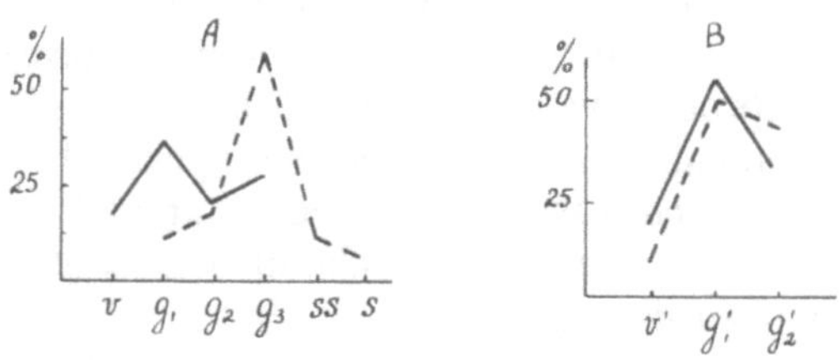

Fig. 8. Age-state composition of coenopopulations of *Agrostis gigantea* derived from (A) seed and (B) particules. Data from two separate plots are shown in each case. **v**, virginile; $\mathbf{g_1}$, $\mathbf{g_2}$, $\mathbf{g_3}$, young, mature and old reproductive; **ss**, subsenile; **s**, senile; **v′**, $\mathbf{g_1'}$, $\mathbf{g_2'}$, virginile, young and mature reproductive particules.

Microevolutionary processes

In most bentgrasses examined (except *A. gigantea*), such phenomena as seed renewal, successive development of seed individuals and dynamics in peaks of individual age-state population density are recorded in coenoses with scarce projective cover. The shift of generations or 'life waves' (Chetverikov 1905) occurs under these conditions. Thus in most of these species the manifestation of microevolutionary processes should be expected in poorly closed coenoses. Seed renewal is impeded in dense coenoses, the population density of particules is subject to slight fluctuation and they can retain territory by means of vegetative renewal for a rather long period of time.

To reveal the morphological diversity of individuals in the test coenoses we thoroughly measured different parameters in tufts of similar age. *A. tenuis* individuals in the coenosis with 50% projective cover were characterized by heterogeneity of external morphological characters and three phenotypes were distinguished (Table 3). The phenotypes differ in leaf colour, blade width and length, in the direction of renewal shoot growing, rhizome length; that is, in the totality of characters that are hereditary according to Vavilov (1922). In the past five years there has been a reduction of the first phenotype which was more widely distributed in late 1960s while the second phenotype has increased. At present tufts of the second phenotype make up 60%, those of the first phenotype 25–30%, while the tufts of the third phenotype are extremely scarce. *A. tenuis* individuals did not differ in shape or colour of blades in dense coenoses, and differed slightly in leaf length and width and in the length of the ligule and rachis.

Hybridization with *A. tenuis* was recorded in an *A. syreistschikowii* coenopopulation on a plot with 50% projective cover (Myachkovo). Apart from individuals with well-pronounced characters of *A. syreistschikowii* (awn projecting from the spikelet glume; narrow leaves pressed to the stalk; thin, rapidly decaying sheaths of dead leaves; short diageotropically directed rhizomes), individuals with *A. tenuis* characters (lost or underdeveloped awns; broad leaves coming off the stalk; broad, compact, persistent sheaths of dead leaves; 3–4 cm long rhizomes) are rather frequent there. The individuals of *A. syreistschikowii* did not differ in morphological characters in dense coenoses with 80% projective cover (Central Chernozem preserve).

Unlike these species, the coenopopulations of *A. gigantea* were characterized by regular seed renewal, in spite of high grass stand density (95–100%) and intensive competitive relations, and the individuals differed in the size of inflorescences, length and width of blades. The formation of hybrids with *A. stolonifera*, with lodging shoots, was observed.

In summary, it can be said that the diversity of morphological indicators, the formation of phenoforms and hybridization phenomena, all considered as manifestations of microevolutionary processes, are observed in most species of coenoses with low grass density.

Table 3. Characters of *A. tenuis* phenotypes

Phenotype	Leaf characters					Shoot thickness at base (mm)
	Colour	Shape	Width of III–V leaves (mm)	Length of III–V leaves (mm)	Ligule length (mm)	
I	Dark green	Thin, narrow with rolled edges	1.5–2.0	10–13	0.8	1.8
II	Light green	Flat	2.5–3.0	10–12	0.5	2.0
III	Intermediate	Flat	5.0–6.0	9–10	1.5	2.0

In dense coenoses, the plants of the test species reveal an external similarity of characters. Since evolution is slower where there is vegetative renewal, and the number of genotypes is considerably lower per unit area than in sexually reproducing generations (Sinskaya 1963), it can be assumed that the dense coenoses are 'gene pool guardians' for most of these bentgrasses.

Favourable microevolutionary conditions can be specified for each of the test species: disturbed plots, fallow ploughlands, felling areas, roadsides and quarries for *A. tenuis*; dry ridges of inundation meadows and fallow lands for *A. syreistschikowii*; moderately moist banks of drying pools and peat fields for *A. canina*; both disturbed and dense grass stands with a high percentage of projective coverage for *A. gigantea*. Under the conditions favourable for microevolutionary processes, the potential variation possibilities are different for each species. Since the rate of evolutionary processes is related to generation frequency, a species with a short life cycle and high generation frequency is likely to evolve rapidly (Timofeyev-Resovsky et al. 1977). Data on the period of tuft development from seedlings to the old reproductive state were presented in Table I. Reproductive individuals, particularly in the state characterised by the highest population density and high quality of seeds are the principal suppliers of genetic diversity. Individuals in subsenile and senile states are of no further evolutionary significance. From Table 1 it appears that the species can be placed in a sequence according to longevity and rate of development: *A. canina* followed by *A. gigantea*, *A. tenuis*, *A. syreistschikowii*, and its two allied species *A. trinii* and *A. marschalliana*. *A. canina* could have been expected to have the highest variability compared to other bentgrass species, but it is restricted by the coenotic factor: within its range, the plot area with seed renewal is not high, and under conditions of overmoistening and high grass stand density *A. canina* undergoes vegetative renewal.

REFERENCES

(English translations of Russian-language titles are given in brackets)

Alekhin, V.V. and Uranov, A.A. (1933) (Methods of steppe investigation.) Sovet. Bot. 18(2).

Chetverikov, S.S. (1905) (Waves of life.) J. Zool. Sect. Imp. Soc. Natur. Ethnogr. 3(6): 102–105.

Dobzhansky, T. (1937) Genetics and the Origin of Species. New York: Columbia University Press. 364 pp.

Gatsuk, L.E., Smirnova, O.V., Vorontzova, L.I., Zaugolnova, L.B. and Zhukova, L.A. (1980) Age states of plants of various growth forms: a review. J. Ecol. 68: 675–696.

Grant, V. (1977) Organismic Evolution. San Francisco: W.H. Freeman. 407 pp.

Khokhlov, S.S. (1947) (Origin centers or geographical tricks.) Bot. Zhur. 32(1): 33–41.

Kurchenko, E.I. (1979) (On the population-ontogenetic approach to studies of bentgrasses.) Bull. Mosk. Obsch. Isp. Prir. Otd. Biol. 84(5): 93–105.

Kurkin, K.A. (1976) (Systematic Studies of Meadow Dynamics.) Moscow: Nauka. 284 pp.

Mayr, E. (1970) Populations, Species and Evolution. Cambridge, Mass: Harvard University Press. 453 pp.

Sinskaya, E.I. (1963) (On variation categories and regularities in higher plant populations.) Problema populyatsii u vysshikh rastenii, 2. Leningrad: USSR Academy of Sciences. 124 pp.
Smirnova, O.V., Zaugolnova, L.B. and Ermakova, I.M. (Eds.) (1976) (Plant Coenopopulations: Basic Concepts and Structure.) Moscow: Nauka. 215 pp.
Stebbins, G.L. (1974) Flowering Plants: Evolution Above the Species Level. Cambridge, Mass: Harvard University Press. 399 pp.
Sukatchev, V.N. (1927) (About competition among biotypes of one species.) Reprinted in Selected Works, Vol. 3, 1975. Moscow: Nauka.
Timofeyev-Resovsky, N.V. (1925) (On phenotypic manifestation of the genotype). Zhur. Exper. Biol. 1: 3–4.
Timofeyev-Resovsky, N.V., Vorontsov, N.N. and Yablokov, A.V. (1977) (Brief Review of Evolution Theory). Moscow: Nauka. 301 pp.
Tsvelev, N.N. (1976) (Grasses of the USSR.) Leningrad: Nauka. 788 pp.
Uranov, A.A. (1975) (Age spectrum of the phytocoenopopulation as a function of time and energetic wave process.) Biol. Nauki 2: 7–34.
Uranov, A.A., Zaugolnova, L.B. and Smirnova, O.V. (Eds.) (1977) (Plant Coenopopulations: Development and Interrelations). Moscow: Nauka. 133 pp.
Vavilov, N.I. (1922) (The law of homologous series in hereditary variability.) Reprinted in *Klassiki sovetskoi genetiki*, pp. 9–50, 1968. Leningrad: Nauka.
Vavilov, N.I. (1926) (Centers of cultivated plant origin.) Reprinted in Selected Works. Vol. 1: 62–87, 1967. Leningrad: Nauka.
Yunatov, A.A. and Lavrenko, E.M. (Eds.) (1969) (Biocomplex Investigations in Kazakhstan.) Leningrad: Nauka. 335 pp.
Zavadsky, K.M. (1968) (The species and species formation.) Leningrad: Nauka. 404 pp.

THE POPULATION BIOLOGY OF EPHEMEROIDS

N.I. SHORINA and O.V. SMIRNOVA

INTRODUCTION

The ephemeroid synusia is well represented in a number of plant formations: in broad-leaved forests, *Stipa* steppes, northern and southern semi-deserts, some meadow and desert communities. The species of this synusia belong to different genera and families and differ in their habit and morphological structure. However their ecological and biological similarities make them phytocoenotically homogeneous (Bykov 1957, Yaroshenko 1961, Shennikov 1964, Rabotnov 1978). Ephemeroid vegetation grows actively in the most favourable (spring) months for moisture and light, when it avoids competition with the dominants and edificators in the upper layers of the phytocoenosis. It is believed (Kleopov 1941, Tolmachev 1960, Fedorov 1952) that in the humid areas of the temperate climatic zones the spring ephemeroid synusia originated in Tertiary deciduous forests and that it appeared much later in steppe and meadow communities, which were established mainly in the Quaternary Period.

The ephemeroid synusia can have a substantial biomass and a well developed photosynthetic surface. For example, in forest steppe oak-groves the overground ephemeroid organs make up 300–400 kg per hectare, and the leaf surface index is 0.8–0.9 (Goryshina 1969). It is clear that ephemeroids are active in coenotic metabolism and in light utilization. During spring snow melting they hinder the outwash of mineral substances from the litter and upper soil horizons (Remezov et al. 1959, Goryshina 1969).

In this paper we shall outline the population biology of this plant group in an attempt to understand the self-maintenance and stability mechanisms of ephemeroid synusiae in various types of phytocoenoses.

SPECIES INVESTIGATED

The prevalent ephemeroids of the broad-leaved forests in the USSR were studied. These include Carpathian beech forests (*Fagus sylvatica* L.) in the Ukraine, the eastern European forests of oak (*Quercus robur* L.) and

White, J. (ed.), The Population Structure of Vegetation.
 ISBN 978-94-010-8927-2

Table 1. Brief characteristics of the species studied, their geographical range and habitat.

No.	Species	Life Form	General range and ecology	Habitats studied	Number of coenopopulations studied	References
	Amaryllidaceae					
1	Galanthus woronowii Lasinsk	Bulbiferous early-spring ephemeroid	Medium-mountain forests of West Transcaucasus	West Transcaucasus forests	3	Shorina 1970 Shorina and Prosvirnina 1971 Shorina and Smirnova 1976
	Brassicaceae					
2	Dentaria bulbifera L.	Rhizomatous early-spring ephemeroid	Broad-leaved forests of Europe, Asia Minor, Caucasus, Iran.	Ukranian forests	5	–
3	D.quinquefolia Bieb.	Rhizomatous early-spring ephemeroid	Broad-leaved forests of Europe, Asia Minor, Caucasus, Iran.	Ukranian forests	5	–
	Fumariaceae					
4	Corydalis bracteata (Steph) Pers.	Tuberous early-spring ephemeroid	Forests and brushwoods of West and East Siberia, Caucasus, Asia Minor, Balkan Peninsula	Linden forests of West Siberia (Kemerovo region)	2	Bezdeleva 1972 Smirnova and Cheremushkina 1976
5	C.caucasica DC.	Tuberous early-spring ephemeroid	Forests of Caucasus	Broad-leaved forests of Stavropol Territory	1	Bezdeleva 1972 Smirnova and Cheremushkina 1976
6	C.cava Schweigg. et Koerte.	Tuberous early-spring ephemeroid	Forests of Europe, Caucasus, Asia Minor (Iran)	Broad-leaved forests of Ukraine, Stavropol Territory, Voronezh preserve	3	Bezdeleva 1972 Smirnova and Cheremushkina 1976
7	C.intermedia (L.) Merat	Tuberous early-spring ephemeroid	Forests of Europe, Asia Minor	Broad-leaved forests of Ukraine, Stavropol Territory, Voronezh preserve	14	Shorina and Smirnova 1976
8	C.halleri Willd.	Tuberous early-spring ephemeroid	Forests of Europe, Asia Minor	Broad-leaved forests of Ukraine, Stavropol Territory, Voronezh preserve	14	Shorina and Smirnova 1976

9	C.marschalliana Pers.	Tuberous early-spring ephemeroid	Forests of Europe, Caucasus, Asia Minor, Iran	Broad-leaved forests of Ukraine, Stavropol Territory, Voronezh preserve	9	Ryberg 1959
	Iridaceae					
10	Crocus scharojanii Rupr.	Bulbo-tuberous autumn-spring hysterantheous ephemeroid	Subalpine and alpine meadows of West Transcaucasaus	High-mountain meadows and waste lands of West Transcaucasus	9	Shorina 1974, Shorina 1979, Shorina and Smirnova 1976
11	C.vallicola Herb.	Bulbo-tuberous autumn-spring hysterantheous ephemeroid	Subalpine and alpine meadows of West Transcaucasus	High-mountain meadows and waste lands of West Transcaucasus	9	Shorina 1974, Shorina 1979, Shorina and Smirnova 1976
	Liliaceae					
12	Allium ursinum L.	Bulbiferous late-spring ephemeroid	Broad-leaved forests of Europe, Asia Minor	Broad-leaved forests of Ukraine	6	Starostenkova 1974, 1978, Ernst 1978
13	A.victorialis L., s.l.	Rhizomatous-bulbiferous late spring summer hemiephemeroid	Forests, meadows of Europe, Caucasus, Asia Minor, India, China, North America	Broad-leaved forests of Ukraine and West Siberia (Kemerovo region)	4	Shorina and Smirnova 1976
14	Colchicum speciosum Stev.	Bulbo-tuberous autumn-spring hysterantheous ephemeroid	Caucasus. From the low-mountain to high-mountain belts	Broad-leaved forests, medium-mountain and subalpine forests and West Transcaucasus	12	Shorina 1967, 1968 Shorina and Smirnova 1976
15	Erythronium sibiricum (Fisch et Mey.) Kryl.	Bulbiferous early-spring ephemeroid	Forests and meadows of West and East Siberia and Mongolia	Linden forests of West Siberia (Kemerovo Region)	2	Shorina and Kuklina 1976
16	Gagea granulosa Turz.	Bulbiferous early-spring ephemeroid	Forests, scrub, open slopes, ploughlands of Europe and USSR, West and East Siberia	Linden forests of West Siberia (Kemerovo Region)	2	Stamm 1954
17	G.erubescens (Bess.) Schult. et Schult. f.	Bulbiferous early-spring ephemeroid	Mid-Europe (steppe-forests and dry slopes), Caucasus, Siberia, Far East	Broad-leaved forests of Ukraine, Stavropol Territory, Voronezh preserve	1	Shorina and Smirnova 1976
18	G. lutea (L.) Ker.-Gawl	Bulbiferous early-spring ephemeroid	Europe, Caucasus, West Siberia, Asia Minor	Broad-leaved forests of Ukraine, Stavropol Territory, Voronezh preserve	12	Shorina and Smirnova 1976

Table 1. Continued

No.	Species	Life Form	General range and ecology	Habitats studied	Number of coenopopulations studied	References
19	G.minima (L.) Ker.-Gawl	Bulbiferous early-spring ephemeroid	Europe, Caucasus, West Siberia Asia Minor	Broad-leaved forests of Ukraine, Stavropol Territory, Voronezh preserve	6	Shorina and Smirnova 1976
20	Tulipa biebersteiniana Schult. et Schult. f.	Bulbiferous early-spring ephemeroid	Forest steppes of Europe, Asia Minor, Iran, Caucasus, West Siberia and Middle Asia	Broad-leaved forests of Voronezh Region	5	Bochantseva 1962
21	Scilla bifolia L.	Bulbiferous early-spring ephemeroid	Broad-leaved forests and bushes of West Europe, western regions of Europe and the USSR, Caucasus	Broad-leaved forests of Ukraine	4	–
22	S.sibirica Haw.	Bulbiferous early-spring ephemeroid	Crimea, Caucasus, Balkan Peninsula, Asia Minor, steppes and forest steppes of the European part of the USSR	Broad-leaved forests of Ukraine, Stavropol Territory, Voronezh Region	12	Smirnova 1967, Shorina and Smirnova 1976
	Ranunculaceae					
23	Anemone altaica Fisch. ex C.A. Mey	Rhizomatous early-spring ephemeroid	Forests of West and East Siberia	Linden forests of West Siberia (Kemerovo region)	2	Starostenkova 1976
24	A.coerulea	Rhizomatous early-spring ephemeroid	Forests of West and East Siberia	Linden forests of West Siberia (Kemerovo region)	1	Starostenkova 1976
25	A.ranunculoides L.	Rhizomatous early-spring ephemeroid	Europe, Ciscaucasia, Asia Minor	Ukrainian forests	12	Starostenkova 1976
26	Ficaria verna Huds.	Root-tuberous, early-spring ephemeroid	Dampish deciduous forests, lawns, short-grass meadows of Europe, Caucasus, West Siberia and Middle Asia	Ukrainian forests	4	–

hornbeam (*Carpinus betulus* L.), the beech forests (*F.orientalis* Lipsky) of West Caucasus, the Transcaucasian mixed forests of Colchis type, with *F.orientalis*, *Castanea sativa* Mill., *Carpinus caucasica* Grossh. and the linden forests (*Tilia sibirica* Bayer) in west and central Siberia. Detailed phytocoenotic descriptions of the communities investigated for each species are given in the appropriate references in Table 1, and a general account is given by Sochava and Semenova (1956). The studies were carried out in conformity with the population-ontogenetic methods developed by Soviet botanists (Gatsuk et al. 1980, Rabotnov 1950, Rabotnov 1969, Rabotnov 1978, Smirnova et al. 1976, Uranov 1975, Uranov et al. 1977). We follow Gatsuk et al. (1980) in interpreting the concepts 'individual', 'age-state', 'particule', 'particulation', 'clone'.

Ten to twenty specimens of each species were analysed for age-state characterisation. Quantitative characteristics of individual plants, such as lifetime, size, productivity and ability to propagate vegetatively were obtained from 20–100 specimens of each species. Each coenopopulation was studied along transects or on quadrats within the confines of a specific plant association (i.e., of defined floristic composition). The transects were established randomly or regularly and samples within the transects were continuous, regular or random. Isolated quadrats were selected mostly at random, less frequently on a regular basis. The numbers of samples within a particular association varied according to their size and ranged from 20 to 30, each 1 m^2, to 50 to 200, each 0.25 m^2.

The composition of the generative plant age-state group was determined without subdivision into $\mathbf{g}_1$, $\mathbf{g}_2$ and $\mathbf{g}_3$. This enabled us to avoid errors and to get comparable results for all species. The samples were collected in the flowering and fruiting stages. In those co-occurring species which are practically indistinguishable in the non-flowering state, two-species populations were studied: this applies to the pairs *Crocus vallicola–C. scharojanii* and *Corydalis halleri–C.intermedia.* Coenopopulations in a state of dynamic equilibrium with the ecotope were predominantly studied, although coenopopulations undergoing unidirectional irreversible changes were also considered in specific cases.

BIOLOGICAL CHARACTERISTICS OF THE SPECIES STUDIED

We shall confine ourselves for brevity to the characteristics of age-state ontogeny in eight species. These represent the diversity of biomorphs and versions in the life cycles of all ephemeroids examined. Figures 1–4 show the changes undergone by the morphological structure of individuals in the course of their ontogeny and the characters used for age-state differentiation, together with the possibilities and characteristics of vegetative reproduction in various species. The age changes of clones (development of supra-organism systems in time) are presented in Fig. 4, using the example of *Colchicum speciosum.* Since the ephemeroids investigated belonged to different genera, families and life forms (Table 1), it was natural to use various basic characters for age-state differentiation of their

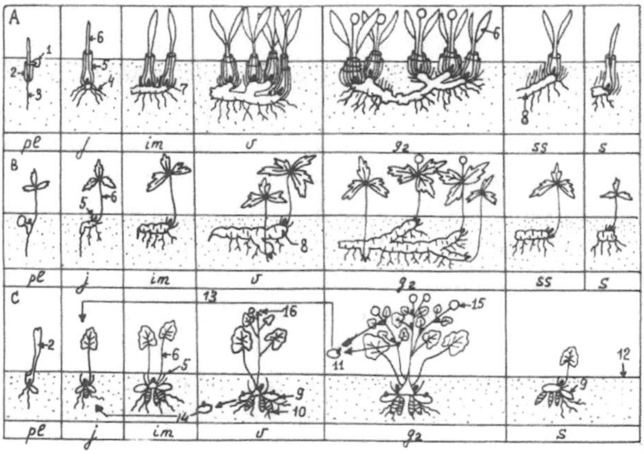

Fig. 1. Schematic representation of age-states in individuals of ephemeroids. A, *Allium victorialis*, a rhizomatous-bulbiferous hemiephemeroid; B, *Anemone ranunculoides*, a short-rhizome ephemeroid; C, *Ficaria verna*, a root-tuberous ephemeroid with bulbils above ground. 1, Remains of seed; 2, cotyledon; 3, main root; 4, adventitious roots; 5, lower leaves; 6, median leaves; 7, regenerative buds; 8, rhizome; 9, living root tubers; 10, dying root tubers; 11, bulbils; 12, surface of soil; 13 and 14 show rejuvenation of the vegetative generation from root tubers and bulbils; 15, flowers; 16, dead top of the shoot.

individuals: the number and shape of median leaves (*Colchicum*), leaf size and leaf blade shape (*Corydalis*, *Anemone*, *Dentaria*), leaf sizes (*Galanthus*, *Scilla*, *Ficaria*), shape and structure of bulbo-tubers (*Crocus*) and so on. The data presented in Figs. 1–4 are supplemented by Table 2 showing a

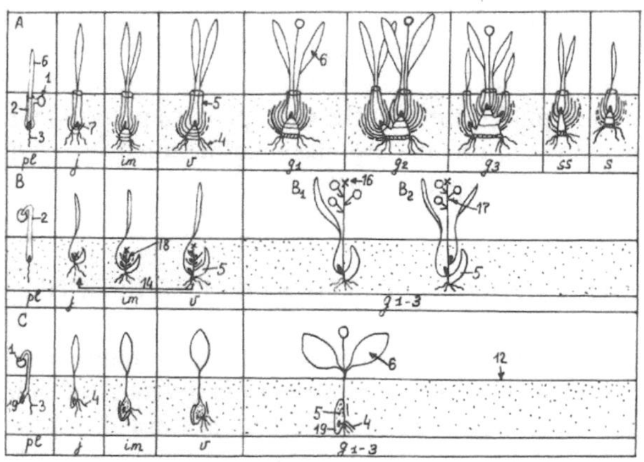

Fig. 2. Schematic representation of age-states in individuals of ephemeroids. A, *Galanthus woronowii*, a monopodial bulbiferous ephemeroid reproducing vegetatively; B, *Gagea*, a sympodial bulbiferous plant with subsoil bulbils and active vegetative reproduction; B_1, *Gagea lutea*; B_2, *G.minima*, *G.erubescens*; C, *Erythronium sibiricum*, a sympodial bulbiferous vegetative subperennial without vegetative reproduction. 17, bract; 18, sub-soil stolons with bulbils; 19, 'spur', a downward outgrowth in the lower part of the bulbotuber on which the regeneration bud is located: the spur is long in young plants, but missing in old plants. Other numbers as in Fig. 1.

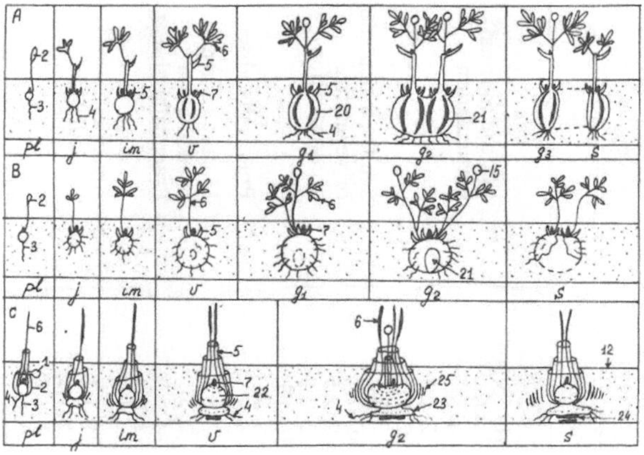

Fig. 3. Schematic representation of age-states in individuals of ephemeroids. A, *Corydalis* sect. *Pes-gallinaceus*, a sympodial tuberous ephemeroid; B, *Corydalis* sect. *Radix-cava*, a monopodial tuberous ephemeroid; C, *Crocus* sp., a tuber-bulbiferous ephemeroid. 20, tuber; 21, dying tissues; 22, tuber-bulb; 23, remainder of previous year's tuber-bulb; 24, remains of dead tuber-bulbs; 25, tunica, an accumulation of dry scales. Other numbers as in Figs. 1 and 2.

number of ephemeroid characteristics essential for self-maintenance of their coenopopulations. As may be seen from Table 2, most of the test ephemeroids are marked by vegetative reproduction, with the exception of *Crocus* and *Erythronium sibiricum* in which it is practically absent in natural populations. Like other herbaceous perennials, ephemeroids show three types of vegetative reproduction (Smirnova et al. 1976):

1. Senile particulation: formation of the ageing progeny at the end of ontogeny; *Corydalis* and *Scilla* are examples. This group of ephemeroids

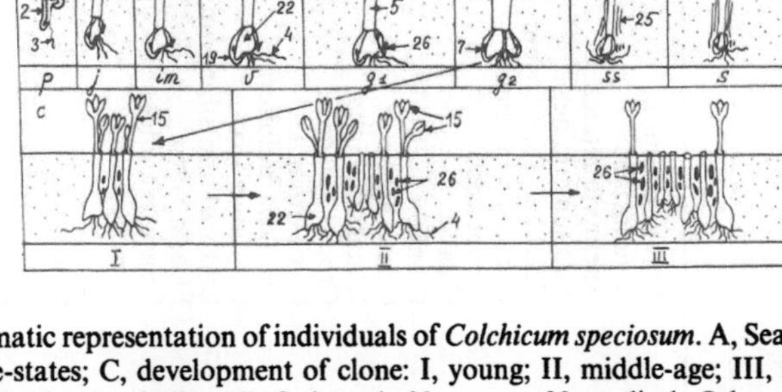

Fig. 4. Schematic representation of individuals of *Colchicum speciosum*. A, Seasonal development; B, age-states; C, development of clone: I, young; II, middle-age; III, senile. 26, The remains of dead tuber-bulbs; 27, fruit-pod; 28, ovary; 29, pedicel. Other numbers as in Figs. 1, 2 and 3.

Table 2. Biological features which are important in coenopopulation maintenance of ephemeroids.

No.	Species	Characters				
		Pregenerative period (years)	Vegetation (months)	Number of reproductive buds in one individual	Duration of the subterranean reproduction organs (years)	Presence and type of vegetative reproduction
1	Allium ursinum	4–5	1.5–2	1–2	1–2	
2	A.victorialis	7–12	3–4	24–28	10–15(20)	
3	Anemone altaica	4–6	1.5–2	2–3	4–6	Mature (normal) particulation
4	A.ranunculoides	4–6	1.5–2	2–3	4–6	
5	Colchicum speciosum	15–18	2.5–3	2	2–2.5	
6	Corydalis bracteata	3–5	1.5–2	1	2–2.5	
7	C.caucasica	3–5	1.5–2	1	2–2.5	
8	C.cava	4–6	1.5–2	1–4	10–15	Senile particulation or vegetative decomposition
9	C.halleri	3–5	1.5–2	1–4	2–2.5	
10	C.intermedia	3–5	1.5–2	1–4	2–2.5	
11	C.marschalliana	4–6	1.5–2	1–4	10–15	
12	Crocus vallicola	5–9	2–2.5	2–5	3–4	Practically absent
13	C.scharojanii	5–9	2–2.5	2–5	3–4	
14	Dentaria bulbifera	4–6	2–2.5	3–5	2–3	Mature (normal) particulation
15	D.quingefolia	4–6	1.5–2	3–5	2–3	
16	Erythronium sibiricum	4–6	1.5–2	1(2)	1.5	Practically absent
17	Ficaria verna	4–6	1.5–2	1–2	1–1.5	
18	Gagea granulosa	4–6	1.5–2	7–10(20)	1–1.5	
19	G.erubescens	4–6	1.5–2	2–5	1–1.5	Juvenile particulation
20	G. lutea	4–6	1.5–2	5–20	1–1.5	
21	G.minima	4–6	1.5–2	2–5	1–1.5	
22	Galanthus woronowii	6–7	1.5–2	2–4	3–4	
23	Scilla bifolia	5–8	1.5–2	2–4	4–8	Mature (normal) particulation
24	S.sibirica	5–8	1.5–2	2–4	4–8	
25	Tulipa biebersteiniana	5–10	1.5–2	1–2	4–8	Juvenile and mature (normal) particulation

is characterized only by particulation of individuals in **g**$_3$, **ss** and **s** age-states and the vegetative progeny is not rejuvenated since its age-state is also estimated as **ss** and **s**. This particulation type is sometimes specified as 'vegetative decomposition' (the notion and term were introduced by Senyaninova-Korchagina 1967) and contrasts with vegetative reproduction in which progeny is rejuvenated.

2. Mature (or 'normal', a term introduced by Rabotnov) particulation: formation of a slightly rejuvenated progeny (**v**, **g**$_1$, **g**$_2$) in the middle of ontogeny (in **g**$_2$, **g**$_3$ individuals). This particulation is characteristic of *Allium*, *Anemone*, *Colchicum*, *Dentaria*, *Galanthus*.

3. Juvenile particulation: formation of highly rejuvenated (**j**, **im**, **v**) progeny by mature (**g**$_1$, **g**$_2$) or young (**j**, **im**, **v**) individuals of *Gagea* species, *Ficaria verna*, *Dentaria bulbifera*, *Tulipa biebersteiniana*.

Juvenile particulation is the most perfect mode of vegetative reproduction. It is often associated with the formation of highly specialised vegetative diaspores in the form of 'brood buds' (*Gagea*, *Ficaria verna*, *Dentaria bulbifera*). Such diaspores can easily break away and spread by running water, animals, etc.: the plants acquire a substantial vegetative mobility. The morphological nature of these diaspores differs in the ephemeroids examined. In *Gagea* species it is a specific vivipary, when a rudimentary inflorescence metamorphoses into a short underground stolon with a group of secondary bulbs. *Ficaria verna* forms the dispersed tuber buds in leaf axils, with a tuber-like hypertrophied adventitious root fulfilling the storage function. *Dentaria bulbifera* is also marked by axillary brood bulb buds, where the reserve substances are localized in scaley leaves. All of these diaspores produce a highly rejuvenated, vegetative progeny (**j**, **im**), thus to a certain extent being equivalent to seeds. However, such juvenile particulation of the ephemeroids is an exception rather than a rule.

Most of the ephemeroids investigated were in fact marked by low mobility or even immobility. An exception is *Tulipa biebersteiniana*, with underground stolons of leaf nature. Vegetative reproduction is most frequent in them in the second half or in late stages of ontogeny. This is phytocoenotically defective since it does not ensure a rapid and active acquisition of space but just enables individuals to retain areas of the coenosis already inhabited.

If one tries to correlate the ontogenies of the investigated ephemeroids with the ontogeny types suggested by Gatsuk et al. (1980), then the species of *Corydalis*, *Crocus*, *Scilla* and *Erythronium* should evidently be drawn together with the vegetatively immobile, tap-rooted polycarpics, and the species of *Anemone*, *Allium*, *Colchicum*, *Dentaria* and *Galanthus* with caespitose polycarpics. In this case the terms 'tap-rooted' and 'caespitose' should not be understood literally since all adult ephemeroids examined had a secondary root system, and as a result of individual branching do not form true intact sods, but specific clones. These clones may be either dense and compact (*Colchicum*, *Galanthus*, *Scilla*, *Gagea*, *Allium ursinum*) or loose and spreading (*Anemone*, *Dentaria*, *Allium victorialis*). The first group can be compared phytocoenotically with dense sods, the second with the loose ones.

EPHEMEROID COENOPOPULATION STRUCTURE

The results of coenopopulation studies are summarized in Figs. 5 and 6 and a comparison shows that ephemeroids may be classified into four groups of species, differing in age-state spectra.

1. Coenopopulations with constant prevalence of juvenile and immature plants over the adult ones **(pl + j + im):(v + g + s)** = 8:2. They show a gradual reduction of the relative number of each age-state with respect to the previous one and typically have a low percentage of generative plants (under 10–15%), particularly in coenopopulations with high density. Coenopopulations with this spectrum occur both in forest and meadow communities. They were found both in plants reproducing only by seeds (*Crocus*, *Erythronium*) and in species with juvenile particulation (*Gagea* spp., *Ficaria verna*, *Tulipa biebersteiniana*). Actually this type represents a convergence of age-state spectra, but with different modes of coenopopulation self-maintenance both in the origin of individuals (seed or vegetative reproduction) and in plant distribution over the coenosis. With seed reproduction the distribution is relatively uniform and diffuse, while with vegetative reproduction it is aggregated in more or less distinct groups.

2. Coenopopulations with an unstable relationship between the fractions of juvenile and adult individuals. The ratio **(pl + j + im):(v + g + s)** varies from 7:3 to 4.5:5.5. These are mainly typical of the forest

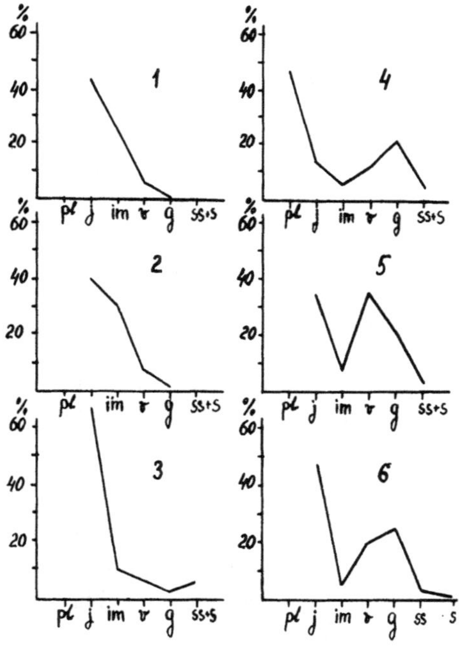

Fig. 5. Age spectra of coenopopulations (1) *Gagea lutea*, (2) *G.minima*, (3) *Crocus scharojanii* and *C.vallicola*, (4) *Allium victorialis*, (5) *Galanthus woronowii*, (6) *Colchicum speciosum*. The abscissa indicates the age-states, the ordinate the proportion of plants in each.

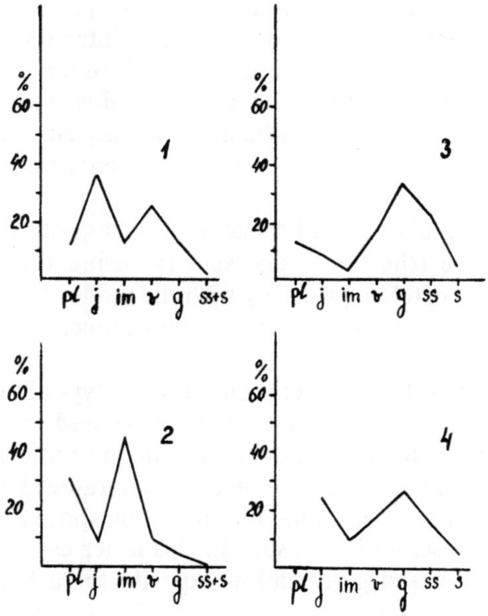

Fig. 6. Age spectra of coenopopulations: (1) *Scilla sibirica*, (2) *Corydalis solida* + *C.intermedia*, (3) *Anemone ranunculoides*, (4) *Dentaria bulbifera*.

ephemeroids having senile particulation and reproducing principally by seeds (*Scilla* sp., *Corydalis* spp.). They have more generative individuals, compared with the former type (15–30%, rarely 6–50%). The age spectra of such coenopopulations are rather dynamic in space and time. They vary strongly among samples of similar associations or even on different plots within one association and also show a rather high variability in different years (Smirnova et al. 1976). In such coenopopulations the distribution of individuals is more or less diffuse.

3. Coenopopulations with bimodal age spectra in which the participation of adult individuals increases to 40–70%. They are typical of the forest and meadow-forest species with mature particulation and mixed (that is, seed and vegetative) modes of coenopopulation self-maintenance (*Allium* sp., *Galanthus*, *Colchicum*, Fig. 5). These bimodal spectra are characterized by gradual decrease of relative abundance in the groups **pl–j–im** and then a distinct increase in **v** and **g** portions. This can be readily explained biologically: the first three age-states are only replenished through seed reproduction, while the two other groups are maintained by both seed and vegetative reproduction. In nature, these coenopopulations can easily be distinguished visually since the diffuse distribution of single individuals of seed origin is always associated with numerous compact clones. The structure of such coenopopulations is relatively stable in space, that is, the bimodal age-state spectra persist in different habitats.

4. Coenopopulations with generative or subsenile individuals prevalent are typical of the forest species (*Anemone* spp., *Dentaria* spp.) and reproduce similarly to the species of group 3. However the later onset of

vegetative reproduction and a longer generative period, compared with those previously mentioned, account for an absolute prevalence of generative and (where there is strong competition with summer herbs) subsenile individuals. The low productivity of seeds (*Anemone* spp., *Dentaria quinquefolia*) and vegetative diaspores (*D.bulbifera*), as well as the poor rooting ability of young plants result in a low representation of the younger age-states.

In summary it should be emphasized that the age-state spectrum types distinguished above (the 'basic age spectra' using the terminology of Zaugolnova 1974) correlate primarily with the modes of coenopopulation self-maintenance and correspond, in general outline, to the types of ephemeroid vegetative reproduction.

At first glance, the difference between spectrum types 1 and 2 is not quite clear. Essentially in both cases reproduction by seeds or disseminules is prevalent. The differences depend mainly on the uniformity of annual coenopopulation replenishment. It may be guaranteed by regular germination of disseminules or abundant annual diaspore formation (this is unlikely) or from reserves in the soil. In this latter case the disseminules must retain viability and germination capacity for a long time: *Crocus* seeds retain their germination capacity for 7–8 years, as do bulbils of *Gagea*. Irregular replenishment of coenopopulations also occurs where seeds, capable of even germination on moist substrates, lose their germination capacity on drying. This results in instability of coenopopulation dynamics, density and age-state structure in space and time. This is observed in coenopopulations of *Scilla* sp. and *Corydalis* sp. The self-reproduction of such coenopopulations is distinctly wave-like and even in stable coenoses they may have a considerable fluctuation in the age spectra dynamics.

The comparison of the age-state spectra shows that most of the ephemeroids investigated, with the exception of *Anemone* and *Dentaria* species, have distinctly left-handed spectra, that is, spectra in which the pregenerative or generative plants are prevalent and the senile ones are extremely rare. One of the reasons for this feature seems to be the long virginile period of life typical of these plants. Another possible reason is that some of these perennials do not repeat seed production frequently (oligocarpous): for example, *Crocus vallicola* bears seeds 3–4 times during a total life-span of 10–12 years.

Finally, most individuals of the coenopopulations do not survive until the senile state, disappearing from the coenosis in earlier age-states. The idea of such an ontogeny type in herbaceous perennials was suggested by Rabotnov (1950).

INTRASPECIFIC VARIATIONS IN THE POPULATION STRUCTURE OF *COLCHICUM SPECIOSUM*

Species such as *Colchicum speciosum* with an extensive ecological and phytocoenotic amplitude are of particular interest for investigating the intraspecific variety of coenopopulation structure. It is a tall (up to 60–70 cm) bulbotuberous ephemeroid, occurring over a wide range of altitudes from 200 m to 2000 m above sea level. It grows in the broad-leaved mountain forests, in mid-altitude meadows created by felling forests, as well as in high mountain areas, in subalpine meadows and in the so-called Colchis tall herbaceous vegetation above the timberline in West Caucasian mountains. Autumn crocus is hysterantheous (Fig. 4): it blossoms in autumn prior to the appearance over-ground of the phytosynthetic leaves, while the assimilating green shoots and fruits develop only next spring. In Abkhazia and the adjacent areas of West Transcaucasus, *Colchicum speciosum* has two ecological races, high-mountain subalpine and medium-mountain forest races (Shorina 1961), referred sometimes to a particular species, *C.liparochyadys* Wor. The two races slightly differ in their population parameters (Table 3).

These differences can be explained by the different modes of coenopopulation self-maintenance: in the forests it is mainly due to seed reproduction, in high-mountain areas to vegetative reproduction. Thus the type, distribution and structure of coenopopulation elements are changed. In forest communities, the basic coenopopulation element of autumn crocus is constituted by single individuals and their distribution is diffuse in the coenosis. In high mountain meadows, the basic coenopopulation element is the compact nest-clone (Fig. 4) (sod analog), in which 18–20 (sometimes up to 30) individuals can be found. In high-mountain populations, the single plants are subordinate and, on the whole, the distribution acquires a diffuse-compact character; that is, the more or less randomly distributed single individuals are associated with numerous compact nest-clones. It is clear that the competitive power of a compact clone is much higher than that of a single individual.

While these population parameters might favour the concept of the taxonomic distinctness of the forest race of autumn crocus, it is not supported by a more detailed analysis. On the medium-mountain compactly-sodded meadows, established in place of eradicated former forests, the autumn crocus coenopopulations become similar to those on the high-mountain meadows. The populations are relicts, remaining as a form of forest inheritance. Seed reproduction practically ceases here and is replaced by extensive vegetative reproduction, so that individuals of vegetative origin are prevalent. Consequently, the ability of the autumn crocus to reproduce vegetatively is a rather plastic character. Potentially it occurs in all coenopopulations, but is actually present only in herbaceous

Table 3. Population differences in the medium and high-mountain races of *Colchicum speciosum*.

Races	%				No. of individuals. m^{-2}		
	im	v	g	s	Juvenile	Adult	Vegetative origin
Medium-mountain	8–23	19–39	38–53	1-5	25–42	4–14	12–40
High-mountain	4–9	40–53	25–38	10–15	2.8–38	12–17	60–85

communities. The compact clones emerging in the dense meadow sod contribute to a higher competitive power of the autumn crocus. These changes in the population structure help it to persist even after forest eradication.

Favourable conditions for both the seed and vegetative reproduction of the autumn crocus are established in the post-forest loose-sodded medium-mountain meadows, as well as in the disturbed high-mountain communities. Coenopopulation density is highest here, with up to 25–30 adult plants per m^2, and it is the dominant species in the peculiar autumn-spring synusia formed by it. Hence, autumn crocus plasticity and its ability to change the mode of morphological behaviour and life history under changed conditions guarantee its persistence.

CONCLUSIONS

1. Ephemeroids present an example of phytocoenotic plant convergence, of species differing in their taxonomic position, morphology, modes of reproduction, which occupy one spatial-time niche and play a similar role in phytocoenoses. The phytocoenotic similarity of ephemeroids is based on the coincidence of their seasonal development rhythms and a relatively lower (as against plants of other coenosis layers) competitive ability.
2. The age spectra of ephemeroid coenopopulations depend on their modes of self-maintenance: the vigour of seed and vegetative reproduction, the place of vegetative reproduction in plant ontogeny, the rate of vegetative progeny rejuvenescence, the ability to form a soil reserve of seeds and vegetative diaspores.
3. Ephemeroids are characterized by full-membered, left-handed age spectra, in which the pregenerative plants predominate over the generative and senile. This can be explained by the rather long pregenerative period and the low levels of vegetative reproduction in most of the species studied.
4. Ephemeroids are marked by high plasticity of biological characteristics and are able to change their population behaviour (for instance, the type of coenopopulation self-maintenance), have a wide ecological and phytocoenotic range and can form a single-species synusia in various plant formations.

REFERENCES

(English translations of Russian-language papers are given in brackets)

Bezdeleva, T.A. (1972) (New mode of vegetative reproduction in *Corydalis* Medic. section *Pes-gallinaceus* Irmisch.) Bot. Zhur. 57(7): 356–361.

Bochantseva, Z.P. (1957) (Tulips: Morphology, Cytology, Biology.) Tashkent. 406pp.

Bykov, B.A. (1957) (Geobotany.) Alma-Ata. 381pp.

Ernst, W.H. (1979) Population biology of *Allium ursinum* in Northern Germany. J. Ecol. 67: 347–362.

Fedorov, A.A. (1952) (The history of high-mountain flora of Caucasus in Quaternary as example of autochtonal development of Tertiary flora.) Materials of Quaternary in USSR 4: 215pp. Moscow.

Gatsuk, L.E., Smirnova, O.B., Vorontzova, L.I., Zaugolnova, L.B. and Zhukova, L.A. (1980) Age states of plants various growth forms: a review. J. Ecol. 68, 675–696.

Goryshina, T.K. (1969) (Spring Ephemeroid Synusia in Oak Forests.) Leningrad: LGU. 234pp.

Kleopov, U.D. (1941) (Development of broad-leaved forest flora in European region of USSR.) pp. 183–256 in History Material About Flora and Vegetation of USSR. Moscow–Leningrad: A.N. SSSR.

Rabotnov, T.A. (1950) (Life cycle of perennial herbaceous plants in meadow coenoses.) Trudy Bot. Inst. A.N. SSSR, ser. III, 1(6): 7–204.

Rabotnov, T.A. (1969) (Some problems of the study of coenotic populations.) Biull. Mosc. Obsh. Ispyt. Prir. Otd. biol. 74(1): 141–149.

Rabotnov, T.A. (1978) (Phytocoenology.) Moscow: MGU. 207pp.

Remezov, N.P., Bykova, L.P. and Smirnova, K.M. (1959) (Consumption and circulation of nitrogen and mineral elements in the forests in European region of USSR.) Moscow: MGU. 282pp.

Ryberg, M.A. (1959) Morphological study of *Corydalis nobilis*, *C.cava*, *C.solida* and some allied species with reference to their underground organs. Acta Hort. Berg. 19(3).

Senjaninova-Korchagina, M.V. (1967) (Geophily and its significance for structure of the phytocoenosis: organism integration in higher plants.) Uchoenyje zapiski LGU. 327: 7–96.

Shennikov, A.N. (1964) (Introduction to Geobotany.) Leningrad: LGU. 315pp.

Shorina, N.I. (1961) (Forms of *Colchicum speciosum* in West Transcaucasus.) Bull. GBS AN.SSSR. 43: 71–78.

Shorina, N.I. (1967) (Life cycle of *Colchicum speciosum* in medium-mountain and high-mountain ranges of West Transcaucasus.) pp. 70–99 in (Ontogeny and Age Composition of Flowering Plants Populations). Moscow: Nauka.

Shorina, N.I.. (1968) (Age spectrum and density of populations of *Colchicum speciosum* in medium-mountain and high-mountain ranges of West Transcaucasus.) pp. 125–154 in (Questions of Morphogenesis Flowering Plants and Structure their Populations). Moscow: Nauka.

Shorina, N.I. (1970) (Pecularities of space disposition of different age individuals in coeno-population *Galanthus woronowii*.) pp. 347–369. in (The Mosaic of Phytocoenosis and its Dynamics). Vladimir.

Shorina, N.I. (1974) (Age biomorphological characters of individual saffrons in coeno-populations of high-mountain range of West Transcaucasus.) pp. 225–237 in (Age Composition of Flowering Plants Populations in Relation to their Ontogeny). Moscow: MGPI.

Shorina, N.I. (1979) (Structure of saffron coenopopulations in high-mountain range of West Transcaucasus.) pp. 50–69 in (Analysis of Composition and Structure of Flora and Vegetation). Moscow: MGPI.

Shorina, N.I. and Prosvirnina, E.A. (1971) (Pecularities of life cycle of *Galanthus woronowii* in forests of West Transcaucasus.) Biolog. Neuki. 4: 65–74.

Shorina, N.I. and Kuklina, A.G. (1976) (Biology of adder's-tongue.) Bull. Main Bot. Gard. AN.SSR. 102: 88–95.

Shorina, N.I. and Smirnova, O.V. (1976) (Age spectra of ephemeroid coenopopulations in connection of pecularities of their ontogeny.) pp. 166–200 in (Plant Coenopopulations). Moscow: Nauka.

Smirnova, O.V. (1967) (Life cycle of *Scilla sibirica* Andr.) Biol. Nauki 2: 76–84.

Smirnova, O.V. and Cheremushkina, V.A. (1976) (Genus fumitory-*Corydalis* Medic.) (Biological Flora of the Moscow Region.) 2: 48–72. Moscow: MGU.

Smirnova, O.V., Zaugolnova, L.B. and Ermakova, I.M. (Eds.) (1976) (Plant Coenopopulations: Basic Concepts and Structure.) Moscow: Nauka. 216pp.

Sochava, V.B. and Semenova-Tyan-Shanskaya, A.M. (1956) (Broad-leaved forests of the USSR.) pp. 365–440 in (The Plant Cover of the USSR: An Explanatory Text to the Geobotanical Map of the USSR), Vol. 1. Moscow–Leningrad: A.N. USSR.

Stamm, V.A. (1954) (Biology of genus *Gagea*.) Bull. Main Bot. Gard AN.SSSR. 19: 90–95.

Starostenkova, M.M. (1974) (Life cycle of *Allium ursinum* L.) Biol. Nauki. 74–82.

Starostenkova, M.M. (1976) (Genus windflower-*Anemone* L.) (Biological Flora of the Moscow Region) 3: 119–133. Moscow: MGU.

Starostenkova, M.M. (1978) (Ramson – *Allium ursinum* L.) (Biological Flora of the Moscow Region). 4: 52–61. Moscow: MGU.

Tolmachev, A.I. (1960) (The role of migration and autochtonal development in the formation of high-mountain floras of the World.) Prob. Bot. 5, 275pp.

Uranov, A.A. (1975) (Age spectrum of phytocoenopopulation as a function of time and energy wave processes.) Biol. Nauki 2: 7–34.

Uranov, A.A., Zaugolnova, L.B. and Smirnova, O.V. (Eds.) (1977) (Plant Coenopopulations: Development and Interaction.) Moscow: Nauka. 133pp.

Yaroshenko, P.D. (1961) (Geobotany.) Moscow–Leningrad: A.N. SSSR.

Zaugolnova, L.B. (1974) (Connection between age state composition of population and biological characters of species: the model of *Acantholimon diapensiodes* Bge). pp. 38–57 in (Age State Composition of Flowering Plants Populations in Relation to their Ontogeny). Moscow: MGPI.

THE DEMOGRAPHY OF COENOPOPULATIONS OF FOREST HERBS

K. FALIŃSKA

ABSTRACT

The diversity of population dynamics among and within species is due to their biology and to the type of ecosystem in which they grow. The demography of coenopopulations of three herbaceous species is described, to illustrate the variety of population dynamics of forest herbs. Changes in demographic properties during the course of regenerative succession from a meadow to forest are discussed.

INTRODUCTION

The herb layer of forest vegetation is composed mainly of perennial herbaceous plants and undershrubs, with a constant but not high participation of annual and biennial plants. The herb layer under stable conditions has a characteristic mosaic spatial structure for a given type of forest ecosystem. Its contiguous spatial structure reflects the commonness of vegetative reproduction among forest herbs, and their morphological and biological properties. Monospecific aggregations in the herb layer result either from a group of individuals developing from seeds or from a clone arising by vegetative diaspores rooting around the maternal plant. They may also be individuals with many shoots and roots, that is *polycormones*. Individuals of perennial herbaceous plants expand their underground organs and multiply their aboveground shoots to form clusters of varying size and permanence.

Evaluation of the dynamics of a population of individuals consisting of many basic units causes numerous census difficulties. Frequently such investigations are based on what is called 'comparative units' which, as a rule, are aboveground shoots or leaf rosettes. In this paper I use the following definitions:

(1) An individual is a morphologically and physiologically uninterrupted organism, irrespective of how many basic units (modules) form it. It is an ontogenetic unit which arises from one generative *or vegetative*

White, J. (ed.), The Population Structure of Vegetation.
© 1985, Dr W. Junk Publishers, Dordrecht. ISBN 978-94-010-8927-2

Fig. 1. (a) One year-old *Filipendula ulmaria* individuals: 1, juvenile; 2, generative. (b) Two year-old *Filipendula ulmaria* individual (polycormone).

diaspore. An example of the development of such an individual is shown in Fig. 1.

(2) For individuals composed of many rooted aboveground shoots, joined together by an extensive system of rhizomes, the name *polycormone* is used. An example based on *Filipendula ulmaria* is shown in Figs. 1b, c and 2.

(3) The term *clone* is reserved here for a set of *independent* individuals of vegetative origin. The interruption of the morphological link between

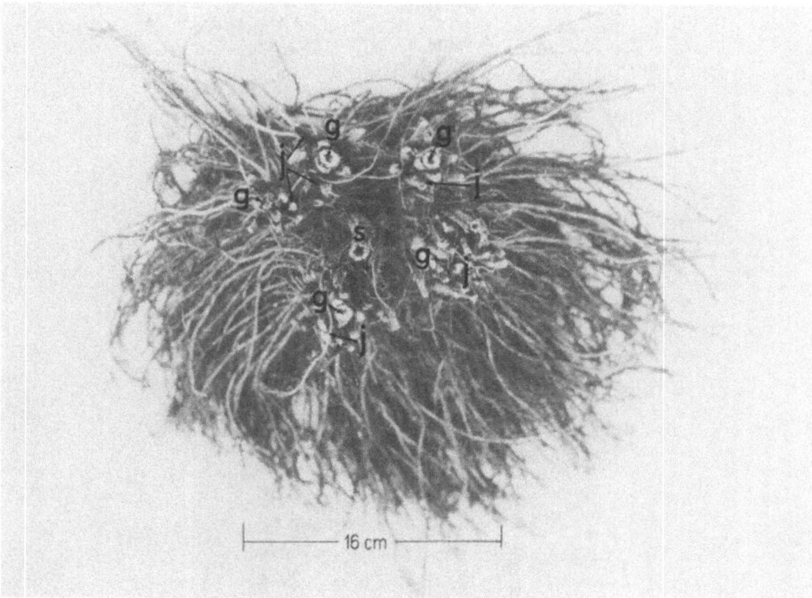

Fig. 1. (c) Underground organs of a 2 year-old *Filipendula ulmaria* individual (polycormone): the centre is occupied by a senile shoot (s) which was reproductive in the previous season; g, generative (reproductive) shoots; j, juvenile leaf rosettes.

new units and the parental specimen serve as a criterion for distinguishing a clone.

(4) In demographic analyses of the populations, both the change in the number of individuals (*N*) in the biological sense and the change in the number of shoots (*n*), that is, units of which they are formed, were taken into account. In the demographic characterisation of the population the numbers of individuals and of shoots were treated as complementary data.

The diversity of population phenomena among and within species is due both to their biology and to the type of ecosystem in which they grow. This diversity is demonstrated in this paper using examples of several species considered to be representative of the population dynamics of forest herbs: *Mercurialis perennis* L., *Filipendula ulmaria* L. and *Geranium robertianum* L.

METHODS

The methodological basis of the present investigations may be stated as follows:

(1) Maximum advantage of the environmental diversity is taken by conducting parallel investigations in all types of ecosystems in which

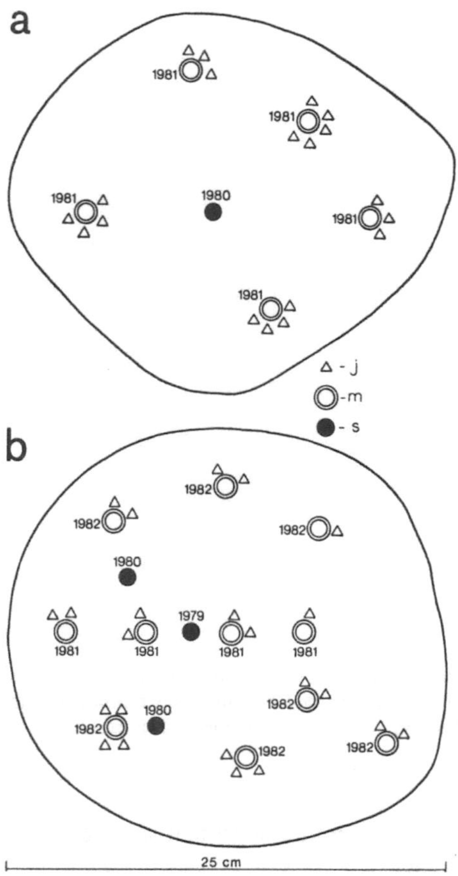

Fig. 2. Distribution and exchange (substitution) of shoots in a 2 year-old (a) and 3 year-old (b) polycormone of *Filipendula ulmaria*. j, juvenile leaf rosettes; m, mature shoots (virginile and generative); s, senile shoots, i.e., those in which renewal buds do not form. The year of appearance is marked on the shoots.

a given species grows; such conditions are available in the Białowieża Forest, northeast Poland.

(2) Analysis of population dynamics is made by following the fate of marked individuals and their progeny over as long a time period as possible.

(3) Individual variability is assessed by growing specimens in uniform conditions in an experimental garden.

The studies were performed during the period 1972–1983 in various forest and shrub communities: oak-linden-hornbeam Tilio-Carpinetum, flood-plain Circaeo-Alnetum, bog-alder Carici elongatae-Alnetum and bog willow scrub Salicetum pentandro-cinereae. Depending on the type of ecosystem and on the biology of the species, plots of various sizes were chosen: 100, 200, 400, 800 m^2 and divided into 1, 2, 4, 16, 25 m^2 squares.

The network of squares was used for mapping the distribution of the polycormones both seasonally and over many years. An example of the mapping procedure is shown in Fig. 3. The demography of populations was investigated by observing marked (tagged) individuals from seedling to vegetative diaspore phases, and by mapping the plots to determine the turnover rate of individuals. In order to become acquainted with the rate of development of some chosen individuals in each season, maps were prepared on a 1 : 10 scale of the number, development phase and distribution of marked shoots, as shown in Fig. 2. Analogous investigations were performed in garden culture using the appropriate forest soil and optimal irrigation of the plots. Observations in the garden also made possible a fuller recognition of the morphological and developmental changes associated with the aging of individuals.

MERCURIALIS PERENNIS COENOPOPULATIONS

M.perennis is one of the few dioecious species in the herb layer of forest communities. It mainly reproduces vegetatively. Its underground parts consist of adventitious roots, renewal shoots and communication rhizomes; the latter link the distant clusters of aboveground shoots. The intensive spread of polycormones with simultaneous vegetative propagation leads in a relatively short time to obliteration of their distinctiveness. Owing to the difficulties in distinguishing individuals, rooted aboveground shoots are frequently treated in natural populations as comparative units (Mukerij 1936, Hutchings and Barkham 1976, Toropova 1977, Falińska 1979a). This species is convenient for investigations on the biology of polycormone development. The purpose of this study was to ascertain the rate of growth and development of individuals and to investigate why they lose their independence. The evaluation was done on the basis of the following observations: (a) the manner in which a new individual arises, its size, and the variability of the area it occupies over a period of several years; (b) dynamics of the number of shoots, with reference to their developmental phases; (c) the durability of the rhizomes joining the single shoots or clusters of aboveground shoots, that is, the permanence of the individual as a polycormone. It was hoped that the observations would reveal both the internal dynamics of discrete polycormones and the overall coenopopulation dynamics.

Individuals of *M.perennis* preserve their distinctiveness above all in the early developmental phases of the population. Therefore, the study plots were established in the period of penetration of this species into the Tilio-Carpinetum herb layer. In a patch of this forest, where in 1974 development of the first individuals of *M.perennis* had been noted, two study plots were established. These study plots of 400 m^2 were divided into 4 × 4 m^2 plots and numbered (Fig. 4).

Over nine seasons (1974–1982) polycormones were mapped in a 1 : 100 scale as they developed on the particular plots (Fig. 4). For detailed investigations on the development and growth of polycormones 30 of

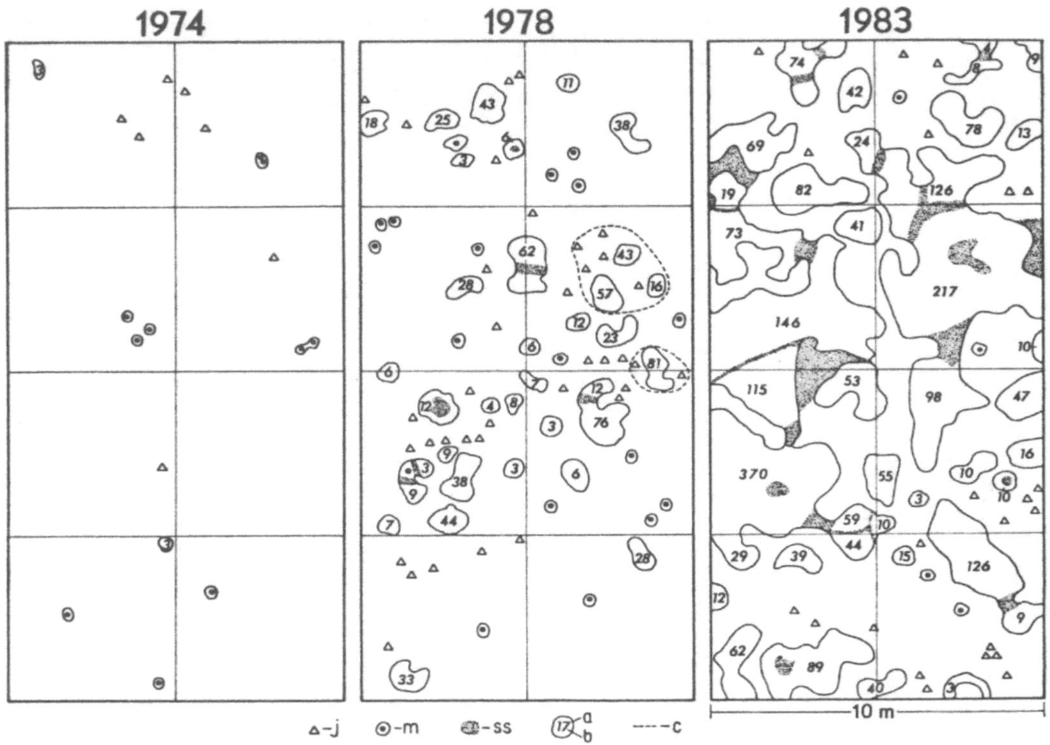

Fig. 3. Changes in the spatial structure of *Filipendula ulmaria* coenopopulations over several years. j, Juvenile individuals in the form of a leaf rosette; m, one year old individuals, with one generative shoot; a, boundary of a polycormone; b, number of generative shoots in a polycormone; c, boundary of clone; ss, senile and dying rhizomes.

them were labelled on the plots. In each vegetative season these individuals were mapped in a 1 : 10 scale and the distribution of juvenile, virginile (mature but not flowering) and generative shoots was marked.

Development and growth of polycormones

The development cycle of *M.perennis* polycormones lasts about 6–8 years. Under natural conditions the development of new individuals starts by detachment of a fragment of the rhizome containing renewal buds or juvenile shoots. Usually this takes place in summer or autumn. When the rhizomes root, the development of aboveground shoots begins, usually in spring. In the first and second seasons the number of shoots in the polycormone doubles whereas in the third and fourth season the ratio of the number of new shoots to those of the previous year is 3 : 1, 4 : 1 and even 6 : 1. In the seventh season after that in which the individual first arose a relative stabilization is observed in the number of shoots in the polycormone: it varies by this time within a range 80–200; generative shoots constitutes 50–70%. Usually polycormones are of spherical or oval shape (Fig. 4). The size of the polycormone area varies within a range 0.5–2 m^2. In this phase the process of dying back of the central part of the polycormone frequently occurs, with a simultaneous increase of the number of juvenile shoots on the periphery. This process leads gradually to a disappearance of the monocentric structure of the polycormones.

In natural populations individuals mainly of vegetative origin were noted. Those developing from seeds are very rare. Over five years on an area of 800 m^2 only 120 seedlings were recorded, of which as few as 8 reached the juvenile phase.

Age structure

The age structure of the *M.perennis* was determined on the basis of a calendar (1974–1982) of development of 92 polycormones (49 male and 43 female) in an area of 800 m^2. The largest group consisted of 4–6 year old polycormones (Fig. 4). In the oldest polycormone age classes these were 19% of the total; these had been noted in the herb layer in 1974 and classified at that time in the group of 1–2 year old individuals. The age pyramid by 1982 is characteristic of a developing population already passing to the stabilisation phase.

According to observations of the development of individuals in garden culture and under natural conditions five age groups of *M.perennis* polycormones were distinguished (Fig. 5).

- I Juvenile polycormones (1–2 year old). They occupy an area of 0.10–0.25 m^2 and are formed of 2–20 shoots. When a new individual arises by detachment of a fragment of a mature polycormone, it reaches the generative phase in the first or second season. If it develops from renewal buds and juvenile shoots, this phase is attained in the second or third year of life.
- II Maturing polycormones (2–3 year old) exhibit a dominance of

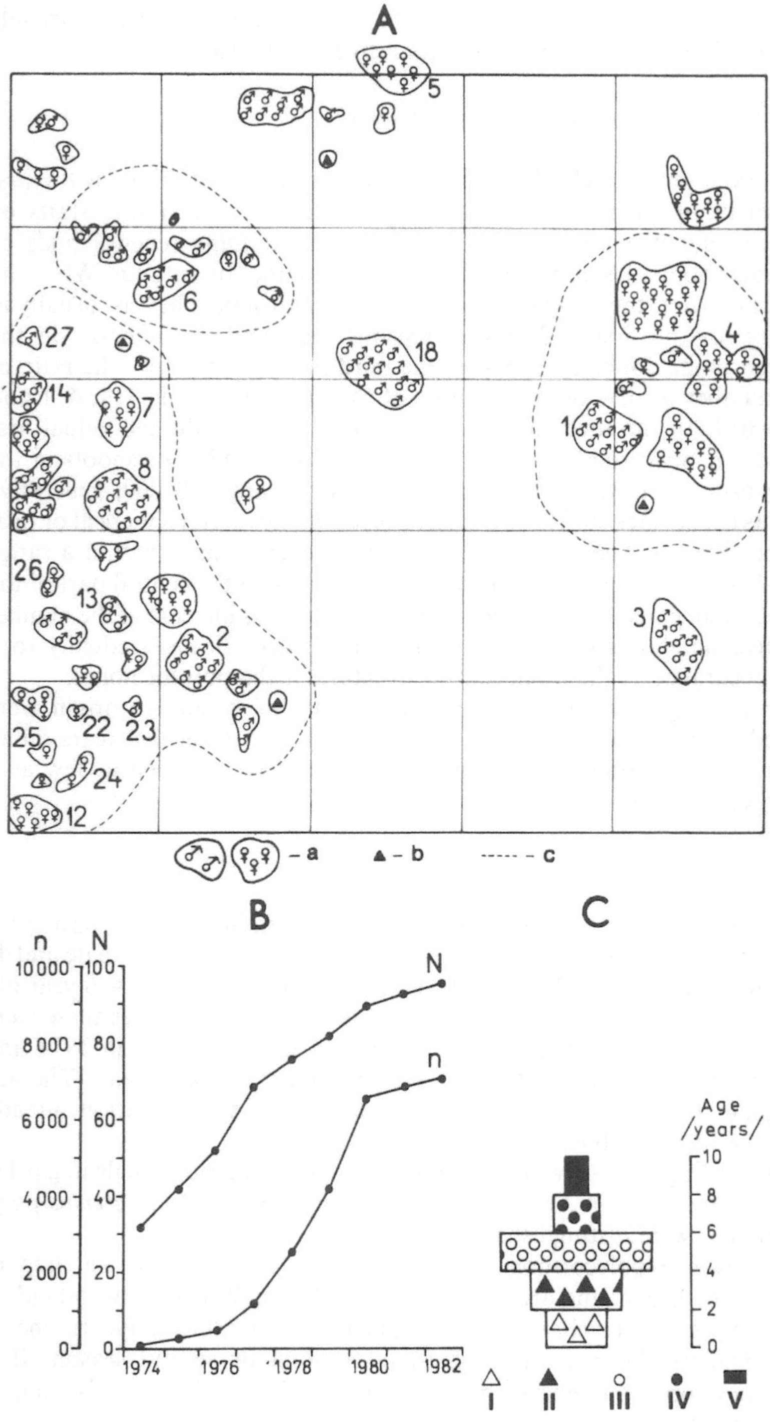
A
5
6
27
18
4
14
7
1
8
20 m
26
13
3
2
25
22
23
24
12
– a
– b
– c
B
C
n
N
10000
100
8000
80
6000
60
4000
40
2000
20
0
1974
1976
1978
1980
1982
Age
/years/
10
8
6
4
2
0
I
II
III
IV
V

juvenile shoots. The area of polycormones varies within the limits of 0.25–1 m^2 and the number of shoots within 20–100.

III Mature polycormones (4–6 year old) occupy an area of 1–3 m^2 and produce 100–200 shoots, including 30–60% of generative ones.

IV Senile polycormones, characterised by a similar area as those in group III, but from many of their parts no further aboveground shoots appear. In the area occupied by polycormones more and more empty spaces appear and the number of juvenile shoots seldom exceeds 10%. In the underground parts the process of dying prevails over renewal.

V Polycormone disintegration.

Spatial structure of polycormones and populations

The polycormones change their shape and size with age, thus affecting the dynamic character of the spatial structure of the *Mercurialis perennis* population. During penetration of this species into the Tilio-Carpinetum herb layer the population exhibits a random distribution of polycormones. On two plots of 400 m^2 32 polycormones began to develop in 1974. They mostly produced 2–3 aboveground shoots. The rapid growth of the polycormones in the first three years led to a contiguous spatial structure. Owing to the intensive development of the polycormones in the successive years, the processes of their disintegration and of coalescence of the newly formed polycormones occurred almost simultaneously (Fig. 4).

According to the biology of polycormone development and earlier studies on this species (Falińska 1979a, 1982) it is concluded that compact one-species patches of *Mercurialis perennis* take ten years to form in forest communities.

FILIPENDULA ULMARIA COENOPOPULATIONS

Filipendula ulmaria is a species with a broad ecological amplitude and a considerable morphological variability. It is a component of meadow, shrub and forest communities. As a species derived from the forest (that is, a forest apophyte) it reaches the greatest abundance beyond its native community which is riverside forest (Circaeo-Alnetum). The highest density of populations of this species is found on the contact zone between forest and meadow (Falińska 1979a). The morphological and biological properties of *F.ulmaria* and its varying participation in the formation of

Fig. 4. Properties of *Mercurialis perennis* coenopopulation in Tilio-Carpinetum coenosis. (A) Spatial structure on a plot of 400 m^2: a, male and female polycormones; b, juvenile individuals arising from seeds; c, polycormone integration (underground organs); numerals, polycormones chosen for observation of development biology. (B) Abundance dynamics of a population on two plots of 400 m^2. N = number of polycormones; n = number of shoots. (C) Age structure of polycormones with their development phase attained by this time: I, juvenile; II, maturating; III, mature; IV, senile; V, senile and dying (polycormone distintegration). The distribution of polycormones in 1980 is shown in A.

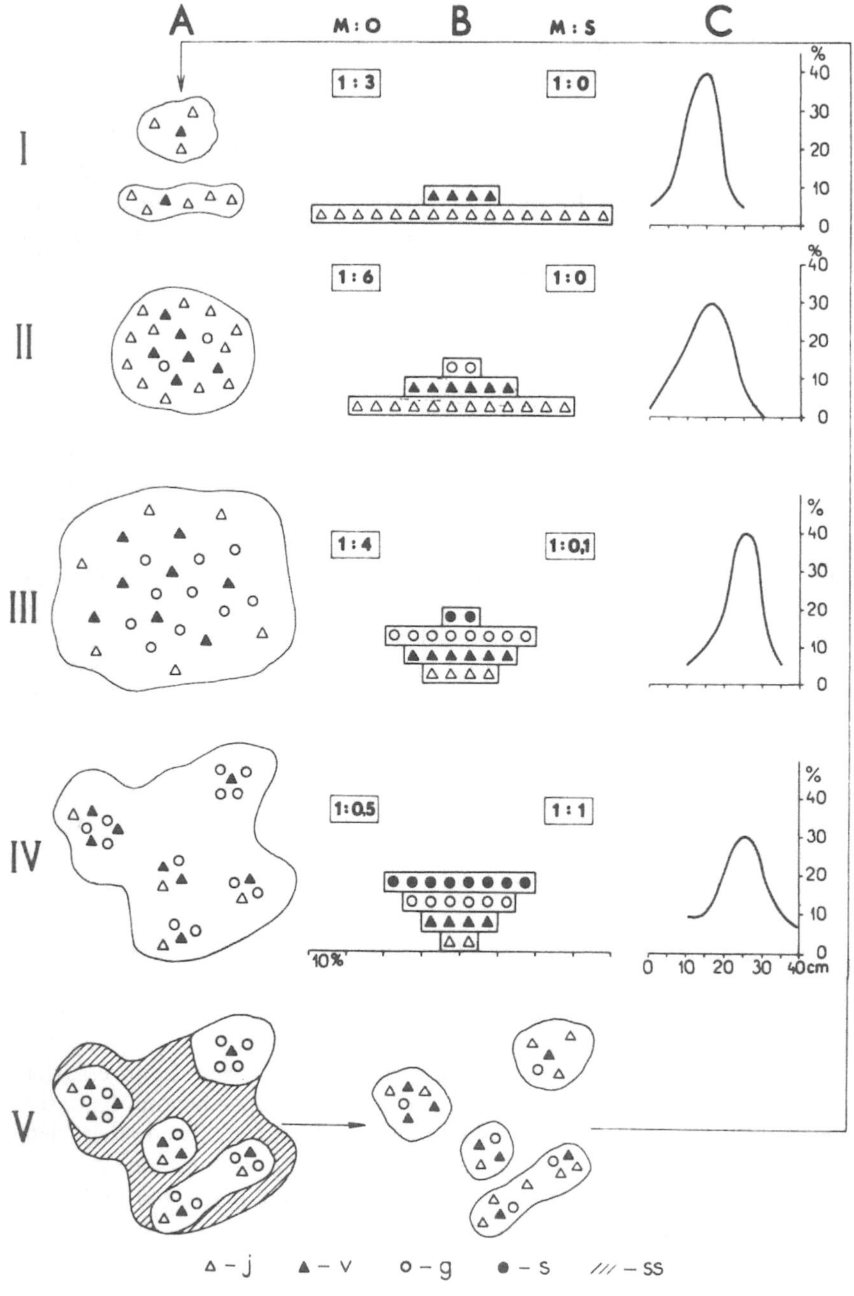

Fig. 5. Model of *Mercurialis perennis* polycormone development. Five phases are shown: I, juvenile; II, maturing; III, mature; IV, senile; V, polycormone disintegration; ss, senile and dying rhizomes. (A) Polycormone area and distribution of above ground shoots and dying parts. (B) Pyramids of developmental phases of shoots: j, juvenile; v, mature but not flowering; g, generative; s, senile. (C) Shoot size structure in polycormone. M : O, ratio of shoots to renewal buds; M : S, ratio of mature (v, g) to senile shoots (Falińska 1982).

meadow and forest phytocoenoses offer a possibility of finding answers to the following questions:

(1) What determines the different sizes of the populations in meadow, scrub and forest, developing on fertile and moist habitats; is it the different life conditions or are these populations in different development phases and, accordingly, differentiated as regards abundance?
(2) What is the fate of a population developing on a particular site on which a sequence of phytocoenoses is observed in the course of regenerative succession of forest?

To answer these questions investigations were undertaken on coenopopulations in neighbouring meadow, shrub and forest communities. The fates of *F.ulmaria* populations were also followed from birth over the successive development phases in the course of regenerative forest succession from moist meadows to the formation of a willow-bog scrub Salicetum pentandro-cinereae (Fig. 6). Spatial structure, age structure and abundance dynamics were monitored. Observations on the biology of polycormones, as a widespread form of individuality in plants, were conducted to elucidate their role in the development and fate of the coenopopulation. The investigations were performed during the period 1974–1983 in the centre of the Białowieża Forest complex in the valley of the River Narewka, deforested for 200 years, in the neighbourhood of the Białowieża National Park.

Biology of polycormone development

Filipendula ulmaria individuals under natural conditions develop mainly from seedlings which appear in May and at the end of the season attain the juvenile phase as a leaf rosette. In the next growth season 20–30% of the plants reach the generative phase and produce one shoot with a small inflorescence. Beginning with the second year of life 80–90% of the individuals produce generative shoots and the numbers of such shoots increase subsequently in a geometrical progression. As a rule this process lasts up to the fourth or fifth year of life, but afterwards wide differences were observed in the course of the life cycle.

In the developmental cycle of the polycormones, which live on the average for 8–10 years, four phases were distinguished:

I juvenile phase, lasting one or two seasons. In this time the seedling develops to a leaf rosette with a well developed root (Fig. 1a);
II phase of maturity, lasting from the second to the fifth year of life; in successive seasons the individuals double or treble the number of generative shoots and new rhizome segments (Fig. 1b, c);
III the subsenile phase begins in the fifth or sixth year of life of the individual. The rate of renewal decreases in the polycormones, while the rate of senile processes increases. Because of this the process of expansion is inhibited;
IV the senile phase is characterised by the prevalence of processes of gradual dying over those of renewal, leading to disintegration of the polycormone and formation of a clone.

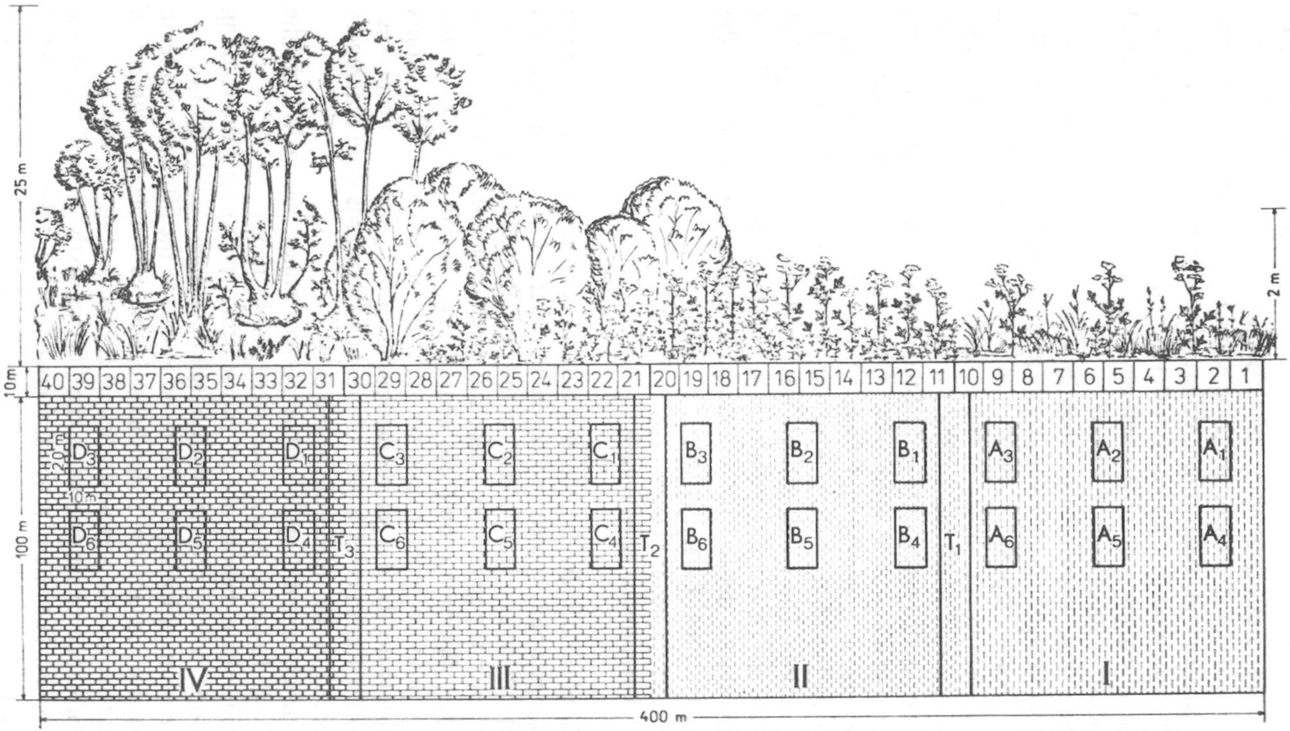

Fig. 6. Plan of study area with permanent plots in communities with various degrees of advancement of regenerative succession from meadow to forest. Each plot (A–D) occupies 200 m^2 and is divided into 5 × 5 m squares; T_1–T_3 are 100 m long transects, 10 m wide, between two neighbouring communities.

The properties of the polycormones decide the spatial (Fig. 7) and age structure of the population. The durability of the underground organs and the ability of polycormones to multiply vegetatively ensure the persistence of the population in the course of regenerative succession from abandoned meadows to forest. The exponential rate of expansion of the individuals rapidly fills the area in 5–7 years. Because of the properties of its polycormones, *F.ulmaria* is a community-forming species, determining or organising the structure of this community in the first phase of regenerative succession towards forest (Fig. 8).

Population phases

The population size of *F. ulmaria* differs depending on the type of community in which it occurs and also on the time for which the meadow has been abandoned (Fig. 9).

On the basis of ten years' observations of the population development on meadows abandoned at various known times, the following phases were distinguished:

I an increase in the number of polycormones close to a logistic model, and of the shoots close to an exponential model. This phase lasts on the average 4–6 years;

II the increase in the number of polycormones is inhibited, but the increase in the number of shoots continues in a geometrial progression. Monospecific patches arise, the formation of which lasts 6–10 years;

III the number of polycormones diminishes rapidly, and the number of aboveground shoots declines at a much slower rate;

IV the population size stabilised at a low level, with changes in the number of shoots only slowly fluctuating.

These changes may be described in more detail as follows:

Phase I – The period of settlement and increase in numbers of the *F.ulmaria* population lasts on average 4–6 years. As a rule in the second year after cessation of mowing the meadow, the first individuals appear. In each subsequent season the number of individuals may either double in relation to the total population size or increase by as little as 30–50%. As a consequence of this the number of polycormones in the fifth year of population development on an area of 200 m^2 ranges from 61 to 121, and the number of shoots from 450 to 843.

Phase II – The second period of population development is characterised by a rapid filling of the settled surface. The increase in shoot number approaches the exponential model with a relatively constant number of polycormones in the population. The number of shoots on particular plots varies within the limits of 3011–5388 per 200 m^2. At this time *F.ulmaria* forms almost pure one-species patches with a high shoot density and a poorly-outlined distinctiveness of polycormones.

Phase III – The third period of development of the population shows a considerable decrease of the number of polycormones and of shoots. At this time the oldest polycormones, constituting 50–70% of the population, die. These are polycormones which developed in the first period after

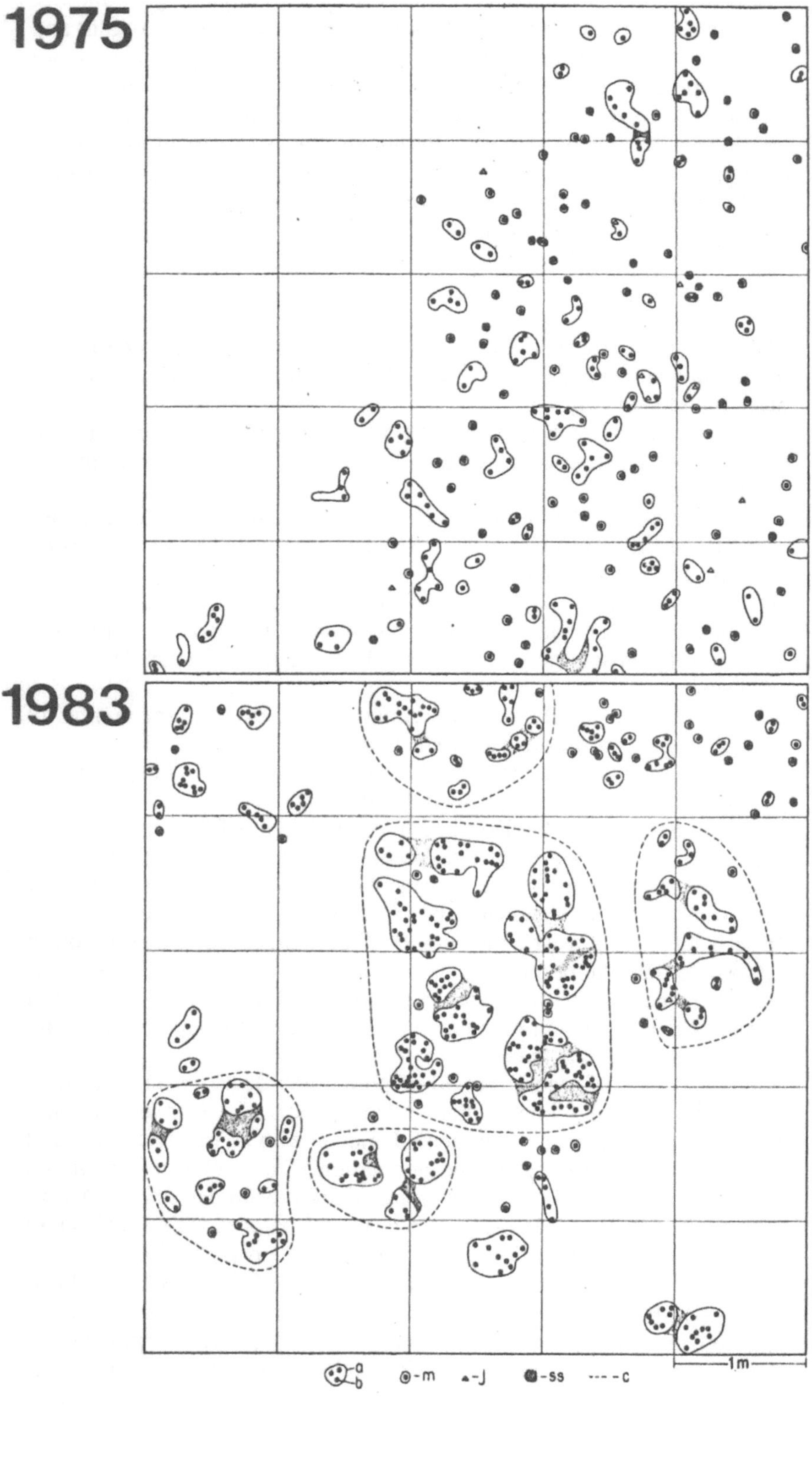
1975
1983
a
b
-m
-J
-ss
----c
1m

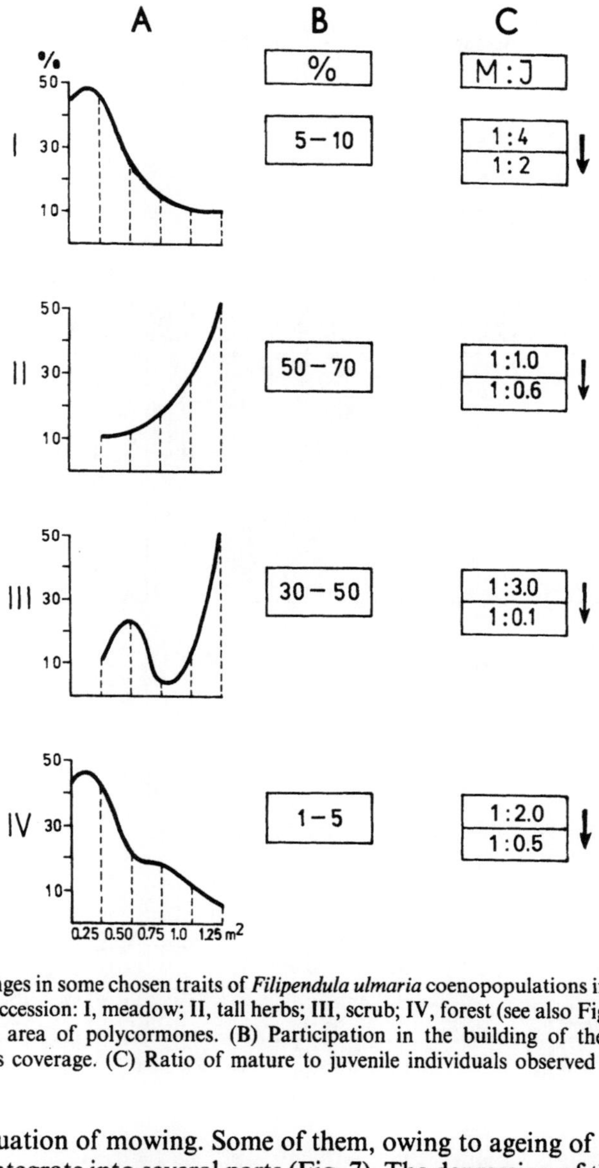

Fig. 8. Changes in some chosen traits of *Filipendula ulmaria* coenopopulations in consecutive phases of succession: I, meadow; II, tall herbs; III, scrub; IV, forest (see also Fig. 6). (A) Size distribution area of polycormones. (B) Participation in the building of the community expressed as coverage. (C) Ratio of mature to juvenile individuals observed in particular phases.

discontinuation of mowing. Some of them, owing to ageing of the central part, disintegrate into several parts (Fig. 7). The depression of the population size is accompanied by an increase in the participation of shrub willows and of other tree species.

Phase IV – This period is characterized by stabilization of the number of polycormones and shoots. On 200 m^2 plots 17–25 polycormones grow at this time, producing jointly 74–143 generative shoots. They form small aggregations distinct in the forest herb layer (Fig. 9).

←

Fig. 7. Size and distribution of *Filipendula ulmaria* polycormones on a plot of 25 m^2. a, Polycormone boundary; b, generative shoots in polycormone; c, clone boundary; j, juvenile individuals; m, mature individuals with one or two generative shoots; ss, dying part.

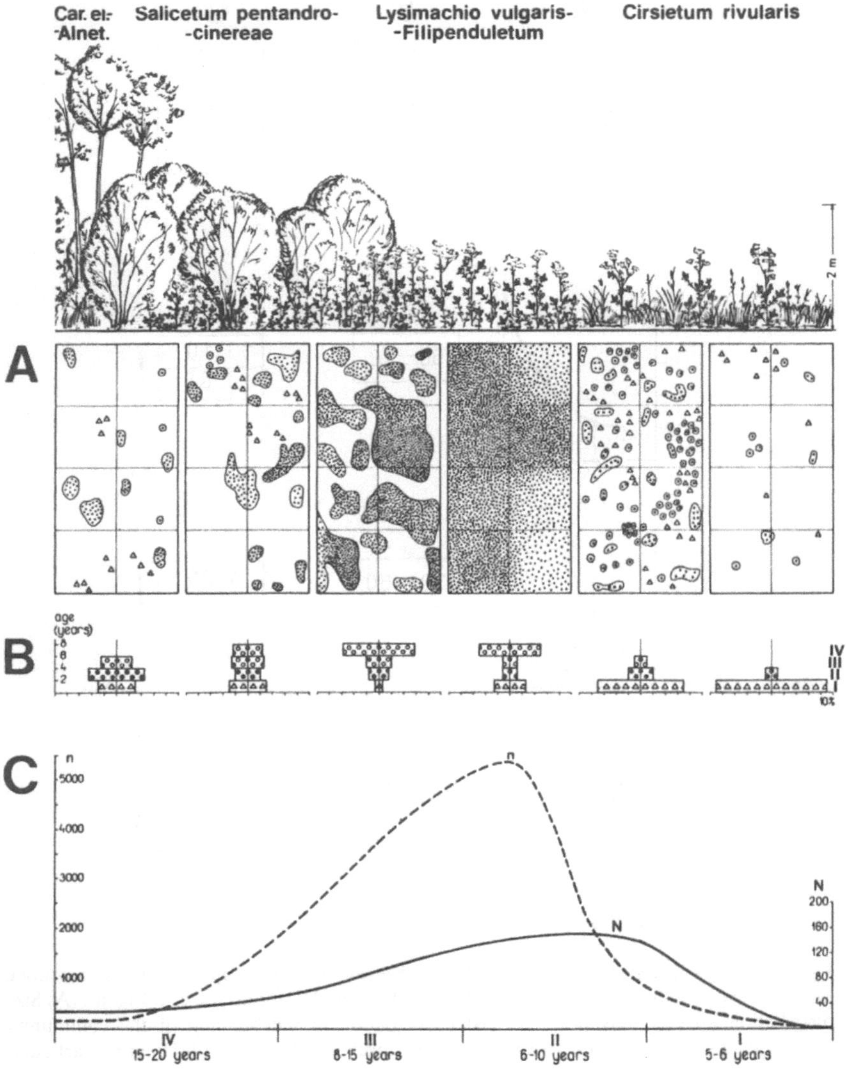

Fig. 9. Coenopopulation development of *Filipendula ulmaria* against the background of vegetation transformations. (A) Spatial structure of populations on 200 m^2 plots. (B) Age structure and development phase attained at this time by polycormones: I, juvenile; II, mature; III, subsenile; IV, senile. (C) Abundance dynamics of a population: N = number of polycormones; n = number of shoots; I–IV, population phases (see text).

The fate of the population, against the background of vegetation transformations occurring after cessation of mowing, indicates that 15–27% of the first individuals settling on the meadows and labelled in 1974 survived to 1983 on particular plots. In this period changes in numbers, age structure and spatial structure of the population (density and distribution of polycormones and of shoots) were observed. The phases of growth, stability and diminution of the *F. ulmaria* populations were accompanied by a succession of consecutively developing communities from moist meadows to the formation of willow-bog scrub.

The question posed at the beginning of the investigations, whether in the course of regenerative forest succession the same *F.ulmaria* population is present may be given a positive answer. Ten years' observations of labelled individuals demonstrated that, at least to the time of formation of the willow-bog scrub, we are dealing with the same population. Comparison of the age structure of the willow shrub population and the riverside forest population of *F.ulmaria* leads, however, to the assumption that, because of the longevity of the individuals and their ability to reproduce themselves vegetatively, the continuity of the population is preserved from the meadow to the formation of forest communities. With the development of the population its role in the building of the consecutive phytocoenoses formed in the course of succession changes (Fig. 9).

GERANIUM ROBERTIANUM COENOPOPULATIONS

Geranium robertianum is a species with a variable life history and proved an interesting object for investigations on the following problems: (1) the causes of variability in the life span of the individual; (2) with prolongation of life of the individual, the changes occurring in reproductive patterns, from mono- to polycarpic; (3) does such variability involve usually all individuals of the population (then characterized as either mono- or polycarpic), or does it concern only some of the individuals, such that particular populations differ in the fractions of mono- and polycarpic individuals and in the participation of annual and biennial plants?; (4) is simultaneous flowering and fruiting a result of more and more individuals reaching the generative phase, or do individuals after reaching the generative phase continue to flower and fruit throughout the season. Another purpose of the study was to determine how the biology of individuals affects the spatial structure and the abundance dynamics in seasonal and yearly cycles in Tilio-Carpinetum vegetation.

Development cycle of individuals of various cohorts

Progeny in *Geranium robertianum* populations appear several times in the course of the vegetation season. Spring, summer and autumn cohorts were distinguished on the basis of three peaks in the seasonal dynamics of the number of seedlings. The spring cohort starts to develop in May (Fig. 10). The juvenile phase lasts to the end of June or the first half of July or August. The plants reach a height of 10–15 cm and develop a three- to five-leaf rosette. The generative phase is attained in May or June of the following year. The flowering and fruiting phases last simultaneously to the end of the season, and then the individuals die. Individuals of the spring cohort are, therefore, biennial. The summer cohort starts develop-

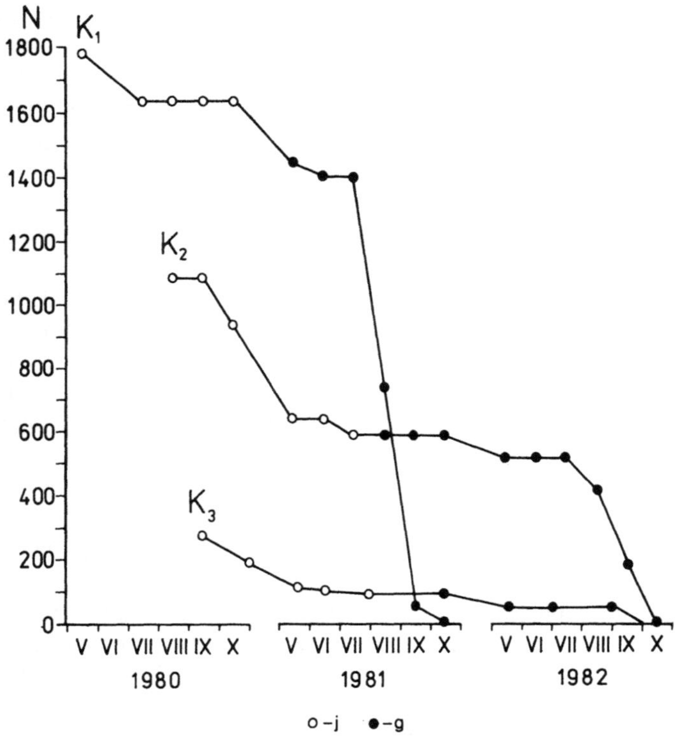

Fig. 10. Development of seasonal cohorts in a *Geranium robertianum* population in a Tilio-Carpinetum community. K_1, Spring cohort; K_2, summer cohort; K_3, autumn cohort; j, juvenile; g, generative individuals.

ment towards the end of July or August and attains the early juvenile phase towards the end of the vegetation season. It overwinters as minute plants with a habit and size similar to seedlings (that is 5–7 cm high) but they have no cotyledons. In the next season the plants start to grow intensively, reaching a height of 15–20 cm in the autumn. Some of them flower and bear fruit in September and October, and the remaining ones attain this phase as late as May or June of the third vegetation season. Plants which attain the generative phase towards the end of the season do not (in general) die after its end, as is the case with plants fruiting in spring or summer. At the beginning of the next year some of them flower and fruit again. These are the small group of two-year or polycarpic individuals of the *Geranium robertianum* populations. Individuals of the autumn cohort overwinter in the seedling phase because they start development in September or October. Some of the individuals, like the summer cohort, have a life span of two years. The proportion of the autumn cohort individuals in the population is small. Moreover seedlings do not appear at the end of the vegetation season every year (Falińska and Pirożnikow 1983).

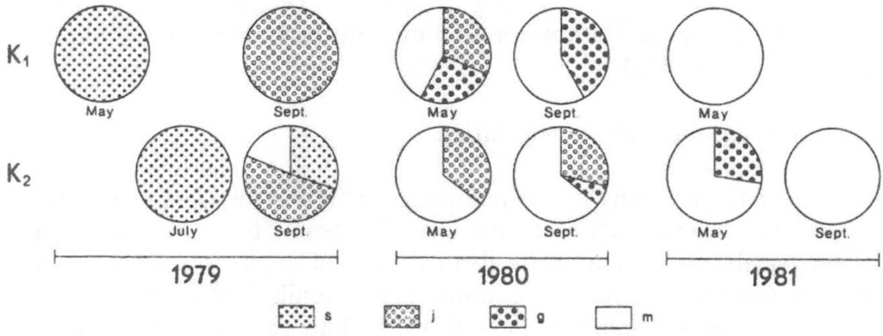

Fig. 11. Percentages of individuals in different developmental phases in the spring (K_1) and summer (K_2) cohorts of *Geranium robertianum*. s, seedlings; j, juvenile individuals; g, flowering and fruiting individuals; m, mortality (Falińska and Pirożnikow 1983).

Since the cohorts start their life cycles at various dates, the successive developmental phases are also reached at different times. Hence in the *Geranium robertianum* populations seedlings, juvenile individuals and flowering and fruiting ones are present at any phenological season. In the course of one vegetation season, however, the fraction of individuals in various development phases changes (Fig. 11).

Structure of age states

The *Geranium robertianum* populations observed in the Tilio-Carpinetum herb layer differ little in age structure. As a rule, individuals aged less than two year dominate, constituting 80–90% of the whole population. However, these populations differ widely in the proportion of individuals in the particular developmental phases. They are mostly characterised by the following age-state structures:

1. juvenile individuals are dominant (70–80%), with a low participation of seedlings and individuals in the generative phase;
2. seedlings and juvenile individuals are present in similar proportions in the population, their joint participation is 80–90%;
3. seedlings predominate (50–60%), and juvenile individuals and those in the generative phase each appear in similar proportions in the population (23–30%).

The age-state spectra change from year to year in the same population when evaluated at the same phenological seasons, as a result of a different number of individuals of the spring, summer and autumn cohorts (Fig. 10). The following regularity was noted: if in one vegetation season the spring cohort is present in highest proportion in the population structure, then in the next season juvenile and generative individuals dominate; when, however, in the preceding season individuals of the summer and autumn cohort prevail, seedlings and juvenile individuals will as a rule dominate in the following season. Hence in relatively stabilized populations in some years a structure of age states characteristic of progressive populations was observed. This is rare, however, because usually individuals of the spring generation are dominant in the populations (Falińska and Pirożnikow 1983).

Abundance dynamics in the population

The seasonal variations in the population numbers of *Geranium robertianum* have the character of fluctuations. The populations are not numerous in April and May and are formed then of scarce seedlings of the autumn generation and more numerous juvenile individuals of the previous spring and summer generations. In June the size of the populations begins to increase and this lasts to July or August. At this time the populations are as a rule most numerous, the decisive factor being the participation of juvenile individuals of the spring generation and numerous seedlings of the summer one. The decrease of population size starts in August and lasts almost to the end of the vegetation season. It is caused by the mortality of seedlings of summer and autumn generations and dying of individuals which have ended their generative phase.

Population size fluctuates in a three- or four-year cycle. For instance, on a 400 m^2 area comprising a fragment of a stabilised population, the size of the population in three successive years was 1958, 1094 and 2547. On an area of similar size in another phytocoenosis of Tilio-Carpinetum an increase in number was observed during 1978–1981 (160, 977, 1157, 1399) and then a decrease in 1982 to 828. The population size depends above all on the spring cohort, and less so on the summer one. *Geranium* individuals require two vegetation seasons to complete their developmental cycle, so that individuals of the spring cohort have in this respect the most favourable conditions. Seedlings develop in spring and attain the juvenile phase in full summer. They start the generative phase at once in the spring of the next vegetation season, producing several score of fruits. The participation of this generation in the population is 60–70% and of the summer generation 20–30%, whereas that of the autumn generation is about 10%. Therefore, the *Geranium* populations differ considerably in the structure of the age states, although they consist as a rule of one and two year-old individuals.

Spatial organisation of populations

In the Tilio-Carpinetum herb layer *Geranium robertianum* individuals form well-outlined aggregations (e.g., Fig. 12). On a 1-ha surface for instance in one vegetation season 60 such aggregations were noted, whereas in the second season there were 45. The smallest aggregations were formed of two or three individuals and occupied 0.10–0.25 m^2, whereas the largest ones were several hundred square metres in extent, 682 m^2 in one instance. About half of the observed aggregations occupied areas of 0.25–5 m^2.

Although the boundaries of the aggregations in particular seasons do not overlap precisely, their situation changes only slightly in the course of

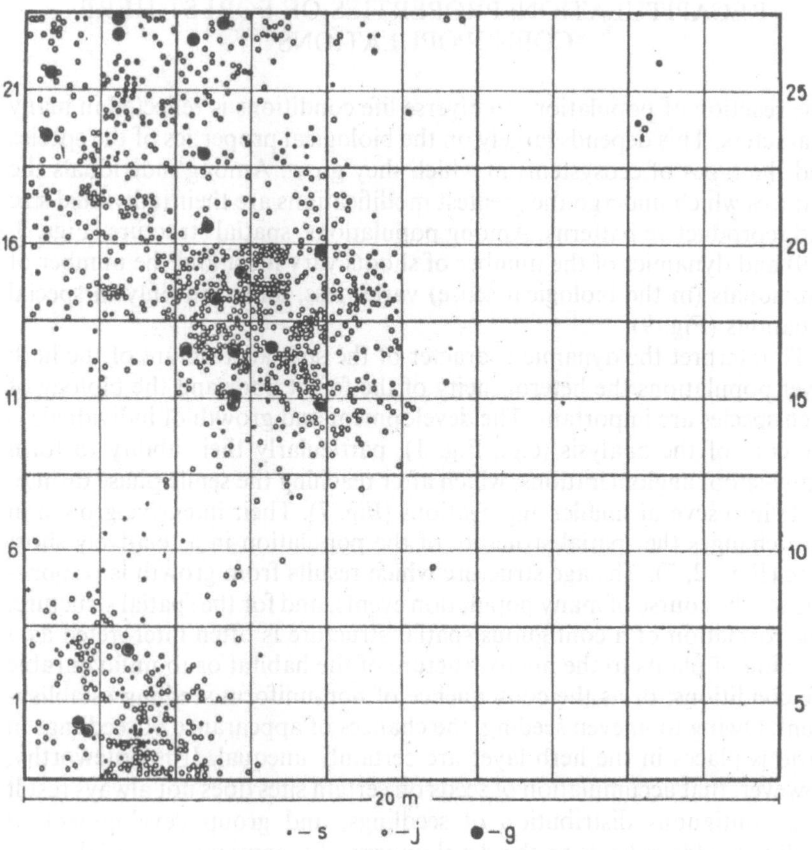

Fig. 12. Spatial organization of *Geranium robertianum* populations in Tilio-Carpinetum ground vegetation in July 1981 on a plot of 400 m^2. s, Seedlings; j, juveniles; g, individuals in generative phase (Falińska and Piroźnikow 1983).

several years. From year to year, however, the number of individuals changes in the aggregations as well as their density (per 1 m^2), for instance 6.33 ± 0.69, 3.19 ± 0.24, 6.67 ± 0.58. The appearance of progeny from May to October decides the seasonal variability of the spatial organisation – every several months the contiguousness in distribution of the individuals increases (e.g., Fig. 12). This is the effect of the heterogeneity of the biotope and of the population processes. As regards *Geranium robertianum*, the spatial organisation of the populations is above all influenced by population processes, particularly by its specific reproductive features.

RECAPITULATION: PROPERTIES OF FOREST HERB COENOPOPULATIONS

The reaction of populations to diverse life conditions is reflected in many characters. This depends largely on the biological properties of the species and the types of ecosystems in which they grow. Among individuals the features which undergo the greatest modifications are their individual size and reproductive patterns. Among populations, spatial structure (Figs. 3, 7, 9) and dynamics of the number of shoots vary most and the number of individuals (in the biological sense) varies less, and then only in special situations (Fig. 9).

To interpret the dynamic character of the spatial structure of the herb layer populations the heterogeneity of the forest floor and the biology of each species are important. The development and growth of individuals is the core of the analysis (e.g., Fig. 1), particularly their ability to form many-shoot agglomerations, which after reaching the senile phase disintegrate into several smaller aggregations (Fig. 7). Their intensive growth in turn changes the spatial structure of the population in a relatively short time (Figs. 2, 7). The age structure which results from growth is responsible for the course of many population events, and for the spatial structure. The formation of a contiguous spatial structure is often interpreted as a reaction of plants to the microstructure of the habitat or to unfavourable life conditions, or as the consequence of non-uniform seedling establishment. Owing to uneven seeding, the chances of appearance of seedlings in various places in the herb layer are certainly unequal. It is noteworthy, however, that accumulation of seeds on certain sites does not always result in a contiguous distribution of seedlings, and group development of seedlings seldom leads to the development of a contiguous spatial structure in populations of perennial plants. This is so because only few seedlings attain the successive developmental phases. The correspondence between contiguous distribution of seeds, seedlings and mature individuals was noted only in *Geranium robertianum* populations in which the seasonal cohorts occupied different sites (Falińska and Pirożnikow 1983). Vegetative reproduction by rooting of diaspores around the maternal plant leads to the formation of clones in the form of large aggregations, for example, in *Caltha palustris* populations (Falińska 1978, 1979b, 1981b).

Although populations in the herb layer exhibit a certain short-term stability of population processes, observations conducted over a longer period indicated that they are dynamic (Figs. 4, 9). The course of population size dynamics in herbaceous plant populations may run in either of two ways, illustrated diagrammatically in Fig. 13.

1. The population starts its development from a small number of individuals, and in successive vegetation seasons increases in numbers. After several years the population size stabilises at a certain level under the given ecological conditions (Fig. 4). Such a type of growth is seen in species where vegetative reproduction dominates, as in many species of the forest herb layer.

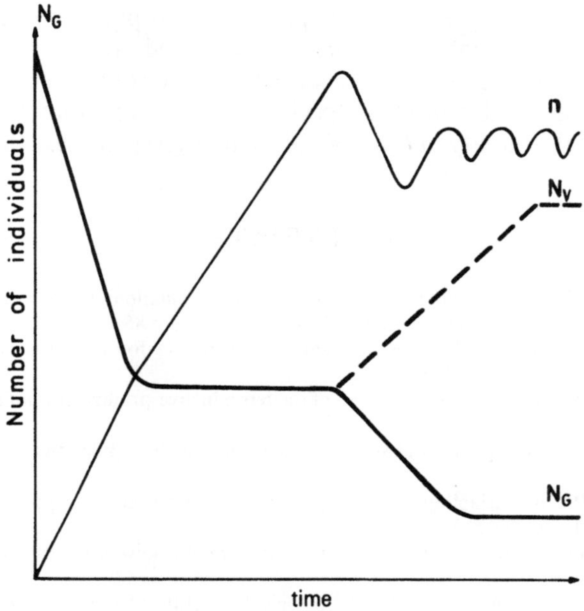

Fig. 13. Model of population size dynamics of perennial herbaceous plants with two modes of propagation. N_G = number of individuals of generative origin; N_V = number of individuals of vegetative origin; n = number of shoots.

2. The population starts its development with a very high number of individuals, with a drastic depression of their numbers in the second or third season. At this time the population size stabilises at a number appropriate to the prevailing life conditions. This type of population growth is characteristic for species with good generative reproduction by high output and germination capacity of seeds.

In numerous perennial species, after a certain stabilisation of the population size, individuals arise mainly by vegetative reproduction, with a smaller participation by individuals of generative origin (Fig. 13). If the species reproduces in both ways, the relations between the vegetative and generative progeny are highly variable in various communities.

In perennial herbaceous plant populations the dynamics of numbers of individuals are not always correlated with the dynamics of the number of shoots. The increase in the shoot number occurs frequently according to the logistic model, or by exponential growth, with a relatively constant number of individuals. This phenomenon is caused by polycormones reaching their highest rate of development in the second to fourth year of life, when, as a rule in forest populations, the number of individuals stabilises. The increase in the number of shoots in individuals of perennial herbaceous plants in the first period of life occurs in a geometrical progress (Fiala 1978, Falińska 1982). Inhibition after a time of individual development is probably caused by ageing and may also be due to the influence of individuals of other species in the herb layer. It is noteworthy that dominance of one species in a given phytocoenosis patch occurs when the herb layer or tree stand structure are disturbed. If no irregularity is observed in the phytocoenosis structure, the mosaic of the herb layer, both as regards the combination of species and their distribution in space, exhibits a high stability and repeatability in a given ecosystem.

REFERENCES

Falińska, K. (1978) Behaviour of *Caltha palustris* L. populations in forest and meadow ecosystems of the Białowieża National Park. Ekol. Pol. 26: 85–109.

Falińska, K. (1979a) Modifications of plant populations in forest ecosystems and their ecotones. Pol. Ecol. Stud. 5: 89–150.

Falińska, K. (1979b) Experimental studies of the reproductive strategy of *Caltha palustris* L. populations. Ekol. Pol. 27: 527–543.

Falińska, K. (1981a) Experimental studies of perennial plant populations. Wiad. bot. 25: 209–230.

Falińska, K. (1981b) Variability of *Caltha palustris* L. populations in garden culture. Acta Soc. Bot. Pol. 50: 493–513.

Falińska, K. (1982) The biology of *Mercurialis perennis* L. polycormones. Acta Soc. Bot. Pol. 51: 127–148.

Falińska, K. and Pirożnikow, E. (1983) Ecological structure of *Geranium robertianum* L. populations. Ecol. Pol. 31: 93–121.

Fiala, K. (1978) Underground organs of *Typha angustifolia* and *Typha latifolia*, their growth, propagation and production. Acta Sc. Nat. Brno, 12: 1–43.

Hutchings, M.J. and Barkham, J.P. (1976) An investigation of shoot interactions in *Mercurialis perennis* L., a rhizomatous perennial herb. J. Ecol. 64: 723–743.

Mukerij, S.K. (1936) Contributions to the autecology of *Mercurialis perennis* L. J. Ecol. 24: 39–81.

Toropova, H.A. (1977) Development of overgrowth of *Mercurialis perennis* L. (Euphorbiaceae). Bot. Zhur. SSSR 62: 1433–1440.

POPULATION STRUCTURE OF PSAMMOPHYTE VEGETATION

E. SYMONIDES

ABSTRACT

Vegetation changes in the inland dune succession and the demography of selected psammophyte populations (*Corynephorus canescens*, *Spergula vernalis*, *Jasione montana*, *Koeleria glauca*, *Festuca psammophila*, *Tragopogon heterospermus*) are described. The roles of these species in fixing sands and in the formation and development of vegetation are discussed. In the initial phase the main roles are played by *C.canescens* and *S.vernalis*: the aggregated spatial structure, high individual fecundity and fast population turnover of these species create habitats which become favourable for those species that immigrate later. The more compact vegetation which then develops is relatively stable: its component coenopopulations are characterized by longevity, slow rates of growth and development, low capacities for vegetative propagation, and high offspring survival. Demographic processes of individual coenopopulations are directly related to the complex of habitat and phytocoenotic relationships.

INTRODUCTION

Psammophytes occur in two different types of ecosystems: they are components of stable, climax biocoenoses of deserts, semideserts and coastal sand dunes; they also colonize unstable biotopes of inland dunes and sands of anthropogenic origin, usually on deforested areas. As highly dynamic systems, the latter are well suited for studying the formation, differentiation and succession of plant communities, as well as the role of populations of individual species in the formations of phytocoenoses. Furthermore, they show relatively simple relationships both between their component species and between the populations and their biotopes, as a result of a small number of species, low vegetation density and infertile, poorly differentiated substrates.

In this paper I shall discuss the population biology of several common psammophyte species in a series of communities which exemplify secondary

White, J. (ed.), The Population Structure of Vegetation.
 ISBN 978-94-010-8927-2

succession, from the initial stages of Spergulo-Corynephoretum (Tx.1928) Libb. 1933 through Festuco-Koelerietum glaucae Klika 1931, to the compact turf formed mainly by *Festuca duriuscula*.

The studies were carried out in the foreland of a large, classic complex of dunes in Toruń Basin, Poland. The dunes are formed of rewinnowed terrace sands of local origin. Genetically they are connected with the younger dune period (postlitorin) and arose as a secondary formation on the area of waste dunes deprived of their forest cover. Large parts of the dunes were completely deprived of soil cover. Small patches of raw soils are encountered in their vicinity which have either an unformed or weakly-formed profile: horizon A_1 lies directly on parent rock. The first level of subsoil water is situated at a depth of 5–6 m. The average annual temperature is 7.6°C; the mean average rainfall has not been higher than 500 mm (Mrózek 1958, Ziembińska 1969). The flora of the study area is fairly varied: c. 90 species of vascular plants and 20 mosses and lichens.

PSAMMOPHYTE INVASION OF BARE DUNES AND SANDY HABITATS

The initial colonization phase of dunes and sandy habitats is relatively slow. Shifting, dry, poorly permeable and easily heated sands are highly unfavourable for seed germination and seedling development. *Corynephorus canescens*, common in secondary dune habitats of Europe, is one of the few species that colonize such areas successfully. Its colonization success is due to those features which enable it to withstand a mobile and dry substrate, and, when weather conditions are favourable, to invade available space quickly: (i) an ability to renew the tussocks layer by layer (Marshall 1967); (ii) a rich secondary root system (Stoutjesdijk 1959); (iii) a typical xeromorphic stem structure, so that its transpiration (usually intensive on a moist substratum) can be almost completely stopped (Rychnovská-Soudkova 1961, Rychnovská 1963, Berger-Landefeldt and Sukopp 1965); (iv) a short pre-reproductive period (Symonides 1979a); (v) a high diaspore output, characteristic of plants of open communities (Salisbury 1942, 1952).

The factors governing germination of the first caryopses and survival of the seedlings that initiate the formation of populations and communities of Spergulo-Corynephoretum are not fully known. However some data indicate that the accumulation of a sufficiently large diaspore density in the soil is very important. A few caryopses germinate, but only when sown in aggregations of at least 30 seeds (Symonides, unpublished data). Germination of caryopses growing in the middle of such aggregation probably occurs because they are protected against extreme temperatures and utilise the moisture retained by the diaspore cluster and perhaps also because it is stimulated by secretions from the mass of germinating diaspores (Linhart and Pickett 1973).

The process of dune colonization starts with the establishment of the pioneer tussocks of *Corynephorus canescens*. These form a natural barrier,

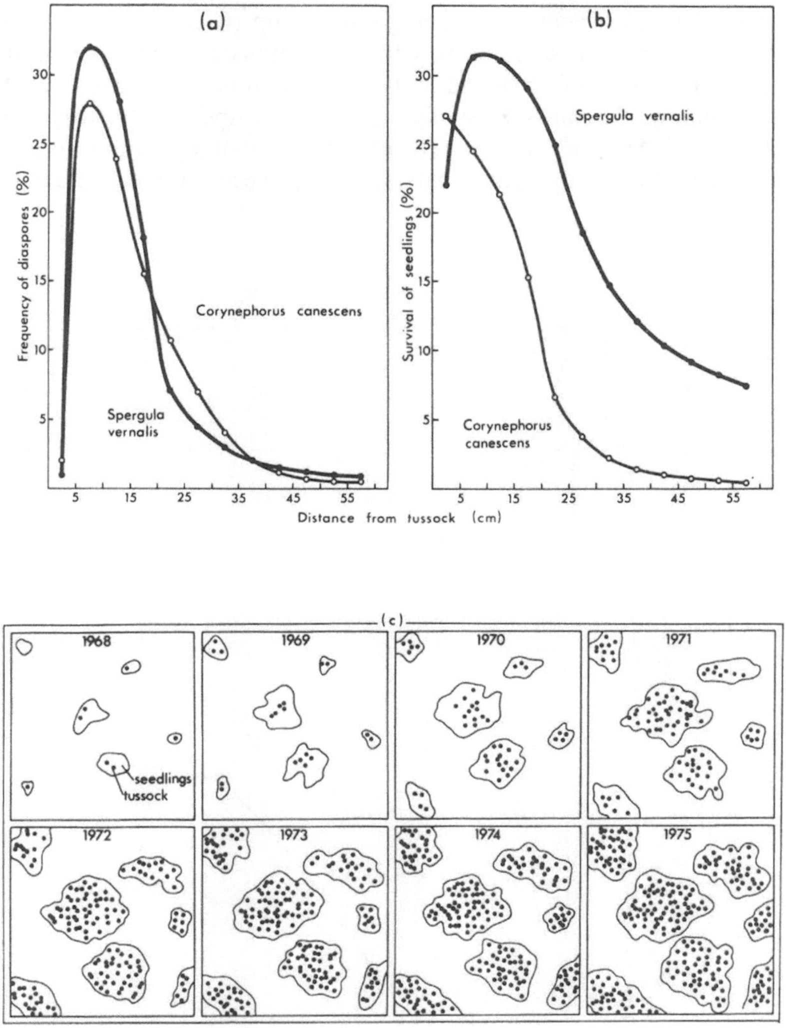

Fig. 1. Effect of the pioneer *Corynephorus canescens* tussocks on (a) diaspore distribution, (b) seedling survival, and (c) formation of initial aggregations of *Spergula vernalis* and *Corynephorus canescens*. Sampling of diaspore distribution was made in 1969 and 1970, at a depth of 2–5 cm and at a distance of every 5 cm from 10 pioneer tussocks. Seedlings were counted within a transect 20 cm wide at 5 cm intervals from the same tussocks (original data).

stopping diaspores carried by a wind. They also fix the substrate and protect it from drying out and overheating. Consequently, seedlings grow and survive mainly in the close vicinity of pioneer tussocks, which results in the aggregated spatial structure of the developing population (Fig. 1). Diaspore dissemination around the tussocks and enhanced survival of seedlings near them cause a rapid increase in the density of aggregations. But on the other hand, there is a slow rate of sand stabilization over a wider area, causing difficulty for settlement of other species which are less resistant than *Corynephorus* to being covered-up (Chadwick and Dalke 1965). Under extreme conditions, when the humus content in the initial soils does not exceed 0.25% and moisture content is less than 2–3%, the initial community consists usually only of two species, *Corynephorus canescens* and *Spergula vernalis*. The changes in such a phytocoenose are purely quantitative for many years, as these species alone fluctuate in

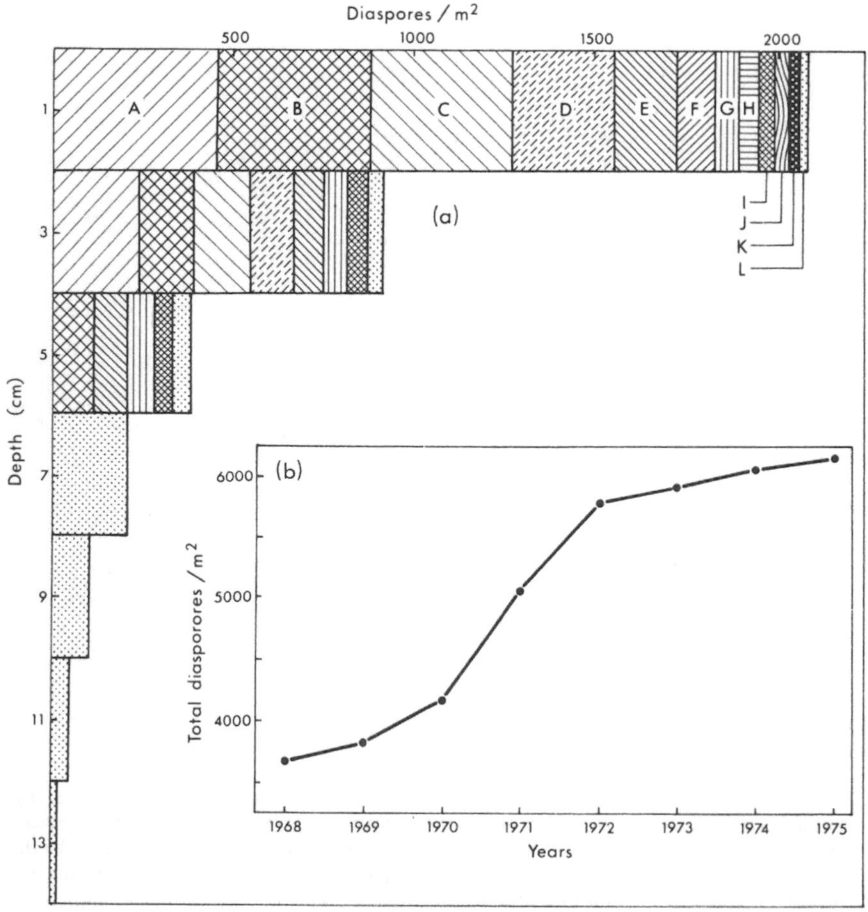

Fig. 2. The number, depth distribution and specific composition of buried viable seeds (a) and changes in the total seed number (b) in the sandy soil beneath initial patches of a community of Spergulo-Corynephoretum. Samples were collected at 50 random sites. A, *Spergula vernalis*; B, *Corynephorus canescens*; C, *Festuca psammophila*; D, *Koeleria glauca*; E, *Helichrysum arenarium*; F, *Scleranthus perennis*; G, *Jasione montana*; H, *Plantago indica*; I, *Hieracium pilosella*; J, *Carex arenaria*; K, the sum of the following species: *Sedum sexangulare, Thymus serpyllum, Veronica dillenii, Potentilla arenaria, Artemisia campestris, Achillea millefolium, Tragopogon heterospermus, Pinus sylvestris, Rumex acetosella*; L, species which were not identified.

Table 1. Changes in the chemical properties of dune soil in the initial patches of Spergulo-Corynephoretum. Samples of soil were taken each September at the same 10 points of aggregation of *Corynephorus canescens* and *Spergula vernalis*.

Year	pH (KCl)	Humus (%)	Exchangeable cations*				P_2O_5**
			Ca^{++}	Mg^{++}	K^+	Na^+	
1968	3.7	0.79	0.05	trace	0.031	0.016	1.9
1969	4.0	1.33	0.05	0.002	0.031	0.017	2.7
1970	4.1	0.70	0.05	0.004	0.031	0.018	2.8
1971	4.2	1.22	0.06	0.006	0.041	0.019	2.9
1972	4.2	·1.33	0.07	0.006	0.041	0.019	3.8
1973	4.2	1.52	0.08	0.008	0.051	0.019	3.9
1974	4.5	0.42	0.08	0.008	0.051	0.022	5.8
1975	4.5	0.39	0.12	0.012	0.051	0.022	6.5
1976	4.5	0.47	0.15	0.025	0.056	0.023	6.6
1977	4.6	1.79	0.34	0.037	0.072	0.023	8.5

*m.e./100 g of soil;
**By Egner's–Riehm's method: mg/100 g of soil.

abundance. At this time all the species typical of the developed Spergulo-Corynephoretum association and many characteristic of Festuco-Sedetalia are represented in the seed bank (Fig. 2).

The populations of pioneer species transform the habitat in their vicinity. This is due mainly to a high biomass production and fast turnover of individuals. Consequently surface sand layers are enriched with organic matter (Table 1), enabling the germination and development of seedlings of other psammophyte species. Individuals of *Festuca psammophila* and *Koeleria glauca* are among the first. Thus, from the beginning of the development of Spergulo-Corynephoretum the direction of successional changes becomes outlined. As a result the next phytocoenosis, Festuco-Koelerietum glaucae is formed (Symonides 1979a, b).

THE STRUCTURE, DIFFERENTIATION AND DYNAMICS OF PSAMMOPHYTE COMMUNITIES

(a) *Spergulo-Corynephoretum*

When a certain threshold of humus content, moisture and coverage of sands is passed in the initial patches of Spergulo-Corynephoretum an increase in the number of its component species occurs. In several years bare dunes are overgrown with vegetation which has a relatively high degree of diversity of growth forms, phenology, biology and reaction on the habitat of its components (Wilkon-Michalska and Symonides 1974). The process of formation of a mosaic structure, which is characteristic of the psammophyte loose sod, and the stabilization of the age and abundance structure of individual populations takes a longer time.

The emergence of the seedlings of perennial species within the primary aggregations of *Corynephorus canescens* and *Spergula vernalis* is of the nature of a single "big bang", to use a metaphor employed by some life

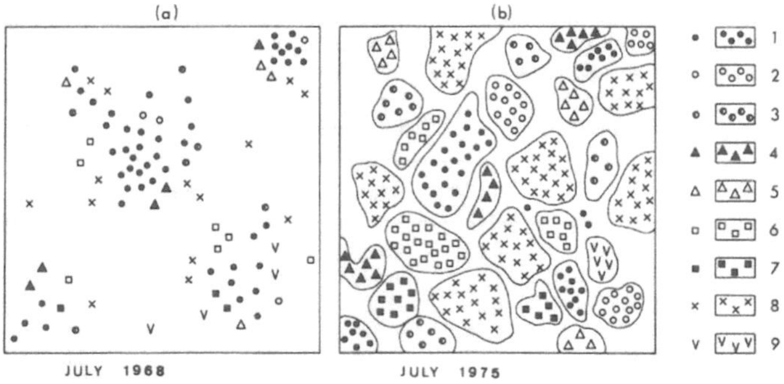

Fig. 3. Changes in the spatial structure of a phytocoenosis of Spergulo-Corynephoretum during the course of succession. Individuals (a) and aggregations (b) of the following species are shown on a 4 × 4 m^2 plot: 1, *Corynephorus canescens*; 2, *Koeleria glauca*; 3, *Festuca psammophila*; 4, *Helichrysum arenarium*; 5, *Sedum sexangulare*; 6, *Polytrichum piliferum*; 7, *Scleranthus perennis*; 8, *Hieracium pilosella*; 9, *Calluna vulgaris*.

history theorists (Gadgil and Bossert 1970). During the following years an increase in vegetative propagation and changes in the age structure of the populations are observed. A many-species mixture of individuals is gradually transformed into a mosaic of one-species patches, which develop at different rates, according to the intensity of vegetative growth and spread of individual populations. Linkage of aggregations, initially separate from each other, takes place, leading to complete occupation of the habitat (Fig. 3). A similar phenomenon has been observed in halophytes (Wilkoń-1961, Michalska 1976).

The quantitative relations between the populations and the spatial structure of the phytocoenose in its optimal phase is determined by the sequence of appearance (or immigration) of particular species and the number, survival and distribution of their seedlings. These are the demographic factors which are pre-eminently responsible for the differentiation of the similar initial patches of Spergulo-Corynephoretum into the various subassociations, variants and facies described in the literature (Fukarek 1961, Krausch 1968). The duration of the optimal phase of the community, as well as the direction and the rate of its further transitions are also determined thus (Symonides 1979b).

If the process of qualitative changes in the initial phytocoenosis is fast and seedlings of new species occur in large numbers and in similar proportions, the typical subassociation of Spergulo-Corynephoretum is formed, when *Corynephorus canescens* and *Spergula vernalis* are dominants. After some time populations of *Koeleria glauca*, *Festuca psammophila*, *Helichrysum arenarium* and *Scleranthus perennis* become subdominant. The small synusiae of mosses (*Polytrichum piliferum*, *Ceratodon purpureus*) and lichens (various species of *Cladonia*) make up a ground layer. The first individuals of *Pinus sylvestris* and *Calluna vulgaris* that settle comparatively early point towards the future climax vegetation.

In the course of the whole optimal phase a gradual decrease in the abundance and cover of the dominant species simultaneously with an increase in the participation of the remaining species takes place. When the populations of *Festuca psammophila* and *Koeleria glauca* become superior in numbers to those of other species a phytocoenosis enters the terminal phase. Then the second stage of qualitative change of the community begins: the encroachment of further species characteristic of Festuco-Koelerietum glaucae and the retreat of original components of Spergulo-Corynephoretum (Symonides 1979a, b).

On the poor acid sands the variant with lichens of the typical subassociation is formed: *Cladonia* species cover 70–80% of the habitat and make free development of herbaceous perennials impossible. However they protect the soil against overdrying under drought conditions (Newman 1967), so that juvenile individuals of *Pinus* and *Calluna* find good conditions for growth. After several years the lichen variant of the association changes probably directly into vegetation of Dicrano-Pinion: this is a facies with *Polytrichum piliferum*, poor in vascular plant species but with a different dynamics, since *Polytrichum piliferum* is usually short-lived. Numerous seedlings of *Festuca psammophila, Koeleria glauca, Helichrysum arenarium* and other psammophile species replace the old, dry and crumbling parts of the moss turf. After several years of stabilization the community is transformed into vegetation similar to Festuco-Koelerietum glaucae.

In the initial patches the mass emergence of *Festuca* seedlings precedes the encroachment of the remaining species. It leads to the formation of the distinct and relatively floristically rich Spergulo-Corynephoretum subassociation with *Festuca psammophila* and such differentiating species as *Jasione montana, Artemisia campestris* and *Hypochoeris radicata*. Here the succession rate is usually faster but the pine seedlings find less favourable conditions.

The subassociation with *Festuca psammophila* can occur both in a typical variant and in the variant with lichens, according to the acidity and the fertility of a substratum, to rainfall and also to the specific composition of the seed bank or to distance from other communities that may supply the habitat with diaspores. The subvariants and facies of both of them contain *Helichrysum arenarium, Cladonia verticillata, Hieracium pilosella* or *Jasione montana* (Wika 1975).

Other developmental tendencies occur in the Spergulo-Corynephoretum patches with *Carex arenaria*, a species well adapted to the dune conditions by its strongly developed rhizomes (Noble, Bell and Harper 1979). A "travelling" sedge sod gradually eliminates almost all of the populations of flowering plants, mosses and lichens within its range of action, except the populations of *F. duriuscula, F. psammophila* and *K. glauca* (Fig. 4). After the retreat of *Carex arenaria*, seedlings of the species new to the community arise: these are mainly *Armeria elongata, Dianthus carthusianorum, D. deltoides* and *Veronica spicata*, characteristic of Diantho-Armerietum Krausch 1959 (Krausch 1968, Symonides 1979b). The transitional dominance of *Carex arenaria*, in other words, causes an intrinsic habitat transformation and changes the direction which the succession of the Spergulo-Corynephoretum would otherwise take.

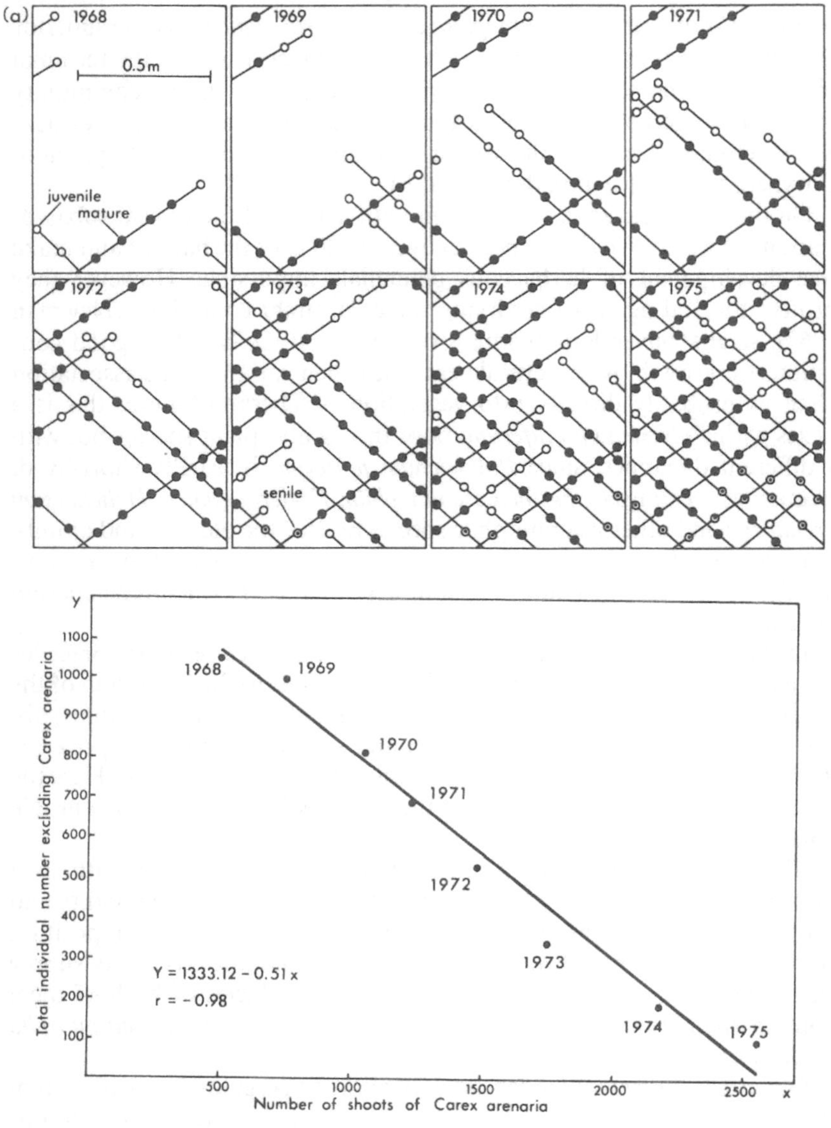

Fig. 4. Development of *Carex arenaria* polycormones (a) and the effect of increasing number of its above-ground shoots on the total number of individuals of other species in a patch ($100\,m^2$) of Sperguló-Corynephoretum (b). The patch was subdivided into $1 \times 1\,m^2$ plots and part of a single plot is shown in (a). (a, Symonides 1979b; b, original.)

(b) *Festuco-Koelerietum glaucae*

The Festuco-Koelerietum glaucae patches are considerably more stable than those of Spergulo-Corynephoretum and transformations in the community mainly lie in small fluctuations of the abundance of individual species. However in the course of several years' investigations a trend towards directional changes can be observed.

The Festuco-Koelerietum glaucae phytocoenoses are relatively stable because of the biological and populational features of the species which constitute the community. Their high competitive power allows them to resist encroachment and their low reproductive potential prevents a fast growth of population abundance (Symonides 1979b–d): both factors enable the community with its high species diversity to function for a long time.

It is essential for the stability of a phytocoenosis which consists mainly of herbaceous perennials and dwarf-shrubs that the component species be characterized by longevity, slow rates of growth and development, low capacities for vegetative propagation and high offspring survival. The

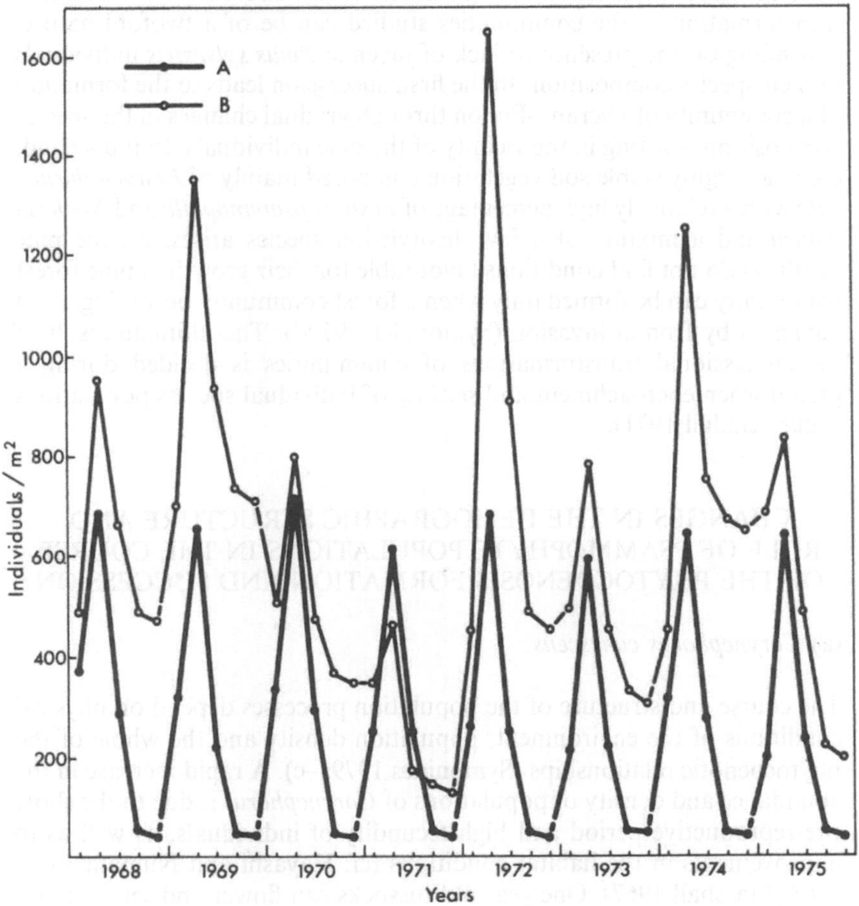

Fig. 5. Seasonal changes in the density of *Cerastium semidecandrum* population in (A) natural Festuco-Koelerietum glaucae patches and in (B) 8 experimental plots (1 m^2) without grass tussocks. Grass tussocks were removed in 1968; single specimens of other species were left. (A, Symonides 1979c, experimental data; B, original).

birth and death processes are usually strictly synchronized; hence intra-populational turnover of individuals is a basic character of their dynamics. The populations of early Spring therophytes which complete their life cycles before the period of active growth of the perennials also show a startling stability (Fig. 5). The abundance and spatial structure of therophyte populations are determined by the surface size of gaps between the aggregations of perennials (Symonides 1983).

The diversity of species is not only a function of the fertility and geographical situation of an area: to a large extent phytocoenosis structure depends on the specific composition and number of seedlings which were the first to establish themselves in the initial patches of Spergulo-Corynephoretum and start the formation of populations. Hence phytocoenoses of this syntaxon are characterized by a high diversity not only in their slightly different structure and species composition but also in their distinct developmental tendencies (Krausch 1962). Indeed the successional transformation of the communities studied can be of a twofold nature, depending on the presence or lack of juvenile *Pinus sylvestris* individuals in their species composition. In the first, succession leads to the formation of a community of Dicrano-Pinion through gradual changes in the species composition, starting in the vicinity of the pine individuals. In the second, compact highly stable sod vegetation composed mainly of *Festuca duriuscula* with a relatively high percentage of *Festuca psammophila* and *Koeleria glauca* and a mixture of a few dicotyledon species arises. As the pine seedlings do not find conditions favourable for their growth, a pine forest community can be formed only when a forest community bordering a sod advances by frontal invasion (Symonides 1979d). The ultimate result of the successional transformations of communities is decided during a period when encroachment and settling of individual species populations occur (Gadgil 1971).

CHANGES IN THE DEMOGRAPHIC STRUCTURE AND ROLE OF PSAMMOPHYTE POPULATIONS IN THE COURSE OF THE PHYTOCOENOSIS FORMATION AND SUCCESSION

(a) *Corynephorus canescens*

The course and structure of the population processes depend on physical conditions of the environment, population density and the whole of the phytocoenotic relationships (Symonides 1979a–c). A rapid increase in the abundance and density of populations of *Corynephorus* is due to the short pre-reproductive period and high fecundity of individuals, as well as to improvements in the habitat conditions (cf. Hayashi and Numata 1967, 1968, Marshall 1967). One year old tussocks can flower and set seed and within three years they attain full reproductive capacity (Table 2). An increase in reproductive output with an increase in the total individual number typically results in a positive net increment of the population, even though 90% or more of seedlings die (Symonides 1979a). Over several

Table 2. Relative distribution of biomass in *Corynephorus canescens* tussocks of different ages. The data are the means of 100 tussocks at each age. Samples were collected from a Spergulo-Corynephoretum community in its optimal phase.

Age (years)	Mean biomass (%)		
	Roots	Vegetative shoots	Caryopses
1	56.2	36.5	7.3
2	52.4	34.4	13.2
3	50.1	22.5	27.4
4	48.2	22.3	29.5
5–6	46.3	25.3	28.4

years in the growing population some vital changes occur: there is an increase in the percentage of the individuals of the older age groups; in very compact aggregations the tussocks attain their reproductive maturity later and death rate increases; the fraction of germinating diaspores decreases fourfold. A slower rate of reproduction eventually results in a decrease in the rate of population increment.

Populations attain stabilization in poor patches of Spergulo-Corynephoretum as a result of (a) an equilibrium between the processes of reproduction and mortality, (b) a constant rate of the maturing of individuals and (c) the stabilized age structure (Fig. 6). Population abundance, considerably lower than at the initial stages, comes under control of the phytocoenotic factors which mark the beginning of the regression phase of a *Corynephorus* population. An increase in the abundance of dominants, transitional to further stages of the succession, is particularly damaging.

In various patches the occurrence of *Polytrichum piliferum*, *Festuca psammophila*, *Koeleria glauca* and *Carex arenaria* can cause similar effects. However the mechanisms of action of these species on *C.canescens* populations are different.

At first *Polytrichum piliferum* forms loose turfs in the vicinity of the *Corynephorus* tussocks. It expands in years of abundant rainfall and spreads easily even on shifting sands (Marsh and Koerner 1972). The moss fixes sands in a short time and forms a hard compact cover that excludes diaspores and makes seedling growth impossible. It causes changes in the age structure and decreases the reproduction and abundance of population of *C.canescens* as well as other psammophyte species that maintain their population by seed (Symonides 1974a, 1979a).

Unlike moss, the proximity of individuals of *Festuca psammophila* causes an increase in the mortality of juvenile as well as mature *Corynephorus* tussocks. *F.psammophila*, like *Koeleria glauca*, has a specific ability to retain moisture in the rhizosphere, thereby overdrying the soil layers close to it (Rychnovská 1963, Rychnovská and Květ 1963). The tussocks of *C.canescens* lack this ability and, sensible to the water loss, wither in masses within range of action of *F.psammophila* or *K.glauca* especially in periods of drought (cf. van der Maarel 1981). Few *C. canescens*

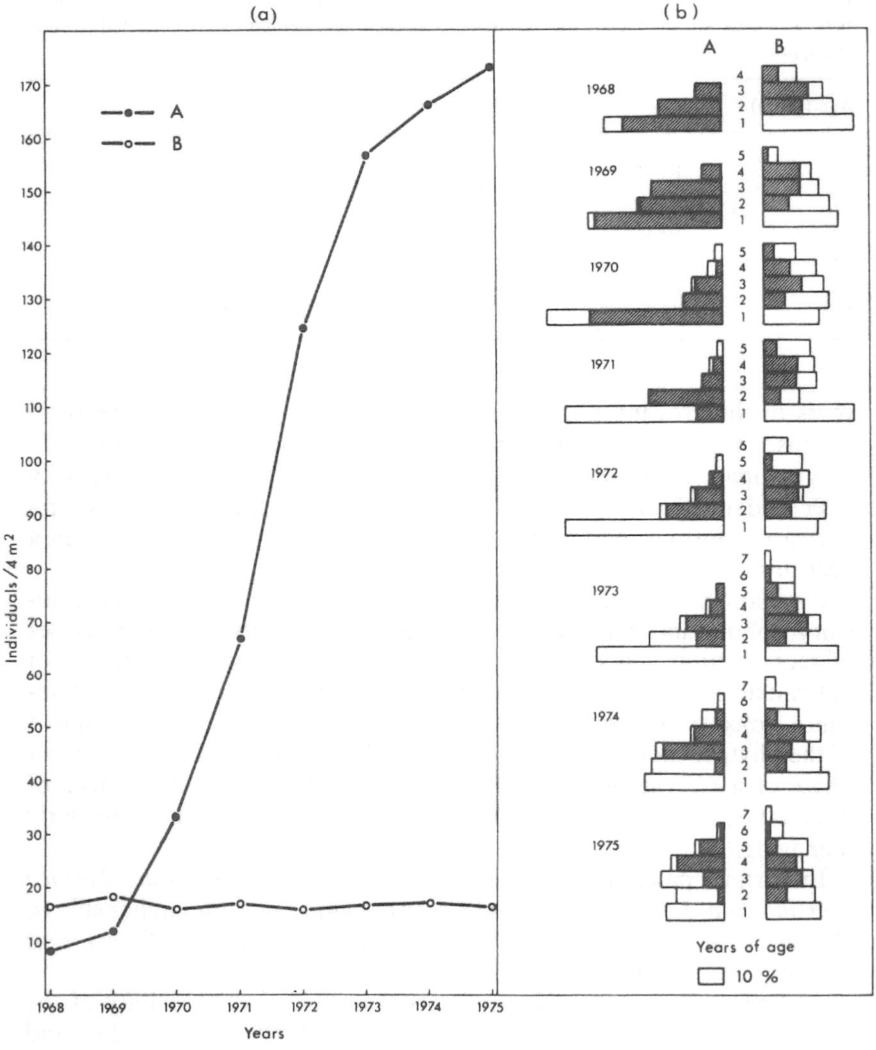

Fig. 6. Changes in density (a) and age structure (b) of *Corynephorus canescens* population in the initial (A) and optimal (B) phases of development of Spergulo-Corynephoretum. The study area was $256\,m^2$, subdivided into 64 plots, each $2 \times 2\,m^2$. Shading indicates the fraction of fruiting individuals (Symonides 1979a).

individuals survive until the pronounced dominance of competitively stronger grasses becomes evident in a phytocoenosis (Symonides 1979b, c).

In the case of *Carex arenaria* the mechanisms of action are different. In the close vicinity of the sedge sod the *Corynephorus* tussocks die, probably as the result of a considerable turgor decrease (Marshall 1965), whereas seedlings do not grow at all or soon perish. Simultaneously a decrease in the survival and reproduction in the local *C. canescens* population results in its complete elimination (Symonides 1979b).

(b) *Spergula vernalis*

The water balance of *S.vernalis* is less elastic than *Corynephorus canescens*. Low values of transpiration and leaf suction pressure, as well as the plant's ability to fall easily into water deficit indicate that in spite of some xeromorphic features in its anatomical structure it is a mesophyte (cf. Abd El Rahman and Batanouny 1965a, b, Symonides 1974b). Its short life span is restricted to autumn (germination and early juvenile stages) and spring (intensive growth and generative phases) when (in the temperate zone) rainfall is high. This is its main adaptation to the living on dry sands, like *Teesdalia nudicaulis*, *Aira praecox*, *Vulpia membranacea* etc. (Newman 1963, 1965, 1967, Symonides 1974c, 1978a, Watkinson and Harper 1978).

S.vernalis invades shifting dunes following the pioneer tussocks of *C.canescens*. The initial patches of Spergulo-Corynephoretum are built by both species. Grass tussocks make it more difficult for the seeds of *S. vernalis* to be blown away and covered-up with a thick sand layer which, as in many psammophytes, inhibits their germination through the lack of light (Symonides 1974a, 1978b, van der Meijden and Waals-Kooi 1979). The seeds germinate more abundantly near *C.canescens* than in an open space, where individuals have a better chance of survival and reproduction. The distribution and abundance of the pioneer *Corynephorus* tussocks determine both the spatial structure and, to a large extent, abundance of *Spergula* populations (Symonides 1979a).

The seedlings of both species emerge in mass almost at the same time and do not compete with each other for environmental resources under extreme abiotic conditions. Furthermore, the survival of *Corynephorus* seedlings is higher in the presence of *Spergula* seedlings than in a one-species system. In this way *S.vernalis* populations indirectly accelerate the process of fixing the dunes, formation of the initial soils and, in consequence, the process of encroachment of further species. On account of their small individual sizes and low productivity *S.vernalis* individuals do not directly play an important role in the formation of a habitat.

In the early colonization phases negative effects of intra-specific competition are not marked, because individuals within large aggregations, with a relatively high crowding index, mainly survive (Symonides 1974a, 1977). At low density 90–95% die, covered up with sand and damaged by frost.

The pioneer *S.vernalis* population is characterized by a highly aggregated spatial structure, a poorly differentiated size and high fecundity of individuals (Fig. 7). The survivorship curve is similar to Deevey type II, and strictly speaking a period of seasonal stabilization does not exist (Fig. 8a). Owing to mass seed production and progressively better conditions for germination and seedling survival, population abundance increases over many years, attaining its peak in the loose Spergulo-Corynephoretum patches (Fig. 8b).

Demographic processes in *Spergula* populations are different to those in many other winter annuals of similar extreme habitats. *Cerastium atrovirens*, *Vulpia membranacea* and *Phleum arenarium* produce only small

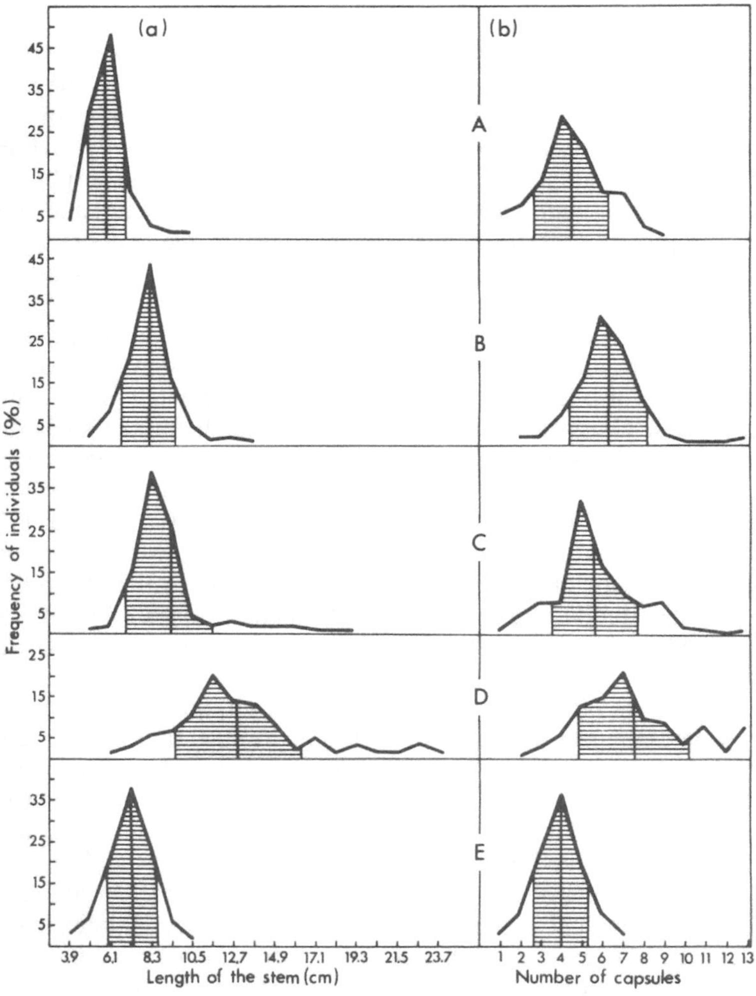

Fig. 7. Diversification of size (a) and fecundity (b) of the individuals of *Spergula vernalis* population in different phytocoenoses. A, Initial aggregations of *S.vernalis* and *Corynephorus canescens*; B–C, phases of Sperguio-Corynephoretum development with 30–50% cover respectively; D, Arctostaphyllo-Callunetum Tx. et Prsg. 1940; E, compact sods with a high content of *Calamagrostis epigeios*. Samples were collected from each phytocoenosis for three years; each sample consisted of 100 randomly chosen individuals. Shaded area shows the value of arithmetical mean and its standard deviation (Symonides 1974a).

numbers of seed, but they have a high germination capacity and their seedlings survive very well (Watkinson and Harper 1978, Mack 1976). About 50% of the biomass production of *S.vernalis* is destined for reproduction, whereas, for example, only 28% in *Oenothera deltoides* (cf. Symonides 1974a, Bell, Hiatt and Niles 1979). Unlike seeds of many dune species which are able to germinate only during one season and then die,

Spergula does not loose its germination capacity for many years (Pemadasa and Lovell 1974, 1975; Symonides in preparation); this is of real importance to a bare sand colonizer.

The development within primary aggregations of *C.canescens* and *S. vernalis* of populations of perennials propagating themselves vegetatively cause changes in the spatial structure of the *S.vernalis* population. Now individuals form small, dispersed groups and are more diversified in respect of size, weight and fecundity than in the pioneer phases. Their mortality is considerably lower and limited mainly to the juvenile stages (Deevey type III). However the better the vegetation (especially turfs of mosses and lichens) covers the sand surface, the worse become conditions for the germination of seeds. Reproduction is reduced and a further increase in the abundance of the next generation is stopped. From the compact sods *S.vernalis* retreats completely, leaving free space for populations of therophytes with even shorter life spans than itself; these are better able to avoid competition with perennials (Figs. 7, 8).

In the successional series of psammophyte communities only in loose, species-poor patches of Spergulo-Corynephoretum does *Spergula* find its ecological optimum. Both severe abiotic conditions in the phases of sand colonization and interactions with other species at a later stage restrain the population; in the first case through high mortality, in the second through a drastic decrease in reproductive output (Symonides 1974a).

(c) *Koeleria glauca and Festuca psammophila*

Populations of both these grass species are tolerant of a wider range of differentiation of psammophyte communities than *Corynephorus canescens*. This is a result of, among others, the following features common to *K.glauca* and *F.psammophila*: (i) capacity for both generative and vegetative reproduction, (ii) high survival of vegetative offspring, (iii) longevity of individuals, (iv) an ability to retain moisture in the rhizosphere (Rychnovská and Květ 1963, Symonides 1979b, d). As with *C.canescens* the formation and development of their populations are accompanied by vital changes in their demographic structure.

Both species are among the first to enrich the initial patches of Spergulo-Corynephoretum. Like *S.vernalis* they grow under the cover of the *Corynephorus* tussocks. Both populations preserve the nature of invasive populations for 5–6 years, although the majority of the seedlings emerge in a single flush of germination. During this time pioneer tussocks reach reproductive capacity.

An increase in the abundance of invasive populations is slow. It is conditional on the immigration of large diaspore numbers, since seeds have low germination capacity (several per cent) and lose their viability within 3–4 years; seedling survival is also low and individuals have a long pre-reproductive period in their life cycle (Symonides 1977, 1978c, 1979b). During this time survivorship curves of populations of both species are similar to Deevey type III. Age and size structures gradually become more complex. The distribution of the *Corynephorus* tussocks, around which

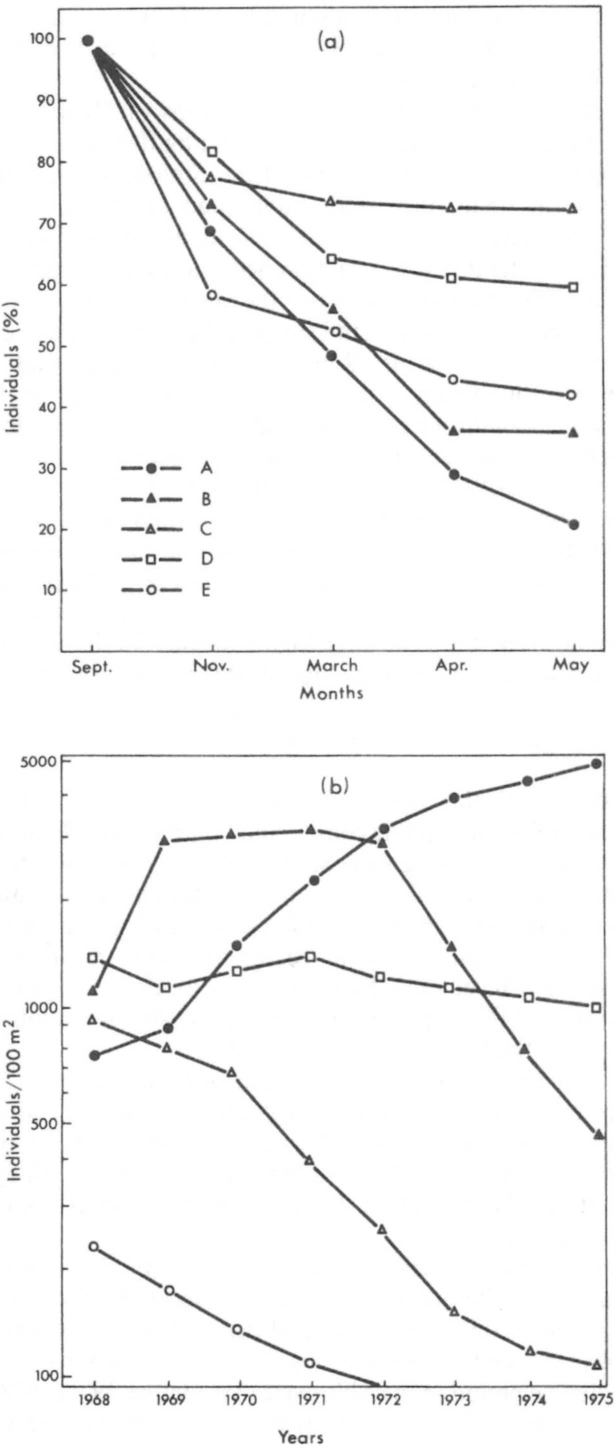
(a)
100
90
80
70
60
50
40
30
20
10
Individuals (%)
A
B
C
D
E
Sept.
Nov.
March
Apr.
May
Months
(b)
5000
1000
500
100
Individuals/100 m²
1968
1969
1970
1971
1972
1973
1974
1975
Years

caryopses and seedlings of *Festuca* and *Koeleria* are crowded, determines the type of the spatial structure in a macroscale. Pioneer tussocks of both species gradually become centres of aggregations of the individuals of vegetative origin. The slow rate of increase in the population abundance of both species results in prolonged dominance of *C.canescens* and a duration of the Spergulo-Corynephoretum association. On the other hand, only conditions of low vegetation density favour the growth of seedlings of *Festuca* and *Koeleria.*

The maturing of the pioneer tussocks, at the same time as the gradual overgrowing of the sandy substrate, causes further changes in the structure and dynamics of populations. Numbers of individuals increase faster than in the invasion period. At first they are of vegetative and generative origin, but gradually the vegetative offspring become predominant. In the fully developed Festuco-Koelerietum patches the tussocks propagate only vegetatively, the degree of the population aggregation grows and correspondingly the rate of spread of individuals within the phytocoenosis declines (Symonides 1979b).

The change in the mode of reproduction results only from worsening conditions for germination of caryopses. The production of generative diaspores increases with the maturing of the population, while germination capacity remains unchanged. A low rate of vegetative propagation coupled with tussock longevity and high survival of the offspring stabilize the population abundance during the year and over many years (Fig. 9). Survivorship curves of the stabilized populations are similar to Deevey type I (*K. glauca*) or type II (*F. psammophila*). As with the age structure, the frequency of individuals of different size class varies insignificantly from year to year (Symonides 1979b).

The *Festuca* and *Koeleria* populations are scarcely susceptible to interference from other species occurring in the phytocoenosis. Their strongly developed compact roots make it difficult for other psammophytes to settle in their vicinity. High competitive ability and a slow rate of expansion of *F.psammophila* and *K.glauca* within a habitat maintain the equilibrium of Festuco-Koelerietum glaucae for many years. Both species populations retreat only under the influence of *Festuca duriuscula* individuals or as a result of significant changes in the abiotic conditions, caused usually by the expansion of *Tragopogon heterospermus* populations or the action of young pine individuals, but then only if they are numerous in the patch.

Changes in the structure of both populations caused by reduced reproduction and a lower survival of the juvenile tussocks are the first signs of their regression (Fig. 10), followed by a considerable reduction in abundance several years later. In compact sods dominated by *Festuca duriuscula* both species have an insignificant presence (Symonides 1979c).

Fig. 8. Survivorship curves (a) and abundance dynamics (b) of *Spergula vernalis* populations in different phytocoenoses. A–E: see Fig. 7. The study area was 10 m^2, subdivided into 100 plots of 0.1 m^2. (a, Symonides 1974a; b, original.)

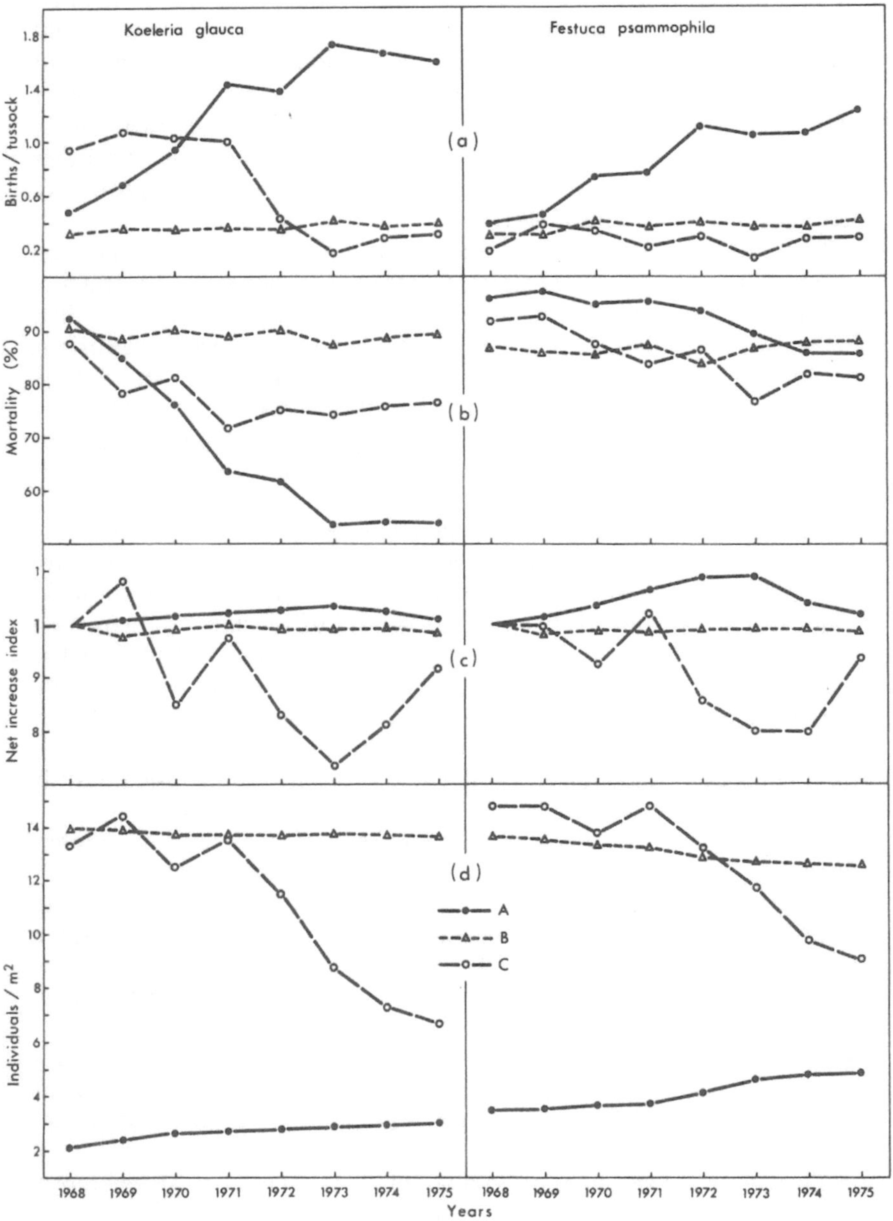

Fig. 9. Demographic indices of (a) reproductive output, (b) mortality, (c) net increase and (d) density in populations of *Koeleria glauca* and *Festuca psammophila* in different phytocoenoses. A, Spergulo-Corynephoretum (initial phase of both populations); B, Festuco-Koelerietum glaucae (optimal phase of both populations); C, Festuco-Koelerietum glaucae during *Tragopogon heterospermus* expansion (regression of both populations). The study area was $256\,m^2$ subdivided into 64 plots, each $2 \times 2\,m^2$. (A, original; B and C, calculated after Symonides 1979b.)

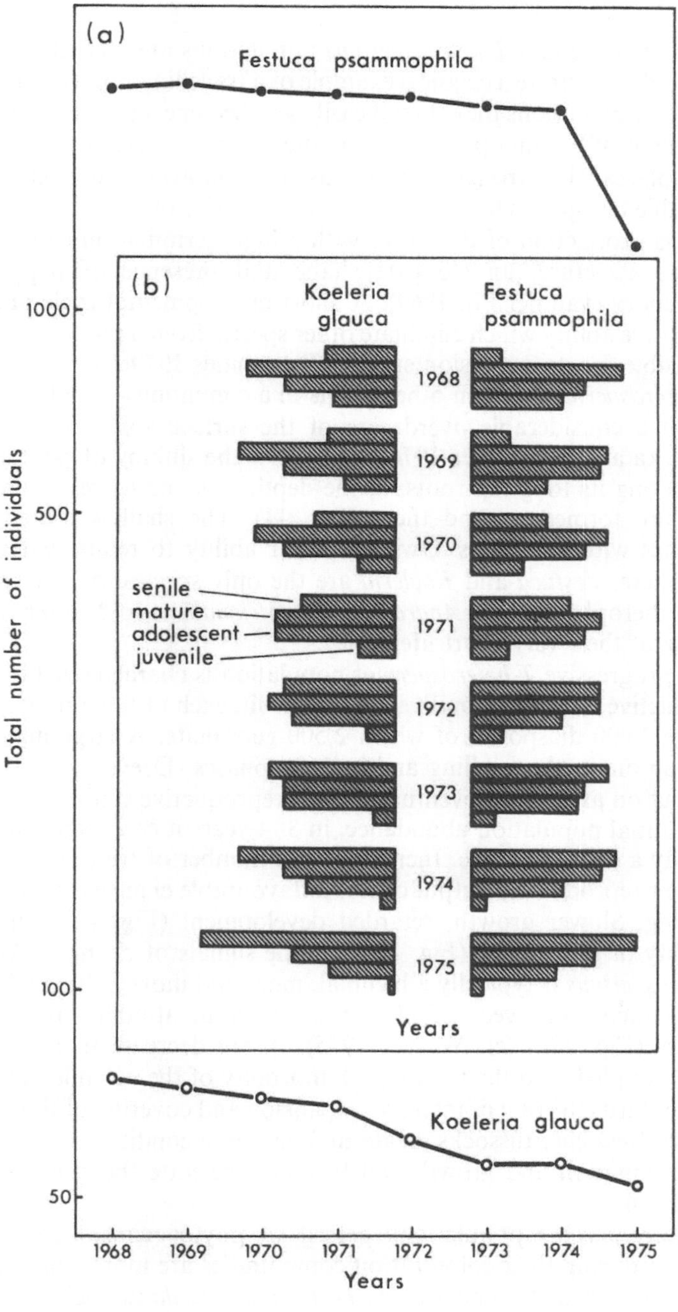

Fig. 10. Changes in abundance (a) and age structure (b) of populations of *Koeleria glauca* and *Festuca psammophila* in Festuco-Koelerietum glaucae patches (256 m^2) as the result of increasing abundance of *Festuca duriuscula* population. (a, Symonides 1979c; b, original.)

(d) *Tragopogon heterospermus*

On the inland dunes *T.heterospermus* populations are immensely expansive and demonstrate a classical example of a travelling population. Under favourable conditions they spread easily and become a transient dominant component of psammophyte communities except on extremely dry sands and in places with strongly turfed sods. The following biological properties enable the species to spread easily: (i) large size of fruiting individuals, (ii) mass production of diaspores with a high germination capacity, (iii) diaspore structure: an elongate shape and presence of pappus, (iv) anemochory (van der Pijl 1969). A short developmental cycle and good competitive ability which eliminate other species from a phytocoenosis are responsible for its expansion success (Symonides 1979b, d).

T.heterospermus acts on other plants in a community mainly indirectly through a considerable overdrying of the surface soil layers (Wilkoń-Michalska and Symonides 1974), a result of the sinking of rainfall water down along its long tap-roots; at the depth of some scores of cm water lenses are formed around them (Fig. 11). The shallow-rooted psammophytes wither in mass. Owing to their ability to retain water in the rhizosphere, *Festuca* and *Koeleria* are the only species to survive. Early Spring therophytes (e.g., *Androsace septentrionalis*) avoid water shortage because of their very short life cycle.

The progressive *T.heterospermus* population is characterized by a high reproductive output: second-year plants fruit, each of them producing on average 3,000 diaspores, of which 2,500 germinate. A large majority of offspring die in the seedling and juvenile phases (Deevey type III) (Fig. 12a), but on average 96 eventually reach reproductive status. Irrespective of the initial population abundance, in 3–4 years it can occupy a habitat and play a dominant role. Increase in the number of fruiting individuals and their reproductive output causes unfavourable changes through overcrowding. Slower growth, retarded development (Fig. 12b) and lower fecundity of individuals (Fig. 12c) are the signals of changes. Although *T.heterospermus* is typically a biennial, more and more individuals flower only in their third year of life, or die without attaining reproductive capacity (Fig. 12a) (cf. Werner 1975). A big decrease in reproductive output coupled with the typical high mortality of the juvenile individuals at first retards the rate of increase; expansion and covering of the patch by *Festuca duriuscula* tussocks create unfavourable conditions for the seedling germination and growth and further accelerate the regression of a population.

T.heterospermus populations persist in phytocoenoses considerably longer, if during their colonization communities are more compact (with 70–80% cover) and dominated by *Festuca duriuscula* tussocks distributed regularly. The interlaced grass roots forming one compact whole make it difficult for *T.heterospermus* seedlings to settle and grow. Under such conditions the population is less abundant and the intensity of its reaction on a phytocoenosis is lower.

Unlike *Carex arenaria*, *T.heterospermus* does not cause the permanent, adverse transformation of a habitat, but changes the competition relationships within a community. It accelerates the colonization of a phytocoenosis with *Festuca duriuscula* and the succession of the Festuco-Koelerietum glaucae association into stable sod vegetation through the elimination and decrease in abundance of most species populations.

(e) *Jasione montana*

J.montana is a common component of the psammophyte vegetation, although owing to its low abundance, cover and productivity its role is apparently unimportant (Wika 1975, Symonides 1979c). Seeds have a low germination capacity: in Petri dishes only 1.1% of seeds germinated and even less than 1% in the dune sands (Symonides 1979c, d). Seeds attain germination capacity in the third year after dissemination and completely

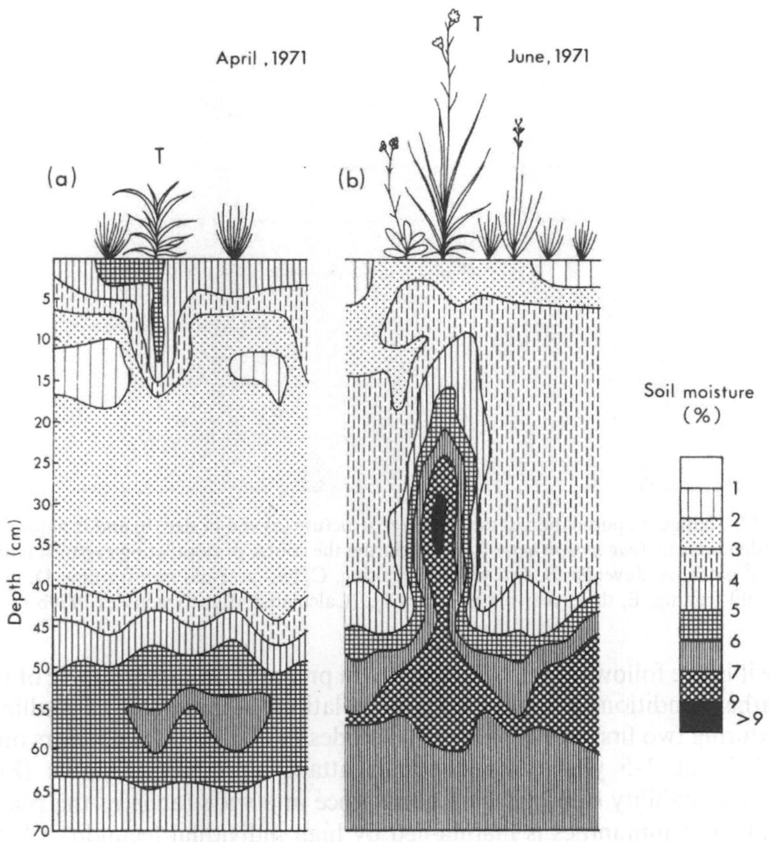

Fig. 11. Effect of (a) juvenile and (b) adult individuals of *Tragopogon heterospermus* (T) on the moisture distribution in dune sands. Mean root length of juvenile individuals is 15.3 cm and 67.9 cm for adults. Transect width shown is 0.5 m (Wilkoń-Michalska and Symonides 1974.)

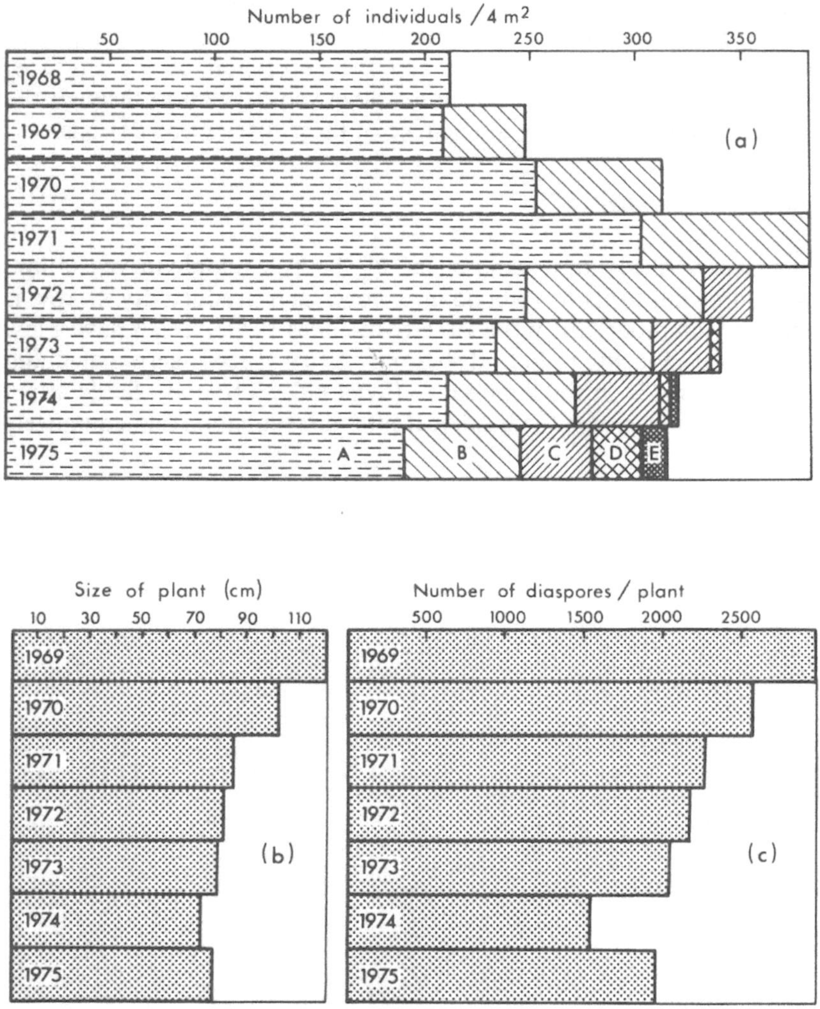

Fig. 12. Changes in population density and age structure (a) and in size (b) and fecundity (c) of individuals in *Tragopogon heterospermus* during the course of population expansion on a $256\,m^2$ patch. A, Juvenile; B, two year-old fruiting; C, two year-old non-fruiting; D, three year-old fruiting; E, three year-old non-fruiting. (Calculated after Symonides 1979b.)

lose it in the following year (Symonides in preparation). Irrespective of the weather conditions and phytocoenotic relations at least 60% of seedlings die during two first weeks of life (Symonides 1977). Among survivors only 20–50% of 2–5 year old individuals attain reproductive status (Fig. 13). The stability of population abundance in a wide range of the psammophyte communities is maintained by high individual fecundity (2,500 seeds on average) and plasticity.

Jasione settles in the loose patches of Spergulo-Corynephoretum, attains maximum abundance in Festuco-Koelerietum glaucae (with a 60% cover) and only retreats from the strongly compacted sods. In different

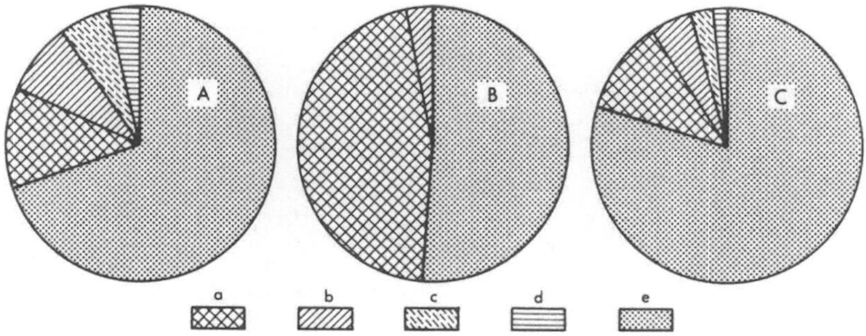

Fig. 13. Proportion of individuals in *Jasione montana* populations which survive the seedling phase attaining reproductive maturity in the (a) second, (b) third, (c) fourth, (d) fifth year of life, and those (e) which do not attain it at all in different phytocoenoses. A, Spergulo-Corynephoretum (40% vegetation cover); B, Festuco-Koelerietum glaucae (60% vegetation cover); C, Festuco-Koelerietum glaucae with *Festuca duriuscula* domination (80% vegetation cover). (A and B, original; C, Symonides 1979c.)

phytocoenoses *Jasione* individuals form small, loose aggregations, each aggregation functioning by itself to a large extent. All of them are characterized by their own size structure and balanced processes of birth and death. The turnover rate and abundance of aggregations oscillate slightly from year to year and these inner dynamics reflect processes in the population as a whole.

In the course of the successional transformations of a habitat *J.montana* populations at first change stepwise, but then attain a high stability in demographic structure and equilibrium between the processes of reproduction and mortality (Fig. 14). In poor Spergulo-Corynephoretum patches *Jasione* individuals grow slowly, mature late and produce few fruits. Only 33% of individuals which survive the seedling phase fruit and only 11–12% in the second year of life. Within the aggregations only the largest individuals attain their reproductive stages in the second year of life; the remainder, in spite of being the same age, one, two or even three years later. Seedling numbers (low) and population mortality vary slightly between years.

In the loose Festuco-Koelerietum glaucae patches a three-fold higher abundance is maintained. Individuals grow more rapidly, attain reproductive maturity and die earlier, behaving as monocarpic plants, than in the initial phases of the population development. Consequently, the age structure and the rate of individual turnover change. The total seed number increases three-fold, caused by a higher fraction of mature individuals. Owing to more favourable conditions the fraction of the seeds which germinate is several times higher. But the reproductive output per one individual decreases four-fold (Symonides unpublished data).

The more compact the vegetation, the lower the population abundance. Like harsh habitat conditions in the initial growth stage of a population, the reaction of other species in the phytocoenosis results in a significantly retarded growth and development of individuals. Only 20% of juvenile

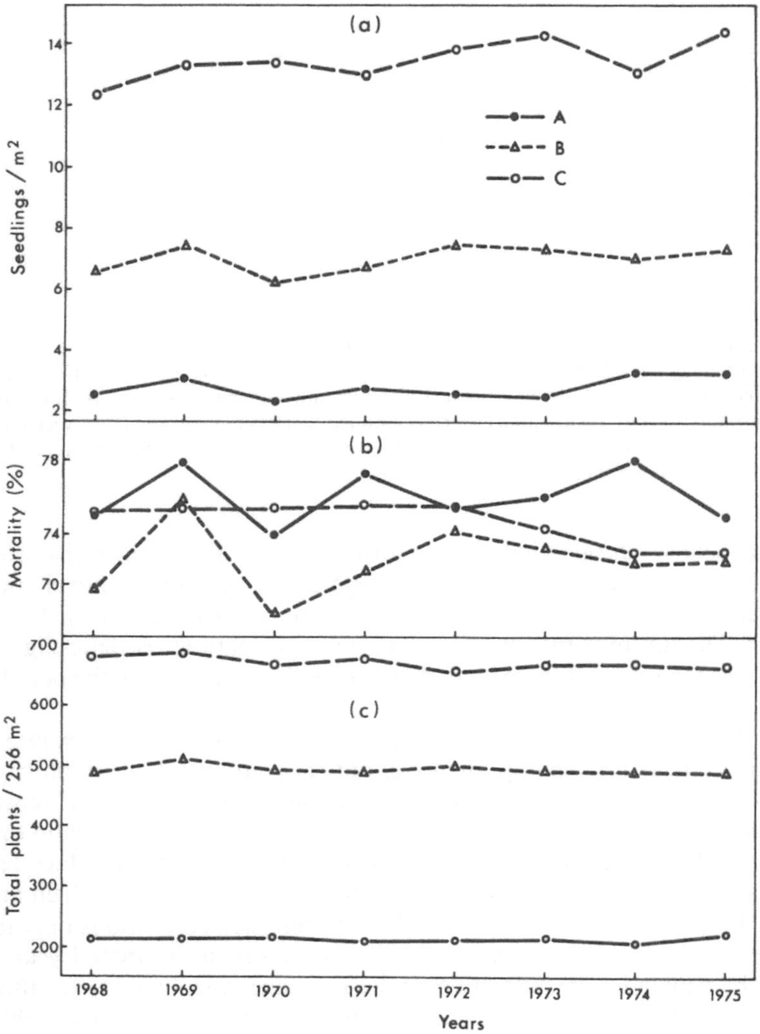

Fig. 14. Demographic indices of (a) seedling numbers, (b) mortality, and (c) abundance of *Jasione montana* populations in different phytocoenoses. A, Spergulo-Corynephoretum (40% vegetation cover); B, Festuco-Koelerietum glaucae (60% vegetation cover); C, Festuco-Koelerietum glaucae with *Festuca duriuscula* domination (80% vegetation cover). The study area in each vegetation type was $256\,m^2$. (A and B, original; C, Symonides 1979c.)

individuals attain reproductive status. The number of seeds produced is also considerably lower on account of the smaller number of fruiting individuals. On the other hand the fraction of seeds which germinate increases: from 0.004 in Spergulo-Corynephoretum, through 0.02 in Festuco-Koelerietum glaucae to 0.99 in compact sods, assuming that emigration and immigration processes of seeds are completely balanced.

Equilibria in the abundance of *Jasione* populations are attained in different phytocoenses and habitats in different ways. They are outcomes of the rate of development and longevity of individuals, as well as the reproductive capacity and mortality of populations. Unfortunately, the mechanisms determining the relative constancy of this type of population in vegetation and their role in the functioning of phytocoenoses have not yet been elucidated.

CONCLUSIONS

A detailed analysis of the structure and dynamics of psammophyte vegetation in the course of the phytocoenosis formation and succession point towards the following conclusions. (i) From the beginning psammophyte communities have the nature of an organized population system. (ii) The formation and succession of communities result from a complex but not chaotic interplay of environmental, population and phytocoenotic factors. (iii) The selection of species is being accomplished in the initial community phases; thus from the beginning the vegetation has a definite ecological nature and constitutes the initial developmental stage of a strictly specified association. (iv) The direction of succession is, in substance, already determined during the formation of initial patches, but the rate and course of changes depend on the course of population phenomena of the individual components of a phytocoenosis. (v) Demographic processes of individual populations are strictly connected with the complex of habitat and phytocoenotic relationships. (vi) In these simple dune systems the abundances of individual populations determine above all their dynamic value and habitat-forming role.

In the initial phases of the development of plant communities the main role is played by those populations, (a) which are capable of spreading fast and increasing their numbers efficiently; (b) whose high fecundity and at the same time fast turnover of individuals determine a fast transformation of the habitat by enriching infertile sands with organic matter and initiating the soil-forming process; (c) which owing to their aggregated spatial structure create in their vicinity a specific microclimate which is more favourable also for those species that enter at a later stage. It is the life activity of the pioneer plants that initiates spatial habitat diversification which becomes intensified with the development of communities.

The diversity of an initial community is manifested only by individual variation. Gradually there also occurs an increasing differentiation in respect of the growth phases and individual age. This is followed by the differentiation of the action on the environment and of the utilization of its resources. The incipient phytocoenotic relationships which are being formed at the stage of a two-species phytocoenosis become more complex with the appearance of further components. The diversity of growth forms, phenology, methods of action on the environment and utilization of its resources leads on to a gradual differentiation of niches and their gradual filling.

As conditions affecting the growth and development of seedlings deteriorate, the populations of perennials, at once able to reproduce vegetatively and generatively, assume a high dynamic value. A low mortality of the offspring in such populations, the possibility of producing offspring many times, as well as longevity of the individuals make it possible for the system to attain stability.

REFERENCES

Abd El Rahman, A.A. and Batanouny, K.H. (1965a) Transpiration of desert plants under different environmental conditions. J. Ecol. 53: 267–272.

Abd El Rahman, A.A. and Batanouny, K.H. (1965b) The water output of the desert vegetation in the different microhabitats of Wadi Hoff. J. Ecol. 53: 139–145.

Bell, K.L., Hiatt, H.D. and Niles, W.E. (1979) Seasonal changes in biomass allocation in eight winter annuals of Mojave Desert. J. Ecol.: 67: 781–787.

Berger-Landefeldt, U. and Sukopp, H. (1965) Zur Synökologie der Sandtrockenrasen, inbesondere der Silbergrasflur. Verh. bot. Ver. Prov. Brandenb. 102: 41–98.

Chadwick, H.W. and Dalke, P.D. (1965) Plant succession on dune sands in Fremont county, Idaho. Ecology 46: 765–780.

Fukarek, F. (1961) Die Vegetation des Darss und ihre Geschichte. Pflanzensoziologie 12: 1–321.

Gadgil, M. (1971) Dispersal: population consequences and evolutions. Ecology 52: 253–256.

Gadgil, M. and Bossert, W.H. (1970) Life historical consequences of natural selection. Amer. Nat. 104: 1–24.

Harper, J.L. (1977) Population Biology of Plants. London: Academic Press. 896pp.

Hayashi, J. and Numata, M. (1967) Ecology of pioneer species of early stages in secondary succession. I. Bot. Mag., Tokyo 80: 11–22.

Hayashi, J. and Numata, M. (1968) Ecology of pioneer species of early stages in secondary succession. II. The seed production. Bot. Mag., Tokyo, 81: 55–56.

Krausch, H.D. (1962) Vorschläge zür Gliederung der mitteleuropaischen Sand- und Silikat-Trockenrasen. Mitt. Flor.-soz. Arbeitsgem.N.F.9.

Krausch, H.D. (1962) Vorschläge zür Gliederung der mitteleuropäischen Sand- und Silikat-Trockenrasen. Mitt. Flor.-soz. Arbeitsgem.N.F.9: 266–269.

Linhart, Y.B. and Pickett, R.A. (1973) Physiological factors associated with density-dependent seed germination in *Boisduvallia glabella* (Onagraceae). Z. Pflanzenphysiol. 70: 367–370.

Maarel van der, E. (1981) Fluctuations in a coastal dune grassland due to fluctuations in rainfall: experimental evidence. Vegetatio 47: 259–265.

Mack, R.N. (1976) Survivorship of *Cerastium atrovirens* at Aberffraw, Anglesey. J. Ecol. 64: 309–312.

Marsh, W.M. and Koerner, J.M. (1972) Role of mosses in slope formation. Ecology 53: 489–493.

Marshall, J.K. (1965) *Corynephorus canescens* (L.) P.Beauv. as a model for the *Ammophila* problem. J. Ecol. 53: 447–463.

Marshall, J.K. (1967) *Corynephorus canescens* (L.) P.Beauv. J. Ecol. 55: 207–220.

Meijden, van der, E. and Waals-Kooi, van der, R.E. (1979) The population ecology of *Senecio jacobaea* in a sand dune system. I. Reproductive strategy and the biennial habit. J. Ecol. 67: 131–153.

Mrózek, W. (1958) Wydmy Kotliny Toruńsko-Bydgoskiej. pp. 7–59 in Wydmy śródladowe Polski, II, R. Galon (Ed.). Warszawa.

Newman, E.I. (1963) Factors controlling the germination date of winter annuals. J. Ecol. 51: 625–638.

Newman, E.I. (1965) Factor affecting the seed production of *Teesdalia nudicaulis*. II. Soil moisture in spring. J. Ecol. 53: 211–232.

Newman, E.I. (1967) Response of *Aira praecox* to weather conditions. I. Response to drought in spring. J. Ecol. 55: 539–556.

Noble, J.C., Bell, A.D. and Harper, J.L. (1979) The population biology of plants with clonal growth. I. The morphology and structural demography of *Carex arenaria*. J. Ecol. 67: 983–1008.

Pemadasa, M.A. and Lovell, P.H. (1974) The mineral nutrition of some dune annuals. J. Ecol. 62: 855–868.

Pemadasa, M.A. and Lovell, P.H. (1975) Factors controlling germination of some dune annuals. J. Ecol. 63: 41–59.

Pijl, L. van der (1969) Principles of Dispersal in Higher Plants. Berlin: Springer-Verlag. 154pp.

Rychnovská, M. (1961) An outpost site of *Corynephorus canescens* in the region between the Danube and the Tiza and its causal explanation. Acta biol. Szeged 14: 57–66.

Rychnovská-Soudkova, M. (1963) *Corynephorus canescens* (L.) P.Beauv. (Physiologisch-ökologische Studie einer Pflanzenart). Rozpr. čsl. Akad. Věd. 71: 1–84.

Rychnovská, M. and Květ, J. (1963) Water relations of some psammophytes with respect to their distribution. pp. 190–198 in The Water Relations of Plants, A.J. Rutter and F.H. Whitehead (Eds.). Oxford: Blackwells.

Salisbury, E.J. (1942) The Reproductive Capacity of Plants. Studies in Quantitative Biology. London: G. Bell and Sons. 244pp.

Salisbury, E.J. (1952) Downs and Dunes. Their Plant Life and its Environment. London: G. Bell and Sons. 328pp.

Stoutjesdijk, Ph. (1959) Heaths and inland dunes of the Veluwe. Wentia 2: 1–96.

Symonides, E. (1974a) Populations of *Spergula vernalis* Willd. on dunes in the Toruń Basin. Ekol. pol. 22: 379–416.

Symonides, E. (1974b) The water economics of *Spergula vernalis* Willd. Ekol. pol. 22: 457–472.

Symonides, E. (1974c) Morphological variability of *Spergula vernalis* Willd. from different dune biotopes of the Toruń Basin. Ekol. pol. 22: 417–440.

Symonides, E. (1977) Mortality of seedlings in natural psammophyte populations. Ekol. pol. 25: 635–651.

Symonides, E. (1978a) Effect of population density on the phenological development of individuals of annual plant species. Ekol. pol. 26: 273–286.

Symonides, E. (1978b) Effect of seed size, density and depth of sowing on the germination and survival of psammophyte seedlings. Ekol. pol. 26: 123–139.

Symonides, E. (1978c) Number, distribution and specific composition of diaspores in the soil of the plant association Spergulo-Corynephoretum. Ekol. pol. 26: 111–122.

Symonides, E. (1979a) The structure and population dynamics of psammophytes on inland dunes. I. Population of initial stages. Ekol. pol. 27: 3–37.

Symonides, E. (1979b) The structure and population dynamics of psammophytes on inland dunes. II. Loose-sod populations. Ekol. pol. 27: 191–234.

Symonides, E. (1979c) The structure and population dynamics of psammophytes on inland dunes. III. Populations of compact psammophyte communities. Ekol. pol. 27: 235–257.

Symonides, E. (1979d) The structure and population dynamics of psammophytes on inland dunes. IV. Population phenomena as a phytocoenose-forming factor (a summing-up discussion). Ekol. pol. 27: 259–281.

Symonides, E. (1983) Population size regulation as a result of intra-population interactions. I. Effect of density on the survival and development of individuals of *Erophila verna* (L.) C.A.M. Ekol. pol. 31(4): 839–881.

Watkinson, A.R. and Harper, J.L. (1978) The demography of a sand dune annual: *Vulpia fasciculata*. I. The natural regulation of populations. J. Ecol. 66: 15–33.

Werner, P.A. (1975) Predictions of fate from rosette size in teasel (*Dipsacus fullonum* L.). Oecologia 20: 197–201.

Wika, S. (1975) Vegetation of sward communities in the surroundings of Kamionna and Dormowo in the district Międzychód. Pozn. Tow. Przyj. Nauk, 40: 5–48.

Wilkoń-Michalska, J. (1976) Structure and dynamics of populations of *Salicornia patula* Duval-Jouve. Uniwersytet Mikotaja Kopernika, Rozprawy, Toruń. 156pp.

Wilkoń-Michalska, J. and Symonides, E. (1974) The influence of vegetation on the dynamics of soil moisture in the patch of Spergulo-Corynephoretum (Tx. 1928) Libb. 1933. Fragm. flor. geobot. 20: 497–528.

Ziembinska, H. (1969) Notes on Toruń's climate. Zesz. nauk. UMK, Toruń, Geografia, 6: 91–108.

VEGETATION DYNAMICS OF PRAIRIE GLACIAL MARSHES

A.G. VAN DER VALK

ABSTRACT

All vegetation change can be reduced to one of three basic phenomena, succession, maturation, and fluctuation, or some combination of these. Each of these phenomena is a result of a change in some attribute of one or more of the plant populations comprising the vegetation of an area. Succession ocurs when different populations are present from time to time. Maturation is an increase in the biomass of an area which is the result of a change in the age/size structure of the populations with time. Fluctuations result from changes in the number of individuals or ramets in the populations of an area from year to year.

The contribution of succession, maturation, and fluctuation to the vegetation dynamics of Eagle Lake, a prairie glacial marsh in Iowa, is examined. In those areas where changing water levels and extensive muskrat damage occur, succession is the most important phenomenon. A knowledge of the life-history characteristics of each species, particularly its establishment requirements, the presence or absence of its seeds in the seed bank, and its life-span, enables successional sequences to be predicted in this marsh. There are short periods where maturation is the major phenomenon causing vegetation change. Fluctuations also occur both in the emergent vegetation and the submerged vegetation.

INTRODUCTION

Vegetation is everchanging. It changes both qualitatively in its floristic composition and quantitatively in its floristic or physical structure over time.

Any qualitative change in the vegetation of an area from year to year is here defined as succession (Fig. 1). This definition of succession, proposed by van der Valk (1981, 1982), is consistent with the reductionist view of vegetation first championed in North America by H.A. Gleason (1917, 1927, 1939). It differs significantly from the definition of succession, the development of climax vegetation, formulated by F.E. Clements

White, J. (ed.), The Population Structure of Vegetation.
 ISBN 978-94-010-8927-2

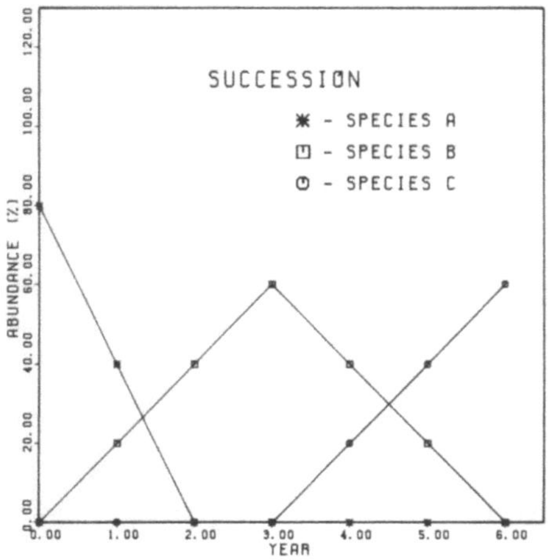
SUCCESSION
✱ - SPECIES A
⊡ - SPECIES B
⊙ - SPECIES C
ABUNDANCE (%)
0.00
20.00
40.00
60.00
80.00
100.00
120.00
YEAR
0.00
1.00
2.00
3.00
4.00
5.00
6.00

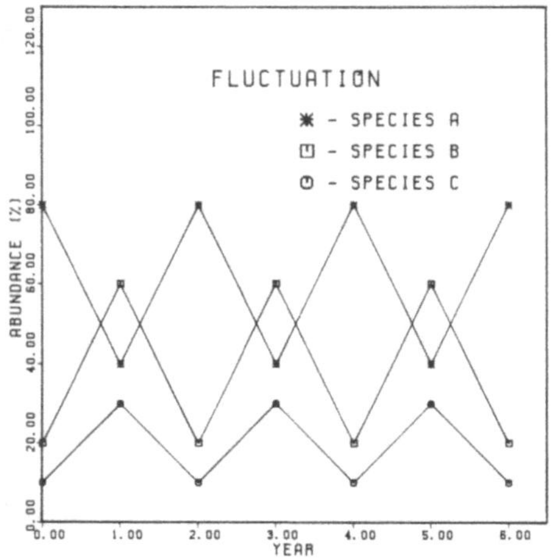
FLUCTUATION
✱ - SPECIES A
⊡ - SPECIES B
⊙ - SPECIES C
ABUNDANCE (%)
0.00
20.00
40.00
60.00
80.00
100.00
120.00
YEAR
0.00
1.00
2.00
3.00
4.00
5.00
6.00

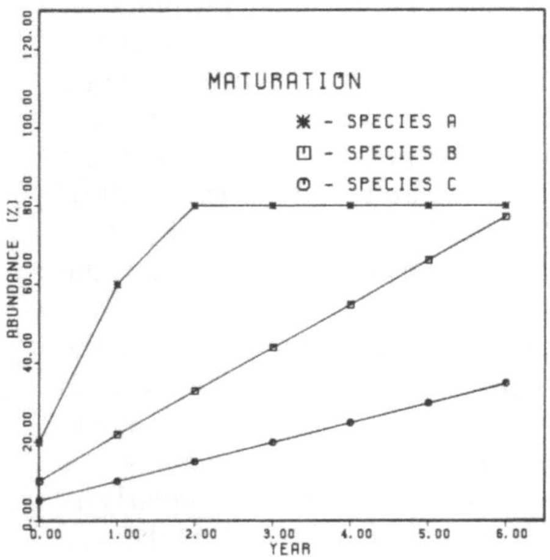
MATURATION
✱ - SPECIES A
⊡ - SPECIES B
⊙ - SPECIES C
ABUNDANCE (%)
0.00
20.00
40.00
60.00
80.00
100.00
120.00
YEAR
0.00
1.00
2.00
3.00
4.00
5.00
6.00

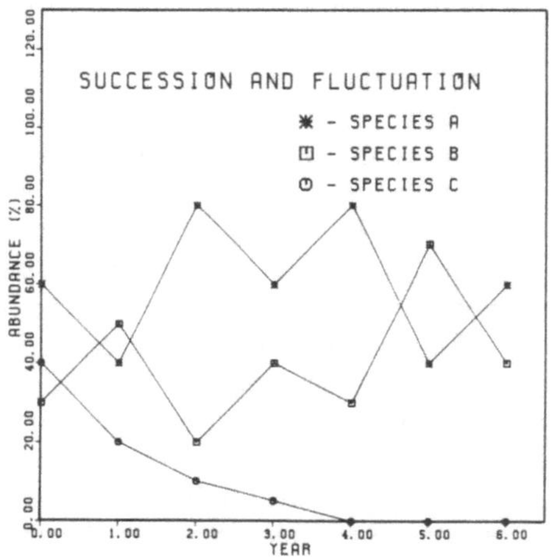
SUCCESSION AND FLUCTUATION
✱ - SPECIES A
⊡ - SPECIES B
⊙ - SPECIES C
ABUNDANCE (%)
0.00
20.00
40.00
60.00
80.00
100.00
120.00
YEAR
0.00
1.00
2.00
3.00
4.00
5.00
6.00

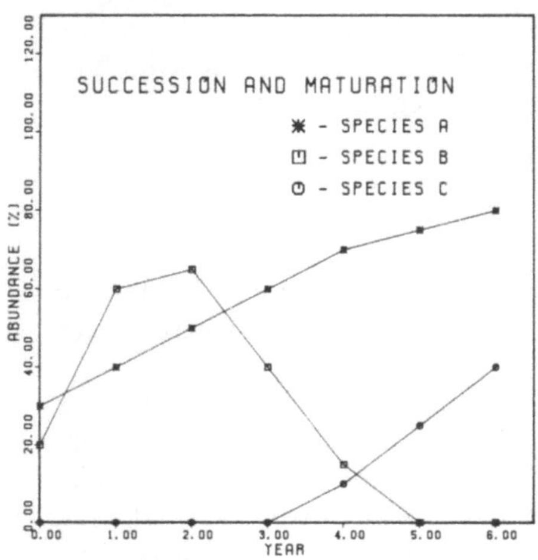
SUCCESSION AND MATURATION
✱ - SPECIES A
⊡ - SPECIES B
⊙ - SPECIES C
ABUNDANCE (%)
0.00
20.00
40.00
60.00
80.00
100.00
120.00
YEAR
0.00
1.00
2.00
3.00
4.00
5.00
6.00

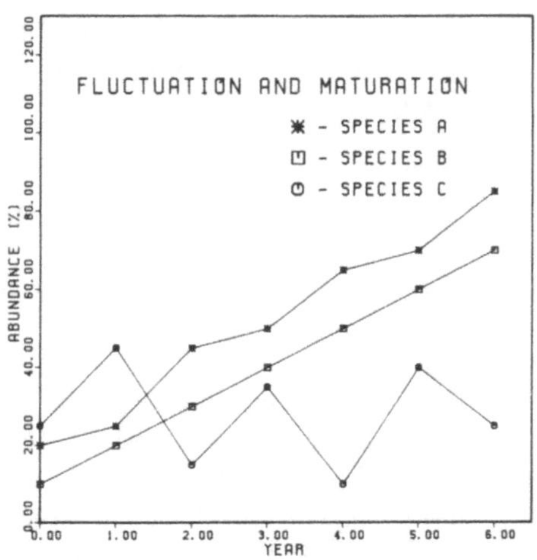
FLUCTUATION AND MATURATION
✱ - SPECIES A
⊡ - SPECIES B
⊙ - SPECIES C
ABUNDANCE (%)
0.00
20.00
40.00
60.00
80.00
100.00
120.00
YEAR
0.00
1.00
2.00
3.00
4.00
5.00
6.00

(1916, 1936, Weaver and Clements 1938). I will not review the impact of Clements' views about the nature of vegetation change on the development of ecological theory, nor the various attempts to modify, update or refute his ideas. There are two excellent reviews of the concept of succession by McIntosh (1980, 1981), and a critical review of successional concepts as applied to wetlands by van der Valk (1982). I have adopted a Gleasonian definition of succession for two basic reasons: it is free of the implicit and explicit assumptions about the ontogenetic nature of vegetation change that underlie Clements' definition, and it provides an unambiguous criterion for recognizing occurrences of succession.

Quantitative changes in the vegetation of an area from year to year are produced by two different phenomena. In vegetation dominated by perennials, particularly woody perennials, there is generally an annual increase in the standing crop of living plants. This accumulation of biomass from year to year is accompanied often by a change in the physiognomy or physical structure of the vegetation. This type of quantitative change is defined as maturation (Fig. 1). It is an irreversible phenomenon that takes a minimum of two years to complete, but usually takes much longer. It is an important phenomenon in forest vegetation and in some types of deserts, but is less important in perennial herbaceous vegetation.

A second type of quantitative change in vegetation is the result of differences in the abundance of species in an area from year to year caused by dissimilar environmental conditions, particularly meteorological and hydrological conditions. A change in the floristic structure of vegetation without a change in its floristic composition is a fluctuation (Rabotnov 1974). These short-term changes are a direct response of plant populations in an area to changes in environmental conditions and are thus reversible. Fluctuations are most common in vegetation dominated by perennial herbaceous species, e.g., grasslands and marshes, where they can cause significant changes in the annual primary production and floristic structure of the vegetation, but have less of an impact on its physical structure. How much structural change occurs in vegetation as a result of fluctuations depends primarily on whether the vegetation is composed of one, or more than one, growth form. If species with different growth forms alternate in abundance, then fluctuations can result in significant structural changes from year to year.

Succession, maturation, and fluctuation alone or in some combination are responsible for all changes in vegetation. Two or three of these phenomena frequently occur simultaneously. For example, in newly established forest vegetation, maturation is accompanied usually by succession. The closing of the canopy as young trees grow and the subsequent environmental changes often result in the loss of ruderal species and in the establishment of understory species. Changes in vegetation caused by interactions of succession, maturation, and fluctuations are illustrated in Fig. 1.

←

Fig. 1. Changes in the abundance of three species over time in a hypothetical community that result in succession, fluctuation, maturation, or some combination of these three phenomena.

The failure to emphasize that vegetation change can be caused by several distinct phenomena has hindered the development of a unified theory of vegetation dynamics. As a consequence, two quite different approaches to vegetation dynamics have developed (McIntosh 1981). The earliest group of ecologists to study vegetation change were community ecologists. They have been concerned traditionally with documenting floristic changes in an area over time, and to discover the reasons for them (Braun-Blanquet 1932, 1964, Weaver and Clements 1938, Oosting 1956, Daubenmire 1968). Later, ecosystem ecologists, influenced by the work of Lindeman (1942), Odum and Pinkerton (1955), Margalef (1963), and Odum (1969) began to study changes both in the primary production of vegetation over time and in the movement of energy from one trophic level to another, in what usually were interpreted to be the different stages of Clementsian succession. These two groups have, in fact, been studying two different phenomena, although they often seem to be unaware of it. Community ecologists are concerned primarily with Gleasonian succession, ecosystem ecologists primarily with maturation. Because studies of these two different phenomena require disparate research techniques and different theoretical frameworks for presenting the results, these two types of ecologist often have little in common and, as a result, frequently ignore each other's work in spite of the fact that they claim to be studying the same phenomenon (McIntosh 1981).

There is an essential underlying uniformity for all vegetation change: vegetation change is always the result of change in the populations of plants that comprise the vegetation of an area. The establishment of new populations or the extirpation of existing populations causes succession. The growth of individuals in established populations, sometimes accompanied by an increase in the size of the populations, causes maturation. The recruitment or loss individuals or ramets and differences in the size of individuals or ramets from year to year within established populations causes fluctuation. Thus all vegetation change can be understood and predicted from a knowledge of the life-history characteristics of the species that compose the vegetation, since these life-history characteristics determine how each population will respond to changes in environmental conditions. However, even with excellent information about the life-history characteristics and the impact of environmental changes on the size and age structure of populations of each species, not all types of vegetation change can be predicted accurately. Maturation and fluctuation can be predicted relatively easily and accurately, since these are, at least in part, the result of ontogenetic changes for a given set of species. Succession always will be more difficult to predict, particularly on a small geographic scale because of the vagaries of local seed dispersal, climate, and disturbance.

The recruitment, growth rate, and loss of individuals in a population is controlled by three basic factors: the physical-chemical environment, interactions among the plant species, and predators and parasites. How much of the vegetation change at a site over a particular period of time is the result of each of these three factors is one of the unresolved questions of plant ecology. These three factors will be referred to as first-order interactions (the impact of the environment on the establishment, growth, and survival of individuals of a species), second-order interaction (interference competition, exploitative competition, and allelopathy), and third-order interactions (disease and herbivory). In this tripartite scheme, disturbance is treated as a temporary, extreme first- or third-order interaction. If succession, fluctuation, and maturation are due to first order interactions then all three phenomena will be relatively easy to understand and to predict. If second- and third-order interactions are major factors producing vegetation change, then significantly more information will be needed before vegetation changes can be either understood or predicted.

I will use information collected in prairie glacial marshes, particularly studies done at Eagle Lake in Iowa, to examine succession, maturation, and fluctuation in such wetlands, and also to illustrate how changes at the population level cause these three types of vegetation change at the community level.

PRAIRIE GLACIAL MARSHES

Prairie glacial marshes, or prairie pothole marshes as they are also called, are found in central North America in an area that was covered both by drift from Wisconsin-Age glaciers during the Pleistocene and by prairie at the time of European settlement. This region has poorly developed drainage and is dotted with numerous small, undrained depressions, locally called "potholes". Geographically the prairie glacial region includes southern Alberta, southern Saskatchewan, and southwestern Manitoba in Canada, and northeastern Montana, northern and eastern North Dakota, eastern South Dakota, western Minnesota, and northwestern Iowa in the United States (Stewart and Kantrud 1971). Most studies of vegetation change in this type of wetland have been done in Iowa near the southern limit of this wetland type. Eagle Lake, a 365 ha marsh in northcentral Iowa, has been one of the most intensively studied prairie glacial marshes in Iowa.

Eagle Lake is a shallow marsh, the deepest sections having a maximum water depth of about 1 m. Water depth in this marsh can be raised or lowered by means of a dam at the northern end. Because Eagle Lake, like all prairie glacial marshes, has a small watershed, its water level at any time is directly dependent on the amount of recent precipitation. As a result of regional droughts, since the early 1950s there have been three periods when water levels at Eagle Lake dropped low enough to expose all or most of the marsh bottom (Bishop et al. 1978). The last of these natural "drawdowns", as they are called, occurred in 1977. There have also been two artificial drawdowns in the last thirty years, one in 1964 and another in 1974. Artificial drawdowns involve draining the marsh by lowering the level of the dam. They were used to re-establish emergent vegetation in the northern section.

Vegetation in the northern, deeper half of Eagle Lake undergoes cyclical changes over a period of 5–30 years. This part of the marsh will be referred to as the cyclic area. Non-cyclic sections of the marsh include all of the shallow southern part of the marsh and a narrow fringe around most of the northern part.

Vegetation cycles in prairie marshes

Prairie glacial marshes undergo cyclical changes in their vegetation that were first described by Weller and Spatcher (1965). There are four stages in an idealized cycle (van der Valk and Davis 1978a): dry marsh, regenerating marsh, degenerating marsh, and lake marsh (Fig. 2). Table 1 outlines the changes of the most abundant species during all four stages of a vegetation cycle at Rush Lake in northwest Iowa.

During the dry marsh stage (Fig. 2) most of the marsh is free of standing water and initially free of vegetation. The absence of standing water is normally the result of a drought. The absence of vegetation, particularly emergent vegetation, is due to the local eradication of emergent species during the previous two stages of the cycle. During the dry marsh stage, the seeds of mud-flat annuals (*Bidens cernua*, *Cyperus* spp., *Polygonum* spp., *Rumex maritimus*, etc.) and some perennial emergents (*Scirpus validus*, *Typha glauca*, *Sagittaria latifolia*) germinate. The seeds of both

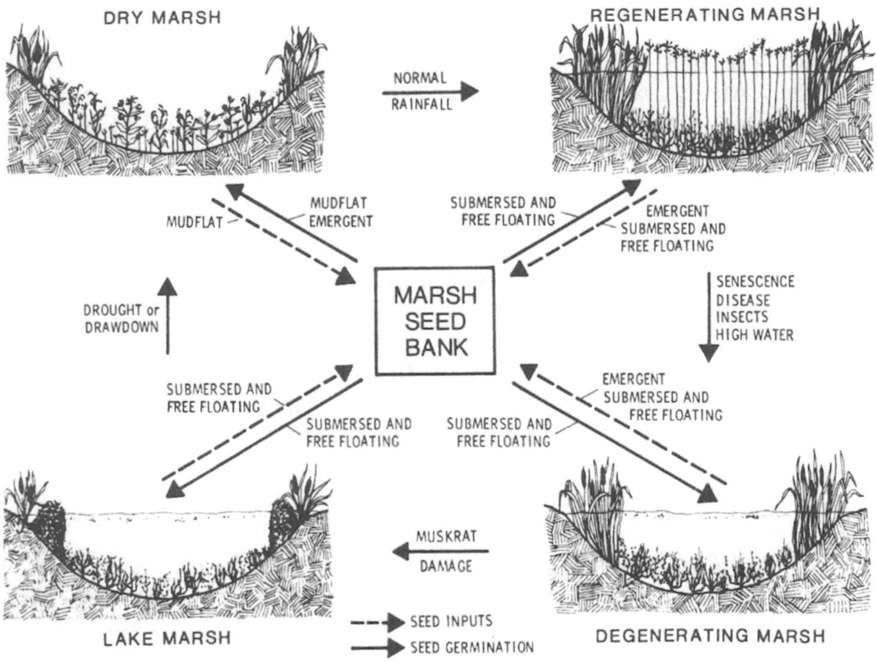

Fig. 2. A generalized outline of a vegetation cycle in a prairie glacial marsh. (Reprinted from van der Valk and Davis 1978a.)

Table 1. The frequency (%) of species present in the four stages (dry marsh, regenerating marsh, degenerating marsh, and lake marsh) of a vegetation cycle at Rush Lake, Iowa. Only species with a frequency of 5% or more at least one year are included. About 6,900 points were sampled each year. (Data from Weller and Fredrickson 1974).

Species	*Dry* 1964	*Regenerating* 1965	1966	1967	*Degenerating* 1968	1969	*Lake* 1970
	Mud-flat annuals						
Bidens cernua	13.2	0.4	0.0	0.0	0.0	0.0	0.0
Cyperus spp.	7.8	0.0	0.0	0.0	0.0	0.0	0.0
Polygonum spp.	7.6	0.1	0.1	<0.1	<0.1	0.0	0.0
	Emergent species						
Sagittaria latifolia	15.7	8.6	4.1	1.2	0.0	0.0	0.0
Scirpus validus	23.0	12.0	4.5	0.3	<0.1	0.0	0.0
Typha glauca	17.1	24.6	36.1	34.1	27.9	6.0	1.0
	Submersed species						
Anacharis sp.	6.7	0.0	<0.1	0.0	0.0	0.0	0.0
Ceratophyllum demersum	2.6	6.4	24.2	6.5	19.7	19.0	1.9
Potamogeton pectinatus	3.5	12.3	17.5	4.9	8.1	20.6	3.9
Utricularia vulgaris	0.0	10.6	12.5	13.8	7.8	6.1	3.1
	Free-floating species						
Lemna minor	2.7	41.4	66.8	71.1	56.8	40.9	18.1
Riccia fluitans	0.0	1.5	0.0	0.0	0.0	27.6	5.3
Spirodela polyrhiza	0.4	6.8	9.8	11.6	34.7	24.3	5.3

types of species are common in the surface sediments of the marsh, i.e., in the seed bank. Other species whose propagules reach the marsh during the dry marsh stage can also become established, e.g., *Salix* spp. and *Populus* spp., but this did not occur at Rush Lake (Table 1). When the drought ends, the marsh quickly refloods. This prevents further germination of the seeds of mud-flat annual and emergent species and quickly eliminates the established annual populations. Seedlings and adults of emergents that become established during the dry marsh stage, however, persist (van der Valk and Davis 1978a).

The reflooding of the marsh marks the beginning of the regenerating stage (Fig. 2). During this stage, the emergent populations spread as a result of their clonal growth, individual culms of the emergent species reach their full size, and submersed and free-floating species become re-established in the marsh. The most common submersed species to reappear during this stage are *Ceratophyllum demersum*, *Najas flexilis*, *Potamogeton pectinatus*, and *Utricularia vulgaris*. *Lemna* spp., *Riccia fluitans*, and *Spirodela polyrhiza* are the most common free-floating species that become established.

After a number of years of high water levels, the emergent populations of these marshes begin to decline. The first emergent species to wane is normally *Scirpus validus*, often one of the most common species during the dry and regenerating marsh stages. *Scirpus validus* populations can not survive more than two or three years of continuous standing water (Harris and Marshall 1963, Weller and Spatcher 1965, and van der Valk and

Davis 1980). At Rush Lake, the *Scirpus validus* population began to decline as soon as the marsh reflooded (Table 1). The decline of the Rush Lake population occurred much earlier after reflooding than did the similar decline at Eagle Lake where the *Scirpus validus* population did not decline for a couple of years after reflooding, but then succumbed all at once. The pattern and rate of decline of *Scirpus validus* seems to be a function of water depth, but this has never been confirmed experimentally. Other emergent species show a similar decline when a marsh refloods, e.g., *Sagittaria latifolia* at Rush Lake (Table 1). The elimination of the *Scirpus validus* population from a marsh may or may not mark the start of the degenerating stage.

A reduction in the overall annual biomass production of the emergent vegetation is the criterion used for recognizing the start of the degenerating stage. At Rush Lake the annual production of the emergent vegetation is estimated to have increased from about 540 metric tonnes (mt) in 1965 to 640 mt in 1966 and then declined to 560 mt in 1967, marking the beginning of the degenerating marsh stage (van der Valk and Davis 1978b). During the regenerating stage at Rush Lake, decreases in the annual primary production of both *Scirpus validus* and *Sagittaria latifolia* populations were more than offset by the increase in the annual production of the *Typha glauca* population (van der Valk and Davis 1978b). Muskrat (*Ondatra zibethica*) populations, which increase both in density and areal coverage from the beginning of the regenerating stage (Weller and Spatcher 1965), reach their maximum size during the degenerating stage. Their feeding and lodge-building activities eventually destroy the emergent vegetation in the deep sections of a marsh. This rapid destruction of the emergent vegetation of a marsh by muskrats is referred to as an "eatout". These eatouts produce the fourth stage of the cycle, the lake stage, during which the marsh resembles a small lake or pond (Fig. 2).

During the lake stage, only a narrow fringe of emergents may remain along the shores of the marsh (Fig. 2). Submersed vegetation often declines during this stage because of the increased turbidity of the water caused by wave action stirring up the sediments or by increased algal growth (Table 1). Blue-green algal blooms are common during the lake stage of many marshes. A marsh will remain in this stage until the next drawdown.

Vegetation change during a cycle in a prairie glacial marsh is caused by first-order interactions (primarily changes in water level) that control the germination of the seeds and the size of the emergent populations, and by a third-order interaction (muskrats destroying the emergent vegetation). Second order interactions seem to be of little or no significance because the vegetation is decimated so quickly after it becomes established. Emergent vegetation during the regenerating and early degenerating stages of marshes that undergo vegetation cycles usually has two or three co-dominant species. In other marshes or in other areas in the same marsh that do not have these cycles (because they are too shallow), emergent vegetation is usually composed of monodominant stands (van der Valk and Davis 1979). Competition and/or allelopathy may very well be important in these latter situations, as is the case in other types of wetlands (Buttery and Lambert 1965, Grace and Wetzel 1981).

Changes in the composition and structure of wetland vegetation caused by water level changes also have been described in a number of other wetlands around the world. Hejny and Husak (1978) outline what they refer to as an ecocycle in Czechoslovakian fishponds that are drained periodically. Changes in the vegetation of these ponds occur during the periods when they are drained and in the following years when water levels return to normal are very similar to those found in prairie glacial marshes in the dry, regenerating, and degenerating marsh stages. Likewise changes in the hydrology of wet meadows along streams result in changes in the composition and location of these wetlands along these streams (Grootjans 1980). Wetlands in monsoonal climates and in other climates with distinct annual wet and dry seasons often undergo yearly changes in their vegetation that in many ways parallel the longer cycles found in prairie glacial marshes and other temperate wetlands (Saxton 1924, Misra 1946).

SUCCESSION

An examination of the life-history characteristics of plant species found in prairie glacial marshes indicates that information about three features of their life histories make it possible to understand and thus predict Gleasonian succession. These three life-history features are life-span, seed or propagule longevity, and seed germination or propagule establishment requirements.

Life-span

Wetland species have only three basic types of life-span: annual, perennial with a fixed life-span, and perennial with an indefinite lifespan. In prairie glacial marshes, annuals (A-species) are found growing on mud-flats and other areas during the dry marsh stage, and a second group composed of free-floating and submersed aquatics is found during the other three stages. Perennial plants with a fixed life-span (P-species) may or may not have clonal growth. Perennial species with clonal growth that have an indefinite life-span are designated V-species; potentially they can survive in a wetland forever. Many species with clonal growth, however, are P-, not V-, species. For example, *Scirpus validus*, an emergent perennial with clonal growth, is eliminated from a prairie glacial marsh by two years or more of continuous flooding.

Propagule longevity

In prairie glacial marshes, seed longevity can be used to distinguish two types of species: species with long-lived seeds (at least 20–30 years) and species with seeds that remain viable for one year or less (van der Valk and Davis 1979). Species with long-lived seed are always represented in the seed bank of a marsh. New individuals of such species become established whenever environmental conditions are suitable for the germination of their seed, even if adults of the species are no longer present in the marsh. Such species are referred to as S-species or seed-bank species. Species with short-lived seed are called D-species or dispersal-dependent species. These species can become (re)established in a marsh, if adult plants are not present, only if their seeds reach the marsh at a time when environmental conditions are appropriate for their germination. The probability that a D-species can become established in a marsh is very much a function of the dispersability of its seeds or other dissemules. Observations on the colonization of new polders in the Netherlands indicate that wind dispersed species such as *Senecio congestus* and *Typha latifolia* are much more rapid colonizers than water dispersed species such as *Scirpus maritimus* (van der Toorn et al. 1969).

Establishment requirements

Field observations and laboratory experiments have shown that the most significant factor controlling the establishment of species from seed or other types of propagules is water depth (Salisbury 1970, Iwata and Ishizuka 1967, Weller 1975, Reinink and van der Toorn 1976, van der Valk and Davis 1978a). Freshwater species can be divided into two groups based on the impact of water depth on seed germination: species whose seeds can germinate under water, and species whose seeds can only germinate when exposed to air or in very shallow water (2 cm or less). Mud-flat annuals and many emergent species have seeds with the latter seed germination requirements, while submersed, free-floating, and some emergents have the former. Species whose seeds germinate only when exposed to air, i.e., on mud flats during the dry marsh stage, are called type-I species. Species whose seeds germinate when covered with water are called type-II species. Seed germination of type-I species often is inhibited if covered by litter or a plant canopy (Reinink and van der Toorn 1976, van der Toorn and Reinink 1978, van der Valk and Davis 1978).

Life-history types

In freshwater wetlands, 12 basic life-history types can now be defined by combining life-span, seed longevity, and seed germination requirements. There are four annual life history types (AS-I, AS-II, AD-I, and AD-II), four comparable perennial (P), and four comparable perennial with an indefinite life-span (V) types. Each of these 12 types has its own unique life-history features (van der Valk 1981). Noble and Slatyer (1980) have developed a similar classification of the life histories of tree species. The terminology used to describe different life histories of wetland species by van der Valk and Davis (1981) is based on that developed by Nobel and Slatyer (1980).

Obtaining the information necessary to assign each species in a marsh to its proper life-history type is relatively straightforward. At least some of the information needed may be already available in the literature. Some information about most species can be obtained from field observations, particularly information about their life-spans. But much of the information needed can be obtained only by examining the seed bank of the marsh. Seed-bank studies involve collecting surface sediment samples from the marsh and exposing subsamples of each sample to conditions similar to those on an exposed mud flat and to those when the seeds are submersed. The number of seedlings of each species whose seeds germinate under simulated drawdown and submersed conditions are recorded after a suitable amount of time has passed (usually several months). Such a seed-bank assay establishes which species in the marsh are S-species and whether these S-species are type-I or -II species.

Table 2 presents the results of a seed-bank study of this type using sediment collected from Eagle Lake. This seed bank contained viable

Table 2. Average number of viable seeds, spores or turions (per m^2) in the upper 5 cm of sediment in the cyclic and non-cyclic sections of Eagle Lake, Iowa. Only species with 100 or more seeds m^{-2} have been included. Each number is an average of four replicates. (Data from van der Valk and Davis 1978a.)

Species	*Cyclic*		*Non-cyclic*		Life-history type
	Open[1]	*Sc.val*	*Carex*	*Sc.flu*	
	Mud-flat annuals				
Bidens cernua	6155	10089	p[2]	p	AS-I
Cyperus odoratus	105	550	421	16	AS-I
Polygonum lapathifolium	1893	1202	1520	3411	AS-I
Polygonum punctatum	0	0	p	723	AS-I
Rorippa islandica	251	1226	1738	2688	AS-I
Rumex maritimus	754	1128	2826	381	AS-I
	Emergent species				
Eleocharis sp.	55	8	3020	8	PS-I
Leersia oryzoides	153	163	703	0	PS-I
Sagittaria latifolia	323	633	p	209	PS-I
Scirpus validus	7221	20890	4281	10678	PS-I
Typha glauca	845	543	1252	612	VS-I
	Submersed and free-floating species[3]				
Ceratophyllum demersum	0	533	p	0	AS-II
Chara sp.	1646	879	2060	p	AS-II
Najas flexilis	880	1135	0	p	AS-II
Spirodela polyrhiza	0	p	p	4213	AS-II
	Meadow species				
Carex sp.	0	0	p	96	VS-I
Mentha arvensis	p	p	1445	462	PS-I
Sium suave	p	0	999	226	PS-I
Stachys palustris	0	0	1405	p	PS-I

[1] Open – area free of emergent vegetation. *Sc.val* – area dominated by *Scirpus validus*. *Sc.flu* – area dominated by *Scirpus fluviatilis*.

[2] A p indicates that less than 100 seeds m^{-2} were present.

[3] *Lemna* spp. were present in Open, *Sc. val*, *Carex* sp., and *Sc.fluv* samples, but they were not enumerated.

seeds, spores, or turions of about 49 taxa (75 to 80% of the taxa recorded for the whole marsh). Twenty taxa had seeds that germinated in the submersed treatment, and 40 taxa germinated in the drawdown treatment. Seeds of all the dominant species in the marsh germinated in one or the other treatment, except those of *Sparganium eurycarpum* and *Scirpus fluviatilis* (van der Valk and Davis 1978a). There were significant differences in the composition of the seed bank in the cyclic and non-cyclic parts of the marsh. For example, *Bidens cernua*, the most abundant mud-flat annual in the cyclic area of the marsh, was almost absent from samples from the non-cyclic area. Seed of four meadow species was found only in samples from the non-cyclic area. Wet-meadow vegetation is found only in shallow, non-cyclic areas of the marsh. All five emergent species whose seeds germinated were present in both areas.

The floristic composition of the samples of the seed bank from the cyclic area is very similar to the floristic composition of the vegetation of this area during a drawdown (van der Valk and Davis 1978a). The only emergent species that was present at Eagle Lake in the cyclic area the seeds of which did not germinate in the seed bank study was *Sparganium eurycarpum*. Viable seeds of this species were present, but did not germinate under the environmental conditions to which these samples were exposed. Field observations at Eagle lake indicate that *Sparganium eurycarpum* is a VS-II species whose seeds germinate in the early spring in flooded areas free of emergents. Annuals, whose seeds were restricted primarily to the seed bank in the non-cyclic area, have seeds that can germinate in existing vegetation when there is no standing water. These species are common understory species in this area in dry years, but are found infrequently in wet years.

From an examination of the seeds banks in this marsh, the life history type of each species in the seed bank was established (Table 2). With this information, it is possible to predict succession in this wetland over time. Predicted successional changes caused by first- (water level changes) and third- (muskrat damage) order interactions that occur during a vegetation cycle are outlined in Fig. 3. Predicted changes (Fig. 3) based on life history features are comparable to actual changes observed at Rush Lake (Table 1) and other Iowa marshes (Weller and Spatcher 1965, Weller and Fredrickson 1974, van der Valk and Davis 1978a). Using this approach, successional changes in these marshes become both comprehensible and easily predictable. A number of studies in other types of wetlands have also found that succession can be predicted from a knowledge of the life-history characteristics of their species (Hall et al. 1946, Gaudet 1977, Leck and Graveline 1979, Pederson 1981).

By increasing our knowledge of the life-history characteristics of wetland species and refining the life-history classifications, more detailed and reliable models of succession can be developed. Currently only successional changes resulting from some first-order interactions can be predicted, but even these predictions need to be improved. For example, specific requirements for the germination of seed of type-I species vary considerably. As a result different type-I species germinate at different times during the

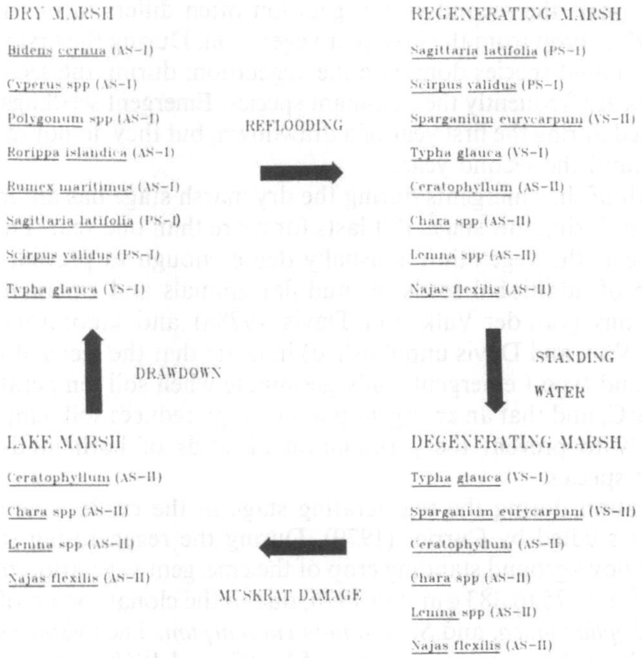

Fig. 3. **Succession at Eagle Lake during a vegetation cycle predicted from the life-history characteristics of its flora (see Table 2). Only the most abundant species are shown. For an explanation of the symbols see the text.**

growing season (Harris and Marshall 1963, Meeks 1969). A drawdown at different times of the year can result in the establishment of different dominant species during the dry marsh stage and possibly for the remainder of the cycle. Currently it is impossible to predict the impact of the timing of a drawback on succession. Similarly, a better understanding of factors that control the life-span of P-species, e.g., *Scirpus validus*, would also increase the reliability of predictions of succession.

Other factors ignored in this life-history model of succession are all second-order and most third-order interactions. These evidently are not of major significance in prairie glacial marshes that have been studied. Nevertheless, these two types of interactions may be significant in other types of wetlands and will have to be understood before this type of successional model can be developed for them.

MATURATION

In prairie glacial marshes, maturation is a considerably less important phenomenon than in forests or deserts. Nevertheless, it does occur and has an impact on the physical structure of the vegetation during both the dry marsh and regenerating marsh stages. If the dry marsh stage lasts for more than one year, the second-year vegetation often differs in physiognomy and standing crop from the first-year vegetation. During the first year, fast growing annual species dominate the vegetation; during the second year, emergents are frequently the dominant species. Emergent seedlings become established during the first year of a drawdown, but they do not reach their full size until the second year.

Growth of the emergents during the dry marsh stage has an impact on succession during this stage, if it lasts for more than one year. During the second year, the vegetation is usually dense enough to prevent the germination of additional seeds of mud-flat annuals and emergents. Field observations (van der Valk and Davis 1978a) and laboratory studies (van der Valk and Davis unpublished) indicate that the seeds of mud-flat annuals and type-I emergents only germinate when soil temperatures are above 30°C, and that an emergent plant canopy reduces soil temperatures sufficiently to prevent the germination of seeds of both mud-flat and emergent species.

Maturation during the regenerating stage in the cyclic zone at Eagle Lake was studied by Currier (1979). During the regenerating stage, the average above-ground standing crop of the emergent vegetation rose from 162 $g\,m^{-2}$ in 1975 to 383 $g\,m^{-2}$ in 1976, due to the clonal spread of *Scirpus validus*, *Typha glauca*, and *Sparganium eurycarpum*. The average standing crop of these three species increased 54, 107, and 106%, respectively, in 1976 (Table 3). Because of dissimilar rates of clonal growth and differences in the average weight of a culm among these three species, the relative contribution of each to the overall primary production of the cyclic zone at Eagle Lake was different in 1975 and 1976. In 1975, *Scirpus validus* accounted for 43% of the total biomass, and *Typha glauca* and *Sparganium eurycarpum* 15 and 13%, respectively. In 1976, *Scirpus validus* accounted for 31% of the total biomass and *Typha glauca* and *Sparganium*

Table 3. Absolute frequency, average biomass ($g\,m^{-2}$), and relative contribution (%) of dominant species to the total biomass of the cyclic section of Eagle Lake during the regenerating stage. Only species that contributed at least 2% to the total biomass of the cyclic section at least one year have been included. (Data from Currier 1979.)

Species	Absolute frequency		Average biomass		Relative contribution	
	1975	1976	1975	1976	1975	1976
			Emergent species			
Scirpus validus	106	105	123	189	43	31
Typha glauca	26	43	177	366	15	24
Sparganium eurycarpum	44	103	87	179	13	28
Sagittaria latifolia	41	28	46	32	6	1
			Submersed species			
Potamogeton pectinatus	101	25	43	34	14	1
Najas flexilis	22	3	58	34	4	< 1
Ceratophyllum demersum	5	43	14	59	< 1	4
Chara sp.	0	8	0	444	0	5

eurycarpum 24 and 28%, respectively. All these changes in primary production were due to changes in the size and density of populations of these two species. In 1976, *Sparganium eurycarpum* and *Typha glauca* were present in 165% and 65% more samples, respectively, than in 1975 (Table 3). Maturation in the cyclic zone during the first two years of the regenerating stage caused a significant change in the physical structure (particularly density and maximum standing crop) of the emergent vegetation.

FLUCTUATION

Water level changes have different consequences for deep and shallow marshes, and the deep and shallow sections of the same marsh. (A marsh is classified as deep if it does not freeze to the bottom during the winter. Because the depth of freezing varies from year to year, depending on snow cover and average winter temperatures, the absolute water depth required for a marsh to be classified as deep varies from year to year. In general, marshes with a fall water depth of 1 m or more are classified as deep marshes, and marshes with a depth of less than 1 m as shallow marshes.) In deep marshes, changes in water levels are an important first-order interaction that controls the rate and nature of the vegetation cycle. In shallow marshes, vegetation cycles do not occur because the water is not deep enough to allow muskrat populations to survive over the winter and they never become large enough to have a significant impact on the emergent vegetation. Water level changes in shallow marshes produce fluctuation, not succession, because of variations in the abundance of species due to differences in water level from year to year.

Changes in the primary production from 1975 to 1979 of the two most abundant emergent species in the non-cyclic area of Eagle Lake are summarized in Fig. 4. This marsh was drawndown in 1974 to re-establish emergent vegetation in its cyclic area. In the non-cyclic area the average primary production of *Typha glauca* and *Sparganium eurycarpum* increased during 1975 and 1976 as their populations recovered from the impact of the 1974 drawdown. A drought from late summer 1976 till the end of the 1977 growing season resulted in the marsh being drawndown again in 1977. The average primary production of populations of both species declined in 1977; that of the *Sparganium eurycarpum* population by 50%, but that of the *Typha glauca* population by only 10%. The decrease in the average primary production of *Typha glauca* in 1977 was, in fact, not statistically significant when compared with the average primary production of the population in 1976 (van der Valk and Davis 1980). In 1978, the average primary production of *Typha glauca* continued to decline while that of *Sparganium eurycarpum* increased significantly. In 1979, *Sparganium eurycarpum* annual production apparently declined while *Typha glauca* annual production increased over 1978 (Fig. 4). The decrease in the primary production of the *Sparganium eurycarpum* population in 1979, however, was not statistically significant.

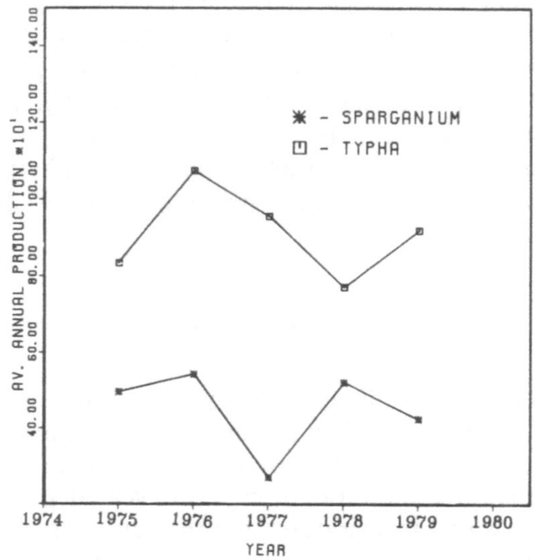

Fig. 4. Average primary production ($g\,m^{-2}$) of *Typha glauca* and *Sparganium eurycarpum* in the non-cyclic section of Eagle Lake from 1975 till 1979.

The average primary production of *Typha glauca* and *Sparganium eurycarpum* populations is a function of culm density and the average weight of a culm, both of which are to a large extent controlled by environmental conditions under which the population is growing (van der Valk and Davis 1980). Table 4 summarizes some changes in the populations of these two species in 1976 (pre-drawdown) and 1978 and 1978 (post-drawdown) in the non-cyclic area of Eagle Lake. In 1978, the total culm density of the *Typha glauca* population was still declining from the impact of the 1977 drawdown, but in 1979, total culm densities had almost returned to pre-drawdown levels. The increase in *Typha glauca*

Table 4. Average maximum culm density, weight, and height of *Typha glauca* and *Sparganium eurycarpum* in the non-cyclic area of Eagle Lake in 1978 and 1979.

Population Feature	*Typha glauca*			*Sparganium eurycarpum*		
	1976	1978	1979	1976	1978	1979
Av. Maximum Culm Density (culms/m^2)						
Flowering Culms	43	10	16	18	6	16
Total Culms	79	45	68	139	138	136
Av. Maximum Culm Weight (g)						
Flowering Culms	32	28	40	14	10	11
Vegetative Culms	29	33	24	7	6	8
Av. Maximum Culm Height (cm)						
Flowering Culms	226	199	219	114	109	123
Vegetative Culms	264	254	271	140	136	143

culm density in 1979 over 1978 was accompanied by a decrease in average weight of vegetative shoots of about 25%. The *Sparganium eurycarpum* population had culm densities in 1978 and 1979 that were almost identical to pre-drawdown densities. The average weight of vegetative and flowering shoots of *Sparganium eurycarpum* increased in 1978 and 1979, but vegetative shoots still weighted less in 1979 than they had in 1976 (Table 4). The average height of both species increased in 1978 and 1979, and in 1979 shoot heights were similar to those in 1976. Adjustments at the population level in shoot density and shoot weight caused by differences in growing conditions from year to year were different for these two species (Table 4). These differences resulted in changes in the relative abundance of emergent species from year to year in the non-cyclic area, i.e., a fluctuation (Fig. 4).

Fluctuations also occur in the submersed vegetation in these marshes. The size of populations of most submersed species can change dramatically from year to year in a marsh (Table 1). At Rush Lake, *Ceratophyllum demersum* frequencies from 1965 to 1970 were 6.4%, 24.2%, 6.5%, 19.7%, 19.0%, and 1.9%. Similar variations in the abundance of other submersed species at Rush Lake can be seen in Table 1; they have also been reported for other Iowa marshes (Currier 1979).

Fluctuation caused by changes in the relative abundance of *Typha glauca* and *Sparganium eurycarpum* at Eagle Lake are less extreme than have been reported in some other studies, but those involving submersed species are as extreme as any reported for other vegetation types (Rabotnov 1974). Fluctuations have an appreciable impact on the floristic structure and primary production of the vegetation of Eagle Lake.

CONCLUSIONS

As illustrated by data from prairie glacial marshes, succession, maturation, and fluctuation are community-level phenomena caused by changes at the population level. As a consequence, all three types of vegetation change can only be understood and predicted from a knowledge of the how first-, second-, and third-order interactions influence the establishment and growth of populations of species that constitute the vegetation of these marshes.

The first step in this approach to vegetation change is to identify critical life-history features of the species that determine when and where these species can become established, and under what conditions their populations can survive and grow. This approach was used to study succession in prairie glacial marshes. Three critical life-history features of species in these marshes were all that were needed to predict succession during a vegetation cycle in them. The utility and predictive power of this approach to succession in wetlands makes it a powerful tool for wetland management. This was demonstrated first by Hall et al. (1946), who used their knowledge of the life-history features of wetland species, particularly the impact of different water levels on the growth of these species, to control the vegetation in Tennessee reservoirs.

Fluctuation and maturation, in comparison with succession, have been little studied in prairie glacial marshes. Nevertheless, information available shows (1) that populations of different species react differently to changes in environmental conditions, and that these differences cause fluctuations; and (2) that there is an increase in the standing crop of emergent species during the regenerating stage of the vegetation cycle, i.e., maturation. Not enough information is available, however, to begin to develop any formal classification of the life-history features of species that could be used to predict fluctuation or maturation in these wetlands.

Fluctuation, succession, and maturation undoubtedly have an impact on rates of mineral cycling and other ecosystem level properties of wetlands, since they change the absolute or relative abundance of species that make up the vegetation. Other features of the marsh which potentially could be affected are invertebrate production, which is dependent on litter quantity and quality (Lee 1980), and vertebrate production, both to the extent that it is dependent on invertebrate production, and to the extent that the physical structure of the vegetation influences the quality of the marsh as vertebrate habitat (Weller 1981). The ability of a marsh to act as a nutrient sink has also been linked to litter quality, patterns of litter fall, and litter decomposition rates (van der Valk et al. 1979). Changes in rates of mineral cycling, nutrient budgets, or invertebrate and vertebrate production have never been investigated during succession, maturation, or fluctuation. These and other changes in wetlands caused by these three phenomena deserve careful study because they are of both great theoretical and practical interest.

REFERENCES

Bishop, R.A., Andrews, R.D. and Bridges, R.J. (1979) Marsh management and its relation to vegetation, waterfowl, and muskrats. Proc. Iowa Acad. Sci. 96: 50–56.

Braun-Blanquet, J. (1932) Plant Sociology. (Trans. and ed. G.D. Fuller and H.S. Conrad). New York, McGraw-Hill. 439pp.

Braun-Blanquet, J. (1964) Pflanzensoziologie. 3rd Ed. Wien, Springer-Verlag. 865pp.

Buttery, B.R. and Lambert, J.M. (1965) Competition between *Glyceria maxima* and *Phragmites communis* in the region of Surlingham Broad. I. The competition mechanism. J. Ecol. 53: 163–181.

Clements, F.E. (1916) Plant Succession. Carnegie Institution of Washington, Publ. 242. 512pp.

Clements, F.E. (1936) Nature and structure of the climax. J. Ecol. 24: 252–284.

Currier, P.J. (1979) Floristic composition and primary production of the postdrawn vegetation of Eagle Lake marsh, Hancock County, Iowa. M.S. Thesis, Iowa State University, Ames, Iowa.

Daubenmire, R.E. (1968) Plant Communities. Harper and Row, New York, 300pp.

Gaudet, J.J. (1977) Natural drawdown on Lake Naivasha, Kenya, and the formation of papyrus swamps. Aquat. Bot. 3: 1–47.

Gleason, H.A. (1917) The structure and development of the plant association. Bull. Torrey Bot. Club 44: 463–481.

Gleason, H.A. (1927) Further views of the succession concept. Ecology 8: 299–326.

Gleason, H.A. (1939) The individualistic concept of the plant association. Amer. Midland Natural. 21: 92–110.
Grace, J.B. and Wetzel, R.G. (1981) Habitat partitioning and competitive displacement in cattails (*Typha*): experimental field studies. Amer. Natur. 188: 453–474.
Grootjans, A.P. (1980) Distribution of plant communities along rivulets in relation to hydrology and management. pp. 143–170. In: O. Wilmanns and R. Tüxen (Eds.), Epharmonie, Bericht Internat. Symp. I.V.V. 1979, Vadug, Cramer Verlag.
Hall, T.F., Penfound, W.T. and Hess, A.D. (1946) Water level relationships of plants in the Tennessee valley with particular reference to malaria control. J. Tennessee Acad. Sci. 21: 18–59.
Harris, S.W. and Marshall, W.H. (1963) Ecology of water-level manipulations on a northern marsh. Ecology 44: 331–343.
Hejny, S. and Husak, S. (1978) Higher plant communities. pp. 22–64. In: D. Dykyjova and J. Kvet, (Eds.), Pond Littoral Ecosystems. Berlin, Springer-Verlag. 464pp.
Iwata, E. and Ishizuka, K. (1967) Plant succession in Hachirigata polder, ecological studies on common reed (*Phragmites communis*) I. Ecol. Review (Sendai) 17: 37–56.
Leck, M.A. and Graveline, K.J. (1979) The seed bank of a fresh-water tidal marsh. Amer. J. Bot. 66: 1006–1015.
Lee, J.J. (1980) A conceptual model of marine detrital decomposition and the organisms associated with the process. Adv. Aquat. Microbiol. 2: 257–291.
Lindeman, R.L. (1942) The trophic-dynamic aspect of ecology. Ecology 23: 399–418.
Margalef, R. (1963) On certain unifying concepts in ecology. Amer. Natur. 97: 357–374.
McIntosh, R.P. (1980) The background and some current problems of theoretical ecology. Synthese 43: 195–255.
McIntosh, R.P. (1981) Succession and ecological theory. pp. 10–23. In D.C. West, H.H. Shugart and D.B. Botkin (Eds.), Forest Succession. Springer-Verlag, New York.
Meeks, R.L. (1969) The effect of drawdown date on wetland succession. J. Wild. Manage. 33: 817–821.
Misra, R. (1946) A study in the ecology of low lying lands. Indian Ecologist 1: 27–46.
Noble, I.R. and Slatyer, R.O. (1980) The use of vital attributes to predict successional changes in plant communities subject to recurrent disturbances. Vegetatio 43: 5–21.
Odum, E.P. (1969) The strategy of ecosystem development. Science 164: 262–270.
Odum, H.T. and Pinkerton, R.C. (1955) Time's speed regulator, the optimum efficiency for maximum output in physical and biological systems. Amer. Sci. 43: 331–343.
Oosting, H.J. (1956) The Study of Plant Communities. 2nd Ed. Freemann, San Francisco, 440pp.
Pederson, R.L. (1981) Seed bank characteristics of the Delta Marsh, Manitoba: applications for wetland management. pp. 61–69. In B. Richardson (Ed.), Selected Proceedings of the Midwest Conference on Wetland Values and Management. Freshwater Society, Navarre, Minnesota.
Rabotnov, T.A. (1974) Differences between fluctuations and successions. pp. 19–24, In R. Knapp (Ed.), Vegetation Dynamics. Junk, The Hague.
Reinink, K. and van der Toorn, J. (1976) Effect of changes in water table on the vegetation development of reed beds in the Zuid Flevoland area. Verhand. Kon. Ned. Akad. Wetensch., Afd. Natuurk., Tweede Reeks, 67: 120–167.
Salisbury, E.J. (1970) The pioneer vegetation of exposed muds and its biological features. Philos. Trans. Roy. Soc. London B 259: 207–255.
Saxton, W.T. (1924) Phases of vegetation under monsoon conditions. J. Ecol. 12: 1–38.
Stewart, R.E. and Kantrud, H.A. (1971) Classification of natural ponds and lakes in the glaciated prairie region. U.S. Fish and Wildlife Service, Resource Publ. 92, 57pp.
van der Toorn, J., Brandsma, M., Bates, W.B. and Penny, M.G. (1969) De vegetatie van Zuidelijk Flevoland in 1968. De Levende Natuur 72: 56–62.
van der Toorn, J. and Reinink, K. (1978) Experiments on the establishment of plant species in a reed field. Verhand. Kon. Ned. Akad. Wetensch., Afd. Natuurk., Tweede Reeks, 71: 292–296.
van der Valk, A.G. (1981) Succession in wetlands: a Gleasonian approach. Ecology 62: 688–696.
van der Valk, A.G. (1982) Succession in temperate North American wetlands. pp. 169–179. In B. Gopal, R.E. Turner, R.G. Wetzel and D.F. Whigham (Eds.). Wetlands: Ecology and Management. International Scientific Publishers, Jaipur, India.

van der Valk, A.G. and Davis, C.B. (1978a) The role of seed banks in the vegetation dynamics of prairie glacial marshes. Ecology 59: 322–335.
van der Valk, A.G. and Davis, C.B. (1978b) Primary production of prairie glacial marshes. pp. 21–37. In R.E. Good, D.F. Whigham and R.L. Simpson (Eds.), Freshwater Wetlands. Academic Press, New York.
van der Valk, A.G. and Davis, C.B. (1979) A reconstruction of the recent vegetational history of a prairie glacial marsh, Eagle Lake, Iowa, from its seed bank. Aquat. Bot. 6: 29–51.
van der Valk, A.G. and Davis, C.B. (1980) The impact of a natural drawdown on the growth of four emergent species in a prairie glacial marsh. Aquat. Bot. 9: 301–322.
van der Valk, A.G., Davis, C.B., Baker, J.L. and Beer, C.E. (1979) Natural fresh water wetlands as nitrogen and phosphorus traps for land runoff. pp. 457–467. In P.E. Greeson, J.R. Clark and J.E. Clark (Eds.), Wetland Functions and Values: the State of our Understanding. American Water Resources Association, Minneapolis.
Weaver, J.E. and Clements, F.E. (1938) Plant Ecology. 2nd ed. McGraw-Hill, New York, 601pp.
Weller, M.W. (1975) Studies of cattail in relation to management for marsh wildlife. Iowa State J. Sci. 49: 383–412.
Weller, M.W. (1981) Freshwater Marshes. Univ. Minnesota Press, Minneapolis, 146pp.
Weller, M.W. and Fredrickson, L.H. (1974) Avian ecology of a managed glacial marsh. Living Bird 12: 269–291.
Weller, M.W. and Spatcher, C.S. (1965) Role of habitat in the distribution and abundance of marsh birds. Iowa Agriculture and Home Economics Experiment Station, Special Report 43, Ames, Iowa.

ANALYSIS AND PREDICTION OF POPULATION AND COMMUNITY CHANGE: A GRASSLAND CASE STUDY

DAVID R. PEART and THEODORE C. FOIN

ABSTRACT

The mechanisms of successional change were analysed in a mediterranean coastal grassland undergoing post-grazing succession at Sea Ranch, northern California, U.S.A. The study included field experiments and observations on the five most abundant species, including four perennial grasses and one annual grass. The results explain the observed trends in the populations, and allow prediction of future change. Patchiness in species distributions, dispersal ability, and localised disturbances strongly affected the outcomes of species interactions. Competitive ability was found to have two distinct components, inhibition and invasion, which were not well correlated. Overall, the analyses show that it is possible to link population and community ecology effectively using appropriate field methods, at least in simple communities.

INTRODUCTION

In this paper, a case study of community analysis will be presented in some detail. The emphasis will be on the approach taken and the empirical results, rather than on review. This research was motivated by the belief that the population-level mechanisms that produce community patterns must be identified and quantified in order to make generalisations or to construct soundly based theoretical models of these mechanisms. This is a view shared in some general treatments, for example those of Foin and Jain (1977), Connell and Slatyer (1977), and Werner (1976).

We will use the term community structure to refer to the distributions and abundances of species in the community; succession will be used in the broad sense to mean change in time of one or both of these aspects of community structure. Two aspects of the complexity of plant communities, spatial heterogeneity and size–asymmetric interactions, are very important to the success of individuals and populations, but have been inadequately studied at the population level, especially by experimental methods. Nutrient and moisture variation in the soil, patchiness in species

White, J. (ed.), The Population Structure of Vegetation.
 ISBN 978-94-010-8927-2

distributions, local disturbances and limited dispersal all contribute to spatial heterogeneity in demographic processes.

With the above consideration in mind, a set of experimental and observational field studies were conducted on the populations of the five most cover-dominant species in a grassland at Sea Ranch, Northern California, U.S.A. These studies have been reported in more detail elsewhere (Peart 1982, Peart 1985). This paper is an overview, and an attempt to integrate the results in the context of the theme of this book. First the general experimental approach will be outlined, followed by a description of the grassland, and the application of the research design to the grassland community.

A FIELD EXPERIMENTAL METHOD OF COMMUNITY ANALYSIS

If reproduction is by seed, there is limited potential for change in cover of the species in a community as a result of growth of existing mature plants. This limitation is especially severe when total cover is already close to 100%, and individuals of a species tend to be aggregated. Then it is only at the edges of patches that individual growth can increase the cover of species. The potential for change in the distributions and abundances of populations is therefore largely determined by species' abilities to invade each type of site encountered by seeds and seedlings.

Whether invasion is successful or not depends also on the abilities of resident adult individuals to inhibit invasion under their canopies. Both inhibition and invasion abilities may be considered as components of competitive ability. Because the process of invasion can be so important for population changes, adult-seedling interactions are a key factor in the analysis of the population basis of community structure and succession.

The invasibility of a site may depend on the species of the resident vegetation (if any), the cover or biomass of that vegetation, substrate or topographic differences, localised disturbances, and possibly other factors as well. In addition to invasion and inhibition abilities, seed production and dispersal ability can strongly influence population dynamics. Generally species distributions are aggregated, and dispersal is localised, so the density of seed fall per unit area (i.e., the "seed rain") for each species is usually quite variable in space. The seed rain in each type of site depends jointly on the spatial distribution of that type of site, patchiness in seed production for each species, and the dispersal abilities of the species.

An experimental design is required which will take account of these complexities of spatial variation and size asymmetry. To do this in a manageable and useful way, sites must be classified into categories according to properties which are likely to affect their invasibility. If site types can be defined in terms of the species identities of the resident vegetation, even where further subdivision below this level is required, the desired community analysis will be greatly facilitated. When such a classification is possible, experiments testing site invasibility simultaneously provide data on the invasion ability of the introduced ("invader") species, and the inhibition ability of the existing ("resident") species.

Once a suitable site-classification scheme has been devised, quite a simple experimental design can be used to measure invasion and inhibition abilities, with seeds of each species introduced at natural densities into each type of site (including disturbed sites), and invasion success measured in terms of establishment, growth and reproduction of the invader.

Finally, the distributions and amounts of each type of site and the dispersal abilities of the species under field conditions must be quantified. Together with the outcomes of the invasion experiments, these studies can provide the data necessary to predict the outcomes of population interactions, project changes in distributions and abundances of populations and, as a result, predict the course of succession. The study to be described in the following sections is an application of the approach outlined above.

THE SEA RANGE GRASSLAND

The northern California coastal grassland is a recognised floristic association (Heady et al. 1977). An annotated flora of the Sea Ranch area, located approximately 200 km north of San Francisco, can be found in Hektner and Foin (1977a). The climate is mediterranean, with 94% of the mean annual rainfall of 103 cm occurring between October and April. Mean monthly temperatures range from 6.5 to 15°C (U.S. Environmental Data and Information Service 1951–1975).

After about 100 years of continuous grazing, sheep were removed from the study area in 1968. The protected section is a 17 km long stretch of coastal terrace, between 1 km and 2 km wide, bounded by the California coast ranges on the east and the Pacific Ocean on the west (Fig. 1). About 90% of this area is free of human disturbance. Sea Ranch is privately owned, with low density housing limited to specific zones. The study sites, which have never been ploughed, were all well removed from areas of human influence. The area is believed to be natural prairie, since there is no evidence of former dominance by woody plants.

Marked plots were monitored for changes in cover of all species over the period 1974–1978 (Hektner and Foin 1977b, Foin and Hektner 1985). The data of Foin and Hektner document quite rapid successional change, with perennial grasses as a group increasing in relative cover at the expense of forbs and annual grasses (Fig. 2). There has been a regular increase in total absolute cover, which by 1978 had reached 100% or more almost everywhere, with bare ground occupying less than 1% of total area. Areas of coastal grassland near Sea Ranch which are still grazed by sheep appear to be dominated by annuals, with the characteristic bunch growth form of the perennial grasses much less prominent, but no sampling data from these areas are available.

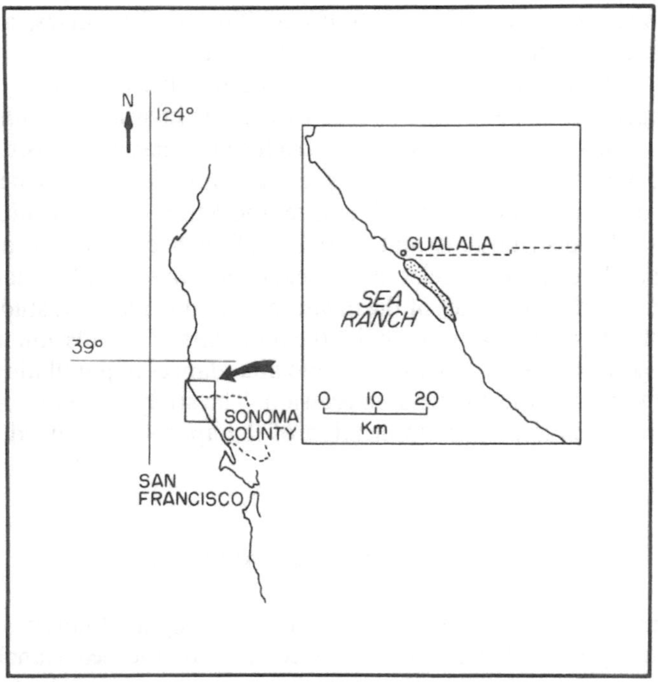

Fig. 1. Map of California coastline, with insert showing Sea Ranch area.

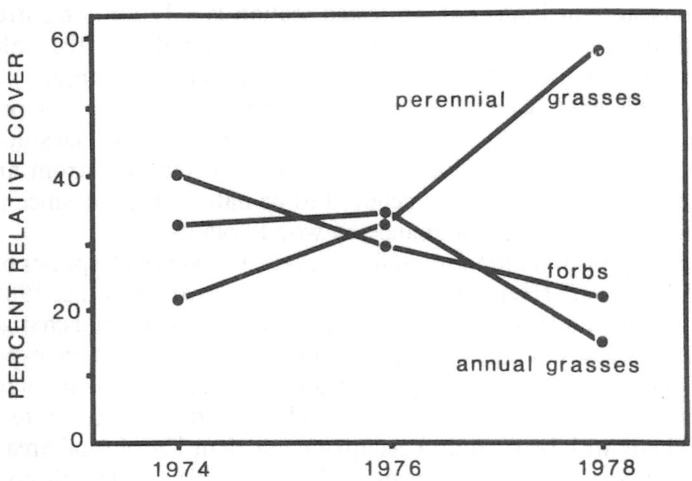

Fig. 2. Changes in percent relative cover of the main growth forms of the Sea Ranch vegetation 1974–1978. Values shown are for those areas in which annuals contributed more than 50% relative cover in 1974. In plots more heavily dominated by perennials, perennial cover also increased over the period (from data of Foin and Hektner 1985).

The most cover-dominant species were included in the research programme. These are *Anthoxanthum odoratum*, *Rytidosperma pilosum*, *Holcus lanatus* and *Deschampsia holciformis* (all perennial bunchgrasses) and *Vulpia bromoides*, an annual grass. Henceforth, species will be referred to by their generic names only, for brevity. Species distributions are very markedly clumped. Three types of patches ("perennial patches") can be defined by dominance by a single perennial species (*Anthoxanthum*, *Deschampsia* or *Holcus*). A fourth type of patch ("annual patch") is dominated by annual grasses and forbs. Unlike the other perennials, *Rytidosperma* does not form its own patches, but its distribution is almost entirely limited to the annual patches, where individuals are interspersed among the annuals and forbs.

Operationally, patches were defined by 85% or greater relative cover, of a single perennial species or of the characteristic *Rytidosperma*/annual grass/forb association. In all but *Deschampsia* patches, this figure was near 100%. Patch sizes are very variable, from 1000 m^2 or more down to the size of a single bunchgrass, which may be up to 30 cm in foliage diameter. In a three hectare area examined for patch sizes, about 75% of the area consisted of patches larger than 9 m^2. Of these larger patches, each patch type had a mean patch size greater than 100 m^2. The patches have been formed by individuals which established from seed. Extensive excavations did not produce evidence of vegetative spread, although *Holcus* apparently does produce rhizomes in mown pastures in England (R. Snaydon, personal communication). Typical canopy heights are 60 cm (perennial patches) and 25 cm (annual patches), and root penetration is about 60 cm in all patch types.

All of the grasses, and most of the forbs, are synchronised in their phenologies, according to the strongly seasonal rainfall pattern. Germination occurs with the autumn rains (usually in October). The soil begins to dry from the surface down in the spring months. Flowering occurs in late spring and early summer, with seed set from May until July. This seasonal pattern is very predictable.

Over the study area, the five species studied account for about 75% of total cover (Table 1). Most of the remaining 25% consists of forbs and other annual grasses which, along with *Vulpia* and *Rytidosperma* , form the annual patches. It can be seen from Table 1 that the four patch types account for over 90% of total cover. The experiments and observations addressed the performance of the five species in the four patch types. The research design therefore dealt with the main occupiers of space in this community.

The term "disturbance" will be used broadly to refer to any factor which alters canopy structure, or otherwise produces marked and discrete changes in the resources available for colonisation and growth. The importance of disturbances for establishment has been shown in several studies (e.g., Platt 1975) and reviewed in Grubb (1977). Three types of naturally occurring disturbances were examined for their effects on the dynamics of populations. These were the mortality of individual bunchgrasses, soil disturbances by gophers, and grazing by non-domestic

Table 1. Relative cover of the five species included in the study. See text for description of patches (from data of Foin and Hektner 1985).

		Percent relative cover
Species of perennial patch types:	*Anthoxanthum*	43
	Holcus	6
	Deschampsia	5
Species of the annual patch type:	*Rytidosperma*	14
	Vulpia	7
	Forbs and other annual grasses	20

animals. The main herbivores, now that sheep have been removed, are deer, hares, gophers, meadow voles and grasshoppers. These animals are heavily concentrated in the annual patches. It is likely that grazers find better forage in annual vegetation. Pocket gophers, for example, prefer forbs (Ward and Keith 1962), which are most abundant in the annual patches. The extensive root systems of the perennials may also inhibit burrowing activity. During the dry summer months, the prevailing winds are from the ocean, and therefore do not tend to spread fire from lightning strikes on the coast ranges. Fire therefore does not seem to be as important a factor as it is in most communities in mediterranean climate. Records of the Sea Ranch fire station support this conclusion. Certainly there have been no uncontrolled fires in the Sea Ranch grassland since 1974.

OBSERVATIONAL STUDIES

Moisture and soil properties and their relationship to patchiness in species distributions

To test whether the pronounced patchiness in the distributions of the grasses was associated with underlying variations in the substrate, soil chemical properties and soil moisture were measured at several sampling locations in each of three areas (patches) belonging to each of the four patch types. Samples were analysed for pH, salinity, organic matter content, cation exchange capacity, exchangeable cations, total phosphorus and total sulphur. To measure soil moisture, gypsum blocks were buried, and readings taken on a Bouyoucos conductance meter throughout the growing season.

The results showed that there was little variation in soil chemical properties, and no association with patchiness in the grasses. A similar lack of correspondence was found for soil moisture, except in one case: *Deschampsia*-dominated patches were found in areas that consistently remained wetter, late in the growing season, than those dominated by other perennial species or by annuals (Fig. 3). The *Deschampsia* patches took about three months longer than other patch types to reach zero soil

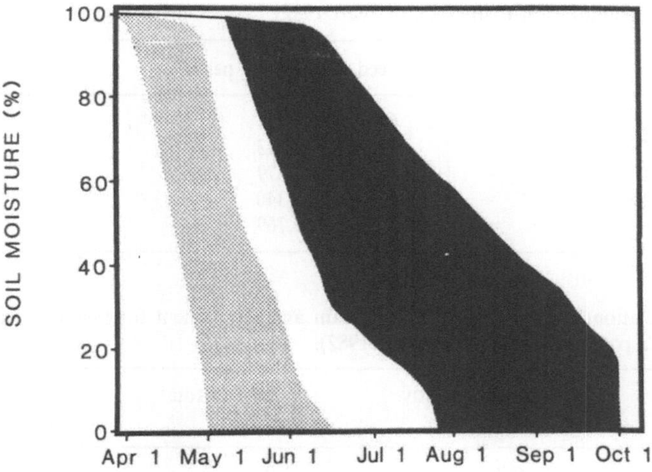

Fig. 3. Soil moisture (percentage of saturation) in *Deschampsia* patches compared to that in the other three patch types. Readings were taken through the growing season. The regions plotted encompass the range of readings for each date for *Deschampsia* patches (dark stippling) and for the other three patch types combined (light stippling). While *Deschampsia* patches were still close to 100% saturation in early June, most soil moisture had been lost from other patch types (from data of Peart 1982).

moisture levels after the end of winter rains. Within the sensitivity of the measurements, the rather surprising conclusion is that the patchiness in other species' distributions is probably not caused by heterogeneity in soil or soil moisture. It is of course possible that the vegetation may itself produce changes in soil properties over time, but the general lack of association between soil properties and species distributions provides no evidence for such effects at this stage of succession. The relatively high soil moisture in *Deschampsia* patches is clearly due to their location in low-lying areas or drainage depressions.

Seed rain, seed bank and recruitment

For each of the five species, measurements were made of the density of seed rain, recruitment, and the contribution of the seed bank to recruitment. Recruitment was measured as the number of individuals establishing and surviving until the end of their first growing season. The contribution of dormant seeds was obtained by preventing entry of seeds to local areas during the period of seed fall, using fine mesh exclosure cages, and then removing the covers and measuring recruitment as above. All measurements were taken at least 2 m inside the more or less well-defined patch boundaries.

The seed rain for each species in the patch type it dominates (or the annual patch type, in the case of *Rytidosperma* and *Vulpia*) is shown in Table 2. Clearly there are substantial differences among the perennials in the numbers of seeds produced per unit area. The mean relative cover of

Table 2. Densities of seed rain produced by each species in its respective patch type. Values are ± one standard error (from data of Peart 1982).

Species	Seed rain density per m^2
Anthoxanthum	64,285 ± 2,331
Holcus	82,389 ± 3,513
Deschampsia	5,290 ± 579
Rytidosperma	2,263 ± 340
Vulpia	29,781 ± 2,769

Table 3. Relationship between cover, seed rain and recruitment for perennial grasses in *Deschampsia* patches (from data of Peart 1982).

Species	Percent relative cover	Percent of total seed rain	Percent of total recruitment
Anthoxanthum	0.7	31.0	73.0
Holcus	3.6	33.9	21.5
Deschampsia	88.2	35.1	5.5

Rytidosperma in the annual patches was only about 35%. The relatively low seed rain density of *Rytidosperma* can therefore be explained in part by its lower tendency to form monospecific stands, compared to the other perennial species.

In annual patches *Vulpia* accounted for more than 50% of the seed community. In all patch types, except those dominated by *Deschampsia*, seed fall of non-resident species was less than 1% of the total, and was zero in most samples. In *Deschampsia* patches the resident species contributed 85% to 95% of relative cover, but *Anthoxanthum* and *Holcus*, in spite of their low relative cover values, produced approximately the same density of seed rain there as did *Deschampsia* (Table 3).

Deschampsia, *Holcus* and *Rytidosperma* seed fall resulted in few (less than five per m^2) recruits in their respective patch types, and recruits from the seed bank numbered less than two per m^2 for each of these three species. Of the perennials, *Anthoxanthum* had by far the greatest numbers of recruits, with a mean of 30 per m^2. The density of *Anthoxanthum* recruits from the seed exclosures was approximately equal to the total, indicating that recruitment within *Anthoxanthum* patches in any one year may be largely independent of the seed rain in that year. *Vulpia* seeds produced many recruits in annual patches (over 200 per m^2) with the seed bank contributing about 3% of this.

Recruitment was observed for non-resident species in only two cases. In *Deschampsia* patches, both *Holcus* and *Anthoxanthum* recruits actually outnumbered those of *Deschampsia* (Table 3). In annual patches, some *Anthoxanthum* recruitment was observed near the borders of *Anthoxanthum* patches. With the exception of *Anthoxanthum* recruitment in annual patches, the following generalisations can be made. Recruits of perennial species were small (less than 1 cm^2 leaf area after one year) and

Table 4. Diaspore ("seed") weights and dispersal abilities of the five species included in the experiments. The 90 percentile dispersal distance is the radial distance within which 90% of the seeds were dispersed. Diaspore weights are ± one standard error (from data of Peart (1982).

Species	Diaspore weight (mg)	90 percentile dispersal distance
Anthoxanthum	0.523 ± 0.008	1.2
Holcus	0.318 ± 0.005	5.2
Deschampsia	0.290 ± 0.005	2.7
Rytidosperma	1.079 ± 0.019	2.1
Vulpia	1.429 ± 0.024	0.6

non-reproductive. Fewer than 50% of them survived to the second growing season in any patch type, and the survivors had apparently not increased in size. Finally, in spite of considerable spatial variation in the densities of seed rain, there was no relationship between local seed rain density and recruitment for any perennial species. For the annual *Vulpia*, however, all recruits reproduced, and a significant positive correlation between seed rain density and recruitment was found on spatial scales of 0.01 m^2 and 1 m^2.

Field measures of dispersal ability

Seeds were marked on the infructescences of field plants, using a fluorescent dye, without disturbing the marked individuals. As marked seeds fell during the summer, they were captured on seed traps, which were arrayed in radial pie-shaped patterns, and positioned in an arrangement that left intact most of the vegetation structure around the marked plants. The 50 and 90 percentile distances for seed dispersal of each species are shown along with diaspore weights in Table 4. The rankings of dispersal ability, based on statistical comparisons of the cumulative probability of dispersal as a function of distance, were *Holcus* > *Deschampsia* > *Rytidosperma* ≃ *Vulpia*. A similar ranking was obtained by dropping seeds by hand from various heights and recording distances travelled. All dispersal curves were strongly skewed, with most seeds falling downwind in the direction of prevailing summer winds.

Anthoxanthum demography in the four patch types

A three-year field study of the demography of mature individuals of *Anthoxanthum* was made. It was hypothesised that individuals would grow faster and survive better when surrounded by annuals or by perennials of low biomass, than when in the vicinity of perennials of high biomass.

A total of 481 individuals, distributed among the four patch types, were permanently marked and measured for base diameter, foliage diameter and fecundity. The species present, and a qualitative index of vegetation

Table 5. Percentage of area disturbed in each patch type. Measurements were made at the beginning of the growing season (from Peart 1985).

Patch type	% area disturbed	
	Resident mortality	Soil disturbance
Anthoxanthum	3.3	–
Holcus	0.4	–
Deschampsia	1.2	–
Annual	0.1*	2.5

**Rytidosperma* mortality.

biomass were recorded for the neighborhood of each plant. The measures of size and fecundity were repeated each year, and mortality noted.

The results confirmed the predictions only for the smallest size-classes. After attaining a moderate adult size, individuals were less strongly affected by neighbors. Mortality was concentrated in the smallest size class (plants with one reproductive culm) which had approximately 50% survival over the period of the study, compared to 80% survival for larger classes. The survival rates suggest that plants which attain a moderate size (> 3 cm bunch diameter) may live many years. Growth rates differed substantially among individuals, largely because small individuals can be suppressed by neighbors for several years. As a result, size was not well correlated with age. However size was a good predictor of fecundity.

Mortality and soil disturbance

At the beginning of the growing season, the percentage of area in each patch type occupied by standing dead individuals and by soil disturbances (gopher mounds) was estimated. Gopher mounds occurred only in the annual patches. The greatest area of standing dead was in *Anthoxanthum* patches, followed by *Holcus*, *Deschampsia* and *Rytidosperma* (Table 5).

FIELD EXPERIMENTS

Application of the experimental design to the Sea Ranch grassland

Sites were categorised first by patch type. Aboveground dry weight biomass was measured within patches, so that sites could be classified by resident vegetation biomass as well as by species. Biomass was estimated in each experimental patch by harvesting all vegetation in three randomly chosen 50 cm × 50 cm quadrats, and obtaining dry weights for each quadrat. Types of disturbed sites within patches included those occupied by standing dead individual bunchgrasses ("resident mortality") in each patch type, and those where gopher mounds were present ("soil disturbances"). Also, exclosures were built to exclude grazers from the annual patch vegetation, forming a "grazer exclusion" site type. Naturally occurring gopher mounds were used in the experiments. However, for the resident mortality treatments, bunchgrasses were killed by application of a low persistence herbicide (trade name Roundup), applied to the leaves of individual plants.

Four sets of invasion experiments were then carried out, involving the introduction of seed into quadrats categorised as (i) unmanipulated sites, (ii) sites containing a standing dead individual, (iii) sites containing a gopher mound, and (iv) sites with grazers excluded. The quadrats of the unmanipulated sites were located in three separate areas (patches) belonging to each patch type (for a total of 12 patches). Experiments involving disturbances and grazer exclusion were performed in one patch of each type. Although most patches of each type appeared similar, there were some which clearly had lower aboveground biomass. One of the three patches of each type was therefore chosen to be of relatively low biomass, so that the effects of biomass on invasion success could be tested.

The treatments included in the design are shown in Table 6. These did not include all possible combinations, but proved adequate for the aims of the study. The abundance of *Anthoxanthum* made it an important focus in the experiments. All possible treatments involving this species, as an invader or as a resident, were included in the design. Also, all species were tested for their abilities to invade annual patches. Each treatment was replicated with five quadrats in each of the patches for which that treatment was done. Each quadrat was 50 cm × 50 cm in size except in the grazer exclusion treatment, where cages were 110 cm × 110 cm. Control quadrats were used to measure natural colonisation and to monitor any year to year changes in cover or seed production of resident species.

Introduced seed densities were calculated as 50% of the mean density of the 10 sampling stations (out of the total of 33 represented in the data of Table 2) that had the highest seed rain densities. This method was used to obtain a high but realistic level of seed input from a species resident in a patch to sites near the edge of a neighbouring patch. An exception to this method was made for *Rytidosperma*. The seed fall data for the other perennial species were from areas of nearly 100% cover of those species. However *Rytidosperma* does not dominate annual patches to this extent as noted earlier. To calculate the seed input for *Rytidosperma*, the Sea Ranch area was surveyed for areas of highest *Rytidosperma* cover and seed production. 50% of the mean seed rain density from a total of 15 samples from such areas was used for *Rytidosperma* introduction. This procedure maximised the similarity, among the species, of the stands from which the estimates of densities were made. However it resulted in an overestimate, relative to the other species, of the natural levels of *Rytidosperma* seed inputs. The calculated values of seed inputs used in the introductions are shown for all species in Table 7. For each quadrat, introduced seed densities were modified to take account of natural seed input from nearby individuals (if any) of the introduced species. The experiments were initiated in the summer and autumn of 1980, and followed through 1981 and 1982.

Table 6. Treatments included in the invasion experiments (see text for explanation).

Invasion without resident mortality or soil disturbance		Invasion with resident mortality		Invasion with soil disturbance		Invasion with grazers excluded	
Invading species	Patch type	Invading species	Patch type	Invading species	Patch type	Invading species	Patch type
Anthoxanthum	Annual	*Anthoxanthum*	Annual	*Anthoxanthum*	Annual	*Anthoxanthum*	Annual
	Holcus		*Holcus*	*Holcus*	Annual		
	Deschampsia		*Deschampsia*	*Deschampsia*	Annual		
				Rytidosperma	Annual		
				Vulpia	Annual		
Holcus	*Anthoxanthum*	*Holcus*	*Anthoxanthum*				
	Annual	*Deschampsia*	*Anthoxanthum*				
Deschampsia	*Anthoxanthum*	*Rytidosperma*	*Anthoxanthum*				
	Annual						
Rytidosperma	*Anthoxanthum*	*Vulpia*	*Anthoxanthum*				
	Annual						
Vulpia	*Anthoxanthum*						
	Annual						

Table 7. Calculated input densities of seeds for each species in the invasion experiments (from Peart 1985).

Species	Introduced density per m^2
Anthoxanthum	38,870
Holcus	49,766
Deschampsia	4,424
Rytidosperma	7,483
Vulpia	21,666

Invasion success was scored in 1981 as the number of individuals per quadrat, leaf area per individual, total leaf area per quadrat, and seed production. In 1982, numbers of flowering individuals and seed production were recorded. For the perennials, only results for total leaf area per quadrat will be reported. The results for other measures showed similar patterns and led to the same conclusions, except where stated in the text. For *Vulpia*, the annual species, only seed production will be reported. The results for the low biomass patches showed the same relative trends (species comparisons) as for the two patches of typical biomass of each type, although the absolute values of the results differed. Results for the low biomass patches will be included in an analysis of the effects of biomass later in the chapter, but omitted from the main results.

Results of the experiments

The perennial species will be considered first. For unmanipulated quadrats, Fig. 4A shows *Anthoxanthum* invasion success compared to natural establishment by the residents in three patch types, and the success of invading species compared to resident establishment in the *Anthoxanthum* patch type. In all perennial patches both invasion success and natural establishment were relatively low for all species, although *Anthoxanthum* and *Holcus* invaded best. Perennials were therefore effective inhibitors.

In Fig. 4B similar comparisons are made for the site condition defined by resident individual mortality. Even though the standing dead individuals remained in these sites, invasion success was greatly increased compared to that in undisturbed canopies. Note the change in scale between Figs. 4A and 4B. There was also a clear difference in the pattern of the results with and without resident mortality.

Relative invasion abilities in undisturbed annual patches were *Anthoxanthum* > *Holcus* > *Rytidosperma* ≃ *Deschampsia* (Fig. 5A). For gopher mounds in annual patches, corresponding results are depicted in Fig. 5B. Invasion success was much increased as a result of soil disturbance (again, note the change in scale from Fig. 5A to Fig. 5B). Here *Anthoxanthum* and *Holcus* invaded about equally well in the first growing season. However *Holcus* survived very poorly and produced no seed in the following season, while *Anthoxanthum* increased its cover and produced

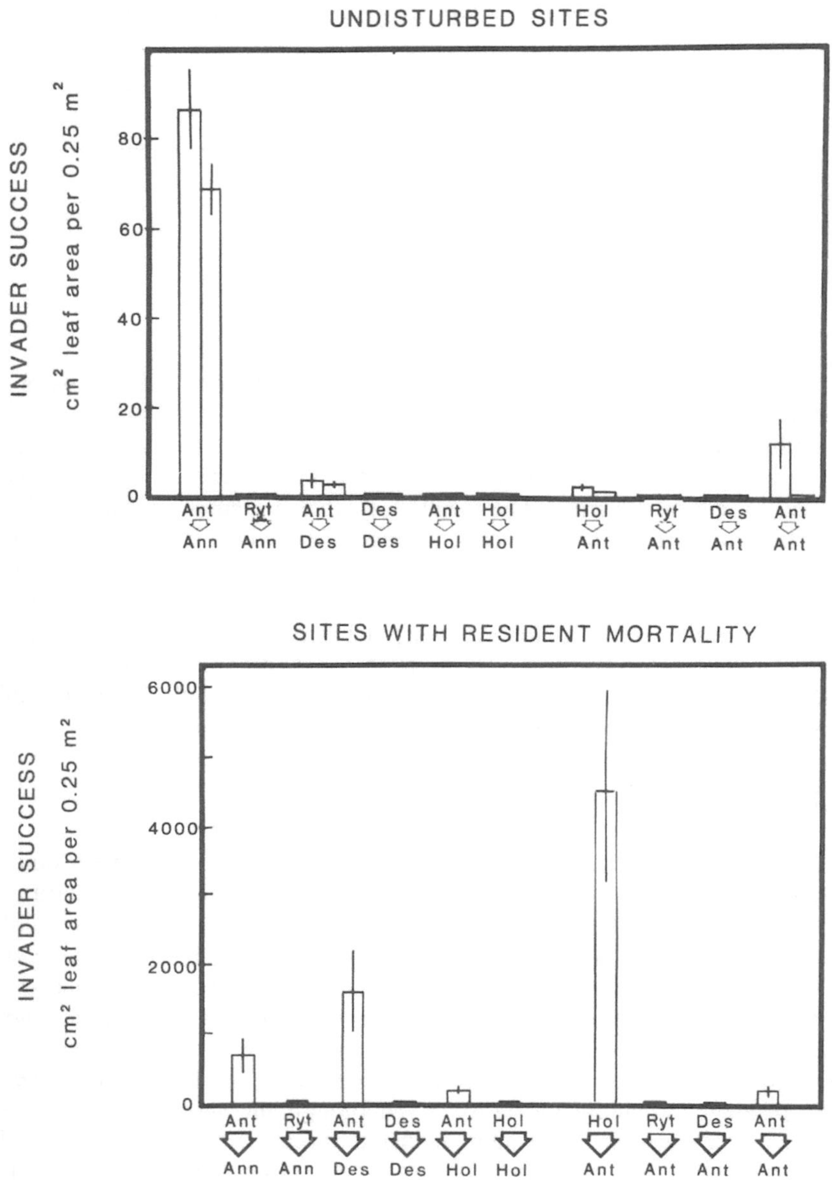

Fig. 4. Invasion success of perennials in unmanipulated quadrats (A) and with resident mortality (B). Results are from two patches of each type for unmanipulated sites, and from one patch of each type for sites with resident mortality. In both figures, the results show, from left to right, *Anthoxanthum* invasion in annual patches compared to natural establishment by the resident *Rytidosperma*, *Anthoxanthum* invasion of *Deschampsia* patches compared to natural establishment by *Deschampsia*, *Anthoxanthum* invasion of *Holcus* patches compared to natural establishment by *Holcus*, and invasion of *Anthoxanthum* patches by *Holcus*, *Rytidosperma*, and *Deschampsia* compared to natural *Anthoxanthum* establishment. Abbreviations are Ant = *Anthoxanthum*, Ryt = *Rytidosperma*, Des = *Deschampsia*, Hol = *Holcus*, Ann = annual patch type. See text for further explanation. Error bars span two standard errors (from data of Peart 1985).

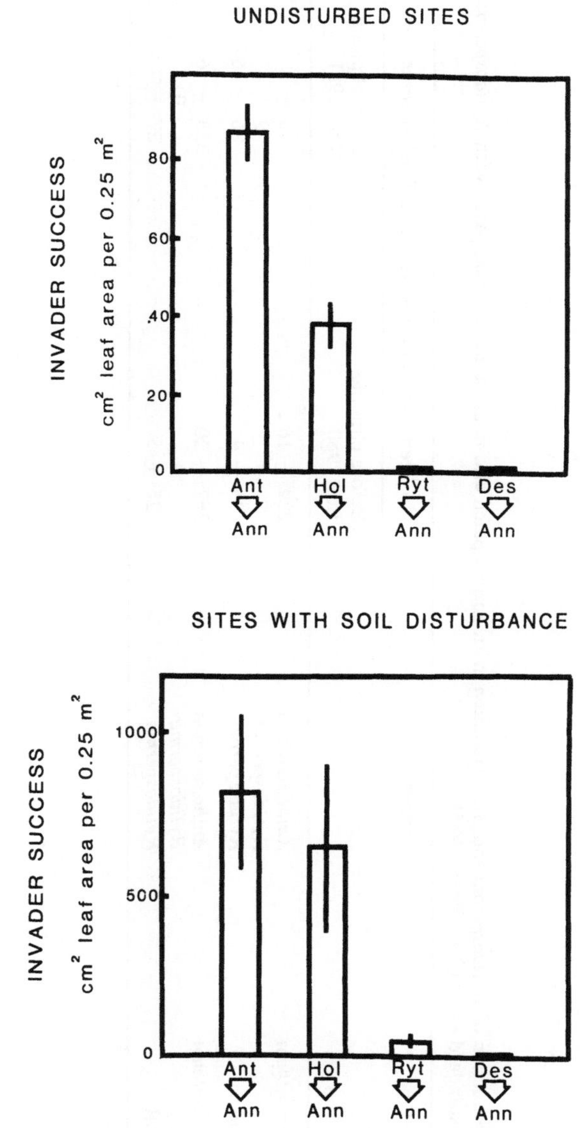

Fig. 5. Invasion success of perennials in annual patches in unmanipulated sites (A) and in sites with soil disturbance (B). Abbreviations as in Fig. 4. Error bars span two standard errors (from data of Peart 1985).

abundant seed. Therefore the rankings again were *Anthoxanthum* > *Holcus* > *Rytidosperma* ≃ *Deschampsia*. *Anthoxanthum* invasion of annual patches was not significantly affected by grazer exclusion, and other experiments indicated that wild grazers had little impact on plant size or biomass accumulation, at least in the first growing season after introductions.

When vegetation in annual patches was clipped for biomass samples, an apparent stimulatory effect on *Rytidosperma* was observed. To quantify

Table 8. Effects of clipping on seed production of *Rytidosperma* and *Anthoxanthum* in an annual patch and in an *Anthoxanthum* patch with *Rytidosperma* understory. Values are means ± one standard error (from Peart 1985).

	Patch type	Species	No. seeds heads produced per 0.25 m^2 quadrat	
			Before clipping July 1980	After clipping July 1981
Experimental quadrats (clipped):	*Anthoxanthum*	*Anthoxanthum*	229 ± 16	5 ± 2
		Rytidosperma	0	9 ± 4
	Annual	*Rytidosperma*	35 ± 4	86 ± 16
Control quadrats (unclipped):	*Anthoxanthum*	*Anthoxanthum*	367 ± 20	333 ± 16
		Rytidosperma	0	0
	Annual	*Rytidosperma*	26 ± 5	27 ± 7

Table 9. Invasion success of the annual *Vulpia*. Also shown is natural seed production in annual patches. Values are means ± one standard error (from Peart 1985).

Site type	*Vulpia* seed production per 0.25 m²	
	First growing season	Second growing season
Anthoxanthum patches, undisturbed (experiment)	94 ± 28	0
Anthoxanthum patches with resident mortality (experiment)	11,997 ± 7,122	189 ± 100
Annual patches with soil disturbance (experiment)	6,278 ± 1,365	2,441 ± 1,026
Annual patches (field measurement without manipulation	7,541 ± 913	Not measured

this, *Rytidosperma* seed production was measured in quadrats before, and one year after clipping to ground level. A similar experiment was done in an *Anthoxanthum* patch with *Rytidosperma* understory. *Rytidosperma* seed production increased as a result of the treatment in both cases, but seed production in control quadrats did not change significantly over the period for either species (Table 8). Stimulation by clipping was not observed for species other than *Rytidosperma*.

The annual *Vulpia* produced seed in the first year in all treatments in which it was tested in the invasion experiments (Table 9). Resident *Anthoxanthum* mortality greatly increased seed production, while soil disturbance in the annual patches had little effect. In the second year, seed production in all invasion experiments, especially those with resident mortality, was markedly reduced, as other species increased their cover.

THE SPECIES POPULATIONS: THEIR ROLES IN THE DYNAMICS OF THE COMMUNITY

Anthoxanthum odoratum

Invasion success in perennial patches without resident mortality was minimal for all species, and standing dead occupy 3% or less of space in those patches. Annual patches account for about 40% of total area (Table 1), and are invasible by both *Anthoxanthum* and *Holcus*. Therefore, short term increases in the abundances of perennials will reflect their abilities to invade annual patches. *Anthoxanthum* was the most effective invader of annual patches, both with and without soil disturbances. As a result, *Anthoxanthum* is predicted to increase relatively rapidly, but at a decreasing rate as annual patches are reduced.

Anthoxanthum lives only a few years on zinc mines in Wales (Antonovics 1972) but has greater longevity in this community, especially when surrounded by annuals, as noted earlier. In the invasion experiments, *Anthoxanthum* increased its cover and seed production in the second year of growth, demonstrating its ability to gain and hold space and to generate further local colonisation in annual patches.

Anthoxanthum also invaded very well with resident mortality in annual and *Deschampsia* patches and, to a lesser extent, in *Holcus* patches. It is therefore an excellent invader in a variety of site types, and it can form stands which are not invasible except where individuals die. *Anthoxanthum* dominates the borders of annual patches. Its stands produce many seeds (Table 2) but its relatively poor dispersal ability (Table 4) results in invasion as an advancing front, and the borders of the annual patches remain distinct. Dwire (1983) has documented the rapid decline in the density of *Anthoxanthum* seed rain with distance from *Anthoxanthum* patch boundaries. *Anthoxanthum* produced more recruits under its canopies than did other perennials, as noted earlier, but these individuals had poor growth and survival, and would require subsequent disturbance to emerge and grow to maturity.

Holcus lanatus

Holcus invaded annual patches less effectively than did *Anthoxanthum* and is less abundant on the borders of annual patches. Therefore, its short-term rate of increase will be lower. However, as annual patches are reduced, rates of change in abundances of perennials will be determined largely by their abilities to invade perennial patches. With its relatively good dispersal ability (Table 4), high seed production (Table 2) and excellent ability to invade *Anthoxanthum* patches with resident mortality, *Holcus* should have several advantages over *Anthoxanthum* as succession proceeds. It is predicted to continue to increase slowly, even when there is little opportunity to invade annual patches. Some of this increase is likely to occur at the expense of *Anthoxanthum*. Note that resident mortality was considerably greater in *Anthoxanthum* patches than in *Holcus* patches (Table 5).

The imbalance in favour of *Holcus* in its reciprocal interactions with *Anthoxanthum* in resident mortality sites is not easily explained. Chemical effects are apparently not responsible: Newman and Rovira (1975) showed that leachates (mostly from roots) were more allo-toxic than auto-toxic for each of these species in a British grassland. Nor were strong herbivore effects observed. Mycorrhizal or pathogenic infections may be involved, but no evidence is available. The interaction will be further investigated in future work. *Holcus* is known as an aggressive coloniser in other habitats, and forms dense stands in some of areas in the Park Grass experiments at Rothamstead, England (Brenchley and Warrington 1958).

Deschampsia holciformis

This native species is almost entirely restricted to wet sites (Fig. 3), where it is invasible by *Anthoxanthum* with resident mortality. It also has relatively low seed production (Table 2). The results recorded above show that *Deschampsia* has few seedlings present in its own stands. This species has relatively good dispersal (Table 4), but its poor invasion ability and habitat restriction prevent successful colonisation in adjacent patches. Isolated *Deschampsia* individuals can sometimes be found in drier areas, but these appear to have a high incidence of mortality (personal observation). All indications are that this species will decline, even in wet sites.

Rytidosperma pilosum

Rytidosperma has some unusual attributes. It is the only species found to survive as a suppressed understory in perennial patches (generally under *Anthoxanthum*, but in one observation, under *Holcus*). In each case, the *Rytidosperma* foliage was sparse, a few cm tall, with a leaf area per m^2 estimated to be less than 0.1% of that of *Anthoxanthum*. The existence of the understory is circumstantial evidence for past invasion of annual patches by *Anthoxanthum*, which has a taller growth form than *Rytidosperma* and could easily overtop it.

The response of the *Rytidosperma* understory to clipping (Table 8) shows that it can recover when the canopy is opened. Heavy grazing may have a similar effect. *Rytidosperma* had poor invasion success in the experimental introductions, even though the seed input levels were somewhat above naturally occurring densities. Furthermore the annual patches which it inhabits are invasible by *Anthoxanthum* and *Holcus*. It is therefore likely to decline in abundance, in the absence of extensive canopy disturbance.

Vulpia bromoides

Vulpia invades well, but cannot maintain reproduction in disturbed sites in perennial patches. Because of its poor dispersal ability (Table 4), it cannot easily reach such disturbed sites. It is therefore unable to persist by a fugitive behavior in perennial-dominated areas. Because of its annual habit, this species has low effective inhibition ability. Since annual patches are invasible by perennials, *Vulpia* must eventually decline in abundance.

TESTS OF THE PREDICTIONS

The predictions can be tested in part with the monitoring data of Foin and Hektner (1985). Changes in relative cover for the perennial species in a 160 m × 80 m plot at Sea Ranch for a four-year monitoring period are shown in Fig. 6. The trends are as predicted from the experimental analysis, with a rapid increase in *Anthoxanthum*, a slower increase in *Holcus*, and declines in *Rytidosperma* and *Deschampsia*. The predicted long-term future trends are shown in the dotted lines.

Vulpia increased in relative cover in the study plot over the period, but this increase was apparently due to an advantage over other annuals. Foin

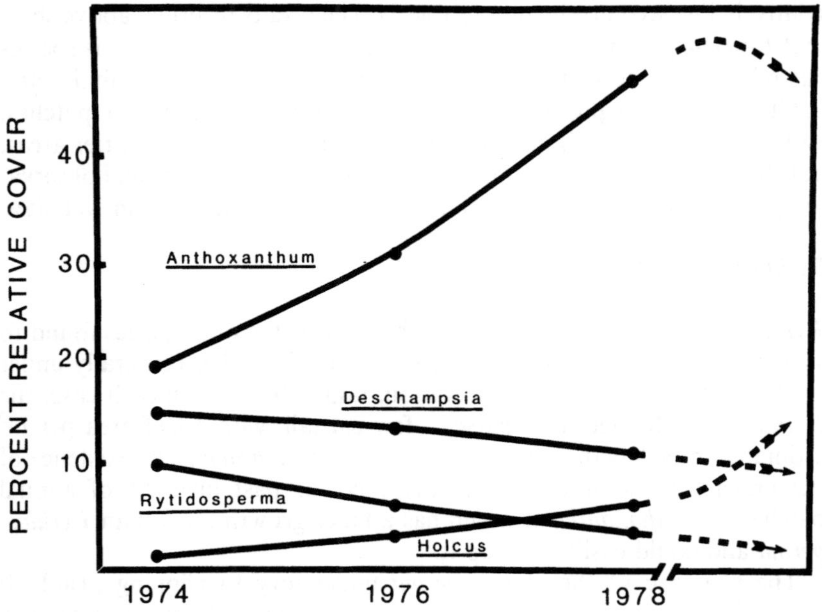

Fig. 6. Changes in percent relative cover over the period 1974–1978 of the perennial species studied. (From data of Foin and Hektner (1985). Also shown are predicted future trends (dotted lines).

and Hektner classified quadrats as "perennial" or "mixed" according to whether perennials (including *Rytidosperma*) contributed more or less than 50% relative cover in 1974. Relative cover of *Vulpia* increased over the sampling period from < 1% to 12% in mixed areas, but from < 1% to only 3% in perennial areas. The internal dynamics of annual patches do not affect the prediction of an eventual decline in *Vulpia*, along with other annuals.

As noted previously, wild grazers are concentrated in the annual areas. A mechanism that could maintain annual patches, at much lower cover than at present, might be intensified grazing by wild herbivores as annual patches decrease in number and size. This is speculative, however, because it depends on the unknown functional and numerical responses of the herbivores, and on the response of the vegetation to increased grazing intensity.

Although year-to-year variations in conditions are generally important for plant populations, there are several sources of evidence that suggest the conclusions from the experiments may be quite robust. The seasonal pattern is a very predictable one, with the main source of variation being the amount of winter rainfall and the timing of first and last rains. The data of Fig. 2 and Fig. 6 include years of normal rainfall and very dry years, but indicate fairly regular trends in abundance (although some of the annual species show more fluctuation). The main experiments were initiated in 1980, a near-average year. A subset were repeated in 1981, a wet year, giving similar results. Seed production in control quadrats in each patch type did not change significantly from 1980 to 1981. Finally, watering treatments late in the season did not change the relative performance of species in the experiments (unpub. data). In contrast to the fairly regular trends exhibited by the perennials, the annual grasses and forbs probably fluctuate somewhat in relative abundance from year to year, despite the overall decline of annuals in the community. Year to year fluctuations in abundances of annuals was documented for an annual grassland in the interior valley of California, which included the dominant annual grasses of the Sea Ranch area (Bartolome 1979).

THE DEPENDENCE OF INHIBITION ON BIOMASS

The large number of treatments involving *Anthoxanthum* introductions into unmanipulated quadrats provided sufficient data to evaluate the relation between *Anthoxanthum* invasion success and established vegetation biomass. A two-parameter negative exponential curve explained 91% of the variance in total leaf area of *Anthoxanthum* seedlings (Fig. 7). For other measures of invasion success – numbers of individuals and sizes of individuals – similar curves explained 79% and 53% of the variance, respectively.

These results suggest that, for an intact canopy, inhibition ability is

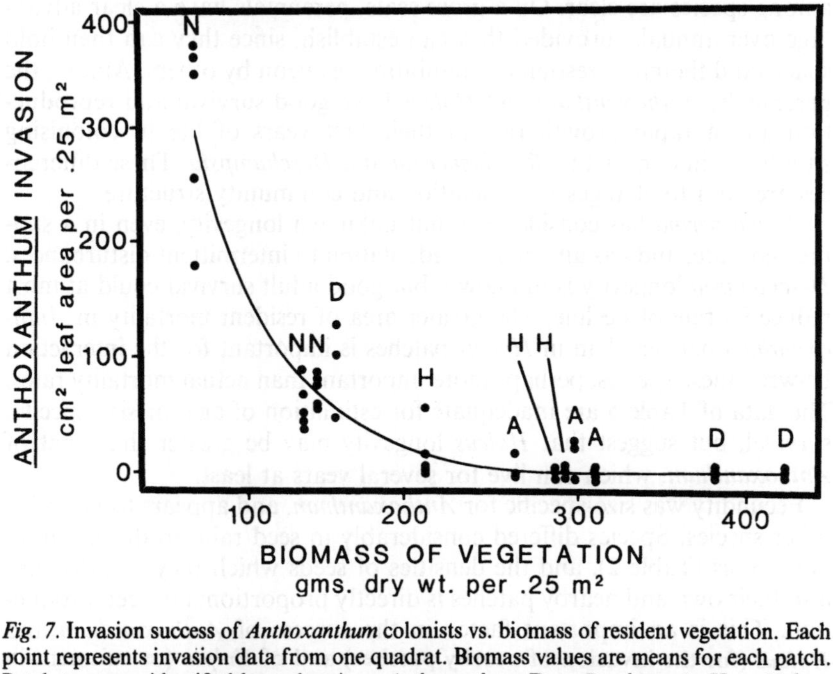

Fig. 7. Invasion success of *Anthoxanthum* colonists vs. biomass of resident vegetation. Each point represents invasion data from one quadrat. Biomass values are means for each patch. Patch types are identified by codes: A = *Anthoxanthum*, D = *Deschampsia*, H = *Holcus*, N = Annual. Points vertically under each letter represent data from one patch. Curve is best fit exponential $y = a \exp(bx)$ (from Peart 1985).

independent of species identity of the vegetation, but increases with above-ground biomass. The presence of annual grasses and forbs in the annual patches did not seem to affect the relationship. For species other than *Anthoxanthum*, invasion success was also negatively related to resident biomass. The dependence of invasion success on biomass rather than species identity may be general, as hypothesised by Goldberg and Werner (1983). The relationship extends many of the experimental results to patch boundaries and other areas where neighbouring plants are of different species.

There are two points that require emphasis in relation to this biomass effect. First, species differed greatly in their invasion abilities in vegetation of given biomass, i.e., invasion ability was strongly species-specific. Second, where disturbances were involved, biomass of surrounding vegetation was *not* a good predictor of invasion success. This is not surprising, since in these cases the canopy itself was altered.

LIFE HISTORY AND POPULATION SUCCESS

Survival and fecundity

Reflecting the great plasticity of plants in general, these attributes were extremely variable for the species examined, and depended on the site type into which their propagules were introduced. However some differences among species are clear. On a gross scale, perennials have a clear advantage over annuals, provided they can establish, since they can then hold space (and therefore resources), inhibiting invasion by others. Among the perennials, *Anthoxanthum* and *Holcus* have good survival and reproduction and a rapid growth rate in their first years of life in colonising situations, in contrast to *Rytidosperma* and *Deschampsia*. These differences are vital to changes in population and community structure.

Rytidosperma has considerable, but unknown longevity, even in a suppressed state, and has an apparent adaptation to intermittent disturbances. *Deschampsia* longevity is unknown, but good adult survival could at most reduce its rate of decline. The greater area of resident mortality in *Anthoxanthum* patches than in *Holcus* patches is important for the interaction between these species, perhaps more important than actual mortality rates. The data of Table 5 are inadequate for estimation of age- or size-specific survival, but suggest that *Holcus* longevity may be greater than that of *Anthoxanthum*, which can live for several years at least.

Fecundity was size-specific for *Anthoxanthum*, and appears to be so for other species. Species differed considerably in seed rain produced on an areal basis (Table 2), and the densities of seeds which they can disperse into their own and nearby patches is directly proportional to seed production. This is an important factor in the success of *Anthoxanthum* and *Holcus* (for the invasion of nearby patches) and of *Vulpia* (recruitment in annual patches). As noted previously, recruitment within the patch of a a resident was dependent on the density of seed rain only for *Vulpia*.

Dispersal

The patchiness in species distributions, inhibition effects of perennial canopies, and the localised nature of disturbances all contribute to the importance of dispersal for population change. Examples are the relatively poor dispersal of *Anthoxanthum* (limiting its rate of invasion of annual patches and access to disturbed sites in perennial patches) and of *Vulpia* (preventing it from surviving as a fugitive in perennial patches) and the relatively good dispersal of *Holcus* (enabling it to reach more disturbances, including sites of *Anthoxanthum* resident mortality).

Dormancy

Although potentially an important aspect of population success, dormancy is apparently not a strong factor in succession in this community. *Anthoxanthum* and *Vulpia* have dormant seed banks, but no evidence was found that these play an important role under present conditions. *Anthoxanthum* is an increasing species, and its seed bank should not be found in abundance in other patch types, with the possible exception of *Holcus* patches. A *Vulpia* seed bank might be expected in *Anthoxanthum* and *Holcus* patches. However, in none of the experiments, even in treatments involving disturbances, were recruits of *Vulpia* or *Anthoxanthum* found where there was no seed rain for these species, except where seeds were experimentally introduced.

Seed weights

From Table 4, the ranking of diaspore weights is *Vulpia* > *Rytidosperma* > *Anthoxanthum* > *Holcus* > *Deschampsia*, with the last two differing only slightly. *Vulpia* showed evidence of seedling-seedling competitive ability, outgrowing perennial seedlings in annual patches, where interactions among juvenile plants are an important factor. However *Rytidosperma* seedlings grew poorly in those patches, and *Deschampsia* seedlings were much less successful than those of *Holcus*, which has similar seed size. Large seeds may be expected to be advantageous in disturbed sites also, but again there is no clear pattern comparable to that shown for old field species, for example, by Gross and Werner (1982) and Gross (1984). The range of variation in seed weights among these species is relatively small, so that any effect may be masked by other aspects of the species' biology. It is possible, although it seems unlikely from the morphologies of the seeds, that a clearer pattern would emerge if the seeds were dissected for endosperm weights.

DISCUSSION AND CONCLUSIONS

The results demonstrate the importance of patchiness, dispersal and disturbance for the interactions among populations and show that the effects of these factors can be evaluated in the field. The invasion experiments show clearly that the early establishment phase is vital for explaining the distribution and abundance of adult plants. This supports the conclusions of a number of recent studies in herbaceous communities (e.g., Platt 1975, Werner 1977, Gross and Werner 1982, Bartolome 1979).

Size-asymmetric interactions were evaluated by considering two components of competition: inhibition and invasion abilities. Inhibition was found to depend on aboveground biomass rather than the species of resident vegetation. However invasion ability differed among species. The best inhibitors were not necessarily the best invaders, and vice versa, so the two components of competitive ability were not well correlated. This lack of correlation and the biomass effect, if they are general, have important implications for both theoretical models and experimental analyses of species interactions in plant communities. If the degree of inhibition exerted by resident plants is not species-specific, models and experimental designs can be simplified considerably. On the other hand it is clear that a single parameter is generally inadequate to describe competition between plant species.

The studies outlined in this paper were designed to analyse the factors affecting population change in the present time. These factors can be put in historical context using the available monitoring data for the community, summarised in Fig. 2 and Fig. 6. The trends in these figures can be explained by mechanisms of interaction among species that were discovered in the experimental and observational studies. The coastal prairie has developed from a predominantly annual community to one dominated by a few species of perennial bunchgrasses. The process of replacement of annuals by perennials is still occurring. However not all perennial species have increased. Biological interactions appear to control the process of invasion, which results in the replacement of individual plants and change in community structure. Each species has a distinct role in succession, and a rich set of interactions is apparent.

The results underscore the importance of site pre-emption by the perennials. *Anthoxanthum*, in particular, was able to occupy sites in previously annual-dominated quadrats, and to resist re-invasion by annuals in the subsequent growing season. The demographic study of *Anthoxanthum* showed that new colonists could survive relatively well in the annual patches. The superiority of *Anthoxanthum* over other perennials in its ability to invade annual vegetation explains its rapid rise to dominance. It is worth noting that the importance of some interactions for vegetation change depends on the relative abundances of species. For example, the rapid rise of *Anthoxanthum* depended on the high cover of annuals early in succession, and future trends in the *Holcus* population will depend on the relative cover of *Holcus* and *Anthoxanthum*. The elucidation of such relationships depends on a community wide approach to the study of populations in succession.

An obvious question is why were *Anthoxanthum* and *Holcus* less abundant in the past? This is not answerable directly from the results of this study, but it seems likely that differences in the nature and intensity of disturbances when sheep were present, were responsible for preventing *Anthoxanthum* and *Holcus* from invading annual vegetation. Disturbances at present either increase invasion by perennials (in the case of gopher mounds or resident mortality), or have little effect (grazing by wild herbivores). The present advantage of *Anthoxanthum* and *Holcus* is due to their ability to hold sites after initial colonisation. If disturbances are such as to reduce perennial survival substantially in the first year, their advantage relative to the annuals disappears. The perennial colonists in annual patches produced relatively few seeds in their first year of growth, and grew more slowly than the annuals surrounding them. It is likely that the sheep, by physical disturbance as well as foliage removal, opened sites for colonisation each year. Under these circumstances, the higher growth rates and reproductive allocations of the annuals are more successful growth patterns than those of the perennials, which rely on longevity and site-holding (i.e., inhibition) ability.

Another question is why are the species' distributions so patchy? Heterogeneity in underlying resources appears not to be the controlling factor, but there are other possibilities. Localised dispersal might produce clumped species distributions, as suggested by simulation studies (in progress). A predominance of allo-toxic chemical interactions among species (Newman and Rovira 1975) could also be responsible. Finally, historical causes, such as management practices which resulted in heterogeneous grazing intensity, may have played a part. The longevity and inhibition abilities of the perennials could maintain historical differences for many years.

The attributes of species quantified in these studies were clearly related to the trends in their populations. However these relationships could not be predicted without detailed analysis. Life history data that are relatively easily obtained, for example seed size and the survival and fecundity of adults, would clearly not be sufficient in the case of this community. It seems there are many more kinds of combinations of traits in real species than are allowed in simple life history theory. Further analyses of communities, involving other populations with different suites of life-history traits, will provide a larger sample size on which to draw reliable generalisations about the roles of various life histories in species interactions. This is clearly desirable as we cannot examine every community experimentally and in detail.

Plant communities differ greatly in number of species, species longevities and reproductive modes, the role of consumers and the heterogeneity of the substrate. Some would be much more difficult to deal with experimentally than the grassland of this study. Nevertheless some plant communities offer great opportunity for experimentation. The effects of spatial heterogeneity, a problem for theorists (Schaffer and Leigh 1976), can be examined, as the heterogeneity is often stable enough in time to observe or manipulate. Site-specific reproductive inputs and outputs can be measured and manipulated, and dispersal quantified in some cases. Finally individual success can be measured in relation to neighbourhood environments. Such opportunities are rare in animal communities.

The experimental design and supporting observational studies outlined in this paper show that a community analysis at the population level can be done, for some communities at least, within reasonable constraints of time and effort. Such an analysis necessarily lacks detailed examination of specific processes. However a whole community study is an effective (perhaps the most effective) means of establishing the relative importances of particular interactions for the success of populations. Once these interactions are identified, the techniques of physiological ecology can be focused on them, helping to forge another vital link in the ecological hierarchy, that between population and physiological ecology.

ACKNOWLEDGEMENTS

C.L. Folt, D.E. Goldberg and K.R. Hopper constructively criticised the manuscript. The Sea Ranch Association generously provided field accommodation and access to field sites. We thank C. Seely for typing.

REFERENCES

Antonovics, J. (1972) Population dynamics of the grass *Anthoxanthum odoratum* on a zinc mine. Journal of Ecology 60: 351–366.

Bartolome, J.W. (1979) Germination and seedling establishment in California annual grassland. Journal of Ecology. 67: 273–281.

Brenchley, W.E. and Warrington, K. (1958) The Park Grass Plots at Rothamstead 1856–1949. Rothamstead Experimental Station, Harpenden.

Connell, J.H. and Slatyer, R.O. (1977) Mechanisms of succession in natural communities and their role in community stability and organization. American Naturalist 111: 1119–1144.

Davidson, E.D. (1975) Demography of *Lupinus arboreus* at Bodega Head, California. Ph.D. dissertation, University of California, Davis.

Dwire, K. (1983) Dispersal and colonization of the invading perennial grass *Anthoxanthum odoratum* in annual patches in a California coastal grassland (Sea Ranch; Sonoma County). M.S. thesis, University of California, Davis.

Foin, T.C. and Hektner, M.M. (1985) Trends in dominance and diversity in a successional coastal prairie. Manuscript.

Foin, T.C. and Jain, S.K. (1977) Ecosystem analysis and population biology: lessons for the development of community ecology. Bioscience 27: 532–538.

Goldberg, D.E. and Werner, P.A. (1983) Equivalence of competitors in plant communities: a null hypothesis and a field experimental approach. American Journal of Botany 70: 1098–1104.

Gross, K.L. (1984) Effects of seed size and growth form on seedling establishment of six monocarpic perennials. Journal of Ecology 72: 369–386.

Gross, K.L. and Werner, P.A. (1982) Colonizing abilities of four "biennial plant species in various vegetation patch types: implications for distributions in a successional sere. Ecology 63: 921–931.

Grubb, P.J. (1977) The maintenance of species richness in plant communities: the importance of the regeneration niche. Biological Reviews 52: 107–145.

Heady, H.F., Foin, T.C., Hektner, M.M., Taylor, D.W., Barbour, M.G. and Barry, W.J. (1977) Coastal prairie and northern coastal shrub. In M.G. Barbour and J. Major (Eds.) Terrestrial Vegetation of California. Wiley, New York.

Hektner, M.M. and Foin, T.C. (1977a) Vegetation analysis of a northern California coastal prairie: Sea Ranch, Sonoma County, California. Madrono 24: 83–103.

Hektner, M.M. and Foin, T.C. (1977b) A flora of the coastal terraces of Sea Ranch, Sonoma Country, California. Wasmann Journal of Biology 35: 26–53.

Newman, E.I. and Rovira, A.D. (1975) Allelopathy among some British grassland species. Journal of Ecology 63: 727–737.

Peart, D.R. (1982) Experimental analysis of succession in a grassland at Sea Ranch, California. PhD thesis, University of California, Davis.

Peart, D.R. (1985) Experimental analysis of grassland succession. Ecological Monographs, in press.

Platt, W.J. (1975) The colonization and formation of equilibrium plant species associations on badger disturbances in a tallgrass prairie. Ecological Monographs 45: 285–305.

Schaffer, W.M. and Leigh, E.G. (1976) The prospective role of mathematical theory in plant ecology. Systematic Botany 1: 209–232.

U.S. Environmental Data Service (1951–1975) Climatological data; California section. U.S. Environmental Data Service, Asheville, N.C.

Ward, A.L. and Keith, J.O. (1962) Feeding habits of pocket gophers on mountain grasslands, Black Mesa, Colorado. Ecology 43: 744–749.

Werner, P.A. (1976) The ecology of plant populations in successional environments. Systematic Botany 1: 246–268.

Werner, P.A. (1977) Colonization success of a "biennial" plant species: experimental studies of cohabitation and replacement. Ecology 58: 840–849.

DEMOGRAPHY OF SHORT-LIVED FORBS IN CHALK GRASSLAND IN RELATION TO VEGETATION STRUCTURE

H.J. DURING, A.J. SCHENKEVELD, H.J. VERKAAR and J.H. WILLEMS

ABSTRACT

The effects of vegetation structure on population processes in five short-lived species occurring in two Dutch chalk grasslands are analyzed on the basis of some laboratory and field experiments. The main factors regulating population numbers appear to be light quantity available for seedlings in dense microsites and water availability, which affects seedling survival, growth and fecundity of established plants in open micro-habitats.

Two broad categories are recognized, "spenders" and "savers". These are based on seed weight, mortality of seedlings, presence of a persistent seed bank and variation in life span. We suggest that coexistence of the short-lived species is due to niche-sharing rather than niche-differentiation.

INTRODUCTION

Grasslands under a constant management regime can be considered as stabilized plagioclimax communities, the composition of which is grossly at equilibrium (Silvertown 1980b). The rank order of the relative abundance of the perennial species involved is more or less fixed (Grubb et al. 1982). Even in such stabilized communities, however, some degree of dynamics is always present (White 1979) leading to gaps in the vegetation. Differentiation of size, nature and timing of such gaps leads to differentiation of the regeneration niche (Grubb 1977). The regular occurrence of gaps offers possibilities for coexistence of plant species with different life histories. Gaps play an important role in maintaining diversity of perennial species (Grubb 1982), but are especially important for short-lived species. Two rather different types of life history patterns have been suggested for short-lived species by During (1979): "fugitives" and "shuttle" species. Fugitives are found in large but rarely and irregularly occurring gaps, are notable for their good dispersal capacity and large diaspore numbers, and usually lack a seed bank (Cohen 1966, 1967).

Shifting or shuttle species are found in smaller gaps which usually occur regularly within the community. For such species a persistent seed bank

White, J. (ed.), The Population Structure of Vegetation.
 ISBN 978-94-010-8927-2

(Thompson and Grime 1979) is often essential, seed size is variable, and seed numbers are high compared to perennials, but low compared to fugitives. Short-lived forbs in grasslands may be considered as shuttle species, though some species on ant hills with high numbers of small seeds are somewhat intermediate (King 1977, Silvertown 1981b).

Recently, Grubb and co-workers (Grubb 1976, 1977, Grubb et al. 1982, Gay et al. 1982) have analyzed the ecology of some short-lived species in detail. Grubb (1976) classified the short-lived plants of chalk downland in England into four groups: 1. species occurring in more or less closed turf ("turf-compatible species", Gay et al. 1982); 2. species characteristic of the tops of active ant hills; 3. species of larger disturbed sites in usually more fertile situations, which presumably cannot grow in a dense turf primarily because of nutrient deficiency (Gay et al. 1982); and 4. species of bare chalk, e.g. around rabbit warrens. Groups 2–4 are called "turf-incompatible" by Gay et al. (1982). All winter annuals appear to be turf-incompatible. This may be related to their small seed size and low competitive ability (Silvertown 1981b).

Gaps originate from the death of individual plants caused by animal activities such as mole heaps, worm casts, ant hills, rabbit scrapes, cattle hoof prints, and from abiotic influences such as wind throws in forests, erosion in grasslands on slopes (Watt 1974, Grubb 1977). The concept of gap is in fact rather imprecise. It may be conceived simply as "new space available" (Miller 1982): for example Gross (1980) showed that gap size so defined was crucial in determining whether *Verbascum thapsus** could complete its life cycle in an old-field. However other habitat factors such as nutrient availability, water, temperature, light intensity and R/FR ratio primarily determined the "ecological space" available for plants limited to gaps. In other words, the plants do not necessarily react to space *sensu stricto*, but to more direct ecological factors. This leads to a gap concept that comes near to the more general concept of niche. The gap concept may still be used for openings in the canopy which provide temporary possibilities for germination and establishment of species locally absent above-ground. It may refer to large-scale gaps such as wind throws in a forest or arable fields, but also to small-scale ones such as the local opening in a grassland turf created by the death of a plant.

Ecological conditions in gaps in grasslands are strongly determined by the structure of the vegetation (Verkaar et al. 1983a), including the bryophyte layer (During et al. 1984).

The structure of grasslands shows strong seasonal fluctuations especially under a mowing regime (Verkaar et al. 1983a), but even under continuous grazing seasonal variation is apparent (Al-Mufti et al. 1977), especially in the bryophyte layer, where cover is usually minimal in summer. As seeds are dispersed in autumn, the bryophyte layer increases strongly, thus covering at least part of the seeds, and reaches maximum cover in early spring. With the onset of summer drought and the growth of the

*Nomenclature follows Van der Meijden et al. (1983).

phanerogam vegetation, the bryophyte cover decreases again quickly. This seasonality will also influence gap quality considerably.

Chalk grasslands

Chalk grasslands of the alliance Mesobromion are widely distributed over W. Europe. Over large parts of this area, the botanical composition is sufficiently uniform to incorporate the stands in one association: Mesobrometum erecti (Willems 1982). Other associations of the alliance are mainly found at the fringes of its area, notably in England and Scandinavia (Willems 1982). Generally, species diversity is high, with often 50 or more phanerogam species and up to 10 bryophytes per m^2, in some area with several lichens as well (Willems 1973, 1978, Rosén 1982).

The characteristically low productivity of chalk grasslands (e.g., Smith 1980) is usually ascribed to mineral nutrient deficiency (Rorison 1960, Grime 1963, Jeffrey 1971, Smith et al. 1971, Bradshaw 1980), mostly of phosphate and nitrate (Lloyd and Pigott 1967, Grime and Curtis 1976), or of phosphate alone (Davison, cited in Read et al. 1976). The levels of these nutrients show strong seasonal variations (Gupta and Rorison 1975, Taylor et al. 1982) and the nutrient economy of several perennials appears to be related to a spring peak in phosphate and nitrate availability (Gay et al. 1982). Whether water shortages are involved is less clear; Anderson (1927) argued that chalk grasslands are ecologically dry, and Perring (1960) found a close correlation between botanical composition of chalk grasslands in England with the ratio between precipitation and saturation deficit. The interactions between drought and mineral nutrient levels probably determine botanical composition and low productivity of chalk grasslands to a large extent (Lloyd and Pigott 1967, Grime and Curtis 1976).

The structure of the grassland depends on the management regime. In heavily grazed sites the turf is very dense and only a few cm high; if the site is mown, as is the case in our sites, the turf is ca 30–80 cm high in summer with a lower shoot density. Usually one or two grass species (often *Brachypodium pinnatum* and/or *Bromus erectus*) are dominant, but a number of other grasses and forbs such as *Leontodon hispidus*, *Centaurea* spp. and *Thymus* spp. are present in fair amounts. In this "matrix" of perennial species several short-lived species – annuals, biennials and pauciennial hapaxanths – occur, often in great quantities.

How do such short-lived species maintain themselves in a dense turf? Some indications of the answer(s) to this question are given by Gay et al. (1982). They investigated the nutrient economy and mycorrhiza of four biennial chalk grassland species: *Linum catharticum*, *Blackstonia perfoliata*, *Centaurium erythraea* and *Gentianella amarella*. Though nutrient concentrations in rosettes at the onset of their first winter were high in comparison with associated perennials in the community, it appeared that such a nutrient store was of little importance for the growth in the next season. The main nutrient uptake took place during the period of fastest growth, in the summer of the second year; there was no evidence of a build-up of a store in spring, when nutrient availability is highest, as happened in the perennials. This suggests that as far as nutrient relations are concerned, the biennials are more or less complementary in time to the perennials. These four biennials are rather shallow-rooted and how they cope with summer drought which affects the upper soil layers is as yet unknown.

Nearly all species in chalk grasslands appear to have vesicular-arbuscular mycorrhiza. The biennials studied by Grubb and co-workers show a high density of mycorrhiza on the roots, but only in *Linum catharticum* was this density appreciably higher than in some perennials. The mycorrhiza of the biennials showed differences in morphology from those in perennials, and *Linum* showed another type from that in the Gentianaceae. The significance of these morphological differences is presently under investigation (Gay et al. 1982).

Grubb et al. (1982) presented some interesting theoretical considerations about the maintenance of species with consistently low relative abundance (explerents *sensu* Rabotnov 1978). In short-lived species in chalk grasslands, actual numbers of individuals show much variation from year to year. Still, a rather consistent rank order of the species was observed from most to least abundant: *Linum catharticum* > *Rhinanthus minor* > *Blackstonia perfoliata* = *Centaurium erythraea* = *Gentianella amarella* > *Medicago lupulina* = *Picris hieraciodes* > *Carlina vulgaris*. They state that interference between the short-lived species themselves is rather limited, but that these species suffer markedly from interference from the perennials in the vegetation, which implies that gap size and duration are important for their success. The rank order mentioned above is tentatively interpreted in terms of plant size in relation to gap size: the species with the largest plants at maturity (*Carlina*, *Picris*, *Medicago*) would need the supposedly scarce relatively large gaps. Also, the larger species tend to need more years for maturation, which implies a slower reaction to a series of favourable years.

In the next sections we shall present some information on the demography of short-lived forbs in Dutch chalk grasslands, and discuss the selective factors operating in the different stages of their life history, with particular attention for the effects of the structure of the vegetation. Finally, we shall discuss the different life history patterns that may be recognized among short-lived forbs in chalk grasslands.

THE STUDY AREA

Sites

The two chalk grasslands studied are situated in two nature reserves in the province of S-Limburg, the Netherlands. The Gerendal site (2 ha) is a NW-facing slope with a soil of ca. 40 cm of colluvial clay over unweathered chalk from the Upper Cretaceous; the Vrakelberg site (4 ha) is a S to SE-facing slope with 10–25 cm clay over chalk. Further site details can be found in Willems (1980) and Verkaar et al. (1983a). At present the vegetation of both sites is a chalk grassland belonging to the Mesobrometum erecti.

The Gerendal site has been a heavily fertilized pasture, but since 1970 part of it has been mown annually without application of fertilizers. Another part of the slope has been grazed by sheep since 1973. The development from heavily fertilized pasture to Mesobrometum has been followed by Willems (1980, 1983a) in a series of experimental plots in which the effects of mowing, mowing and fertilizing, grazing without fertilizing and leaving untouched have been compared. The vegetation has a peak standing crop of 500–600 g/m^2. In this site, short-lived forbs are rather rare. *Daucus carota* is the only species present in some quantities, and *Linum* occurs occasionally. In 1980, the first few individuals of an annual, *Rhinanthus minor*, were found on the mown part of the slope. In 1982 already ca. 200 individuals were present.

The Vrakelberg site has long been an arable field. It has been mown annually since 1961, without application of fertilizers. At present *Brachypodium pinnatum* is dominant over large parts of the site; locally patches with dominance of *Carex flacca* or *Briza media* can be found. In this site, the vegetation structure is more variable. A large part of the site is covered by a dense vegetation reaching a peak standing crop of 400–500 g/m^2. Locally the vegetation is rather open, with a peak standing crop of ca. 200 g/m^2. These open patches are usually 2.5 m in diameter, but a few long stretches of ca. 2 m width are present, too. Patches with intermediate vegetation density (peak standing crop ca. 250–400 g/m^2) are present as well. With only few exceptions, such differences in vegetation density were maintained throughout the year. In this site quite a number of short-lived species occur. Winter annuals are rare: only *Arenaria serpyllifolia* is occasionally found. *Euphrasia stricta* is the only summer annual occurring in large quantities; other summer annuals include *E. rostkoviana*, *Rhinanthus minor* and *R. alectorolophus*. *Linum catharticum* and *Gentianella germanica*, both very common in the site, represent the strict biennials. *Satureja acinos* and *Teucrium botrys* are mainly biennial in this site, too. Predominantly monocarpic perennials include *Anthyllis vulneraria*, *Daucus carota*, and small quantities of *Carlina vulgaris*, *Echium vulgare*, *Inula conyza*, and *Picris hieracioides*. Since we expected *Scabiosa columbaria* to belong to this category as well we included this species in our investigations. Occasionally, a few plants of arable weeds such as *Euphorbia exigua* can still be found, reminiscent of the time that part of the slope was an arable field. In the seed bank, several more such species are present (Verkaar and Schenkeveld 1981). For brevity most of these species will be indicated only by their genus name subsequently.

The bryophyte layer is well-developed in Gerendal and parts of the Vrakelberg, especially in winter and early spring; at Gerendal, peak standing crop of this layer may reach 100–350 g/m^2 dry weight in March. The bryophyte flora is rather rich, with 42 and 44 species for Gerendal and Vrakelberg, respectively. In Gerendal and Vrakelberg-dense, large pleurocarps dominate, but in Vrakelberg-open turf-forming acrocarps are more abundant (During and ter Horst 1983). In the soil, a large diaspore population of seeds, bryophyte and fern spores and vegetative diaspores of bryophytes is present (Schenkeveld and Verkaar 1983, Willems 1983b, During and ter Horst 1983).

Microclimate

Effects of the structure of the vegetation on the performance of short-lived species are likely to be due at least partly to microclimatological factors. Many features of the vegetation structure influence the microclimate at ground level, such as leaf inclination, and distribution of LAI and biomass over vegetation layers (Barkman 1979). In our sites, differences in microclimatological factors appear to be chiefly attributable to differences in vegetation density, expressed as standing crop of dry matter (Verkaar et al. 1983a).

Transmission of light at ground level is strongly reduced at higher densities, reaching very low values at a standing crop of 400 g/m^2 or more (Fig. 1). The R/FR ratio also decreases with increasing standing crop, reaching very low levels at approximately 500 g/m^2 (Fig. 1).

Chalk grasslands can become very dry in summer (Anderson 1927). In our sites, soil moisture also fluctuates markedly over the seasons, being continuously high in winter and showing short-term fluctuations depending on rain fall distribution in summer, especially in the upper soil layer (0–1 cm). Soil moisture is consistently higher in the Gerendal site: winter values fluctuate around 1 g water/g dry soil, in summer a lowest value of 0.65 g/g was found. At Vrakelberg, winter values fluctuate around 0.5 g/g (open) and 0.7 g/g (dense). In summer, the top soil below open vegetation desiccates more strongly than that below dense vegetation in periods of

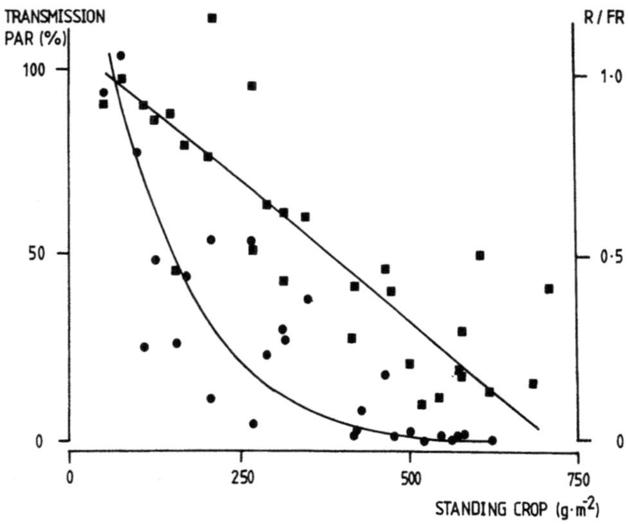

Fig. 1. **Transmission of light (PAR) (dots) in % of incoming radiation and R/FR ratio of transmitted light (squares) at ground level in relation to above ground phytomass. Data from Gerendal and Vrakelberg chalk grasslands combined.**

dry weather (lowest values 0.18 and 0.35 g/g, respectively), but in March and April – the main period of seedling emergence for most species – the differences are less pronounced.

The temperature of the upper soil layer does not fluctuate very much on the Gerendal slope, due to the north-west exposure and the dense vegetation in summer. On the S-exposed Vrakelberg slope, fluctuations throughout the year are stronger. At 1 cm depth, we found temperature ranges of −4 to 43°C below open and −2 to 35°C below dense vegetation.

Wind profiles in dense and open vegetation on the Vrakelberg did not differ much. Air movement is strongly reduced below a height of 50 cm, in the lowest layers more so in dense than in open vegetation (Verkaar et al. 1983b).

DEMOGRAPHY OF SOME SHORT-LIVED FORBS

In the demographic studies reported upon here, attention was focused on the age-state of the plants along with their calendar age (Gatsuk et al. 1980). The following categories were distinguished: seed; seedlings – plants with cotyledons only; juveniles – plants with one or more leaves, but with the cotyledons still present and green; virginiles – plants without cotyledons or reproductive organs; mature plants – plants with reproductive organs. The senile stage does not apply to monocarpic plants (Gatsuk et al. 1980).

In a preliminary study, correlations between frequency at different age-states of short-lived forbs and vegetation density were investigated (Verkaar et al. 1983a). Though variation in the data was very large, the highest frequency of mature plants of all species was found at some optimal vegetation density, which ranged from 200 g/m^2 peak standing crop for *Euphrasia* to 450 g/m^2 for *Daucus*. In several species frequency of seedlings tended to peak at a deviating vegetation density. In *Euphrasia* a double shift occurred; seedlings and flowering plants had their highest frequency in low-productive sites, but the highest frequency of virginile plants was at medium vegetation density.

Now what is the relation between these observations and the actual course of population numbers? Only for *Anthyllis* have demographic data been published earlier (Sterk 1975, Sterk et al. 1982), also from the Vrakelberg reserve. For the other species, we can only present data of individually marked plants in numerous small permanent plots for about three years of observations (1980–1983). Obviously this is a very short period in view of the importance of strong weather fluctuations (Watt 1981b), but even so, interesting results emerge and some tentative conclusions can be drawn.

Synoptic diagrams (after Sagar and Mortimer 1976, slightly modified) of the fate of year classes of plants in open, intermediate and dense vegetation at the Vrakelberg site have been constructed for *Euphrasia stricta*, *Linum*, *Gentianella*, *Daucus* and *Scabiosa*; some examples are

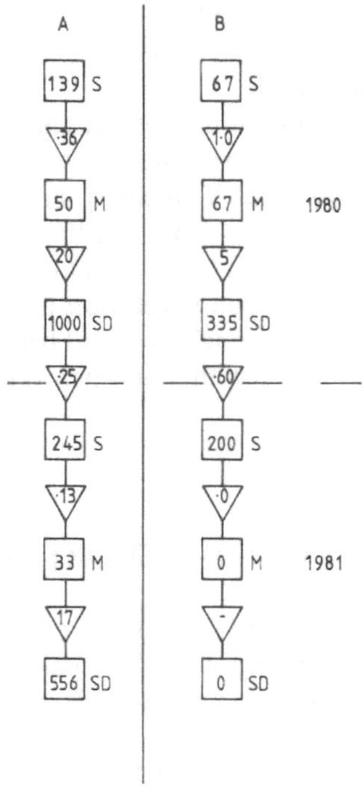

Fig. 2. Fate of year classes of seedlings emerged in 1980 and in 1981 of seedlings of *Euphrasia stricta* in open (A) and dense (B) vegetation in the Vrakelberg chalk grassland. Numbers per m^2, survival rates in summer and mean number of seeds per mature plant are indicated. S = seedlings, M = mature plants, SD = seeds.

presented in Figs. 2–4. Because of large variation in the data, the transition values between the periods are coarse approximations only.

The *Euphrasia* population (Fig. 2) is strictly annual. Salient features in the diagrams are the high fraction of plants not reaching the mature state and the small seed production per plant. Apparently, seed losses in the soil are moderate. In dense vegetation, frequency of plants and seed production per plant are distinctly lower than in open vegetation.

A small fraction of the *Linum* population is annual, but the major part of it is strictly biennial (Fig. 3). Mortality is highest in the first summer, moderate in winter and rather low in the second summer. Seed production per plant is rather higher than in *Euphrasia*, especially in open vegetation. Losses between seed dispersal and seedling emergence appear to be considerable. *Gentianella* (not shown), strictly biennial in this site, shows a similar pattern but mortality is high in the second summer as well. *Daucus* (not shown) locally is a monocarpic pauciennial. The plants occur at low frequencies, and mortality rates are low.

In *Scabiosa* (Fig. 4) the diagram becomes more complicated. Mature plants are usually several years old, and some of them survive after flowering. The density is remarkably high. Mortality rates and seed

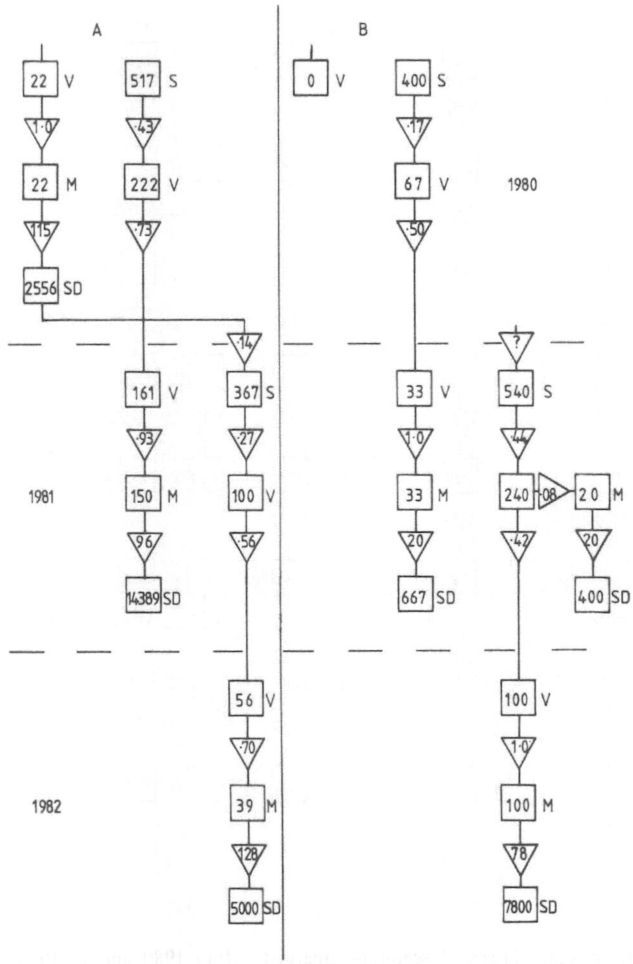

Fig. 3. Fate of year classes of seedlings emerged before 1980, in 1980 and in 1981 of *Linum catharticum*. Numbers per m^2, survival rates in each summer and winter half-year, and mean number of seeds per mature plant are indicated. S = seedlings, V = virginile plants, M = mature plants, SD = seed. A: open; B: dense vegetation.

production per plant are low. Many flowering stems are eaten by rabbits before seed set.

Availability of seeds

The seeds of most species are dispersed in late summer and autumn, *Carlina* (spring) and *Anthyllis* (early summer) being the main exceptions. The seeds of *Gentianella*, *Euphrasia*, *Linum*, and some of the seeds of *Daucus* (Dale 1970) and *Scabiosa* need a chilling period (Verkaar et al.

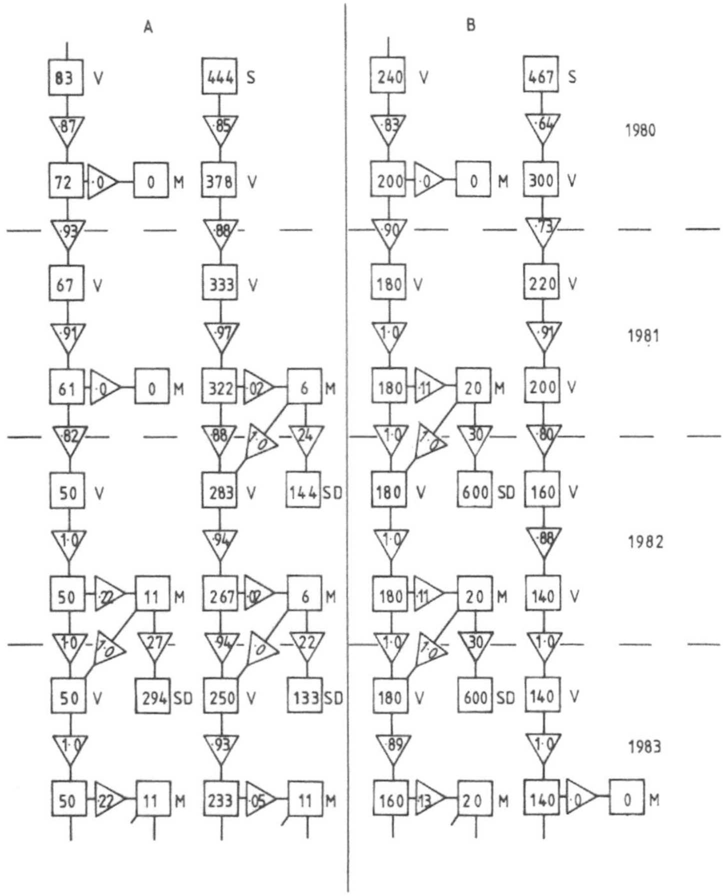

Fig. 4. Fate of year classes of seedlings emerged before 1980 and in 1980 of *Scabiosa columbaria*. Numbers per m^2, survival rates in each summer and winter half-year, and mean number of seeds per mature plant are indicated, S = seedlings, V = virginile plants, M = mature plants, SD = seed. A: open; B: dense vegetation.

subm.). In both sites, the seed bank was sampled every six weeks through 1.5 years (Fig. 5).

Seeds of *Gentianella* and *Euphrasia* which are capable of germinating are present in the soil for only a short period in winter and early spring; whether all seeds germinate or some enter secondary dormancy and remain as viable seeds in the soil is as yet unknown. In February 1980, mean seed density of *Euphrasia* in the soil was 128 and 54 per m^2 below open and dense vegetation, respectively; for *Gentianella*, these figures are 293 (open) and 180 (dense). Since the number of seedlings emerged in spring 1980 (Fig. 2) lies in the same order of magnitude, we suppose that practically all seeds germinate. This is also indicated by the fact that all germinable seeds were found in the upper 3 cm of the soil profile, whereas in the other species a fraction (usually small) is also found in the

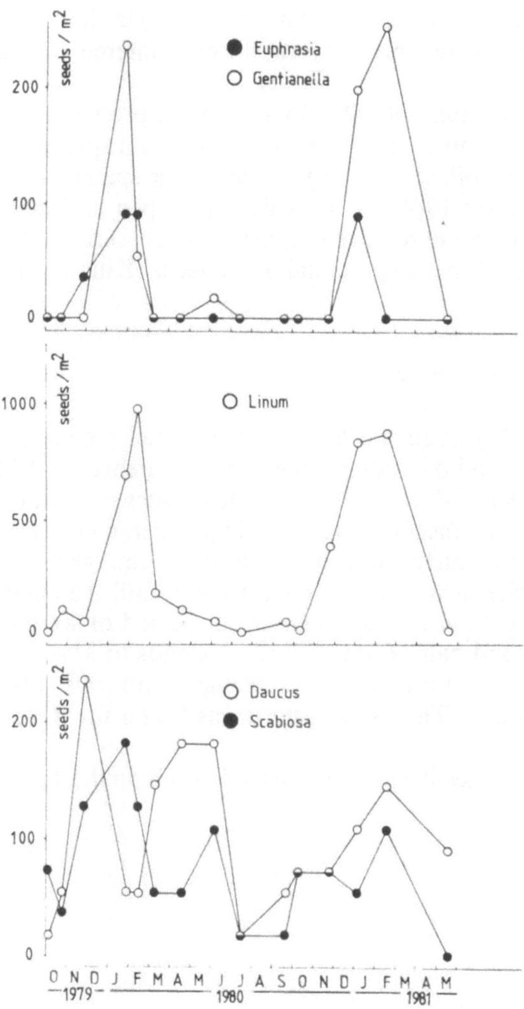

Fig. 5. Seasonal pattern in availability of germinable seed of *Euphrasia stricta*, *Gentianella germanica*, *Linum catharticum*, *Daucus carota*, and *Scabiosa columbaria* in the Vrakelberg site.

3–6 cm layer. Since the seeds of these other species are no more suitable morphologically for rapid burial than those of *Euphrasia* and *Gentianella*, it would seem that such seeds remained viable in the soil for several years.

Germinable seeds of *Linum* are present mainly from December to March, but also during summer and autumn a few occur. The seeds of *Scabiosa* and *Daucus* are mainly available in the soil from the time of dispersal to the spring, and also during the summer they are present in small amounts. From data of Sterk et al. (1982) it appears that the seeds of *Anthyllis* become available only very gradually after dispersal, and that a small proportion of the seeds will remain viable in the soil for 2–3 years. Seeds of *Carlina* were only very rarely encountered in the seed bank studies.

In the Gerendal soil, only *Daucus* seeds were present in some quantity (peak ca. 200 m^{-2}); no seeds of the other short-lived species were met with. In the Vrakelberg soil, seed density showed large spatial variation (Schenkeveld and Verkaar 1984). Mean values for open and dense microsites correlated well with the relative frequencies of mature plants, which may reflect the limited dispersal potential of the seeds (King 1976, Greig-Smith 1979).

Seedling and juvenile stage

Gentianella and *Euphrasia*, the first species of which seedlings emerge, have a rather short period of seedling emergence in (February–) March–April, with a peak in early March (Fig. 6). In dense sites, the peak is about one month later for *Gentianella. Linum* seedlings emerge in March–May, with a peak in March or early April. A few seedlings emerge as late as June. In these three species, juvenile plants are present until the beginning of July (Figs. 7a–c). For *Daucus* and *Scabiosa*, the period of seedling emergence starts in the second half of March and it extends to May, some seedlings emerging as late as June or July. Juvenile plants can still be found in August (Figs. 7d, e). The same pattern was found for *Anthyllis* by Sterk et al. (1982).

The length of the seedling period partly depends on the date of emergence:

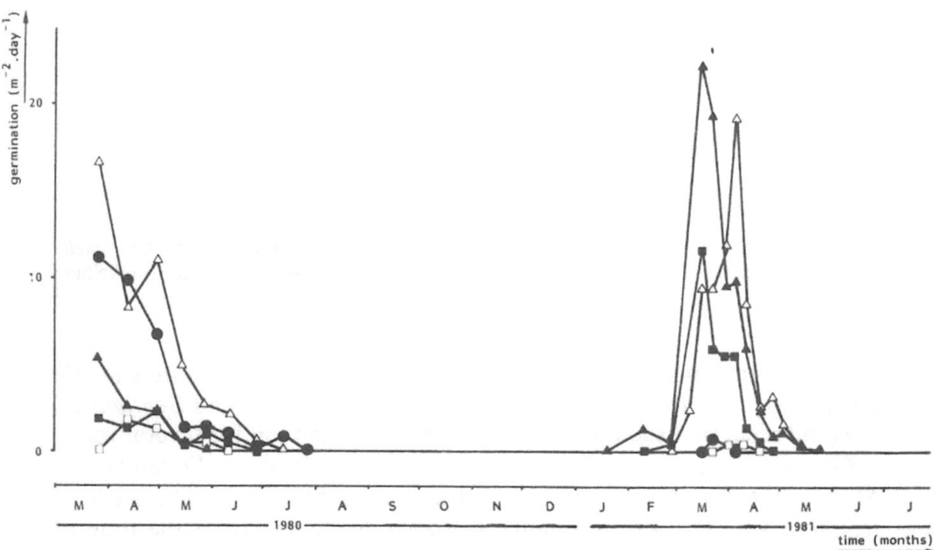

Fig. 6. Seedling emergence through time in *Euphrasia* (■—■), *Gentianella* (▲—▲), *Linum* (△—△), *Daucus* (□—□), and *Scabiosa* (●—●) in the Vrakelberg chalk grassland.

generally, the first seedlings remain in the stage longest (Fig. 8). A short seedling stage together with a long juvenile stage are characteristic of *Euphrasia* (Fig. 7a). The reverse is true for *Gentianella* (Fig. 7b).

The actual density of seedlings is extremely variable, both in space and from year to year. The variability in space can be explained partly by the large variation in density of the seed bank mentioned earlier. Variation between the years 1980 and 1981 is particularly clear in *Daucus* and *Scabiosa*, which in 1980 were present with mean densities of 85 and about 450 per m^2, respectively, and in 1981 with only 11 per m^2 each. This is probably related to the great difference in density of flowering plants in the previous year: both species flowered abundantly in 1979, but only very sparely in 1980 on the Vrakelberg. For *Gentianella*, *Euphrasia* and *Linum*, such differences are much less pronounced; *Gentianella* and *Euphrasia* were respectively 3 and 2 times more abundant in 1981 than in 1980, but *Linum* was 1.5 times more abundant in 1980 in open vegetation.

In *Linum* and *Gentianella*, the first seedlings are most liable to die, while survival is relatively good in the later cohorts (Table 1). This suggests that density-dependent mortality within a species at least does not play a significant role here. Again, *Daucus* and *Scabiosa* behave differently from the other species, showing low mortality (0–20%) which is not concentrated in a particular period or age-state. Seedling mortality is not correlated to vegetation density in *Daucus*. In *Scabiosa*, mortality is very low except in the most open plots (vegetation cover 10% or less; Fig. 9). For some unknown reason, animal activity (e.g., scratching by rabbits) is highest in such very open spots, and many seedlings die because of burial or erosion.

Table 1. Percentage surviving plants of cohorts of seedlings emerged between the recording dates indicated until 15 October of their first year, in relation to date of emergence in some short-lived chalk grassland forbs. In *Euphrasia*, the proportion of plants that reached the mature state is used. In brackets, original size of the cohorts.

Recording dates	*Euphrasia*		*Daucus*		*Scabiosa*	
1980						
19–20/3	50	(4)	–	(0)	43	(13)
28/3–2/4	56	(9)	100	(5)	80	(44)
22–23/4	33	(15)	85	(13)	91	(55)
6–7/5	27	(11)	100	(5)	90	(31)
22–23/5	?	(2)	67	(3)	57	(7)
	Euphrasia		*Gentianella*		*Linum*	
1981						
18/2	–	(0)	0	(8)	0	(1)
4/3	100	(2)	33	(3)	–	(0)
12–13/3	32	(22)	19	(43)	14	(7)
17–18/3	19	(21)	15	(40)	19	(16)
25–26/3	12	(14)	13	(56)	34	(26)
1–2/4	0	(14)	25	(24)	33	(30)
6/4	11	(9)	31	(16)	31	(32)
13–14/4	?	(4)	27	(15)	28	(25)
23–24/4	?	(4)	13	(8)	56	(9)

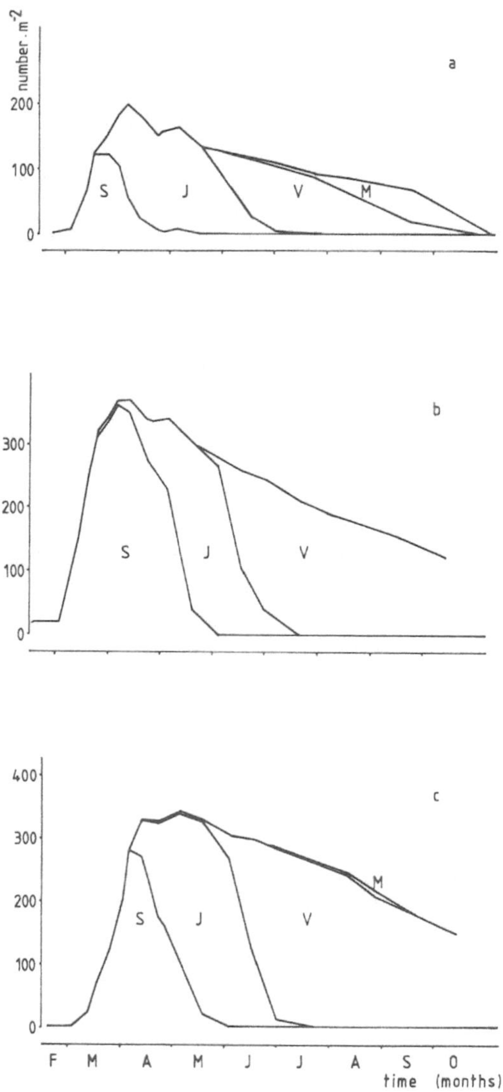

Seedlings and juveniles of *Linum*, *Gentianella* and *Euphrasia* suffer high mortality, especially as young seedlings. For *Euphrasia* seedlings and juvenile plants, mortality is highest at low vegetation cover.

In *Gentianella* the relation between mortality percentage and vegetation cover shows a peak at intermediate vegetation density (Fig. 9). In *Linum* no correlation was found, which is in accordance with the wide range of densities at which mature plants were found in earlier observations. *Anthyllis* is more or less intermediate between the two groups, showing a mortality of ca. 40%, with a peak in August (Sterk et al. 1982). Mortality of seedlings in this species is much higher than in *Scabiosa* according to Silvertown and Dickie (1981).

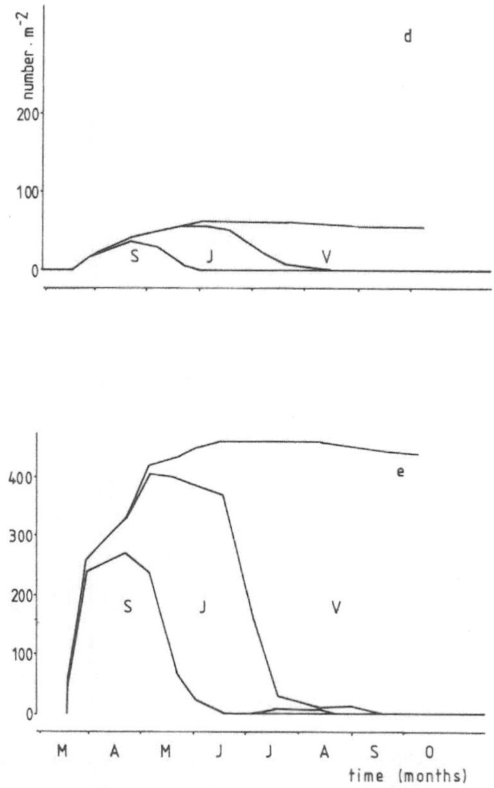

Fig. 7. Seasonal development of age-state composition of one year class of seedlings of *Euphrasia* (a), *Gentianella* (b), *Linum* (c), *Daucus* (d), and *Scabiosa* (e) through their first summer. S = seedlings, J = juvenile plants, V = virginile plants, and M = mature plants.

Virginile plants

The virginile state of *Euphrasia* lasts ca. 2 months, during which 40–50% of the plants die, especially in dense vegetation. The remainder enters the flowering stage during summer. The plants of *Gentianella* and *Linum* remain vegetative until the next summer, except ca. 2% of *Linum* plants which flower in their first summer, and may belong to the annual ecotype described by Zijlstra (1974). Especially in *Gentianella*, a considerable proportion of the plants die before flowering, mainly in the second summer; this mortality is highest in dense vegetation. For *Linum* no correlation between mortality and vegetation density was found. In the second winter, all plants of both species that did not flower died.

In *Anthyllis, Daucus* and *Scabiosa* the virginile period lasts at least until the next summer in the field, and for a large proportion of the plants much longer, up to 4 (*Anthyllis*) or more years. *Anthyllis* shows considerable mortality in the first winter, but afterwards mortality of virginile plants is low (Sterk 1975). In *Daucus* and *Scabiosa* mortality is very low in summer,

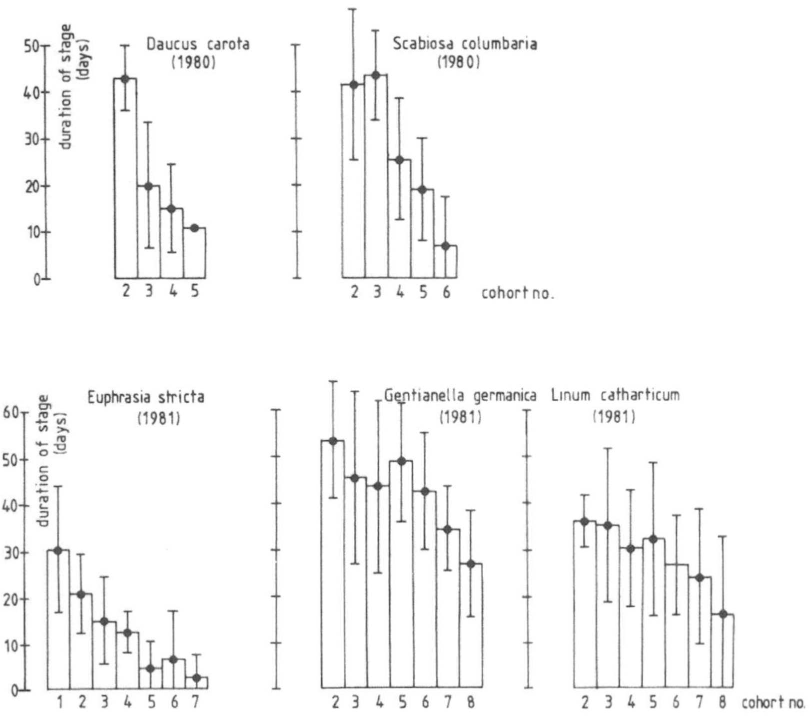

Fig. 8. Duration of the seedling stage in relation to date of emergence in some short-lived forbs. Each cohort consists of the seedlings emerged between two consecutive recording dates (see Table 1).

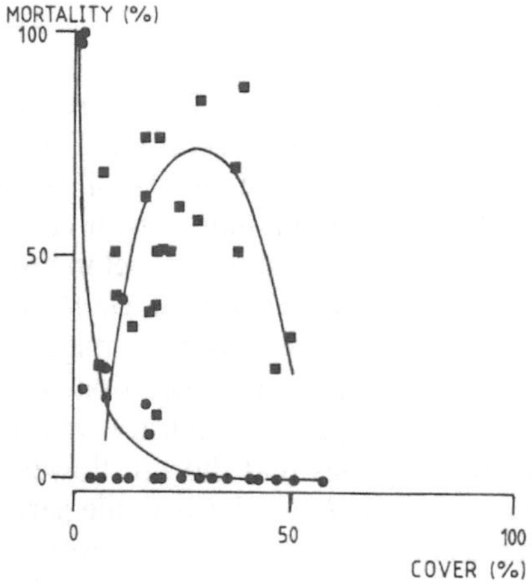

Fig. 9. Mortality of *Gentianella* (squares) and *Scabiosa* (dots) seedlings in relation to vegetation cover.

and slightly higher but yet low in winter. There is hardly any correlation with vegetation cover.

Though predation may be a significant cause of mortality in *Anthyllis* (Sterk 1975), in other species it seems to be less important.

Gentianella is probably not very palatable to grazing animals, as Watt (1981a) found for *G. amarella*, and as far as our observations go the same holds for *Euphrasia*. In *Linum* some mortality is due to predation by slugs (Kelly 1982). In some years flowering stems of *Scabiosa*, and probably the leaves as well, suffer severe predation by rabbits and the same is true for *Daucus*, but even these have comparatively low mortality rates.

Flowering and fruit setting

In *Euphrasia* all surviving virginile plants enter the flowering stage in the first summer, after which they die. *Gentianella* and *Linum* plants (except the few annual plants) surviving in the second summer all flower; again, all plants die after flowering. We could not find any correlation between rosette size and dying or flowering of the virginile plants. Since the nutrient store in the rosettes after their first summer is very small compared to the amounts taken up in the second summer – the rosettes of *Gentianella* even die off above-ground in winter (Gay et al. 1982) – this is not surprising.

In *Anthyllis* 30% of the plants surviving in the second summer are recruited to the flowering stage; in the third, fourth and fifth summer this proportion is 43%, 35% and 1%, respectively. About 90% of the plants on the Vrakelberg flower only once (Sterk 1975). We do not know whether or not the initiation of the flowering stage of an individual is related to a certain critical plant size.

Daucus is strictly monocarpic. In contrast to Lacey et al. (1983) we did not observe flowering in the first and second summer in the field, though in the garden a few plants flowered in their first summer. About 20% of the plants in the third and following summers are recruited to the flowering stage; probably this ratio will be higher in favourable years. Plants with thicker tap roots, and thus greater nutrient store, will have better chances to enter this stage (Gross 1981).

Most *Scabiosa* plants are able to produce new shoots after flowering and are thus polycarpic (Verkaar and Schenkeveld 1984c). Plants grown at low nutrient levels in the garden often died after flowering, however. Flowering in the first summer was found in the garden in some plants that had been well supplied with water and nutrients, but has never been observed in the field. Only very few plants started flowering in the second summer. In the third and following summers, about 10–20% of the plants enter the flowering stage; again, it may well be that this ratio is much higher in favourable years. In 1979 we observed that a large proportion of the plants flowered. Chances of entering the flowering stage are correlated with the dry weight of the individual rosettes in the previous autumn (Fig. 10). We have the impression that in open sites the minimal weight of above-ground parts necessary for flowering is lower than in densely vegetated sites, but the number of observations is yet rather low. Since we

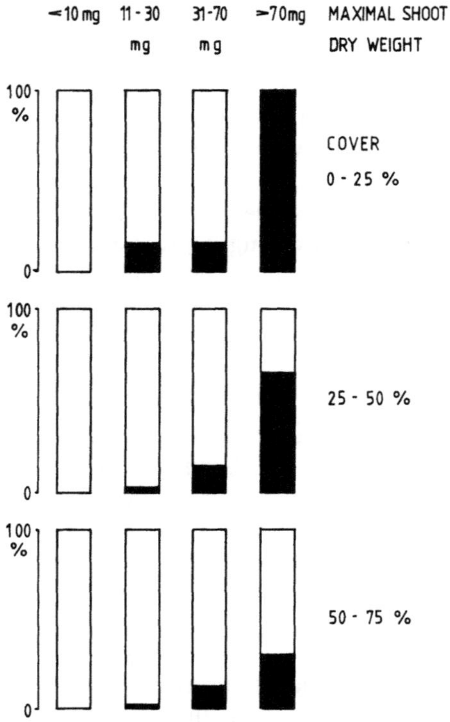

Fig. 10. Percentage of *Scabiosa* plants entering the flowering stage (as indicated by elongation of a flowering stem) in open, intermediate and dense vegetation (maximum cover 0–25, 25–50, and 50–75%, respectively) in relation to rosette dry weight in the previous autumn (calculated from number and size of the rosette leaves, from Verkaar and Schenkeveld 1984b).

also found a retarded growth of rosettes in dense sites, this would mean that it takes several more years for plants in dense sites to reach the flowering stage than for plants in open sites.

Seed production per plant is highly variable, but a general pattern is that in the strict biennials fecundity is distinctly higher than in the perennials. In *Linum*, seed production per flower is very constant (ca. 10), while seed production per individual is more variable: 20–250 (mean 75) per plant (Fig. 3). Those few plants that flower in their first year produce only 20–30 seeds per plant. In *Euphrasia*, number of seeds per flower is also rather constant (ca. 5). Again, the number of flowers is rather variable, resulting in a mean seed output per plant of ca. 15 (Fig. 2). Generally, fecundity tends to be higher in open sites. In *Gentianella* there is much more variation in these values. The number of seeds per flower is 0–70, and the number of flowers per plant is 1–80! Average seed production per plant is 570; most seeds are produced in rather dense sites.

By contrast, seed production of *Scabiosa* plants is rather constant and limited: 20–60 seeds per plant (Fig. 4). It should be noted that many *Scabiosa* seeds are predated while still in the flower head. Seed production per plant is highest in vegetation of intermediate density. Average seed production of *Daucus* is ca. 80 per plant; most seeds are produced in dense sites. For *Anthyllis*, Sterk et al. (1982) mention a mean of 50 seeds per plant in favourable years – implying that this value usually is rather lower.

We did not estimate reproductive effort, even though this measure is an important trait in life history studies (e.g., Stearns 1976), because we do not have data on below-ground production, and because interpretation of data relating to allocation of biomass are difficult to interpret (Thompson and Stewart 1981, Abrahamson and Caswell 1982).

INFLUENCE OF THE VEGETATION STRUCTURE

In the following paragraphs we shall consider the effects of some ecological factors on the demographic processes described, using the results of some laboratory and field experiments. We will try to indicate in which ways the variation in vegetation structure affects these processes through its influence on the microclimate, though often it will appear that our knowledge is far from complete, and that the enormous variability of plants and environment in the field almost defies analysis. Such variation may be caused by plasticity of the genotype or by genetic polymorphism (Harper 1977, Jain 1979). Nothing is known about the species involved in this respect. Ultimately, multiple cause analysis will be necessary as indicated by Silvertown (1981b) and Hilborn and Stearns (1982).

Dispersal

Seeds of most of the species discussed do not possess specialized structures to enhance or ensure long-distance dispersal, and will be dispersed over short distances only (Sheldon and Burrows 1973, Levin and Kerster 1974, ter Borg 1979). In view of the large variation in seed bank density of the species involved, we investigated dispersal distances in the field by three methods: with seed traps around plants in 9 m^2 squares cleared from fruiting neighbours, with marking-recovery techniques, and with a dispersal model for such seeds based on measured wind profiles, wind speed and direction data over the fruiting period, terminal seed velocity, and average height of the inflorescence (Verkaar et al. 1983b). The values predicted by the model proved to be very consistent with the values observed in the two other methods. Since both wind profile and average height of inflorescence differ in dense and open vegetation, we compared dispersal possibilities in relation to vegetation structure. In dense vegetation, wind speed is more strongly reduced, but inflorescences are higher in all species as compared with open vegetation. Maximum trajectories of some species in open and dense vegetation were calculated for a high wind speed of 20 m/s at 1 m height (Table 2).

In the species of small stature, the higher inflorescence does not compensate for the lower wind speed in dense vegetation, resulting in a smaller

Table 2. Average height of inflorescences and calculated maximum dispersal trajectories in dense and open chalk grassland vegetation (Vrakelberg) for some short-lived forbs. The calculations are made assuming a wind speed of 20 m/s at 1 m height.

	Average height of the inflorescence (cm)		Maximum dispersal trajectory (cm)	
	Open	Dense	Open	Dense
Euphrasia	8.3 ± 1.9	13.1 ± 4.7	70	48
Gentianella	10.5 ± 3.8	17.7 ± 3.1	78	64
Linum	12.1 ± 3.3	17.3 ± 3.3	81	55
Daucus	32.8 ± 10.1	50.5 ± 8.7	234	296
Scabiosa	23.5 ± 3.3	45.0 ± 11.9	149	262

dispersal trajectory in dense compared to open vegetation, while in the larger species such as *Daucus* and *Scabiosa*, the reverse holds. This may partly explain the pattern in variation of density in the seed bank of these species.

The model does not apply to species with plumed seeds (such as *Carlina*), for which a model was developed by Burrows (1973). Dispersal capacity might be expected to be larger in such species. Dispersal in *Carlina* is strongly hampered, however, by the fact that the seeds remain attached to the capitulum for most of the autumn and winter, and usually fall on the ground together. Even if wind-dispersed, the seeds of this species have a dispersal trajectory of only a few metres (Sheldon and Burrows 1973).

The numbers of seeds produced per m^2 are considerably higher than the densities of seeds in the seed bank. In this stage, quantitatively the most significant reduction in population numbers takes place. For *Anthyllis*, Sterk et al. (1982) estimated that 40–60% of the seeds is lost through death and predation, largely by the snout beetle *Tychius schneideri*. We have recorded considerably higher values for the species we studied: for *Gentianella*, *Linum*, *Scabiosa*, and *Carlina* only 12, 21, 16 and 43% respectively, of the seeds sown in autumn gave rise to emerged seedlings (During et al. 1984). We know only very little about the causes of these losses. Predation will certainly be important. Field observations show that seeds of *Scabiosa* are already predated upon in the flower heads (loss ca. 12% in 1983); part of the seeds remain undeveloped (in 1983 ca. 38%). *Carlina* seed is heavily predated by small animals (Greig-Smith and Sagar 1981).

Germination, growth and mortality

After dispersal the seeds will at least be partially buried and become incorporated into the seed bank. Recruitment from this bank in spring partly depends on the capability of germination in the dark (this capability may be influenced again by burial, Baskin and Baskin 1980). Seeds of *Linum* do not germinate in the dark (Grime et al. 1981). Germination tests in complete darkness of some other species, with inspection only at the end of the experiment, resulted in the following figures: *Carlina*, 2% germination; *Scabiosa*, 17%; *Daucus*, 30%; and *Euphrasia*, 100%. Since in the field we observed emergence of seedlings of *Gentianella* from seeds buried 1–2 cm deep in the soil, we suppose that this species is not light-dependent for germination either.

In our demographic plots, seedling emergence was only slightly influenced by phanerogam vegetation density, which is generally low in spring. This means that light quality below the canopy is not liable to be a selective factor in this stage, though we found differences in germination response to the R/FR ratio among the species studied (see also King 1975, Silvertown 1980a, 1981a). For some species, notably *Euphrasia* and *Carlina*, emergence was strongly negatively correlated with bryophyte cover, however.

From germination experiments with different temperature regimes the start of the period of seedling emergence in the field could be predicted fairly well. In open microsites the end of this period could be predicted fairly well from the course of the water potential in the upper 1 cm soil layer in combination with germination experiments involving different water potentials. Below a dense vegetation water content of the soil is suitable for a longer period, which may account for the extended emergence period of species such as *Scabiosa* and *Daucus* here. For *Gentianella* and *Euphrasia* this period apparently ends because all seeds have germinated, as is indicated by the pattern in availability from the seed bank in the soil (Fig. 5).

Mortality of the seedlings takes place mainly in March and April in *Euphrasia*, *Gentianella*, and *Linum*, and is usually highest in open vegetation. Soil erosion, predation and extremely low temperatures seem to be among the main causes. A positive correlation between survival and bryophyte cover was found for *Linum*, whereas *Gentianella* showed no correlation, and too few seedlings of *Euphrasia* survived in the plots to allow a conclusion for this species. We have no experimental data to explain these patterns.

Daucus and *Scabiosa* germinate later and these species as well as *Carlina* suffer less mortality which is not strongly concentrated in the early age stages, in contrast to *Anthyllis* (Sterk et al. 1982). Mortality patterns could be related to vegetation density: *Carlina* and *Scabiosa* seedlings are more sensitive to low light intensities (below 1 W/m^2 P.A.R.) and less sensitive to drought than are *Daucus* seedlings (Verkaar and Schenkeveld 1984a).

Growth and mortality of virginile plants are strongly influenced by water deficits and in dense vegetation by low light intensities and a low R/FR ratio below the canopy. The reactions to low water levels of the different species can largely be explained from their different rooting systems (Anderson 1972). The morphology of *Daucus* rosettes, with their erect leaves, might be related to the light climate in dense stands (Gross and Werner 1982). Removal of perennials in the field and addition of nutrients to plants in the greenhouse resulted in strikingly larger plants producing much more seeds, which indicates that nutrient shortages may play a significant role in the field as well.

In the chalk grasslands studied by us, many discrete, small-sized gaps occur as a result of animal activities and erosion. In such gaps the soil is very loose, and soil moisture content is extremely low periodically. Even very small gaps are hardly colonized by the short-lived species we studied, but (if at all) by other species such as *Teucrium botrys*. The distribution pattern of our species in the field is very fine-grained and not directly related to any such pattern in the vegetation structure; the plants usually are interwoven with plants of other, perennial, species. They may thus be regarded as "turf-compatible" species *sensu* Gay et al. (1982).

While the "gap approach" is clearly very suitable for the study of turf-incompatible species (Grubb 1976, 1977, Gross 1980, Miller 1982), we think that establishment and abundance of turf-compatible species should be investigated first in relation to more general effects of spatial and temporal variation in the vegetation structure. In dense vegetation micro-variation at ground level, e.g., in bryophyte cover, appears to be smothered by the overall effect of the phanerogams on light quality and quantity. Whether such small-scale variations in the more open sites cause micro-patterns in the species distribution remains a topic for further research. To apply the gap concept even on this scale would perhaps stretch its meaning too much.

CONCLUSION

Within the group of short-lived chalk grassland forbs the "fine-tuning" to the environment in different age-states of each species has led to different sets of traits. Silvertown (1981b) views seed size, life span and germination data as a set of co-adapted features and he recognized two principal patterns among 75 chalk grassland species: annuals, especially winter annuals, with seeds of 0.01–1.0 mg, and biennials, pauciennial monocarps and perennials, with seed weight of 1.0–3.0 mg. It may be noted in passing that the strict biennials we studied fall in the former category. Whereas the large-seeded species usually germinate in spring, germination of the small-seeded species generally takes place in autumn, when competition for water and nutrients is reduced; this may be related to the small energy reserve in the seed, giving the species considerable disadvantage compared to the faster-growing perennials and seedlings from larger seeds in spring. Autumn germination, as a consequence, entails heavy mortality in winter, which is counterbalanced by the large numbers of seeds produced. In Silvertown's view, the life history pattern of winter annuals is due to "ecological displacement" for habitats where seed size is a constraint. The fact that our only small-seeded annual, *Euphrasia*, is a summer annual may be related to its hemiparasitic character – it needs a well-growing host!

A similar relation between seed size and survival in dune (winter) annuals has been observed by Grubb et al. (1982). Gross and Werner (1982) also note the correlation between seed size and survival of the seedlings, and they state that large-seeded species have better opportunities of establishment in a dense vegetation. In the species studied by us, the general tendency of small-seeded species to show higher seedling mortality is apparent; in the large-seeded species the correlation between seed weight and mortality is not so evident, however. The correlation between high seed weight and preference for dense vegetation is not always present, *Linum* (also abundant in rather dense sites) and *Carlina* and *Scabiosa* (more common in open vegetation) being the main exceptions.

Among the winter annuals in a Californian annual grassland, Bartolome (1979) recognized five different groups of plants differing mainly in regeneration characteristics. At one extreme, the grasses *Aira praecox* and *Festuca* spp. produce large, though highly variable, quantities of seeds, of which only a few remain viable in the seed bank for more than a year; germination extends over a long period, and mortality of the seedlings is high. In contrast to this, the grass *Taeniatherum asperum* and *Erodium* spp. produce only few, large seeds, in rather constant quantities per year, with complex dormancy patterns. Mortality of the seedlings is low, though higher in a later age-state for *Taeniatherum*.

Similar patterns can be found in the species we have studied, though there is an added complication for some species in which there is a shift in "optimal" vegetation density for the survival of the different age-states. This leads to a considerable loss of individuals. Such species are termed "spenders" by Schenkeveld and Verkaar (1981). These are different from species referred to as "savers", in which maximal population density is consistently at the same vegetation density throughout the life cycle. When we consider some relevant characteristics (Table 3), each species appears to have its own set of traits. Among the short-lived plants studied two broad categories can be recognized.

Spenders

These are shallow-rooted, very short-lived plants of small stature, producing many small seeds. Plants of one year class all flower in one year. A persistent seed bank is absent or small. Mortality in seedling, juvenile and virginile state is high, often with a marked shift among the different age-states between open and dense sites and with a preference for open vegetation in the mature state. Maximal growth occurs in summer or early autumn. The annual *Euphrasia* is the most extreme example of this pattern. The biennials *Gentianella* and *Linum* also belong here, though *Gentianella* does not show a distinct shift between the age-states, and *Linum* occurs preferentially in denser vegetation than the other two. Most of the species in this group are classified as stress-tolerant ruderals in the system of Grime (1979).

Savers

These are deeper-rooted and longer-lived plants than spenders, attain a higher stature and produce few large seeds. Flowering in plants of one year class is spread over several years. A persistent seed bank is usually present.

Table 3. Characteristics related to the life history patterns of some short-lived forbs.

	Seed weight (mg)	Average number of seeds per plant	Life span (years)	Mortality of seedlings up to August (%)	Presence of 'shift' between age-states in relation to cover of	
					phanerogams	bryophytes
'Spenders'						
Euphrasia	0.12	15	1	90	+	+
Gentianella	0.18	570	2	70	(+)	–
Linum	0.15	75	(1–)2	80	(+)	+
Carlina	1.3*	105	⩾2	20	–	+
'Savers'						
Scabiosa	1.3	30	⩾2	10	–	–
Daucus	0.9	80	⩾2	5	–	?
Anthyllis	3.4	50	⩾2	40	(+)	?

*Seed weight without plumes.

Mortality in seedling, juvenile and virginile states is low (less so in *Anthyllis*), without a shift between open and dense sites. They have a preference for dense vegetation, except *Scabiosa*. Maximal growth presumably occurs in spring. This pattern is "conservative" (Bartolome 1979). The risk of local extermination is minimized by the persistent (though often small) seed bank and by the spread of flowering over several years (after which the monocarpic plants die). *Daucus* is a clear example; *Anthyllis* and *Scabiosa* also belong here. *Anthyllis* and *Daucus* are classified as C–S–R strategists by Grime (1979), and *Scabiosa* fits in this category as well. The place of *Carlina* is more or less in between; it has mostly characteristics of a saver, but it exhibits a marked shift in frequency at low and high bryophyte cover among the age-states, and in the mature state it is most common in open sites. Moreover, it is rather shallow-rooted. Grime (1979) classifies it as a stress-tolerant ruderal, to which category also our spenders belong.

The relative abundance of savers tends to be lower than that of spenders, which is generally in accordance with the observations of Grubb et al. (1982). In some years, the frequency of *Scabiosa* is exceptionally high.

What, briefly, is the place of spenders and savers in the framework of life history systems? Both occur in habitats in which spots with suitable environmental characteristics occur regularly and frequently. They are contrasted with species which exploit (usually large) gaps which occur irregularly and infrequently: these are either fugitives *sensu* During (1979), with emphasis on diaspore production and high dispersal capacity (extremely *r*-selected species *sensu* MacArthur and Wilson 1967) or species with an emphasis on a very persistent seed bank, with viable seeds in the soil for many years. As shown by Bartolome (1979) and others, a continuum of specific sets of traits may be found, but if we restrict ourselves to spenders and savers the following hypothesis may be put forward.

In both groups dispersal capacity is very limited and the numbers of diaspores produced are small compared to fugitives. In these respects, they are comparable to "annual shuttle species" among bryophytes (During 1979). However, whereas most "shuttle" bryophytes depend on gaps (cf. Furness and Hall 1982) and thus may be compared to turf-incompatible monocarpic phanerogams, the turf-compatible spenders and savers are more dependent on spatial and temporal variation in the general vegetation structure and perhaps in the ecosystem as a whole including predators.

Spenders react more than savers to spatial variation; their emphasis is not on safeguards to survive highly unfavourable years, but on finding suitable sites within the ecosystem each year. This involves high losses of individuals before the reproductive stage, but each year a number of individuals flower and produce sufficient seeds. The low palatability of several of these species discussed earlier may be important in preventing local extinction in years with exceptionally many predators. It is not yet known on what scale the spatial variation operates, but in our areas we think it would be on a scale of centimeters in the open sites in which most species occur.

Savers, on the other hand, are better adjusted than spenders to temporal variations. Several safeguards – persistent seed bank, lower mortality rates resulting in higher mean age of individuals, extended period of flowering of one year class – exist. These enable them to survive unfavourable periods, such as years with hardly any flowering, perhaps due to excessive predation or bad weather conditions, as was shown for *Anthyllis* by Sterk et al. (1982). A reaction to spatial micro-variation is improbable for most species because of the smothering effect on such variation of the dense vegetation in which they are found. Thus, the five patterns discussed might be seen as bridging unfavourable periods ("gaps" in time) of different length and regularity of occurrence. Spenders usually do not bridge periods longer than one winter or summer, and spatial variation is more prominent. Savers are able to bridge periods of one to a few years. Gap-dependent species may occur for several years in the same gap such as ant-hills, but often have to bridge periods of ten or more years. In extreme examples, some species occur in gaps often separated in time by a hundred years or more (e.g., storm gaps in forests). The fugitives have discarded the bridging of time gaps altogether and their emphasis is on bridging spatial distances.

To explain the fact that a number of short-lived species may co-exist, mechanisms of niche-differentiation are generally invoked (e.g., Silvertown 1981b), such as micro-distribution of suitable germination sites (Harper et al. 1965, Oomes and Elberse 1976) or neighbour effects (Mack and Harper 1977). This viewpoint, starting from the applicability of Gause's principle of competitive exclusion (attacked by Alley 1982, among others), assumes a critical role for competition, as exemplified by the interesting models for co-existence presented by Newman (1982). However, Grubb et al. (1982) provide evidence that competition within the guild of annuals and biennials is rather limited, and Fowler (1981) argues that within guilds in a North Carolina grassland there is rather weak, diffuse competition (*sensu* MacArthur 1972). In our view because of the mowing regime which prevents accumulation of litter and establishment of shrubs and trees, "ecological space" (the amount of which is determined by the peak standing crop, among other features) is available in the Vrakelberg vegetation continuously and is filled partially by the short-lived species. Even though each species has a different set of traits, we think that co-existence of such species in grasslands with low to moderate productivity needs to be interpreted not in terms in niche-differentiation, but in terms of "niche-sharing" (During and Willems 1984). The selective factors imposed by the abiotic environment in combination with the perennial vegetation keep the levels of each species rather low; the life history patterns of the species are not directed towards attaining maximal density or competitive power, but to prevent local extinction. Thus, the maximal number of species in the corridor model of Grime (Al-Mufti et al. 1977) would at least partly depend on the number of suitable species able to reach the site, as has been argued by Whittaker (1977) and Grime (1979). A possible test of this hypothesis might be deliberately to introduce some short-lived species in such a vegetation, or preferably in an experimental garden, and to follow their fate together with that of the other species present.

In view of the considerations of Huston (1979), the same ideas might apply to the perennials in very species-rich communities. In such communities niche-sharing may be more important than niche-differentiation and analysis should therefore be directed more towards mortality causes and maintenance of a minimum level of subsistence, rather than towards competitive power in the face of other species.

ACKNOWLEDGEMENTS

Many studies have contributed to our insight into the chalk grassland ecosystem. We wish to thank especially D. van de Born, F. Kranenburg, M. Kropff, H. Smit and F. Stouthart (water relations); H. van Osch and A. van Rossum (temperature responses); M. Verstegen and E. Veneklaas (germination in relation to water potentials); C. Huurnink (predation of seeds); and P.J. Keizer and B.F. van Tooren (role of the bryophyte layer).

REFERENCES

Abrahamson, W.G. and Caswell, H. (1982) On the comparative allocation of biomass, energy, and nutrients in plants. Ecology 63: 982–991.

Alley, T.R. (1982) Competition theory, evolution, and the concept of an ecological niche. Acta Biotheor. 31: 165–179.

Al-Mufti, M.M., Sydes, C.L., Furness, S.B., Grime, J.P. and Band, S.R. (1977) A quantitative analysis of shoot phenology and dominance in herbaceous vegetation. J. Ecol. 65: 759–791.

Anderson, V.L. (1927) Studies of the vegetation of the English chalk V. The water economy of the chalk flora. J. Ecol. 15: 72–129.

Barkman, J.J. (1979) The investigation of vegetation texture and structure. In M.J.A. Werger (Ed.), The Study of Vegetation, pp. 123–160. Junk, The Hague.

Bartolome, J.W. (1979) Germination and seedling establishment in California annual grassland. J. Ecol. 67: 273–281.

Baskin, J.M. and Baskin, C.C. (1980) Ecophysiology of secondary dormancy in seeds of Artemisia artemisifolia. Ecology 61: 475–480.

Borg, S.J. ter (1979) Some topics in plant population biology. In M.J.A. Werger (Ed.), The Study of Vegetation, pp. 11–55. Junk, The Hague.

Bradshaw, A.D. (1980) Mineral nutrition. In J.H. Rorison and R. Hunt (Eds.), Amenity Grassland, an Ecological Perspective. Wiley, Chicester.

Burrows, F.M. (1973) Calculation of the primary trajectories of plumed seeds in steady winds with variable convection. New Phytol. 72: 647–664.

Cohen, D. (1966) Optimizing reproduction in a randomly varying environment. J. Theor. Biol. 12: 119–129.

Cohen, D. (1967) Optimizing reproduction in a randomly varying environment when a correlation may exist between the conditions at the time a choice has to be made and the subsequent outcome. J. Theor. Biol. 16: 1–14.

Dale, H.M. (1970) Germination patterns in *Daucus carota* (L.) ssp. *carota*. Variations in the 1967 collections. Canad. J. Bot. 48: 413–418.

During, H.J. (1979) Life strategies of bryophytes: a preliminary review. Lindbergia 5: 2–18.

During, H.J. and ter Horst, B. (1983) The diaspore bank of bryophytes and ferns in chalk grassland. Lindbergia 9: 57–64.

During, H.J., Keizer, P.J. and van Tooren, B.F. (1984) The influence of bryophytes on germination and establishment of phanerogams in a chalk grassland. In J.M. Géhu (Ed.), Vegetation of Chalk Grasslands and Conservancy, Colloques Phytosoc. 11: 457–469.

During, H.J. and Willems, J.H. (1984) Diversity models applied to a chalk grassland. Vegetatio 57: 103–114.

Fowler, N. (1981) Competition and coexistence in a North Carolina grassland II. The effects of the experimental removal of species. J. Ecol. 69: 843–854.

Furness, S.B. and Hall, R.H. (1982) An explanation of the intermittent occurrence of *Physcomitrium sphaericum* (Hedw.) Brid. J. Bryol. 11: 733–742.

Gatsuk, L.E., Smirnova, O.V., Vorontzova, L.I., Zaugolnova, L.B. and Zhukova, L.A. (1980) Age states of plants of various growth forms: a review. J. Ecol. 68: 675–696.

Gay, P.E., Grubb, P.J. and Hudson, H.J. (1982) Seasonal changes in the concentrations of nitrogen, phosphorus and potassium, and in the density of mycorrhiza, in biennial and matrix-forming perennial species of closed chalkland turf. J. Ecol. 70: 571–593.

Greig-Smith, J. and Sagar, G.R. (1981) Biological causes of local rarity of *Carlina vulgaris*. In H. Synge (Ed.), The Biology of Rare Plant Conservation, pp. 389–400. Wiley, Chichester.

Greig-Smith, P. (1979) Pattern in vegetation. J. Ecol. 67: 755–779.

Grime, J.P. (1963) An ecological investigation at a junction between two plant communities in Coombsdale on the Derbyshire limestone. J. Ecol. 51: 391–402.

Grime, J.P. (1979) Plant Strategies and Vegetation Processes. Wiley, Chichester.

Grime, J.P. and Curtis, A.V. (1976) The interaction of drought and mineral nutrient stress in calcareous grassland. J. Ecol. 64: 975–988.

Grime, J.P. et al. (1981) A comparative study of germination characteristics in a local flora. J. Ecol. 69: 1017–1059.

Gross, K.L. (1980) Colonization by *Verbascum thapsus* (mullein) of an old-field in Michigan: experiments on the effects of vegetation. J. Ecol. 68: 919–927.

Gross, K.L. (1981) Predictions of fate from rosette size in four "biennial" plant species: *Verbascum thapsus*, *Oenothera biennis*, *Daucus carota* and *Tragopogon dubius*. Oecologia (Berl.) 48: 209–213.

Gross, K.L. and Werner, P.A. (1982) Colonizing abilities of "biennial" plant species in relation to ground cover: implications for their distributions in a successional sere. Ecology 63: 921–931.

Grubb, P.J. (1976) A theoretical background to the conservation of ecologically distinct groups of annuals and biennials in the chalk grassland ecosystem. Biol. Conserv. 10: 53–76.

Grubb, P.J. (1977) The maintenance of species-richness in plant communities: the importance of the regeneration niche. Biol. Rev. 52: 107–145.

Grubb, P.J. (1982) Control of relative abundance in roadside Arrhenatheretum: results of a long-term garden experiment. J. Ecol. 70: 845–861.

Grubb, P.J., Kelly, D. and Mitchley, J. (1982) The control of relative abundance in communities of herbaceous plants. In E.I. Newman (Ed.), The Plant Community as a Working Mechanism. Spec. Pap. B.E.S. 1: 79–97.

Gupta, P.L. and Rorison, I.H. (1975) Seasonal differences in the availability of nutrients down a podsolic profile. J. Ecol. 63: 521–534.

Harper, J.L. (1977) Population Biology of Plants. Academic Press, London.

Harper, J.L., Williams, J.T. and Sagar, G.R. (1965) The behaviour of seeds in soil I. The heterogeneity of soil surface and its role in determining the establishment of plants from seed. J. Ecol. 53: 273–286.

Hilborn, R. and Stearns, S.C. (1982) On inference in ecology and evolutionary biology: the problem of multiple causes. Acta Biotheor. 31: 145–164.

Huston, M. (1979) A general hypothesis of species diversity. Amer. Nat. 113: 81–101.

Jain, S. (1979) Adaptive strategies: polymorphism, plasticity, and homeostatis. In O.T. Solbrig, S. Jain, G.B. Johnson and P.H. Raven (Eds.), Topics in Plant Population Biology, pp. 160–187. MacMillan Press, London.

Jeffrey, D.W. (1971) The experimental alteration of a Kobresia-rich sward in Upper Teesdale. In E. Duffey and A.S. Watt (Eds.), The Scientific Management of Animal and Plant Communities for Conservation, pp. 78–89. Blackwell, Oxford.

Kelly, D. (1982) Demography, population control and stability in short-lived plants of chalk grasslands. Ph.D. Thesis, Cambridge.

King, T.J. (1975) Inhibition of seed germination under leaf canopies in *Arenaria serpyllifolia, Veronica arvensis* and *Cerastium holosteoides*. New Phytol. 75: 87–90.

King, T.J. (1976) The viable seed contents of ant-hill and pasture soil. New Phytol. 77: 143–147.

King, T.J. (1977) The plant ecology of ant-hills in calcareous grasslands III. Factors affecting the population sizes of selected species. J. Ecol. 65: 279–315.

Lacey, E.P., Real, L., Antonovics, J. and Heckel, D.G. (1983) Variance models in the study of life histories. Amer. Nat. 122: 114–131.

Levin, D.A. and Kerster, H.W. (1974) Gene flow in seed plants. Evol. Biol. 7: 139–220.

Lloyd, P.S. and Pigott, C.D. (1967) The influence of soil conditions on the course of succession on chalk of southern England. J. Ecol. 55: 137–146.

Mack, R. and Harper, J.L. (1977) Interference in dune annuals: spatial pattern and neighbourhood effects. J. Ecol. 65: 345–363.

MacArthur, R.H. (1972) Geographical Ecology. Harper and Row, New York.

MacArthur, R.H. and Wilson, E.O. (1967) The Theory of Island Biogeography. Princeton Univ. Press, Princeton.

Meijden, R. van der, Weeda, E.J., Adema, F.A.C.B. and de Joncheere, G.J. (1983) Heukels–Van der Meijden flora van Nederland, 20th ed. Wolters-Noordhoff, Groningen.

Miller, T.E. (1982) Community diversity and interactions between the size and frequency of disturbance. Amer. Nat. 120: 533–536.

Newman, E.I. (1982) Niche separation and species diversity in terrestrial vegetation. In E.I. Newman (Ed.), The Plant Community as a Working Mechanism. Spec. Pap. B.E.S. 1: 61–77.

Oomes, M.J.M. and Elberse, W.Th. (1976) Germination of six grassland herbs in micro-sites with different water content. J. Ecol. 64: 745–755.

Perring, F. (1960) Climatic gradients of chalk grassland. J. Ecol. 48: 415–442.

Rabotnov, T.A. (1978) On coenopopulations of plants reproducing by seeds. In A.H.J. Freijsen and J.W. Woldendorp (Eds.), Structure and Functioning of Plant Populations, pp. 1–26. North Holland Publ. Comp., Amsterdam.

Read, D.J., Koucheki, H.K. and Hodgson, J. (1976) Vesicular-arbuscular mycorrhiza in natural vegetation systems I. The occurrence of infection. New Phytol. 77: 641–653.

Rorison, I.H. (1960) Some experimental aspects of the calcicole-calcifuge problem I. The effects of competition and mineral nutrition upon seedling growth in the field. J. Ecol. 48: 585–599.

Rosén, E. (1982) Vegetation development and sheep grazing in limestone grasslands of south Oeland, Sweden. Acta Phytogeogr. Suec. 72.

Sagar, G.R. and Mortimer, A.M. (1976) An approach to the study of the population dynamics of plants with special reference to weeds. Appl. Bio. 1: 1–47.

Schenkeveld, A.J. and Verkaar, H.J. (1981) Hapaxanthous species in limestone grasslands: concious savers or big spenders? Acta Bot. Neerl. 30: 134.

Schenkeveld, A.J. and Verkaar, H.J. (1983) The distribution in space and time of viable seeds of some short-lived forbs on two chalk grasslands in S. Limburg. Acta Bot. Neerl. 32: 244–245.

Schenkeveld, A.J. and Verkaar, H.J. (1984) On the ecology of short-lived forbs in chalk grasslands: Distribution of germinative seeds and its significance for seedling emergence. J. Biogeogr. 11: 251–260.

Sheldon, J.C. and Burrows, F.M. (1973) The dispersal effectiveness of the achene-pappus units of selected Compositae in steady winds with convection. New Phytol. 72: 665–675.

Silvertown, J. (1980a) Leaf-canopy-induced dormancy in a grassland flora. New Phytol. 85: 109–118.

Silvertown, J. (1980b) The dynamics of a grassland ecosystem: botanical equilibrium in the Park Grass Experiment. J. Appl. Ecol. 17: 491–504.

Silvertown, J. (1981a) Micro-spatial heterogeneity and seedling demography in species-rich grassland. New Phytol. 88: 117–128.

Silvertown, J. (1981b) Seed size, life span, and germination date as co-adapted features of plant life history. Amer. Nat. 118: 860–864.

Silvertown, J. and Dickie, J.B. (1981) Seedling survivorship in natural populations of nine perennial chalk grassland plants. New Phytol. 88: 555–558.

Smith, C.J. (1980) Ecology of the English Chalk. Academic Press, London.

Smith, C.J., Elston, J. and Bunting, A.H. (1971) The effects of cutting and fertilizer treatments on the yield and botanical composition of chalk turf. J. Br. Grassld. Soc. 26: 213–219.
Stearns, S.C. (1976) Life history tactics: a review of the ideas. Quart. Rev. Biol. 51: 3–47.
Sterk, A.A. (1975) Demographic studies of *Anthyllis vulneraria* L. in the Netherlands I. Fluctuations and survivorship in natural populations. Acta Bot. Neerl. 24: 315–337.
Sterk, A.A., van Duikeren, A., Hogervorst, J. and Verbeek, E.D.M. (1982) Id. II. Population density fluctuations and adaptations to arid conditions, seed populations, seedling mortality, and influence of the biocoenosis on demographic features. Acta Bot. Neerl. 31: 11–40.
Taylor, A.A., De-Felice, J. and Havill, D.C. (1982) Seasonal variation in nitrogen availability and utilization in an acidic and calcareous soil. New Phytol. 92: 141–152.
Thompson, K. and Grime, J.P. (1979) Seasonal variation in the seed banks of herbaceous species in ten contrasting habitats. J. Ecol. 67: 893–921.
Thompson, K. and Stewart, A.J.A. (1981) The measurement and meaning of reproductive effort in plants. Amer. Nat. 117: 205–211.
Verkaar, H.J. and Schenkeveld, A.J. (1981) Kiemkrachtige zaden van enkele akkeronkruiden gevonden in de bodem van een voormalige akker in Zuid-Limburg. Gorteria 10: 225–227.
Verkaar, H.J., Schenkeveld, A.J. and Brand, J.M. (1983a) On the ecology of short-lived forbs in chalk grasslands: micro-site tolerances in relation to vegetation structure. Vegetatio 52: 91–102.
Verkaar, H.J., Schenkeveld, A.J. and Klashorst, M.P. van der (1983b) The ecology of short-lived forbs in chalk grassland: dispersal of seeds. New Phytol. 95: 335–344.
Verkaar, H.J. and Schenkeveld, A.J. (1984a) On the ecology of short-lived forbs in chalk grasslands: seedling development under low photon flux density conditions. Flora 175: 135–141.
Verkaar, H.J. and Schenkeveld, A.J. (1984b) On the ecology of short-lived forbs in chalk grasslands: life history characteristics. New Phytol. 98: 659–672.
Verkaar, H.J. and Schenkeveld, A.J. (1984c) On the ecology of short-lived forbs in chalk grasslands: semelparity and seed output of some species in relation to various levels of nutrient supply. New Phytol. 98: 673–682.
Verkaar, H.J., Schenkeveld, A.J., van Osch, H.H. and van Rossum, A.C. (subm.) On the ecology of short-lived forbs in chalk grasslands: germination responses of fresh and buried seeds to temperature. New Phytol.
Watt, A.S. (1974) Senescence and rejuvenation in ungrazed chalk grassland (grassland B) in Breckland; the significance of litter and moles. J. Appl. Ecol. 11: 1157–1171.
Watt, A.S. (1981a) A comparison of grazed and ungrazed grassland A in East Anglian Breckland. J. Ecol. 69: 499–508.
Watt, A.S. (1981b) Further observations on the effects of excluding rabbits from grassland A in East Anglian Breckland: the pattern of change and factors affecting it (1936–1973). J. Ecol. 69: 509–536.
White, P.S. (1979) Pattern, process, and natural disturbance in vegetation. Bot. Review 45: 229–299.
Whittaker, R.H. (1977) Animal effects on plant species diversity. In R. Tüxen (Ed.), Vegetation und Fauna. Ber. Int. Symp. Int. Ver. Vegetationskunde, Rinteln 1976: 409–425.
Willems, J.H. (1973) Limestone grasslands in the central part of the French Jura, South of Champagnôle (dept. Jura). Proc. Kon. Ned. Ac. Wet. Ser. C, 76: 231–244.
Willems, J.H. (1978) Phytosociological and ecological notes on chalk grasslands of southern England. Vegetatio 37: 141–150.
Willems, J.H. (1980) Observation on North-West European limestone grassland communities V. An experimental approach to the study of species diversity and above-ground biomass in chalk grassland. Proc. Kon. Ned. Ac. Wet. Ser. C, 83: 279–306.
Willems, J.H. (1982) Phytosociological and geographical survey of Mesobromion communities. Vegetatio 48: 227–240.
Willems, J.H. (1983a) Species composition and above-ground phytomass in chalk grassland with different management. Vegetatio 52: 171–180.
Willems, J.H. (1983b) The seed bank as a part of the vegetation. Acta Bot. Neerl. 32: 243.
Zijlstra, G. (1974) Variabiliteit bij *Linum catharticum* L. in Nederland. Contactblad Oecologen 10(1): 10–14.

LIFE CYCLE ECOLOGY OF ANNUAL PLANT SPECIES OF CEDAR GLADES OF SOUTHEASTERN UNITED STATES

JERRY M. BASKIN and CAROL C. BASKIN

ABSTRACT

In southeastern United States, a region where the zonal vegetation is deciduous forests, shallow soils over limestone and dolomitic bedrock support edaphic climax herbaceous plant communities known as cedar glades. The flora includes over 400 species of vascular plants; 79 of these are winter annuals and 47 are summer annuals. This chapter reviews the life cycle ecology of both groups of therophytes in the cedar glade habitat in relation to their tolerances, requirements and adaptations. Temperature, through its influence on seed dormancy, dormancy break and germination, is the single most important environmental factor regulating the timing of the life cycle of annual plants of cedar glades.

INTRODUCTION

In the eastern United States deciduous forest vegetation is the climatic climax (Braun 1950), but in some areas where the soils are too shallow to support trees, natural, long-persisting communities dominated by herbaceous vascular plants have developed. When these nonforested areas occur where limestone or dolomite rock is at or near the surface, they are called glades (Fig. 1A, B). The herbaceous plant communities developed on outcrops of Ordovician limestone in the Central Basin of Tennessee were observed by Galloway (1919) and Quarterman (1950a), and they wrote definitions of the word glade. According to Galloway (1919) the word originally meant "... an open space in a wood," but "... now it is applied to any rocky place whether the trees still surround it or have been removed." Quarterman (1950a) defined glades as "... open spaces occasioned by the presence of rock at or near the surface. They may be simply exposed rock surfaces or gravelly or grassy expanses surrounded by cedar woods." The term "cedar glades" comes from the fact that *Juniperus virginiana* (commonly called redcedar) is often the most frequent woody species growing in the deeper soils adjacent to the glades. (Unless given, authorities of scientific names are found in Fernald 1950.) The primary

White, J. (ed.), The Population Structure of Vegetation.
ISBN 978-94-010-8927-2

Fig. 1. Cedar glades in central Tennessee. A. Glade developed on thin-bedded limestone. **B.** Glade developed on thick-bedded limestone.

focus of this review is on the life cycle ecology of annual plants of cedar glades of the southeastern United States and most especially those in the cedar glades of Tennessee and Kentucky.

DISTRIBUTION, SUBSTRATE AND CLIMATE

Within the Southeast, well-developed cedar glades occur in Kentucky, Tennessee, Alabama and Georgia with the most numerous, extensive and floristically rich ones occurring in the Central Basin of Tennessee (Fig. 2). In the southeastern United States, cedar glades are developed mostly on gently sloping topography on thin- and thick-bedded limestone and dolomite rock formations by the processes of superficial solution and erosion.

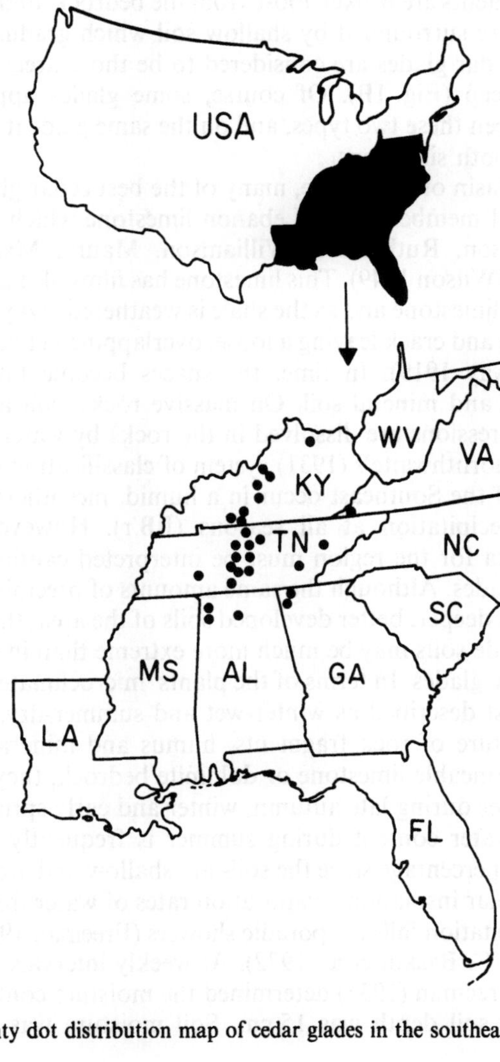

Fig. 2. County dot distribution map of cedar glades in the southeastern United States.

As the soluble part of the rocks is dissolved, it is carried away leaving a small amount of insoluble material, mainly clay. However, most of the clay also is removed by rain water (Galloway 1919). Cedar glade soils do not develop the normal A to C soil profile; instead, they are azonal soils (lithosols) composed of a mixture of rock fragments, humus and mineral soil. Cedar glades developed on thin-bedded formations have more rock fragments in the plant rooting zone than those developed on thick-bedded formations. Also, on thin-bedded formations, rocks are broken loose from the layers of bedrock and fragments of varying sizes are scattered over much of the surface area of the cedar glades forming a rock pavement (Fig. 1A). Soil depths in cedar glades on thin-bedded formations range from 0–25 cm and occasionally up to 50 cm in deep soil-filled crevices, and a transect across the glade shows a mosaic of soil depths rather than a gradually increasing or decreasing gradient. On thick-bedded formations relatively few fragments are broken loose from the bedrock, and exposures of solid bedrock are surrounded by shallow soil which gradually grades into deeper soil; cedar glades are considered to be those areas where the soil is 0–25 cm deep (Fig. 1B). Of course, some glades appear to be intermediate between these two types, and on the same glade it sometimes is possible to see both situations.

In the Central Basin of Tennessee, many of the best cedar glades occur on the thin-bedded member of the Lebanon limestone which is exposed in Wilson, Davidson, Rutherford, Williamson, Maury, Marshall and Bedford Counties (Wilson 1949). This limestone has films of shale between the layers of dense limestone and as the shale is weathered away, the layers of limestone slump and crack leaving a loose, overlapping network of rock fragments (Galloway 1919). In time, the spaces become filled with a mixture of humus and mineral soil. On massive rocks, soil and humus accumulate as depressions are dissolved in the rocks by water.

According to Thornthwaite's (1931) system of classification of climate, the cedar glades of the Southeast occur in a humid, mesothermal region with adequate precipitation at all seasons (BB′r). However, general meteorological data for the region must be interpreted cautiously when applied to cedar glades. Although the same amounts of precipitation may fall on glades as on deeper, better developed soils of the area, the moisture relations of the glade soils may be much more extreme than in the deeper soils adjacent to the glades. In terms of the plants' microclimate, the glade environment is best described as winter-wet and summer-dry. Since the soils (i.e., the mixture of rock fragments, humus and mineral soil) are underlain by impermeable limestone or dolomite bedrock, they often are saturated with water during late autumn, winter and early spring. On the other hand, soil water content during summer is frequently below the permanent wilting percentage since the soils are shallow and are subjected to high intensity solar insolation, evaporation rates of water from the soil are high and precipitation falls as sporadic showers (Freeman 1933, Baskin and Baskin 1971a, C. Baskin et al. 1972). At weekly intervals during the summer of 1931, Freeman (1933) determined the moisture content of soil cores taken where soil depth was 15 cm. Soil moisture was below the wilting coefficient 11 times between 13 June and 24 October, and it was below the wilting coefficient for up to 5 weeks at a time.

VEGETATION AND FLORA

Quantitative studies have been published on the plant communities of the cedar glades of middle Tennessee (Freeman 1933, Quarterman 1950a, b, Baskin and Baskin 1977b), but no such studies have been done for the cedar glades in Kentucky, Alabama and Georgia. For the latter states, only floristic information has been published. In the only published study on the successional relations of the cedar glade plant communities, Quarterman (1950a) concluded that primary succession beginning on bare rock leads to herbaceous communities and then to shrubs, red-cedar and finally to hardwood forest. She recognized that the intermediate stages of succession, which are dominated by herbaceous plants, are long-persisting.

Various published floristic lists and published notes on specific species were combined to make a composite list of all the herbaceous vascular plants that have been collected in cedar glades in Kentucky, Tennessee, Georgia and Alabama. The total number of species is 414, and this includes endemic species, characteristic (but not endemic) widely distributed species, prairie species and native and introduced weeds. Cedar glade endemics are plant taxa that are restricted to glades or glade-like areas such as small, isolated limestone outcrops. Approximately 25 plant taxa are endemic or near-endemic to the cedar glades of southeastern United States (Baskin and Baskin unpub.). Characteristic widely distributed cedar glade species grow in cedar glades, certain disturbed sites and in other kinds of plant communities such as dry prairies. Prairie species are widely distributed in the prairies and/or plain regions of the United States and have disjunct populations in the cedar glades and other xeric habitats of the Southeast. Many native and introduced perennial and annual weeds have invaded disturbed glades from nearby roadsides, fields and waste places. Of the 414 species recorded on glades, 20% of them are nonnative, according to the information given in Fernald (1950). The flora of the cedar glades consists of about 68% polycarpic perennials, 2% monocarpic perennials and 30% annuals. There are 79 species of winter annuals and 47 of summer annuals; 49 of the winter annuals and 35 of the summer annuals are native species. A few winter annuals also can behave as summer annuals, and these will be discussed in more detail later.

In the life cycles of both kinds of annuals, the same events occur (Fig. 3), but the time of germination, and in most species the time of seed set, is different. In winter annuals, seeds germinate in late summer or autumn, and plants produce a rosette or semi-rosette of leaves during autumn and/or early winter. Plants overwinter as rosettes or semi-rosettes, and shoot production, flowering and seed set occur the following spring or summer, depending on the species. In most species seed dispersal immediately follows seed set, and seeds lie on, or in, the soil until summer

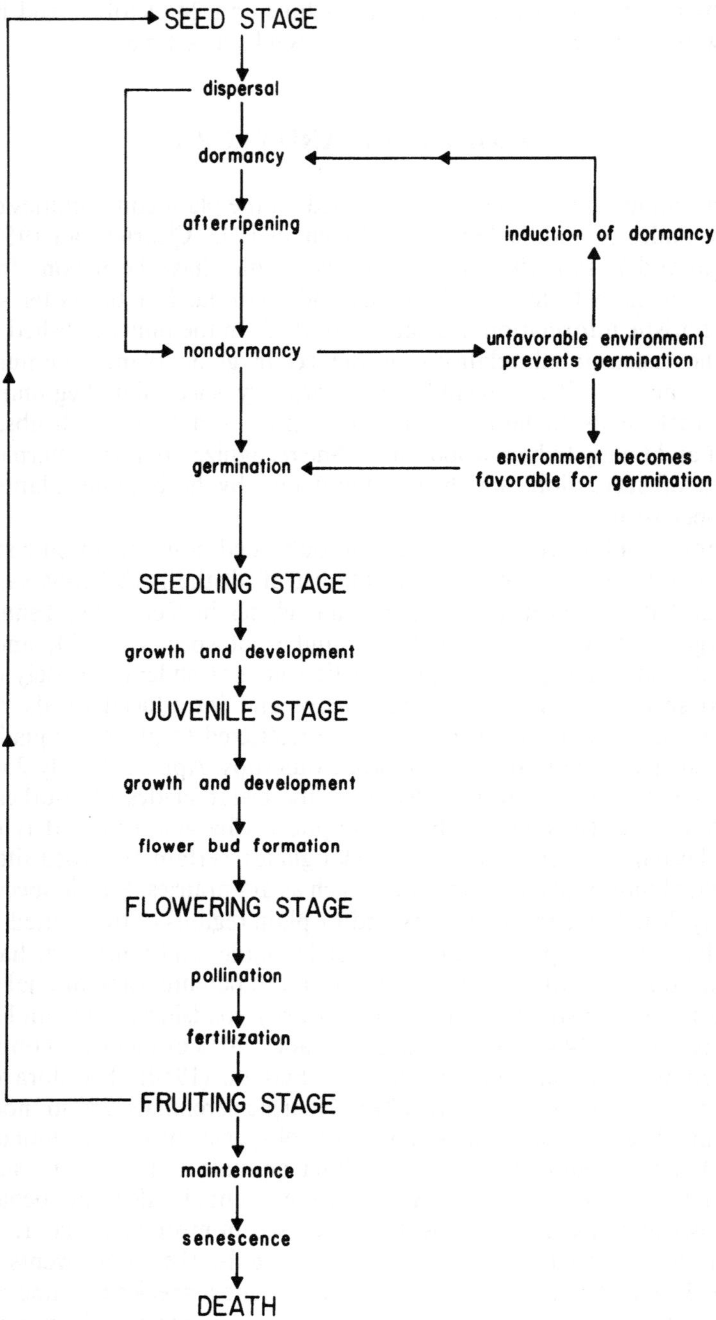

Fig. 3. Stages in the life cycle of annual plant species.

has passed. In a few species seeds are retained on the dead, upright plants until autumn. In summer annuals seeds germinate in late winter, spring and/or in summer in some species, and plants usually produce shoots without first forming a rosette of leaves. Flowering and seed set are completed by the end of the growing season, and seeds lie on, or in, the soil until winter has passed.

WINTER ANNUALS

Glade habitats and species

Although cedar glades are dominated by herbaceous vascular plants, the various species are not uniformly distributed within a given glade because of the variation in soil depth. Winter annuals grow in glades developed on thin- as well as thick-bedded calcareous rocks, and they are most abundant in those areas of the glades where soils are 1–5 cm deep. Due to the shallowness of the soil, perennials are almost nonexistent where winter annuals grow. The only perennial of any importance in soils 1–5 cm deep is the small, shallow-rooted, succulent, *Talinum calcaricum* Ware, which flowers and produces seeds in summer. Other perennials grow in areas of the glades where soil depth ranges from 5–25 cm.

The only endemic, strictly winter annual species in the glades belong to the genus *Leavenworthia* (Fig. 4). This genus consists of seven species, and all of them are endemic to cedar glades or glade-like habitats. All the

Fig. 4. Leavenworthia stylosa in the shallow-soil zone of the cedar glade habitat.

species, except *L. uniflora* and *L. aurea* Torr., are restricted to glades in the Southeast (Rollins 1963). According to Rollins (1963) the cedar glades have been available to *Leavenworthia* for a long period of time, and species and varieties have evolved during this long occupation of the habitat. He believes that northern Alabama is the area of origin of the present day species of *Leavenworthia* and that *L. crassa* Rollins, which is endemic to the glades of northern Alabama, is the basic species of the genus.

Our observations in glades throughout the Southeast indicate that there are few, if any, glades that have been left completely undisturbed; most of them have been used at some time for pasture. In central Tennessee and northern Alabama *Leavenworthia* spp. have spread from undisturbed glades into overgrazed pastures and cultivated fields that are not plowed until after the seeds are shed (Lloyd 1965, Rollins 1981). *Leavenworthia* spp. are present in relatively undisturbed as well as badly disturbed glades, and with an increase in disturbance the number of species of winter annuals increases. In relatively undisturbed glades *Leavenworthia* may be the only winter annual present, but this is a rare occurrence. In lightly to moderately disturbed glades in central Tennessee, *Leavenworthia* frequently grows in association with *Sedum pulchellum* and *Arenaria patula*. In badly disturbed glades the list of associated winter annuals also includes *Helenium amarum* (Raf.) H. Rock, *Plantago virginica* (Baskin and Baskin 1972a), *Arenaria serpyllifolia, Draba verna, Lepidium virginicum, Lithospermum arvense, Veronica arvense* (Baskin and Baskin 1981a) and many others.

Since *Sedum pulchellum* and *Arenaria patula* occur on limestone outcrops in semi-shaded situations in open woods and on the ledges of limestone cliffs of rivers and since the numbers of individuals of these two species appear to increase dramatically after a glade has been disturbed, there is some question as to whether or not these species were present in great abundance in glades prior to the time of European settlement of the region. In those glades where *Leavenworthia* spp., *A. patula* and *S. pulchellum* occur together, *Leavenworthia* spp. flower first, followed by *A. patula* and then *S. pulchellum*. Seed maturation in *Leavenworthia* spp. occurs in early to mid May, *A. patula* in late May to early June and *S. pulchellum* in mid June to early July. Thus, in most years *S. pulchellum* completes its life cycle during a period when soil moisture in its shallow-soil glade habitat may be below the permanent wilting point. *Sedum pulchellum* is a succulent and, therefore, is more drought tolerant than *Leavenworthia* spp. and *A. patula* which are nonsucculent. Plants of *S. pulchellum* subjected to soil water potentials of −15 to −40 bars for 80 days and to −50 to −85 bars for 50 days still had a net CO_2 uptake during a 24-hour period (Smith and Eichmeier 1983).

Life cycle ecology

Germination

Most of the winter annuals in the cedar glades set seeds and die in mid to late spring or early summer. At maturity seeds are in a state of dormancy

Table 1. Dormancy state of freshly-matured seeds of cedar glade winter annuals.

Species	Dormancy state at maturity	References
Arabidopsis thaliana	Conditional	Baskin and Baskin 1972c
Arenaria patula	Conditional	Caudle and Baskin 1968, Baskin and Baskin 1982
A. serpyllifolia	Conditional	Baskin and Baskin unpubl.
Cardamine hirsuta	Innate	Baskin and Baskin unpubl.
Draba verna	Innate	Baskin and Baskin 1970
Holosteum umbellatum	Innate	Baskin and Baskin 1973b
Leavenworthia crassa	Conditional	Caudle and Baskin 1968
L. exigua Rollins	Innate	Baskin and Baskin 1972a
L. stylosa Gray	Innate	Baskin and Baskin 1971a
L. torulosa	Innate	Baskin and Baskin 1971a
L. uniflora	Innate	Baskin and Baskin 1971a
Phacelia dubia (L.) Trel. var. *dubia* McVaugh	Conditional	Baskin and Baskin 1971b
Plantago virginica	Innate	Baskin and Baskin unpubl.
Sedum pulchellum	Innate	Caudle and Baskin 1968, Baskin and Baskin 1977a
Thlaspi perfoliatum	Innate	Baskin and Baskin 1979a
Valerianella radiata	Innate	Baskin and Baskin unpubl.
Veronica arvensis	Innate	Baskin and Baskin 1983a
Viola rafinesquii	Innate	Baskin and Baskin 1972b

and, thus, do not germinate immediately following dispersal, even if the soil is moist. Seeds of some winter annuals are innately dormant and none, or only a low percentage, will germinate in either light or darkness at either low (5, 10, 15, 20, 15/6 and 20/10°C) or high (25, 30, 25/15, 30/15 and 35/20°C) temperatures (Table 1). Seeds of other winter annuals are conditionally dormant (*sensu* Vegis 1964) and will germinate to moderate to high percentages if incubated at low, but not at high, temperatures (Table 1). Also, seeds usually germinate to higher percentages in light than in darkness (Figs. 5 and 6). Conditionally dormant seeds do not germinate following dispersal in spring or early summer because temperatures of the habitat are above those required for germination.

During summer the seeds of winter annuals undergo certain physiological changes (afterripening) that result in changes in their germination responses. Innately dormant seeds gradually pass into a state of conditional dormancy, and with additional afterripening they become completely nondormant. Seeds which are conditionally dormant at maturity also afterripen and become completely nondormant. An example of the pattern of changes in germination responses of seeds during afterripening is illustrated by *Leavenworthia uniflora* (Figs. 5 and 6). Most of the seeds were innately dormant at maturity, and less than 5% of them germinated in light or darkness. During afterripening, (1) the optimum and maximum temperatures for germination increased, (2) rates and final percentages of germination increased and (3) seeds became less dependent on exposure to light for germination. Thus, by September and October seeds are fully afterripened, and temperatures of the habitat have declined so that there

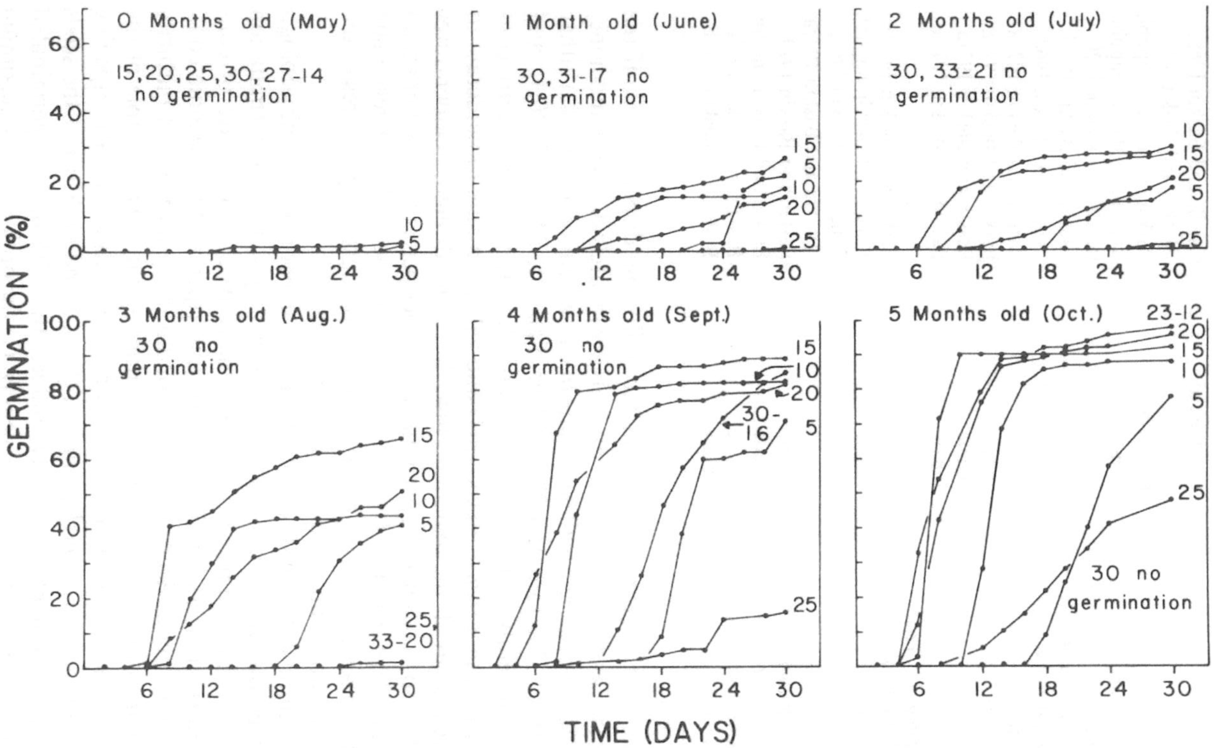

Fig. 5. Germination of *Leavenworthia uniflora* at a 14-hr photoperiod at various temperatures. Ages of the seeds are given in the figure. (Baskin and Baskin, 1971a, permission of *The American Midland Naturalist*.)

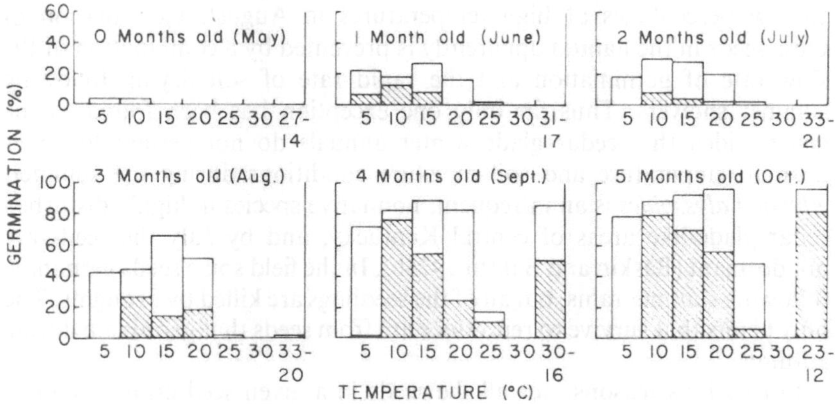

Fig. 6. Germination of *Leavenworthia uniflora* in constant darkness (hatched area) and at a 14-hr photoperiod after 30 days. Ages of the seeds are given in the figure. (Baskin and Baskin, 1971a, permission of *The American Midland Naturalist.*)

is an overlap between the temperatures required for germination and those prevailing in the habitat.

In some winter annuals low temperatures inhibit and high temperatures promote afterripening of the seeds (Baskin and Baskin 1972b, 1976a, Roberts and Neilson 1982). Thus, although dormancy is considered to be an adaptation of winter annuals to habitats that are hot and periodically or continuously dry during summer (Ratcliffe 1961, Newman 1963, Thompson 1970, 1974, Baskin and Baskin 1971a, Hájková and Krekule 1972, Janssen 1973), high temperatures are required to break seed dormancy, in at least some species, and, therefore, are necessary for the completion of the life cycle.

Although the temperature requirements for germination of seeds of winter annuals increase during the summer afterripening period, seeds generally do not germinate during summer in the habitat. In many species, e.g., *Leavenworthia* spp. (Baskin and Baskin 1971a, 1972a) and *Viola rafinesquii* (Baskin and Baskin 1972b), germination is prevented because temperatures of the habitat are above those required for germination. Thus, temperature and not soil moisture is the overriding environmental factor controlling germination from mid to late summer. This would preclude seeds from germinating during the short periods of time following summer showers when the soil is moist. However, if low temperatures were to occur during a time when the soil was moist, the seeds of many species would be able to germinate. This is exactly what happened in July 1970 in the cedar glades in middle Tennessee. From 21–23 July the soil was moist, and mean daily maximum and minimum air temperatures were 22.3 and 15.7°C, respectively; these were 10.7 and 5.3°C, respectively, below normal (U.S.D.C. 1965). On 25 July we found newly-germinated seedlings of 17 species of winter annuals, but by early September only a few seedlings of three species were still alive (Baskin and Baskin 1971c).

In a few species such as *Arenaria patula* (Baskin and Baskin 1982) and

Phacelia dubia var. *dubia* (Baskin and Baskin 1971b), the seeds germinate to low percentages at high temperatures in August. Germination of these seeds in the habitat apparently is prevented by a combination of the slow rate of germination and the rapid rate of soil drying following summer showers. Thus far only one exception has been found to the general idea that cedar glade winter annuals do not germinate under normal temperature and soil moisture conditions during the summer. *Alyssum alyssoides* is an infrequent, nonnative species in highly disturbed cedar glade-like areas of central Kentucky, and by July the seeds are nondormant (Baskin and Baksin 1974b). In the field some seeds germinate following summer rains, but all of the seedlings are killed by droughts. The only plants that survive to reproduce are from seeds that germinate during autumn.

For various reasons, not all the seeds in a given seed crop germinate during the first autumn following dispersal. Lack of germination may be due to (1) soil too dry in early autumn when temperatures are favorable for germination, (2) light-requiring seeds are buried, (3) temperatures too low in late autumn although soil moisture is not limiting, (4) lack of afterripening during summer and thus seeds are still innately dormant in autumn and (5) seeds may not be dispersed until after temperatures are too low for germination (Baskin and Baskin 1974a, 1975a, 1977a). Exposure of nondormant seeds of winter annuals such as *Phacelia dubia* var. *dubia* (Baskin and Baskin 1973c) and *Sedum pulchellum* (Baskin and Baskin 1977a) to low winter temperatures induces them into dormancy. These dormant seeds become nondormant again when exposed to actual or simulated summer temperatures. In *P. dubia* var. *dubia*, seeds induced into dormancy by low winter temperatures require high temperatures to afterripen (Baskin and Baskin 1978a). No doubt, seeds of some species go into and out of dormany many times before microhabitat conditions become favorable for germination in autumn. Induction of dormancy in seeds of winter annuals during winter means that they can not germinate in spring. This could be of possible survival value to the species because it would prevent germination at a time of the year when subsequent periods of summer drought and/or competition from established plants might kill the seedlings.

In winter annuals, such as the cedar glade endemic *Lobelia gattingeri* Gray (Baskin and Baskin 1979b), exposure of nondormant seeds to low winter temperatures does not include them into dormancy, and they can germinate in early spring. If seeds germinate in microsites which remain moist until early to mid summer, the plants flower and set seeds and thus exhibit a summer annual type of life cycle.

Seeds in the soil

Seeds that do not germinate the first autumn after maturity become part of the viable, ungerminated seed pool in and on the soil and may germinate at some time in the future. In a study of the germination phenology of 57 species of cedar glade winter annuals, seeds of all species germinated the first autumn after they were planted, and seeds of 49 and 36 species germinated the second and third autumns, respectively (Baskin and Baskin unpubl.). It long has been known that many weedy species of winter annuals may have large populations of buried viable seeds, especially in arable soils (Brenchley 1918, Brenchley and Warington 1930, Champness and Morris 1948, Kropáč 1966, Roberts and Stokes 1966). More recently, some of the nonweedy winter annuals in cedar glades, including *Sedum pulchellum* (Baskin and Baskin 1977a) and *Leavenworthia stylosa* (Baskin and Baskin 1978b) have been found to have large seed reserves at their population sites.

According to Cohen (1966), seed pools at population sites of annual species growing in habitats with a high probability of an occasional total failure of the plants to set seeds are thought to be necessary for the long-term survival of a species at a given site. The cedar glades constitute one of those "risky" habitats, and in 1977 we made a direct observation on the consequences of the presence of a seed pool at a population site of *Sedum pulchellum* (Baskin and Baskin 1980a). During May all plants of *S. pulchellum* at a population site were killed by drought, and no seeds were produced. However, in autumn seeds in the seed pool germinated, and a population of 5.6 plants/dm^2 was established. Thus, the presence of a seed reserve in the soil is an adaptation which allows *S. pulchellum* to persist at a site without immigration, even if plants do not produce seeds every year.

Another advantage of seed reserves for species like *Leavenworthia* spp. which, before disturbance of cedar glades, occurred in small relatively isolated populations (Lloyd 1965) is that they may increase the genetic variability and stability of the population. In *L. stylosa* a portion of the reserve seeds present in the soil in the spring of 1971 germinated in the autumn of each of 7 subsequent years, and a total of 11,261 seeds/m^2 germinated (Baskin and Baskin 1978b). Thus, the total gene pool of a population of *L. stylosa* at a particular site might be considerably larger than that represented by the number of individuals growing at the site in any given year. In the desert annuals *Linanthus parryae* (Gray) Greene (Epling et al. 1960) and *Stephenomeria exigua* ssp. *coronaria* (Greene) Gottlieb (Gottlieb 1974), the presence of large seed reserves is believed to give genetic stability to the populations. In contrast, desert annuals in the genus *Clarkia* do not have seed reserves, and this is believed to account partly for the great genetic instability of *Clarkia* populations (Lewis 1962, Bartholomew et al. 1973).

Environmental requirements for flowering

Since winter annuals overwinter as rosettes or semi-rosettes, is exposure to low winter temperatures (vernalization) a prerequisite for flowering? Bangma (1966) found that neither low winter temperatures nor a specific photoperiod or temperature was required for flower bud initiation and anthesis in *Leavenworthia stylosa.* In a study of 31 native and 19 nonnative species of winter annuals that occur in cedar glades, none, except possibly

Galium cruciata (L.) Scop. and *Thalspi perfoliata* (both nonnative), had an absolute low temperature requirement for flowering (Baskin and Baskin 1974c). In 38 species, including *Arenaria patula*, *Leavenworthia exigua*, *L. stylosa*, *L. torulosa*, *L. uniflora* and *Sedum pulchellum*, plants kept in a nonheated greenhouse flowered later in spring than those in a heated greenhouse. Thus, photoperiod *per se* probably does not control flowering, and in the field flowering is delayed until late winter and/or spring by low temperatures.

Breeding system and population structure of Leavenworthia

On the basis of chromosome number, the seven species of *Leavenworthia* can be divided into three groups: *L. alabamica* Rollins, *L. crassa* and *L. exigua*, $n = 11$; *L. stylosa*, *L. torulosa* and *L. uniflora*, $n = 15$; *L. aurea*, $n = 24$. Species in the first two groups are diploid, and *L. aurea* is a polyploid (Rollins 1963). Of the seven species, only *L. crassa* and *L. alabamica* can produce fertile hybrids (Rollins 1963). *Leavenworthia sylosa* is self-incompatible, *L. exigua*, *L. uniflora*, *L. torulosa* and *L. aurea* are self-compatible, and *L. alabamica* and *L. crassa* have both self-incompatible and self-compatible races (Rollins 1963, Lloyd 1965, 1968a, b). Lloyd (1967) studied the genetics of self-incompatibility in *L. crassa* and showed that the incompatibility system is of the sporophytic type and is controlled by multiple alleles at a single locus. Self-compatible species grow in drier microsites on the glades and occupy wider geographical areas than self-incompatible species (Rollins 1963, Solbrig and Rollins 1977).

Solbrig and Rollins (1977) believe that, "There is strong evidence for evolutionary shifts from self-incompatibility to self-compatibility, both at the specific and intraspecific levels." They suggest that the strongest selective force for the evolution of self-compatibility in *Leavenworthia* is the timing of emergence of native pollinators in relation to drying of the shallow-soil glade habitat in spring. This hypothesis is supported by their observations that self-incompatible species and races (in *L. alabamica* and *L. crassa*) occupy wetter sites and flower later in the season than self-compatible species and races. Thus, self-compatible plants can mature seeds earlier, when there is a paucity of insect visitors, than self-incompatible plants which can not be pollinated until temperatures are favourable for insect flight. Consequently, in a dry spring and in dry microhabitats inbreeders would be more likely to produce seeds than outbreeders.

In a pollination study of *Leavenworthia crassa* and *L. stylosa*, Lloyd (1965) found that 57% of the insects visiting the flowers were nonnative honeybees, 28% were native nonsocial bees and 15% were native non-Hymenoptera. Lloyd concluded that honeybees and native bees are the only significant pollinators of these two species because the foraging activities of non-Hymenoptera on flowers of *Leavenworthia* are such that they do not effect pollination.

Both self-incompatible and self-compatible *Leavenworthia* populations

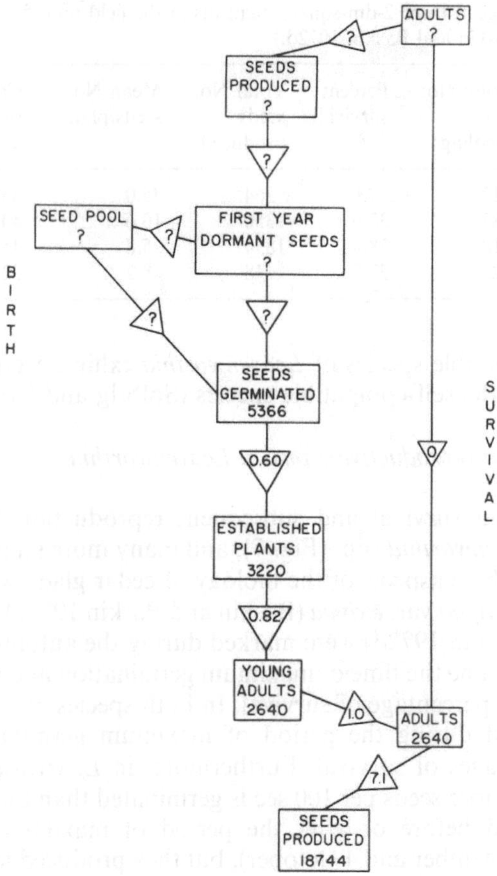

Fig. 7. Diagrammatic life table for *Leavenworthia stylosa*. Population sizes are per m². (Outline is according to method of Begon and Mortimer 1981.)

have a clumped distribution, but the number of plants per population is much greater in self-incompatible than in self-compatible taxa and races (Lloyd 1965, Solbrig and Rollins 1977). Frequencies of size classes within populations of both inbreeders and outbreeders are log normal (Solbrig and Rollins 1977), but like many other annual plants *Leavenworthia* is phenotypically plastic. Plants growing in favorable microhabitats are much larger and produce many more seeds than those growing in less favorable microhabitats.

Interpopulation variability in *Leavenworthia* occurs in the breeding systems (*L. alabamica* and *L. crassa*), flower color and pattern, flower and fruit morphology and isozyme pattern and frequency (Rollins 1963, Lloyd 1965, 1969, Solbrig 1972, Solbrig and Rollins 1977). In fact, enough morphological variation occurs in *L. alabamica*, *L. crassa* and *L. exigua* that Rollins (1963) designated varieties within these species. On the whole,

Table 2. Survival and seed production of *Leavenworthia stylosa* plants marked as newly-germinated seedlings in three 2-dm-square quadrats in the field on different dates in 1970. (Modified from Baskin and Baskin 1972d.)

Date seedlings were marked	Total No. new seedlings	Percent survival	Total No. seeds produced	Mean No. seeds/plant	Calculated No. of seeds produced/100 seeds germinated
28 July	117	7.6	364	45.0	342.0
13 September	87	37.9	350	10.6	401.7
4 October	418	78.4	1894	5.8	454.7
18 October	22	27.2	48	8.2	223.0

the self-incompatible species of *Leavenworthia* exhibit a greater range of variation than the self-compatible species (Solbrig and Rollins 1977).

Survivorship and reproductive effort in Leavenworthia

Studies on plant survival and subsequent reproduction have centered mostly on *Leavenworthia* spp. (Fig. 7), and many more studies need to be conducted on these aspects of the biology of cedar glade winter annuals. Cohorts of *L. exigua* var. *exigua* (Baskin and Baskin 1972a) and *L. stylosa* (Baskin and Baskin 1972d) were marked during the autumn germination season to determine the time of maximum germination and which cohorts had the highest percentage of survival. In both species, plants from seeds that germinated during the period of maximum germination had the highest percentages of survival. Furthermore, in *L. stylosa* these plants also produced more seeds per 100 seeds germinated than those from seeds that germinated before or after the period of maximum germination (between 13 September and 4 October), but they produced fewer seeds per individual plant than those in the other cohorts (Table 2). Although the number of seeds produced per plant was lower, plants from seeds that germinated during the period of maximum germination added more seeds to the population than did plants in all other cohorts combined.

Solbrig and Rollins (1977) determined the reproductive effort (weight of seeds/total weight of plant) in field-grown plants of four species of *Leavenworthia*. The average reproductive effort for several populations of each species was: *L. exigua*, 36.4; *L. alabamica*, 26.9; *L. crassa*, 21.1 and *L. stylosa*, 33.2.

WINTER ANNUALS AND THE CEDAR GLADE COMMUNITY

With the exception of *Sedum pulchellum*, none of the cedar glade winter annuals, either native or nonnative species, behaves as a biennial or perennial in some other type of community. Although Baldwin (1943) studied *S. pulchellum* throughout its geographical range and knew it only as a winter annual, Clausen (1975) reported a population of perennial plants of this species on a limestone ledge in northern Alabama. Some of the weedy species such as *Helenium amarum*, *Lamium amplexicaule* and *Veronica arvensis* (Baskin and Baskin 1973a, 1981b, 1983a) and the endemic *Lobelia gattingeri* (Baskin and Baskin 1979b) have the ability to behave as both winter and summer annuals.

Understanding the physiological reasons for the occurrence of winter annuals in cedar glade habitats requires a knowledge of (1) the plant and environmental factors controlling the timing of events in the life cycle, particularly seed germination and reproduction and (2) the environmental tolerances and requirements of plants during all phases of their life cycle. The physiological responses and germination requirements of the seeds as well as requirements for flowering have been discussed already. The only characteristic, native cedar glade winter annual in which photosynthesis has been studied is *Sedum pulchellum*. This succulent species fixes CO_2 primarily *via* the Calvin cycle (C_3 species), but it also is capable of taking up CO_2 at night (Crassulacean acid metabolism) when soil moisture content is very low (Smith and Eichmeier 1983). It is expected that the other cedar glade winter annuals have the C_3 pathway of carbon fixation as has been demonstrated in, for example, winter annuals of deserts of southwestern United States (Mulroy and Rundel 1977, Kemp 1983) and unshaded granite outcrops of southeastern United States (Martin et al. 1982). Further, it is suspected that the optimum temperature for photosynthesis in cedar glade winter annuals is relatively low (especially during winter and spring) and that their light requirements are relatively high as has been shown in oldfield (Regehr and Bazzaz 1976) and desert (Mooney et al. 1976, Ehleringer et al. 1979) winter annuals. Since the soils in the cedar glade habitat are subject to waterlogging from late autumn to early spring, winter annuals must be tolerant of flooding. Plants of *L. uniflora* grown with their roots under flooded conditions did not show an increase over the nonflooded control in alcohol dehydrogenase activity, implying that ethanol did not accumulate and that *Leavenworthia* is metabolically adapted to waterlogged soils (Baskin and Baskin 1976b).

Winter annuals are excluded from many habitats because they apparently are poor competitors with taller species of plants that shade them. A survey of the various types of winter annual habitats shows that most receive full, or near full, sunlight at canopy level. In the cedar glades and in other shallow-soil habitats, the summer dryness of the shallow soils prevents the establishment of large, shade-producing plants. When winter annuals do grow in habitats in which the soils are deep and not droughty during summer, plants that could compete with the winter annuals are destroyed by plowing, mowing or some other form of disturbance. Those few winter annuals that grow in thickets and woodland habitats also receive near full sunlight during the active phase of their life cycle. They germinate after the leaves of the woody species have fallen in autumn, and the plants mature at or slightly after the time of canopy closure (Baskin and Baskin 1983b).

The competitive abilities of cedar glade winter annuals are an almost untouched area of research. Rollins (1963) observed that plants of *Leavenworthia* spp. were killed by *Poa annua* in an unweeded portion of an experimental plot, whereas those in the weeded portion of the plot grew normally and produced seeds. In northern Alabama, Lloyd (1965) observed that *Leavenworthia* spp. competed poorly with the invading weedy flora in abandoned agricultural fields. Turner and Quarterman (1975) grew *Arenaria patula* with the summer annual *Sporobolus vaginiflorus*. At all levels of shoot competition the growth of *A. patula* was inhibited, but root competition had no effect.

SUMMER ANNUALS

Glade habitats and species

There are two general types of summer annual plant communities in the cedar glades. One occurs in the main winter annual habitat where soils are 1–5 cm deep and the other where soils are 5–25 cm deep. Depending on soil depths in a given glade, one community can gradually grade into another, or the two communities may occur as a mosaic. In soils 1–5 cm deep, the most important species are the natives *Cyperus inflexus*, *Euphorbia supina*, *Panicum capillare* and the nonnative *Portulaca oleracea*. The three native species are widely distributed in North America, and *P. oleracea* is found in tropical and temperate regions throughout the world (Gleason 1952). In soils 5–25 cm deep, *Sporobolus vaginiflorus* is the most important species in terms of numbers and biomass. In glades on thin-bedded rocks it is most abundant where soils are 5–15 cm deep, while in glades on thick-bedded rocks it is most abundant where soils are 10–25 cm deep (Baskin and Baskin 1973d). Summer annuals with > 10% frequency that grow in association with *S. vaginiflorus* in soils 5–25 cm deep include: *Aristida longespica* (15%), *Croton capitatus* (31%), *C. monanthogynus* (18%), *Heliotropium tenellum* (Nutt.) Torr. (42%), *Isanthus brachiatus* (55%) and *Panicum flexile* (13%). Other summer annuals frequently found growing in this community include *Aristida oligantha*, *Diodia teres*, *Euphorbia dentata* and *E. maculata* (Baskin and Baskin unpubl.). All of these species, including *S. vaginiflorus*, are native and occur in other open, nonforested habitats, including disturbed sites, over a large portion of eastern and/or southern United States (Gleason 1952). No summer annuals are endemic or near endemic to cedar glades.

Life cycle ecology

Germination

The biology of cedar glade summer annuals has not been studied in as much detail as the biology of winter annuals. We have done germination studies on the summer annual species of the shallow-soil community and on five of the most frequent species in the deep-soil community (Table 3). Our results show that seeds of all these species, except *Euphorbia supina*, are dormant at maturity in autumn and will not germinate in light or

Table 3. Dormany state of freshly-matured seeds of cedar glade summer annuals.

Species	Dormany state at maturity	Reference
Shallow-soil community		
Cyperus inflexus	Innate	Baskin and Baskin 1971d
Euphorbia supina	Conditional	Baskin and Baskin 1979c
Panicum capillare	Innate	C. Baskin and J. Baskin 1969
Portulaca oleracea	Innate	Baskin and Baskin unpubl.
Deep-soil community		
Aristida longespica	Innate	Baskin and Caudle 1967
Heliotropium tenellum	Innate	Baskin and Baskin unpubl.
Isanthus brachiatus	Innate	C. Baskin and J. Baskin 1969, Baskin and Baskin 1975b
Panicum flexile	Innate	Baskin and Baskin unpubl.
Sporobolus vaginiflorus	Innate	Baskin and Caudle 1967, Baskin and Baskin 1973d

darkness over a range of alternating temperatures. Seeds of *E. supina* (Baskin and Baskin 1979c) are conditionally dormant and will germinate only in light at high (30/15 and 35/20°C) temperatures. This is in contrast to seeds of winter annuals which will germinate only at low (15/6, 20/10°C) temperatures while in a state of conditional dormancy. Although freshly-matured seeds of *Portulaca oleracea* collected from a cedar glade population were innately dormant, seeds collected from different habitats in other geographical locations are nondormant or conditionally dormant (Miyanishi and Cavers 1980, Baskin and Baskin unpubl.). Exposure of imbibed seeds to natural or simulated winter temperatures (5°C) conditions causes both innately and conditionally dormant seeds to become nondormant. When seeds come out of dormancy, they first gain the ability to germinate at high and then at low temperatures, just the reverse of what happens in winter annuals.

Nondormant seeds of *Panicum capillare* (Baskin and Baskin unpubl.), *Cyperus inflexus* (Baskin and Baskin 1971d, 1976c), *Euphorbia supina* (Baskin and Baskin 1979c) and *Portulaca oleracea* (Baskin and Baskin unpubl.) require light for germination. Thus, the four major species in the shallow-soil community require light for germination. In the deep-soil community, *Isanthus brachiatus* requires light for more than about 10% germination (Baskin and Baskin unpubl.), *Sporobolus vaginiflorus* and *Panicum flexile* do not require light (Baskin and Baskin unpubl.) and the light requirement for *Aristida longespica* and *Heliotropium tenellum* is unknown.

Since seeds of all the members of the deep-soil community studied thus far germinate only in spring, it is conceivable that ungerminated, nondormant seeds of some of them enter secondary dormancy in late spring and remain in this state until the following winter. In *Ambrosia artemisiifolia*, a native weed which occasionally occurs in the deep-soil habitat, nondormant seeds that do not germinate enter secondary dormancy by late spring. These seeds become nondormant during the following winter and

Table 4. Cumulative germination percentages of seeds of *Isanthus brachiatus* and *Heliotropium tenellum* planted in a nonheated greenhouse in various years.

Year seeds germinated	Year seeds were planted				
	Isanthus			Heliotropium	
	1969	1970	1971	1969	1973
1970	31.5			11.5	
1971	69.3	30.8		31.3	
1972	75.0	65.0	4.1	35.5	
1973	77.0	71.4	29.0	36.8	
1974	81.8	74.6	45.6	41.5	7.5
1975	84.5	77.3	54.2	43.3	15.3
1976	86.3	78.8	58.5	44.3	31.2
1977	87.3	79.9	59.5		31.9
1978	87.5	79.9	59.7		

will germinate in spring if exposed to light. Seeds of *A. artemisiifolia*, and probably other summer annuals in the glades, can go into and out of secondary dormancy each year for many years (Baskin and Baskin 1980b).

Although seeds of *Cyperus inflexus*, *Panicum capillare*, *Portulaca oleracea* and *Euphorbia supina* germinate in spring, they also can germinate in summer, whenever soil moisture and light are nonlimiting. In late spring and early summer, nondormant seeds of these species lose the ability to germinate at low but not at high temperatures. Consequently, they can germinate during summer because temperatures are within the range of those required for germination, but they do not germinate in autumn because temperatures are too low (Baskin and Baskin 1978c, 1979c, unpubl.). Thus, species in the shallow-soil habitat may have several opportunities for seed germination and seedling establishment during a single growing season, whereas those in the deep-soil habitat have only one.

Seeds in the soil

Intact soil samples (6 dm^2 in area) were collected in the cedar glades and placed in a nonheated greenhouse where germination of various species was monitored for up to 7 years. Seed pools exist for the four species in the shallow-soil community, and for *Heliotropium tenellum* and *Isanthus brachiatus* in the deep-soil community. We found no evidence of seed pools for *Sporobolus vaginiflorus*, *Aristida longespica* or *Panicum flexile*. The largest seed pools were for *P. capillare*, and over a 7-year period 26,361 seeds/m^2 germinated (Baskin and Baskin unpubl.). Additional information on seed pools comes from planting experiments in the nonheated greenhouse. Some seeds of *I. brachiatus* and *H. tenellum* germinated for 9 and 7 years, respectively, after planting (Table 4) (Baskin and Baskin 1975b, unpubl.), indicating that seeds in a given seed crop germinate over a long period of time. One reason for the spread of seed germination in *I. brachiatus* through time is the physiological polymorphism of the seeds with respect to the conditions necessary to overcome dormancy. The seeds are dormant at maturity in autumn and require moist, low temperature treatment (stratification) before any of them will germinate. However, stratification breaks dormancy only in a portion of the seed crop, and three or more additional stratification treatments, separated by periods of warm pretreatment (i.e., summer) are needed to break dormancy in the remainder of the seed crop (Baskin and Baskin 1975b). Field observations indicate that if all the plants of *I. brachiatus* at a population site are killed by drought before seeds are produced, a population is established the following spring from seeds in the soil (Baskin and Baskin unpubl.).

Environmental requirements for flowering

As far as we know, information on the flowering response to length of day is available for only one of the important summer annuals that grows in cedar glades. Plants of *Portulaca oleracea* will flower in both long and short photoperiod regimes and thus are day-neutral (Zimmerman 1976, Vengris and Dunn 1972). We speculate that *Panicum capillare* is a qualitative short day plant. Plants of this species that emerged in late August produced seed heads in 35 days, whereas those that emerged in May required 80 days to produce seed heads (Vengris and Damon 1976). In *Isanthus brachiatus* and *Sporobolus vaginiflorus*, flowering is delayed until late summer, and therefore short days may be the environmental cue for flowering. On the other hand, *Heliotropium tenellum* begins flowering during the long days of late spring and early summer, and thus long days may stimulate flowering. Much research needs to be done before we can understand the community-wide pattern of flowering, and its significance, in summer annuals of cedar glades.

Survivorship and reproduction

Preliminary studies have been done on plant survival from seed germination to reproductive maturity in *Sporobolus vaginiflorus*, *Heliotropium tenellum* and *Isanthus brachiatus* in the cedar glades of Tennessee and/or Kentucky. As would be expected, soil depth of a given microsite and summer rainfall greatly influence the survival rate of seedlings and the number of seeds produced per plant. In *S. vaginiflorus* survival has been determined only where soils are 3–10 cm deep, and there 60% of the plants that germinated survived to maturity (Fig. 8). Survival of *H. tenellum* was 30 and 60% at a glade site in two different years (Baskin and Baskin unpubl.), and in *I. brachiatus* all the plants at some sites in some years died. However, 48% survival has been recorded in the latter species with the number of seeds produced per plant ranging from 1 to 140 and averaging 8.1 (Baskin and Baskin unpubl.).

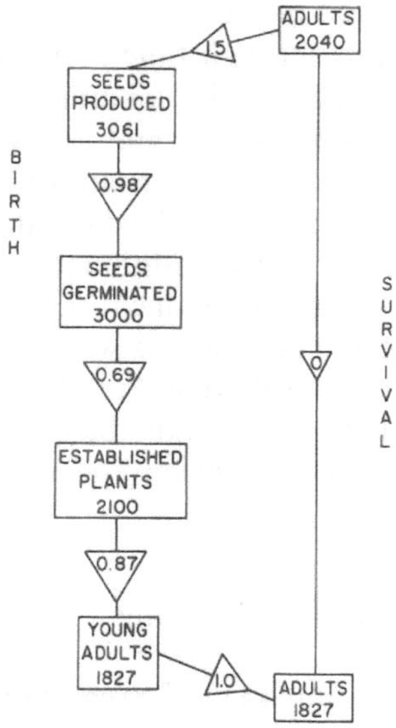

Fig. 8. Diagrammatic life table for *Sporobolus vaginiflorus* in shallow (3–10 cm) soil. Population sizes are per m^2. (Outline is according to method of Begon and Mortimer 1981.)

SUMMER ANNUALS AND THE CEDAR GLADE COMMUNITY

Although the most abundant summer annuals in both the shallow- and deep-soil habitats in the cedar glades are native species, none of them is restricted to cedar glades. As previously mentioned, they also grow in other unshaded habitats, where they also behave only as summer annuals. This type of life cycle would seem to be a response to a climate in which temperatures are at or near freezing for part of the year, rather than a response to the local environmental conditions of the cedar glades *per se*. Although our observations indicate that the summer annuals are intolerant of long periods of temperatures below freezing, low temperatures are required to break dormancy in innately dormant seeds of some summer annuals and to lower the temperature requirement in the conditionally dormant seeds of others. Thus, while summer annuals are intolerant of low winter temperature conditions, the life cycle cannot be completed in species such as *Isanthus brachiatus* and *Heliotropium tenellum* unless the seeds are stratified.

Just as winter annuals are excluded from habitats where taller species of plants can shade them, most summer annuals (exceptions being such species as *Impatiens capensis*, *I. pallida* and *Pilea pumila* which grow in forests) are also intolerant of shaded situations and, thus, compete poorly in vegetation that overtops them. As noted in the discussion of winter annuals, the summer dryness of the cedar glade habitat prevents growth of large shade-producing plants and near 100% insolation reaches the soil surface. Thus, plants growing in cedar glades during summer must tolerate

Table 5. Kranz syndrome in characteristic cedar glade summer annuals.

Species	Kranz anatomy*
Aristida longespica	+
A. oligantha	+
Croton capitatus	−
C. monanthogynus	−
Cyperus inflexus	+
Diodia teres	−
Euphorbia maculata	+
E. dentata	−
E. supina	+
Heliotropium tenellum	−
Isanthus brachiatus	−
Leucospora multifida	−
Linum sulcatum	−
Panicum capillare	+
P. flexile	+
Polygala verticillata	−
Portulaca oleracea	+
Sporobolus vaginiflorus	+

*+ = with Kranz anatomy; − = without Kranz anatomy.

not only very low soil moisture but also intense solar radiation. Much work remains to be done to understand how summer annuals are adapted to the cedar glade habitat; however, it is known that some of them are C_4 plants and all of them have relatively small leaves. The leaf anatomy of 18 species of cedar glade summer annuals was examined to determine whether or not they have Kranz or non-Kranz anatomy (Table 5). Nine species have Kranz anatomy (C_4 plants), and nine do not have it (C_3 plants) (Baskin and Baskin 1981c, unpubl.). Studies on plants of *Heliotropium tenellum* (Baskin and Baskin 1978d), *Isanthus brachiatus*, *Diodia teres*, *Euphorbia dentata* and *Croton capitatus* (Baskin and Baskin unpubl.) growing in the cedar glades revealed that temperatures of their small leaves remain near air temperatures during periods of intense solar radiation. Thus, small leaves and/or C_4 photosynthesis are adaptations to the hot, dry summer environment of cedar glades. Adaptation to drought in *Sporobolus vaginiflorus* (Shinners 1941) and *Panicum capillare* (Weaver and Albertson 1943, Savage 1937) is demonstrated further by their invasion into and persistence in over-grazed pastures and other dry habitats during the great drought of the 1930's. Additionally, in studies of the water requirements of certain weeds, it has been shown that growth of *P. capillare* (Dillman 1931) and *Portulaca oleracea* (Briggs 1914, Shantz 1927) is highly efficient with regard to the amount of water used.

CONCLUSIONS

Although winter and summer annuals occur in the same habitat, their niches are very different. From late autumn until early- to mid-spring, species with winter annual life cycles dominate those portions of the cedar glades where the soils are only 1–5 cm deep. The winter annuals are active during the period when summer annuals are in the seed stage, and most of the winter annuals complete their life cycles while the summer annuals are either small seedlings or still in the seed stage. The most important environmental factor regulating the timing of events in the life cycles of both groups of annuals is temperature, and the most important event in the life cycle is germination. Germination must occur at a time when environmental conditions are suitable for seedling establishment and eventual maturation of the plant. There are some exceptions, but, in general, the temperatures required to break dormancy in one type of annual will cause seeds of the other type to re-enter dormancy or, in the case of the summer annuals in the shallow-soil habitats, cause a decrease in the temperature range for germination. These differences in responses of seeds to temperature ensure that seeds of species of the two life cycle types germinate in different seasons of the year.

We speculate that the most important selective force acting on the life cycle of winter annuals is summer drought, whereas low winter temperatures are the most important selective force acting on the life cycle of summer annuals. In other words, since winter annuals are drought intolerant, germination of seeds in spring or summer would result in death of the seedlings; and since summer annuals are intolerant of winter temperature conditions, any plants from autumn-germinating seeds would not survive. The presence of endemic species in a given habitat may be the result of evolution that has occurred in response to strong selective forces unique to that habitat. Why are there endemic winter annuals in the glades and no endemic summer annuals? Summer drought is more intense in the cedar glades than it is in other habitats in the same geographical region, while temperatures in winter are about the same in the cedar glades as they are in other habitats within the region. Thus, the selective force in the cedar glades (summer drought) acting on winter annuals is of greater intensity than that of other habitats in the region, while the selective force in the glades acting on summer annuals (winter cold) is of the same intensity as that in other habitats in the region.

REFERENCES

Baldwin, J.T., Jr. (1943) Polyploidy in *Sedum pulchellum* – I. Cytogeography. Bull. Torrey Bot. Club 70: 26–33.

Bangma, G.S. (1966) Temperature and related factors affecting germination and flowering of *Leavenworthia stylosa* (Cruciferae). M.A. thesis. Vanderbilt University, Nashville, Tennessee.

Bartholomew, B., Eaton, L.C. and Raven, P.H. (1973) *Clarkia rubicunda*: A model of plant evolution in semiarid regions. Evolution 27: 505–517.

Baskin, C.C. and Baskin, J.M. (1969) Germination and dormancy in cedar glade plants. IV. *Isanthus brachiatus*, *Panicum capillare*, *Cyperus inflexus*, *Eragrostis spectabilis* and *Ruellia humilis*. J. Tenn. Acad. Sci. 44: 69–70.

Baskin, C.C., Baskin, J.M. and Quarterman, E. (1972) Observations on the ecology of *Astragalus tennesseensis*. Amer. Midl. Natur. 88: 167–182.

Baskin, J.M. and Baskin, C.C. (1970) Germination eco-physiology of *Draba verna*. Bull. Torrey Bot. Club 97: 209–216.

Baskin, J.M. and Baskin, C.C. (1971a) Germination ecology and adaptation to habitat in *Leavenworthia* spp. (Cruciferae). Amer. Midl. Natur. 85: 22–35.

Baskin, J.M. and Baskin, C.C. (1971b) Germination ecology of *Phacelia dubia* var. *dubia* in Tennessee cedar glades. Amer. J. Bot. 58: 98–104.

Baskin, J.M. and Baskin, C.C. (1971c) Germination of winter annuals in July and survival of the seedlings. Bull. Torrey Bot. Club 98: 272–276.

Baskin, J.M. and Baskin, C.C. (1971d) Germination of *Cyperus inflexus*. Bot. Gaz. 132: 3–9.

Baskin, J.M. and Baskin, C.C. (1972a) The ecological life cycle of the cedar glade endemic *Leavenworthia exigua* var. *exigua*. Can. J. Bot. 50: 1711–1723.

Baskin, J.M. and Baskin, C.C. (1972b) Physiological ecology of germination of *Viola rafinesquii*. Amer. J. Bot. 59: 981–988.

Baskin, J.M. and Baskin, C.C. (1972c) Ecological life history and physiological ecology of seed germination of *Arabidopsis thaliana*. Can. J. Bot. 50: 353–360.

Baskin, J.M. and Baskin, C.C. (1972d) Influence of germination date on survival and seed production in a natural population of *Leavenworthia stylosa*. Amer. Midl. Natur. 88: 318–323.

Baskin, J.M. and Baskin, C.C. (1973a) Ecological life cycle of *Helenium amarum* in central Tennessee. Bull. Torrey Bot. Club 100: 117–124.

Baskin, J.M. and Baskin, C.C. (1973b) Studies on the ecological life cycle of *Holosteum unbellatum*. Bull. Torrey Bot. Club 100: 110–116.

Baskin, J.M. and Baskin, C.C. (1973c) Delay of germination in seeds of *Phacelia dubia* var. *dubia*. Can. J. Bot. 51: 2481–2486.

Baskin, J.M. and Baskin, C.C. (1973d) Observations on the ecology of *Sporobolus vaginiflorus* in cedar glades. Castanea 38: 25–35.

Baskin, J.M. and Baskin, C.C. (1974a) Some aspects of the autecology of *Bupleurum rotundifolium* in Tennessee cedar glades. J. Tenn. Acad. Sci. 49: 21–24.

Baskin, J.M. and Baskin, C.C. (1974b) Germination and survival in a population of the winter annual *Alyssum alyssoides*. Can. J. Bot. 52: 2439–2445.

Baskin, J.M. and Baskin, C.C. (1974c) Influence of low winter temperatures on flowering of winter annuals. Castanea 39: 340–345.

Baskin, J.M. and Baskin, C.C. (1975a) Ecophysiology of seed dormancy and germination in *Torilis japonica* in relation to its life cycle strategy. Bull. Torrey Bot. Club 102: 67–72.

Baskin, J.M. and Baskin, C.C. (1975b) Seed dormancy in *Isanthus brachiatus* (Labiatae). Amer. J. Bot. 62: 623–627.

Baskin, J.M. and Baskin, C.C. (1976a) High temperature requirement for afterripening in seeds of winter annuals. New Phytol. 77: 619–624.

Baskin, J.M. and Baskin, C.C. (1976b) Evidence for metabolic adaptation to flooding in *Leavenworthia uniflora*. J. Chem. Ecol. 2: 441–447.

Baskin, J.M. and Baskin, C.C. (1976c) Effect of photoperiod on germination of *Cyperus inflexus* seeds. Bot. Gaz. 137: 269–273.

Baskin, J.M. and Baskin, C.C. (1977a) Germination ecology of *Sedum pulchellum* Michx. (Crassulaceae). Amer. J. Bot. 64: 1242–1247.

Baskin, J.M. and Baskin, C.C. (1977b) An undescribed cedar glade community in middle Tennessee. Castanea 43: 140–145.

Baskin, J.M. and Baskin, C.C. (1978a) Temperature requirement for afterripening of seeds of a winter annual induced into secondary dormancy by low winter temperatures. Bull. Torrey Bot. Club 105: 104–107.

Baskin, J.M. and Baskin, C.C. (1978b) The seed bank in a population of an endemic plant species and its ecological significance. Biol. Conserv. 14: 125–130.

Baskin, J.M. and Baskin, C.C. (1978c) Seasonal changes in the germination response of *Cyperus inflexus* seeds to temperature and their ecological significance. Bot. Gaz. 139: 231–235.

Baskin, J.M. and Baskin, C.C. (1978d) Leaf temperatures of *Heliotropium tenellum* and their ecological implications. Amer. Midl. Natur. 100: 488–492.

Baskin, J.M. and Baskin, C.C. (1979a) The ecological life cycle of *Thlaspi perfoliatum* and a comparison with published studies on *Thlaspi arvense*. Weed Res. 19: 285–292.

Baskin, J.M. and Baskin, C.C. (1979b) The ecological life cyle of the cedar glade endemic *Lobelia gattingeri*. Bull. Torrey Bot. Club 106: 176–181.
Baskin, J.M. and Baskin, C.C. (1979c) Timing of seed germination in the weedy summer annual *Euphorbia supina*. Bartonia 46: 63–68.
Baskin, J.M. and Baskin, C.C. (1980a) Role of seed reserves in the persistence of a local population of *Sedum pulchellum*: A direct field observation. Bull. Torrey Bot. Club 107: 429–430.
Baskin, J.M. and Baskin, C.C. (1980b) Ecophysiology of secondary dormancy in seeds of *Ambrosia artemisiifolia*. Ecology 61: 475–480.
Baskin, J.M. and Baskin, C.C. (1981a) Geographical distribution and notes on the ecology of the rare endemic *Leavenworthia exigua* var. *laciniata*. Castanea 46: 243–247.
Baskin, J.M. and Baskin, C.C. (1981b) Seasonal changes in the germination responses of buried *Lamium amplexicaule* seeds. Weed Res. 21: 299–306.
Baskin, J.M. and Baskin, C.C. (1981c) Photosynthetic pathways indicated by leaf anatomy in fourteen summer annuals of cedar glades. Photosynthetica 15: 205–209.
Baskin, J.M. and Baskin, C.C. (1982) Comparative germination responses of the two varieties of *Arenaria patula*. Trans. Ky. Acad. Sci. 43: 50–54.
Baskin, J.M. and Baskin, C.C. (1983a) The germination ecology of *Veronica arvensis* L. J. Ecol. 71: 57–68.
Baskin, J.M. and Baskin, C.C. (1983b) Germination ecology of *Collinsia verna*, a winter annual of rich deciduous woodlands. Bull. Torrey Bot. Club 110: 311–315.
Baskin, J.M. and Caudle, C. (1967) Germination and dormancy in cedar glade plants. I. *Aristida longespica* and *Sporobolus vaginiflorus*. J. Tenn. Acad. Sci. 42: 132–133.
Begon, M. and Mortimer, M. (1981) Population Ecology. Sinauer Associates, Inc., Sunderland, Massachusetts.
Braun, E.L. (1950) Deciduous Forests of Eastern North America. Hafner Publ. Co., New York.
Brenchley, W.E. (1918) Buried weed seeds. J. Agric. Sci. 9: 1–31.
Brenchley, W.E. and Warington, K. (1930) The weed seed population of arable soil. I. Numerical estimation of viable seeds and observations on their natural dormancy. J. Ecol. 18: 235–272.
Briggs, L.J. (1914) Relative water requirement of plants. J. Agric. Res. 3: 1–64 + 7 plates.
Caudle, C. and Baskin, J.M. (1968) The germination pattern of three winter annuals. Bull. Torrey Bot. Club 95: 331–335.
Champness, S.S. and Morris, K. (1948) The population of buried viable seeds in relation to contrasting pasture and soil types. J. Ecol. 36: 149–173.
Clausen, R.T. (1975) *Sedum* of North America North of the Mexican Plateau. Cornell University Press, Ithaca, New York and London.
Cohen, D. (1966) Optimizing reproduction in a randomly varying environment. J. Theor. Biol. 12: 119–129.
Dillman, A. (1931) The water requirement of certain crop plants and weeds in the northern Great Plains. J. Agric. Res. 42: 187–238.
Ehleringer, J., Mooney, H.A. and Berry, J.A. (1979) Photosynthesis and microclimate of *Camissonia claviformis*, a desert winter annual. Ecology 60: 280–286.
Epling, C., Lewis, H. and Ball, F.M. (1960) The breeding group and seed storage: A study in population dynamics. Evolution 14: 238–255.
Fernald, M.L. (1950) Gray's Manual of Botany. 8th ed. American Book Co., New York.
Freeman, C.P. (1933) Ecology of cedar glade vegetation near Nashville, Tennessee. J. Tenn. Acad. Sci. 8: 143–228.
Galloway, J.J. (1919) Geology and natural resources of Rutherford County, Tennessee. Tenn. Geol. Surv. Bull. 22: 1–81.
Gleason, H.A. (1952) The new Britton and Brown illustrated flora of the northeastern United States and adjacent Canada. 3 Vols. Hafner Publ. Co., New York and London.
Gottlieb, L.D. (1974) Genetic stability in a peripheral isolate of *Stephanomeria exigua* ssp. *coronaria* that fluctuates in population size. Genetics 76: 551–556.
Hájková, L. and Krekule, J. (1972) The developmental pattern in a group of therophytes. I. Seed dormancy. Flora 161: 111–210.
Janssen, J.G.M. (1973) Effects of light, temperature and seed age on the germination of the

winter annuals *Veronica arvensis* L. and *Myosotis ramosissima* Rochel ex. Schult. Oecologia 12: 141–146.
Kemp, P.R. (1983) Phenological patterns of Chihuahuan Desert plants in relation to the timing of water availability. J. Ecol. 71: 427–436.
Kropáč, Z. (1966) Estimation of weed seeds in arable soil. Pedobiologia. 6: 105–128.
Lewis, H. (1962) Catastrophic selection as a factor in speciation. Evolution 16: 257–271.
Lloyd, D.G. (1965) Evolution of self-compatibility and racial differentiation in *Leavenworthia* (Cruciferae). Contr. Gray Herb. 195.
Lloyd, D.G. (1967) The genetics of self-incompatibility in *Leavenworthia crassa* Rollins (Cruciferae). Genetica 38: 227–242.
Lloyd, D.G. (1968a) Pollen tube growth and seed set in self-incompatible and self-compatible *Leavenworthia* (Cruciferae) populations. New Phytol. 67: 179–195.
Lloyd, D.G. (1968b) Partial unilateral incompatibility in *Leavenworthia* (Cruciferae). Evolution 22: 382–393.
Lloyd, D.G. (1969) Petal color polymorphisms in *Leavenworthia* (Cruciferae). Contr. Gray Herb. 198.
Martin, C.E., Lubbers, A.E. and Teeri, J.A. (1982) Variability in Crassulacean acid metabolism: A survey of North Carolina succulent species. Bot. Gaz. 143: 491–497.
Miyanishi, K. and Cavers, P.B. (1980) The biology of Canadian weeds. 40. *Portulaca oleracea* L. Can. J. Pl. Sci. 60: 953–963.
Mooney, H.A., Ehleringer, J. and Berry, J.A. (1976) High photosynthetic capacity of a winter annual in Death Valley. Science 194: 322–324.
Mulroy, T.W. and Rundel, P.W. (1977) Annual plants: Adaptations to desert environments. BioScience 27: 109–114.
Newman, E.I. (1963) Factors controlling the germination date of winter annuals. J. Ecol. 51: 625–638.
Quarterman, E. (1950a) Major plant communities of Tennessee cedar glades. Ecology 31: 234–254.
Quarterman, E. (1950b) Ecology of cedar glades. I. Distribution of glade flora in Tennessee. Bull. Torrey Bot. Club 77: 1–9.
Ratcliffe, D. (1961) Adaptation to habitat in a group of annual plants. J. Ecol. 49: 187–203.
Regehr, D. and Bazzaz, F. (1976) Low temperature photosynthesis in successional winter annuals. Ecology 57: 1297–1303.
Roberts, H.A. and Neilson, J.E. (1982) Role of temperature in the seasonal dormancy of seeds of *Veronica hederifolia* L. New Phytol. 90: 745–749.
Roberts, H.A. and Stokes, F.G. (1966) Studies of the weeds of vegetable crops. VI. Seed populations of soil under commercial cropping. J. Appl. Ecol. 3: 181–190.
Rollins, R.C. (1963) The evolution and systematics of *Leavenworthia* (Cruciferae). Contr. Gray Herb. 192.
Rollins, R.C. (1981) Weeds of the Cruciferae (Brassicaceae) in North America. J. Arnold Arb. 62: 517–540.
Savage, D. (1937) Drought survival of native grass species in the central and southern Great Plains. U.S.D.A. Tech. Bull. No. 549.
Shantz, H.L. (1927) The water requirements of plants at Akron, Colorado. J. Agric. Res. 34: 1093–1190.
Shinners, L. (1941) Notes on Wisconsin grasses–II. *Muhlenbergia* and *Sporobolus*. Amer. Midl. Natur. 26: 69–73.
Smith, T.L. and Eichmeier, W.G. (1983) Limited photosynthetic plasticity in *Sedum pulchellum* Michx. Oecologia 56: 374–380.
Solbrig, O.T. (1972) Breeding system and genetic variation in *Leavenworthia*. Evolution 26: 155–160.
Solbrig, O.T. and Rollins, R.C. (1977) The evolution of autogamy in species of the mustard genus *Leavenworthia*. Evolution 31: 265–281.
Thompson, P.A. (1970) Changes in germination responses of *Silene secundiflora* in relation to the climate of its habitat. Physiol. Plant. 23: 739–746.
Thompson, P.A. (1974) Correlations between climatic features of the plant's environment and germination responses to temperature. Proc. 12th Brit. Weed Cont. Conf. 3: 1109–1120.

Thornthwaite, C.W. (1931) The climates of North America according to a new classification. Geogr. Rev. 21: 633–655.
Turner, B.H. and Quarterman, E. (1975) Allelochemic effects of *Petalostemon gattingeri* on the distribution of *Arenaria patula* in cedar glades. Ecology 56: 924–932.
United States Department of Commerce (1965) Climatography of the U.S. 86–35. Decennial census of the U.S. climate. Climatic summary of the U.S.–Supplement for 1951 through 1960. Weather Bureau, Washington, D.C.
Vegis, A. (1964) Dormancy of higher plants. Ann. Rev. Pl. Physiol. 15: 185–224.
Vengris, J. and Damon, R.A., Jr. (1976) Field growth of Fall panicum and Witchgrass. Weed Sci. 24: 205–208.
Vengris, J. and Dunn, S. (1972) Life history studies as related to weed control in the Northeast. 7–Common purslane. Univ. Massachusetts Agric. Exp. Sta. Res. Bull. 598.
Weaver, J. and Albertson, F. (1943) Resurvey of grasses, forbs and underground plant parts at the end of the great drought. Ecol. Monogr. 13: 63–117.
Wilson, C.W., Jr. (1949) Pre-Chattanooga statigraphy of central Tennessee. Bull. 56. State of Tennessee, Dept. of Conservation. Div. of Geology.
Zimmerman, C.A. (1976) Growth characteristics of weediness in *Portulaca oleracea* L. Ecology 57: 964–974.

POPULATION STRUCTURE AND PROCESSES OF TUNDRA PLANTS AND VEGETATION

T.V. CALLAGHAN and U. EMANUELSSON

ABSTRACT

Selective forces of the physical environment are particularly strong in the tundra: positive plant interactions tend to replace the competition or self-thinning typical in warmer latitudes. Tundra areas are young and evolution is slow because of long life spans, low reproductive rates and a predominance of vegetative proliferation. However, recruitment from seed is important in open habitats such as fell-fields, but early mortality rates are high and recruitment is intermittent. Age class distributions resulting from vegetative reproduction show low recruitment levels but high survival rates due to the physiological interdependence of modules. The dynamic equilibrium of forest tundra plant populations is regulated by interactions between fluctuating populations of animals and plants.

INTRODUCTION

Tundra is derived from a Finnish word to describe the treeless plateaus of Lapland (Warming 1909) but its meaning has become vague (Dagon 1966). It has been used to describe the treeless vegetation of the Arctic/Antarctic climatic zonobiome IX (Walter 1979) and, more specifically, a geobotanical region of the Arctic which is characterised by hekistothermic plants aggregated into communities of hyparctic, arctic and arctic-alpine shrubs, dwarf shrubs, herbaceous perennials, mosses and lichens (Aleksandrova 1980). In the present paper, however, tundra is used in its widest sense and that used by Murray (1978), i.e., landscapes of vegetation which are treeless due to predominantly cold climates. Tundra vegetation is found therefore, beyond the latitudinal limits of tree growth (polar tundra vegetation) and above the altitudinal limits of tree growth (alpine tundra vegetation). Detailed discussions of classifications of tundra subdivisions and vegetation are given in Aleksandrova (1980) and Bliss (1981a, b). The present paper concentrates on population processes and life-history phenomena within the tundra vegetation of the Arctic and

White, J. (ed.), The Population Structure of Vegetation.
 ISBN 978-94-010-8927-2

sub-Arctic region, and the taiga, or forest tundra ecotone between the tundra vegetation and the boreal forest.

Tundra areas cover 17% of the earth's land surface (Bliss 1971) and are characterised by short, cool growing seasons, long, cold winters and primitive impoverished humus soils which consist of a lower permanently frozen layer – the "permafrost" – and an upper "active" layer which thaws in summer. The southern limit of tundra in the northern hemisphere is associated with the mean July isotherm of 10°C, the southern limit of permafrost, the southern limit of auroral activity, the isoline for areas with air temperatures above 0°C for less than 140 days and a decrease in annual absorbed radiation which is less than 2100 MJ m^{-2} (Lewis and Callaghan 1976). The summer photoperiod increases with latitude in these areas until the midnight sun effect occurs.

Within the tundra area, however, there are considerable differences in climate along an oceanic/continental gradient. An oceanic location such as South Georgia in the sub-Antarctic region has minimum monthly mean temperatures in winter of only −2°C and maximum monthly mean temperatures in summer of +5°C with mean monthly temperatures above 0°C for 8 months of the year (Rosswall and Heal 1975). At the other extreme, Devon Island in the High Arctic region has mean monthly winter temperatures down to −36°C and mean monthly summer temperatures of 9°C with mean monthly temperatures above 0°C for only 3 months of the year (Rosswall and Heal 1975). The variation in climate between locations tends to form a continuum between these extremes. There also exists a considerable diversity of tundra habitats resulting from small differences in microclimate related to local variations in topography. These variations affect solar radiation absorption, snow accumulations, moisture retention, air and soil temperatures, soil conditions etc.

TUNDRA HABITATS

The plant communities of the tundra are numerous and many classifications exist (e.g., Wielgolaski 1972, Böcher 1975, Webber 1978, Bliss 1981b, French 1981, Wielgolaski et al. 1981). Unfortunately, the range of classifications and terminology of plant communities is great and varies from author to author.

(a) *Forest tundra communities* are characterised by a tree layer with an open canopy of scattered individuals which are diminished in height and which may be polycormic in growth form.

(b) *Scrub communities* are dominated by erect shrubs which are often species of *Salix*.

(c) *Herb meadow communities* are found on sheltered slopes with moist fertile soils and species diversity is high.

(d) *Fell-field communities* are open communities of scattered individuals aggregated into islands. The plants are dwarfed and prostrate and the habitat is characterised by little winter snow cover, water shortage, primitive soils, and frequent soil movement due to freeze-thaw cycles.

(e) *Snow-bed communities* are found in sheltered depressions. This often results in an extremely short growing season.

(f) *Dwarf shrub-heath communities* are found on poor soils with moderate winter snow cover. They are often dominated by clonal Ericaceous species and cryptogams.

(g) *Wet meadow communities* are found on poorly drained areas, often with permafrost. Many of the communities are dominated by tussock forming graminoids.

(h) Other communities occur on salt marshes, beaches and dunes. Polar desert, a particularly extensive zone, may be regarded physiognomically as an extreme fell-field community.

SELECTIVE FORCES IN THE TUNDRA AND ADAPTATION

It is perhaps the nature of the forces involved in the evolution of tundra plants which contrasts most strikingly with the population processes of mesic, temperate and tropical areas. The forces which select genotypes capable of growing in the tundra are connected with soil conditions and the climate near to the ground surface. Plant competition and herbivory are probably only important in increasing the probability of plant mortality in areas of closed vegetation in more favourable sub-arctic climates and soils. Indeed, it is the *positive* interactions of the biotic environment (e.g., commensalism and mutualism) which are particularly important in the harsher environments of the tundra. The survival probabilities of individuals are *increased* by the close association of individuals of the same population or other species. This situation is the opposite of the competitive exclusion and self-thinning typical of closed vegetation of other geographical regions (e.g., Grime 1979). Plants which climb, sting, poison or possess spines and thorns are absent from the tundra, with the implication that this protection is not necessary (Porsild 1951), although recent work in the sub-Arctic suggests that chemical defence mechanisms against grazing occur (Haukoja 1980).

Mechanisms have evolved in many plants which reduce the harsh influence of the tundra environment and therefore increase the probability of survival. The environments of the Arctic are however relatively young and evolution has not proceeded very far. Savile (1972) suggested that the Arctic flora represents "a depauperate miscellany from various regions", as wholly Arctic genera are unknown and endemic sections and species groups are scarce. The high frequency of vegetative reproduction and vivipary, together with considerable longevity, in tundra plants have probably decreased the rate of evolution significantly in the tundra. Savile (1972) suggested that some of the first species to colonise Arctic areas were species pre-adapted to cold environments, having arisen (possibly from cold-night desert species) in alpine regions of more southerly latitudes. Other species, such as those of cool temperate marshes and bogs, may have had sufficient phenotypic plasticity to have become morphologically and/or physiologically acclimated on exposure to arctic conditions. Within these populations the initial stages of evolution may have produced distinct ecotypes along environmental gradients. Unfortunately there are few quantitative assessments of the increased probability of survival of individuals which possess these hypothetical adaptations.

The soil environment

The main soil conditions limiting plant growth in the tundra are concerned with the primitive nature of the soil, the extremes of water content and temperature, and the low levels of plant nutrients (Russell 1940, Haag 1974). Fell-field soils are often mobile as a result of freeze-thaw cycles and they represent perhaps the harshest environment, with 'high stress and high disturbance' (Grime 1977). Plants of the dry fell-field soils are often xerophytic (e.g., *Diappensia lapponica*, Petersen 1908), they often have deep tap-roots (e.g., *Papaver radicatum*, Bell and Bliss 1978) and often occur as tight cushions (e.g., *Diappensia lapponica, Silene acaulis*).

Low nutrient availability has been overcome to some extent by species such as *Eriophorum angustifolium* which can grow and absorb nutrients efficiently at low temperatures (Chapin 1974). Some non-woody plants have roots which may be active for up to 12 years (Headley et al., in press). In general, belowground biomass increases in proportion to aboveground biomass as latitude or environmental severities increase (Callaghan and Collins 1981) and tundra plant species have a larger root volume than temperate species (Chapin 1974). Once inside the plant, mineral nutrients may be conserved in various ways. Plant groups typical of tundra areas appear to be physiologically active with lower tissue concentrations of nutrients than other groups (Rodin and Bazilevich 1967, Wielgolaski et al. 1975). Species with indeterminate apical growth and distal senescence (e.g., *Lycopodium annotinum*, Callaghan 1980 and *Empetrum hermaphroditum*, Malmer and Nihlgård 1980) may recycle nutrients from senescing structures to growing points. Other growth forms such as mats of *Dryas integrifolia* may accumulate dead leaves which serve as a supply of nutrients during senescence and decomposition (Svoboda 1977). Tussock plants (e.g., *Poa flabellata* and *Eriophorum vaginatum*, Chapin et al. 1979) also accumulate dead leaves and the roots of young tillers are active exclusively in the attached litter of older generations within the same tussock. The rhizomatous growth of many tundra plant species allows a "foraging strategy" whereby parts of a clone may be subsidized in terms of energy, nutrients and water by older established parts of the clone until areas favourable for new root production are found (Callaghan 1980). In addition to physiological adaptations, life history phenomena such as longevity, indeterminate growth patterns and vegetative "tillering" of many tundra plant species are thus important mechanisms which increase the survival potential of species growing in polar soils.

The aerial environment

Plant growth in the aerial environment of tundra is restricted by extremely low temperatures, desiccating conditions, abrasion of plant parts by wind blown particles during winter, together with xeric conditions during summer in the fell-field habitat; by extremely short and late growing seasons, with low solar angles, in the snow-bed habitat; and by generally cool short growing seasons in meadows and heath habitats (Lewis and Callaghan 1976). Low summer temperatures limit the period available for growth and developmental processes, such as the flowering cycle. Low temperatures also affect flowering indirectly by limiting the availability and activity of insect pollinators.

Tundra plants are obviously winter-hardy but do not appear to be very different from temperate species in this respect, although lichens can withstand −198°C (Ahmadjian 1970). The impacts of the adverse aerial environments are reduced mainly by "avoidance strategies" (Savile 1972) whereby dwarf growth forms exploit sheltered microhabitats. Elaborate mechanisms to withstand harsh conditions appear to be lacking (Savile 1972).

Most tundra plant species, particularly those of oceanic environments, must be able to tolerate sudden freezing in summer and they must be able to resume growth rapidly once favourable conditions return. *Braya humilis* (Sørensen 1941) and *Epilobium latifolium* (Porsild 1951) may be frozen during any stage of flowering and yet continue to complete the cycle on thawing. Morphological adaptations to low summer temperatures include dwarf, compact growth which increases the differential between ambient and plant temperatures (Salisbury and Spomer 1964, Cernusca and Seeber 1981), deep colour pigmentation, a protective screen of dead tissues and the possession of transparent hairs (Krog 1955). Such adaptations may enable temperature differentials of up to 32°C to be achieved (Saville 1972, Larcher 1980, Mølgaard 1982). Limited growth in short growing seasons is enhanced by the presence of storage organs and the ability to produce a photosynthetic surface rapidly at the beginning of the season; winter-green leaves are common. The reduced length of time for flowering is overcome by many tundra plant species which have the ability to extend the flowering cycle over two or more years (Sørensen 1941, Hodgson 1966).

The main physiological adaptations of tundra plant species appear to be an ability to photosynthesise and translocate materials at low temperatures (40% of the maximum rate of photosynthesis of wet meadow grasses occurs at 0°C (Tieszen 1973) and translocation occurs at 0°C (Allessio and Tieszen 1975a), a general ability for high physiological activity at low temperatures and therefore high respiration rates (Billings 1974) and the possession of erectophilic leaves to intercept low angle radiation efficiently (Miller and Tieszen 1972). Chabot (1979), however, suggests that there are no qualitative differences between tundra and non-tundra species in their adaptations for low temperature metabolism.

The selective forces of the aerial environment have their greatest effect by killing young individuals. It is not surprising therefore that vegetative reproduction which allows a period of "postnatal care" is common in tundra plants; this will be discussed later.

The biotic environment

Too little quantitative data exist to characterise the biotic environment of tundra plants. It may be postulated however that the major biotic selective pressures follow a gradient associated with the severity of the physical environment or the successional stages of the vegetation. In the fell-field environment, positive interaction (i.e., commensalism and mutualism) between plants would probably be important in contributing to survival. As the vegetation closes in more favourable physical environments, negative plant interactions (i.e., competition) occur and herbivory becomes more important.

Positive plant interactions

In the most extreme physical environments or at the earliest stages of colonisation, the vegetation is characteristically sparse, scattered and aggregated into islands: the vegetation does not generally occur as isolated monospecific areas. It is reasonable to assume, therefore, that competition caused by the aggregation of species is relatively unimportant when compared with the mutual benefits of aggregation. These benefits are mainly shelter from wind exposure, provision of moisture for germination and seedling establishment, provision of organic rooting substrates, and an increased temperature locally. Shelter is afforded by large cushions or tussocks: Polunin (1936) and Whitehead (1951) have observed seedlings of various species growing in the shelter of large cushions of *Silene acaulis*.

Moisture is often limiting in fell-field situations and tends to be more freely available within plant cushions in fell-field habitats: existing cushions and mats of pioneer species support the establishment of many other species. Griggs (1956) recorded 93 species of vascular plants developing within cushions of 5 pioneer species and claimed that in most cases there was no evident damage suffered by the host plant other than loss of ground; indeed the roots of the host cushion of *Silene acaulis* and the invader species *Festuca ovina* survived in separate areas. Bonde (1968) found that after three years, half of the surviving seedlings of *Trifolium nanum* occurred in mats or cushions of vegetation and Bell and Bliss (1980) found that most seedlings in a high Arctic environment become established in moss or lichen mats.

At the individual plant level, the cushion or tussock growth form would appear to represent a situation of intense competition between shoots for light and space and for water and nutrients supplied by a single tap-root system. The predominance of these life forms in fell-fields must however indicate a positive benefit of the aggregation of shoots; survival without such close aggregation may be impossible. In *Hylocomium splendens*, for example, shoot densities of over 6000 per m^2 were recorded (Callaghan et al. 1978). Individual shoots could not control their own water balance and were susceptible to water loss. Only by tight packing with other shoots could water loss be reduced and a mutualism therefore exists between individual shoots. At the tree line in Keewatin, spruce trunks reduce ice crystal abrasion by causing eddies: high density is essential for the survival of individuals (Bryson et al. 1965). The increased survival probabilities conferred by close packing of shoots or of species, in contrast to the density-dependent mortality of temperate species, is an area of tundra plant biology which requires much greater attention than it has received to date. Commensalism may even occur in closed areas of vegetation in the sub-Arctic. Evergreen *Vaccinium vitis-idaea* co-exists with deciduous *V. myrtillus* and *V. uliginosum* over vast areas. Preliminary observations suggest that the deep-rooting deciduous species exploit deep soil reserves and return nutrients to the soil surface where they are available to the shallow-rooting evergreen species. The different canopy structures of the two groups result in the minimisation of competition for light as well as nutrients between them.

Negative plant interactions

As the vegetation becomes closed and resources become more abundant commensalism and mutualism become replaced by competition and density-dependent mortality increases in importance. Wijk (1980), for example, showed that *Salix herbacea* in a snow bed habitat was limited by the climate where snow cover is longest and by competition from *Vaccinium myrtillus* at the outer limits of the snow bed. Thus the dominance of some plants in a harsh environment may result from an inability of more competitive species to tolerate that environment. Reciprocally, the paucity of species tolerant of harsh environments in sites with more abundant resources may result from their poor competitive ability (e.g., Klikoff 1965, Jolls 1982).

Differentiation between 'stress-tolerant' and 'competitive' (Grime 1979) biotypes may exist within one species. Fell-field populations of tetraploid *Phleum alpinum* were genetically distinct and were characterised by dwarf stature, a tufted growth form (resulting from lack of rhizome growth) and short wide leaves (Callaghan and Lewis 1971, Callaghan 1974). A biotype from dense meadow vegetation had pronounced rhizome growth, fewer leaves which were 5 times longer than those of the fell-field population, higher reproductive capacities and greater biomass. When grown under the same conditions, both biotypes had similar relative growth rates but these were achieved by high assimilation rates and low leaf area ratios in the fell-field type and *vice versa* in the meadow ('competitive') biotype.

Collins (1976) showed that *Polytrichum alpestre* growing in a pure turf had a much slower turnover than when growing mixed with *Chorisodontium aciphyllum*. In a pure turf, an individual apex could grow for at least 9 years whereas 4 years appeared to be the maximum longevity within the mixed population. Data on density-dependent mortality among plants of the tundra are few but Collins (1976) showed that the relationship between weight of the annual increment of *P. alpestre* and the spatial density approximated to the density-effect law, well established in crop plants (White 1980). Lindeberg-Johnson, quoted in Jolls (1982), showed that an increase in the total plant density of *Caltha leptosepala* produced a decrease in the number of leaves per individual, average leaf area, the length of the primary root and the number of lateral roots in plants raised from seed.

Within closed areas of vegetation, invasion and colonisation by rhizomatous species becomes important (Griggs 1956). Species with this growth form have considerable competitive advantages. They may avoid competition by developing roots and shoots in unoccupied microsites as a result of a "foraging strategy" (e.g., *Lycopodoium annotinum*, Callaghan 1980) or they may out-compete existing occupants of a microsite by the development of roots and shoots subsidised (in terms of energy, nutrients and water) from a firmly established parent plant. In some species (e.g., *Carex aquatilis*, Shaver and Billings 1975 and *Rostkovia magellanica*, Callaghan 1977) aggressively competitive pioneer rhizomes and tillers may grow with subsidies from non-disseminating assimilatory tillers. However such rhizomatous species can only show limited competitive dominance because, by nature, they are mobile and will eventually vacate their microsites on death of their tillers.

Some graminoid species may show competition for internal parental resources between vegetatively produced offspring. In *Carex bigelowii* tillers produced during the middle period of the parent tillers' life-span are smaller than either those produced earlier when competition is lacking, or later when they are nearest to the source of energy and nutrients from senescing leaves and floral structures of the parent (Callaghan 1976). Similarly, tillers of *Phleum alpinum* in the final flowering year of the parent may flower themselves many years before older sibling tillers. This effect may be due to the close proximity (i.e., a relative competitive advantage) of the youngest tillers to substrates remobilised from senescing parent structures (Callaghan 1977).

Plant–animal interactions

Plant–animal interactions may be both positive and negative. Positive interactions include pollination whereas negative interactions are mainly direct (grazing) and indirect (disturbance) effects of herbivores.

Insects pollinate many Arctic species which have large conspicuously coloured flowers: a review of insect-pollinated flowers and their vectors is given by Bliss (1971).

Herbivory by rodents is extensive from the coastal tundra of Barrow, Alaska (Batzli et al. 1980) to the sub-Arctic forests of Fennoscandia (Oksanen 1982, Emanuelsson 1984). Whereas herbivory from rodents tends to follow cycles (Batzli et al. 1980), grazing by the reindeer of North America and Fennoscandia is relatively stable over time. In the birch forests of sub-Arctic Fennoscandia, the defoliation of vast areas of trees by moth larvae is a cyclical phenomenon (Tenow 1972, Kallio and Lehtonen 1973).

Although morphological defense mechanisms are absent in tundra plants, many species are resistant to grazing (e.g., *Lycopodium* species) because of chemical defence mechanisms. Some species, such as *Betula pubescens* ssp. *tortuosa*, show an induction of polyphenols in new leaves as a response to insect grazing (Haukoja 1980).

LIFE CYCLES OF TUNDRA PLANTS

The basic patterns of life cycle followed by tundra plants are not dissimilar from those found in other regions.

"Ruderal" life cycle

The word "ruderal" in terms of long-lived tundra plants is not inappropriate because there are parallels between some of the long-lived plants of open ground (such as fell-fields and moraines) and the weeds of disturbed ground in temperate regions. Plants following this life cycle show a long vegetative period but, unlike their temperate region monocarpic equivalents, flowering can take place repeatedly after some minimum developmental state has been reached. Vegetative reproduction is absent and death is not associated with flowering. *Oxyria digyna* and mat-forming dwarf shrubs such as species of *Dryas* which do not produce axillary roots would be examples of this category. This life cycle predominates in the High Arctic.

Viviparous life cycles

Vegetative proliferation or vivipary is the formation of leafy shoots (plantlets or bulbils) within the inflorescences of normally seed-bearing plants (Harmer and Lee 1978a, 1978b). The life cycles of these species (e.g., *Polygonum viviparum*) resemble the pattern discussed above but differ in that all reproduction is vegetative. Successful establishment of new genotypes by sexual reproduction in the present polyploid races of *P. viviparum* has not been demonstrated (Engell 1973), although one observation has been recorded of a seed germinating (Law personal communication). Although *P. viviparum* is polycarpic (if this term can be applied to bulbil production), whereby one corm may produce flowering culms for many years (Callaghan 1973, Petersen 1981), viviparous grasses are monocarpic. The viviparous grasses also possess the ability to reproduce vegetatively by tillering whereas *P. viviparum* bulbils are usually the sole means of reproduction. It is surprising therefore that *P. viviparum* is an extremely successful tundra species occurring throughout the circumpolar area (Polunin 1959) in most habitats from severe fell-fields in the High Arctic to tall dense herb meadows in the sub-Arctic (Callaghan and Collins 1981). It is one of the few species in which young individuals are common in closed vegetation (Petersen 1981).

Rhizomatous/stoloniferous graminoids

Many grasses and sedges of the tundra belong to this category, particularly those of waterlogged areas and meadows. Their life cycle is characterised by vegetative reproduction in the early stages of growth and later sexual reproduction which terminates the life of the parent tiller. Although seeds may germinate, it is unusual for seedlings to survive in closed swards and recruitment to the population is usually by tillering.

Tussock graminoids

In tussock graminoids both sexual and vegetative reproduction are successful but differ in relation to the site of colonisation. Vegetative reproduction by tillering increases the size of the parent colony (i.e. the tussock or tuft) whereas seed produced from sexual reproduction is dispersed over wider areas and germinates to form new colonies. Species with this life cycle are found typically in open heathland and fell-fields but sometimes occur as extensive areas (e.g., *Eriophorum vaginatum*) in the Arctic (Miller 1982) and (e.g., *Poa flabellata* and *P. foliosa*) in the sub-Antarctic (Warming 1909).

Cushion plants

Cushion plants follow a life cycle closely resembling that described above. Throughout the period of growth of an individual, it reproduces vegetatively to form other modules (White 1980) which, because of lack of dispersion, remain close to the parent so that successive generations form a dense cushion of closely-packed modules. Like tussock graminoids, sexual reproduction serves to disperse the species and establish new colonies but the flowering of cushion plants is not associated with death. Cushion plants are found in open heaths and on fell-fields. They reach their peak of development in terms of height in austral and sub-Antarctic regions, e.g., *Bolax* species (Moore 1968).

"Ericaceous" species

This group, consisting for example of species of *Vaccinium*, *Empetrum*, *Lycopodium* etc., has a life cycle in which sexual and vegetative reproduction take place throughout life (Flower-Ellis 1971, 1980a, 1980b) but seedling establishment is usually unsuccessful in closed swards. It would appear that the role of sexual reproduction both in this group and in the rhizomatous/stoloniferous graminoids is to establish new colonies of new genotypes outside the closed community in areas where there are extensive gaps in the vegetation. Once colonies arising from sexual reproduction have developed into a closed canopy, maintenance of the population is by vegetative reproduction in which large propagules subsidised in terms of energy, water and nutrients by the parent can out-compete small fragile, self-sufficient seedlings. Sometimes, extensive areas may be covered by a single clone (Oinonen 1967, 1968). The different ecological requirements of the two modes of reproduction are highlighted by cryptogams such as *Lycopodium* species in which the sporophyte and gameteophyte generations differ vastly in terms of morphology and physiology, the sporophyte being a leafy autotrophic organism whereas the gametophyte is a subterranean heterotroph.

Differentiation of life cycles within a population

Vegetative reproduction is theoretically a process whereby offspring are produced which are genetically identical to the parent. Large morphological and physiological differences may, however, exist between parent and offspring and between siblings (Callaghan 1977, Callaghan and Collins 1981). Tillers of *Phleum alpinum* produced in the vegetative years of a parent tillers' life cycle follow the parents' life cycle by developing long rhizomes over a number of years before flowering. Sibling tillers produced by the same parent in its final flowering year have little rhizome and flower after one year, many years before their older siblings. At the simplest level, tillers of *Carex bigelowii* produced by a parent tiller in the middle of its life span (Fig. 1) have statistically significant lower dry weights throughout their life cycles than sibling tillers produced earlier or later in the parents' life span (Callaghan 1976). In the sedges *Carex aquatilis* (Shaver and Billings 1975) and *Rostkovia magellanica* (Callaghan 1977), two classes of tiller are produced vegetatively. One class develops long rhizomes with a

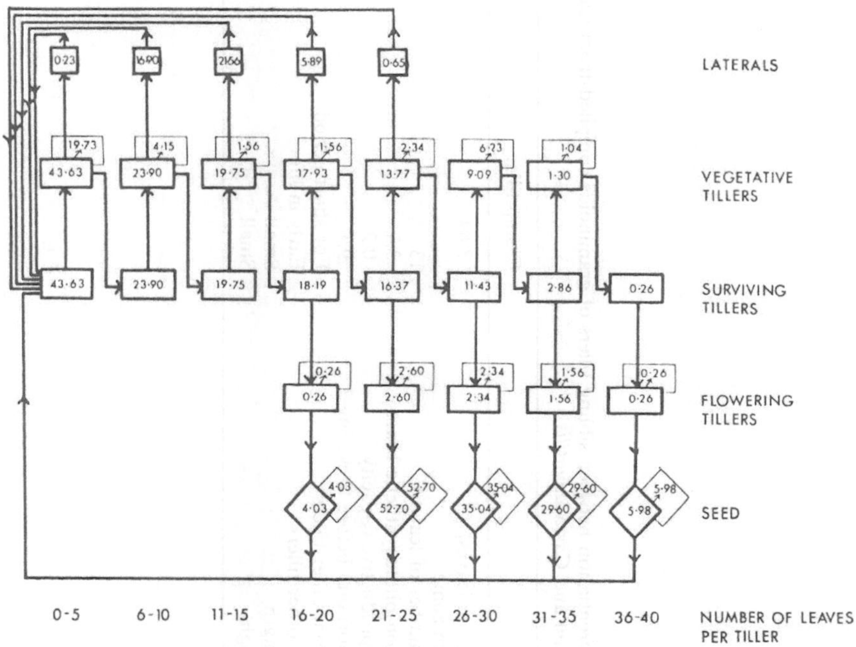

Fig. 1. Life cycle pathways of *Carex bigelowii* (Callaghan 1976). Rows represent types of tiller or reproductive propagules while columns represent age-classes. Thin-walled boxes represent dead tillers. Values are density per m^2 and present an integrated pattern for *C. bigelowii.*

Table 1. Differentiation between sibling tillers of graminoids compiled from Callaghan (1977) and Shaver and Billings (1975) for *Rostkovia magellanica*, *Phleum alpinum* and *Carex aquatilis*.

Parameter	Clump tiller	Pioneer tiller	Species
Dry weight green leaves/ dry weight rhizome	3.69	0.89	*Rostkovia magellanica*
Longevity (number of leaves)	13	19	*Rostkovia magellanica*
Number of non-photosynthetic bracts	3.7	8.4	*Rostkovia magellanica*
Vegetative reproductive capacity	0.2	2.4	*Rostkovia magellanica*
Percentage mortality before flowering	9.1	7.7	*Rostkovia magellanica*
C^{14} import from other tillers	For short period	For long period	*Rostkovia magellanica*, *Phleum alpinum*
C^{14} export to other tillers	Early in life	Late in life	*Rostkovia magellanica*, *Phleum alpinum*
Total root length	Small	Great	*Carex aquatilis*
Root dry weight	Small	Great	*Carex aquatilis*

large root complement and only develops green leaves late in its life span, after being subsidised in carbon by the parent tiller (Table 1). This tiller type has a long life span and reproduces by tillering and by seed. The second class of tiller has a short life span in which a large leaf complement is produced at an early stage and rhizomes and roots are poorly developed. The assimilatory 'colonising' or 'clump' tillers have two functions; they expand the existing colony and they subsidise the growth of sibling 'pioneer' or 'spreading' tillers which disperse the population.

THE MODULAR GROWTH OF TUNDRA PLANTS

The growth of many plants may be interpreted as the production of a population of modules which may vary in morphology to a greater or lesser degree (White 1980, Harper 1981). Several modules may often be produced within a growing season. Mattheis et al. (1976) have described the growth of *Dupontia fischeri* from Alaska in terms of the production of phytomers, each phytomer consisting of a leaf, leaf sheath and internode. Perhaps the simplest form of modular growth is shown by the moss *Hylocomium splendens* from Abisko (Swedish Lapland). The vertical shoot consists of a sympodium of feathery segments, or modules. Young sub-apical buds branch in their second year to produce a photosynthetic "frond" and the next generation of sub-apical buds. A shoot consists therefore, of a series of annually produced modules which are generated at the top of the shoots while older modules at the base of the shoots are senescing, dying and decomposing. Each segment has a short period of determinate growth (two years) whereas each shoot potentially has an indeterminate period of growth; the formation of new shoots by sexual reproduction is rare (Callaghan et al. 1978). Within the shoot each module appears to be a physiologically independent individual, as translocation has not been demonstrated (Callaghan et al. 1978) and the shoot may be analogous to a 'colony' of individuals.

Lycopodium annotinum is more complex and consists of a series of annually produced modules (Callaghan 1980). Some modules are long, have relatively high dry weight and possess relatively few microphylls. These 'horizontal segments' (Fig. 2) can produce a similar segment apically to continue the growth of the main axis; they can produce similar segments from lateral buds which form side branches; they can produce roots (up to a maximum of three) and they can produce buds from which segments grow vertically. 'Vertical segments' (Fig. 2) have lower dry weight, a shorter life span (Callaghan 1980) and have many more microphylls than horizontal segments. They may reproduce to continue the upward growth of the vertical branch or to produce new vertical branches. After two or more modules have been produced within a vertical branch, a vertical module may produce a strobilus which is an annual structure terminating the growth of the whole branch (Fig. 2).

These morphological (and functional) differences between the basic module types are paralleled by differences within each module type. For

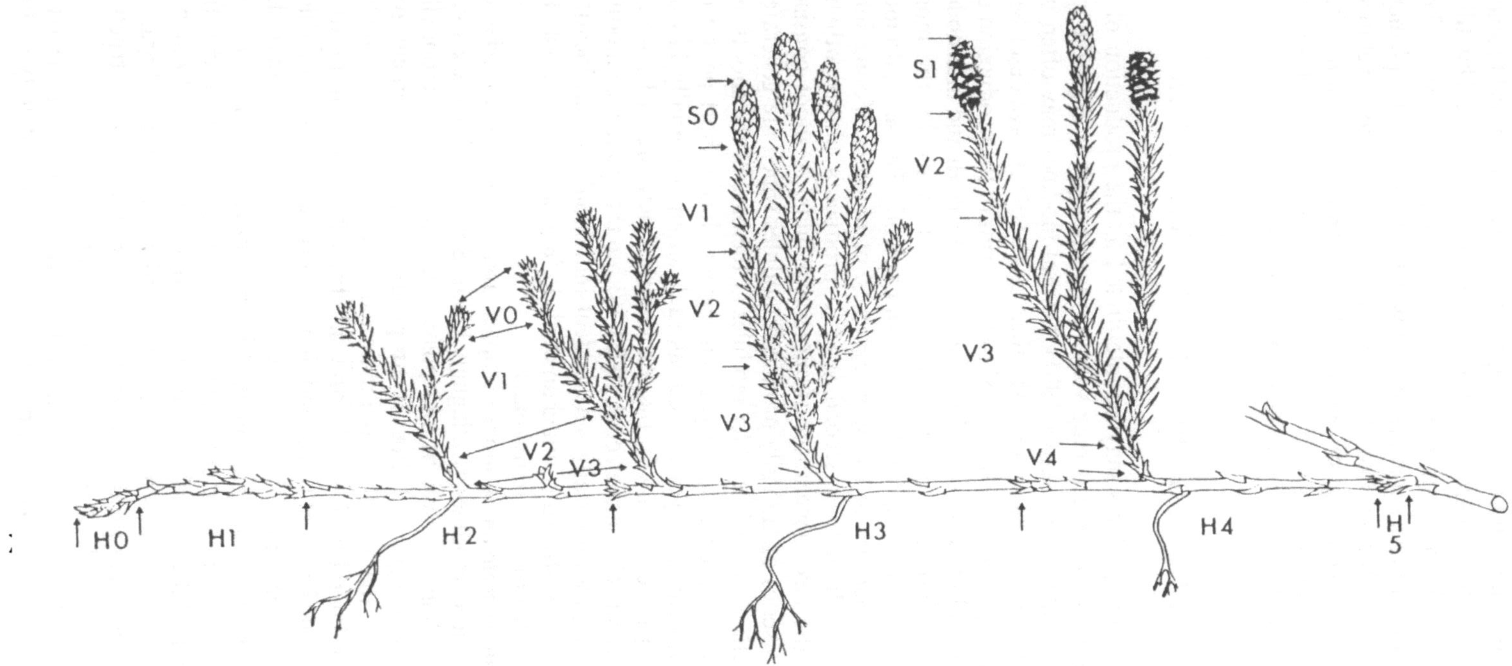

Fig. 2. Growth form of *Lycopodium annotinum.* H = Horizontal segments; V = Vertical segments; S = Strobili. Numbers refer to ages in years while arrows denote morphological markers of annual growth (taken from Callaghan 1980).

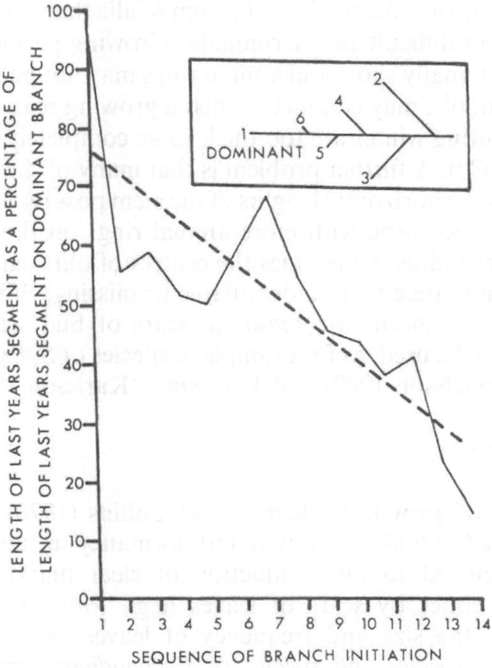

Fig. 3. The effect of apical dominance on the length of growth modules of *Lycopodium annotinum*. All modules are the same age but belong to branches of various orders of sub-dominance (see inset). Points are means, while the regression line is: $y = 76 - 3.88x$; $r^2 = 0.64$ with 224 degrees of freedom.

example, the first vertical segment to be produced within a vertical branch has the longest life span and the smallest dry weight of the vertical segments. Reproduction (production of lateral buds) is associated with the first-formed modules within a vertical branch. Within horizontal branches, great differences in length are associated with apical dominance effects (Fig. 3). Thus genetically controlled patterns of growth of modules are the prime determinants of the growth form of *L. annotinum* while climate just modifies growth by controlling numbers and size of modules, but not the interrelationships between them.

AGE-CLASS DISTRIBUTIONS OF TUNDRA PLANTS

Age determinations of tundra plants

Woody species

Annual rings of wood are usually counted to determine the age of shrubs and dwarf shrubs (see Warren Wilson 1964, Flower-Ellis 1971). The generally slow growth of tundra plants is evidenced by the narrowness of annual rings which vary from 2.73 to 0.7 mm (Callaghan 1973). The rings are therefore often difficult to discriminate. Growing periods may occasionally be exceptionally short and annual rings may not be formed. *Salix herbacea*, for example, may completely miss a growing season when snow accumulations during winter are too thick to be completely melted in the summer (Wijk 1980). A further problem is that many of the dwarf shrubs are prostrate and the horizontal lengths of the stem possess reaction wood, i.e., the stems are eccentric with more annual rings on the lower radius than on the upper radius. Sometimes the centres of old stems decompose and the oldest rings are either unidentifiable or missing. When age determinations of young shoots are required, scars of bud scales are clear markers which can be used, as for example in species of *Salix* (Wijk 1980), *Empetrum* (Emanuelsson 1980) and *Vaccinium* (Karlsson 1980).

Herbaceous species

Annual markers of growth. Callaghan and Collins (1976) stressed that clearly demarcated periods of growth and dormancy in tundra summers and winters often led to the production of clear markers of annual growth, formed either by scars of leaves (e.g., *Polygonum viviparum*) or a decrease in the size and frequency of leaves along a main axis (e.g., *Polytrichum* species and species of *Lycopodium* (Primack 1973)). In some species, particularly mosses, apical buds may be killed each winter and renewed sub-apical bud growth results in a sympodium of annual segments (Callaghan and Collins 1981, Callaghan et al. 1978).

Leaf production rates. In the early stages of the growth of dicotyledons and throughout the life span of graminoid tillers, the number of leaves produced from birth may be used with annual leaf production rates to determine age (Wager 1938, Callaghan 1976, 1977, 1984, Shaver and Billings 1975). In many tundra habitats, particularly on fell-fields, slow decomposition rates result in the preservation of entire leaf complements on graminoid tillers. Some species, such as *Phleum alpinum* (Callaghan 1977, Heilbronn and Walton 1984) and *Saxifraga oppositifolia* (Wager 1983) appear to have stable annual leaf production rates whereas other species such as *Unicinia meridensis* (Callaghan 1984) and *Luzula spicata* (Wager 1938) have an annual leaf production rate which varies with tiller age and age determinations must be made with care.

Size relationships and plant development. Wager (1938) found that the longest leaf of *Trisetum spicatum* increased in size as tillers aged and used this parameter to determine age. Petersen (1981) assigned *Polygonum viviparum* plants to age classes depending upon the presence of bulbil remains, the size of the rhizome, the shape of the rhizome and the accumulation of dead scapes and leafstalks. In *Oxyria digyna*, one group of xylem vessels is produced each year in the tap root and this pattern of growth has been used by Humlum (1981) to determine the ages of older individuals.

Longevity of tundra plants

Many authors (e.g., Johnson 1969) have recognised that the short harsh growing seasons of tundra environments have selected plants which grow slowly over many years. In habitats with high 'stress' and low disturbance it may be expected that life spans would be long and reproductive effort low (Grime 1977). This assumption tends to be confirmed by the distribution of tundra plants between the growth forms devised by Raunkiaer (1934): only 2% of tundra plants are therophytes (annuals) compared with the maximum of 73% in desert floras (Whittaker 1975). Most of the plants of tundra (60%) are hemicryptophytes (perennial herbs with their perennating tissues at the soil surface), 22% are chamaephytes (dwarf shrubs), 15% are geophytes (perennial herbs with underground perennating tissues) and 1% are phanerophytes (erect shrubs and trees) (Whittaker 1975).

Annuals in the tundra are often either small in stature and associated with aquatic environments (e.g., *Koenigia islandica* and *Montia fontana*) or are hemiparasites (e.g., *Euphrasia frigida*, *Rhinanthus minor* and several species of *Pedicularis*). Some species of *Gentiana*, *Gentianella* and *Lomatogonium* may behave either as annuals or biennials.

In contrast to the annuals, the greatest life spans in the tundra are achieved by the lichens and ages of 4500 years have been calculated (Table 2). It must be stressed however that these ages refer really to colonies or clones and not to an individual part of a thallus. The mountain birch (*Betula pubescens* ssp. *tortuosa*) can survive for up to 150 years in Swedish Lapland and the dwarf shrubs of harsh environments may survive over 100 years (Table 2) but annual ring widths may be as small as 0.07 mm which means that a 100 year old plant would have a stem diameter of only 14 mm. In general, longer life spans are attained on dry fell-fields than in wetter areas.

Lycopodium annotinum possesses units which may survive for over 20 years, but like the ericaceous species (e.g., *Vaccinium* and *Empetrum*) the plant grows apically and dies distally so that the plant – or clone – has indeterminate growth. Where the production of adventitious roots is impossible in dry areas (e.g., the tap-root forms of *Empetrum hermaphroditum* (Emanuelsson 1984) and *Vaccinium vitis-idaea*) an individual plant shows a finite life span.

Fewer data are available on the life-spans of other growth forms in the tundra because of difficulties associated with determining age. However, it is apparent that several herbs survive for between 10 and 50 years whereas individual tillers of graminoids may survive for as long as 8 years before flowering and dying (Table 2). Tufts or tussocks of graminoids apparently survive for particularly long periods but data are lacking. Mosses show shorter life-spans with the few observations ranging from a pseudo-annual cycle to 12 years.

Age-class distributions of tundra plants

The distribution of individuals between age-classes within a population gives information for one point in time. However, current age-class

Table 2. The maximum recorded ages achieved by tundra plant groups and species excluding annuals (G = genet, R = ramet, ? = uncertainty).

Species	Location	Age (yr)	Reference
Trees			
Betula pubescens	Abisko, Sweden, at treeline	120	Sonesson and Hoogesteger 1983
ssp *tortuosa* (G)	Abisko, Sweden, in forest	150	Sonesson and Hoogesteger 1983
Juniperus communis (G)	Kola, Russian Lapland	> 400	Kihlman 1890
Sorbus aucuparia (G)	Kola, Russian Lapland	112	Kihlman 1890
Dwarf Shrubs			
Acaena magellanica (R)	South Georgia	> 5	Walton 1976
A. tenera (R)	South Georgia	> 5	Walton 1976
Arctostaphylos alpina	Kola, Russian Lapland	> 84	Kihlman 1890
A. uva-ursi	Kola, Russian Lapland	> 80	Kihlman 1890
Betula nana	Kola, Russian Lapland	> 75	Kihlman 1890
Dryas integrifolia (G?)	Devon Island, fell field	80–120	Svoboda 1977
D. octopetala (G?)	Kola, Russian Lapland	> 108	Kihlman 1890
E. hermaphroditum (R?)	Abisko, Sweden	72	Emanuelsson 1984
Empetrum nigrum (G?)	Kola, Russian Lapland	> 79	Kihlman 1890
Loiseluria procumbens (G?)	Kola, Russian Lapland	64	Kihlman 1890
Phyllodoce caerulea (G?)	Kola, Russian Lapland	> 35	Kihlman 1890
Salix artica (G)	Devon Island	65	Callaghan 1973
S. arctica (G)	Cornwallis, fell field	56	Warren Wilson 1964
S. arctica (G)	Cornwallis, bog	27	Warren Wilson 1964
S. glauca (G)	N.E. Greenland, fell field	> 10	Wager 1938
S. glauca (G)	W. Greenland, forelands	70	Beschel 1963
S. glauca (G)	W. Greenland, fell field	65	Callaghan 1973
S. glauca (R)	W. Greenland, herb slope	16	Callaghan 1973
S. glauca (?)	Kola, Russian Lapland	26	Kihlman 1890
S. hastata (?)	Kola, Russian Lapland	14	Kihlman 1890
S. herbacea	Vassijaure, Sweden	35	Wijk ms.
S. lanata (G?)	Kola, Russian Lapland	> 40	Kihlman 1890
S. lapponum (R)	Stigstuv, Norway, wet meadow	25	Callaghan 1973

S. lapponum (?)	Kola, Russian Lapland	39	Kihlman 1890
S. myrsinites (G?)	Kola, Russian Lapland	99	Kihlman 1890
S. planifolia (R)	Niwot Ridge, USA, meadow	50	Callaghan 1973
S. pulchra (G)	Barrow, Alaska, meadow	57	Callaghan 1973
S. reticulata (R)	Stigstuv, Norway, dry meadow	65	Callaghan 1973
Vaccinium uliginosum (?)	Kola, Russian Lapland	59	Kihlman 1890
V. vitis-idaea (R)	Kevo, Finland, heath	109	Callaghan 1973
Vascular Cryptogams			
Lycopodium annotinum (R)	Abisko, Sweden, forest tundra	21	Callaghan 1980
L. annotinum (G)	Finland, forest	250	Oinonen 1968
Herbs			
Alchemilla glomerulans (R)	W. Greenland, herb slope	16	Lewis et al. 1972
Cerastium cerastoides (?)	N.E. Greenland, fell field	> 10	Jolls 1982
Oxyria digyna (G)	W. Greenland	18–50	Humlum 1981
O. digyna (G)	Norway, moraines	13–17	Humlum 1981
Polygonum viviparum ("G")	Devon Island, meadow	26	Callaghan and Collins 1981
Saxifraga nivalis ("G")	N.E. Greenland, fell field	> 10	Wager 1938
S. oppositifolia (G?)	N.E. Greenland, fell field	> 10	Wager 1938
S. rivularis (G)	N.E. Greenland, fell field	> 10	Wager 1938
S. tricuspidata (G)	W. Greenland, forelands	97	Beschel 1963
Sedum lanceolatum (?)	Rocky Mountains USA	> 5	Jolls 1982
Silene acaulis (G)	Rocky Mountains USA	23	Griggs 1956
S. acaulis (G)	Tatra Mountains, Poland	100	Schroeter 1926
S. acaulis (G)	Disko Island, Greenland	56	Beschel 1963
Graminoid tillers			
Carex aquatilis (R)	Barrow, Alaska, meadow	7	Shaver and Billings 1976
Carex bigelowii (R)	Stigstuv, Norway, heath	4	Callaghan 1976
Dupontia fischeri (R)	Barrow, Alaska, meadow	5	Shaver and Billings 1976
Eriophorum angustifolium (R)	Barrow, Alaska, meadow	8	Shaver and Billings 1976
Phleum alpinum (R)	South Georgia, meadow	8	Callaghan 1977
Rostkovia magellanica (R)	South Georgia	10	Callaghan unpublished
Uncinia meridensis (R)	South Georgia, meadow	6	Callaghan 1984

Table 2. Continued

Species	Location	Age (yr)	Reference
Graminoid tufts and tussocks			
Luzula spicata (G)	N.E. Greenland, fell field	> 10	Wager 1938
Phleum alpinum (G)	South Georgia, fell field	20	Heilbronn and Walton 1984
Trisetum spicata (G)	N.E. Greenland, fell field	> 20	Wager 1938
Mosses			
Hylocomium splendens (R)	Abisko, Sweden, forest tundra	1	Callaghan et al. 1978
Polytrichum alpestre (R)	Signy Island, moss banks	12	Collins 1976
P. commune (R)	Abisko, Sweden, forest tundra	6	Callaghan et al. 1978
Lichens			
Rhizocarpon spp. (G)	West Greenland	4500	Beschel 1958
Rhizocarpon tinei (G)	Disko Island, Greenland	1550	Beschel 1963

distributions have been determined by past events of natality and mortality and, with considerable caution, these patterns may in some cases enable predictions to be made about future likely events and general fluctuations in population composition (Hett and Loucks 1976, Kawano et al. 1982). These predictions may be made by inferring probabilities of survival and death of individuals of each age-class *assuming* that proportions of individuals entering and leaving the population are constant from year to year. In tundra areas this method of modelling population flux is particularly important as few observations of actual survival and mortality have been made (Wager 1938, Bonde 1968). An intermediate method of analysing population change was developed by Callaghan (1976, 1984) whereby individuals which died at a known age were recorded from preserved specimens in the field, during the past twelve-year period.

Distributions arising from sexual reproduction

Age-class distributions arising from sexual reproduction have been documented by Wager (1938) for several species from East Greenland, by Bonde (1968) for *Trifolium nanum* from Colorado, by Humlum (1981) for several populations of *Oxyria digyna* from Disko Island, West Greenland and Norway, and for *Ledum palustre*, *Eriophorum vaginatum* and *Empetrum nigrum* from Alaska (McGraw and Shaver 1982). Typically, there is a very high frequency of the youngest age-class individuals and then a constant but high rate of decrease in frequency (Fig. 4a) which, assuming a steady state, is equivalent to mortality. On unstable and sparsely vegetated glacial deposits nearly 70% of the individuals of *Oxyria digyna* were in the year 0 age-class whereas only 14% were in the year 1 age-class; maximum longevity was 13 to 17 years. In contrast, *Oxyria digyna* had herb-field populations with only 35% of individuals in the year 0 age-class and ca. 3% were in the 18 to 50 year age-class (Humlum 1981). An intermediate stable snow-bed site with 50 to 60% vegetation cover had ca. 10% of individuals in each of the year 0 to year 4–5 age-classes. Thereafter, there was a constant but low rate of mortality until the 18 to 50 year age-class. Humlum (1981) concluded that *Oxyria digyna* was an 'arctic-alpine weed' of open spaces which could not become established in closed vegetation.

Luzula spicata (Wager 1938) and *Ledum palustre* (McGraw and Shaver 1982) also showed the trend of age-class distribution typical of sexual reproduction but seedlings of *Chionochloa rigida* from New Zealand showed high survival of 42–50% after 3 years (Mark 1965).

Distributions arising from vegetative reproduction

Tillering. Few observations have been published on the age specific survival of tillers produced by tundra graminoids (Callaghan 1976, 1977, 1984, Lawrence et al. 1978, Chapin et al. 1980). However, the growth of new branches of some moss species (e.g., *Polytrichum* spp.) may be regarded as tillering (Collins 1976, Callaghan et al. 1978, Callaghan and Collins 1981).

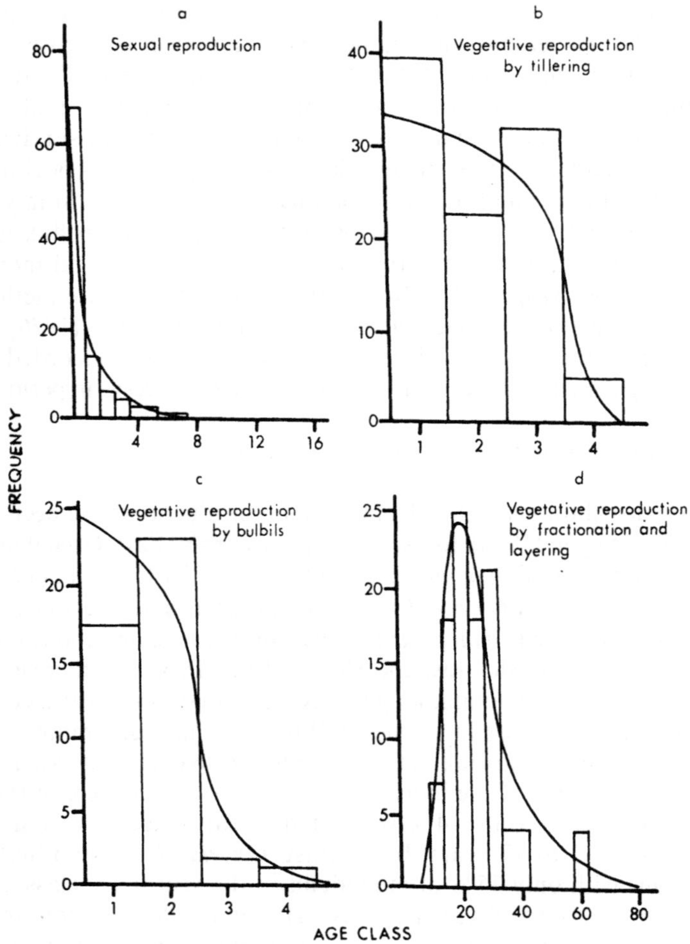

Fig. 4. Types of age class distribution associated with various methods of reproduction. (a) *Oxyria digyna* from Fanneraki Breen, Norway (Humlum 1981). (b) *Carex bigelowii* from Hardangervidda, Norway (Callaghan 1976). (c) *Polygonum viviparum* from Group 6 sites (dense vegetation), Greenland (Petersen 1981). (d) *Salix arctica* from Devon Island (Callaghan 1973). The generalised curves are fitted by hand.

The age-class distributions arising from tillering typically suggest a low mortality of young individuals and a short period of high mortality of old individuals associated with natural senescence (Fig. 4b). In *Carex bigelowii* from a lichen heath in Norway approximately 30% of the individuals occurred in each of the first three annual age-classes with a decrease to 5% in year 3 (Callaghan 1976). Similarly in *Dupontia fischeri* from Alaska between 18 and 30% of the individuals were found in each of the first 4 age-classes and only 3% were found in the last age-class (Chapin et al. 1980). *Alchemilla glomerulans*, from a herb-field on Disko Island, West Greenland reproduces from buds on the rhizome and it showed a slow linear decrease in the % frequency of individuals over the first 5 years followed by a negligible decrease until natural senescence became important in year 11 (Lewis et al. 1972). These patterns are typical of individuals with a well-determined life-span (e.g., man) (Pearl 1928, Deevey 1947).

Within the basic pattern, the biotic and abiotic environment affects rates of mortality. Rates of mortality of the moss *Polytrichum alpestre* from the maritime Antarctic were lower in pure turf than in a mixture with *Chorisodontium aciphyllum* (Collins 1976). Grazing had little effect on the rate of mortality of *Dupontia fischeri* (Lawrence et al. 1978).

Bulbils. *Polygonum viviparum* is a ubiquitous arctic-alpine species which colonises a wide range of habitats from severe fell-fields to dense herb-meadows (Callaghan and Collins 1981, Petersen 1981). The production of viable seed from this species is extremely rare.

Age-class distributions for *Polygonum viviparum*, which reproduces entirely from bulbils, have been presented by Petersen (1981). Bulbils, which are intermediate in size between seeds and lateral buds, show some age-class distributions which are intermediate between seedlings and lateral tillers (Fig. 4c). In general, a large percentage of individuals survive until the older age-classes: e.g. 60% of the individuals from a competitive situation in the dense vegetation of herb-fields survived after the first two years (Fig. 4c, after Petersen 1981). As site severity increases there appears to be a more even distribution of individuals between age-classes, until at the most extreme sites the recruitment of individuals into the population becomes intermittent (see below).

Fractionation and layering. Many dwarf shrubs (Callaghan and Collins 1981), *Lycopodium annotinum* (Callaghan 1980), *Empetrum hermaphroditum* (Emanuelsson 1980), *Andromeda polifolia* and *Vaccinium vitis-idaea* (Flower-Ellis personal communication) and *Juncus scheuchzerioides* (Fig. 9.8 in Callaghan and Collins 1981) reproduce by fractionation or layering, a process whereby side-branches develop axillary roots and become detached from the main axis when the old tissues at the junction of the side and main branch die. Consequently, the side branch is 'old at birth', that is, it has attained considerable age before it can be classed as an independent unit or new individual. This process leads to a gap in the age-class distribution as young individuals are either completely absent or occur at lower frequencies than older individuals (Fig. 4d). In *Salix pulchra* from Point Barrow, Alaska and *S. arctica* from Devon Island no individuals younger than 10 years were found (Callaghan 1973) and Warren Wilson (1964) was unable to find individuals of *S. arctica* younger than 4 years on Cornwallis Island.

A similar anomaly arises in the age-class distributions of *Calluna vulgaris* from moorland in the north of England (Forrest 1971, Forrest and Smith 1975), and *Andromeda polifolia* and *Vaccinium vitis-idaea* from Swedish Lapland (Flower-Ellis personal communication).

This type of age-class distribution emphasises the need for caution in extrapolating from age-class distributions to probabilities of survival and death. A cursory appraisal of the age-class distribution illustrated in Fig. 4d would lead to a prediction of the failure of recruitment and the imminent extinction of the population. As shown above, this is not the case.

After the statistical mode of frequency of individuals within a particular age-class has been attained by this group of species, there is a constant risk of mortality characteristic of many plant species (Harper 1967).

Distributions resulting from severe environmental conditions

Age-class distributions resembling the one just discussed above are common in tundra areas, particularly distributions showing low or zero frequencies in the youngest age-class. Most of the species investigated by Wager (1938) showed this pattern during 1936 (Fig. 5b). In the previous year (1935) there was however a greater abundance of the youngest age

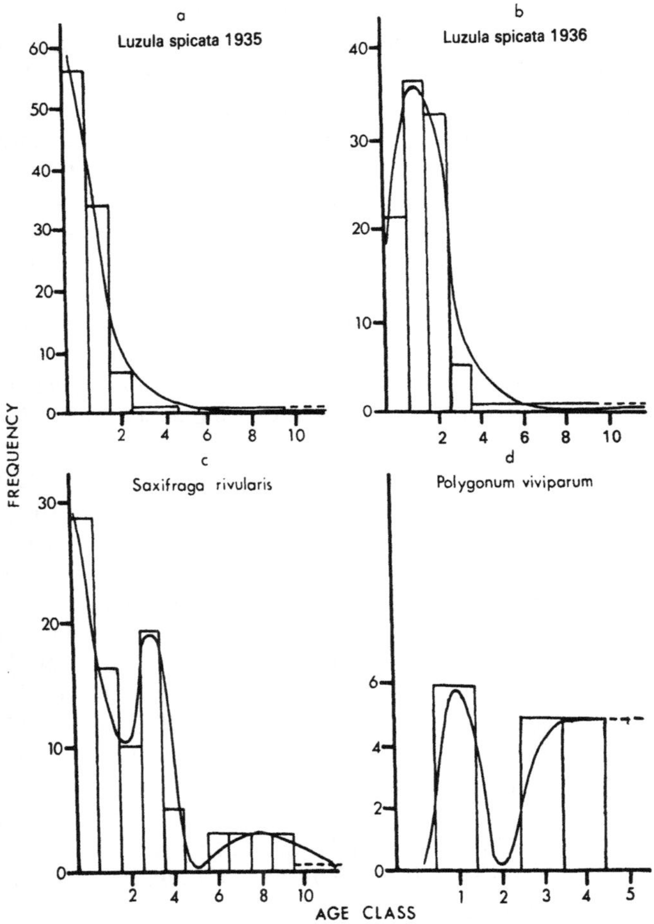

Fig. 5. Types of age-class structure, showing intermittent recruitment, associated with environmental extremes. (a) and (b) *Luzula spicata* and (c) *Saxifraga rivularis* from East Greenland (Wager 1938), (d) *Polygonum viviparum* from Disko Island, West Greenland, Site 10 (Petersen 1981).

class of *Luzula* (Fig. 5b). It would appear, therefore, that extreme environmental conditions exert their main effects on the recruitment of young individuals into a population either by reducing the frequency of flowering and seed set or by killing the youngest individuals in the population. In the case of *Luzula*, the low frequency of the youngest individuals in 1936 was the result of an excessive scattering of seed in the very dry snow-free autumn of 1935 (Wager 1938). Out of the 110 plants present in 1935 (all age-classes) 48 individuals had either died by July 1936 or had been blown away or buried – 34 of these individuals were from the year 0 age-class of 1935. Wager concluded that death resulted from carbohydrate starvation in slow-growing individuals whose small leaf areas were particularly susceptible to wind and ice damage. Low frequencies of the youngest seedlings of *Eriophorum vaginatum* and *Empetrum nigrum* were also recorded by McGraw and Shaver (1982) while Heilbronn and Walton (1984) showed that young graminoid plants are more susceptible to death from frost heave in solifluction areas compared with older and deeper rooted plants.

Indications of past unfavourable growing seasons can sometimes be detected from an age-class distribution (Fig. 5c) such as that of *Saxifraga rivularis* (Wager 1938), where two periods of low recruitment are evident. In particularly severe environments, recruitment may not occur in certain years (Bell and Bliss 1980) and the age class-distribution is disjunct (Fig. 5d). This pattern is shown by *Polygonum viviparum* in a gravelly depression on Disko Island which is subject to solifluction (Petersen 1981).

DEMOGRAPHY AND ECO-PHYSIOLOGY

The basic physiological processes within a plant vary according to the patterns of development within a growing season and over a life-span. Relationships between physiology and seasonal development are well documented (e.g., Tieszen et al. 1980, Chapin et al. 1980, Wielgolaski et al. 1975). However, physiological processes can be studied over much longer time scales and their interpretation within a life cycle framework should enable trends – or even reasons – for survival, mortality and natality to be understood.

Age related trends of root growth and nutrient uptake

Up to 98% of the living biomass of wet graminoid tundra areas may be below ground (Billings et al. 1978) and the processes within this part of the vegetation are particularly important. Root and rhizome growth of tundra graminoids at Barrow, Alaska takes place in soils where mean daily temperatures rarely exceed 7°C in the upper 10 cm (Billings et al. 1978) and roots of *Eriophorum angustifolium*, *E. vaginatum* and *Arctagrostis latifolia* grow at the surface of the permafrost as this thaws during summer (Bliss 1956, Shaver and Billings 1975, Billings et al. 1976, 1977). Roots of *E. angustifolium* survive for only one year whereas those of *Carex aquatilis* may survive for up to 10 years, which is longer than the life-span of the shoot. *Dupontia fischeri* roots produce 90% of their laterals in the first year and live for 5–6 years (Shaver and Billings 1975). In the polar semi deserts of the Canadian High Arctic Islands only 10–52% of live biomass was found underground (Bell and Bliss 1978a). *Phippsia algida* possesses roots which are tightly spiralled and can stretch during soil movement. In the sub-arctic birch forests of Abisko, Sweden Headley et al. (in press) identified living roots of *Lycopodium complanatum* up to 17 years old while 12 year old roots of *L. annotinum* showed active uptake and translocation of P^{32}. Three to four year-old roots of *Lycopodium* showed the greatest activity in taking up P^{32} and nitrate and translocating them both apically and distally within the plant (Headley et al., in press).

Age related trends of nutrient allocation

Once inside plant tissues, nutrients tend to accumulate according to their function and the physiological activity of the tissue. In *Lycopodium annotinum* nitrogen, phosphorus and potassium concentrations were highest in the youngest individual growth modules and they increased as the total life spans of the segment type decreased (Callaghan 1980). Thus highest concentrations were generally found in the annual strobili where protein synthesis, energy transfer and general physiological activity would be greatest and lowest concentrations were in modules of the same age-class on horizontal branches with a life-span of over 20 years. In contrast, calcium, a non-mobile element which accumulates in cell walls, increases in concentration as the module age increases and highest concentrations occur in horizontal segments over 20 years old (Callaghan 1980). Similar trends have recently been recorded for *L. selago* (Headley et al., in press) and various tundra mosses and dwarf shrubs (Malmer and Nihlgård 1980).

Age related trends of energy fixation and storage

Surprisingly few data are available on the relationship between photosynthetic rates and the age of the photosynthesising tissues. Karlsson (1982) showed that the photosynthetic rate of one year-old leaves of *Vaccinium vitis-idaea* was 66% of that during the first year; in the subsequent three years the rate decreased by 10% per year and at the age of five years the rate was 25% of current year leaves. In a comparison between the evergreen *V. vitis-idaea* and the deciduous *V. uliginosum*, which co-exist in much of the Fennoscandian forest tundra, Karlsson (1982) found that similar photosynthetic capabilities were achieved by two contrasting patterns: in *V. vitis-idaea* low momentary photosynthetic rates were compensated by a long life-span and the ability to use marginal parts of the growing season when *V. uliginosum* is leafless.

Early season photosynthesis in Alaskan mosses is carried out by tissues from the previous year, or the year before this, but their rates are lower than those of current year tissues (75 and 40% respectively: Collins and

Oechel 1974). Initial reductions in photosynthetic rates with age were even greater in *Hylocomium splendens* and *Polytrichum commune* from Swedish Lapland as rates of one year old tissues were 55 and 58% of the rate of current year's tissues respectively (Callaghan et al. 1978).

Once inside the interconnected tiller systems of tundra graminoids, photo-assimilate is preferentially allocated to the main growing points, i.e. root tips, young tillers and flowering structures (Allessio and Tieszen 1975a, 1975b, Callaghan 1977) but there may be an important inter-dependence between vegetative offspring and inflorescences whereby inflorescences export up to 23% of their fixed carbon to vegetative offspring (Callaghan 1977, 1984). The interdependence of graminoid tillers in the tundra is important (Allessio and Tieszen 1975b), since the probability of survival of an inflorescence or vegetative offspring may be enhanced by subsidies of energy (Callaghan 1984).

Some of the energy fixed in photosynthesis is stored in the form of non-structural carbohydrates. The concentration of these are highest in the youngest tissues of *Carex aquatilis*, *Eriophorum angustifolium* and *Dupontia fischeri* (Shaver and Billings 1976) and in *Lycopodium annotinum* (Callaghan 1980). Long-lived structures, such as rhizomes of *Carex aquatilis*, stem bases of *Eriophorum angustifolium* (Shaver and Billings 1976) and horizontal modules of *Lycopodium annotinum* (Callaghan 1980) tend to invest dry weight in structural tissues but shorter-lived units maintain high concentrations of non-structural carbohydrates useable in new growth. Fats are also an important form of energy stored in tundra plants (Bliss 1962, Karunen and Kallio 1976) which represent a greater concentration of energy (35.6 KJ g^{-1} dry weight) than soluble carbohydrates (15.7–17.5 KJ g^{-1} dry weight) (Verduin 1972). In *Lycopodium annotinum* fats show the highest concentration and total weights in mature segments (Callaghan 1980).

REPRODUCTION

Sexual reproduction

Many authors (e.g., Porsild 1951, Bliss 1971, Billings and Mooney 1968) have discussed the dominance of vegetative reproduction in the tundra in general. However, seed production is the main form of reproduction in the open vegetation of the High Arctic, on fell-fields and on the fore-fields of retreating glaciers. On King Christian Island (High Arctic), 50% of 27 species of vascular plants, out of a total flora of 35, were observed by Bell and Bliss (1980) to produce viable seed while an additional 25% contained well-developed but unripe seed during 1974 and 1975. Only 33% of the 27 species studied show vegetative reproduction.

In such areas, the vegetation consists of isolated individuals surrounded by large areas of bare ground and colonisation in these areas is mainly by seed. *Silene acaulis* reproduces by seed and 80–90% of the colonising individuals in the Rocky Mountain National Park were seedlings of this

Table 3. Reproduction in tundra plants

Species	Sexual reproduction						
	Inflorescence weight (mg)	Investment of weight in inflorescence (%)	Seed weight (mg)	Potential reproductive capacity (flowers/plant)	Actual reproductive capacity (seeds/plant)	Seed/m²	Probability of survival %
Hylocomium splendens	–	–	–	–	44**	–	0
Polytrichum commune	–	–	–	–	754**	–	>0
Lycopodium annotinum	18**	ca. 10	–	–	400,000***	1.8×10^8	>0
Eriophorum vaginatum	91	3.5	0.51	–	–	11	–
Carex bigelowii	93	2.8	0.97	–	–	43	–
Rostkovia magellanica							
Pioneer tillers	–	–	–	–	} 65	–	–
Colonizing tillers	–	–	–	–		–	–
Uncinia meridensis	17.5	6	< 1.6	–	11	712	0
Carex bigelowii	40	7.7	< 2.2	–	18	169	0
Phleum alpinum							
Fell field	19.4	39	–	77	22	275	>0
Meadow	241	29	–	786	0	0	0
Festuca contracta							
Exposed	–	2.7	0.71	–	61	9514	–
Sheltered	–	5.2	0.94	–	82	39565	2.3
Oxyria digyna							
Fell field	–	–	–	–	200	200	7
Meadow	–	–	–	–	305	336	2
Polygonum viviparum							
Arctic meadow (Barrow)	–	–	–	>0	0	0	0
Alpine meadow (Norway)	–	–	–	>0	0	0	0
Betula pubescens ssp. *tortuosa* (70 yr. reproductive output)	–	–	–	–	4×10^7	34,000	>0
Betula nana	22	0.4	0.15	–	–	< 1	–
Vaccinium uliginosum	–	–	0.24	–	–	25	–
Rubus chamaemorus	–	–	7.98	–	–	5	–
Vaccinium vitis-idaea	–	–	0.17	–	–	158	–
Empetrum nigrum	–	–	0.97	–	–	7	–
Ledum palustre	–	–	0.02	–	–	650	–

*This value represents the production of more than 1 generation of offspring.
**These values refer to sporophytes.
***Spores per strobilus.
NB. Probability of survival is calculated until sexual maturity.

species (Griggs 1956). The colonising species *Potentilla tridentata*, *Diapensia lapponica*, *Juncus trifidus* and *Arenaria groenlandica* produce between 2,500 and 60,000 seed/m^2. In similar situations, some dwarf shrubs such as *Salix arctica* (Beschel 1963) and *Empetrum hermaphroditum* (Emanuelsson unpublished) reproduce exclusively by seed whereas in closed vegetation adventitious roots are produced from lateral branches and fractionation results in vegetative reproduction.

Although seed output may occur extensively in some species and habitats, it is generally low but the longevity of polycarpic species compensates for this (Jolls 1982). Monocarpic species such as the sedges often have particularly low seed output, as low reproductive capacities (Table 3) may combine with many years of development of tillers (see Table 2) before flowering takes place (e.g., Fig. 1). Species at their distributional limit (e.g., *Phleum alpinum* on Disko Island, West Greenland: Callaghan

Vegetative reproduction					Reference
Vegetative offspring wt. (mg)	Offspring wt. as % of fully grown shoot	Vegetative reproductive capacity (No./plant)	Vegetative offspring/m²	Probability of survival %	
1.1	10	0.1	–	< 100	Callaghan et al. 1978
–	–	0.3	–	–	Callaghan et al. 1978
–	–	–	–	–	Plotnikov 1977, Callaghan 1980
–	–	–	–	–	Chester and Shaver 1982
–	–	–	–	–	Chester and Shaver 1982
					Callaghan 1977
39	26	2.4	–	92	
4	3	0.2	–	91	
12	10	2.3	191	79	Callaghan 1977, 1984
–	–	1.8	34	43	Callaghan 1976
–	–	2.03	50	97	Callaghan and Lewis 1971
–	–	3.97	30	44	Callaghan 1974
–	–	7*	–	–	Tallowin 1977a, b
–	–	14*	–	–	
–	–	–	0	–	Humlum 1981
–	–	–	0	–	
0.9 (1.5)	13 (2)	11.5 (14)	18.2 (14)	< 58	Callaghan and Collins 1981
2.1 (3.9)	30 (18)	40.2 (21)	728 (17870)	< 14	(Petersen 1981)
–	–	> 0	> 0	> 0	Nordell unpublished
–	–	–	–	–	Chester and Shaver 1982
–	–	–	–	–	Chester and Shaver 1982
–	–	–	–	–	Chester and Shaver 1982
–	–	–	–	–	Chester and Shaver 1982
–	–	–	–	–	Chester and Shaver 1982
–	–	–	–	–	Chester and Shaver 1982

1974) or in particularly severe habitats may set seed only infrequently. Bell and Bliss (1980) recorded complete failure of seed-set during 1974 from the 26 flowering species which they studied in the High Arctic. Savile (1972) recorded only 2 to 10 fruiting stalks of *Carex misandra* and *C. membranacea* per 100 m² on Somerset Island and observed that *Cardamine pratensis* never set seed in the High Arctic. Tree-line spruces at Keewatin also produced no seed and have existed as a clone since the end of the Hypsithermal Interval (Bryson et al. 1965). Fell-field or exposed populations of *Phleum alpinum* (Callaghan 1974), *Oxyria digyna* (Humlum 1981) and *Festuca contracta* (Tallowin 1977a) showed much lower reproductive capacities than more sheltered populations (Table 3), while seed size of *Festuca contracta* in the exposed habitat was 25% lower than in the sheltered situation (Tallowin 1977b; Table 3) and viability may also decrease with site severity (Tallowin 1977b, Mark 1965).

One of the main limitations to seed production in the tundra is the shortness of the growing season. *Phleum alpinum* can produce viable seed if the growing season is about 170 days (Callaghan 1974). Consequently, successful flowering species show pre-formation of flowering primordia (Sørensen 1941, Hodgson 1966, Bell and Bliss 1980, Mark 1970). *Puccinellia vaginata* may initiate flowers at least 3 years before their seed mature (Bell and Bliss 1980). The fast development of the fruiting cycle also combats the short tundra growing season in higher plants (Porsild 1951) and mosses (Clarke and Greene 1970) while seeds generally lack dormany mechanisms and are opportunists (Billings and Mooney 1968).

In general, the investment of dry weight into flowering structures is low (Table 3) but the actual cost of sexual reproduction may be very high because of low survival probabilities. Chapin et al. (1980) calculated that the carbon cost of producing a tiller of *Dupontia fisheri* from seed at Point Barrow, Alaska, was 10,000 times greater than that of producing one vegetatively, whereas it can be calculated from data of Nordell (pers. comm.) that a single tree of *Betula pubescens* ssp *tortuosa* in Swedish Lapland would produce 40,000,000 seeds throughout its life span, only one of which is necessary to replace the original tree for a stable population.

In closed tundra vegetation where both sexual and vegetative reproduction occur, the role of sexual reproduction is often difficult to interpret. This is because sexual reproduction may be completely unimportant at the site of seed production or at the time when they are produced. The vast output of seed from the tundra birch forests is probably only important in the colonisation of new areas outside the forest or after disturbance such as fire within the forest. Large inflorescences of *Phleum alpinum* at its geographical limit (Table 3) appear wasteful as no seed is set in most years. Large potential reproductive capacities are however essential for sufficient seedling recruitment in those (infrequent) growing seasons which are long enough to allow seed set (Callaghan 1974).

Some species of the tundra exist as clones which may be hundreds (e.g. *Lycopodium* species: Oinonen 1968) or even thousands of years old. It can be roughly calculated from data on spore production (0.4×10^6/strobilus and $1.88 \times 10^6/\mathrm{dm}^2$: Plotnikov 1977) and the rate of spread of clones of *Lycopodium annotinum* (an increase in radius of 22 cm/year; Oinonen 1968) that a 250 year-old clone will have produced 1.5×10^{14} spores but only one asexual spore together with one sexual spore are required to initiate these long-lived and extensive clones. As successful recruitment from sexual reproduction may happen so infrequently, it is obviously important to interpret sexual reproduction in an extremely wide context. Sexual reproduction in many circumstances may be adapted to the exceptional and not to the normal circumstances. Thus the environmental conditions which select for sexual reproduction may be very different from those which select for vegetative growth: the sporophyte and gametophyte generations of the Pteridophytes, for example, are completely different morphologically, physiologically and ecologically.

Vegetative reproduction (excluding vivipary)

In many areas, particularly in closed vegetation, successful establishment from seed is negligible (Table 3). Vegetative reproductive capacities are usually low when compared with sexual reproductive capacities but the chance of a vegetative propagule reaching sexual maturity is often very high (Table 3) resulting from a period of 'post natal care' which is absent in physically and physiologically isolated seeds.

In the rhizomatous sedges in particular, a lateral tiller may develop to 39% of its parent tiller's dry weight (Table 3) before initiating an independent root and leaf system. The investment of dry weight into winter buds of *Rubus chamaemorus* was 4 times that into flowers (Flower-Ellis 1980). During the initial period of ramet growth, the water, nutrients and carbon required by the offspring are provided exclusively by the parent, sometimes even by the parent's inflorescence (see above) and other related tillers.

Division of labour may exist between sibling tillers (Table 1) which may show different survival probabilities and it is likely that seeds within an inflorescence will also show different survival probabilities, independently of genotype, as seed weight may vary by 33% according to the position of the seed in the inflorescence (Tallowin 1977b). As selection in the tundra environment acts mainly on the youngest individuals of a population, the evolution of large propagules subsidised by parent tillers is understandable.

Dispersal to safe microsites may occur as a result of subsidised growth until rooting or leaf production is possible. If a propagule becomes shaded or defoliated after energy subsidies from the parent have ceased, then these subsidies may, presumably, be reinstated as in some temperate grasses (Gifford and Marshall 1973). The survival of the apices of horizontal branches of *Lycopodium annotinum* (see Fig. 2) is dependent upon the possibility for roots to be initiated (Fig. 6A). After apical death, lateral branches change the direction of the clone which develops in the direction where roots may grow (Fig. 6B). When few constraints to root development exist, the direction of the clone is determined by microtopography (Svensson and Callaghan unpublished; Fig. 6C).

Many generations of ramets may often remain in physical and physiological contact in tundra areas. This enables the formation of extensive clones which have a well-developed transport system for nutrients and carbon. Indeed, in some graminoids, below ground parts survive far longer than shoots (Callaghan 1976, Billings et al. 1978) and translocations into these shootless rhizomes often occurs (Callaghan 1977, 1984).

The well-developed transport system between generations within a clone allows the translocation of energy and nutrients from senescing individuals to growing points, thus by-passing loss in the decomposition pathway (Callaghan 1980). This conservation of nutrients and energy by recycling between generations is particularly important in the low nutrient and energy environment of the tundra, and obstacles such as rock outcrops may be traversed by young generations of species such as *Lycopodium annotinum* and *Empetrum hermaphroditum* dependent upon established older generations.

Rates of vegetative reproduction, like those of sexual reproduction, depend upon aerial and soil environments. The vegetative reproductive capacity of *Phleum alpinum* varied from 1 tiller per tiller in native soil on

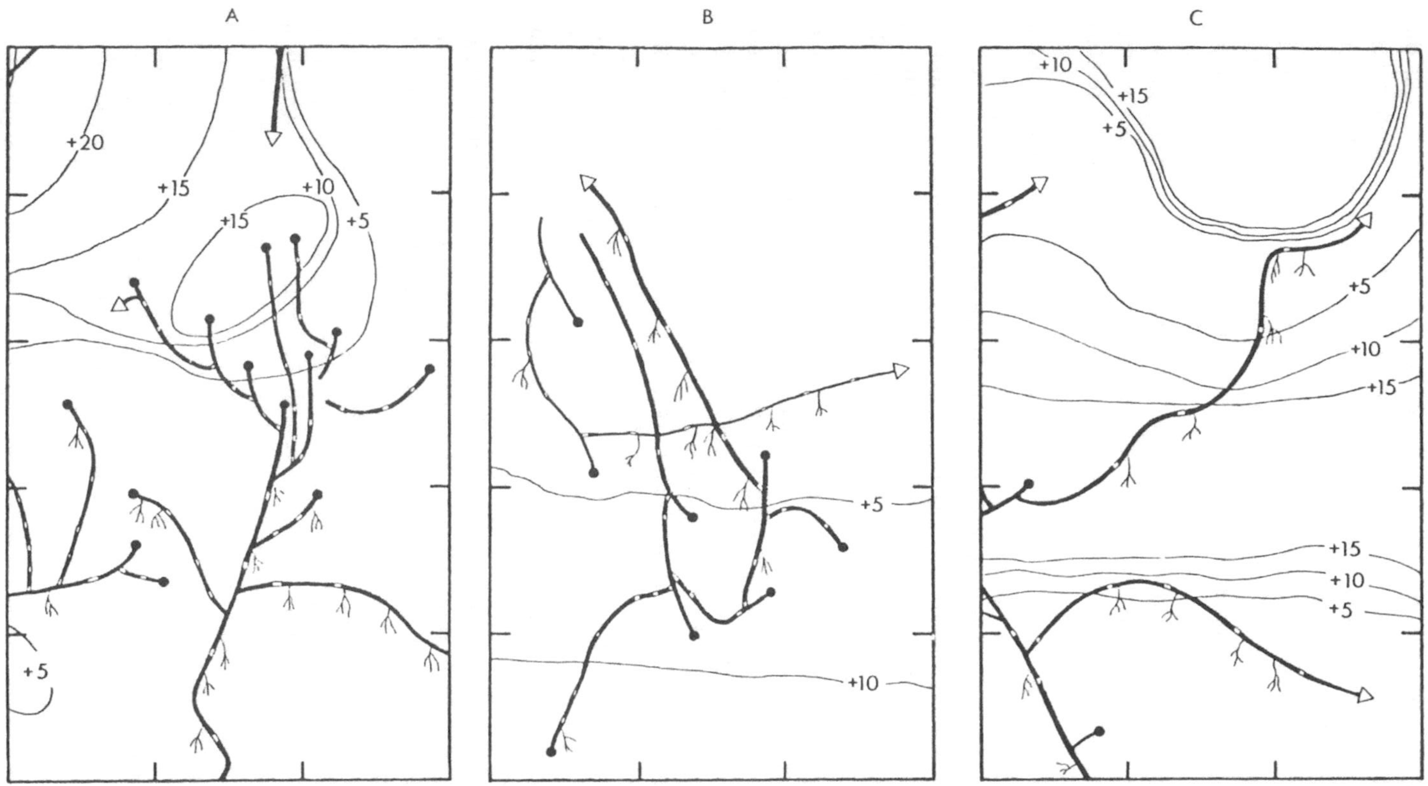

Fig. 6. Map of horizontal rhizomes and roots of *Lycopodium annotinum* in Swedish Lapland. Grid squares are 10 × 10 cm, and the contours (thin lines) are relative. Gaps in the branches denote annual markers of growth, round solid circles denote dead apices and open arrows denote living apices. A shows death of apices and subsequent increased branching in relation to failure of root production. B shows apical death due to root failure and changed direction over an 11 year period. C shows successful development, apical survival and direction related to microtopography.

South Georgia to 6.5 in a non-limiting soil medium, whereas an altitudinal gradient of 218 m from dense vegetation to the barren fore-field of a glacier resulted in a difference of only 7.2 to 4.7 tillers per tiller (Callaghan and Lewis 1971).

Vivipary

Vivipary is a mode of reproduction intermediate between sexual and the vegetative reproduction discussed above. The propagules are intermediate between seed and tillers in size, number and survival potential (Table 3, Fig. 4). Plantlets of *Festuca vivipara* contained between 3 and 4 times the amounts of nutrients and carbohydrates as the seed of the related species *Festuca ovina* as a result of larger size (Harmer and Lee 1978a). Plantlets were only visible at the end of the growing season but they began growth immediately on contacting a moist surface and could survive the cold and high humidity of a winter in upland Britain (Harmer and Lee 1978b). In contrast, seed of *Festuca ovina* did not germinate until the following spring.

This opportunism, combined with the lack of prolonged development required in the tundra flowering cycle makes vivipary particularly successful in harsh and disturbed environments. On King Christian Island in the High Arctic, 2 out of 27 vascular species were viviparous and bulbils of *Saxifraga cernua* were the only propagules to be successfully produced during the short summer of 1974 (Bell and Bliss 1980). In the fore-field vegetation of the Kårsa Glacier in Swedish Lapland, 20% of the vascular species were viviparous and they were often associated with habitats of high 'stress' and high disturbance – indeed vivipary may be adapted to this combination of environmental features. It is not surprising, therefore, that viviparous species, particularly *Polygonum viviparum*, are ubiquitous in tundra areas (Callaghan and Collins 1981, Petersen 1981) although the failure of sexual reproduction (Engell 1973) must render this existence precarious in the long term.

FOREST TUNDRA – A CASE STUDY OF INTERACTIONS BETWEEN POPULATIONS

The structure of any ecosystem is the result of complex interactions between its component plant and animal populations. Tundra ecosystems have been considered relatively simple in structure because of their low species diversity, but this makes them more amenable to study. The forest tundra of the Abisko area in Swedish Lapland is such an example. But this ecosystem of the sub-Arctic with its closed vegetation is dependent far more on its animal populations than more extreme tundra ecosystems.

The forest tundra of Abisko is dominated by the polycormic form of *Betula pubescens* ssp. *tortuosa* which has a cover abundance of 2–20% and four types of dwarf shrub heaths (Sonesson and Lundberg 1974) including *Empetrum hermaphroditum* and *Vaccinium vitis-idaea* heaths. The area

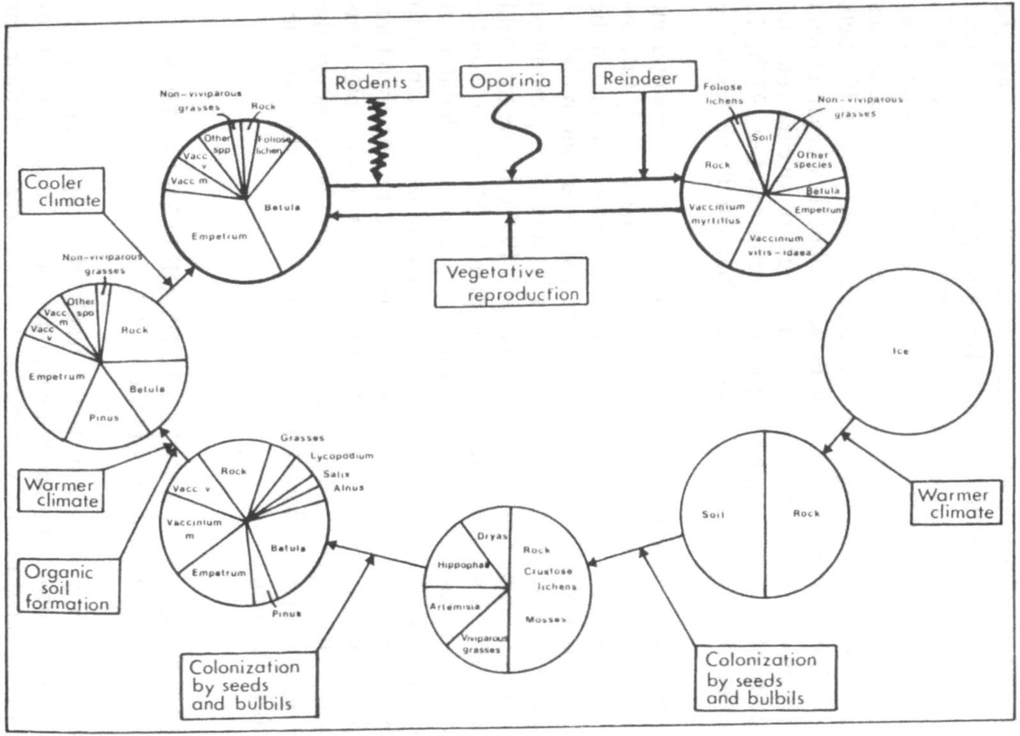

Fig. 7. Model of the development (thin circles) and fluctuations (2 thick circles at top) of the forest tundra ecosystem at Abisko, Swedish Lapland. Data are taken from various sources quoted in the text but the cover of individual species is approximate.

has been deglaciated for about nine thousand years (Sonesson and Lundberg 1974) and the development of the present vegetation (Sonesson 1974) is illustrated in Fig. 7. The present forest structure is, however, constantly varying. The birch tree population varies at the extreme of its distribution in that the altitude of the tree line varies according to mild or severe weather conditions (Sonesson and Hoogesteger 1983). More importantly however, the populations of birch are susceptible to infestations of larvae of the moths *Oporinia autumnata* and *Operophthera brummata* (Fig. 7). These larvae defoliate the birch trees and when their populations are particular high (about every 40–150 years) they kill vast areas of birch forest (Tenow 1975). Following a defoliation event populations of *Deschampsia flexuosa* and *Linnaea borealis* increase (Sandberg 1958). If the defoliation is not too severe, the birch forest may become re-established by the vegetative proliferation of shoots from the base of the trees: alternatively long term devastation may occur (Kallio and Lehtonen 1973).

The size of the populations of foliose lichens (particularly species of *Cladonia* in the ground vegetation of the forests is mainly controlled by stable populations of reindeer (Fig. 7).

One of the surprising features of forest tundra is the vast extent of three dwarf shrubs (*Empetrum hermaphroditum*, *Vaccinium vitis-idaea* and *V. myrtillus*) which co-exist. On careful examination, however, there are continuous fluctuations in their populations and permanent plots show that plants do not have permanent positions year after year (Emanuelsson 1984). Indeed, many of the species, e.g., *Empetrum hermaphroditum*, *Vaccinium vitis-idaea*, *Lycopodium annotinum*, *Linnaea borealis* and *Calamagrostis lapponica* show rhizomatous or stoloniferous growth and 'travel' around the area (see Fig. 6). In addition to the continual movement of individuals, the size of the dwarf shrub population is controlled by the grazing of voles and lemmings (*Cletrynomos ruficanus* and *Lemmus lemmus*). Populations of these two rodents reach high levels approximately every four to five years.

Empetrum populations are most affected by rodent grazing because growth occurs via lateral meristems produced on one year shoots. The only method of regeneration by *Empetrum* in grazed areas is by shoot production and spread is in the order of 3–5 centimetres per year. It may require six to eight years for an area of grazed *Empetrum* to be recolonized. In contrast, *Vaccinium vitis-idaea* can recolonize quickly from deep rhizomes and *V. myrtillus* can regenerate even faster as it posseses buds on shoot bases just below the soil surface. *Festuca ovina* and *Deschampsia flexuosa* populations also increase after severe winter grazing, but unlike the species listed above, their abundance is due to sexual reproduction.

With increasing time after a severe winter grazing event the prolific vegetative growth of *Empetrum* suppresses the growth of *Vaccinium* species and the grasses. Indeed it is probably the rodent grazing which prevents *Empetrum* from totally dominating the forest floor vegetation.

ACKNOWLEDGMENTS

This paper was written at the Abisko Scientific Research Station in Swedish Lapland on funding from the Royal Swedish Academy of Sciences. We are grateful to the Academy and Director of the Research Station, Professor Mats Sonesson, for his hospitality and for commenting on the manuscript. Dr. R.S. Callow also commented on the manuscript. Our thanks also to many colleagues who gave access to material unpublished, or in press and provided valuable discussion. In particular we thank Dr. J.G.K. Flower-Ellis, Dr. S. Wijk, Dr. S. Karlsson, B. Svensson, A. Headley and Dr. J.A. Lee.

REFERENCES

Ahmadjian, V. (1970) Adaptations of antarctic terrestrial plants. In Antarctic Ecology, Vol. 2, M.W. Holdgate (Ed.), pp. 801–811. New York: Academic Press.

Aleksandrova, V.D. (1980) The Arctic and Antarctic: their Division into Geobotanical Areas. Cambridge: Cambridge University Press. 247pp.

Allessio, M.L. and Tieszen, L.L. (1975a) Patterns of carbon allocation in an arctic tundra grass, *Dupontia fischeri* (Gramineae) at Barrow, Alaska. Am. J. Bot. 62: 797–807.

Allessio, M.L. and Tieszen, L.L. (1975b) Leaf age effect on translocation and distribution of C-photoassimilate in *Dupontia fischeri* at Barrow, Alaska. Arct. Alp. Res. 7: 3–12.

Batzli, G.O., White, R.G., Maclean, S.F. Jr., Pitelka, F.A. and Collier, B.D. (1980) The herbivore-based trophic system. In An Arctic Ecosystem. The Coastal Tundra at Barrow Alaska, J. Brown, P.C. Miller, L.L. Tieszen and F.L. Bunnel (Eds.), pp. 335–410. Stroudsberg: Dowden, Hutchinson and Ross Inc.

Bell, K.L. and Bliss, L.C. (1978) Root growth in a polar semi desert environment. Can. J. Bot. 56: 2470–2490.

Bell, K.L. and Bliss, L.C. (1980) Plant reproduction in a high arctic environment. Arct. Alp. Res. 12: 1–10.

Beschel, R.E. (1958) Lichenometrical studies in West Greenland. Arctic 11: 254.

Beschel, R.E. (1963) Observations on the time factor in interactions of permafrost and vegetation. Ottawa. Permafrost Tech. Mem. 76: 43–56.

Billings, W.D. (1974) Adaptations and origins of alpine plants. Arct. Alp. Res. 6: 129–142.

Billings, W.D. (1975) Arctic and alpine vegetation: plant adaptations to cold summer climates. In Arctic and Alpine Environments, J.D. Ives and R.G. Barry (Eds.), pp. 403–443. London: Methuen.

Billings, W.D. and Mooney, H.A. (1968) The ecology of arctic and alpine plants. Biol. Rev. 43: 481–529.

Billings, W.D., Peterson, K.M., Shaver, G.R. and Trent, A.W. (1977) Root growth, respiration, and carbon dioxide evolution in an arctic tundra soil. Arct. Alp. Res. 9: 127–135.

Billings, W.D., Peterson, K.M. and Shaver, G.R. (1978) Growth turnover and respiration rates of roots and tillers in tundra graminoids. In: Vegetation and production ecology of an Alaskian arctic tundra, L.L. Tieszen (Ed.), pp. 415–434. New York: Springer-Verlag.

Billings, W.D., Shaver, G.R. and Trent, A.W. (1976) Measurement of root growth in simulated and natural temperature gradients over permafrost. Arct. Alp. Res. 8: 247–250.

Bliss, L.C. (1956) A comparison of plant development in microenvironments of arctic and alpine tundras. Ecol. Monogr. 26: 303–307.

Bliss, L.C. (1962) Caloric and lipid content in alpine tundra plants. Ecology 43: 753–757.

Bliss, L.C. (1971) Arctic and alpine plant life cycles. Annu. Rev. Ecol. Syst. 2: 405–438.

Bliss, L.C. (1981a) North American and Scandinavian tundras and polar deserts. In Tundra Ecosystems: a Comparative Analysis, L.C. Bliss, O.W. Heal and J.J. Moore (Eds.), pp. 38–46. Cambridge University Press.

Bliss, L.C. (1981b) Summary. In Tundra Ecosystems: a Comparative Analysis, L.C. Bliss, O.W. Heal and J.J. Moore (Eds.), pp. 8–24. Cambridge: Cambridge University Press.

Böcher, T.W. (1975) Det Grønne Grønland. Rhodos, Copenhagen: 256pp.

Bonde, E.K. (1968) Survival of seedlings of an alpine clover (*Trifolium nanum* Torr.). Ecology 49: 1193–1195.

Bryson, R.A., Irving, W.N. and Larsen, J.A. (1965) Radiocarbon and soil evidence of former forest in the southern Canadian tundra. Science 147(3633): 46–48.

Callaghan, T.V. (1973) A comparison of the growth of tundra plant species at several widely separated sites. Institute of Terrestrial Ecology, Merlewood Research and Development Paper No. 53. 52pp.

Callaghan, T.V. (1974) Intra-specific variation in *Phleum alpinum* L. with specific reference to polar populations. Arct. Alp. Res. 6: 361–401.

Callaghan, T.V. (1976) Growth and population dynamics of *Carex bigelowii* in an alpine environment. Strategies of growth and population dynamics of tundra plants, 3. Oikos 27: 402–413.

Callaghan, T.V. (1977) Adaptive strategies in the life cycles of South Georgian graminoid species. In Adaptations Within Antarctic Ecosystems. G.A. Llano (Ed.), pp. 981–1002. Houston, Texas: Gulf Pub. Co.

Callaghan, T.V. (1980) Age-related patterns of nutrient allocation in *Lycopodium annotinum* from Swedish Lapland. Strategies of growth and population dynamics of tundra plants, 5. Oikos 35: 373–386.

Callaghan, T.V. (1984) Growth and translocation in a clonal southern hemisphere sedge, *Uncinia meridensis*. J. Ecol. 72: 529–546.

Callaghan, T.V. and Collins, N.J. (1976) Introduction. Strategies of growth and population dynamics of tundra plants, 1. Oikos 27: 383–388.

Callaghan, T.V. and Collins, N.J. (1981) Life cycles, population dynamics and the growth of tundra plants. In Tundra Ecosystems: a Comparative Analysis, L.C. Bliss, O.W. Heal and J.J. Moore (Eds.), pp. 257–284. Cambridge: Cambridge University Press.

Callaghan, T.V. and Lewis, M.C. (1971) The growth of *Phleum alpinum* L. in contrasting habitats at a Sub-Antarctic station. New Phytol. 70: 1143–1154.

Callaghan, T.V., Collins, N.J. and Callaghan, C.H. (1978) Photosynthesis, growth and reproduction of *Hylocomium splendens* and *Polytrichum commune* in Swedish Lapland. Strategies of growth and population dynamics of tundra plants 4. Oikos 31: 73–88.

Cernusca, A. and Seeber, M.C. (1981) Canopy structure, microclimate and the energy budget in different alpine communities. In Plants and Their Atmospheric Environment, J. Grace, E.D. Ford and P.G. Jarvis (Eds.), pp. 75–81. Oxford: Blackwell.

Chabot, B.F. (1979) Metabolic and enzymatic adaptations at low temperature. In Comparative Mechanisms of Cold Adaptation, L.S. Underwood, L.L. Tieszen, A.B. Callahan and G.E. Folk (Eds.), pp. 282–301. New York: Academic Press.

Chapin, F.S. III (1974) Morphological and physiological mechanisms of temperature compensation in phosphate absorbtion along a latitudinal gradient. Ecology 55: 1180–1198.

Chapin, F.S. III, Tieszen, L.L., Lewis, M.C., Miller, P.C. and McCown, B.H. (1980) Control of tundra plant allocation patterns and growth. In An Arctic Ecosystem. The Coastal Tundra at Barrow, Alaska, J. Brown, P.C. Miller, L.L. Tieszen and F.L. Bunnell (Eds.), pp. 140–185. Stroudsberg: Dowden, Hutchinson and Ross Inc.

Chapin, F.S. III, Van Cleave, K. and Chapin, M.C. (1979) Soil temperature and nutrient cycling in the tussock growth form of *Eriophorum vaginatum*. J. Ecol. 67: 169–189.

Chester, A.L. and Shaver, G.R. (1982) Reproductive effort in cotton grass tussock tundra. Holarct. Ecol. 5: 200–206.

Clarke, G.C.S. and Greene, S.W. (1970) Reproductive performance of two species of *Pohlia* at widely separated stations. Trans. Br. bryol. Soc. 6: 114–128.

Collins, N.J. (1976) Growth and population dynamics of the moss *Polytrichum alpestre* in the maritime antarctic. Strategies of growth and population dynamics of tundra plants, 2. Oikos 27: 389–401.

Collins, N.J. and Oechel, W.C. (1974) The pattern of growth and translocation of photosynthate in a tundra moss, *Polytrichum alpinum*. Can. J. Bot. 52: 355–363.

Dagon, R.R. (1966) Tundra – A definition and structural description. Polar Notes. 6: 22–35.

Deevey, E.S. (1947) Life tables for natural population of animals. Q. Rev. Biol. 22: 283–314.

Emanuelsson, U. (1980) Mechanical impact on vegetation in the Tornetrask area. (Swedish with English summary). Fauna Flora Upps. 75: 37–42.

Emanuelsson, U. (1984) Ecological effects of grazing and trampling on mountain vegetation in northern Sweden. Ph.D. dissertation, Univ. Lund, Sweden. 157pp.

Engell, K. (1973) A preliminary morphological cytological and embryological investigation in *Polygonum viviparum*. Bot. Tidsskr. 67: 305–316.
Flower-Ellis, J.G.K. (1971) Age structure and dynamics of stands of bilberry (*Vaccinium myrtillus* L.) Department of Forest Ecology and Forest Soils, Royal College of Forestry, Stockholm. Research notes Nr. 9, 108pp.
Flower-Ellis, J.G.K. (1980a) Diurnal dry weight variation and dry matter allocation of some tundra plants. 1. *Andromeda polifolia* L. In Ecology of a Subarctic Mire, M. Sonesson (Ed.), pp. 139–162. Stockholm: Ecol. Bull. 30.
Flower-Ellis, J.G.K. (1980b) Diurnal dry weight variation and dry matter allocation of some tundra plants. 2. *Rubus chamaemorus* L. In Ecology of a Subarctic Mire, M. Sonesson (Ed.), pp. 163–179. Stockholm: Ecol. Bull. 30.
Forrest, G.I. (1971) Structure and production of North Pennine blanket bog vegetation. J. Ecol. 59: 453–479.
Forrest, G.I. and Smith, R.A.H. (1975) The productivity of a range of blanket bog vegetation types in the Northern Pennines. J. Ecol. 63: 173–202.
French, D.D. (1981) Multivariate comparisons of IBP tundra biome site characteristics. In Tundra Ecosystems: a Comparative Analysis, L.C. Bliss, O.W. Heal and J.J. Moore (Eds.), pp. 47–75. Cambridge: Cambridge University Press.
Gifford, R.M. and Marshall, C. (1973) Photosynthesis and assimilate distribution in *Lolium multiflorum* following differential tiller defoliation. Aust. J. biol. Sci. 26: 517–526.
Griggs, R.F. (1956) Competition and succession on a rocky mountain fell field. Ecology 37: 8–20.
Grime, J.P. (1977) Evidence for the existence of three primary strategies in plants and its relevance to ecological and evolutionary theory. Amer. Nat. 111: 1169–1194.
Grime, J.P. (1979) Plant Strategies and Vegetation Processes. Chichester, New York: John Wiley and Sons Ltd. 222pp.
Hagg, R.W. (1974) Nutrient limitations to plant production in two tundra communities. Can. J. Bot. 52: 103–116.
Harmer, R. and Lee, J.A. (1978a) The growth and nutrient content of *Festuca vivipara* (L) S.M. plantlets. New Phytol. 80: 99–106.
Harmer, R. and Lee, J.A. (1978b) The germination and viability of *Festuca vivipara* (L.) S.M. plantelets. New Phytol. 81: 745–751.
Harper, J.L. (1967) A Darwinian approach to plant ecology. J. Ecol. 55: 247–270.
Harper, J.L. (1981) The concept of population in modular organisms. In Theoretical Ecology, 2nd Edn, R.M. May (Ed.), pp. 53–77. Oxford: Blackwells.
Haukoja, E. (1980) On the role of plant defences in the fluctuation of herbivore populations. Oikos. 35: 202–213.
Headley, A.D., Callaghan, T.V. and Lee, J.A. (In press). Phosphorus economy throughout the life span of *Lycopodium annotinum* from Swedish Lapland. Oikos.
Heilbronn, T.D. and Walton, D.W.H. (1984) Plant colonization of actively sorted stone stripes in the subAntarctic. Arct. Alp. Res. 16: 161–172.
Hett, J.M. and Louks, O.L. (1976) Age structure models of balsam fir and eastern hemlock. J. Ecol. 64: 1029–1044.
Hodgson, H.J. (1966) Floral initiation in Alaskan Gramineae. Bot. Gaz. 127: 64–70.
Humlum, C. (1981) Age distribution and fertility of populations of the arctic-alpine species *Oxyria digyna*. Holarct. Ecol. 4: 238–244.
Johnson, P.L. (1969) Arctic plants, ecosystems and strategies. Arctic 22: 341–355.
Jolls, C.L. (1982) Plant population biology above timberline: biotic selective pressures and plant reproductive success. In Ecological Studies in the Colorado Alpine. A Fetschrift for John W. Marr, J.C. Halfpenny (Ed.), pp. 83–95. Univ. Colorado, Inst. Arct. Alp. Res. Occ. Pap. No. 37.
Kallio, P. and Lehtonen, J. (1973) Birch forest damage caused by *Oporinia autumnata* (Bkh.) in 1955–66 in Utsjoki, N. Finland. Rep. Kevo Subarct. Res. Stat. 10: 55–69.
Karlsson, S. (1980) Evergreen or deciduous – what does this mean to a plant in the mountains? (Swedish with English summary). Flora Fauna Upps. 75: 25–31.
Karlsson, S. (1982) Ecology of a deciduous and an evergreen dwarf shrub: *Vaccinium uliginosum* and *Vaccinium vitis-idaea* in subarctic Fennoscandia 2. Photosynthesis and carbon economy of leaves. Ph.D. dissertation, Univ. Lund, Sweden. 20pp.

Karunen, P. and Kallio, P. (1976) Seasonal variation in total lipid content of sub-Arctic *Dicranum elongatum*. Rep. Kevo Subarct. Res. Stat. 13: 63–70.

Kawano, S., Hiratsuka, A. and Hayashi, K. (1982) Life history characteristics and survivorship of *Erythronium japonicum*. The productive and reproductive biology of flowering plants. Oikos 38: 129–149.

Kihlman, A.O. (1890) Pflanzenbiologische Studien aus Russichen Lappland. Acta Soc. Fauna Flora fenn. 6(3): 263pp.

Klickoff, L.C. (1965) Microenvironmental influence on vegetational pattern near timberline in central Sierra Nevada. Ecol. Monogr. 35: 187–211.

Krog, J. (1955) Notes on temperature measurement indicators of special organisation in arctic and subarctic plants by utilization of radiated heat from the sun. Physiologia Pl. 8: 836–839.

Larcher, W. (1980) Physiological Plant Ecology. Berlin Heidelberg, New York: Springer-Verlag. 303pp.

Lawrence, B.A., Lewis, M.C. and Miller, P.C. (1978) A simulation model of population processes of Arctic tundra graminoids. In Vegetation and Production Ecology of an Alaskan Arctic Tundra, L.L. Tiezen (Ed.), pp. 599–619. New York: Springer-Verlag.

Lewis, M.C., Callaghan, T.V. and Jones, G.E. (1972) International biological programme, tundra biome, bipolar botanical project, Arctic research programme phase II. Report to the Royal Society, London, 34pp.

Lewis, M.C. and Callaghan, T.V. (1976) Tundra. In Vegetation and the Atmosphere, Vol. 2. J.L. Monteith (Ed.), pp. 399–433. London: Academic Press.

Malmer, N. and Nihlgård, B. (1980) Supply and transport of mineral nutrients in a subarctic mire. In Ecology of a Subarctic Mire, M. Sonesson (Ed.), pp. 63–95. Stockholm: Ecol. Bull. 30.

Mark, A.F. (1965) Flowering, seeding and seedling establishment of narrow-leaved snow tussock, *Chionochloa rigida*. N.Z. J. Bot. 3: 180–193.

Mark, A.F. (1970) Floral initiation and development in New Zealand alpine plants. N.Z. J. Bot. 8: 67–75.

Mattheis, P.J., Tiezen, L.L. and Lewis, M.C. (1976) Responses of *Dupontia fischeri* to simulated lemming grazing in an Alaskan arctic tundra. Ann. Bot. 40: 179–197.

McGraw, J.B. and Shaver, G.R. (1982) Seedling density and seedling survival in Alaskan cotton grass tussock tundra. Holarct. Ecol. 5: 212–217.

Miller, P.C. and Tiezen, L.L. (1972) A preliminary model of processes affecting primary production in the arctic tundra. Arct. Alp. Res. 4: 1–18.

Miller, P.C. (1982) The availability and utilization of resources. In Tundra ecosystems. Proceedings of a workshop held in San Diego, California 9–12 October 1978, P.C. Miller (Ed.). Holarct. Ecol. 5: p.83.

Mølgaard, P. (1982) Temperature observations in high arctic plants in relation to microclimate in the vegetation of Peary Land, North Greenland. Arct. Alp. Res. 14 (2): 105–115.

Moore, D.M. (1968) The vascular flora of the Falkland Islands. Scient. Rep. Br. Antarct. Surv. No. 60. 202pp.

Murray, D.F. (1978) Vegetation, floristics and phytogeography of Northern Alaska. In Vegetation and Production Ecology of an Alaskan Arctic Tundra, L.L. Tieszen (Ed.), pp. 19–36. New York: Springer-Verlag.

Oinonen, E. (1967) Sporal regeneration of bracken. (*Pteridium aquilinum* (L.) Kuhn) in Finland in the light of the dimensions and the age of its clones. Acta For. Fenn. 83: 96 pp.

Oinonen, E. (1968) The size of *Lycopodium clavatum* L. and *L. annotinum* L. stands as compared to that of *L complanatum* L. and *Pteridum aquilinum* (L.) Kuhn. stands, the age of the tree stand and the dates of fire on the site. Acta Forest. Fenn. 87: 53pp.

Oksanen, L. (1982) Trophic exploitation in relation to primary productivity. Reports from the Department of Biology. Univ. Turku No. 5.

Pearl, R. (1928) The Rate of Living. London: University of London Press. 185pp.

Petersen, H.E. (1908) *Dipensiaceae. Diapensia lapponica* L. In The Structure and Biology of Arctic Flowering Plants, E. Warming (Ed.), pp. 138–154. Meddr Grønland. (1912) 36.

Petersen, P.M. (1981) Variation of the population structure of *Polygonum viviparum* L. in relation to certain environmental conditions. Meddr om Grønland Bioscience 4: 1–19.

Plotnikov, V.V. (1977) Sporophyte ontogenesis of *Lycopodium annotinum* L. and its population structure (in Russian) Bot. Zh. SSSR 62: 1196–1200.

Polunin, N. (1936) Plant succession in Norwegian Lapland. J. Ecol. 24: 372–391.
Polunin, N. (1959) Circumpolar Arctic Flora. Oxford: Oxford University Press. 514pp.
Porsild, A.E. (1951) Plant life in the Arctic. Canadian Geogrl. J. 42: 120–145.
Primack, R.B. (1973) Growth patterns of five species of *Lycopodium*. Am. Fern J. 63: 3–7.
Raunkiaer, C. (1934) The Life Forms of Plants and Statistical Plant Geography. Translated by Carter, Fausball and Tansley. Oxford: Oxford University Press. 632pp.
Rodin, L.E. and Bazilevich, N.I. (1967) Production and Mineral Cycling in Terrestrial Vegetation. Edinburgh and London: Oliver and Boyd. 288pp.
Rosswall, T. and Heal, O.W. (1975) The IBP. Tundra Biome – an Introduction. In Structure and Function of Tundra Ecosystems, T. Rosswall and O.W. Heal (Eds.), pp. 7–16. Stockholm: Ecol. Bull. 20.
Russell, R.S. (1940) Physiological and ecological studies on an arctic vegetation. III. Observations of carbon assimilation, carbohydrate storage, and stomata movement in relation to the growth of plants on Jan Mayen Island. J. Ecol. 28: 289–309.
Salisbury, F.B. and Spomer, G.G. (1964) Leaf temperatures of alpine plants in the field. Planta 60: 497–505.
Sandberg, G. (1958) Fjällens vegetationsregioner, der vegetationsserier och viktigaste växtekologiska faktorer. In Renbeten och deras gradering, F. Skuncke (Ed.), pp. 36–40. Uppsala. Lappräsendet-Renforskningen, Meddr 4.
Savile, D.B.O. (1972) Arctic adaptations in plants. Can. Dep. Agric. Monogr. 6: 81pp.
Schroeter, C. (1926) Das Pfanzenleben der Alpen. Zurich: Albert Baustein. 1288pp.
Shaver, G.R. and Billings, W.D. (1975) Root production and root turnover in a wet tundra ecosystem, Barrow, Alaska. Ecology 56: 401–409.
Shaver, G.R. and Billings, W.D. (1976) Carbohydrate accumulation in tundra graminoid plants as a function of season and tissue age. Flora 165: 247–267.
Sonesson, M. and Lundberg, B. (1974) Late Quaternary forest development of the Torneträsk area, North Sweden. 1. Structure of forest ecosystems. Oikos 25: 121–123.
Sonesson, M. (1974) Late Quaternary forest development of the Torneträsk area, North Sweden. 2. Pollen analytical evidence. Oikos 25: 288–307.
Sonesson, M. and Hoogesteger, J. (1983) Recent tree line dynamics (*Betula pubescens*, f. *tortuosa* (Ledeb.) Nyman in northern Sweden. In Tree Line Ecology. Proceedings of the Northern Quebec Treeline Conference, P. Morisset and S. Payette (Eds.), pp. 49–54. Québec: Université Laval.
Sørensen, T. (1941) Temperature relations and phenology of the North East Greenland flowering plants. Meddr Grønland. 125: 305pp.
Svoboda, J. (1977) Ecology and primary production of raised beach communities. In Truelove Lowland, Devon Island, Canada: a high Arctic ecosystem, L.C. Bliss (Ed.), pp. 185–216. Edmonton, Alberta: University of Alberta Press.
Tallowin, J.R.B. (1977a) Vegetative proliferation in *Festuca contracta* T. Kirk on South Georgia. Br. Antarct. Surv. Bull. 45: 13–17.
Tallowin, J.R.B. (1977b) The reproductive strategies of a subantarctic grass *Festuca contracta* T. Kirk. In Adaptations within Antarctic Ecosystems, G.A. Llano (Ed.), pp. 967–980. Houston, Texas: Gulf Publishing Co.
Tenow, O. (1972) The outbreaks of *Oporinia autumnata* Bkn. and *Operophthera* spp. (Lep., Geometridae) in the Scandinavian mountain chain and northern Finland 1862–1968. Zool. Bidr. Upps. Suppl. 2: 107pp.
Tenow, O. (1975) Topographical dependence of an outbreak of *Oporinia autumnata* Bkh. (Lep. Geomitridae) in a mountain birch forest in northern Sweden. Zoon 3: 85–110.
Tieszen, L.L. (1973) Photosynthesis and respiration in arctic tundra grasses: Field light intensity and temperature responses. Arct. Alp. Res. 5: 239–251.
Tieszen, L.L., Miller, P.C. and Oechel, W.C. (1980) Photosynthesis. In An Arctic Ecosystem. The Coastal Tundra at Barrow, Alaska, J. Brown, P.C. Miller, L.L. Tieszen and F.L. Bunnell (Eds.), pp. 102–139. Stroudsberg: Dowden, Hutchinson and Ross Inc.
Verduin, J. (1972) Caloric content and available energy in plant matter. Ecology 53: 982.
Wager, H.G. (1938) Growth and survival of plants in the Arctic. J. Ecol. 26: 390–410.
Walter, H. (1979) Vegetation of the Earth and Ecological Systems of the Geobiosphere 2nd Edition. New York: Springer-Verlag. 274pp.
Walton, D.W.H. (1976) Dry matter production in *Acaena* (Rosaceae) on a subantarctic island. J. Ecol. 64: 399–415.

Warming, E. (1909) Oecology of Plants. Oxford: Clarendon Press. 422pp.
Warren Wilson, J. (1964) Annual growth of *Salix arctica* in the High Arctic. Ann. Bot. 28: 71–76.
Webber, P.J. (1978) Spatial and temporal variation of the vegetation and its production, Barrow, Alaska. In Vegetation and Production Ecology of an Alaskan Arctic Tundra, pp. 37–112. New York: Springer-Verlag.
White, J. (1980) Demographic factors in populations of plants. In Demography and Evolution in Plant Populations, O.T. Solbrig (Ed.), pp. 21–48. Oxford: Blackwell.
Whitehead, F.H. (1951) Ecology of the Altiplano of Monte Maiella, Italy. J. Ecol. 39: 330–335.
Whittaker, R.H. (1975) Communities and Ecosystems, 2nd edition. New York: Macmillan. 385pp.
Wijk, S. (1980) Plants and the snow cover (Swedish with English summary). Fauna Flora, Upps. 75: 32–36.
Wielgolaski, F.E. (1972) Productivity of World Ecosystems, P.E. Reiche, J.F. Franklin and D.W. Goodall (Eds.), pp. 1–12. Washington, D.C.: National Academy of Sciences.
Wielgolaski, F.E., Kjelvik, S. and Kallio, P. (1975) Mineral content of tundra and forest tundra plants. In Fennoscandian Tundra Ecosystems. Part 1. Plants and Microorganisms, F.E. Wielgolaski (Ed.), pp. 316–332. Berlin: Springer-Verlag.
Wielgolaski, F.E., Bliss, L.C., Svoboda, J. and Doyle, G. (1981) Primary production of tundra. In Tundra Ecosystems: a Comparative Analysis, L.C. Bliss, O.W. Heal and J.J. Moore (Eds.), pp. 187–225. Cambridge: Cambridge University Press.

DEMOGRAPHIC, GENETIC AND ECOPHYSIOLOGICAL VARIATION IN *PLANTAGO MAJOR* AND *P. LANCEOLATA* IN RELATION TO VEGETATION TYPE

P.J.M. VAN DER AART

INTRODUCTION

Plant species occur in different types of vegetation under different biotic and abiotic conditions. The regular occurrence of a plant species in a given type of vegetation implies adaptation to the characteristics of that type of vegetation. It is intriguing that some species can perform in a wide range of vegetation types and that other species are more restricted. The aim of our present research program is to obtain a better understanding of the functioning of plant species in different types of vegetation. In our view there are two different approaches for such a study. First is the study of taxonomically closely related species, or preferably ecotypes, in different environments. Ecological characteristics that are shared by such species or ecotypes from different environments can be attributed to the genetic similarity of the species. Characteristics which are different in such species, however, may hint at adaptation to local environmental conditions. The second approach is to study phylogenetically remote species in the same environment. In view of their distant taxonomic relationships convergent properties may point towards adaptations to characteristic local conditions. We have chosen the first approach.

After screening plant genera occurring in the Netherlands, as to the number of species, distribution, commonness, co-occurrence of species and experimental handiness, we selected the genus *Plantago*. This genus is represented by five species in the Netherlands, i.e., *P. coronopus* L., *P. lanceolata* L., *P. major* L., *P. maritima* L. and *P. media* L. From phytosociological studies it is known that these species differ in their ecological amplitude. *P. coronopus*, *P. maritima* and *P. media* are more strictly bound to specific habitats compared to *P. lanceolata* and *P. major*. The latter two species have a cosmopolitan distribution being weeds (Holm et al. 1977) and can be found over a wide range of vegetation types.

In this paper, the apparent success of *P. lanceolata* and *P. major* in exploiting a wide range of vegetation types will be analyzed in terms of their demographic, genetic and physiological properties. In the project reported here, approximately thirty research workers are involved. Both published and unpublished data of these research workers are presented here in an integrated form.

White, J. (ed.), The Population Structure of Vegetation.
 ISBN 978-94-010-8927-2

OCCURRENCE OF *P. LANCEOLATA* AND *P. MAJOR* IN DIFFERENT VEGETATION TYPES IN THE NETHERLANDS

Although *P. lanceolata* and *P. major* have a worldwide distribution, they occur in specific situations. *P. major* is a low rosette-forming species and as such is restricted to vegetation types with sufficient light on ground level. *P. major* is often found on sites disturbed by trampling, grazing or flooding (Blom 1979). It is found in agricultural lands in all climatic zones. *P. lanceolata* with its long, slender and often erect leaves is more suited to grow in dense herb vegetation than *P. major*. The roots of both species do not develop aerenchyma under anaerobic conditions, and are unable to penetrate soils with a low oxygen concentration. In contrast to *P. lanceolata* (Oberdorfer 1949), *P. major* is unpalatable to horses (Kerner von Marilaun 1891). Both species grow on a wide range of soil types over the world (Holm et al. 1977).

Haeck et al. (1982, 1985) have made a detailed phytosociological characterization of the habitats of *P. lanceolata* and *P. major* in the Netherlands. They analysed 361 phytosociological tables, describing 181 plant communities (associations and subassociations) from the 278 plant communities that have been documented by Westhoff and den Held (1969) for the Netherlands. Of the remaining 97 communities insufficient tables were available. The vegetation types not covered are mainly found among swamp and water vegetation. In addition 152 relevés of various vegetation types especially made for this study, all with *Plantago*, were used. For 105 of these relevés a physico-chemical analysis of the soil was made.

To arrange the Dutch plant communities the phytosociological tables were ordinated, using species composition as criterion. The reciprocal averaging technique (Hill 1973) was used. For our purpose this technique was more appropriate than principal component analysis, since extended floristic gradients were involved. The arrangement of the tables (Fig. 1) reveals an overall moisture gradient as the main source of variation in species composition in the Netherlands (axis A). On this axis plant communities are arranged from very dry to wet environments. Salinity (axis B) was found to be the factor second in importance in influencing plant-species composition in the Netherlands. To investigate further environmental factors correlated with the distribution of *Plantago* species, the phytosociological data have been re-arranged omitting the very wet and saline plant associations (Fig. 2). This second arrangement reveals two further gradients, one correlated with "soil fertility" (axis C) and another related to structural complexity of the vegetation (axis D). The ecological amplitude of *P. major* and *P. lanceolata* was then estimated by scoring their occurrence in the arranged plant communities.

Figure 3 shows the frequency of occurrence of *P. lanceolata* and *P. major* along the major lines of species variation. As expected, both species display very wide ecological amplitudes which overlap considerably. However their maxima are different. *P. major* occurs in predominantly open vegetation types (axis D), which are also nutrient-rich

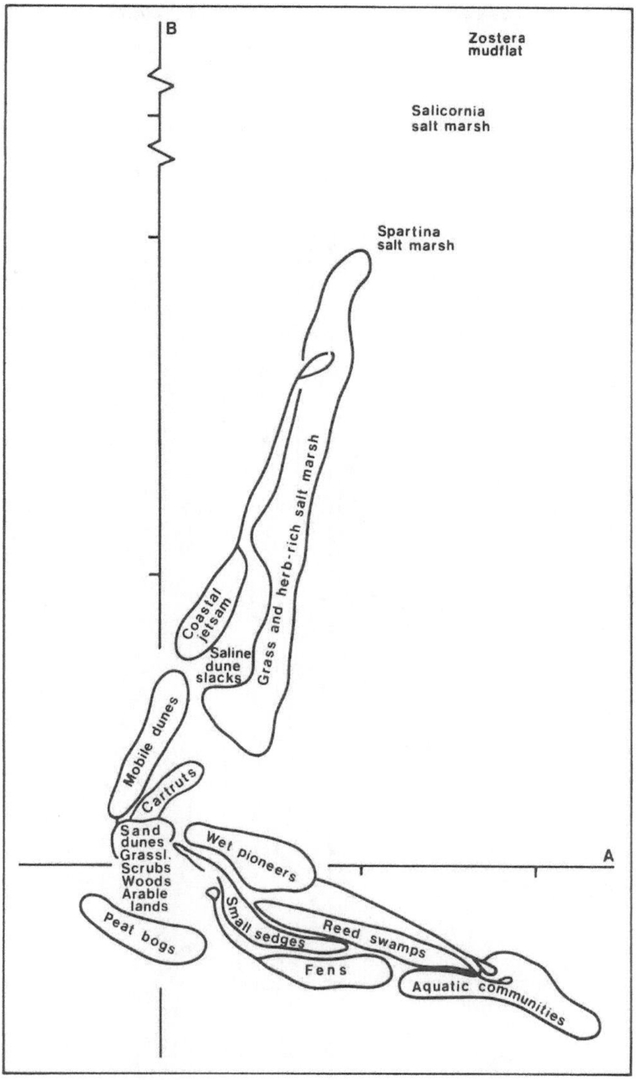

Fig. 1. Arrangement, according to species composition, of the vegetation types occurring in the Netherlands, using the Reciprocal Averaging Technique. Data from Haeck et al. (1982); for further information see text.

(axis C). It also can be found under slightly saline conditions. *P. lanceolata* occurs on soil types of intermediate fertility and in a denser type of vegetation than *P. major*. It occurs neither in saline conditions nor in reed swamps or aquatic communities (Figs. 1 and 3).

Table 1 lists the plant-alliances (groups of floristically related plant associations) in which *P. lanceolata* and *P. major* regularly occur. Both species are associated with an equal number of plant-alliances, though

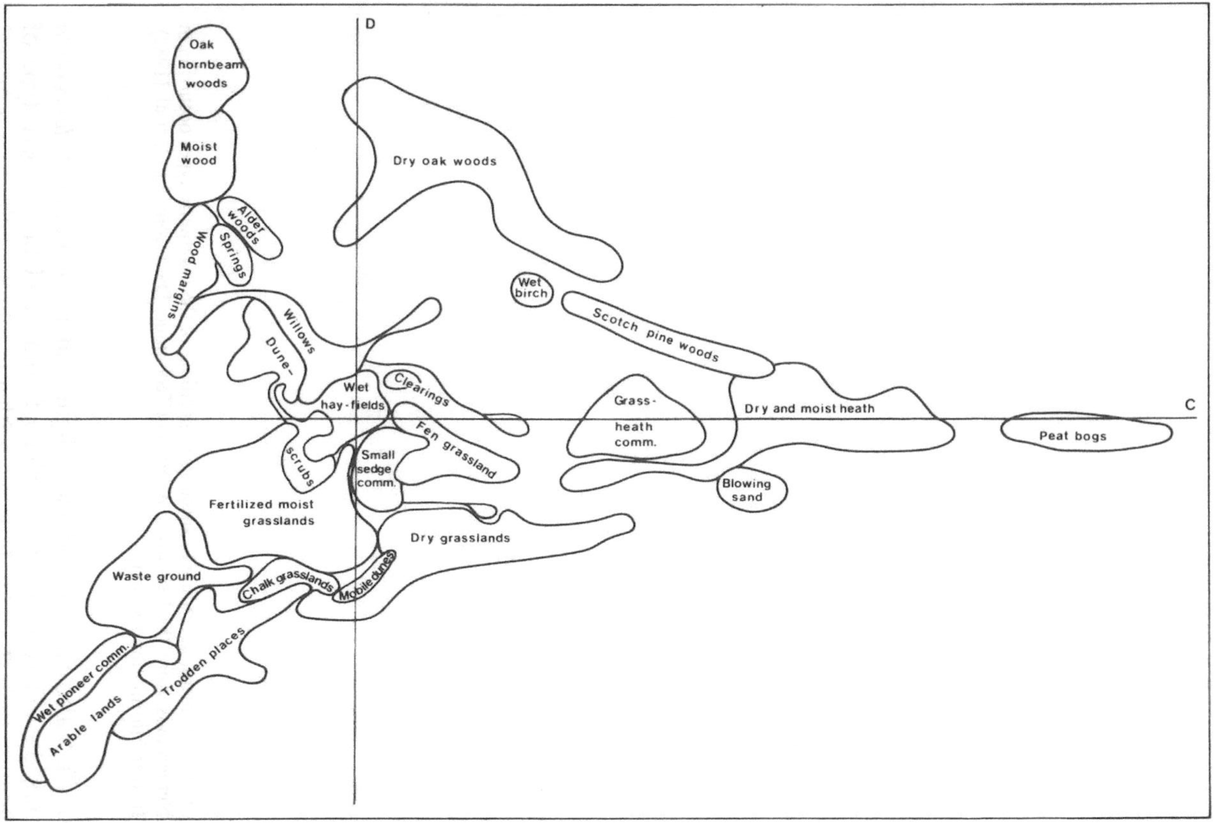

Fig. 2. Ordination of plant communities as for Fig. 1, but without communities from aquatic and saline environments. Data from Haeck et al. (1982).

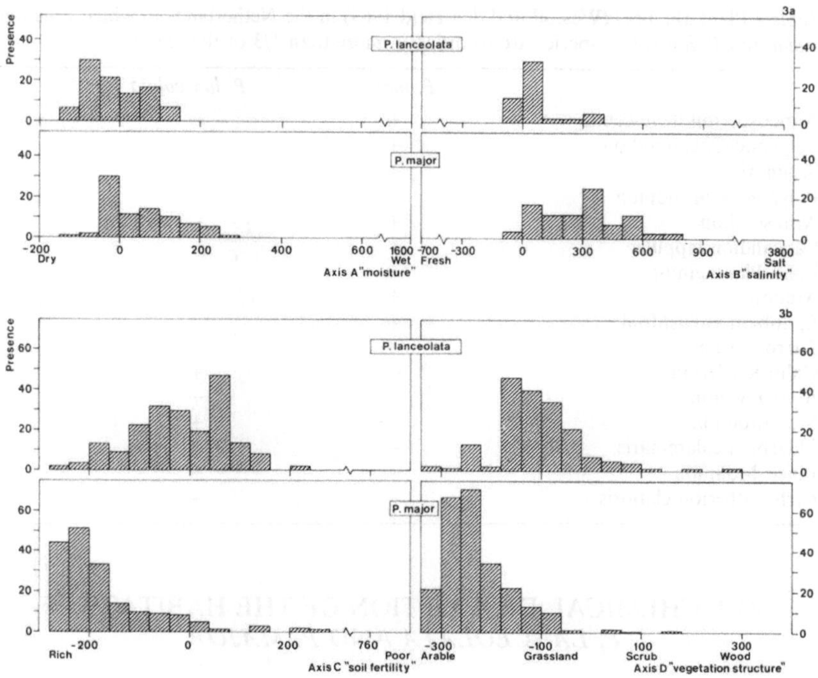

Fig. 3. Average presence of *Plantago lanceolata* and *P. major* in ordinated vegetation types: axis A "moisture" and B "salinity" (in this ordination all vegetation types are included), axis C "soil fertility" and D "vegetation structure" (in this ordination communities from aquatic and saline environments are excluded).

they are nearly all different. This means that the two species are rather distinct phytosociologically. *P. major* occurs on arable fields (Polygono-Chenopodion, Arnoseridion, Caucalidion lappulae, Polygono-Coronopion), ruderal and disturbed sites (Lolio-Plantaginion, Arction) and ephemeral habitats (Epilobion angustifolii). The common feature of these habitats besides being relatively nutrient-rich, is their open character caused by harvesting, trampling, grazing, wood clearing, fire or flooding. *P. lanceolata* also occurs in open nutrient-rich sites, that are of not-too-short longevity (Sisymbrion, Arction). In addition, it often occurs in low, often closed, permanent vegetation on nutrient-poor sites (Thero-Airion, Galio-Koelerion, Mesobromion, Junco-Molinion), or in closed permanent vegetation on nutrient-rich sites where vegetation growth is hampered by heavy metals (Thlaspion calaminaris), haying and/or grazing (Arrhenatherion elatioris), or by extreme exposure and solid substrate (Alysso-Sedion). The common feature of these habitats is their more permanent character and (in closed vegetation) their restriction in height.

Table 1 Plant alliances (Westoff and den Held 1969) in the Netherlands in which *Plantago major* and *P. lanceolata* species are recorded in more than 1/3 of the relevés

	P. major	*P. lanceolata*
Nanocyperion flavescentis	+	−
Polygono-Chenopodion	+	−
Sisymbrion	+	+
Polygono-Coronopion	+	−
Arnoseridion	+	−
Caucalidion lappulae	+	−
Lolio-Plantaginion	+	−
Arction	+	+
Epilobion angustifolii	+	−
Thero-Airion	−	+
Galio-Koelerion	−	+
Alysso-Sedion	−	+
Mesobromion	−	+
Thlaspion calaminaris	−	+
Junco-Molinion	−	+
Arrhenatherion elatioris	−	+

SOIL CHEMICAL DESCRIPTION OF THE HABITATS OF *P. LANCEOLATA* AND *P. MAJOR*

It is generally recognized that *P. major* and *P. lanceolata* may inhabit a wide range of soil types. Most authors state that *P. major* frequents more fertile soils than *P. lanceolata* (e.g., Oberdorfer 1949). Table 2 summarizes the mean and the observed ranges of several soil properties of the natural habitats of *P. lanceolata* and *P. major*. Both species occur over a wide range of pH, organic matter and macro-nutrient contents. The ranges observed show substantial overlap: the macro-nutrient concentrations are hardly different for the two species, except for Olsen-P, a good estimate of P-availability to plants (Abbott 1978). Both species also occur over a wide range of textural classes (Table 3). The mean fractional composition of a soil sample (0–10 cm) of both species is similar but in a given sample, for both species the proportion of a given grain-size class may vary considerably. These large ranges are compatible with their distribution over a wide variety of plant communities. However to detect differences in the soil ecology of these two species the chemical analyses may need to be performed on soil much nearer to the *Plantago* plant root system. Our method was based on soil samples taken from a site where *Plantago* occurred, irrespective of whether or not a *Plantago* individual plant was exploiting the particular soil sample taken. This way of sampling may obscure soil chemical preferences of the species on a very local scale. Besides, "nutrient availability" to plants is a very complex phenomenon in which (apart from soil chemistry) many factors play a role, such as soil physical factors, root morphology, plant competition and mycorrhizal interference.

In the ordination of plant communities (Fig. 2), axis C was interpreted as a soil fertility gradient. In Fig. 3 it can be seen that *P. major* and

Table 2. Mean values and ranges of some selected properties of the 0–10 cm soil layer of habitats of *P. lanceolata* (72 habitats) and *P. major* (17 habitats) in the Netherlands (after Troelstra et al. 1981).

	P. lanceolata		*P. major*	
	Mean value	Range	Mean value	Range
pH–H_2O	6.1	4.3 – 7.8	6.9	5.3 – 8.4
pH–KCl	5.4	3.7 – 7.5	6.2	4.4 – 8.0
% $CaCO_3$	2.3	0 –52.6	2.9	0 – 9.9
% Organic matter	11.3	1.25–74.0	5.0	0.94–11.3
Chloride (mg.100 g^{-1})	0.9	0 – 9.6	0.9	0 – 4.5
Conductivity 1 : 5 extract ($\mu S.cm^{-1}$)	130	16–428	134	46–310
Total N ($\mu g.g^{-1}$)	3480*	350–23590	2220*	280–4380
Total P ($\mu g.g^{-1}$)	506*	105–1367	837*	229–2184
Total organic P ($\mu g.g^{-1}$)	276*	20–1115	229*	49–344
Olsen-P ($\mu g.g^{-1}$)	13*	2.8 –44	49*	3.1 –110
Labile organic P ($\mu g\, g^{-1}$)	24*	3.7 –70	28*	3.5 –39
K (meq 100 g^{-1})	0.28**	0.03– 0.81	0.490**	0.08– 1.81
Na (meq 100 g^{-1})	0.17**	0.01– 0.90	0.16**	0.02– 0.47
Ca (meq 100 g^{-1})***	10.3**	0.73–49.7	4.73**	1.82–10.2
Mg (meq 100 g^{-1})	1.3**	0.10– 5.18	1.42**	0.20– 2.88
CEC (NH_4OAc pH7) (meq 100 g^{-1})***	13.1**	0.97–52.9	7.40**	3.58–11.7

*in $\mu g.cc^{-1}$.
**in meq.100 cc^{-1}.
***Calcareous soils not included.
CEC Cation exchange capacity.

Table 3. Fractions (in percentages) of the 0–10 cm soil layer collected from various natural habitats of *Plantago lanceolata* and *P. major* in the Netherlands (data from Troelstra et al. 1981).

	P. lanceolata		*P. major*	
	Mean value	Range	Mean value	Range
$CaCO_3$	2.3	0–53	2.9	0–10
Organic matter	11.3	1–74	5.0	1–11
Inorganic particles				
$> 53\,\mu$	58	0–95	53	13–93
$16–53\,\mu$	11	1–58	18	1–56
$2–16\,\mu$	8	0–31	10	1–27
$< 2\,\mu$	10	0–54	10	1–34

P. lanceolata occupy about the same range on this soil fertility gradient, their optima however are clearly different. These identical ranges are to a large extent also apparent from the soil-parameter values given in Tables 2 and 3. Correlation analysis showed a close correspondence only between the ordinated vegetation relevés on axis C (Fig. 3) and Olsen-P values (Fig. 4). It may tentatively be concluded that phosphate availability is a critical factor in the distribution of *P. major*. This hypothesis was confirmed in water-culture experiments (Troelstra, unpublished).

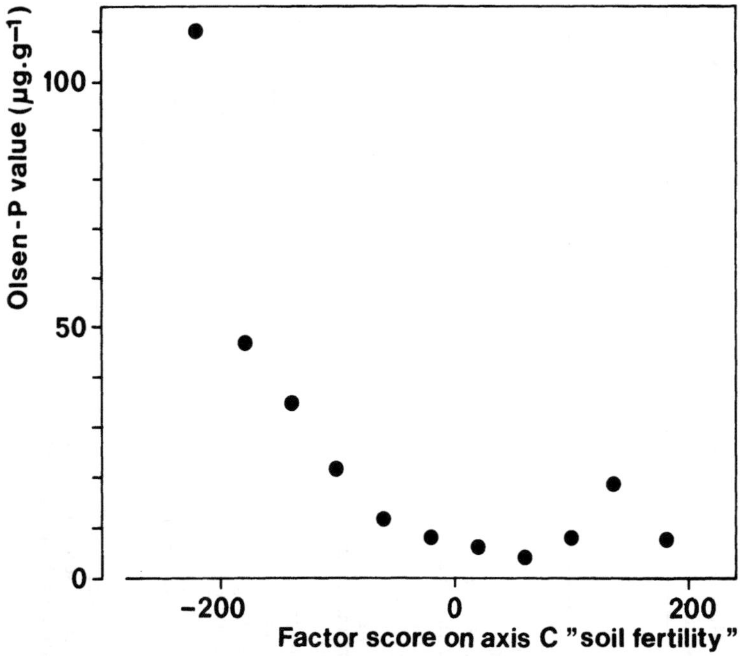

Fig. 4. The relation between phosphate availability (Olsen-P) and the "soil fertility" axis from Fig. 2.

DEMOGRAPHIC AND ECOLOGICAL CHARACTERISTICS OF *P. LANCEOLATA* AND *P. MAJOR* IN RELATION TO VEGETATION TYPE

P. lanceolata

In Fig. 5 the population dynamics of *Plantago* species are given. As can be seen from this Figure *P. major* shows more marked fluctuations than *P. lanceolata* at the two locations where both occur. This proved to be a general feature. The population dynamics of *P. lanceolata* in six types of vegetation are different. The populations in dry open vegetation show comparatively large fluctuations, whereas the populations in closed vegetation have reduced recruitment and mortality curves, which are reflected in small fluctuations. Some populations increased in number and others decreased during the study interval. These reduced recruitment and mortality curves correlate with density of the vegetation in summer (estimated as light transmission through the vegetation) and are mainly due to differences in survival of seedlings and small plants (Fig. 6).

To study the ecological characteristics in relation to vegetation type in more detail, two *P. lanceolata* populations in contrasting vegetation types were chosen. The first area Westduinen is an open, and in summer a dry, coastal dune grassland (Galio-Koelerion, Thero-Airion) on sandy soil which has been grazed by cattle for the last 400 years (Blom et al. 1979). The sward is low (less than 10 cm) and biomass production is restricted by summer droughts. The other area Merrevliet, is a hayfield on former fen

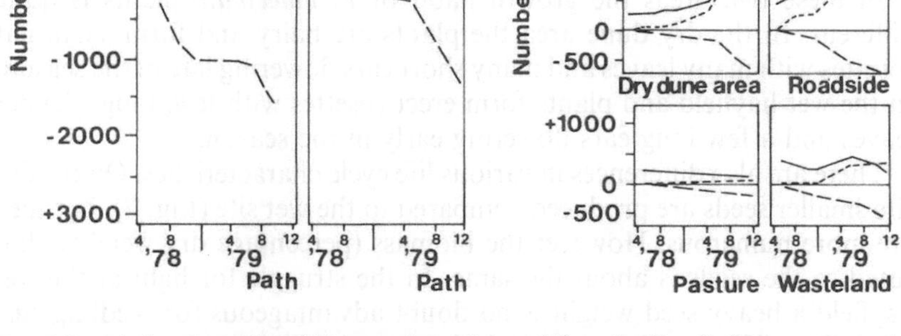

Fig. 5. Population dynamics of *Plantago lanceolata* and *P. major*. Populations were recorded four times a year. Plot size 30 × 40 cm. Per population several plots were recorded and the number of individuals per m^2 calculated. The solid lines represent the number of rosettes present, the upper dashed lines represent the recruitment and the lower dashed lines represent the mortality. Data from van der Toorn et al. (1980).

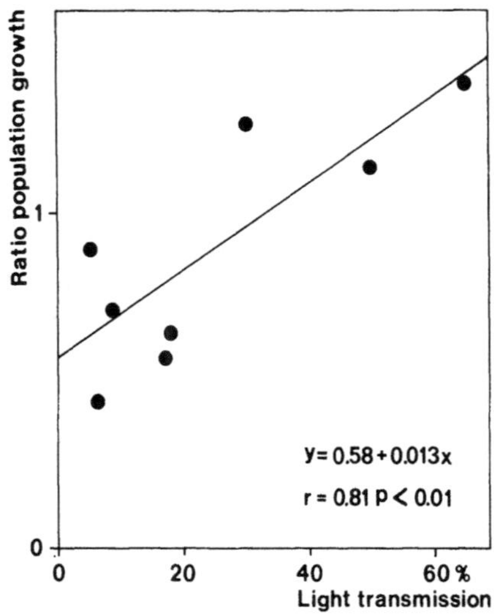

Fig. 6. Ratios of population growth of *Plantago lanceolata* in relation to light transmission of the vegetation in summer (May–August). Data from Mook et al. (1981). The ratio of population growth is defined by the number of rosettes present in 1979 divided by the number in 1978.

land (Calthion). The water table is a few centimeters below soil surface. Fresh water overlays brackish groundwater to a depth of less than 1 m. This brackish water prevents the development of shrubs and luxuriant reed. Vegetation height increases to 40 cm during the growing season. The area is mown in mid July.

The growth conditions for *P. lanceolata* are quite different in the two areas. Water supply and consequently nutrient mineralization are unpredictable in the dry dune area. Light in the vegetation at the height where most plantain leaves are present is approximately 80 per cent of full daylight. In the wet hayfield the water supply is constant. Light intensity at the height with most *Plantago* leaves is reduced to 5% of full daylight in summer.

In these two areas the growth habit of *P. lanceolata* plants is quite different. In the dry dune area the plants are hairy and form small flat rosettes with many leaves and many short ears flowering late in the season. In the wet hayfield and plants form erect rosettes with few, long, slender leaves and a few long ears flowering early in the season.

There are also differences in various life cycle characteristics. On the dry site smaller seeds are produced compared to the wet site (Fig. 7), but they are more numerous. However the biomass (percentage dry weight) allocated to the seeds is about the same. In the struggle for light at the wet hayfield a heavy seed weight is no doubt advantageous for seedling survival. The difference in seed weight between both sites appeared to be at least partly genetically fixed (van der Toorn and ten Hove 1982). Selection for heavier starting weight during seedling stage in the wet hayfield is likely to occur.

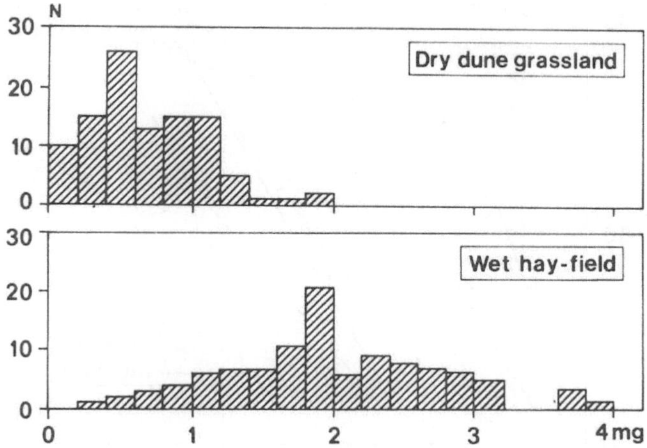

Fig. 7. Seed-weight distribution of *Plantago lanceolata*. Data from van Groenendael (personal communication). N = number of seeds.

P. lanceolata seeds do not need light for germination (Blom 1978). In the light germination is hardly affected by spectral composition (Pons, personal communication). On the dry dune site the main germination period is in spring when only a fraction of the seeds germinates. There appears to be a substantial seed bank. On the wet site the main germination period is in autumn, when nearly all seeds germinate so that the remaining seed bank is very small. The survival of these seedlings in spring is low due to competition for light in the growing vegetation. Van Groenendael (personal communication) found 26 and 28% survival of these seedlings for two successive years, compared with 59 and 45% in the dry dune grassland. Due to summer drought adult survival in dune grassland is low: the half-life time expectancy for adults was 1.8 and 1.2 years. In the wet area the corresponding figures were 2.5 and 3.4 years. We conclude that the population-dynamic behaviour of *P. lanceolata* in the dry dune grassland and in the wet hayfield is different.

Differences in the time of flowering may also have a survival value. Figure 8 shows the percentage of flowering plants of *P. lanceolata* in the course of the growth season. A comparison of three hayfields mown at the end of June, in mid July and early in August, showed a correlation between the time of mowing and the time of flowering. *P. lanceolata* plants in grazed meadows flower relatively late in the season. Transfer of adult plants from these fields to an experimental garden did not completely remove these differences in time of flowering in subsequent years. This suggests that differences in flowering time are at least partly genetically fixed.

The variability of leaf characteristics of *P. lanceolata* has been studied by van der Toorn and ten Hove (1982). They measured the angle between the third and fourth leaf as a measure of the degree of erectness and leaf length and leaf width of cloned plants made from roots grown

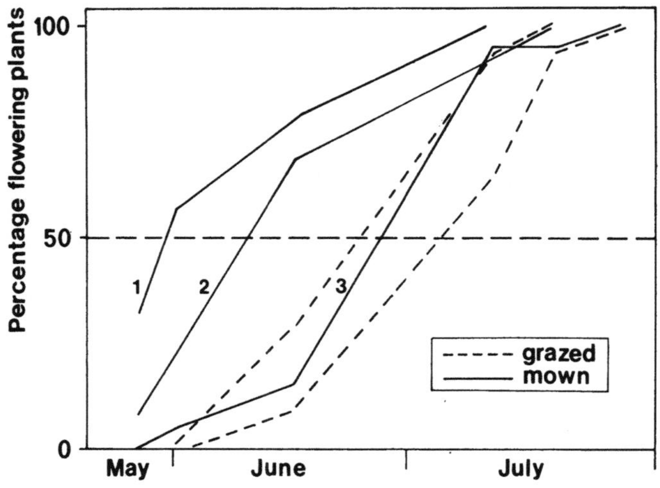

Fig. 8. Percentage flowering plants of *Plantago lanceolata* in the course of the growing season, for hay-fields mown end of June (1), mid-July (2) and start of August (3) and two grazed meadows.

under controlled conditions. Parent plants were collected from hayfields and pastures. Figure 9 shows the relation between vegetation height and various leaf characteristics. Leaf length is correlated positively with vegetation height, whilst a negative correlation exists between leaf angle and height of vegetation. The use of cloned material allows the conclusion that the variability in leaf characteristics is at least partly genetically determined. The correlations are likely to result from natural selection favouring erect, long-leaved types in tall vegetation and flat-lying, short-leaved types in short vegetation (Fig. 10).

In one situation a hayfield and a pasture were adjacent to each other. Both fields were extensively grazed until twenty-two years ago; thereafter

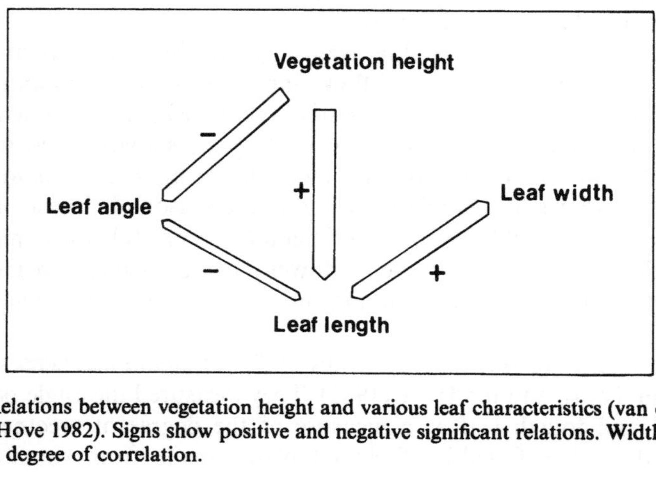

Fig. 9. Relations between vegetation height and various leaf characteristics (van der Toorn and ten Hove 1982). Signs show positive and negative significant relations. Width of arrow indicates degree of correlation.

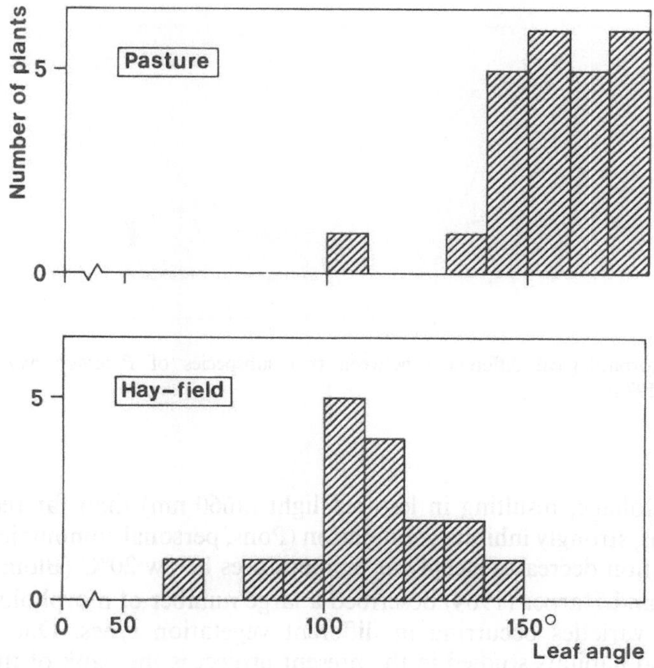

Fig. 10. Distribution of leaf-angles (angle between third and fourth leaf) in *Plantago lanceolata* in a dry dune grassland and a wet hay-field population.

one of the fields was managed as a hayfield. In this case selection for erect long-leaved plants was realized within twenty-two years at a distance of ca. 30 m (van der Toorn and ten Hove 1982). This is remarkable for a species whose pollen is spread by wind.

P. major

P. major has a wide ecological amplitude. The species consists of a number of ecotypes (van Dijk and van Delden 1981). In the Netherlands the subspecies *P. major* spp. *pleiosperma* Pilger and *P. major* ssp. *major* form the extremes of a continuum of ecotypes. The species occurs on arable fields, compacted soils, in brackish environments and ephemeral habitats. These vegetation types all have an open structure. Skellam (1952) found that the distance to nearest neighbours of *P. major* is greater than statistically can be expected in field situations. Blom (1979) has shown a reduction in root biomass of *P. major*, when its root system is sharing the soil volume with *P. lanceolata*. Its ability for root competition seems to be low. Germination of *P. major* is inhibited in the dark. Light passing

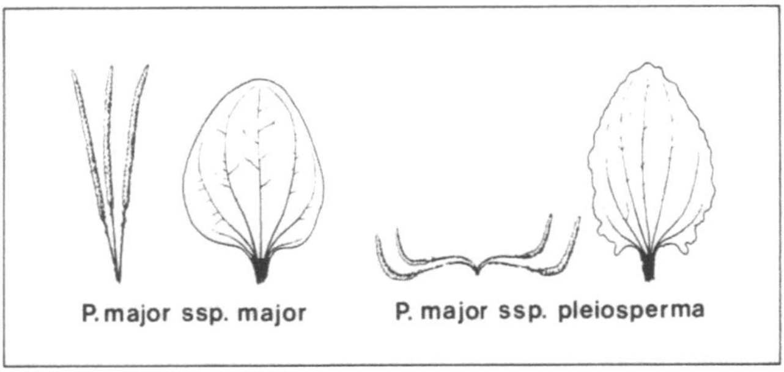

Fig. 11. Morphological differences between two subspecies of *Plantago major* (after Mølgaard 1976).

through foliage, resulting in less red light (λ660 nm) than far red light (λ730 nm), strongly inhibits germination (Pons, personal communication). Germination decreases steeply at temperatures below 20°C (Blom 1979).

Sagar and Harper (1964) described a large number of morphologically different varieties occurring in different vegetation types. One of the ephemeral habitats studied in the present project is the bank of the river Rhine, where *P. major* spp. *pleiosperma*, formerly known as *P. intermedia* auct. non Gilib., occurs. This is a short-lived subspecies: it germinates and establishes in spring and produces seeds before the river floods the banks in the autumn. *P. major* ssp. *pleiosperma* is also found as a weed in crops (Mølgaard 1976). In trampled situations the subspecies *P. major* ssp. *major* is found (Blom 1979).

The two subspecies of *P. major* show morphological differences (Fig. 11). The main differentiating characteristic is in the number of ovules per capsule ranging from 14 or more in *P. major* ssp. *pleiosperma* and from 4–12 in *P. major* ssp. *major* (Mølgaard 1976). The spikes of *P. major* ssp. *major* from roadsides are erect, which may not seem to be very suitable to withstand trampling. However after pressing the spike on the ground, it straightens itself immediately (Blom 1979). A different ecotype of *P. major* ssp. *major* with a prostrate growth form occurs on mown and grazed sites. The life expectancy of *P. major* spp. *major* is several years. Overground the plants in our area die off each winter. The growing point retracts just

Table 4. Demographic properties of *Plantago major* and *P. lanceolata* (means for 2 or 3 years)

	P. major ssp. *major*	*P. major* ssp. *pleiosperma*	*P. lanceolata*
Mean weight per 1000 seeds (g)	0.20–0.32	0.11–0.15	1.2–2.0
Seed production per plant	400–2500	30–300	30–500
Seedling survival (%)	3–4	26–34	3–60
Adult survival (%)	68–75	27–36	40–88

below the soil surface. In spring the plants reappear. Some of the demographic characteristics of *P. lanceolata* and *P. major* are summarized in Table 4.

GENETIC VARIABILITY WITHIN *PLANTAGO MAJOR* AND *P. LANCEOLATA*

As shown above, both *P. major* and *P. lanceolata* show marked morphological and demographic differences in relation to their environment. Theoretically there are two main possibilities for a species to function in divergent habitats. One is by genetic differentiation (formation of ecotypes), the other by phenotypic plasticity of the individuals. A combination of both may also exist.

The data on flowering time, leaf morphology and seed weight in *P. lanceolata* and on number of seeds per capsule in *P. major*, indicate genetic differentiation within these species. Also Groot and Boschhuizen (1970) and Warwick and Briggs (1980) have collected evidence indicating genetic differentiation in *P. major*. Böcher (1943) and Primack (1976, 1978) demonstrated genetic differences for *P. lanceolata*.

Most ecotypic differences are due to genes affecting morphology or physiology. Genetic differences for such genes are unfortunately not easy to study. Genes coding for specific enzymes are much easier in this respect. The allozyme technique (van Dijk and van Delden 1981) can be used to estimate genetic characteristics such as the degree of polymorphism and of inbreeding, the size of the genetic neighbourhood and the intensity of the gene flow. The enzymes can be used as parameters to quantify genetic variation in and between populations and to analyse the genetic structure of populations. Subsequently, the minimal degree of genetic specialization can be defined.

Genetic variability (Table 5) as expressed in the values for heterozygosity is close to the mean level reported by Hamrick (1979) in *P. lanceolata* and very low in *P. major*. The low value in the latter is largely due to the high amount of self-fertilization. Conversion into a "polymorphic index" makes the heterozygosity values independent of the mating system. The value of *P. lanceolata* remains unchanged, but *P. major* populations are in

Table 5. Genetic variability for *Plantago major* and *P. lanceolata*. Data from van Dijk et al. (1985).

	P. major	*P. lanceolata*
Degree of polymorphism	0.25	0.37
Degree of heterozygosity	0.005	0.100
Percentage self fertilization	85	0

Table 6. Linkage between allozymes and morphological characteristics for *Plantago major* as demonstrated by van Dijk (1984).

Allozymes	Morphological characteristics
1. Pgm-1	leaf form
	length of petiole
	rate of development
	number of seeds per capsule
	inflorescence position
2. Got-1	spike/scape ratio

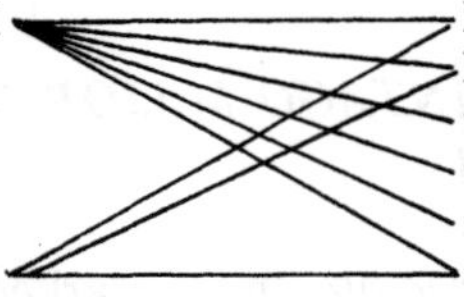

the range 0.03–0.06. The mean heterozygosity for herbaceous perennials is 0.116 ± 0.032 (Hamrick 1979).

In *P. major* two out of seven loci with intermediate allele frequencies proved to have alleles restricted to only one subspecies. Linkage of these two loci, i.e., phosphoglucomutase-1 (= Pgm-1) and Glutamate-oxaloacetate-transaminase-1 (= Got-1), with a number of loci which control various ecologically important characters explains their subspecies specificity (Table 6). These characters are leaf morphology, inflorescence morphology and generative development and are discriminative on the subspecific level.

In *P. lanceolata* no discriminative differences in allozyme variation were found between populations, although the morphologically most extreme populations were also investigated.

P. major clearly shows difference in presence and frequency of enzyme alleles between populations in divergent habitats. It is striking, however, that the overall allozyme variation between populations of *P. major* is very similar in similar habitats. Detailed sampling in the field revealed that populations of *P. major* are composed of subpopulations, each of some square decimeters (van Dijk et al. 1985). Each subpopulation consists of genetically strongly related individuals, which are the progeny of only one (often selfing) or a few parent plants. Adjacent subpopulations may be very different in allele types and/or frequencies.

P. lanceolata being a wind-pollinated and partly insect-pollinated species (Stelleman and Meeuse 1976) is self-incompatible (Ross 1973). Consequently, the degree of heterozygosity in this species is high. The structure of genetic neighbourhoods, which is very pronounced in *P. major*, is obscured in *P. lanceolata*. The high degrees of heterozygosity and polymorphism in *P. lanceolata* also hamper the detection of genetic differences between populations. The variation between populations in terms of allozymes is about 4 per cent of the total variation. However, experiments on the inheritance of leaf-morphology characteristics, time of flowering, and seed weight, show distinct genetic differences between populations.

The results of the allozyme analysis, which reveal only minor differences between populations of *P. lanceolata* seem to contradict those obtained by the experimental ecologists on flowering time, leaf characteristics and seed size. A likely explanation for this paradox is that allozymes may in general be neutral with respect to natural selection. Cross-fertilization ensures a high rate of gene flow and the neutral variants are consequently easily spread. Characteristics with selective value basically have the same rate of gene flow as those which are neutral, but their spread into populations in alien habitats is opposed by natural selection. The characteristic differences between populations of *P. lanceolata* very likely have local survival value, though this remains to be investigated.

Van Valen (1965) suggested a positive relationship between the amount of genetic variation and the ecological amplitude of a species (his niche-width variation hypothesis). In *P. lanceolata* the degree of polymorphism as well as heterozygosity is higher than in *P. major*. However for both species the ecological ranges occupied are of comparable size, so that the hypothesis is not supported by these data.

ECOPHYSIOLOGICAL ADAPTATION TO HABITAT TYPE IN *P. LANCEOLATA* AND *P. MAJOR*

Both *P. lanceolata* and *P. major* are able to grow under a very wide range of soil-chemical, soil-textural and phytosociological conditions. This raises the question whether each individual of these species is capable of performing under a wide range of conditions, or whether ecotypes have evolved with specific adaptations for particular localised conditions.

To investigate this problem *P. lanceolata* and *P. major* were grown under a wide range of conditions of pH, nutrient availability (Kuiper and Kuiper 1979c, Stulen 1981a/b), nitrate availability (Freijsen and Otten 1979), nitrate versus ammonium supply (Troelstra 1982), and different light intensities (Smit unpublished).

These investigations indicated that both *P. major* and *P. lanceolata* individuals can grow under a wide range of stationary environmental circumstances, and no pronounced differences were found between the species.

When plants were grown in a fluctuating, rather than a constant environment, fundamental ecophysiological differences between the two species were found. Kuiper and Kuiper (1979a–c) carried out experiments in which individual plants under experimental conditions were subjected to a switch in the nutrient level.

After a switch of plants to lower nutritional conditions *P. lanceolata* changed its shoot-root ratio, whereas *P. major* did not respond (Fig. 12). After a switch from low to high nutrient level the same is observed. A similar response to a change in the nutrient level, in the case of *P. lanceolata* was also found for other parameters such as the concentration of nitrogen in the plant and the activity of enzymes like nitrate reductase and glutaminesynthetase, enzymes which play a role in nitrogen metabolism (Stulen et al. 1981a) and some aspects of respiratory metabolism (Lambers et al. 1981a). *P. lanceolata* also shows changes in membrane-lipid composition and activity of ion-stimulated ATP-ases

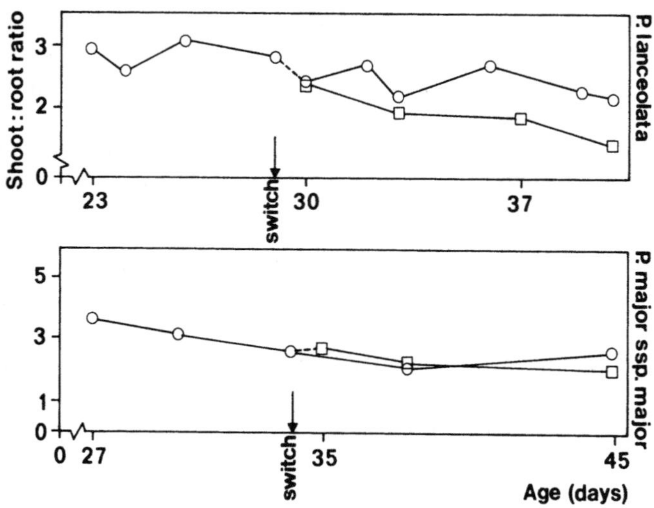

Fig. 12. Reaction in shoot-root ratio after a switch to lower nutrient availability (Kuiper and Kuiper 1979c). Round symbols 25% Hoagland nutrient solution, square symbols 0.5% Hoagland nutrient solution.

after a switch to a different nutritional level (Kuiper and Kuiper 1979b, c). For all of these parameters *P. major* was unable to adjust. After a change in conditions, *P. major* shows no reaction in the first days after that switch (Kuiper and Kuiper 1979a, Stulen et al. 1981b, Lambers et al. 1981b).

Lambers (1979, 1982) investigated a phenomenon which might prove to be one of the basic physiological processes underlying the ability to respond to fluctuating conditions. Under constant environmental conditions part of the respiration in the plant does not generate ATP. This respiration involves an 'alternative' pathway. Upon a change in conditions *P. lanceolata* is able to change the ratio of ATP-generating and 'alternative' respiration in response to energy requirements. In *P. major* this ratio is not flexible under changing conditions and hence no adjustment in energy supply can be made in order to cope with the change. This ability to adjust the intensity of its 'alternative' respiration appears to be linked to the ability of a species to react to environmental changes and hence may be a basic process in physiological plasticity as a reaction to short term changes.

DISCUSSION

The distribution of a plant species is an ever-intriguing problem. *P. lanceolata* and *P. major* both have a cosmopolitan distribution. Within the Netherlands they occur over a wide range of environmental conditions as is apparent from the ordination analysis of the Dutch plant communities (Figs. 1–3) and from soil chemical and textural analyses (Tables 2 and 3). The two species are almost completely overlapping in their abiotic environment.

However, the phytosociological distributions of these species are quite distinct. Of the 16 plant-alliances in the Netherlands in which the species regularly occur, there are only two in which they are both regularly present (Table 1).

As both *Plantago* species are rosetteforming, they are confined to relatively low structured vegetation types. These conditions are brought about by grazing, mowing, trampling, flooding, burning or other habitat disturbances. It transpires that such disturbances affect species composition more than general soil conditions (Troelstra et al. 1981). With this in mind it is not surprising that *P. major* and *P. lanceolata* do not react very differently to a wide range of soil-chemical conditions. Extreme soil conditions may have an overruling effect.

A major difference in habitat preference between *P. lanceolata* and *P. major* is the openness of the vegetation, e.g., the proximity of neighbouring plant individuals. *P. major* is restricted to sites which are more open than those of *P. lanceolata*. In competition experiments Brouwer (personal communication) has shown that *P. lanceolata* outcompetes *P. major* by light interception. This suggests that other tall-growing species may do the same. Blom (1979) showed that the root system of *P. major* is heavily restricted in the presence of *P. lanceolata* as a neighbouring plant, even when the shoots do not compete for light. Root competition ability in *P. major* proved to be inferior to that of *P. lanceolata*. The data on competition offer an explanation for the restriction of *P. major* to more open sites.

It is hypothesized that in a closed vegetation of the type preferred by *P. lanceolata*, nutrient availability and further conditions for growth will be more influenced by neighbouring plants than by soil-chemical conditions. The composition and vitality of neighbouring plants in vegetation is continuously changing and this may require a flexible response of any constituent species. For *P. major* the interference with neighbouring species is less, due to the more open character of the habitat. Thus growth conditions for *P. major* are likely to be more stable, and competition for nutrients is likely to be less severe.

This hypothesis that a plant in a closed vegetation needs to be more flexible compared to species of open vegetation is based on the ecophysiological research done on *P. major* ssp. *major* and *P. lanceolata*. Under stable, noncompetitive experimental conditions both species perform equally well. Only slight differences were detected under a broad range of conditions tested. Under fluctuating conditions, however, *P. lanceolata* is superior to *P. major* ssp. *major*. *P. lanceolata* quickly adjusts both morphologically and physiologically. For *P. major* a stable nutrient supply is ensured in arable field vegetation (Polygono-Chenopodion, Arnoseridion, Caucalidion lappulae and Polygono-Coronopion). In ruderal and disturbed vegetation (Lolio-Plantaginion, Arction, Epilobion angustifolii) on nutrient-rich soil competition for nutrients will be of minor importance to *P. major*.

P. lanceolata is much more able to respond to changes in nutritional conditions. It persists through changing competitive conditions during the growing season (Thero-Airion, Galio-Keolerion, Mesobromion, Junco-Molinion) and under changing conditions of haying and/or grazing (e.g., Arrhenatherion elatioris).

For both species adaptations to the different environmental conditions in divers habitats is realized by genotypic differentiation. The restricted gene flow in *P. major*, however, allows for a much more precise genetic differentiation than in *P. lanceolata*.

ACKNOWLEDGEMENTS

Approximately thirty research workers are involved in the project reported here. C.W.P.M. Blom and J.H. Mook (Institute for Ecological Research, resp. Oostvoorne and Heteren, the Netherlands) supervise the demographic and ecological research. W. van Delden (Groningen University, the Netherlands) is in charge of studies on population genetics. Plant physiological research is carried out under supervision of P.J.C. Kuiper (Groningen University, the Netherlands) and R. Brouwer (Utrecht University, the Netherlands). Microbiological research is guided by J. Woldendorp (Institute for Ecological Research, Heteren, the Netherlands). The chemical analyses of soil and plant material are carried out by S.R. Troelstra (Institute for Ecological Research, Oostvoorne, the Netherlands). I gratefully recognize the use of data from Blom, Brouwer, van Damme, van Doorn, van Dijk, Freijsen, van Groenendael, Haeck, ten Hove, Kuiper, Lambers, Mook, Soekarjo, Stulen, van der Toorn, and Troelstra. I want to thank Beeftink, van Delden, Kuiper, Lambers, White and Woldendorp for the many valuable suggestions to improve the manuscript.

REFERENCES

Abbott, J.L. (1978) Importance of organic phosphorus fraction in extracts of calcareous soil. Soil Sci. Soc. Am. J. 42: 81–85.

Blom, C.W.P.M. (1978) Germination, seedling emergence and establishment of some *Plantago* species under laboratory and field conditions. Acta Bot. Neerl. 27: 257–271.

Blom, C.W.P.M. (1979) Effects of trampling and soil compaction on the occurrence of some *Plantago* species in coastal sand dunes. Thesis Nijmegen University: 135 pp.

Blom, C.W.P.M., Husson, L.M.F. and Westhoff, V. (1979) Effects of trampling and soil compaction on the occurrence of some *Plantago* species in coastal sand dunes. IVa. The vegetation of two dune grasslands in relation to physical soil factors. Proc. Kon. Ned. Akad. Wetensch., Series C, 82: 245–273.

Böcher, T.W. (1943) Studies on variations and biology in *Plantago lanceolata* L. Dansk Bot. Ark. Band II. (3): 1–18.

Freijsen, A.H.J. and Otten, H. (1979) The growth and nitrate uptake of *Plantago lanceolata* and *P. major* in experiments with continuously and intermittently supplied nitrate concentrations of 750 and 25 uM. Verh. Kon. Ned. Akad. Wetensch., afd. Natuurk. II, 73: 364–367.

Groot, J. and Boschhuizen, R. (1970) A preliminary investigation into the genecology of *Plantago major* L. J. Exp. Bot. 21: 835–841.

Haeck, J., van der Aart, P.J.M., Dorenbosch, H., van der Maarel, E. and van Tongeren, O.

(1982) The occurrence of *Plantago* species in ordinated Dutch plant communities. Verh. Kon. Ned. Akad. Wetensch., afd. Natuurk. II, 79 Progress Report I.O.O. 1981: 26–33.

Haeck, J., van Tongeren, O. and van der Maarel, E. (1985) Phytosociological amplitudes of some Dutch coastal species and their ecological interpretation. Vegetatio (in press).

Hamrick, J.L. (1979) Genetic variation and longevity. In: Topics in Plant Population Biology, O.T. Solbrig et al. (Eds.), pp. 84–113. Columbia University Press, New York.

Hill, M.D. (1973) Reciprocal averaging: An eigenvector method of ordination. J. Ecol. 61: 237–249.

Holm, L.G., Plucknett, D.L., Planche, J.V. and Herberger, J.P. (1977) The World's Worst Weeds. The University of Hawaii.

Kerner von Marilaun, A. (1891) Pflanzenleben. Bd. 1. Bibliographisches Institut Leipzig and Wien. 734 pp.

Kuiper, D. and Kuiper, P.J.C. (1979a) Ca^{2+} and Mg^{2+} stimulated ATPases from roots of *Plantago major* and *Plantago maritima*: response to alterations of the level of mineral nutrition and ecological significance. Physiol. Plant. 45: 1–6.

Kuiper, D. and Kuiper, P.J.C. (1979b) Ca^{2+} and Mg^{2+} stimulated ATPases from roots of *Plantago lanceolata*, *P. media* and *P. coronopus*: response to alterations of the level of mineral nutrition and ecological significance. Physiol. Plant. 45: 240–244.

Kuiper, D. and Kuiper, P.J.C. (1979c) Comparison of *Plantago* species from nutrient-rich and nutrient-poor conditions: Growth response, ATPases and lipids of the roots, as affected by the level of mineral nutrition. Physiol. Plant. 45: 489–491.

Lambers, H. (1979) Energy metabolism in higher plants in different environments. Thesis Groningen University: 176 pp.

Lambers, H. (1982) Cyanide-resistant respiration: a nonphosphorylating electron transport pathway acting as an energy overflow. Physiol. Plant. 55: 478–485.

Lambers, H., Posthumus, F., Stulen, I., van de Dijk, S.J. and Hofstra, R. (1981a) Energy metabolism of *Plantago lanceolata* L. as dependent on the supply of mineral nutrients. Physiol. Plant. 51: 85–92.

Lambers, H., Posthumus, F., Stulen, I., Lanting, L., van de Dijk, S.J. and Hofstra, R. (1981b) Energy metabolism of *Plantago major* L. ssp. *major* as dependent on the supply of mineral nutrients. Physiol. Plant. 51: 245–252.

Mølgaard, P. (1976) *Plantago major* ssp. *major* and ssp. *pleiosperma*. Morphology, biology and ecology in Denmark. Bot. Tidskr. 71: 31–56.

Mook, J.H., Haeck, J. and van der Toorn, J. (1981) Survival of *Plantago lanceolata* in vegetations of varying structure. Verh. Kon. Ned. Akad. Wetensch., afd. Natuurk. II, 77, I.O.O. Progress Report 1980: 32–35.

Oberdorfer, E. (1949) Pflanzensoziologische Exkursionsflora für Südwestdeutschland und die angrenzenden Gebiete. Ulmer, Stuttgart. 411 pp.

Primack, R.B. (1976) The evolutionary basis of population dynamics in the genus *Plantago*. Ph.D. thesis. Duke University, Durham, North Carolina.

Primack, R.B. (1978) Evolutionary aspects of wind pollination in the genus *Plantago* (Plantaginaceae). New. Phytol. 81: 449–458.

Ross, M.D. (1973) Inheritance of self-incompatibility in *Plantago lanceolata*. Heredity 30: 169–176.

Sagar, G.R. and Harper, J.L. (1964) Biological Flora of the British Isles. *Plantago major* L., *P. media* L. and *P. lanceolata* L. J. Ecol. 52: 189–221.

Skellam, J.G. (1952) Studies in statistical ecology. I. Spatial patterns. Biometrika 39: 346–362.

Stelleman, P. and Meeuse, A.D.J. (1976) Anthecological relations between reputedly anemophilus flowers and syrphid flies. I. The possible role of syrphid flies as pollinators of *Plantago*. Tijdschr. Entom. 119: 15–31.

Stulen, G., Lanting, L., Lambers, H., Posthumus, F., van de Dijk, S.J. and Hofstra, R. (1981a) Nitrogen metabolism of *Plantago lanceolata* L. as dependent on the supply of mineral nutrients. Physiol. Plant. 51: 93–98.

Stulen, G., Lanting, L., Lambers, H., Posthumus, F., van de Dijk, S.J. and Hofstra, R. (1981b) Nitrogen metabolism of *Plantago major* L. ssp. *major* as dependent on the supply of mineral nutrients. Physiol. Plant. 52: 108–114.

Troelstra, S.R. (1982) The growth and chemical composition of *Plantago lanceolata* on a NO_3/NH_4 nutrient solution; the estimation of nitrate and ammonium utilization from ionic balance aspects. Verh. Kon. Ned. Akad. Wetensch., afd. Natuurk. II, 79, Progress Report I.O.O. 1981: 55–65.

Troelstra, S.R., Sluimer, L., Smant, W., Wagenaar, R. and van der Meulen, M.A. (1981) On the soil chemistry of natural habitats of *Plantago* species and *Hypochaeris radicata* in various parts of the Netherlands in relation to chemical composition of plants. Verh. Kon. Ned. Akad. Wetensch., afd. Natuurk. II, 77: Progress Report I.O.O. 1980: 20–32.

van der Toorn, J., Haeck, J. and Mook, J.H. (1980) Some remarks on the demography of *Plantago* species. Verh. Kon. Ned. Akad. Wetensch., afd. Natuurk. II, 75, Progr. Rep. I.O.O. 1979: 19–23.

van der Toorn, J. and ten Hove, H.J. (1982) Variability of some leaf characters in *Plantago lanceolata*. Verh. Kon. Ned. Akad. Wetensch., afd. Natuurk. II, 79, Progr. Rep. I.O.O. 1981: 321–327.

van Dijk, H. (1984). Genetic variability in *Plantago* species in relation to their ecology. II. Quantitative characters and allozyme loci in *P. major*. Theor. Appl. Genet. 68: 43–52.

van Dijk, H. and van Delden, W. (1981) Genetic variability *Plantago* species in relation to their ecology. I. Genetic analysis of the allozyme variation in *P. major* subspecies. Theor. Appl. Genet. 60: 285–290.

van Dijk, H., Wolff, K. and de Vries, A. (1985) Genetic variability in *Plantago* species in relation to their ecology. III. Genetic structure of populations of *P. major*, *P. lanceolata* and *P. coronopus*. In: Genetic variability in *Plantago* species in relation to their ecology. Thesis, University of Groningen, 157pp.

Van Valen, L. (1965) Morphological variation and the width of the ecological niche. Amer. Nat. 103: 531–535.

Warwick, S.I. and Briggs, D. (1980) The genecology of lawn weeds. V. The adaptive significance of different growth habit in lawn and roadside populations of *Plantago major* L. New Phytol. 85: 289–300.

Westhoff, V. and den Held, A.J. (1969) Plantengemeenschappen in Nederland. Ed. Thieme and Cie, Zutphen, 324 pp.

POPULATION BIOLOGY AND HABITAT RELATIONS OF SOME HEMIPARASITIC SCROPHULARIACEAE

SINY J. TER BORG

ABSTRACT

Recent information is reviewed on the population biology and habitat relations of some hemiparasitic Scrophulariaceae (*Pedicularis*, *Rhinanthus*, *Melampyrum*, *Odontites*, *Euphrasia* and *Parentucellia* species). The species occur in a wide range of habitats; they are not host specific, except perhaps for *Melampyrum sylvaticum* in field conditions. A wide range of hosts may be parasitized, but a high inter- and intra-specific variation of 'host quality' was observed. There is also intra-specific variation of the parasitic ability of the parasites. These may reduce the development of their hosts, and so may affect the composition of a vegetation.

The demographic processes vary with field conditions – both in space (soil conditions, height above the water table) and time (position in a succession sequence) – and between ecotypes. The wide morphological variation of the annuals with respect to node number and flowering time can be related to the length of the growing season; the variation with respect to growth habit and leaf morphology is related to vegetation structure. Such relations may exist because of high light requirements, or because microclimate affects the balance between the hydrostatic pressures of host and hemiparasite, the driving force behind the transport system between the two.

INTRODUCTION

In the plant kingdom parasitism has evolved, probably along eight lines, in about 15 dicotyledon families, but in none of the monocots; it has been claimed for one gymnosperm, *Parasitaxus* (*Podocarpus*) *ustus* (Kuyt 1969, Ozenda and Capdepon 1979, Wilkinson and Musselman 1979).

Parasitism is not limited to a certain life form: there are parasitic trees (e.g., *Santalum album*) as well as shrubs (e.g., *Ximenia*, *Pyrularia*), and within herbaceous species both annuals like *Odontites*, *Orobanche*, and *Cuscuta* as well as long-lived perennials like most *Pedicularis* species can be found; some extreme forms live underground except for the flower

White, J. (ed.), The Population Structure of Vegetation.
 ISBN 978-94-010-8927-2

(*Rafflesia*). All of them, holoparasites without chlorophyll as well as the green hemiparasites are connected with roots and shoots of host plants by means of a haustorium developed on roots, stem or even leaves (Visser et al. 1978) of the parasite. Water, minerals and a wide variety of organic substances are transported through this organ. It is always a one-way flow as was shown by using radio-isotopes or dyes (Piehl 1963, Okonkwo 1966, Govier et al. 1967). The degree of dependence varies: some species can be grown to flower and set seed without a host, whereas others do not even germinate without a host stimulus. The effect on the host is variable too; it can be dramatic as in the case of *Orobanche* in *Vicia faba* (Pieterse 1979) and *Arceuthobium* in pine plantations (Kuyt 1969), or hardly measurable and difficult to detect in other cases (ter Borg and Bastiaans 1973). Strict host specificity does not seem to exist (Kuyt 1979).

The following review will be concerned with hemiparasitic Scrophulariaceae, in particular the taxa occurring in the Netherlands and some related ones. They are root parasites, connected to their hosts by xylem–xylem connections via the haustoria. Most may complete the life cycle without a host, i.e., they are "facultative autotrophic parasites" (Tsivion 1978). I will review the information on their population biology and habitat relations, in particular the part which became available since Kuyt (1969) and Hartl (1974) published their extensive reviews. In the final discussion attention is paid primarily to the influence of air humidity and of the structure and species composition of the vegetation on the distribution of the hemiparasites.

Nomenclature is based on *Flora Europaea* or follows the authors cited; in those cases the authorities are given in the text. Nomenclature of syntaxa is based on Oberdorfer (1979) and Westhoff and den Held (1969).

PERENNIAL TAXA

Pedicularis

The genus *Pedicularis* is the only one of the perennial hemiparasitic Scrophulariaceae which is represented in the Netherlands. It is one of the largest genera in the family, with several hundred species, all parasitic, and mainly occurring in arctic and alpine areas of the northern hemisphere. Some of these have been studied in more detail by Sprague (1962), Piehl (1963, 1965), McInnes (1972), Macior (1980), Lackney (1981) and others. Two taxa, *P. palustris* and *P. sylvatica*, occur in the Netherlands and will be discussed here.

P. palustris occurs all over Europe, except for the south, in marshy habitats, wet meadows and dune slacks, which are mown more or less irregularly or extensively grazed; they are irregularly flooded, with inundations during several months of the year. The common biennial ssp. *palustris* prefers the wetter habitats, the incidentally occurring annual ssp. *opsiantha* was found in the slightly drier places (ter Borg et al. 1980).

P. palustris is a character species of the Scheuchzerio-Caricetea fuscae (Oberdorfer 1979) or Parvocaricetea (Westhoff and den Held 1969). Monocots as well as dicots, herbs and sometimes shrubs were reported as hosts in the field and in a few growth experiments (Heinricher 1924, Weber 1976, Kwak unpublished).

Unlike all other species discussed here its fresh seeds germinate best at high temperatures (25–30°C); stratification widens this range. This holds also for the annual taxon (ter Borg et al. 1980). The seeds do not germinate under water.

Demographic data collected in a coastal dune system (de Noo unpublished) and an inland marsh area (Brouwer unpublished) (Figs. 1 and 2) and later scattered field observations showed that the plants behaved as strict biennials. Germination was concentrated in April/May, but in low sites could be postponed by inundations. One year massive germination was observed in July (Fig. 2A), probably of fresh seeds induced to germinate by the shock of rather early mowing in a warm period. Late seedlings have little chance to produce a bud which is large enough to overwinter and to produce a flowering plant the next year. There is no substantial seed bank. Therefore the populations are divided into two alternating generations, fairly strictly separated from each other.

The observations showed that the fluctuations in time and the variation in space as observed in many *P. palustris* populations can be related to a large extent to the fluctuating water table in their habitats. Since there is

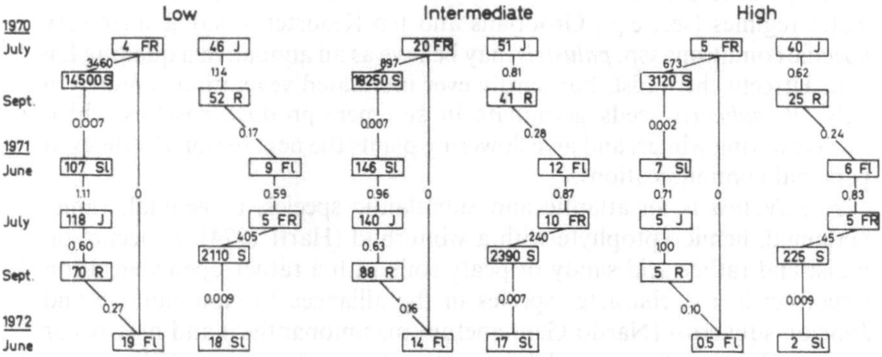

Fig. 1. Demographic data of a population of ***Pedicularis palustris*** at Kapeglop, an older part of the dune system of Schiermonnikoog, one of the Frisian Islands in 1 m² plots at different heights above the ground water table, all with species of the class Parvocaricetea. 'Low': 5 plots at 0–10 cm height, differentiated by species of Littorellion, Phragmition and by several species of Agropyro-Rumicion crispi communities; 'Intermediate': 8 plots at 10–25 cm height, differentiated by species of Phragmition and by some species of Agropyro-Rumicion crispi communities; 'High': 8 plots at 25–50 cm height, differentiated by species of Junco-Molinion, Arrhenatheretalia and by some species of Violion caninae communities. Heights are relative, above the lowest point. Data on seed production are based on counts of fruits. The number of seeds per fruit was counted in two samples, with ca. 16 seeds per fruit. Data of de Noo (unpublished). S, seeds, S^1 before, S^2 after predation, S^3 seeds on the soil; G, germinated; Sl, seedling, J, juvenile; V, vegetative; Pl, plants; Fl, flowering; FR, fruiting and ripening; fr, fruits; R, rosettes; B, winter bud (mm diameter); †: dead. Numbers.m^{-2}.

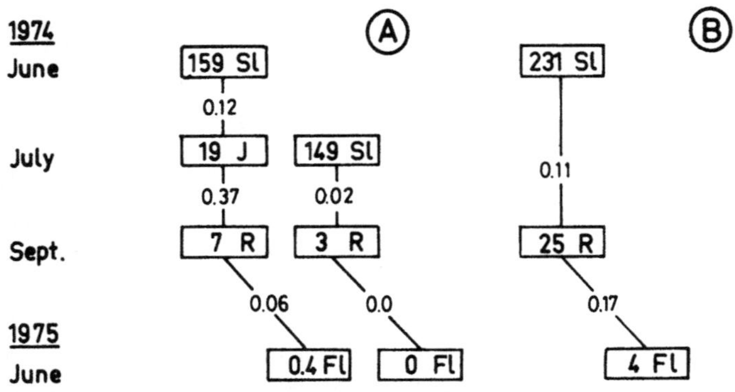

Fig. 2. Demographic data of a population of *Pedicularis palustris* in an inland marsh area, near Vries, N. Netherlands, with species of the Parvocaricetea, Magnocaricion, Agropyro-Rumicion crispi and Molinio-Arrhenatheretea communities. A: mown in July, data from 2.4 m^2; B, undisturbed, data from 4.9 m^2. Letter symbols as in Fig. 1. Data of Brouwer (unpublished). Numbers.m^{-2}.

no seed bank to take over after catastrophes such as late inundations or early mowing, a water table below ground level should occur yearly, early in the year, at least locally, to allow seedlings to germinate. Once established the plants are able to withstand long periods of flooding, even during the period of growth. These conditions apparently are fulfilled in the Scheuchzerio-Caricetea and the Parvocaricetea for which *P. palustris* is a character species, but less so in other marsh habitats, each with slightly different water regimes (see, e.g., Grootjans and ten Klooster 1980). Under very specific conditions ssp. *palustris* may behave as an annual: in a quaking fen near Utrecht the moist, but hardly ever inundated vegetation is mown in July. *P. palustris* seeds germinate in summer, produce rosettes which survive during winter, and give flowering plants the next season (Verhoeven personal communication).

P. sylvatica is an atlantic and subatlantic species, a biennial, rarely perennial, hemicryptophyte with a winterbud (Hartl 1974). It occurs on moist and rather acid sandy or peaty soils with a rather open vegetation structure; it is a character species in the alliances Violion caninae and Juncion squarrosi (Nardo-Gentianetum pneumonanthes) and may occur also in Caricion fuscae and Ericion tetralicis communities (Oberdorfer 1979; Westhoff and den Held 1969). In the northern part of the Netherlands it was either found in Ericetum tetralicis (Ericion) communities kept open by a strongly fluctuating water table and irregular inundations in winter, or in Nardo-Gentianetum pneumonanthes (Violion caninae) communities, where anthropogeneous effects like mowing, grazing, treading or path clearing cause a certain degree of openness and enrichment (Everts unpublished).

When plants were excavated haustoria were noticed on several monocot and dicot herbs, but also on *Picea abies*. In an Ericion community large plants with thick haustoria were found in particular around *Molinia caerulea* (Weber 1976, Everts unpublished). No growth experiments have

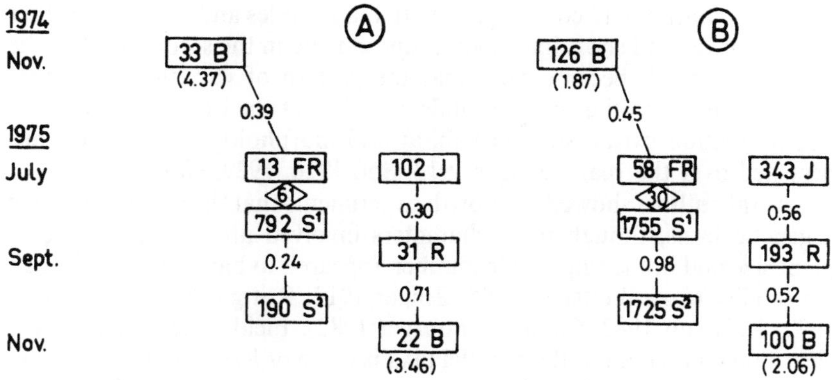

Fig. 3. Demographic data of two populations of *Pedicularis sylvatica*. A: a small population in an Ericion tetralicis community, near Zeegse, N. Netherlands. Data from 3.25 m^2. B: a population in the verge of a sandy lane, in a Violion caninae community, near Anlo, N. Netherlands. Data from 3.75 m^2. Letter symbols as in Fig. 1. Numbers.m^{-2}. Data of Everts (unpublished).

been reported for this species, probably because its seeds are difficult to germinate: they require ca. 5 months chill treatment (Weber 1981, ter Borg unpublished).

Demographic data were collected in two populations, one in each of the habitat types (Fig. 3, Everts unpublished). Some differences can be observed: plants in the Ericion were larger, but few; they suffered intense seed predation from larvae of *Phytomyza spec.* (*Agromyzidae*). In the Violion habitats the plants were smaller, but the numbers higher, and no seed predation was noticed. Note however, that the data concern one year only! Plant size and survival of the first year individuals appeared to be positively correlated with openness of the plant cover. The effect of this factor was proven directly in a test when the overlying vegetation was bent back: this resulted in higher seedling survival and larger winter buds.

In both *Pedicularis* species survival of the largest winterbuds was below that of the medium-sized ones (ter Borg 1979). Scattered field observations suggested that in *P. palustris* the bigger ones had been eaten by water fowl, whereas in *P. sylvatica* the largest were in the lower sites where growth conditions were best, but winter inundations might cause mortality.

ANNUAL TAXA

Several of the annual genera, e.g., *Euphrasia*, *Rhinanthus*, *Melampyrum*, are notorious for the taxonomic problems they offer, both at the specific and the infraspecific level. They gave rise to the concept of seasonal dimorphism, introduced by Wettstein (1895, 1899). He distinguished sets of parallel taxa: an early flowering one (the "vernal" or "aestival") with few nodes below the inflorescence, few branches, rather long internodes, relatively broad leaves, producing seed before hay-making; and a late flowering "autumnal" counterpart, with many nodes and branches, rather narrow leaves and short internodes, reproducing in the second part of the summer. Later it became clear that the pattern of variation was more complex and has other backgrounds as well, but the original idea that a strong relation exists between habitat and morphological variation has persisted over the years (Zinger 1912, Soó 1926, 1929, Hartl 1974).

Several authors showed by growth experiments that the differences have a genetic basis, though most characters involved are very plastic, e.g., plant size and branching. Node number appeared to have a relatively high heritability (e.g., Wettstein 1895, Zinger 1912, Rottgardt 1956, ter Borg 1972, Karlsson 1982, Campion-Bourget 1982). Field observations made clear that emergence of the parallel taxa is more or less simultaneous, and for *Rhinanthus serotinus* (Schönh.) Oborny (syn. *R. angustifolius*) (*cf.* Gutermann 1975) the relative growth rates were shown to be similar (ter Borg 1972). Therefore it must be primarily the different patterns of development which cause the variation.

The differences can be placed into three main sets: one is the number of nodes of the main stem below the inflorescence; this determines the time required between emergence and first seed production, as well as the potential number of branches and hence of seeds. The second concerns growth habit, i.e., length of the internodes and pattern and shape of branching. A third minor group includes size and shape of leaves. For node number and the resulting onset of seed production, it is clear that the length of the growing period is decisive. It may be determined by haymaking, as observed in *Euphrasia* (Wettstein 1895) and *Rhinanthus* (e.g., Zinger 1912, ter Borg 1972). It may also vary along a geographical gradient: node numbers increase from north to south in the *Euphrasia stricta* group (Karlsson 1976) and *Rhinanthus montanus* Saut. (Vasilçenko 1955). In the arctic-alpine taxon *Melampyrum pratense* ssp. *alpestre*, with only 1–3 nodes, the short growing season, either by geographical position or altitude, may play a role. And low node numbers may be due also to a short growing season as effected by summer drought: in N.E. Belgium on top of a slope *R. minor* plants with ca. 7 nodes were found in a dry Thero-Airion/Violion caninae vegetation; lower down was a more mesic Mesobromion where *R.minor* had ca. 9 nodes (Hillegers 1983).

The second set of characters, concerning growth habit, varies with flowering time to some degree, but it may also vary widely between populations with similar flowering times. Again a relation with mowing regime was observed, within autumnal *R. serotinus*. One ecotype grows in meadows which are mown for hay twice a year, a rare type of management. When young, the plants are rosette-like with very short lower internodes and upward bending branches. They were found near Meppel, in the slightly higher parts of alluvial meadows in a Calthion vegetation. When the fields are mown for the first time, just the tops of the branches are cut. After that the plants may regenerate very quickly, are in full flower in July/August, and set seed before the second mowing period (late August). As shown experimentally, other autumnals lack this regeneration capacity (Table 1, Zinger 1912, ter Borg 1972).

Table 1. The effect of June mowing on regeneration and production of fertile branches in three autumnal *Rhinanthus serotinus* ecotypes. Seeds sown in a meadow in autumn, plants counted in June, regeneration and reproduction measured in August. Data from twelve 1 m^2 plots per ecotype, each sown with 2–3000 seeds. After ter Borg (1972).

Provenances of the ecotypes	Dry, heath with grass; not mown	Moist, ditch sides and road verges, mown irregularly	Wet meadow, mown twice yearly
Plants.m^{-2}, 12 June	44.8	22.5	11.3
Fraction surviving	*0.07*	*0.48*	*1.16*
Plants.m^{-2}, 11 August	3.3	10.8	13.1
Branches.$plant^{-1}$	2.0	4.3	5.4
Fraction fertile branches	*0.75*	*0.06*	*0.60*
Fertile branches.$plant^{-1}$	1.5	0.3	3.2
Fertile branches.m^{-2}	5.0	3.0	42.8

In general, variation of the growth habit seems to have adaptive value, as was most clearly illustrated for *Euphrasia* species by Karlsson (1974) (Fig. 4). The causal factors will be discussed later.

Rhinanthus

The genus *Rhinanthus* has its centre of diversity in Europe, with some extension into Asia; one section (*Minores*) has reached N. America, possibly introduced by man. As weeds of winter grain and grassland some species have received considerable attention (Fürst 1931, Tolwinska 1963, Mizianty 1975). Some of the ca. 25 species have a wide distribution, e.g., *R. minor*, others can almost be called endemics, e.g., *R. antiquus*. Some species are well isolated, others easily form hybrids and hybrid swarms (Campion-Bourget 1980, Kwak 1980).

All *Rhinanthus* species are summer annuals. The seeds germinate only after 1–3 months chilling at 5°C; up to ca. 10% may persist as a seed bank. Growth experiments and field observations showed a wide range of potential hosts, monocots and dicot herbs as well as some trees (*Salix*, *Populus*). The performance varies with host species and growth conditions; a host is not strictly required (Heinricher 1924, ter Borg 1972, Weber 1976, 1981, de Hullu 1984). All *Rhinanthus* species prefer grass vegetation; meadows, road verges, dune grasslands, usually moist, never wet, sometimes dry, from sea level (rarely in a salt marsh) up to alpine areas; they do not stand intensive grazing. The distribution of some taxa in Eurasia was correlated with climate and the relation was tested by plotting the isolines for temperature and precipitation on the N. American map (Fig. 5). A clear similarity between the expected and the observed distribution of *R. minor* was found, i.e., between distribution and macroclimate (Prins unpublished).

Three species occur in the Netherlands, *R. serotinus*, *R. minor* and *R. alectorolophus*. Studies of their distribution in S. Netherlands (Hillegers 1981) and the North Holland dunes (Drent unpublished) showed that *R. minor* prefers lower and more open turf (Mesobromion and Koelerio-Corynephoretea), whereas *R. alectorolophus* and *R. serotinus* grow in more

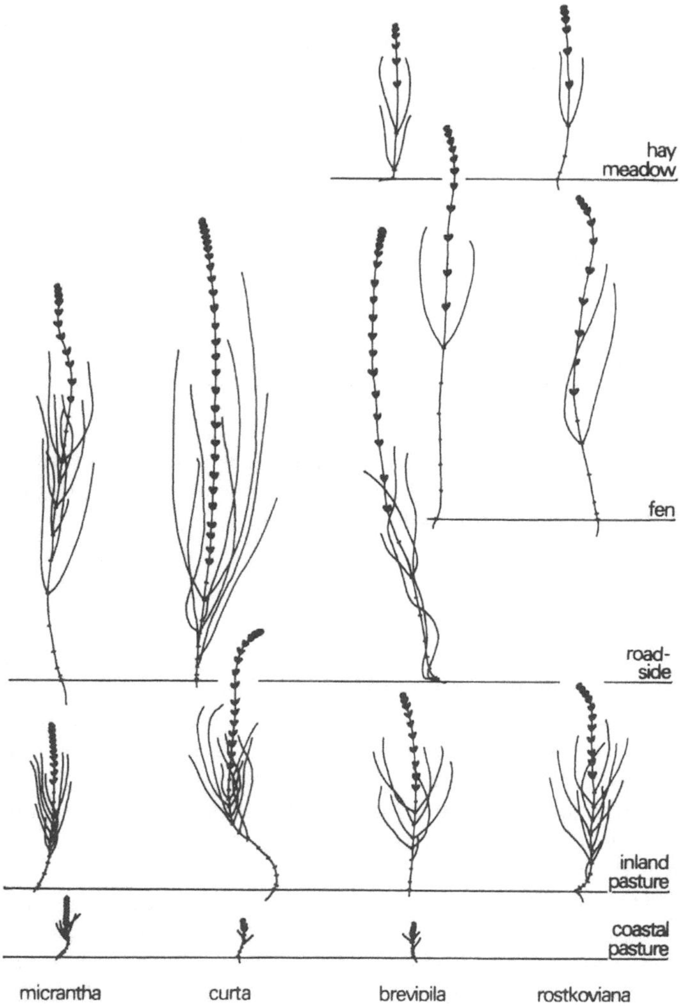

Fig. 4. Parallel, habitat correlated variation of growth habit within four species of *Euphrasia*. Nodes, branches and flowers are represented for average specimens, collected from natural habitats in Southern Sweden. The characteristic features to a great extent remain constant under cultivation. From Karlsson (1974), with kind permission of the author and the publishers of Botaniska Notiser.

mesic conditions, sometimes under shrubs (with more species of Trifolion medii and Prunetalia communities) (Table 2). There is some experimental evidence that the latter have a higher capacity to etiolate (Boudewijn and Blacquière, unpublished).

Densities of *Rhinanthus* populations may vary widely (e.g., Grubb et al. 1982). For meadows in the N. Netherlands the variation could be related to a colonization process which started when management was changed.

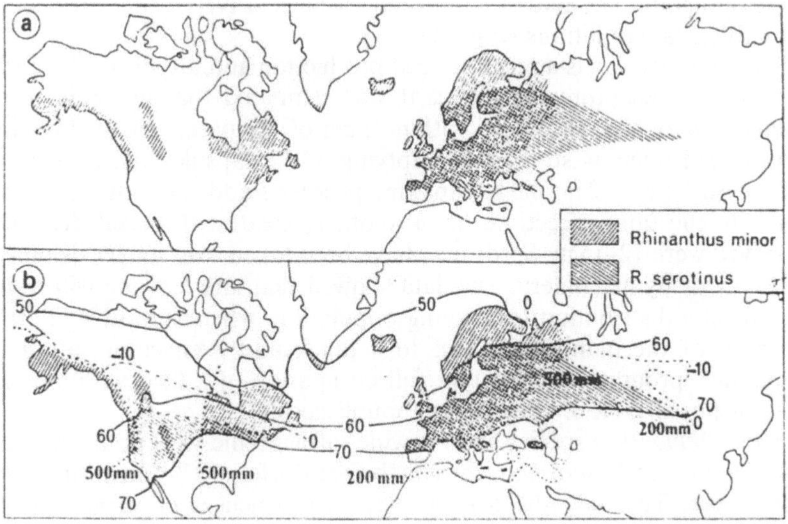

Fig. 5. The distribution of ***Rhinanthus minor*** and the large flowered ***Rhinanthus*** species (mainly *R.serotinus*) related to climate. (a) The distribution in Eurasia and N. America. (b) The same, with the 200 and 500 mm isohypses included and the isotherms for January (– 10 and 0°F) and July (50, 60 and 70°F). Further explanation in text. Maps produced by Prins (unpublished).

These fields were under more or less intensive agricultural use until one after the other was brought under a regime of mowing without fertilization (extensive management). Then the vegetation changes from Agropyro-Rumicion crispi and Cynosurion to Calthion and Junco-Molinion communities (van Duuren et al. 1981). Within eight years *Rhinanthus* populations (*R. serotinus* mixed with *R. minor*) may increase from zero to very high densities (up to $1000/m^2$; '3' on the scale of Braun-Blanquet) and then

Table 2. Habitat differences of the three *Rhinanthus* species in the Netherlands. % of associated species in various syntaxa are shown, calculated from relevés collected in S. Netherlands and adjacent Belgium for *R.minor* and *R.alectorolophus* (Hillegers 1981) and in the coastal dunes between Bergen and IJmuiden for *R.minor* and *R.serotinus* (Drent unpublished).

	S. Netherlands		Coastal dune area	
	R.alectorolophus	R.minor	R.serotinus	R.minor
Number of relevés	19	20	30	31
Species of				
Mesobromion	27.1	33.8		
Trifolion medii	26.7	18.2	1.7	0.9
Arrhenatherion	37.5	40.4	10.3	9.8
Koelerio-Corynephoretea			21.2 }	26.8 }
idem and Festuco-Brometea			14.6 }	20.3 }
Prunetalia			23.9	16.6
Others	8.8	7.6	27.1	25.6

decrease again. The pattern appeared to be relatively little affected by climatological conditions (Fig. 6).

This colonization is a result of seed production and dispersal. By mark, replace and recapture techniques it was estimated that in undisturbed plants 90% of the seeds were within 25 cm of the parent plant. Mowing during seed ripening stimulated emptying of the capsules and resulted in a dispersal of over 2 m; the hay-making processes added another 6 to 7 m, either in the same direction, or in another; greatest dispersal distances observed were 12–15 m. Seeds may have been taken over longer distances with hay or by machinery. The data showed that late mowing will result in a patchy distribution in a young population, whereas mowing during ripening of the plants may lead to a quick and even colonization. A simulation program showed that full occupation of a field within a few years is possible (K.J. Nanninga, unpublished).

Most *Rhinanthus* species show a wide infraspecific variation. This certainly holds for *R. serotinus*, for which Oberdorfer (1979) mentions eight sub-species. To study the demographic consequences of variable node numbers, and the resulting differences in potential seed production, counts were made in densely populated sites in aestival populations growing in meadows and autumnal ones in road verges (ter Borg 1972, Table 3). The data showed that generally plant density is higher in early than in late populations, but that the latter may produce far more fruits per plant. This results in similar numbers of seeds per square meter, as were also reported by Silakova (1975; 2416–30,950 m^{-1}). To see how the difference works out over the whole life cycle demographic data were collected in a vernal and an autumnal population, the first growing in a meadow mown for hay (Calthion and Cynosurion communities), the latter in an extensively

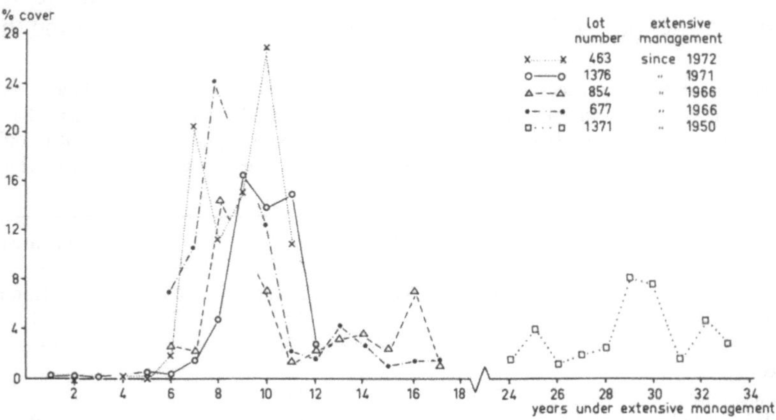

Fig. 6. Changes of densities of *Rhinanthus serotinus* populations with time. % cover is plotted against the number of years since extensive management has been introduced. Data are the average coverages of *Rhinanthus* in five to eight relevés taken from 1972 through 1983 (1975 excepted) in five meadows near each other in the valley of the Drentsche A, N. Netherlands. (cf. van Duuren et al. 1981). Data of J.P. Bakker (Dept. Plant Ecology, Univ. of Groningen).

Table 3. Reproduction and establishment in aestival and autumnal populations of *Rhinanthus serotinus* in N. Netherlands. Data on seed production are based on numbers of fruits.plant^{-1}; the maximum number of seeds per fruit is 18, the actual number was ca. 10 in aestivals and ca. 8 in autumnals. After ter Borg (1972).

	Fruits.pl^{-1}	Plants.m^{-2}	Seeds.m^{-2}	% Established
Aestivals	5.3	250	12,300	2.0
(Range of pop. means)	(3.2–7.7)	(60–600)	(4,600–25,000)	(1.3–3.1)
(Overall range)	(0–33)			
Autumnals	20.1	57	8,000	0.7
(Range of pop. means)	(7.4–43.2)	(5–200)	(1,600–29,000)	(0.3–1.7)
(Overall range)	(0–203)			

grazed pasture (Agropyro-Rumicion crispi) (Leemburg–van der Graaf unpublished, Fig. 7). It can be noticed that here the fruit numbers of autumnal plants were less than twice those of the aestivals. The loss in the autumnals was higher over the whole life-span (seedling establishment, smaller second cohort, lower proportion of seeds producing a fertile plant); their persistent seed bank was largest, as was also found in other populations (ter Borg 1972). The data suggest that the vernals require a rather constant, reliable type of habitat, whereas the autumnals with a high maximum potential are adapted to less stable and uniform conditions.

Melampyrum

The genus *Melampyrum* includes 20–30 species, with two main areas of distribution, one in Europe extending into Asia, and a second one in E. Asia, extending into tropical ranges; one species, *M. lineare* Desr., has a wide distribution in N. America. Interspecific hybridization is rare, if it occurs at all. In some species a wide infraspecific variation has been described (Soó 1926, Hartl 1974).

The further discussion here will concern some Eurasian species and *M. lineare*. All are summer annuals, with seed germination in autumn and emergence in spring, regulated by a complex system of double dormancy. Some 20–80% of the seeds are dormant the first season, and account for the presence of a substantial persistent seed bank (Curtis and Cantlon 1963, Horrill 1972, Oesau 1973, 1975, Masselink 1980). The proportion of dormant seeds appears to be at least partially due to climatic conditions during seed ripening (Masselink 1980).

The species occur from sea level up to ca. 2000 m in Trifolio-Geranietea, Vaccinio-Piceetea and Querco-Fagetea communities, usually along paths, woodland margins etcetera. In two species, *M. arvense* and *M. barbatum*, weedy subspecies have developed, formerly common in winter grain (Caucalidion) communities (Gislén 1949, Hartl 1974, Benkov 1978). All species require a host to complete the life cycle; some have a more or less strong preference for woody species, as shown by excavations and growth experiments (Heinricher 1924, Smith 1963, Bitz 1970, Horrill 1972, Weber 1976, Masselink unpublished). Barsukova and Pyatkovskaja (1977) noticed a

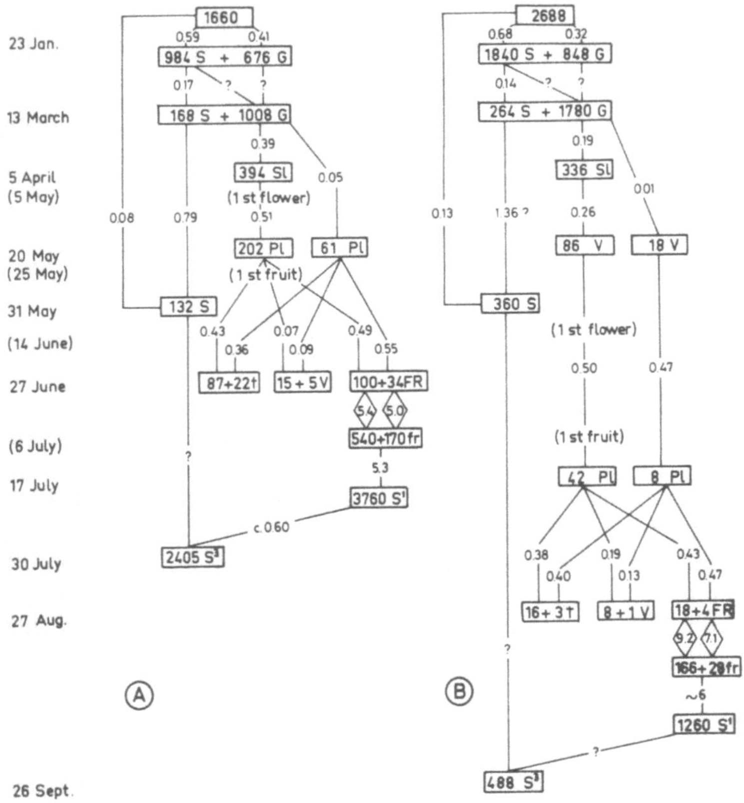

Fig. 7 Demographic data of two populations of *Rhinanthus serotinus*, a vernal and an autumnal one, both growing in grasslands in N. Netherlands. The first was in a meadow with Calthion and Cynosurion communities, the latter in a pasture extensively grazed by ponies and horses (Agropyro-Rumicion crispi). Data are means of 11 and 7 plots resp., and are in numbers.m^{-2}. Data on seeds are based on five 5 cm diameter soil samples per plot, those on plants come from $\frac{1}{4}m^2$ plots. Data of Leemburg–van der Graaf (unpublished). Letter symbols as in Fig. 1.

decrease of *M. pratense* when roots of nearby *Picea abies* were cut. A close relation between the distribution of *M. sylvaticum* and *Picea abies* was described by Albrecht (1973). She excavated many plants which appeared to be connected to *Picea* only, in spite of the nearby presence of several other species which had functioned as host plants in growth experiments. She also indicated the variation of the distribution of *M. sylvaticum* with altitude along a geographical gradient (Fig. 8), which might be related to the distribution of *Picea*. This example illustrates the existence of a system of 'principal' and 'minor' hosts in the hemiparasitic Scrophulariaceae, as was described for mistletoes by Atsatt (1983).

Variation of host preference at the infraspecific level was indicated by Bitz (1970). He studied *M. pratense*, with ssp. *vulgatum* (Pers.) Ronniger on base-rich soils, and ssp. *pratense* in acid habitats. Cultivation experiments

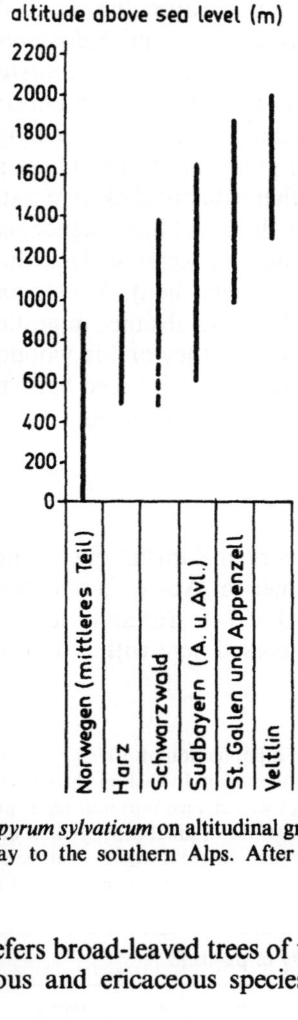

Fig. 8. Distribution of *Melampyrum sylvaticum* on altitudinal gradients along a geographical gradient from central Norway to the southern Alps. After Albrecht (1973), with kind permission of the author.

showed that the first prefers broad-leaved trees of the richer soils, and the second *Betula*, coniferous and ericaceous species, occurring in poorer habitats.

Extensive studies of the population dynamics of two *Melampyrum* species indicated that they may be strongly affected by animals. In a pine–oak–aspen forest in N. Michigan the grasshopper *Atlanticus testaceus* (Scudder) appeared to exert a considerable effect on *M. lineare*, first by feeding on seedlings, later by eating leaves and branches. Major effects were also noticed from ants and beetles, transporting and eating the seeds. After an insecticide treatment the *Melampyrum* populations increased at high rates: two- and three-fold after two and three years respectively. Moreover the pattern of distribution of the plants changed from even to more or less patchy (Cantlon 1969, Manley et al. 1975).

Masselink (1980) studied *M. pratense* in an open Querceto-Betuletum in N. Netherlands and constructed a life-table for its seed population. He compared the processes in different microhabitats, and so could explain why *M. pratense* plants in this habitat with little undergrowth are concentrated around scattered tufts of *Deschampsia flexuosa*: not because the grasses serve as hosts, but because the seeds find shelter there from predating mice (*Mus musculus* (L.) and *Sylvaemus sylvaticus* (L.)). In the same area data on the growth phase of *M. pratense* were collected. Some small scale experiments indicated that both naturally and experimentally different densities affected mortality in the seedling and vegetative phases. Within a few years the original densities had been reached again (Table 4). Effects of grazing and other catastrophes were tested by removal of plants and plant parts. At a low degree of interference seed production was even stimulated, but the population seems to lack the capacity to withstand strong attacks (Bijpost, unpublished). *M. pratense* plants produce few seeds and a mechanism for long distance migration is lacking. Therefore a rather stable habitat, such as they find in wooded areas, seems to be an important prerequisite. The persistent seed bank may be indispensable to overcome temporarily adverse conditions.

Euphrasia

The genus *Euphrasia* occurs in Eurasia, the northern part of N. America, and some parts of S. America, Australia and New Zealand, always in a temperate climate (Hartl 1974). Certainly it is the taxonomically most complex genus to be discussed here with a total of between 100 and 250

Table 4. Demographic data of a population of *Melampyrum pratense* in an open oak-birch woodland near Annen, N. Netherlands. In each of three sites with low, medium and high population density. A 1 m^2 plot was divided into four segments and weeded down to lower densities (25, 50, 75 and 100% of the original density). The data are arranged according to weeded densities (left) and original densities (right) (Numbers.m^{-2}). Data on seed production are based on counts of fruits.pl^{-1}, the maximum number of seeds per fruit is 4, the actual number was 2.9. Data of Bijpost (unpublished).

	Density of plants.m^{-2}						
	25%	50%	75%	100%	Low	Medium	High
1976							
28 May	314	529	806	1016	232	648	2180
Fraction surviving	*0.95*	*0.84*	*0.81*	*0.75*	*0.81*	*0.82*	*0.72*
1 July	299	436	656	762	188	528	1576
Fraction surviving	*0.72*	*0.85*	*0.80*	*0.79*	*0.83*	*0.64*	*0.84*
1 Sept.	215	372	522	604	156	336	1325
Seeds. $plant^{-1}$	*11.2*	*10.5*	*8.6*	*6.1*	*7.5*	*11.4*	*4.5*
Seed production	2415	3925	4506	3709	1171	3944	5962
1977							
26 May	387	492	527	325	32	460	484
1978							
25 May	536	806	668	646	256	812	876
Fraction surviving	*0.66*	*0.65*	*0.58*	*0.78*	*0.78*	*0.80*	*0.76*
5 July	355	524	384	504	200	648	668

taxa. Extensive descriptive and experimental work on the ca. 50 European species is gradually leading to consensus on their taxonomic treatment (Yeo 1978b). It is supposed that the complexity has been caused by migrations induced by glacial and human history; these lead to variable selection and changing contacts between taxa, resulting in hybrids and hybrid swarms, introgression, incipient species, and more or less clearly distinct ecotypes. The genus includes diploid and tetraploid species, some of which might hybridize (Karlsson 1976, 1982, Yeo 1978a, b).

The European *Euphrasia* species are all summer annuals and can be found in relatively dry habitats with a low and open vegetation, from coastal to alpine areas (Hartl 1974, Yeo 1978b, Oberdorfer 1979). As an illustration of the habitat differentiation the information given by Oberdorfer (1979) is summarized in Table 5. The plants can be grown autotrophically, but may profit from a wide range of hosts (e.g., Wilkins 1963). Their seeds require low temperatures to germinate. There is no seed bank (Karlsson 1982, Schenkeveld and Verkaar 1984).

Studies of the chalkgrassland vegetation in S. Netherlands included *E. stricta*. They showed clustering of seeds, due to limited seed dispersal, and

Table 5. Habitat data for *Euphrasia* species, summarized from Oberdorfer (1979). x: the species is a character species in the syntaxon; o: the species occurs preferentially in the syntaxon.

	E. rostkoviana	*E. kerneri*	*E. picta*	*E. pectinata*	*E. stricta*	*E. salisburgensis*	*E. hirtella*	*E. minima*	*E. nemorosa*	*E. drosocalyx*	*E. micrantha*
Molinio-Arrhenatheretea	x										
Molinietalia											
Molinion		o									
Arrhenatheretalia	o										
Poion alpinae			x								
Festuco-Brometea											
Brometalia				x							
Mesobromion					x	o					
Seslerietea											
Seslerietalia						x	o				
Caricion ferrugineae			o								
Salicetea herbaceae											
Salicetalia herbaceae											
Salicion herbaceae								x			
Nardo-Callunetea											
Nardetalia					x		o		x		
Nardion			o					o	o	o	
Violion caninae	o								o		
Vaccinio-Genistetalia											
Genistion											o

site preferences changing with age: seedlings were most abundant in open sites, but suffered high mortality; juveniles were mainly in vegetation of medium density, and reproductive plants appeared to prefer vegetation of relatively lower density (Verkaar et al. 1983, Schenkveld and Verkaar 1984, see also During et al. 1985).

Infraspecific variation of *E. rostkoviana* was studied by Schaeftlein and Karlsson. Schaeftlein (1967) in particular studied the variation of the numbers of leaf glands and noticed a decrease with altitude. Karlsson (1982) related the very restricted distribution of the subspecies in Southern Sweden to the habitat but in particular to dispersal caused by human migrations.

Odontites

The red flowering *Odontites* species occurring in Europe, Asia and eastern N. America, are sometimes taken together as *Odontites rubra* (Baumg.) Opiz (e.g., Hartl 1974) or *O. verna* (Bell.) Dum. (e.g., *Flora Europaea*). I will follow the classification used by Oberdorfer (1979) and clearly described by Snogerup (1982a, 1983), who distinguish three species:

O. verna (Bell.) Dum., $2n = 40$, a weed of winter grain in M. Europe and southern Scandinavia, in Aperion and Caucalidion communities;

O. litoralis Fries, $2n = 18$, an early flowering species of the N. European salt marshes, with ssp. *litoralis* in the Armerion alliance and the eastern ssp. *fennica* along the Gulf of Finland, between broad leaved forbs;

O. vulgaris Moench (= *O. serotina* Dum.), $2n = 18$, a variable species, from mid- to late flowering, with a wide distribution, occurring in a broad range of habitats: salt marshes, inland grasslands, road verges, waste places etc.; some authors distinguished separate ecotypes – an aestival, an autumnal and a salt tolerant one (e.g., Rottgardt 1956).

O. litoralis and *O. vulgaris*, when occurring together in salt marshes, take slightly different positions: *O. litoralis* lower down with *Juncus gerardii*, and *O. vulgaris* higher up with *Festuca rubra* (Snogerup 1983; Meyering unpublished). The species may hybridize and form hybrid swarms. *O. verna* and *O. vulgaris* when in the same arable field are also ecologically separated, the first being in the middle of the field, the latter on the edges (Shmidt 1965, cited by Snogerup 1983), or the first on the drier, the latter on the wetter places (Snogerup 1983). The three species have been reported for the Netherlands, but only the latter is common. Both others are extinct or nearly so.

All species are summer annuals. Chancellor (1976) studying seed germination of a road-side population of *O. vulgaris* concluded that the moment of germination seems to be solely regulated by temperature; the main prerequisite is a cold period below 15°C. Brenchley and Warington (1936) reported that dormancy in *O. vulgaris* from arable land may last for at least nine years.

Odontites species can be grown without hosts, but are favoured by their presence, particularly by the presence of a host mixture (Govier et al. 1967). Snogerup (1982b) reported effects of hosts on some of the taxonomically important characters (e.g., plant size and growth habit, but also leaf succulence). Based on her data she suggested the existence of variable resistance of hosts on the one hand and adaptation to particular host species by the populations of the parasite on the other.

Population variation along gradients was studied in various areas. Van Tooren et al. (1983) recorded the 'shuttling' of *O. vulgaris* along a dune slope in a flooded beach plain on Schiermonnikoog over a 10 year period. The shifting lower margin could be related to high salinity caused by flooding and a successive dry spring period, and to fresh water inundations after late spring storms when the seedlings had already emerged. The fluctuations higher up were related to summer drought on the dune slope. In this area plants with up to 1000 fruits were collected from gulls' nesting sites (Meyering unpublished).

Schmidt (1963) studied the variation of node numbers along gradients, and found them to be ca. 8 (7–15) at the lower end and ca. 13 (8–20) in the higher parts, both along a river and in a salt marsh. Differences at this scale probably have a genetic basis and may at least partly be due to local selection processes, but at the moment the reasons can only be guessed at.

Parentucellia

Little has been published about the genus *Parentucellia*, with two species in S. and W. Europe. One of them, *P. viscosa* reaches its northern limit in S.W. Netherlands, though sporadic, non-persistent populations may occur in more northern sites, probably imported with hay or grass seed (de Jongh 1962, Hartl 1974, Rehmböck 1975, Sunesson 1976). The same means of transport may have resulted in its introduction into N. America (Oregon and N. California, Atsatt and Strong 1970). It is usually in rather open vegetation types in dry habitats; in its northern areas it occurs on sandy soils in a vegetation of grassland species mixed with Nanocyperion elements (van Haperen and de Kogel 1981).

Table 6. The effect of temperature on germination of *Parentucellia viscosa*. 350–1200 seeds scattered on filter paper in petri dishes and tested at various temperatures and 16 h light per day, after 0, 2, 4 and 6 weeks stratification respectively. Data are germination percentages and – in brackets – the number of days required to reach half of the final value (total number of days after the start of the experiment, and the number after the transfer from chilling). (Original data.)

Temperature °C	5	10	20	30
Weeks of stratification				
0	85.5 (42)	57.1 (27)	0	0
2	–	94.0 (28; 14)	78.2 (23; 9)	0
4	–	97.5 (34; 6)	88.3 (31; 3)	37.1 (30; 2)
6	–	97.4 (45; 3)	84.1 (43; 1)	69.0 (43; 1)

The seeds are small, with sizes similar to those of *Striga* (0.25 × 0.40 mm), but unlike those of the latter they do not require a host stimulus to germinate. Tested on filter paper, fresh seeds germinated best at 10°C; after 0–6 weeks stratification the temperature range widened quickly (Table 6).

Parentucellia can be grown autotrophically; the five hosts tested by Atsatt and Strong (1970) only had a small positive effect. No intraspecific variation has been reported for the species. Karlsson (1974) ascribed this to the fact that its habitat does not display such a wide variation as that of the hemiparasites of the manmade forestless habitats in temperate Europe.

DISCUSSION

Hemiparasitic Scrophulariaceae can be found in all but aquatic and marshy, and open and unstable habitats (Table 7). Many authors have noticed that annuals in general are rare in closed vegetation, except for the hemiparasitic species. Karlsson (1974) supported and quantified both notions when he compared the habitat data given in N.W. European floras on hemiparasitic Scrophulariaceae with those on monocarpic species in general (Table 8). He discussed the reasons for the aberrations in some detail. He also pointed to the fact that the hemiparasites occur in more habitats than the autotrophic monocarps. Similarly Fresco (1980) found a wider amplitude for *Rhinanthus* than for other meadow species when he calculated response curves. The parasitism in both cases is the obvious reason.

What is the influence of the species composition of the vegetation on the distribution of the hemiparasites? For some taxa, e.g., *Melampyrum* spp. data are available which suggest a rather close relation between the hemiparasites and particular hosts. For others, long lists of hosts are given, but for several reasons little can be really said about their role. In the first place the presence of haustoria does not imply automatically their functioning. In fact several haustoria observed could not be functional at all: on sand grains, on a seed, on dead material (e.g., Piehl 1965, and unpublished data). Moreover, host lists based on excavations often primarily mention the species with haustorial connections, not those without. Thus they leave half the question unanswered. And finally, information from growth tests can be very contradictory. Species may act as good hosts in one test, and be poor in the next. This may be due to genetic variation in host and parasite (see Snogerup 1982b; Atsatt and Strong 1970) or to growth conditions. This makes it extremely difficult to indicate to what extent species composition of a host vegetation influences distribution of hemiparasites, the more so since information on some hosts does not yet cover a complete vegetation, which may include both good and bad host species. The data presented by Snogerup (1982b) and de Hullu (1984) indicate that effects do exist. A further quantitative analysis must give more information on their relative importance.

It is clear that the parasites depend on a host vegetation to some extent,

Table 7. The distribution of the Rhinanthoideae over the syntaxonomic system. Table based on data presented by Hartl (1974) and Oberdorfer (1979), supplemented with data provided by Dr. J. Haeck. x, a (sub)species is a character species in the class; o a (sub)species occurs preferentially in the class; +, a (sub)species occurs in the class; each symbol denotes a species or subspecies.

	Odontites	*Pedicularis*	*Rhinanthus*	*Euphrasia*	*Melampyrum*	*Parentucellia*
Saginetea maritimae	+					
Asteretea tripolii	x					
Thlaspietea rotundifolii		x				
Secalinetea	x		x		x	
Chenopodietea	+					
Artemisietea	+					
Agrostietea stoloniferae	o		o	+		
Isoeto-Nanojuncetea	+	+	+	+		o
Littorelletea		+				
Phragmitetea		+				
Montio-Cardaminetea		o		+		
Scheuchzerio-Caricetea fuscae	+	xx	o	+		
Molinio-Arrhenatheretea	x	o	xxx	xx		
Violetea calaminariae				+		
Sedo-Scleranthetea	o		+	+		
Festuco-Brometea	x	o	o	xx	o	
Seslerietea		7x	x	x		
Salicetea herbaceae		o		x		
Juncetea trifidi		x				
Nardo-Callunetea		xx	o	xx	o	
Oxycocco-Sphagnetea		o			x	
Trifolio-Geranietea			+	+	xx	
Betulo-Adenostyletea		o			x	
Alnetea glutinosae		+	+			
Erico-Pinetea		o	+			
Vaccinio-Piceetea						xx
Querco-Fagetea					+	xxx

but what about the opposite? In which way do they affect the vegetation? Agriculturalists have noticed for a long time that vegetation is low and open in places with *Rhinanthus* (Fürst 1931, Tolwinska 1963, Gançev 1973). But is the vegetation open because of the presence of the parasite, or is the parasite present because the vegetation is rather low and open? The data on the role of vegetation structure indicate that the latter certainly has some effect. Some authors have presented data which proved a direct effect of hemiparasites such as *Rhinanthus* spp. on yield and composition of vegetation (Rabotnov 1959, ter Borg and Bastiaans 1973, Mizianty 1975, ter Borg 1979). Rabotnov's data showed that leguminous species in particular decreased in the presence of *Rhinanthus*. Other authors (ter Borg 1972, Campion-Bourget 1982) mentioned Leguminosae as

Table 8. A comparison of the distribution of autotrophic and hemiparasitic monocarpic species. Table based on the records given in N.W. European floras, compiled by Karlsson (1974). ++, +, −, and −− point to (high) positive and negative divergence between observed and expected numbers. 'Expected number' was calculated from the ratio of hemiparasitic (31) to all (454) species in all habitats, i.e. 0.68. Information reproduced with kind permission of the author and the publishers of Botaniska Notiser.

General character of the habitat	All monocarps	Hemiparasitic monocarps		
	Observed	Expected	Observed	Divergence
Open	379	25.9	11	−
Dry, semiclosed	95	6.5	3	−
Dry, closed	25	1.7	11	++
Moist to wet, closed	37	2.5	18	++
Shady	39	2.7	8	+
Northern	28	1.9	7	+
Wet to submerged, open	81	5.5	2	−
Open seashores	52	3.5	0	−−
Shore meadows	33	2.3	7	+
Total number of notations above	769	52.5	67	
Total number of actual species	454		31	
Number of $habitats.species^{-1}$	1.69		2.16	

particularly good hosts. As shown by Atsatt and Strong (1970) good hosts are the first to be 'overparasitized' and killed; this might explain the vegetation changes observed. These processes might also affect species diversity, if a dominant one were the first to be reduced.

The demographic data have shown that population behaviour may vary within and between populations belonging to the same species. The reasons are clear when aestivals and autumnals are concerned (Table 3, Fig. 7), but they may be more obscure when morphologically similar populations are involved. Then the differences may be site-induced, as they probably were in *Pedicularis palustris* (Figs. 1 and 2), *Rhinanthus* (Fig. 6) and *Melampyrum pratense* (Table 4). They might be due to local selection to some extent, as suggested by Schmidt's data on *Odontites vulgaris* (Schmidt 1963), or be induced by genetically different host-parasite interactions as indicated for the same species by Snogerup (1982b). There is no reason to suppose that the differences between the two *Pedicularis sylvatica* populations (Fig. 3) had a genetic basis, but this can not be stated with certainty.

All annual species of hemiparasites appear to follow more or less the same life history pattern, with a Deevey type III curve of mortality. The life-cycles vary in details, but the species differ widely with respect to reproductive pattern. Whereas *Parentucellia viscosa* produces several thousand seeds per capsule, *Melampyrum pratense* achieves a maximum of four seeds per fruit, each weighing ca. 20 mg (Hartl 1974); the others are in between, with some dozens of seeds per capsule as a maximum. The large seed size of *Melampyrum* can be considered as an adaptation to cover the distance between the host roots and the surface of the layer of litter and old leaves in woodland vegetation. The tiny seeds of *Parentucellia* should find the host roots and the soil surface within short distance from each other, and may do so in their habitat.

It is beyond the scope of this paper to cover all factors which influence the distribution of hemiparasites, but one which directly relates hemiparasites with their surrounding vegetation will be discussed, i.e., water relations, especially air humidity which is influenced by vegetation structure.

Several species prefer to grow along woodland edges and paths, which allows them subtle phenotypic adjustments to variation of environmental conditions; in other species a close relation was observed between vegetation and genotypic variation of the growth habit (Fig. 4). The negative effect of a dense vegetation was shown for *Pedicularis sylvatica* as well as for *Rhinanthus* (de Hullu 1985). This is usually attributed to high light requirements of the species. However, vegetation structure also affects microclimate, which could be an even more important habitat factor. Microclimate (including air humidity) together with soil water availability and processes within the plant determine the transport through the xylem system. In hemiparasitic Scrophulariaceae this system not only regulates the uptake of water, but also the uptake of minerals and organic compounds. Lower hydrostatic pressures in shoots of hemiparasites compared with those in the host appear to be the driving force (Scholander et al. 1965, Klaren and van de Dijk 1976, Hansen 1979). Therefore a strict adjustment of the hemiparasites to water relations in the habitat, including those of the host plant, must be very important and may explain part of the distribution pattern.

Some data on the relation between macro-climate and the distribution of hemiparasites may support this point. The first is the relation of the distribution of *Rhinanthus minor* with climate, as illustrated in Fig. 5. The second is concerned with the number of leaf glands. According to Govier et al. (1968) these may play a role in transpiration and excretion processes. Therefore the decrease of their numbers with altitude as observed by Schaeftlein (1967) for *Euphrasia rostkoviana* might be interpreted in this way. The third concerns the leaf dissection of *Pedicularis* spp., which appeared to be related to habitat. Savile (1977), describing its variation in American *Pedicularis* spp., thought that light conditions and temperature were the main selective forces; he found more finely dissected leaves in areas with high insolation, where they may act as heat dissipators. Tabulating the degree of leaf dissection of 20 Central European *Pedicularis* species against the openness of their habitats leads to the suggestion that a higher degree of leaf dissection might compensate for higher air humidity, i.e., low transpiration capacities (Table 9). Moreover the fact that hemiparasitic Scrophulariaceae in temperate Europe are summer rather than winter annuals may be relevant: their growth has to take place in periods of sufficient water transport through the plant, and would probably be restricted in the cool part of the year. And finally, hemiparasitic Scrophulariaceae do not occur in tropical rainforests (Kuijt 1969). This might be a matter of light shortage, but the fact that the achlorophyllous and closely related Orobanchaceae are missing also

Table 9. The relation between leaf dissection and habitat of 20 *Pedicularis* species in Central Europe. Data taken from Hartl (1974).

	Open, low grass	Open, shrubs nearby	Marshy, high herb layer	Similar, and open woodland
Leaves pinnatisect	4	3	–	–
Leaves pinnate, leaflets serrate	–	4	1	–
Leaves pinnate, leaflets serrate to pinnatisect	1	–	4	–
Leaves pinnate-bipinnate, leaflets serrate	–	1	–	2

justifies the hypothesis that lack of light is not the only reason for their absence.

ACKNOWLEDGEMENTS

Many of the original data included in this paper were derived from unpublished reports produced by students of the Department of Plant Ecology, University of Groningen. Their contributions in the practical work and the development of ideas form a major basis of this paper. *Pedicularis palustris* was studied by Frans de Noo, Johan Brouwer and Manja Kwak; *Pedicularis sylvatica* by Henk Everts; *Rhinanthus* spp. by Tejo Boudewijn, Tjeerd Blacquière, Piet Drent, Henk Hillegers, Corrie Leemburg-van der Graaf, Henk Nanninga, Klaas Nanninga, Herbert Prins; *Melampyrum pratense* by Ab Masselink and Siem Bijpost, and *Odontites* spp. by Marguerite (Sweers)-Meyering.

The work could not have been completed without the technical assistance of Kor Beukema, Fia Brussen, Jouke Franke, Jakob Hogendorf, Herman Klees, Thea Linders, Syze Nijdam, Louis de Nijs, Ali Ormel and the late M. Robertus.

Dr. Wim Beeftink, Dr. Jan van Groenendael, Ella de Hullu and Dr. Manja Kwak commented on various versions of the manuscript.

The contributions of all of them are gratefully acknowledged.

REFERENCES

Albrecht, B. (1973) Untersuchungen über Lebensweise und Verbreitung von *Melampyrum sylvaticum* L. (Scrophulariaceae). Diplomarbeit Mainz. (cited by Hartl 1974).

Atsatt, P.R. (1983) Host-parasite interactions in higher plants. Encylopedia Pl.Phys. N.S. 12C: 519–535.

Atsatt, P.R. and Strong, D.R. (1970) The population biology of annual grassland hemiparasites. I. The host environment. Evolution 24: 278–291.

Barsukova, A.V. and Pyatkovskaya, V.P. (1977) Influence of spruce root system on dynamics of abundance and vitality of some plants of herb undershrub layer in Picetum hylocomiosum. Vestn. Mosk. Univ. XVI Biol. (4): 56–64. (in Russian).

Benkov, B. (1978) Chemical control of *Melampyrum arvense* in winter wheat. Rasteniev dni Nauki 15: 126–131. (in Russian).

Bitz, W. (1970) *Melampyrum pratense* L., sein Wurzelwerk und seine Wirte. Staatsprüfungsarbeit Mainz. (cited by Hartl 1974).

Borg, S.J. ter (1972) Variability of *Rhinanthus serotinus* (Schönh.) Oborny in relation to the environment. Diss. RU Groningen.

Borg, S.J. ter (1979) Some topics in plant population biology. In M.J. Werger (Ed.), The Study of Vegetation pp. 11–55. Den Haag: Junk.
Borg, S.J. ter and Bastiaans, J.C. (1973) Host-parasite relations in *Rhinanthus serotinus*. I. The effect of growth conditions and host; a preliminary report. Symp. Parasitic Weeds. Eur. Weed Res. Counc. pp. 236–246. Malta Univ. Press.
Borg, S.J. ter, Janse, A. and Kwak, M.M. (1980) Life cycle variation in *Pedicularis palustris* L. Acta Bot. neerl. 29: 397–406.
Brenchley, W.E. and Warington, K. (1936) The weed seed population of arable land. III. The reestablishment of weed species after reduction by fallowing. J. Ecol. 24: 479–501.
Campion-Bourget, F. (1980) L'hybridisation interspécifique expérimentale chez les cinq principales espèces de *Rhinanthus* L. de France. Revue Cytol. Biol. vég. 3: 221–280.
Campion-Bourget, F. (1982) Influence de l'hémiparasitisme du *Rhinanthus alectorolophus* (Scop.) Pollich sur son dimorphisme saisonnier. Revue gén. Bot. 89: 3–77.
Cantlon, J.E. (1969) The stability of natural populations and their sensitivity to technology. Brookhaven Symp. Biol. 22: 197–205.
Chancellor, R.J. (1976) The effects of temperature on seed germination of *Odontites verna* (Bell.) Dum. Proc. 4th intern. Coll. on Ecology and Biology of Weeds, Marseille 1973, pp. 309–319.
Curtis, E.J.C. and Cantlon, J.E. (1963) Germination of *Melampyrum lineare*: Interrelated effects of afterripening and gibberellic acid. Science 140: 406–408.
During, H., Schenkeveld, A.J., Verkaar, H.J. and Willems, J.H. (1985) Demography of short-lived forbs in chalk grassland in relation to vegetation structure. In: J. White (Ed.) The Population Structure of Vegetation, pp. 341–370. Dordrecht: Dr. W. Junk.
Duuren, L. van, Bakker, J.P. and Fresco, L.F.M. (1981) From intensively agricultural practices to hay-making without fertilization (Effects on moist grassland communities). Vegetatio 47: 241–258.
Fresco, L.F.M. (1980) Ecological response curves of *Rhinanthus serotinus*: a synecological study. Acta Bot. neerl. 29: 533–540.
Fürst, F. (1931) Untersuchungen ueber Bau, Verbreitung, Wachstumsverhältnisse und Bekämpfung der in Deutschland als Acker- und Wiesenunkraüter auftretenden Arten der Pflanzengattung *Alectorolophus*. Arch. Pfl. Bau 6: 30–141.
Gançev, I. (1973) Influence of *Rhinanthus* on the composition and dynamics of meadow communities. Izv. bot. Inst., Sof. 23: 88–118.
Gislén, T. (1949) Problems concerning the occurrence of *Melampyrum arvense* in Sweden. Oikos 1: 208–234.
Govier, R.N., Brown, J.G.S. and Pate, J.S. (1968) Hemiparasitic nutrition in angiosperms. II. Root haustoria and leaf glands of *Odontites verna* (Bell.) Dum. and their relevance to the absorption of solutes from the host. New Phytol. 67: 963–972.
Govier, R.N., Nelson, M.D. and Pate, J.S. (1967) Hemiparasitic nutrition in angiosperms. I. The transfer of organic compounds from host to *Odontites verna* (Bell.) Dum. (Scrophulariaceae). New Phytol. 66: 285–297.
Grootjans, A.P. and Klooster, W.P. ten (1980) Changes of ground-water regime in wet meadows. Acta Bot. Neerl. 29: 541–554.
Grubb, P.J., Kelly, D. and Mitchley, J. (1982) The control of relative abundance in communities of herbaceous plants. In: The Plant Community as a Working Mechanism, E.I. Newman (Ed.), pp. 79–97. British Ecological Society, Special Publication No. 1.
Gutermann, W. (1975) Notulae nomenclaturales 1–18 Zur nomenklatur einiger Arten der mitteleuropäischen Flora. Phyton (Austria) 17: 31–50.
Hansen, D.H. (1979) Physiology and microclimate in a hemiparasite *Castilleja chromosa* (Scrophulariaceae). Am. J. Bot. 66: 477–484.
Haperen, A.M.M. van and Kogel, T.J. de (1981) Het voorkomen van *Hordeum jubatum* L. en *Parentucellia viscosa* (L.) Caruel in Zuidwest Nederland. Gorteria 10: 159–168.
Hartl, D. (1974) Scrophulariaceae. In G. Hegi, Illustrierte Flora von Mitteleuropa VI(1): 1–469.
Heinricher, E. (1924) Methoden zur Aufzucht und Kultur der parasitischen Samenpflanzen. In E. Abderhalden, Handbuch der biol. Arbeitsmethoden XI, Teil 2. pp. 237–350.
Hillegers, H.P.M. (1981) Het vegetatiepatroon van de drie Nederlandse Rhinanthussoorten (Ratelaar) in Zuid-Limburg en het Belgische Jekerdal. Uitgave SBB, Consulentschap Natuurbehoud in de Provincie Limburg.

Hillegers, H. (1983) Ratelaars in het Mergelland. Uitgave SBB, Consulentschap Natuurbehoud in de Provincie Limburg.
Horill, A.D. (1972) Biological flora of the British Isles. *Melampyrum cristatum* L. J. Ecol. 60: 235–244.
Hullu, E. de (1984) The distribution of *Rhinanthus angustifolius* in relation to host plant species. In: Proceedings of the Third Symposium on Parasitic Weeds, Aleppo, Syria, 1984. C. Parker et al. (Eds.). pp. 43–52.
Hullu, E. de (1985) The influence of sward density on the population dynamics of *Rhinanthus angustifolius* in a grassland succession. Acta Bot. Neerl. 34: 23–32.
Jongh, S.E. de (1962) Over het inburgeren van adventieven. Gorteria 1: 38–41.
Karlsson, T. (1974) Recurrent ecotypic variation in Rhinantheae and Gentianaceae in relation to hemiparasitism and mycotrophy. Bot. Notiser 127: 527–539.
Karlsson, T. (1976) *Euphrasia* in Sweden: hybridization, parallelism and species concept. Bot. Notiser 129: 49–60.
Karlsson, T. (1982) *Euphrasia* rostkoviana i Sverige. Växtekologiska studier 15. Svenska Växtgeografiska Sällskapet. Uppsala.
Klaren, C.H. and Dijk, S.J. v.d. (1976) Water relations of the hemiparasitic *Rhinanthus serotinus* before and after attachment. Physiologia Pl. 38: 121–125.
Kuijt, J. (1969) The biology of parasitic flowering plants. Univ. of California Press, Berkeley.
Kuijt, J. (1979) Host selection by parasitic angiosperms. Symb. Bot. Ups. XXII(4): 194–199.
Kwak, M.M. (1980) Artificial and natural hybridization and introgression in *Rhinanthus* (Scrophulariaceae) in relation to bumblebee pollination. Taxon 29: 613–628.
Lackney, V.K. (1981) The parasitism of *Pedicularis lanceolata* Michx., a root hemiparasite. Bull. Torrey Bot. Club 108: 422–429.
MacInnes, K.L. (1972) Reproduction in Arctic populations of *Pedicularis*. Ph.D. Thesis Univ. Western Ontario.
Macior, L.W. (1980) Population ecology of the Furbish lousewort, *Pedicularis furbishiae* S. Wats. Rhodora 82: 105–111.
Manley, G.V., Butcher, J.W. and Cantlon, J.E. (1975) Relationship of insects to distribution and abundance of *Melampyrum lineare* (Scrophulariaceae). Pedobiologia 15: 385–404.
Masselink, A.K. (1980) Germination and seed population dynamics in *Melampyrum pratense* L. Acta Bot. Neerl. 29: 451–468.
Mizianty, M. (1975) Influence of *Rhinanthus serotinus* (Schönh.) Oborny on the productivity and floristic composition of the meadow plant association. Fragm. flor. geobot. XXI: 491–505.
Oberdorfer, E. (1979) Pflanzensoziologische Exkursionsflora. Stuttgart: Ulmer.
Oesau, A. (1973) Keimung und Wurzelwachstum von *Melampyrum arvense* L. (Scrophulariaceae). Beitr. Biol. Pfl. 49: 73–100.
Oesau, A. (1975) Untersuchungen zur Keimung und Entwicklung des Wurzelsystems in der Gattung *Melampyrum* L. (Scrophulariaceae). Beitr. Biol. Pfl. 51: 121–147.
Okonkwo, S.N.C. (1966) Studies on *Striga senegalensis* (= *S. hermonthica*). II. Translocation of C^{14}-labelled photosynthate, urea-C^{14}, and sulphur-35 between host and parasite. Am. J. Bot. 53: 142–148.
Ozenda, P. and Capdepon, M. (1979) L'appareil haustorial des phanérogames parasites. Revue gén Bot. 86: 235–343.
Piehl, M.A. (1963) Mode of attachment, haustorium structure and hosts of *Pedicularis canadensis*. Am. J. Bot. 50: 978–985.
Piehl, M.A. (1965) Studies of root parasitism in *Pedicularis lanceolata*. Mich. Bot. 4: 76–81.
Pieterse, A.H. (1979) The broom rapes – a review. Abstr. on Trop. Agr. 5: 9–35.
Rabotnov, T.A. (1959) The effect of *Rhinanthus major* Ehrh. upon the crops and the composition of the floodland herbage. Byull. mosk. Obshch. Ispŷt. Prir. 64: 105–107.
Rehmböck, O. (1975) *Parentucellia viscosa* (L.) Caruel – ein seltener Fremdling unserer Flora. Göttinger Floristische Rundbriefe 9: 8, 10.
Rottgardt, K. (1956) Morphologische, cytologische und physiologische Untersuchungen von Oekotypen in Schleswig-Holstein. Beitr. Biol. Pfl. 32: 225–278.
Savile, D.B.O. (1977) Leaf form and evolutionary patterns in *Pedicularis*. Proc. Indian Natn. Acad. Sci. Pt. B 43: 223–227.
Schaeftlein, H. (1967) Beiträge zur Kenntnis einiger mitteleuropäischer Euphrasien. Der taxonomischen Wert der drüsigen Behaarung. Phyton (Austria) 12: 48–90.

Schenkeveld, A.J. and Verkaar, H.J. (1984) The ecology of short-lived forbs in chalk grasslands. Ph.D. Thesis Univ. Utrecht.

Schmidt, V.M. (1963) The evidence relating to the natural selection in the populations of *Odontites serotina* Dum. (Russian with English summary). Bot. Zh. SSSR 48: 989–1000.

Scholander, P.F., Hammel, H.T., Bradstreet, Edda D. and Hemmingsen, E.A. (1965) Sap pressure in vascular plants. Science 148: 339–346.

Silakova, V.M. (1975) Multiplication by seed of yellow rattle (*Rhinanthus vernalis* Schischk. et Serg.) Nauch Trudy, Kurskii Ped. Inst. 35: 85–89. (in Russian).

Smith, A.J.E. (1963) Variation in *Melampyrum pratense*. Watsonia 5: 336–367.

Snogerup, B. (1982a) *Odontites litoralis* Fries subsp. *litoralis* in the British Isles. Watsonia 14: 35–39.

Snogerup, B. (1982b) Host influence on north western European taxa on *Odontites* (Scrophulariaceae). Ann. Bot. Fennici 19: 17–30.

Snogerup, B. (1983) North west European taxa of *Odontites* (Scrophulariaceae). Acta Bot. Fennica 124: 1–62.

Soó, R.v. (1926) Systematische Monographie der Gattung *Melampyrum* II. Repert. Spec. nov. Regn. veget. 23: 385–397.

Soó, R.v. (1929) Die mittle- und südosteuropäischen Arten und Formen der Gattung *Rhinanthus* und ihre Verbeitung in Südosteuropa. Repert. Spec. nov. Regn. veget. 26: 179–219.

Sprague, E.F. (1962) Pollination and evolution in *Pedicularis* (Scrophulariaceae). Aliso 5: 181–209.

Suneson, S. (1976) *Parentucellia viscosa* i naturlig svensk vegetation. Svensk bot. Tidskr. 70: 61–64.

Tolwinska, M. (1963) The absorption of food, growth and development of *Alectorolophus glaber* (Lam.) Beck and the attempt of its control on meadows. Roczn. Naukro ln. 75: 1–23.

Tooren, B.F. van, Schat, H. and Borg, S.J. ter (1983) Succession and fluctuation in the vegetation of a Dutch beach plain. Vegetatio 53: 139–151.

Tsivion, Y. (1978) Physiological aspects of the association between parasitic angiosperms and their hosts. A review. Isr. J. Bot. 27: 103–121.

Vassilzenko, I.T. (1955) Rhinanthus. In V.L. Komarov, Flora USSR. Part XXII: 659–686.

Verkaar, H.J., Schenkeveld, A.J. and Brand, J.M. (1983) On the ecology of short-lived forbs in chalk grasslands: microsite tolerances in relation to vegetation structure. Vegetatio 52: 91–102.

Verkaar, H.J., Schenkeveld, A.J. and Klashorst, M.P. van de (1983) The ecology of short-lived forbs in chalk grasslands: Dispersal of seeds. New Phytol. 95: 335–344.

Visser, J.H., Weber, H.Chr. and Kuijt, J. (1978) *Hyobanche sanguinea* L. (Scrophulariaceae) ein Blattparasit. Naturwissenschaften 65: 209.

Weber, H.C. (1976) Ueber Wirtspflanzen und Parasitismus einiger mitteleuropäischer Rhinanthoideae (Scrophulariaceae). Plant Syst. Evol. 125: 97–107.

Weber, H.C. (1981) Untersuchungen an parasitischen Scrophulariaceen (Rhinanthoideen) in Kultur. I. Keimung und Entwicklungsweise. Flora 171: 23–38.

Westhoff, V. and Held, A.J. den (1969) Plantengemeenschappen in Nederland. Zutphen: Thieme.

Wettstein, R. von (1895) Der Saison-Dimorphismus als Ausgangspunkt für die Bildung neuer Arten im Pflanzenreiche. Ber. dt. bot. Ges. 13: 303–313.

Wettstein, R. von (1899) Descendenztheoretische Untersuchungen. I. Untersuchungen über den Saison-Dimorphismus im Pflanzenreiche. Denkschr. Akad. Wiss., Wien 70: 305–346.

Wilkins, D.A. (1963) Plasticity and establishment in *Euphrasia*. Ann. Bot. 27: 533–552.

Wilkinson, E. and Musselman, L.J. (1979) The parasitic gymnosperm, *Parasitaxus ustus*. Proc. 2nd Symp. Paras. Weeds. Suppl. L.J. Musselman, A.D. Worsham and R.E. Eplee (Eds.).

Yeo, P.F. (1978a) *Euphrasia*: a taxonomically critical group with normal sexual reproduction. In H.E. Street (Ed.), Essays in Plant Taxonomy, pp. 144–162. London: Academic Press.

Yeo, P.F. (1978b) A taxonomic revision of *Euphrasia* in Europe. Bot. J. Linn. Soc. 77: 223–334.

Zinger, N.V. (1912) Variability of *Alectorolophus major* Rchnb., growing on sites influenced by agriculture, and their evolution. (in Russian). Trudij tiflis. bot. Sada 12: 179–190.

DEMOGRAPHY OF *BROMUS TECTORUM* IN *ARTEMISIA* COMMUNITIES

JAMES A. YOUNG and RAYMOND A. EVANS

ABSTRACT

Bromus tectorum has successfully invaded sagebrush (*Artemisia*) grasslands in western North America within the past century. The population dynamics of this species under drought conditions and during relatively wet seasons are described. Persistence of populations is ensured by means of dormant seed reserves in the soil. *Bromus tectorum* has a flexible life history though it typically behaves as a winter annual. The frequency and intensity of natural wildfires regulates its population size, but it can recover rapidly after fire and quickly dominate the vegetation by its high vegetative vigor and reproductive output. The genetic consequences of the large fluctuations in population size which follow fire are considered.

INTRODUCTION

The demography of plant species in many pristine shrub/grassland communities of the Intermountain region of the western United States differs from what now occurs in communities that have been degraded by a century of improper grazing. Plant succession has been revolutionized in these communities by alien plant species. The pristine communities were dominated by relatively long-lived perennials where recruitment of seedlings to the population may have occurred at irregular intervals in response to extraordinary environmental conditions: We have observed this pattern of recruitment in relict plant communities, and similar observations have been reported by Harris (1967). In contrast, the alien plant communities are dominated by annuals with life cycles that require establishment each year. These profound changes raise several questions: (a) whether the pristine environment lacked the selective pressure to evolve competitive annuals; (b) whether the current dominance by annuals is a relatively short-lived phenomenon; or (c) whether the introduction of man

White, J. (ed.), The Population Structure of Vegetation.
 ISBN 978-94-010-8927-2

and his domestic animals to the *Artemisia* (sagebrush) grasslands has changed the environment sufficiently to perpetuate the dominance of annuals (Young et al. 1972, Young and Evans 1976, Young et al. 1978). Evaluation of these or other appropriate hypotheses may be made by studying the dynamics of recruitment, reproduction, and death of plants in these communities.

The overriding factor in competition among plants in the sagebrush grasslands is available soil moisture for plant establishment and growth (Evans et al. 1970). *Bromus tectorum* is the dominant seral species of many of the *Artemisia* grasslands and the persistence and dynamic response of this species through droughts and moisture abundance provide a striking illustration of the mechanisms necessary for establishment, survival, and perpetuation of colonizing species (Harris 1967). Year-to-year variation of total and seasonal distribution of precipitation necessitates much flexibility in life cycle characteristics for an annual species to persist and dominate on the large scale represented by *B. tectorum* distribution.

HISTORICAL BACKGROUND

Livestock were introduced in large numbers to the *Artemisia* grasslands of the Great Basin in the late 1860s and early 1870s (Young et al. 1978). By the late 1880s, the dominant perennial grasses, species of *Agropyron*, *Stipa*, *Festuca* and *Elymus* (Young et al. 1976), were severely overgrazed and largely depleted from ranges located near water sources on most slopes not too steep for grazing. There were many native annual species in the *Artemisia* grasslands, but none aggressive or competitive enough to preempt the environmental potential released by the destruction of the perennial grasses (Young et al. 1972). The dominant species of *Artemisia*, and especially *A. tridentata*, were not preferred by cattle. As a result, the shrubs increased in density (Stewart 1936).

The near void in the herbaceous community continued until 1900 (Young et al. 1972). At that time, the alien annuals *Salsola iberica*, *Erodium cicutarium* and *Sisymbrium altissimum* were found in western Nevada (Kennedy and Doten 1901) whereas only *Erodium cicutarium* was established in north-eastern Nevada (Kennedy 1903). *Bromus tectorum*, a very important weed which later came to dominate millions of hectares of *Artemisia* rangeland, was not detected in the Great Basin until the turn of the century (Young et al. 1976).

Many of the alien annuals that proved to be such successful colonizers of the *Artemisia* grasslands appear to be native to central Asia (West 1983) although they have shadowed the spread of domestic livestock in many worldwide environments (Young and Evans 1976). *Salsola iberica*, for instance, was probably first introduced into the United States in 1873 or 1874 in flax (*Linum* sp.) seed brought from Russia and sown near Scotland, South Dakota (Robbins et al. 1952). By 1900, it was well established on degraded *Artemisia* rangelands. The portion of California with a Mediterranean-type climate was invaded by a host of alien species at the time the earliest Spanish Missions were built during the early 1800's (Burcham 1956). The spectacular spread of *Bromus tectorum* in northern Nevada, however, probably occurred after 1906 (Young et al. 1976, Mack 1981).

If plant communities dominated by woody *Artemisia* species are disturbed, usually by wildfire or plowing, the initial seral stage is usually dominated by *Salsola iberica* followed successionally by *Sisymbrium altissimum* and finally *Bromus tectorum* (Piemeisel 1951). With moderate disturbance from grazing animals, *Bromus tectorum* will remain the dominant species for a long time. *Bromus tectorum* will even invade the interspaces among perennial grasses in remnant examples of pristine plant communities of the *Artemisia* grasslands (Daubenmire 1940).

This seral continuum of alien annuals coupled with low level disturbances by herbivores, including rodents, rabbits, and domestic livestock, combine virtually to close the disturbed plant communities to the successful reestablishment of seedlings of herbaceous perennials (Robertson and Pearse 1946), although seedlings of semi-woody and woody species of *Artemisia*, *Chrysothamnus*, etc., do become established in communities dominated by the alien annuals (Young and Evans 1974).

Bromus tectorum and *Salsola iberica* dominated communities have proven susceptible to invasion by more recently introduced alien annuals. The new arrivals generally do not colonize a broad spectrum of plant communities, so they do not become prominent on the landscape. The more recent introductions tend to be highly localized in specific environments. For example, sometime before 1935 the toxic, broadleaf plant *Halogeton glomeratus* was introduced to the Great Basin. By the early 1950s this poisonous plant had spread spectacularly across the Intermountain area and east into Wyoming and Colorado (see Cronin 1965 for details). Only a portion of the *Artemisia* grasslands were susceptible to invasion by *Halogeton*. Its spread was generally in salt desert vegetation. Similarly, the alien grass *Taeniatherum asperum* suddenly preempted the dominance of *Bromus tectorum* in many low seral plant communities (McKell et al. 1962). In the *Artemisia* grasslands, this invasion was largely restricted to sites with clay soil (Young and Evans 1970). Within the last decade *Salsola paulsenii* has replaced *Salsola iberica* as the dominant alien species, particularly in arid, degraded *Artemisia* communities along the margin of the salt deserts in the Great Basin (Evans and Young 1980).

POPULATION DYNAMICS UNDER DROUGHT CONDITIONS

It is somewhat of a misnomer to discuss the dynamics and demography of populations of annuals in the *Artemisia* grasslands in relation to drought because moisture for plant growth is more or less limiting every year. Moisture is almost completely out of phase with temperatures conducive to plant growth (Evans et al. 1970). Germination, establishment, and reproduction of annuals occur between the time seedbed temperature and available soil moisture permit germination in the fall or spring and soil moisture is exhausted in early summer. Some species, for example those of *Salsola* and *Halogeton*, partially avoid these moisture-temperature restrictions by inherently being able to utilize soil moisture at low matric and osmotic potentials. *Bromus tectorum*, the dominant annual, seems to be restricted to what is generally considered to be the normal range of water potentials (0.03–1.5 M Pascals) for establishment and growth (Evans 1961).

During a season when precipitation for the period from spring germination to maturity of *Bromus tectorum* averaged one-third of normal, seed production of this species averaged 1.38 seeds per established plant with production below 1.0 seeds per plant at three of the five locations where it was measured (Young et al. 1969). Obviously, no population of annuals can survive with a negative return in seeds produced per established plant. *Bromus tectorum* maintains a large reserve of viable dormant seeds in the litter and surface soil, and these seeds are capable of renewing the population for two or possibly three years without markedly reducing plant density (Young et al. 1969).

The development and environmental sensitivity of this reserve of seeds in the soil is an example of adaptation of an annual to a seasonally erratic moisture supply. Soon after harvest, the seeds of *Bromus tectorum* are highly germinable (90–95%) (Young et al. 1969). Some authors have reported after-ripening requirements for seeds of this species (see Klemmedson and Smith 1964), but we have found no evidence of such requirements (Young et al. 1969). The germinability of *Bromus tectorum* seeds at a wide range of constant and alternating temperatures (Table 1) exceeds that observed for seeds of most cultivars of perennial grass growing in the *Artemisia* grasslands. When these highly germinable seeds are dispersed, they tend to accumulate in natural depressions or with litter accumulations on the surface of the seedbed (Evans and Young 1970).

Over winter the *Bromus tectorum* seeds are exposed to repeated cycles of wetting at temperatures too low for germination, and then desiccation by freezing. In the spring, with warmer temperatures, germination is usually not simultaneous for the entire population. Germination is conditioned by a series of moisture events of varying intensity. Each storm brings varying portions of the seedbed within the potential of germination of *Bromus tectorum*. Microtopography of the soil surface and litter accumulations modify extremes of temperature and increase relative humidity allowing seeds to germinate (Evans and Young 1972). Those seeds that by accidents of dispersal occur in unfavorable situations for germination, acquire dormancy (Young et al. 1969). This environmentally-induced acquired dormancy is an excellent example of the conversion from simultaneous to continuous germination of a seed population (Newman 1963).

The dormancy acquired by *Bromus tectorum* seeds last for several years and breaks down at erratic intervals. The breaking of this acquired dormancy is in response to stimuli from the environment. If the amount of nitrate-nitrogen is enriched in the seedbed, the dormant *Bromus tectorum* seeds will begin to germinate (Young and Evans 1975). Artificial addition of gibberellic acid (GA_3) to the seedbed reacts synergistically with the

Table 1. Percentage germination and confidence interval (in parentheses, $p = 0.01$) of seeds of *Bromus tectorum* in relation to constant and alternating temperatures. Seeds incubated for 4 weeks in dark germinators (Young and Evans 1982).[a]

Cold period temperature °C 16 hr	Warm period temperature °C–8 hr									
	0	2	5	10	15	20	25	30	35	40
0	26(9)	35(8)	48(6)	65(6)	78(6)	87(6)	92(6)	92(6)	87(7)	79(10)
2		40(8)	52(6)	69(5)	82(5)	91(6)	95(5)	95(5)	91(6)	82(9)
5			57(7)	74(4)	87(4)	95(5)	99(5)	99(4)	94(5)	85(7)
10				79(7)	92(5)	100(4)	100(4)	100(4)	97(5)	87(7)
15					92(7)	99(5)	100(5)	100(5)	95(5)	85(7)
20						95(8)	98(5)	96(5)	90(5)	79(8)
25							89(7)	86(5)	80(5)	69(7)
30								73(7)	66(5)	54(7)
35									47(9)	36(9)
40										13(12)

[a]Underlined means are not significantly different from maximum and one half its confidence interval, our definition of optimum germination.

nitrate to further enhance germination (Young and Evans 1975), but we do not know whether gibberellin occurs naturally in *Bromus tectorum* seedbeds. When temperatures warm in the spring (permitting germination), nitrification is also enhanced. In dry years nitrification is limited by available moisture so this enhancement of germination of the *Bromus tectorum* seeds does not occur.

The undisturbed density of *Bromus tectorum* plants in degraded *Artemisia* grassland communities is about 10,000 to 13,000 plants per m^2 (Young et al. 1969). With this population, 10,000 to 15,000 viable, but dormant, seeds per m^2 of *Bromus tectorum* are present in litter and surface soil. This reserve of seeds would appear to be enough to restock the stand for one year only, but two factors interact to ensure renewal of the stand for two or even three years. First, the dormancy of all the seed reserve will not break down the first year, especially if soil moisture availability is below normal. Secondly, the normal stand density far exceeds the optimum density for herbage or seed production. Even after two years of severe drought, there will be sufficient *Bromus tectorum* to stock the stand with sufficient plants to exploit the environmental potential fully.

Diagrammatically, we can follow a population of downy brome through a growing season with below average and inadequate distribution of precipitation (Fig. 1). Starting in late winter with roughly equal populations of dormant and germinable caryopses, each consisting of about 6,500 seeds per m^2, there is a loss of germinable seeds as successive waves of germination occur under inadequate seedbed moisture conditions. Many of the germinable seeds are converted by acquired dormancy to the dormant category. This is possible even if the radicle has emerged and then died, because adventitious root growth from a lateral bud is possible at some future germination episode (Young et al. 1971). Few dormant seeds become germinable because the environmental stimuli for this reversal, including nitrate nitrogen, are lacking. A portion of the dormant seeds lose viability, mainly as a result of soil-borne pathogens, especially *Podosporiella verticillata* (Kreitlow and Bleak 1964).

Because of inadequate moisture during the spring growing season of a dry year, seed production averages less than one seed per established plant (Fig. 1). The low seed production results in the *Bromus tectorum* population entering winter with a comparatively low reserve of germinable seeds. Further reduction occurs over winter by germinable seeds acquiring dormancy. If the following spring had adequate amounts and favorable distribution of precipitation to permit plant establishment, then the low number of germinable seeds is partially compensated for by the breakdown of dormancy in response to stimuli from the environment. The first growing season after the drought is one of high risk for the population because both the germinable and dormant seed resources are relatively depleted. Once the *Bromus tectorum* plants are established during a season with adequate moisture, there is a good chance of returning some viable seed to the system. Even extremely heavy grazing will not completely inhibit flowering and seed production although it may confine seed production to late-developing, nearly decumbent, tillers. The current

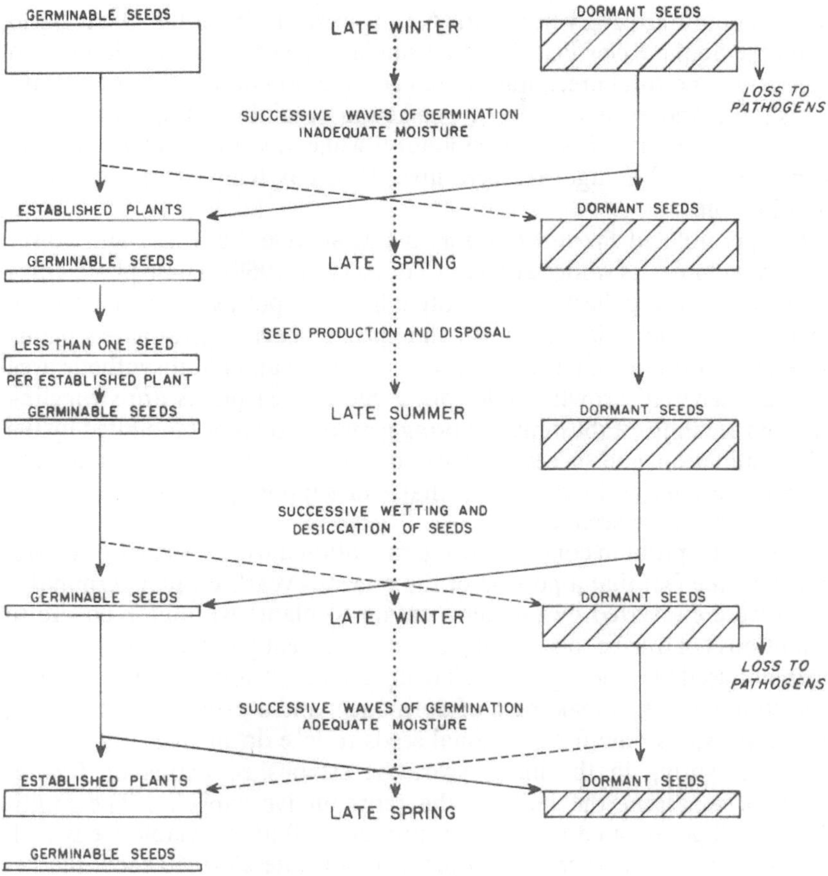

Fig. 1. Diagrammatic presentation of dynamics of *Bromus tectorum* plants and seed populations during a growing season with inadequate moisture for plant growth. Width of horizontal bars indicates relative size of seed reserves. Thickness of arrows indicates relative volume of change from one category to the next.

year's seed crop will not be eliminated by wildfires except in portions of the community where woody fuels contribute to the intensity and duration of the fire (Young et al. 1976). Probably the most significant factor in the dry-season dynamics of *Bromus tectorum* is the continued maintenance of the dormant seed reserve and the immunity of these reserves to the drought because stimuli that promote germination are lacking.

POPULATION DYNAMICS IN A WET YEAR

Fall germination is not an absolute requirement for luxuriant establishment and herbage production by *Bromus tectorum*, but if fall germination occurs, it almost always leads to a good growth and reproduction for this species. In the northern Great Basin, fall germination of *Bromus tectorum* occurs about once every five years. On the western rim of the Great Basin or on the Snake River Plains, with a similar type of *Artemisia*-dominated vegetation and total precipitation, fall germination occurs four out of five years (Hull and Pechanec 1947). The key to obtaining fall germination is the occurrence of sufficient precipitation while seedbed temperatures are warm enough (2°C night temperatures, 15°C day temperatures) to allow germination.

Fall-germinated *Bromus tectorum* plants survive the winter and flower the next summer as winter annuals (Young et al. 1969). These plants form flat rosettes during the winter and often have symptoms of severe nitrogen deficiency during mid-winter. In the spring, the fall-germinated plants have the advantage of an established root system and photosynthetic area to begin renewed growth while spring germinated plants are struggling through seedling establishment. Spring germination is not inhibited by the fall germination, and successive waves of germination occur in response to moisture events as long as sites that can support germination remain unoccupied in the seedbed.

The near-optimum condition for germination during the spring of a wet year plus the fact that a portion of the previous year's crop of germinable seeds were converted by fall germination to plants without exposure to conditions conducive to acquiring dormancy, greatly reduce the reserve of dormant seeds the next year. Relatively abundant nitrate from nitrification stimulates the breakdown of dormancy, and a prolonged wet spring period provides time for additional seeds to lose dormancy (Fig. 2).

In late spring in the mesic year, the established plants of *Bromus tectorum* represent the bulk of the reproductive capacity. The multi-tillered fall-germinated plants often produce 250 to 500 viable seeds each and the small spring-germinated plants contribute 25 to 50 each; the net result is a huge increase in germinable seeds after the current year's seed crop is dispersed to the seedbed.

During seasons when the seed production of *Bromus tectorum* is very prolific, the quality of seed is often severely damaged by infestations of smut (*Ustilago bromiviar*) (Stewart and Hull 1949). The large reserve of germinable seeds is further reduced by predation by rodents and birds (Latourrette et al. 1971, Young 1981). Not all *Bromus tectorum* seeds collected by rodents are consumed, because many are cached by *Heteromyid* rodents and the seeds in the caches eventually germinate and reenter the reproductive cycle (Latourrette et al. 1971). We did not consider predation in the dry season dynamics because in the years of low seed production, not enough seeds are supplied to support minimal predation, populations of predators crash, and the remainder seeks alternate food supplies. Abundance of seeds results in proportionately more losses of seeds to predation.

There are two important factors that affect the huge reserve of germinable *Bromus tectorum* seeds resulting from a season with abundant moisture. First, an abundance of germinable seeds does not necessarily result in a dense population of plants the next season. Plant density is not related

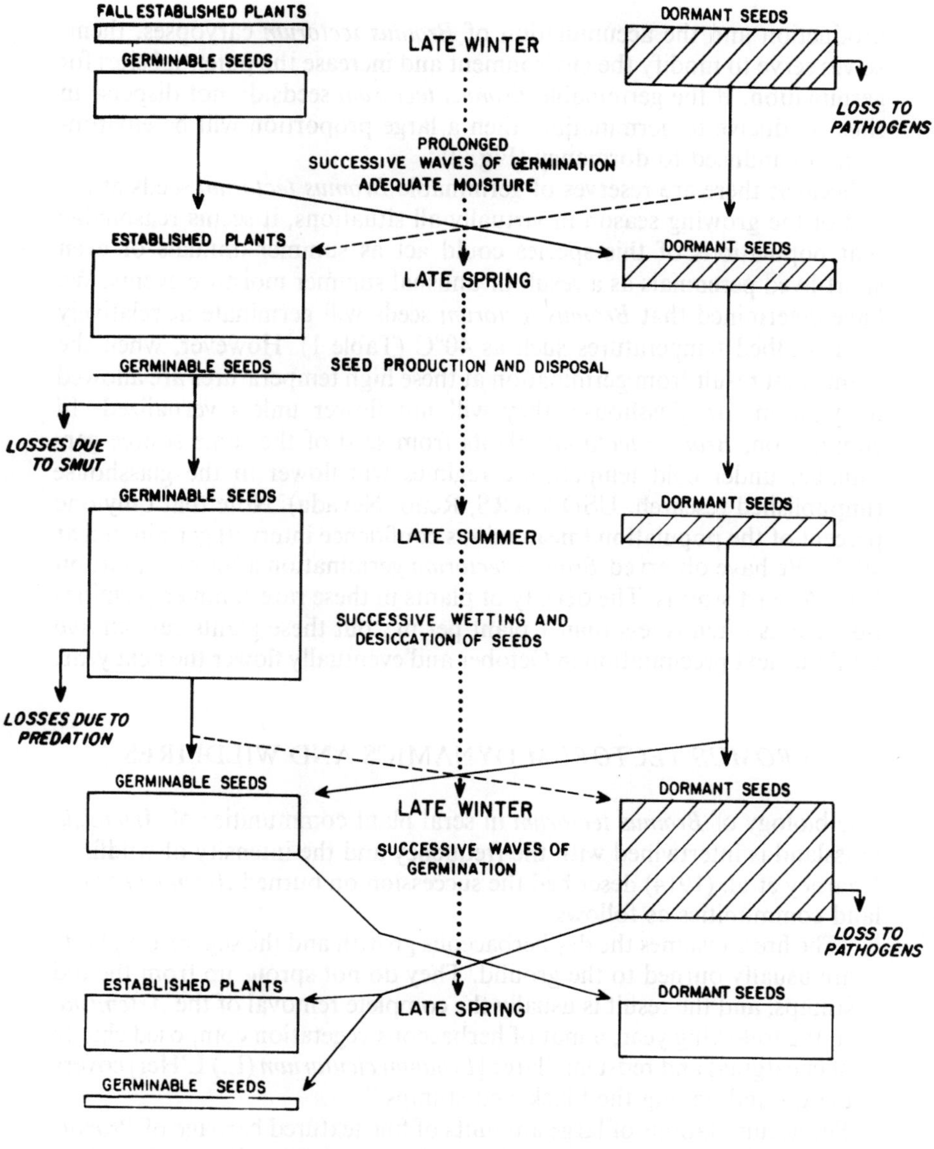

Fig. 2. Diagrammatic presentation of dynamics of *Bromus tectorum* plants and seed populations during a growing season with abundant moisture for plant growth. Width of horizontal bars indicates relative size of seed reserves. Thickness of arrows indicates relative volume of change from one category to the next.

directly to the number of seeds present, but to the number of available sites in the seedbed capable of supporting germination of this species (Harper et al. 1965). Sites that support germination are those with accumulations of litter or microtopography that modify temperature extremes and moisture relations and allow seeds to take up moisture from the substrate faster than they lose it to the atmosphere (Evans and Young 1970, 1972). The carryover of additional litter from a year of higher-than-average herbage production and the accumulation of *Bromus tectorum* caryopses, themselves serve to modify the environment and increase the potential sites for germination. If the germinable *Bromus tectorum* seeds do not disperse in sites conducive to germination, then a large proportion will be environmentally induced to dormancy (Fig. 2).

Because there are reserves of germinable *Bromus tectorum* seeds at the end of the growing season in virtually all situations, it seems reasonable that populations of this species could act as summer annuals or even short-lived perennials as a result of unusual summer moisture events. We have determined that *Bromus tectorum* seeds will germinate at relatively high seedbed temperatures such as 40°C (Table 1). However, when the plants that result from germination at these high temperatures are allowed to grow in the glasshouse, they will not flower unless vernalized. In comparison, *Bromus tectorum* plants from seed of the same source germinated under cold temperature regimes will flower in the glasshouse (unpublished research, USDA/ARS, Reno, Nevada). Note that only one percent of the population (mean minus confidence interval) germinated at 40°C. We have observed *Bromus tectorum* germination after precipitation from August storms. The density of plants in these rare summer germination events often is less than 1 plant per m^2, but these plants can survive until the next precipitation in October and eventually flower the next year.

BROMUS TECTORUM DYNAMICS AND WILDFIRES

The biology of *Bromus tectorum* in seral plant communities of *Artemisia* grassland is intertwined with the frequency and the intensity of wildfires. Kearney et al. (1914) described the succession on burned *Artemisia* grassland communities as follows:

> "The fire consumes the dry herbaceous growth and the sagebrush plants are usually burned to the ground. They do not sprout up from the old stumps, and the result is usually the complete removal of the *Artemisia*. In the following year, a mat of herbaceous vegetation composed chiefly of cheatgrass and redstem filaree [*Erodium cicutarium* (L.) L'Her] covers the ground among the blackened stumps."

The accumulations of large amounts of fine textured herbage of *Bromus tectorum*, following springs with good moisture conditions for plant growth, predisposes the communities to being burned in mid-summer wildfires. Because of the non-sprouting characteristics of the shrubs, repeated wildfires result in herbaceous and eventually *Bromus tectorum* dominance.

The bulk of the reproductive reserve of germinable and dormant *Bromus tectorum* seeds is located under the canopy of *Artemisia tridentata* plants (Young and Evans 1975). The accumulation of litter and nutrients and the more favorable microenvironment result in more plants and greater growth of herbaceous species under the shrubs than in the interspace (Schlatterer and Hironaka 1972). Virtually the only cycling of nitrate-nitrogen in these communities occurs in the litter and surface soil under the shrubs (Charley and West 1975). When these woody shrubs and the subcanopy litter are completely consumed in wildfires, they generate heat of sufficient intensity and duration to destroy *Bromus tectorum* seeds. In contrast, seeds in the interspaces are not completely consumed (Young and Evans 1978).

In the typical *Artemisia*/*Bromus tectorum* community, which contains about 30 herbaceous species, population dynamics following burning are rather stagnant for most species except *Bromus tectorum*. There are few pronounced changes in the number of plants per unit of area of the species. The density of *Bromus tectorum* plants the first season following burning is usually less than 10 plants per m^2 with the plants located in the interspaces between the burned shrubs (Young and Evans 1978). By three years after burning, the density may be greater than 10,000 plants per m^2. This tremendous preemption of environmental potential dampens the expression of dynamics of other species: there are so many *Bromus tectorum* plants per unit of area that moisture and nitrogen are not available for other species.

Although the density of *Bromus tectorum* plants is very low the first year after burning, the vigor and size of individual plants is tremendous. In both numbers of tillers and caryopses produced per plant, the populations of *Bromus tectorum* are far more productive on burned areas than on adjacent unburned areas. In number of seeds per plant, the populations do not overlap. In the case of one burn we sampled, the lowest seed production per plant on the burn was 960 (maximum 5500 seeds per plant), while the highest in the unburned area was 250 (Young and Evans 1978). The number of tillers per plant is important because it extends the flowering period and enhances the chances for hybridization among plants. The reproduction response of the reduced population of *Bromus tectorum* established after the wildfires is an expression of phenotypic plasticity. By greatly increasing the number of tillers per plant and florets per tiller, the number of seeds produced per plant are markedly increased. This plasticity is an obvious response to reduced competition, but the phenotypic response paves the way for changes in breeding systems that have genotypic consequences.

Bromus tectorum plants are largely self-pollinated, and in unburned communities, hybridization probably rarely occurs. The release in environmental potential, more water and nutrients available to *Bromus tectorum* plants, brought on by the destruction of the dominant woody component of the community and the partial reduction in *Bromus tectorum* reproduction potential, pave the way for hybridization and recombination. The phenotypic plasticity of *Bromus tectorum* that produces multiple tillers also prolongs the period of time when anthers are exerted and stigmas are receptive. This is an application of the genetic system for predominantly self-pollinated species established by Stebbins and Allard (e.g., Allard 1965).

We carry the concepts of Stebbins and Allard a logical step further and suggest that the progeny of the hybrids express population heterosis in the F_1 generation, i.e., the second year after burning. The population density of *Bromus tectorum* plants the second year following burning is approximately 2,500 to 3,500 plants per m^2. This population is approximately one-quarter to one-third of the population density observed the third and subsequent years following burning. The competition from this increasing, but still reduced population on the second year, when we postulate an expression of heterosis, completely preempts other annual herbaceous species in the community (Young and Evans 1978).

The second year after burning, when density the *Bromus tectorum* population increases, the plants invade the haloes of ash from the burned shrub canopies occupied by native annuals the season before. The burned shrubs provide a relatively nutrient-rich environment, especially in terms of nitrogen (Schlatterer and Hironaka 1972), for full expression of the potential of *Bromus tectorum*. The nitrogen resulted from nitrification of decaying accumulations of shrub and herbaceous litter under the shrubs and possible cryptogamic fixation (Snyder and Hullstein 1973, Charley and West 1975). The organic nitrogen in the litter under the shrubs is mostly lost through volatilization when burned, but there is a relatively large accumulation of nitrate-N in the soil under the burned canopies compared with the interspaces between the shrubs. The nitrogen was made available for *Bromus tectorum* by the fire in the sense that the shrub competition was removed, intraspecific competition reduced, and a suitable seedbed prepared.

The net result of *Bromus tectorum* populations being destroyed in wildfires is a genetically reconstituted population that takes advantage of environmental potential released by the destruction of the dominant shrubs to assure continued dominance of the site.

THE FUTURE OF *BROMUS TECTORUM* IN *ARTEMISIA* GRASSLAND COMMUNITIES

Considering the many adaptive advantages that can be demonstrated for this species and, most importantly, its apparent ability to evolve rapidly to meet the challenges of a changing environment, it certainly appears that *Bromus tectorum* will be a permanent component of many of the *Artemisia* grasslands in western North America. This species seems to be able to adapt to the alternative possibilities in the harsh and variable environment of the *Artemisia* grasslands. For example, when drought occurs, the seeds in dormant seed reserves fail to receive stimuli to germinate. The inherent requirements for germination of viable seeds make total germination of germinable reserves unlikely. The phenotypic plasticity of the species allows a dynamic rebound in seed production even when population densities are 0.1% of normal.

Because so many of the requirements for growth and stimuli to begin physiologic functions of this species are concerned with the available nitrate-nitrogen, the availability of this compound may be the Achilles' heel of its prolonged survival. Based on past history in this environment, the chance introduction of a new species could negate the competitive advantages of **Bromus tectorum** and reduce its dominance of seral communities.

REFERENCES

Allard, R.W. (1965) Genetic systems associated with colonizing ability in predominantly self-pollinated species. In H.G. Baker and G.L. Stebbins (Eds.), The Genetics of Colonizing Species, pp. 49–76. Academic Press, New York.

Burcham, L.T. (1956) Historical backgrounds of rangeland use in California. J. Range Manage. 9: 81–86.

Charley, J.L. and West, N.E. (1975) Plant-induced soil chemical patterns in some shrub-dominated semi-desert ecosystems of Utah. J. Ecol. 63: 945–964.

Cronin, E.H. (1965) Ecological and physiological factors influencing chemical control of *Halogeton glomeratus*. U.S. Dep. Agric. Bull. 1325. 65 pp.

Daubenmire, R.F. (1940) Plant succession due to overgrazing in the *Agropyron* bunchgrass prairie of southeastern Washington. Ecology 21: 55–64.

Evans, R.A. (1961) Effects of different densities of downy brome (*Bromus tectorum*) on growth and survival of crested wheatgrass (*Agropyron desertorum*) in the greenhouse. Weeds 9: 216–223.

Evans, R.A., Holbo, H.R., Eckert, Jr., R.E. and Young, J.A. (1970) Functional environment of downy brome communities in relation to weed control and revegetation. Weed Sci. 18: 154–162.

Evans, R.A. and Young, J.A. (1970) Plant litter and establishment of alien annual species in rangeland communities. Weed Sci. 18: 697–703.

Evans, R.A. and Young, J.A. (1972) Microsite requirements for establishment of annual rangeland weeds. Weed Sci. 20: 350–356.

Evans, R.A. and Young, J.A. (1980) Establishment of barbwire Russian thistle in desert environments. J. Range Manage. 33: 169–173.

Harper, J.L., Williams, J.T. and Sagar, G.R. (1965) The behavior of seeds in soil. I. The heterogeneity of soil surfaces and its role in determining the establishment of plants. J. Ecol. 53: 273–286.

Harris, G.A. (1967) Some competitive relationships between *Agropyron spictatum* and *Bromus tectorum*. Ecol. Monogr. 37: 89–111.

Hull, A.C., Jr. and Pechanec, J.F. (1947) Cheatgrass – A challenge to range research. J. Forestry 45: 555–564.

Kearney, T.H., Briggs, L.J., Shantz, J.H., McLane, J.W. and Piemeisel, R.L. (1914) Indicator significance of vegetation in Tooele Valley, Utah. J. Agric. Res. 1: 365–417.

Kennedy, P.B. (1903) Summer ranges of eastern Nevada sheep. Agric. Exp. Sta. Bull. 55. Univ. of Nevada, Reno, Nevada. 53 pp.

Kennedy, P.B. and Doten, S.B. (1901) A preliminary report on summer ranges of western Nevada sheep. Nev. Agric. Exp. Sta. Bull. 51. Univ. of Nevada, Reno, Nevada. 53 pp.

Klemmedson, J.O. and Smith, J.G. (1964) Cheatgrass (*Bromus tectorum* L.) Bot. Rev. 30: 226–262.

Kreitlow, K.W. and Bleak, A.T. (1964) *Podosporiella verticillata*, a soil-borne pathogen of some western *Gramineae*. Phytopathology 54: 353–357.

Latourrette, J.E., Young, J.A. and Evans, R.A. (1971) Seed dispersal in relation to rodent activities in seral big sagebrush communities. J. Range Manage. 24: 118–120.

Mack, R.M. (1981) Invasion of *Bromus tectorum* L. into western North America: An ecological chronicle. Agro-Ecosystems 7: 145–165.

McKell, C.M., Robinson, J.P. and Major, J. (1962) Ecotypic variation in medusahead, an introduced annual grass. Ecology 43: 686–698.

Newman, E.I. (1963) Factors controlling the germination date of winter annuals. J. Ecol. 51: 625–639.

Piemeisel, R.L. (1951) Causes affecting change and rate of change in a vegetation of annuals in Idaho. Ecology 32: 53–72.

Robbins, W.W., Crafts, A.S. and Raynor, R.N. (1952) Weed control. McGraw-Hill Book Company, Inc. 503 pp.

Robertson, J.H. and Pearse, C.K. (1946) Give your range a chance – reduce sagebrush and cheatgrass. American Cattle Producer, May 62–63.
Schlatterer, E.F. and Hironaka, M. (1972) Some factors influencing tolerance to moisture stress of three range grasses. J. Range Manage. 25: 364–367.
Snyder, J.M. and Hullstein, L.H. (1973) The role of desert cryptogams in nitrogen fixation. Am. Midl. Natur. 90: 257–265.
Stewart, George (1936) History of range use. In The Western Range, Senate Doc. No. 199. 74th Congress, 2nd Session. U.S. Govern. Printing Office, Wash. DC.
Stewart, G. and Hull, A.C. (1949) Cheatgrass (*Bromus tectorum* L.) – An ecologic intruder in southern Idaho. Ecology 30: 58–74.
West, N.E. (Ed.) (1983) Temperate Desert and Semi-deserts. Elsevier Science Publ. Co. Amsterdam. 418 pp.
Young, J.A. (1981) Chukar partridge. Rangelands 3: 166–168.
Young, J.A., Eckert, Jr., R.E. and Evans, R.A. (1978) Historical perspectives regarding the sagebrush ecosystem. In Proc. Symposium on Sagebrush Ecosystem, pp. 1–13. Utah State Univ., Logan, Utah.
Young, J.A. and Evans, R.A. (1970) Invasion of medusahead into the Great Basin. Weed Sci. 18: 89–97.
Young, J.A. and Evans, R.A. (1974) Population dynamics of green rabbit-brush in disturbed big sagebrush communities. J. Range Manage. 27: 127–132.
Young, J.A. and Evans, R.A. (1975) Germinability of seed reserves in a big sagebrush community. Weed Sci. 23: 358–364.
Young, J.A. and Evans, R.A. (1976) Responses of weed populations to human manipulations of the natural environment. Weed Sci. 24: 186–190.
Young, J.A. and Evans, R.A. (1978) Population dynamics after wildfires in sagebrush grasslands. J. Range Manage. 31: 283–289.
Young, J.A. and Evans, R.A. (1982) Temperature profiles for germination of cool season range grasses. ARR-W-27; U.S. Dept. Agric., Agric. Res. Serv.
Young, J.A., Evans, R.A. and Eckert, Jr., R.E. (1969) Population dynamics of downy brome. Weed Sci. 17: 20–26.
Young, J.A., Evans, R.A. and Kay, B.L. (1971) Germination in caryopses of annual grasses in simulated litter. Agron. J. 63: 551–555.
Young, J.A., Evans, R.A. and Major, J. (1972) Alien plants in the Great Basin. J. Range Manage. 25: 194–201.
Young, J.A., Evans, R.A. and Tueller, P.T. (1976) Great Basin plant communities – pristine and grazed. In Robert Elston (Ed.), Holocene Environmental Change in the Great Basin, pp. 187–215. Nevada Arch. Survey-Res. Paper No. 6. Reno, NV.
Young, J.A., Evans, R.A. and Weaver, R.A. (1976) Estimating potential downy brome competition after wildfires. J. Range Manage. 29: 322–325.

TOWARDS A FUNCTIONAL DESCRIPTION OF VEGETATION

J.P. GRIME

INTRODUCTION

If the description of vegetation is to provide a basis for causal analysis, informed management and effective conservation it must progress beyond the listing of species and the estimating of their relative abundance. Since the writings of Raunkiaer (1934) the desirability has been recognized of devising a system which by reference to attributes of the component plants allows the major formative influences upon the vegetation to be identified. Raunkiaer's concept of life-forms has proved to be of durable value in the attempt to analyse the influence of macroclimate upon vegetation composition and has found modern expression in the comprehensive system of Box (1981). This approach has the very considerable advantage that it relies upon morphological criteria which are readily available for all plant species. However, many important influences upon vegetation are not detectable by reference to plant morphology and studies such as those of Grime (1974) and Noble and Slatyer (1979) bear witness to the continuing effort to recognize additional plant characteristics which are predictably related to habitat and ecology. The purpose in this paper is to review progress in this field of research. First, however, it is necessary to respond to those who have argued that such effort is premature or ill-conceived.

PUTTING FROM THE TEE

The increasing volume of research in plant ecology over the last decade has been associated with specialization. This trend is nowhere more evident than in the present dichotomy between plant population biology and physiological ecology. However, as the following quotations illustrate, there is one point upon which some exponents in these two fields can be found to agree.

First, a plant population biologist:

> "... the search for generalities in ecology has been disappointing – more so in plant than in animal ecology. The few generalities that have emerged come from studies of stands of single species."

White, J. (ed.), The Population Structure of Vegetation.
 ISBN 978-94-010-8927-2

"... it is from the work of many individuals working scattered over a variety of parts of the world, but concentrating their attention over long periods on the behaviour of individual plants, that the development of ecology as a generalizing and predictive science may be possible."

"The detailed analysis of proximal ecological events is the only means by which we can reasonably hope to inform our guesses about the ultimate causes of the ways in which organisms behave."

Harper (1982)

Next, a plant physiologist:

"There is another way to study adaptation, which is to proceed from measurements of physiological responses to the environment to an analysis of the underlying biochemistry and from these to the genetic control of the biochemical characteristics and from the biochemistry back to classical population genetics. This must inevitably be a painstaking and slow approach which will probably, in the short term, have to take second place to the grand designs for the rationalization of a mass of field data afforded by the *a posteriori* and *a priori* methods; but it is an approach with an honourable history ... and offers a sure route to a proper answer to the 'why they work in the way they do' sort of question.

Woolhouse (1981)

These authors are united in their rather bleak assessment of current and future progress in our understanding of vegetation and each advocates a long, narrow and rather stoney path as the only reliable way forward. Particularly since the two recommended paths lead in quite different directions the uncommitted plant ecologist would do well to reflect on whether such philosophies are adjusted more closely to the current divisions within Academia than to efficient problem-solving in vegetation science. Problems in ecology are so numerous and their practical implications so considerable that it is doubtful whether any major sources of evidence, however imperfect, can be ignored. Whilst acknowledging the essential contribution of detailed studies of the kinds suggested by Harper and Woolhouse it seems prudent to bear in mind instances in the earlier history of plant ecology where progress has been limited by excessive devotion to particular methodologies. In plant ecology as in golf there is a time for precision and a time for progression. Only in fog or cases of acute myopia can the hazards of driving justify putting from the tee.

PLANT ECOLOGY AS A JIGSAW PUZZLE

The weakness of golf as a metaphor for ecological research is that it is an individual activity with a well-defined objective. A closer model involves the construction of a large jigsaw at a family party. Initially the task for all participants is to find individual pieces which can be connected and at this stage there is scope for special skills, some members of the family concentrating on edge-pieces, others assembling sections with conspicuous colours, shapes or textures. Soon this activity results in a chaotic state of affairs in which the floor is littered with irregular islands growing at different rates and without reference to each other.

It is at this stage that relationships within the family are tested and begin to determine how quickly the jigsaw is completed. At one extreme is the family in which overall progress is kept under review so that connections between developing islands are established as early as possible allowing the completed picture to be visualized and the remaining gaps to be filled with minimal delay. Not unknown, however, is the family for which the jigsaw provides a longer and more enjoyable diversion in which the construction of each island becomes an absorbing activity in its own right with individual logic, rules and rivalries with neighbouring islands.

It is my submission that progress in developing generalities in plant ecology is limited at the present time by ineffective communication between specialisms. In particular, opportunities to establish connections between plant population biology and comparative physiology have been neglected. This has delayed the recognition of recurrent patterns of plant specialization which could form the basis for a functional description of vegetation.

THE SEARCH FOR PATTERN

As explained in the Introduction, the effort to devise a functional description of vegetation involves the search for plant characteristics, variation in which has a specific indicator value. Success depends not only upon finding attributes which are easily measured on large numbers of plants but also, and more crucially, upon establishing the ecological significance of variation in particular attributes. Opinions differ concerning the procedures needed to attain this second objective. Gould and Lewontin (1979) have criticised what they believe to be a widespread tendency among biologists to accept naive adaptive explanations without adequate proofs and a more specific attack upon "facile guesswork" by plant ecologists has been made by Harper (1982). These two papers contain useful warnings for the unwary but they are polemical in their disregard of the more sophisticated approaches adopted by many ecologists. It seems timely therefore to restate some of the principles which have proved useful in the task of identifying the ecological and evolutionary significance of particular traits in plants or animals.

1. Mechanisms excluding an organism or reducing its abundance in a particular type of habitat may be suggested on the basis of differences in requirements or in tolerance which distinguish it from other organisms which are more successful in the habitat.*

2. Comparisons between species of constrasted ecology reveal many differences and it is difficult to determine which, if any, are of ecological

*Except where stated, points 1–6 are based upon Grime (1965).

significance. This may be resolved to some extent by confining attention to the more consistent differences between large numbers of species successful or unsuccessful in the habitat under study. The advantage of this research strategy, as explained by Clutton-Brock and Harvey (1979), is that "as more species are considered, it becomes progressively more difficult to fit several adaptive hypotheses to the empirical facts."

3. Where available, populations of the same species drawn from contrasted habitats may provide opportunities to examine variation with respect to a smaller number of potentially critical characteristics. However, many of the features which differentiate between species are not adequately reflected in microevolutionary change within species. It seems advisable to review evidence from intra-specific studies within the context provided by broader inter-specific comparisons.

4. It is rarely profitable to examine variation in a single attribute without reference to other characteristics of the organisms under study. Ecological specialization is usually associated with correlated changes in a set of traits (Hutchinson 1951, MacArthur and Wilson 1967, Pianka 1970).

5. In both plants and animals there is need for care in the choice of attributes to be measured and in the sampling of organisms for study. In some plants development is extremely plastic and there is no guarantee that individuals examined under specific experimental conditions will provide data which can be extrapolated to other conditions of study. However, comparisons in which the rooting medium and climate are simplified have the advantage that it is easier to ensure that differences noted between plants are occurring in response to the same known variable or group of variables. Experiments in which field conditions are more closely simulated (e.g., by the use of natural soils) may fractionally increase the probability of eliciting the critical differences in response but "they transfer to the laboratory much of the environmental complexity which has defied analysis in the field" (Grime and Hodgson 1969).

6. Studies involving large numbers of species provide many opportunities for comparison and interpretation (Salisbury 1942, Hutchinson 1967, Grime et al. 1968, 1981, Harper et al. 1970, Seilacher 1970, Baker 1972, Grime and Hunt 1975, Southwood 1976, Clutton-Brock and Harvey 1979). Analysis may extend beyond attempts to identify traits which are (or have been) the subject of particular kinds of natural selection in certain habitats. This further possibility lies in recognition of components which have a limited capacity for phenotypic adjustment. Where such components occur in fundamental aspects of the anatomy, physiology or biochemistry of plants and animals it seems likely that they will have provided a recurrent focus for conflicting selection pressures during the evolution of many different taxa. Using the comparative approach it may be possible therefore to interpret the evolution and present ecology of plants by reference to the "internal constraints" (Grime and Hodgson 1969) which limit the potentiality of particular taxonomic groups.

THE TRIANGULAR MODEL

From various sources, including large-scale comparative studies, there is now for both animals and plants a considerable fund of information relating to the characteristics of organisms of similar or contrasted distribution in the field. These data reveal the existence of recurrent sets of traits, some conspicuously associated with particular ecologies. Information of this kind has prompted animal ecologists to propose 'strategy' theories of which the most widely known is that of r- and K-selection (MacArthur and Wilson 1967, Pianka 1970). Adoption of the term 'strategy' by plant ecologists has attracted some criticism; with its teleological implications the term is not ideal. However, provided that a clear definition* is applied the term can be used with precision. Here strategy is retained as a mark of respect for the pioneers who first used the term and whose achievement was to recognize that organisms exhibit sets of co-adapted traits which are predictably related to their ecology. This simple concept has more than any other allowed ecological theory to begin its escape from a morass of parochial and undigested observations; against this the semantic objections to 'strategy' pale into insignificance.

Despite its seminal influence upon ecology, the theory of r- and K-selection has two major weaknesses (Greenslade 1972a, b, Wilbur et al. 1974, Gill 1978, Grime 1979) which can be summarized as follows:

1. The concept fails to recognize a third strategy (stress-tolerance) of general occurrence in stable but unproductive habitats.

2. The theory does not take into consideration the fact that organisms may exhibit quite different strategies during regenerative (juvenile) and established phases of their life-cycles.

In recent years, additional evidence has supported the suggestion first evident in the writings of Ramenskii (1938) and Greenslade (1972a, b) that there are three primary strategies which can be represented in a triangular model (Grime 1974). This is to suggest that Darwin's "struggle for existence" (1859) can be dissected by recognizing three distinct threats to existence coinciding with particular habitat conditions and resulting in three characteristic types of evolutionary specialization – ruderals, competitors and stress-tolerators. The extreme conditions favouring these three primary strategies are only part of the spectrum of habitats exploited by animals and plants; the full range can be represented by an equilateral triangle (Fig. 1). This model has been applied to herbaceous plants (Grime 1974, 1977, Chapin 1980, Leps et al. 1982), algae (Raven 1981, Shepherd 1982, Coesel 1982, Dring 1982), fungi (Pugh 1980, Cooke and Raynor in press) and corals (Rosen 1981). An account of the theoretical basis of the model has been presented elsewhere (Grime 1979). Various opportunities and problems for vegetation analysis and description are suggested by the model; they include the following:

1. The triangular model provides a compact framework in which to

*Here a strategy is defined as a grouping of similar or analogous genetic characteristics which recurs widely among species or populations and causes them to exhibit similarities in ecology.

Table 1. Some characteristics of competitive, stress-tolerant and ruderal plants

		Competitive	Stress-tolerant	Ruderal
(i)	*Morphology*			
1.	Life-forms	Herbs, shrubs and trees	Lichens, bryophytes, herbs, shrubs and trees	Herbs, bryophytes
2.	Morphology of shoot	High dense canopy of leaves. Extensive lateral spread above and below ground	Extremely wide range of growth forms	Small stature, limited lateral spread
3.	Leaf form	Robust, often mesomorphic	Often small or leathery, or needle-like	Various, often mesomorphic
4.	Canopy structure	Rapidly-elevating monolayer	Often multilayered. If monolayer not rapidly-elevating	Various
(ii)	*Life-history*			
5.	Longevity of established phase	Long or relatively short	Long-very long	Very short
6.	Longevity of leaves and roots	Relatively short	Long	Short
7.	Leaf phenology	Well-defined peaks of leaf production coinciding with periods of maximum potential productivity	Evergreens, with various patterns of leaf production	Short phase of leaf production in period of high potential productivity
8.	Phenology of flowering	Flowers produced after (or, more rarely, before) periods of maximal potential productivity	No general relationship between time of flowering and season	Flowers produced early in the life-history
9.	Frequency of flowering	Established plants usually flower each year	Intermittent flowering over a long life-history	High frequency of flowering
10.	Proportion of annual production devoted to seeds	Small	Small	Large
11.	Perennation	Dormant buds and seeds	Stress-tolerant leaves and roots	Dormant seeds
12.	Regenerative* strategies	V, S, W, B_s	V, B_{sg}, W	S, W, B_s
(iii)	*Physiology*			
13.	Maximum potential relative growth-rate	Rapid	Slow	Rapid

14.	Response to stress	Rapid morphogenetic responses (root-shoot ratio, leaf area, root surface area) maximizing vegetative growth	Morphogenetic responses slow and small in magnitude	Rapid curtailment of vegetative growth, diversion of resources into flowering
15.	Photosynthesis and uptake of mineral nutrients	Strongly seasonal, coinciding with long continuous period of vegetative growth	Opportunistic, often uncoupled from vegetative growth	Opportunistic, coinciding with vegetative growth
16.	Acclimation of photosynthesis, mineral nutrition and tissue hardiness to seasonal change in temperature, light and moisture supply	Weakly developed	Strongly developed	Weakly developed
17.	Storage of photosynthate mineral nutrients	Most photosynthate and mineral nutrients are rapidly incorporated into vegetative structure but a proportion is stored and forms the capital for expansion of growth in the following growing season	Storage systems in leaves, stems and/or roots	Confined to seeds
(iv)	*Miscellaneous*			
18.	Litter	Copious, often persistent	Sparse, sometimes persisent	Sparse not usually persistent
19.	Palatability to unspecialized herbivores	Various	Low	Various, often high
20.	Genome size	Usually small	Various	Small-very small

*Key to regenerative strategy: V, vegetative expansion; S, seasonal regeneration in vegetation gaps; W, numerous small wind-dispersed seeds or spores; B_s, persistent seed bank; B_{sg}, persistent seedling bank.

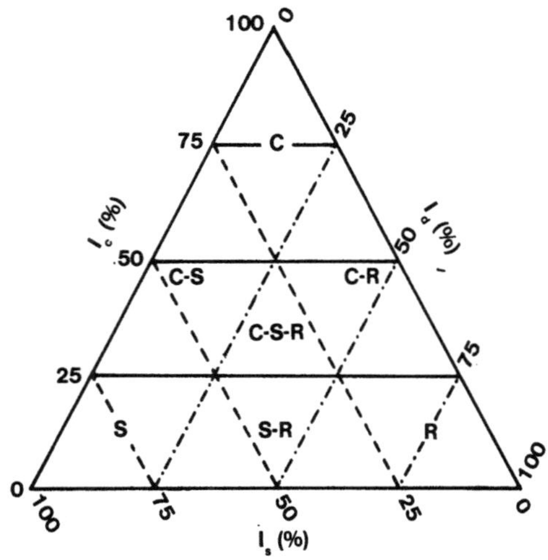

Fig. 1. Model describing the various equilibria between competition, stress and disturbance in vegetation and the location of primary and secondary strategies. C, competitor; S, stress-tolerator; R, ruderal; C–R, competitive-ruderal; S–R, stress-tolerant ruderal; C–S, stress-tolerant competitor; C–S–R, "C–S–R strategist". I_c, relative importance of competition (—); I_s, relative importance of stress (---); I_d, relative importance of disturbance (-·-·-).

connect some disparate threads of ecological information drawn from population biology, comparative physiology and ecosystem analysis. Reference to the lists in Table 1 suggests quite a number of plant characteristics (e.g., life-span, potential relative growth-rate, capacity for physiological acclimation, palatability to generalized herbivores) which may be expected to change in a predictable manner as we move from one type of habitat to another. The model also provides a basis on which to predict the direction and rate of floristic response to alterations in the intensities of stress* or disturbance.

2. A central assertion of the triangular model is that the intensity of competition for resources (C) declines progressively with increasing intensities of stress (S) and/or disturbance (D). It follows that the equilibrium between C, S and D is a major determinant of vegetation structure and species composition at any site. Opportunities for analysis and description may be expected therefore wherever it proves possible to recognize measurable plant characteristics which vary in relation to the prevailing intensities of C, S or D. This approach has been tested in triangular ordinations involving plant morphology (Grime 1974, Shepherd 1981), life-span (Shepherd 1981) and relative growth-rate (Grime 1974). In temperate herbaceous vegetation, seasonal change in shoot biomass (Fig. 2) appears to be predictably related to the C, S, D equilibrium and deserves to be explored further as a potential basis for the recognition of plant strategies.

*Henceforward, use of the terms *stress*, *disturbance* and *competition* is in accordance with the definitions in Grime (1979).

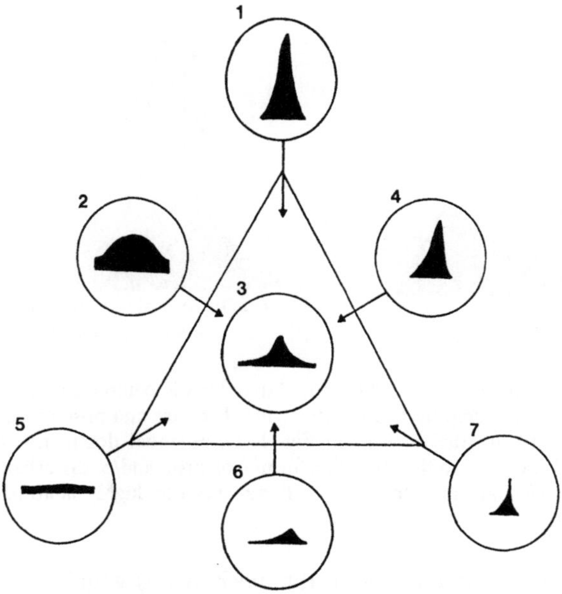

Fig. 2. Scheme relating pattern of seasonal change in shoot biomass to strategy. 1, competitor; 2, stress-tolerant competitor; 3, C–S–R strategist; 4, competitive ruderal; 5, stress-tolerator; 6, stress-tolerant ruderal; 7, ruderal.

3. In any analysis of vegetation it is necessary to distinguish between the selection forces which have determined its essential characteristics (ultimate determinants) and those which are operating at the present time to modulate species composition (proximal determinants). The triangular model suggests a particular need for such distinctions in considering the vegetation types which correspond to the extremities of the triangle. Thus it may be found that although continuously low productivity (stress) dictates that the vegetation on a granitic boulder is composed exclusively of stress-tolerant lichens it is quite feasible that inconspicuous levels of competition and/or disturbance could determine *which* lichen species are present or most abundant. Similar arguments apply to the fine-tuning of species composition in vegetation composed exclusively of ruderal or competitive plants. Here we glimpse a fascinating paradox whereby, contrary to the argument of Harper (1982), the detailed analysis of proximal ecological events may not reliably "inform our guesses about the ultimate causes of the ways in which organisms behave."

4. Except in the most extreme conditions corresponding to the corners of the triangular model it is unlikely that plant communities will be composed exclusively of species of similar strategy. Spatial heterogeneity within a stand of vegetation can arise from differences in topography, soil and microclimate and may be imposed by the vegetation, fauna and microflora. In addition, plant habitats change seasonally and on a longer (successional) time scale. One major effect of this variation will be to cause spatial and temporal changes in the C, S, D-equilibrium. It seems likely therefore that an analysis of the strategic diversity within a plant

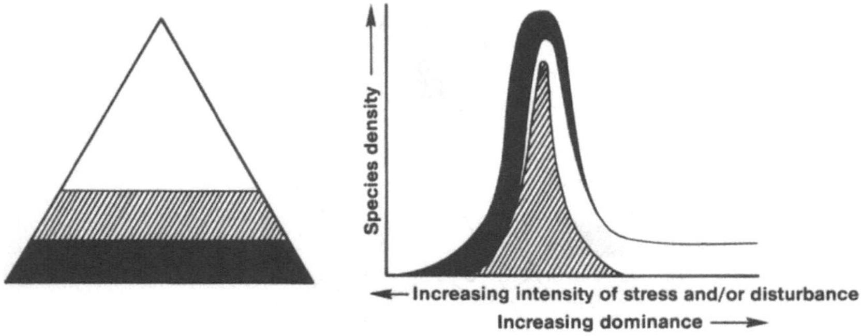

Fig. 3. Scheme describing the distribution of three floristic components in the triangular model (left) and in the hump-backed model (right). The axes and positions of strategies in the triangular model are the same as in Fig. 1. □, potential dominants; ■, species or populations highly adapted to the prevailing form(s) of stress and/or disturbance; ▨, species or populations which are neither potential dominants nor highly adapted to stress or disturbance.

community may provide a clue to the mechanisms which permit co-existence and control the relative abundance of the constituent species.

5. Further insights into the control of species composition and relative abundance are available from the hump-backed model (Grime 1973) which, as shown in Fig. 3, is a close relative of the triangle of primary strategies. Two main propositions, discussed in detail in Grime (1979), are incorporated in the hump-backed model:

(a) Vegetation remains species-poor in habitats subjected to high intensities of stress and/or disturbance (corresponding to the left-hand side of the model) and also in the range of quite different circumstances where dominance* is permitted (on the right-hand side).

(b) Vegetation which is species-rich (in the central corridor) contains a majority of plants which are neither potential dominants nor capable of surviving in extreme habitats. Additional but minor contributions to species-rich vegetation occur in the form of debilitated and transient specimens of both potential dominants and plants of extreme habitats.

REFINEMENTS

Strategy concepts can achieve wide applicability in plant and animal communities by focussing on attributes which vary in relation to the quality and duration of the opportunities for resource capture, growth and reproduction afforded by particular types of habitats (Southwood 1977). A framework is provided on which to base a functional description of vegetation and it has been suggested also (Grime 1979, Leps et al. 1982) that the generality of the concepts allows prediction with respect to

*Dominance is used here in accordance with the definition in Grime (1979).

community characteristics such as resistance and resilience under perturbation. However, if the strategy approach is to be of practical value in the description and management of vegetation it must be refined by the development of additional criteria which can be used to recognize the mechanisms which control the "fine-structure" of plant communities. In this endeavour it will be useful to devise indices which relate to regenerative capacity and, in particular, enable predictions to be made with respect to the dispersal and dormancy of propagules (Noble and Slatyer 1979). It is also essential that objective criteria are found with which to identify niche differentiation in relation to variation in the intensity and seasonal distribution of particular types of stress and disturbance. A specific development here is the recent use of genome size (Grime and Mowforth 1982) as an indicator of plant growth responses to seasonal changes in climate.

REFERENCES

Baker, H.G. (1972) Seed weight in relation to environmental conditions in California. Ecology 53: 997–1010.

Box, E.O. (1981) Macroclimate and Plant Forms: an Introduction to Predictive Modelling in Phytogeography. The Hague: Junk.

Chapin, F.S. (1980) The mineral nutrition of wild plants. Ann. Rev. Ecol. Syst. 11: 233–260.

Clutton-Brock, T.H. and Harvey, P.H. (1979) Comparison and adaptation. Proc. R. Soc. Lond. B. 205: 547–565.

Coesel, P.F.M. (1982) Structural characteristics and adaptation of desmid communities. J. Ecol. 70: 163–178.

Cooke, R.C. and Rayner, A.D.M. (1984) The Ecology of Saprotrophic Fungi: Towards a Predictive Approach. London: Longman.

Darwin, C. (1859) The Origin of Species by Means of Natural Selection or the Preservation of Favoured Races in the Struggle for Life. London: Murray.

Dring, M.J. (1982) The Biology of Marine Plants. London: Arnold.

Gill, D.E. (1978) On selection at high population density. Ecology 59: 1289–1291.

Gould, S.J. and Lewontin, R.C. (1979) The spandrels of San Marco and the Panglossian paradigm: a critique of the adaptationist programme. Proc. R. Soc. Lond. B. 205: 581–598.

Greenslade, P.J.M. (1972a) Distribution patterns of *Priochirus* species (Coeleoptera: Staphylindae) in the Solomon Islands. Evolution 26: 130–142.

Greenslade, P.J.M. (1972b) Evolution in the staphylinid genus *Priochirus* (Coeleoptera). Evolution 26: 203–220.

Grime, J.P. (1965) Comparative experiments as a key to the ecology of flowering plants. Ecology 45: 513–515.

Grime, J.P. (1973) Competitive exclusion in herbaceous vegetation. Nature 242: 344–347.

Grime, J.P. (1974) Vegetation classification by reference to strategies. Nature 250: 26–31.

Grime, J.P. (1977) Evidence for the existence of three primary strategies in plants and its relevance to ecological and evolutionary theory. Am. Nat. 111: 1169–1194.

Grime, J.P. (1979) Plant Strategies and Vegetation Processes. Chichester: Wiley.

Grime, J.P., MacPherson-Stewart, S.F. and Dearman, R.S. (1968) An investigation of leaf palatability using the snail *Cepaea nemoralis* L. J. Ecol. 56: 405–420.

Grime, J.P. and Hodgson, J.G. (1969) An investigation of the ecological significance of lime-chlorosis by means of large-scale comparative experiments. In I.H. Rorison (Ed.), Ecological Aspects of the Mineral Nutrition of Plants, pp. 67–99. Oxford: Blackwell.

Grime, J.P. and Hunt, R. (1975) Relative growth-rate: its range and adaptive significance in a local flora. J. Ecol. 63: 393–422.

Grime, J.P., Mason, G., Curtis, A.V., Rodman, J., Band, S.R., Mowforth, M.A.G., Neal,

A.M. and Shaw, S. (1981) A comparative study of germination charcteristics in a local flora. J. Ecol. 69: 1017–1059.
Grime, J.P. and Mowforth, M.A. (1982) Variation in genome size – an ecological interpretation. Nature 299: 151–153.
Harper, J.L. (1982) After description. In E.I. Newman (Ed.), The Plant Community as a Working Mechanism. Special Pub. No. 1 BES, pp. 11–25. Oxford. Blackwell.
Harper, J.L., Lovell, P.H. and Moore, K.G. (1970) The shapes and sizes of seeds. Ann. Rev. Ecol. Syst. 1: 327–356.
Hutchinson, G.E. (1951) Copepodology for the ornithologist. Ecology 32: 571–577.
Hutchinson, G.E. (1959) Homage to Santa Rosalia or why are there so many kinds of animals? Am. Nat. 93: 145–159.
Hutchinson, T.C. (1967) Comparative studies of the ability of species to withstand prolonged periods of darkness. J. Ecol. 55: 291–299.
Lepš, J.J., Osbornová-Kosinová and Rejmánek, K. (1982) Community stability, complexity and species life-history strategies. Vegetatio 50: 53–63.
MacArthur, R.H. and Wilson, E.D. (1967) The Theory of Island Biogeography. Princeton N.J.: Princeton University Press.
Noble, I.R. and Slatyer, R.O. (1979) The use of vital attributes to predict successional changes in plant communities subject to recurrent disturbances. Vegetatio 43: 5–21.
Pianka, E.R. (1970) On r- and K-selection. Am. Nat. 104: 592–597.
Pugh, G.J.F. (1980) Strategies in fungal ecology. Trans. Br. mycol. Soc. 75: 1–14.
Ramenskii, L.G. (1938) (Introduction to the geobotanical study of complex vegetations). Moscow: Selkhozgiz. 620 pp. (Russian).
Raunkiaer, C. (1934) The Life Forms of Plants and Statistical Plant Geography; being the collected papers of C. Raunkiaer, translated into English by H.G. Carter, A.G. Tansley and Miss Fansboll. Oxford: Clarendon.
Raven, J.A. (1981) Nutritional strategies of submerged benthic plants: the acquisition of C, N and P by rhizophytes and haptophytes. New Phytol. 88: 1–30.
Rosen, B.R. (1981) The tropical high diversity enigma – the corals'-eye view. In P.L. Forey (Ed.), Chance, Change and Challenge. The evolving biosphere, pp. 103–129. British Museum and Cambridge University Press.
Salisbury, E.J. (1942) The Reproductive Capacity of Plants. London: Bell.
Seilacher, A. (1970) Arbeitskonzept zur Konstruktionsmorphologie. Lethaia 3: 393–396.
Shepherd, S.A. (1981) Ecological strategies in a deep water red algal community. Botanica Marina XXIV: 457–463.
Southwood, T.R.E. (1976) Bionomic strategies and population parameters. In R.M. May (Ed.), Theoretical Ecology, Principles and Applications. Oxford: Blackwell.
Southwood, T.R.E. (1977) Habitat, the templet for ecological strategies? J. Anim. Ecol. 46: 337–365.
Wilbur, H.M., Tinkle, D.W. and Collins, J.P. (1974) Environmental certainty, trophic level, and resource availability in life history evolution. Am. Nat. 108, 805–817.
Woolhouse, H.W. (1981) Aspects of the carbon and energy requirements of photosynthesis considered in relation to environmental constraints. In C.R. Townsend and P. Calow (Eds.), Physiological Ecology: an Evolutionary Approach to Resource Use. Oxford: Blackwell.

LIFE HISTORY CHARACTERISTICS OF TEMPERATE WOODLAND PLANTS IN JAPAN

SHOICHI KAWANO

ABSTRACT

The life history features of over thirty woodland plants in Japan are reviewed. The seasonal growth cycles of various species are differentiated into several types of evergreen and deciduous phenologies. Their energy allocation patterns are diverse. *Erythronium japonicum*, *Disporum smilacinum* and *D. sessile* are discussed in detail as examples of the contrasting life histories found among woodland species.

INTRODUCTION

In a series of studies since 1968 on the comparative biology of temperate woodland plants in Japan, we have continuously concentrated our efforts to investigate the life history phenomena of temperate elements of Arcto-Tertiary origin (e.g., Kawano et al. 1968, Kawano 1970, 1975, 1981, Kawano and Nagai 1975, Kawano et al. 1982a). In this paper an attempt is made to describe and summarize all the evidence thus far accumulated and to discuss the modes of evolution in their life history patterns. Although this paper is primarily concerned with species indigenous to Japan, the so-called Arcto-Tertiary elements have very closely related counterparts in North America or Europe (Li 1952, Axelrod 1966), and consideration will be given also to them whenever appropriate (see also Bierzychudek 1982).

LIFE HISTORY PARAMETERS

As a matter of course, life history studies must cover various aspects of the life cycle of the species in question. Any characteristics that the species show have adaptive significance, both in connection with the survival or maintenance of the individual organism itself in the biotic communities with which it is associated and in relation to the maintenance of its future generations. The implication of all such life history traits observed must

White, J. (ed.), The Population Structure of Vegetation.
 ISBN 978-94-010-8927-2

however, be considered in terms of selective forces that have operated in the past. The life history traits of any plant species must be regarded as an outcome of its past ecological and evolutionary history. The following significant life history parameters of plants must at least be carefully analyzed and considered (Kawano 1970, 1975, Solbrig et al. 1979, Solbrig 1980, Silvertown 1982):

(1) Phenology or seasonal growth cycles;

(2) Age of first reproduction;

(3) Reproductive life span, whether it is continuous or intermittent, whether monocarpic or polycarpic;

(4) Various aspects of reproductive biology, including sex expression, pollination system, breeding system and fecundity;

(5) Assimilation behavior and energy allocation to growth and to maintenance activities;

(6) Energy investment in reproduction, sexual as well as asexual (most woodland herbs are perennials or pseudo-annuals (Salisbury 1942, Kawano 1975) and possesses both systems of reproduction);

(7) Reproductive capacity and size of propagules; the role and efficiency of asexual reproduction as a means of recruitment of individuals must also be evaluated;

(8) Dispersal mechanisms, dormancy and germination mechanisms of propagules;

(9) Survivorship and mortality factors;

(10) Population structure, ecological distribution and niche.

PHENOLOGY AND LIFE CYCLES

The phenology or seasonal growth cycle which a particular species exhibits is established as a result of adaptive responses to particular environments where they grow, although numerous cases of convergence or parallelism are known in such characteristics. The life cycle of a plant is to some extent predetermined; it is a function of its genotype and doubtlessly an outcome of its ecological and evolutionary history. A comparison of phenological characteristics of a number of warm- and cool-temperate plants so far examined reveals that woodland plants with seasonal growth cycles have differentiated as a result of adaptive response to woodland habitats where conspicuous periodicity in various physical and biotic regimes predominates. In studies dealing with the phenology or seasonal growth cycles of plants, it is most important to pay attention not only to the active phase in growth when aerial organs appear above ground, but also to the entire process of growth throughout the year, since significant changes in the partitioning of matter into aerial and underground parts (especially storage organs) may be expressed in some form.

Figure 1 is a summary diagram demonstrating the phenology or seasonal growth cycles of various woodland plants studied by us, together with the seasonal changes in thermal and light regimes on the deciduous forest floor. It is noteworthy that there are several different groups of evergreen perennials in terms of their dry matter economy, and also in the manner of the replacement of their aerial organs, although all these plants are usually described simply as 'evergreen' (e.g., Ohwi 1965, Kitamura et al. 1967). The plants referred to as evergreen I normally maintain aerial organs over at least three years and new shoot sprouting takes place either once in spring (e.g., *Pachysandra terminalis*: Kawano et al. 1983) or twice a year, once in spring and again in autumn (e.g., *Heloniopsis orientalis*: Kawano and Masuda 1980), although the extent of the replacement in autumn is slight and gradual. The relative partitioning of dry matter into the component organs of *Pachysandra terminalis* throughout a year is shown in Fig. 2: this species possesses aerial organs over the three years and new aerial organs produced attain approximately 25–30% of the total biomass. Another peculiar and more complex process of replacement of aerial organs is found in *Chamaele decumbens* (Umbelliferae), an evergreen or semi-evergreen perennial. In this species, sprouting of new shoots is somewhat unsynchronized and occurs more or less continuously from March to November. This behavior varies in different regions of the distribution range of this species, especially in northern areas (e.g., in Hokkaido), where the favourable season for its growth is much shorter and limited: its seasonal growth cycle becomes much simpler and somewhat similar to that of typical deciduous perennials.

The plants referred to as evergreen IIa–c (Fig. 1) are conspicuously distinct from the groups mentioned above in the manner of their replacement of aerial organs, which occurs regularly only once a year, beginning in spring to late summer. As shown in Fig. 1, three different types can be distinguished according to the manner of the shoot replacement. In evergreen IIa, represented by species such as *Ophiopogon planiscapus*, a very gradual replacement of aerial organs occurs regularly once a year, initiated in April and completed in November. The ratio of subterranean organs to aerial shoots in this species is approximately 1 : 1. In *Coptis japonica*, which can be referred to as evergreen IIb, the replacement is also yearly, but is more rapid and requires only three months, beginning in early April and completed in late June. The ratio of underground organs to aerial shoots is about 3 : 2 throughout the year in this species. In *Asarum takaoi*, which is referred to evergreen IIc, the replacement is very rapid and brief, being completed within a month from March to April (Fig. 3). The ratio of underground organs to aerial shoots in this species varies from 4 : 1 to 7 : 3, which contrasts markedly with those found in *Heloniopsis orientalis* or *Pachysandra terminalis*, in which the ratios range from 1 : 4 to 3 : 7. To this type *Hepatica nobilis* var. *japonica* may also be referred (Kawano et al. 1980). This evergreen IIc is very close to or almost identical with 'deciduous shade' plants (Fig. 1): for instance, if the senescence of aerial organs occurs slightly earlier than the usual, its phenology precisely matches those of 'deciduous shade' woodland perennials. It is of interest to note that the photosynthetic potential of foliage leaves and the types of light-photosynthetic curves in such evergreen species cited above are very similar to those of typical shade plants in their general patterns, although photosynthetic rates of foliage leaves of these evergreen species showed the

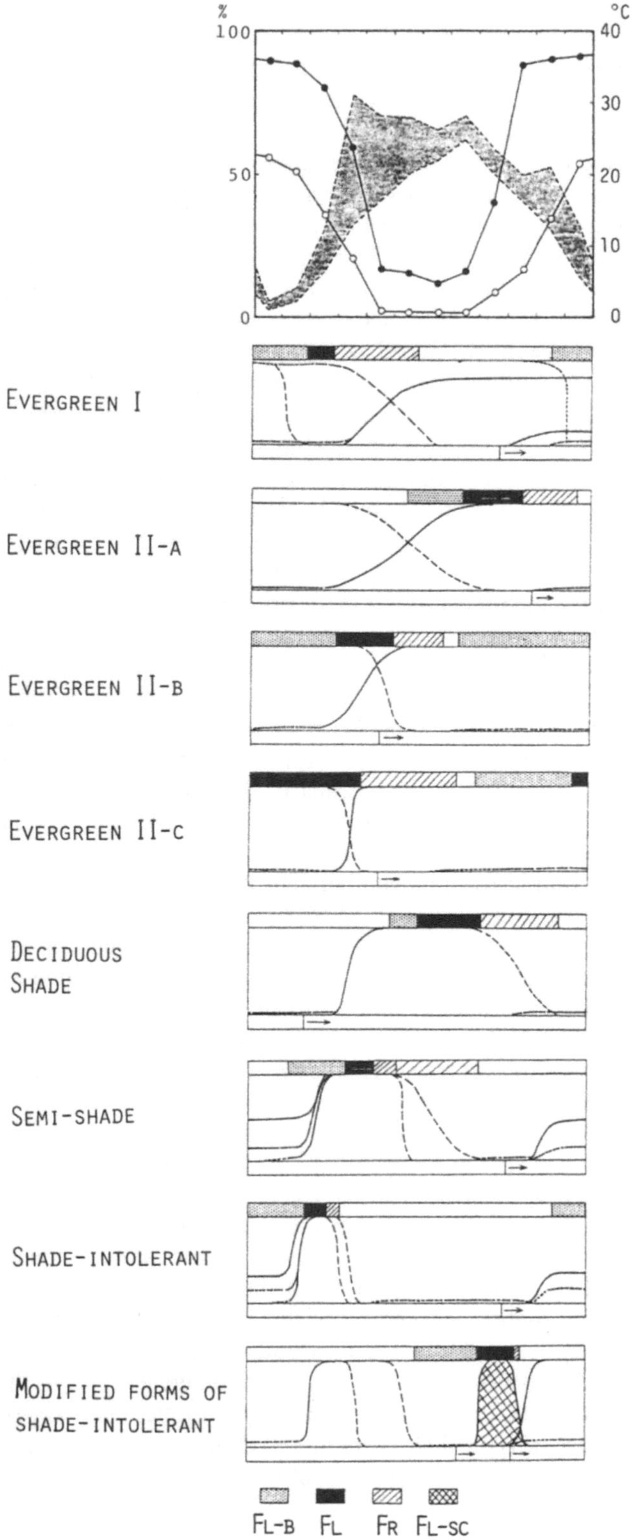
%
100
50
0
°C
40
30
20
10
0
Evergreen I
Evergreen II-a
Evergreen II-b
Evergreen II-c
Deciduous
Shade
Semi-shade
Shade-intolerant
Modified forms of
shade-intolerant
Fl-b
Fl
Fr
Fl-sc

lowest values in dark, shady summer conditions, without exceptions (Kawano 1970, Kawano and Masuda 1979, Kawano et al. 1980, Kawano et al. 1982b, Kawano et al. 1983). In short, all these evergreen herbs or semi-shrub species, despite the differences in their dry matter economy, exhibit a clear seasonality in their growth, leaf replacement and reproductive activity: this is their synchronized response to the annual cyclical changes in light and temperature regimes in woodland environments. Among other typical temperate woodland herbs, four clearly different deciduous perennials shown in Fig. 1 and Table 1 could be distinguished according to their phenology or seasonal growth cycles and also by their assimilation behavior:

(i) Typical shade plants (Lundegårdh 1921, 1924) or green-shade plants (Seybold and Eagle 1937), in which leaf development occurs usually after the almost complete expansion of the canopy in the upper tree layer;

(ii) 'semi-shade' plants (Sparling 1967) or red-green shade plants (Seybold and Eagle 1937) which initiate the expansion of aerial shoots almost simultaneously with the typical spring or shade-intolerant plants but persist for considerably longer periods of time until the forest canopy casts heavy shade, that is, until the end of June or occasionally the beginning of July;

(iii) typical spring plants which appear above ground only for a brief period of one month or slightly more in early spring (mostly from March to April) and are referred to as shade-intolerant plants (Lundegårdh 1924, Seybold and Eagle 1937, Sparling 1967); among them, however, are several species in which some leaves often extend above the ground in autumn depending upon the thermal conditions on the forest floor;

(iv) a modified form of shade-intolerant plants, in which the productive (photosynthetic) phase and reproductive phase are completely separate in the course of a year: that is, the leaves expand above ground only for about a month during March to April or for several months from October to May when there are no foliage leaves in the upper tree layers; and then at the end of July to September only reproductive organs (scapes) are produced.

The examples given exhibit the phenology and dry matter economy of mature individuals of selected woodland species sampled from the fields, but if we examine the entire process of their development and growth throughout the life cycle the characteristics of their life-history will become much more evident. As an example, Fig. 4 exhibits the phenology and seasonal changes in proportional distribution of dry matter into

Fig. 1. Yearly changes (January to December) in relative light intensities on the forest floor (filled circles, the maximum; open circles, the minimum), temperature amplitude between the monthly maximum and minimum (in °C) and seasonal growth cycles of the eight major groups of woodland herbs among temperate forest plants in Japan and North America.

Fl-b, flower buds; Fl, flowers; Fr, fruits or seeds; Fl-sc, scapes lacking leaves. Solid lines in the diagram indicates the formation of new aerial organs, and broken lines the senescence and fall of old aerial organs. Arrows specify new root sprouting (after Kawano 1970, Kawano et al. 1982b).

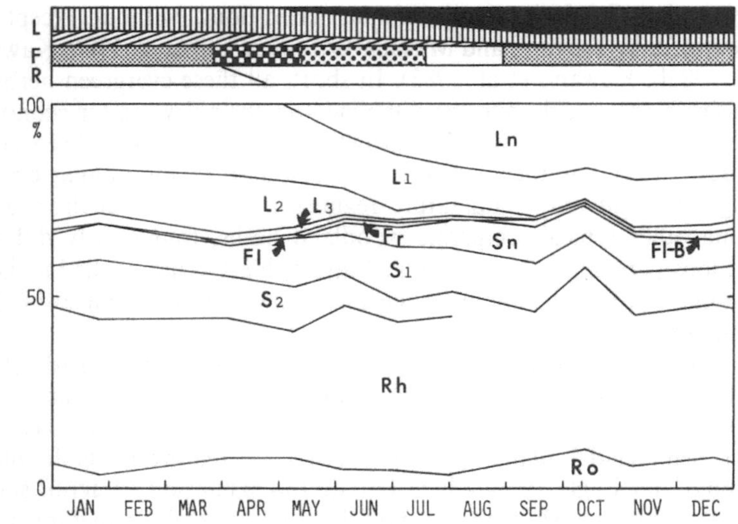

Fig. 2. Phenology and seasonal changes in the partitioning of organs in *Pachysandra terminalis* (Buxaceae).

L, leaf expansion; F, flowering; R, root sprouting; Ln and Sn, new leaf and stem; L_1 and S_1, one year old leaf and stem; L_2 and S_2, two years old leaf and stem; L_3, three years old leaf; Fl-b, flower beds; Fl, flower; Rh, rhizome; Ro, root. The sampling of materials was made at monthly intervals from its natural habitats (Kamiichi, Toyama Prefecture).

The standing crop (both aerial and subterranean) in a 50 × 50 cm quadrat, more or less uniformly covered by this species, on the forest floor was carefully harvested each time, dismembered into each component organ, dried in an oven for more than 48 hr at 80°C, and then weighed. Since this species possesses exceedingly creeping underground organs, it was not possible to distinguish an individual plant.

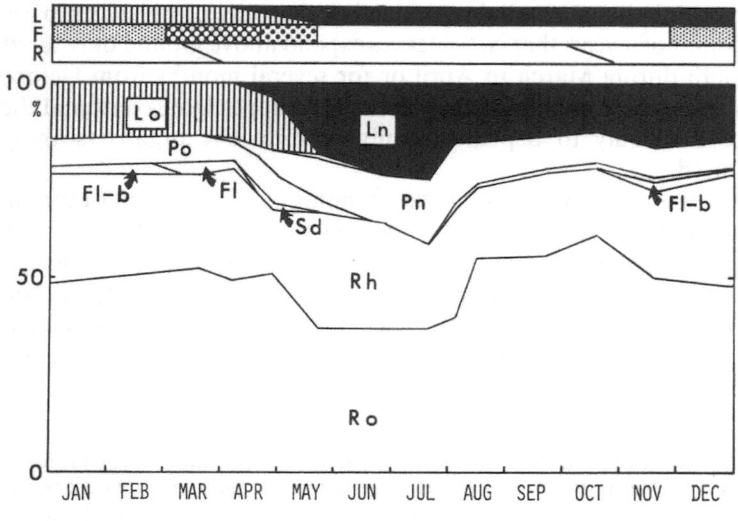

Fig. 3. Phenology and seasonal changes in the partitioning of dry matter into component organs of *Asarum takaoi* (Aristolochiaceae).

Ln, new leaf; Lo, one year old leaf; P, petiole; Fl-b, flower buds; Fl, flower; Sd, seed; Rh, rhizome; Ro, root. For other symbols, see Fig. 2.

Table 1. Patterns of seasonal growth cycles, life cycles and reproductive biology of selected temperate woodland plants in Japan.

Types of seasonal growth cycles (See Fig. 1)	Taxa	Life[1*] cycle	SRA[2*] (Fl. stage)	SRA[2*] (Fr. stage)		No. of sexual propagules /plant	Weight of single propagule (mg)	VRA[3*]	No. of asexual propagules /plant	Weight of single propagule (mg)	Reproductive system[6*]	
				Sexual propagules	Total sexual repro. organs			Asexual propagules			Sexual reproduction	Asexual reproduction
Evergreen:												
Type I	*Chamaele decumbens*[7*,1]	p, p	29.3	17.9	53.6	.[4*]	.	–[5*]	–	–	+++	–
	Chionographis japonica[1]	p, p	11.3	4.4	15.5	308	0.4	–	–	–	+++	–
	Heloniopsis orientalis[4]	p, p	6.8–9.3	2.0–2.6	10.0–25.5	1134–2718	0.049–0.058	?	several	?	++	+
	Mitchella undulata[1]	p, p	.	.	.	.	.	–	–	–	+++	–
	Pachysandra terminalis[1]	p, p	1.0	.	1.5	.	.	–	–	–	+++	–
Type IIa	*Liriope platyphylla*[1]	p, p	.	12.0	22.3	23.9	36.2	–	–	–	+++	±
	Liriope spicata[1]	p, p	3.6	2.2	2.9	76.0	71.0	–	–	–	+++	±
	Ophiopogon japonicus[1]	p, p	1.6	6.7	7.6	9.5	110.0	–	–	–	+++	±
	Ophiopogon ohwii[1]	p, p	2.4	3.4	3.9	19.8	89.9	–	–	–	+++	–
	Ophiopogon planiscapus[1]	p, p	7.8	11.7	14.3	13.7	65.6	–	–	–	+++	–
Type IIb	*Coptis japonica*[6]	p, p	3.1	1.7	11.4	101	0.4	–	–	–	+++	–
	Epimedium sempervirens[1]	p, p	.	.	.	.	.	–	–	–	+++	–
	Oxalis griffithii[1]	p, p	.	.	.	.	.	–	–	–	+++(++)[10*]	–
Type IIc	*Asarum kooyanum* v. *nipponicum*[1]	p, p	5.9	.	7.1	.	.	–	–	–	+++	–
	Asarum takaoi[1]	p, p	3.0	.	2.0	.	.	–	–	–	+++	–
	Hepatica nobilis v. *japonica*[5]	p, p	13.3	2.3	4.7	104	2.4	–	–	–	+++	±
Deciduous:												
Shade	*Adenocaulon himalaicum*[1]	p, p	.	2.7	10.8	65.1	2.0	–	–	–	+++	–
	Ainsliaea acerifolia v. *subapoda*[1]	p, p	.	1.9	8.3	27.8	1.6	–	–	–	+++	
	Arisaema japonicum[8*,8]	p, p	.	23.0	41.1	305	32.0	–	–	–	+++	–
	Arisaema serratum[8*,8]	p, p	.	31.3	34.2	248	20.0	–	–	–	+++	–
	Arisaema urashima[8*,9]	p, p	4.8 15.8	14.9	43.9	161	19.0	7.2 10.4	.	.	++	++

Table 1 (Continued)

Types of seasonal growth cycles (See Fig. 1)	Taxa	Life[1*] cycle	SRA[2*] (Fl. stage)	SRA[2*] (Fr. stage)		No. of sexual propagules /plant	Weight of single propagule (mg)	VRA[3*]	No. of asexual propagules /plant	Weight of single propagule (mg)	Reproductive system[6*]	
				Sexual propagules	Total sexual repro. organs			Asexual propagules			Sexual reproduction	Asexual reproduction
	Asarum caulescens[9]	p, p	.	.	.	.	.	.	.	.	+	+++
	Asarum sieboldii[1]	p, p	.	.	.	.	.	–	–	–	+++	–
	Clintonia udensis[2]	p, p	.	5.9	13.0	31.1	7.6	–	–	–	+++	–
	Macroclinidium rigidulum[1]	p, p	.	2.5	6.9	98.3	4.3	–	–	–	+++	–
	Sanicula chinensis[1]	p, p	.	8.7	11.1	223	2.7	–	–	–	+++	–
	Streptopus streptopoides ssp. *japonicus*[2]	p, p	.	2.5	3.4	28.0	2.2	–	–	–	+++	–
	Smilacina hondoensis[2]	p, p	.	14.1	14.7	139	15.4	–	–	–	+++	–
	Smilacina japonica[2]	p, p	.	6.7	7.2	24.9	14.7	–	–	–	+++	
	Tovara filiformis[1]	p, p	2.8	3.6	8.5	258	1.5	–	–	–	+++	–
	Tricyrtis hirta[1]	p, p	.	.	.	.	.	–	–	–	+++	–
Semi-shade	*Adoxa moschatellina*[1]	p-a, m	1.1	.	.	.	.	27.7	1–3	18.1	±	+++
	Allium victorialis ssp. *platyphyllum*[3]	p, p	19.8	2.6	19.6	34.8	6.4	–	–	–	+++	+
	Cardiocrinum cordatum[1] (incl. v. *glehni*)	p, m	7.5	19.7	71.0	2121	2.9	3.0	2.2	231	++	++
	Disporum sessile[10]	p-a, m	.	1.5	2.5	4.7	21.7	47.6	1.5	1548	+	+++
	Disporum smilacinum[10]	p-a, m	0.8	3.2	5.2	1.5	12.8	32.7	1.6	174	±	+++
	Epimedium cremeum[1]	p, p	.	.	.	.	.	–	–	–	+++	–
	Maianthemum dilatatum[1]	p, p	0.1[11*]	1.1	1.3	.	.	.	.	.	++	++
Shade-intolerant	*Allium monanthum*[9*,3]	p-a, m	16.9	4.0	20.3	2–4	3.3	52.1–72.1	1–4	9.3–33.0	±	+++
			.		7.7							
	Amana edulis[1]	p, p	4.6	.	.	.	.	–	–	–	+++	–
	Anemone debilis[1]	p, p	.	6.9	8.1	12.7	0.5	–	–	–	+++	–
	Anemone flaccida[1]	p, p	2.5	.	.	.	.	–	–	–	+++	±
	Anemone nikoensis[1]	p, p	5.3	.	.	.	.	–	–	–	+++	
	Anemone pseudo-altaica[1]	p, p	6.0	0.4	1.1	26.2	0.2	–	–	–	+++	±
	Anemone raddeana[1]	p, p	6.0	1.7	3.1	17.1	1.3	–	–	–	+++	–
	Corydalis decumbens[1]	p-a, m	3.2	.	.	.	.	27.7	1	113	++	±

	Corydalis incisa[11*,1]	wa, m	5.8	6.4	15.4	1161	0.6	–	–	–	+++	–
	Corydalia remota[1]	p-a, m	5.1	.	.	.	.	58.2	1	270	++	±
	Erythronium japonicum[7]	p, p	12.5	5.0	10.9	21.9	7.1	–	–	–	+++	±
	Fritillaria koidzumiana[1]	p, p	13.7	4.5	15.3	16.6	0.5	–	–	–	+++	–
Modified form	*Lycoris sanguinea*[1]	p, p	21.6	5.0	29.5	17.8	13.8	–	–	–	+++	+
of shade-	*Lycoris radiata*[1]	p, p	25.7	0	3.2	0	.	76.1	2–3	?	–[12*]	+++
intolerant	*Scilla scilloides*[1]	p, p	.	.	.	.	.	–	–	–	++	++

[1*]Life cycle: p, p, perennial, polycarpic; p, m, perennial, monocarpic; p-a, m, pseudo-annual, monocarpic; and wa, m, winter-annual, monocarpic.
[2*]SRA, sexual reproductive allocation (reproductive organs/individual biomass × 100[in g dry weight]).
[3*]VRA, vegetative reproductive allocation (reproductive organs/individual biomass × 100[in g dry weight]).
[4*]No data available.
[5*]Lacking.
[6*]The effectiveness of reproduction: + + +, very effective; + +, effective; +, least but moderately effective; ±, accidental; –, lacking.
[7*]This species is "deciduous, shade" in its northernmost range.
[8*]Sequentially hermaphrodite, functionally dioecious.
[9*]Hermaphrodite or dioecious.
[10*]Cleistogamous.
[11*]Only winter annual species found in temperate woodlands in Japan.
[12*]Triploid (3×) and thus sterile. All figures in the Table indicate mean values.
[1]Kawano et al. unpublished; [2]Kawano 1975; [3]Kawano and Nagai 1975; [4]Kawano and Masuda 1979; [5]Kawano et al. 1980; [6]Kawano and Masuda 1980; [7]Kawano et al. 1982a; [8]Kawano and Takasu unpublished; [9]Takasu unpublished; [10]Kawano and Hiratsuka unpublished.
Note: The figures included here are sometimes based on a single study of a single population and often during only one year, and therefore represent only part of the range of possible values.

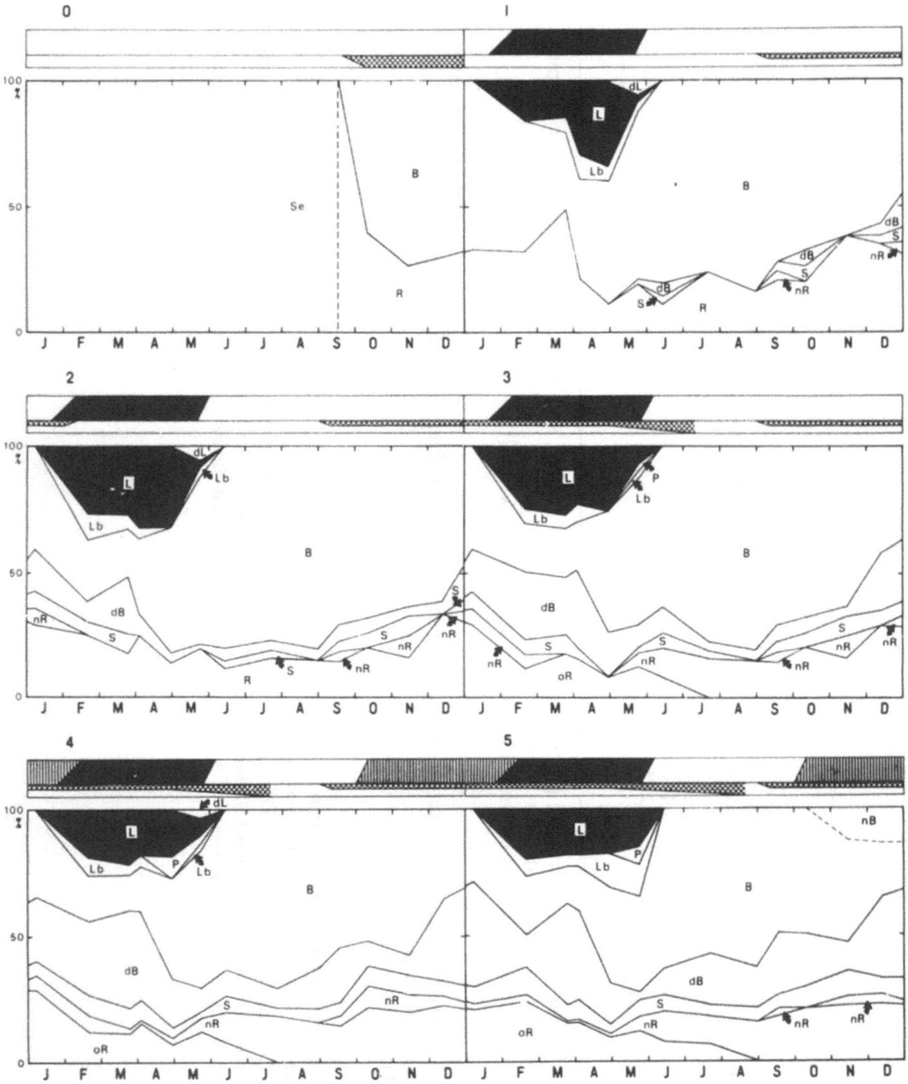

component organs from seedlings to sexually mature individuals in *Allium victorialis* ssp. *platyphyllum*.

There is no doubt that the longevity and expression of the life cycle, the development of unique production systems, reproductive behavior (i.e., the length and timing of the sexual and asexual reproductive phases) and reproductive efficiency of temperate woodland herbs have all differentiated in intimate relation with this rather regular yearly, cyclical change, in environmental factors in temperate woodlands. If we carefully examine reproductive systems in woodland plants, the development of *diverse* perennation patterns can be found. Although all these have been described

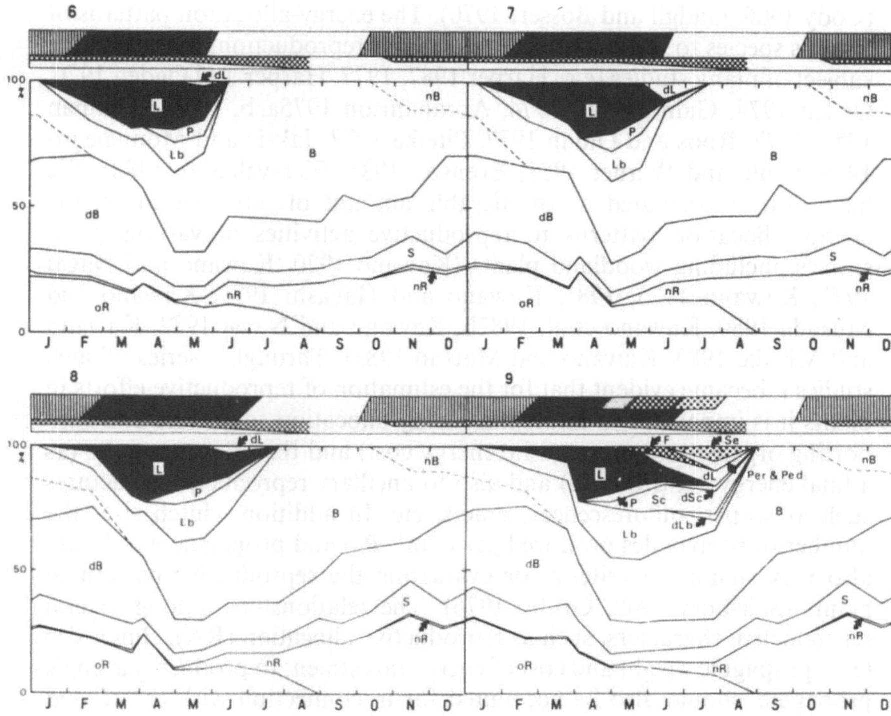

Fig. 4. Phenology and seasonal changes in proportional distribution of dry matter into component organs from seedling to sexually mature individuals of *Allium victorialis* ssp. *platyphyllum* (Kawano and Nagai 1975).

L, leaf; Lb, basal part of leaf; P, petiole; dL, dead leaf; Sc, scape; dSc, decayed scape; Ped, pedicel; Per, perianth; Se, seed; B, bulb; nB, new bulb; dB, decayed bulb; S, stem or axis within the bulb; R, root; nR, new root; oR, old root.

exclusively as perennial, some of them are equivalent to annuals as far as their dry matter economy is concerned (Kawano 1975, Kawano and Nagai 1975), and indeed these plants have previously been called 'pseudo-annual' (Salisbury 1942). Various species such as *Adoxa moschatellina*, *Disporum sessile*, *D. smilacinum*, *Allium monanthum*, *Corydalis decumbens*, and *C. remota*, could be cited as examples (Table 1). A number of species referrable to the 'pseudo-annual' category are also known from North American temperate woodlands, e.g., *Medeola virginiana*, *Uvularia perfoliata* and *U. sessilifolia* (Bell 1974, Whigham 1974, Abrahamson 1980).

REPRODUCTIVE EFFORTS IN TEMPERATE WOODLAND PLANTS

An optimal allocation of the finite resources to various life activities in life history processes of each organism is regarded as the result of selection

(Cody 1966, Gadgil and Bossert 1970). The energy-allocation patterns of various species to life activities, especially to reproduction, have been the subject of many studies (e.g. Harper 1967, 1977, Harper and Ogden 1970, Ogden 1974, Gaines et al. 1974, Abrahamson 1975a, b, 1979, Hickman 1975, 1977, Roos and Quinn 1977, Pitelka 1977, Jaksic and Montenegro 1979, Soule and Werner 1981, Evenson 1981, Bierzychudek 1982). We have also accumulated a considerable amount of information on the energy allocation patterns to reproductive activities in various plant species, including woodland plants (Kawano 1970, Kawano and Nagai 1975, Kawano 1975, 1981, Kawano and Hayashi 1977, Kawano and Musuda 1980, Kawano et al. 1982b, Kawano and Nagai 1982, Kawano and Miyake 1983, Kawano and Matsuo 1983). Through a series of such studies it became evident that for the estimation of reproductive efforts in plants it is important to know the energy allocation to flowers and supporting organs (as a prerequisite energy cost) and then to propagules (as a final energy contribution) and also to ancillary reproductive structures such as scapes, inflorescences, bracts, etc. In addition, clutch size (the number of propagules produced per plant: P_N) and propagule weight are also very significant criteria for evaluating the reproductive pattern of plants (Salisbury 1942, Grubb 1976). The relationships among several reproductive characters, such as reproductive allocation (RA), clutch size (P_N), propagule weight and cost of energy investment to producing a single propagule, should also be accounted for in connection with the related reproductive mechanisms, such as the pollination system and breeding system of the plants studied.

Table 1 compiles the figures for sexual (at both flowering and fruiting stages) as well as asexual reproductive allocation, reproductive output and propagule weight in 35 woodland species thus far examined. It is evident from this table that, even among woodland species, there are diverse reproductive patterns. Although several life history theories make predictions for demographic specializations of woodland plants, and suggest that their life history traits are consequences of adaptive differentiation for a stable, closed and predictable environment (e.g., Pianka 1970, Grime 1979), studies to date clearly indicate that temperate woodland plants encompass a diversity of life history features and reproductive patterns, reflecting the diverse evolutionary history of each species. As shown in Table 1, all such characteristics are well exhibited in the patterns of reproductive allocation (both the SRA and VRA), reproductive output, the weight of propagules, dispersion systems of propagules, and the modes of perennation of these species.

Several noteworthy aspects of the reproductive characteristics of woodland plants are summarized below.

(a) *Relationship between propagule weight (P_W) and propagule output per plant (P_N)*

According to the classic studies in birds (Lack 1954), organisms with large clutch size bear small eggs (or propagules), whereas those with small clutch

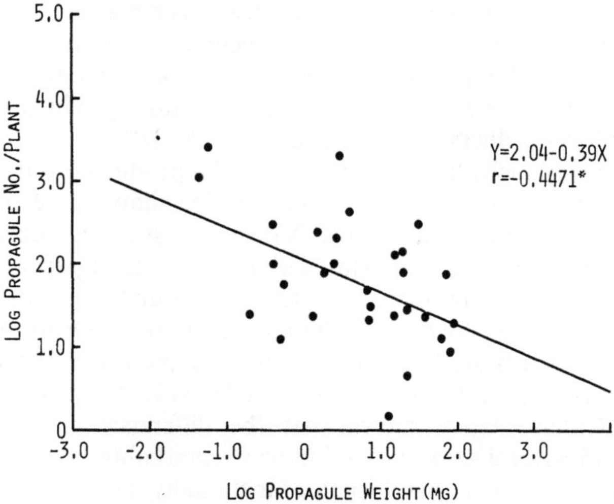

Fig. 5. Relationship between the weight of single propagule (seed, achene or caryopsis) and the number of propagules produced per plant.

sizes of single or few eggs per individual hatch large eggs (or propagules). Although this was suggested by Harper et al. (1970) as a general phenomenon in plants as well, only a weak trade-off relationship was recognized in the woodland plants examined. Figure 5 shows a weak reciprocity found between propagule weight and seed output per plant in 30 Japanese temperate woodland species.

(b) *Relationship between reproductive allocation (RA) and propagule output per plant* (P_N)

Kawano (1981) recently demonstrated that there are three conspicuously distinct types of plants with respect to their patterns of proportional dry matter distribution to total reproductive organs (RA) and propagule output per plant (P_N). The total reproductive structures are considered here to include not only propagules, but also organs ancillary to reproduction such as pedicels, peduncles, inflorescences, bracts and scapes, etc.

One conspicuous trend recognized among numerous perennials, including woodland plants, is that the higher the energy allocation to total reproductive structures per plant, the larger the number of propagules per plant becomes, and *vice versa.* Another notable (but exceptional) type which shows a reverse trend to this was found in an evergreen liliaceous herb *Heloniopsis orientalis* which possesses an exceedingly broad ecological amplitude. (It occurs from the evergreen or deciduous lowland and montane forests consisting of *Carpinus laxiflora*, *Quercus serrata*, *Fagus crenata* and *Cryptomeria japonica*, etc. to the alpine grasslands and high moor (covered by extremely dwarf *Sasa kurilensis*, *Moliniopsis japonica*, *Carex blepharicarpa*, *Sieversia pentapetala*, etc.) developed above 2,500 m in elevation above sea level.) In this species the *higher* the proportionate energy allocation to total reproductive organs per plant, the *lower* the number of propagules borne per plant becomes, and *vice versa*. This suggests that populations in harsher habitats with limited available resources may invest higher relative energy to total reproductive organs (RA), as well as producing a single propagule (R_A).

The reason why such allocation patterns and reproductive output occur in plants was not initially clear, and inexplicable simply in terms of dry matter economy, but it soon became evident that such high correlations noted between the two variables stated above are at least partially superficial phenomena: they are due to the levels of fecundity or efficiency of pollination caused in turn by pollinator limitation and/or breeding systems of the plants studied. Most of the species which show such a tendency are typical outbreeders. Bierzychudek (1981) and Kawano and Nagai (1982) have lately shown that the difference between hand-pollinated and natural seed production was remarkably large in populations of a sequentially hermaphrodite, functionally dioecious, species of Jack-in-the-Pulpit, *Arisaema triphyllum* (Araceae) and in a typical xenogamous spring plant, *Erythronium japonicum* (Liliaceae), respectively: this difference was doubtlessly due to pollinator limitation under field conditions (see also Kawano et al. 1982a, Schemske et al. 1978, Schemske 1980, Willson et al. 1979, Weller 1980, Janzen et al. 1980, Bernhardt 1977). This indicates that the energy translocation into propagules occurs only when the ovules in pistillate flowers are fertilized. Only in such cases is reproduction being limited, not by energy levels, but by pollinator activity (Bierzychudek 1981, Kawano and Nagai 1982).

Evidence to support this interpretation was also obtained from a comparative study of several typical *inbreeders* of annual fox-tail species (*Setaria*, Gramineae) (Kawano and Miyake 1983), which clearly demonstrated that in all these species RA is more or less constant and attains the level of saturation in fecundity, despite great differences in individual biomass and in number of the propagules produced per plant. Recently this was further confirmed in a series of cultivation experiments using *Helianthus annuus*, *Glycine max*, *Coix ma-yuen* and *Oryza sativa* (unpublished data).

(c) *Relationship between proportional energy allocation to a single propagule (R_A) and propagule output per plant (P_N)*

The trends mentioned above in the patterns of reproductive allocation and output of various plant species were verified by an additional analysis. When the values of the average propagule number per plant (P_N) and the RA to a single propagule (R_A) (*not* the same as the weight of a propagule) of various species are plotted on log–log coordinates, a conspicuous reciprocity or 'trade-off' was evident between them, such that $\log R_A = \log C - b \log P_N$, or $R_A = C \cdot P_N^{-b}$, where C is a constant, varying with the particular value of b. Three different ranges of b-values, i.e., $b > -1.0$, $b < -1.0$ and $b = -1.0$ were recognized, each corresponding to the three different types in allocation pattern and reproductive

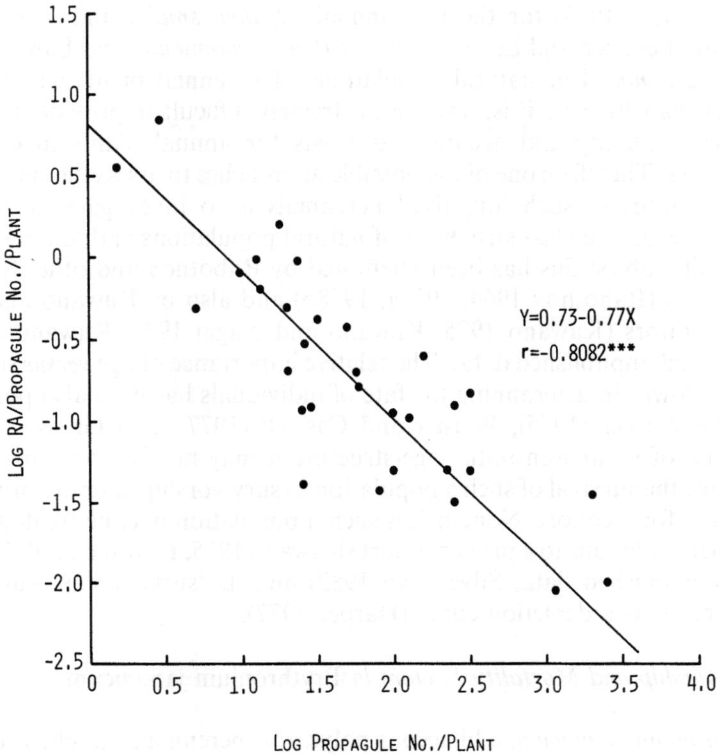

Fig. 6. **Relationship between the number of propagules produced per plant and the reproductive allocation (RA) to a single propagule (Kawano 1981).**

output described above (see also Kawano 1981). According to our recent studies, the values on the ordinate (RA/P_N) evidently indicate the relative cost of producing a single propagule, but the ranges of slopes (*b*) proved to reflect the efficiency of pollination or fecundity in a population of each particular species (Kawano and Miyake 1983, Kawano and Nagai 1982, Kawano and Matsuo 1983, and unpublished data).

Figure 6 demonstrates the relationship between reproductive output per plant and the RA to a single propagule (R_A) in 30 woodland species. The values of RA to a single propagule indicate the differences in the relative cost of energy investment to produce a single propagule in different species. It is interesting to note that the cost of producing a single propagule is high in species with very small 'clutch size', such as *Allium monanthum*, *Disporum smilacinum* and *Lycoris sanguinea* (Table 1).

SURVIVORSHIP PATTERNS AND POPULATION DYNAMICS IN WOODLAND PLANTS

Several workers have provided actuarial life-tables and survivorship curves for natural populations of herbaceous plant species: Sharitz and

McCormick (1973) for the two annuals, *Sedum smallii* and *Minuartia uniflora*, Leverich and Levin (1979) for *Phlox drummondii* and Law (1975) for *Poa annua*. For natural populations of perennial plants which live longer than 10 years it is, however, extremely difficult to provide data of the same quality and accuracy as exists for annual plants in cohort life-tables. Therefore one of the possible approaches to assess the survivorship patterns in such long-lived perennials is to investigate the stage structures or size-class structures of natural populations and to construct static life tables: this has been attempted by Rabotnov and other Soviet ecologists (Rabotnov 1964, 1978a, 1978b) and also by Kawano and his collaborators (Kawano 1975, Kawano and Nagai 1975, Kawano et al. 1982a and unpublished data). The relative importance of age *versus* size of plant growth in determining the fate of individuals has been also pointed out by Werner (1975), Werner and Caswell (1977), and Gross (1981). Because of its uneven initial age-structure it may not be appropriate to describe the survival of such a population as survivorship, a term normally reserved for a cohort. Nonetheless such a population may be treated in a manner analogous to a proper cohort (Kawano 1975, Kawano et al. 1982a and unpublished data, Silvertown 1982) and its 'survivorship' curve is referred to as a depletion curve (Harper 1977).

Survivorship and Mortality Factors in Erythronium japonicum

Erythronium japonicum, a liliaceous polycarpic perennial, was chosen as a characteristic example showing quite distinct survivorship and life history traits among the woodland plants examined. This is one of the most representative spring plants of Japanese temperate broad-leaved deciduous forests consisting of *Carpinus tschonoskii*, *Quercus acutissima*, *Q. serrata*, etc., and often covers the entire forest floor in early spring. If one carefully observes all growing individuals within a population, it can be noticed that numerous sterile and fertile plants of varying size occur. There is no doubt, however, that such differences in individual size reflect different growth stages or possibly age differences of each individual to a certain extent. In this study, sterile plants have been sorted arbitrarily into 13 different size-classes, and fertile plants into 9 size-classes based on their leaf areas and biomasses. All sterile classes bear only single leaf, whereas fertile plants bear always two leaves (Kawano et al. 1982a). Figure 7 demonstrates the size-class structure of *Erythronium japonicum* sampled in 1979. The number of propagules produced per population was also estimated by product of the number of flowering individuals present and the mean seed output per plant.

What is the mortality of each size-class, and the relationship between the size of a plant and its chronological age? In most herbaceous plants, there is actually no way of knowing the exact chronological age. Therefore in order to clarify the extent of age-overlaps among the various size-classes of *E. japonicum* discriminated, an attempt was made to examine the changes in size distribution of populations for over two years. Measurements were taken from ten permanent quadrats (1 × 1 m) established on

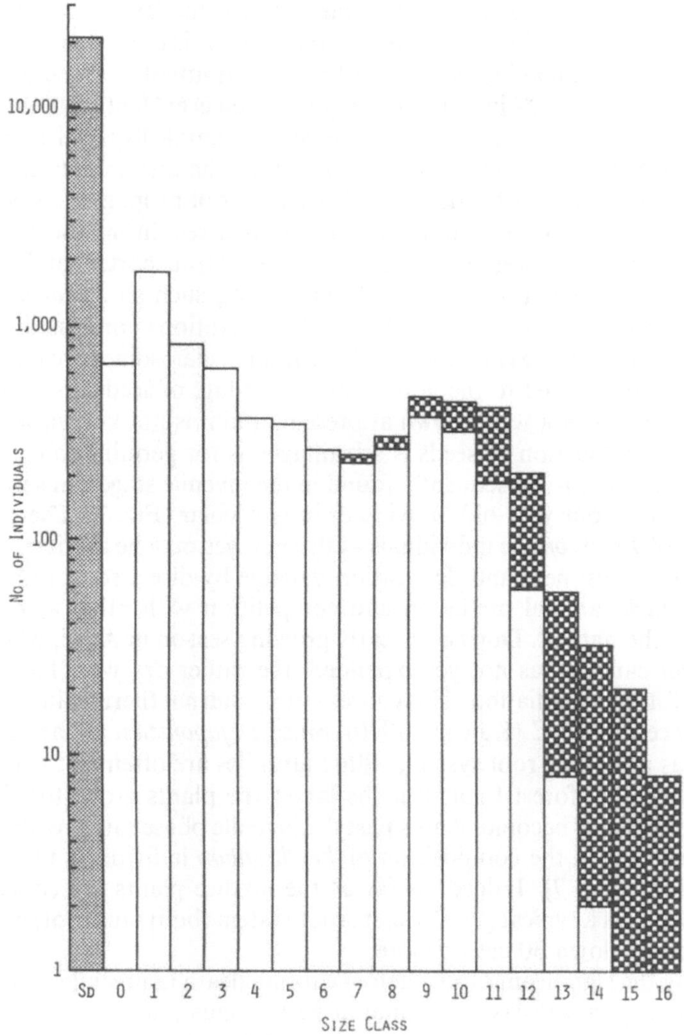

Fig. 7. Size-class structure of *Erythronium japonicum* populations sampled from sixty 1 × 1 m quadrats on the deciduous forest floor in Yatsuo, Toyama Prefecture.

All measured individuals from sixty 1 × 1 m^2 quadrats were sorted into 16 different categories previously discriminated (Kawano et al. 1982a). The number of propagules produced (Sd) was estimated by the number of flowering individuals present times the mean number of seeds produced per plant. 0–14; seedling to single-leaved vegetative individuals; 7–16: flowering individuals. Small peaks noticed at the stage of size-class 2 to 4, and size-class 8 to 11 are due to considerable overlaps of the individuals belonging to different generations.

the forest floor of the *Erythronium* site to determine the rates of growth or changes in leaf size from one year to another. In the young sterile size-classes, growth or increase of leaf areas is very conspicuous and more or less consistent. The behaviors of individuals larger than size class 7 or 8 are quite diverse, and the extent of overlaps is very large. Some individuals remain sterile, some sterile ones became fertile, some fertile ones became sterile, but many produced flowers continuously. It is clear that flowering is continuous in some individuals, but it is intermittent in some others.

Loss of mature seeds in *Erythronium japonicum* is evidently attributable to two major causes; animal predation or accidental dispersal to places unsuitable for germination and establishment. The dissemination of the propagule itself is a stochastic event. The number of propagules produced per cohort must fully compensate for the loss incurred during the dispersal and establishment processes. The most important transporters of *Erythronium* seeds are known to be several ant species, such as *Aphaenogaster famelica* and *Formica japonica*. Our field observations confirmed that a large fraction of the *Erythronium* seeds with small elaiosomes are regularly picked up and carried to the nest by ants. The fate of seeds dispersed by ants is, however, not well known at present, and it is not known whether or not such relocation of seeds is advantageous for germination.

High mortality is subsequently found in the juvenile stages, in seedlings and in two to four year-old individuals in particular (Fig. 7). The drastic decrease of *Erythronium* individuals at these stages may be ascribed to the following causes: heat and desiccation damage by direct solar radiation, fungal attack, animal predation and competition with other associated species in the habitat. During the early growing season in April, when the forest leaf canopy has not yet expanded, the rather dry woodland floor receives full solar radiation. This causes strong diurnal thermal insolation, often exceeding 10°C (Kawano 1970). Since *E. japonicum* at the juvenile stages has a shallow root system, wilted juveniles are often met with here and there in the forest floor. But the larger the plants grow, the deeper their root systems become. Plants past the juvenile phase rarely wither and this is reflected in the composition of *Erythronium* individuals in natural populations (Fig. 7). Indeed, as far as the mature plants are concerned *E. japonicum* is a typical geophyte, its root system (bulb plus root) eventually reaching down 30 cm or more.

Among the significant biotic factors causing death in juvenile plants are *Uromyces erythronii* Pass., a common rust fungus parasite which often kills the host plant at young stages, and a well-known predator of *Erythronium*, the leaf-beetle, *Sangariola punctatostriata*. Another significant biotic factor causing the decrease of *Erythronium* individuals appears to be competition with other plants in the habitat: since the juveniles of *E. japonicum* are very small, and their growing period is brief (confined to only about a month in early spring before canopy closure of the tree layer), this no doubt represents a major factor in the decrease of the juvenile stages. Among the plants associated with *E. japonicum*, those with robust or caespitose growth habits are the most important competitors. Senescence seems to be a primary cause of death in individuals exceeding 13 to 15 year-old (Fig. 7). The death of individuals is also possibly due in part to the consumption of reserved food in storage organs after continuous sexual activity over one season. The death of monocarpic perennials after their 'big-bang reproduction' is a clear indication of high energy consumption by reproductive activity and may be a primary cause of senescence in polycarpic perennials as well (Inouye and Taylor 1980, Beattie et al. 1973, Threadgill et al. 1981, Baskin and Baskin 1979a, b, Schaffer 1974, Schaffer and Gadgil 1975, Schaffer and Schaffer 1977). Often *Erythronium japonicum* does not produce reproductive organs continuously after flowering and production of seeds with a high fecundity in one season (Kawano et al. 1982 and unpublished data).

The mortality schedule or survivorship pattern found in *E. japonicum* is thought to be characteristic of those woodland perennials which occupy environments with high environmental predictability (Kawano and Nagai 1975, Kawano 1975). In any case, there is no doubt that the species with a life history pattern similar to *Erythronium japonicum* are typical members of stable, closed and predictable environments. This is also shown clearly in the population dynamics of this species recorded for over four years (Table 2): in spite of new cohorts of seedlings within the population every year, it maintains a more or less equilibrium state; this is partly due to a rather high mortality in seedling and subsequent juvenile stages. It also seems clear now that *Erythronium japonicum* has a certain critical size or threshold biomass before producing flowers: the size of the plants is more significant or critical than its chronological age (Werner 1975, Werner and Caswell 1978, Gross 1981, Thompson and Beattie 1981). This species takes at least 7 to 8 years to become sexually mature.

Examples similar to *Erythronium japonicum* are found in *Allium victorialis* ssp. *platyphyllum* (Kawano and Nagai 1975) (Fig. 8), *Arisaema japonicum*, *A. serratum* (unpublished data), *Fritillaria koidzumiana*, and also Japanese and North American species of *Trillium* (Lilliaceae) (unpublished data). All these species display regular annual recruitment from seedlings. This behavior is perhaps optimal within an environment of high stability and predictability in its abiotic and biotic factors: there is a clear division in time of growth and reproduction, separated by a full switch-over from vegetative growth to reproduction (Zeide 1978).

Survivorships of Disporum smilacinum *and* D. sessile

Another group of woodland herbs chosen here as a contrasting example of population structure and survivorship patterns is composed of the typical pseudo-annual species of the genus *Disporum* (Liliaceae) (Salisbury 1942; also Table 1).

Disporum smilacinum and *D. sessile* are species which often grow at the marginal parts or clearings of the broad-leaved deciduous forests (e.g., *Carpinus tschonoskii*, *Quercus acutissima*, *Q. serrata*, etc.), often on gently inclined slopes within the woodlands. *Disporum smilacinum* also occurs on the dark shady floor of forests of the Japanese Cedar, *Cryptomeria japonica*. Both species produce one to three (rarely six) underground runners every year (Fig. 9), and at the tips of the runners small buds for the next season are formed. At the end of the season in November to December, all the stored materials in the runners are consumed and independent asexual propagules are produced underground (Fig. 10). The sequence of

Table 2. Population flux in *Erythronium japonicum* sites (each site 1 × 1 m^2)

Quadrat No.	Q-1	Q-2	Q-4	Q-5	Q-6	Q-7	Q-8	Q-9	Q-10
(a) No. of plants/m^2 (1979)	27	7	26	29	65	75	76	131	29
(b) No. of plants/m^2 (1983)	60	8	25	47	85	76	106	195	59
(c) Net change (b − a)	33	1	−1	18	20	1	30	64	30
(d) Rate of increase (b/a)	2.22	1.14	0.96	1.62	1.31	1.01	1.39	1.49	2.03
(e) No. of plants arrived between 1979 and 1983	54	3	20	33	61	34	90	147	48
(f) Total no. of plants lost between 1979 and 1983	21	2	21	15	41	33	60	83	18
(g) Plants present 1979 arrived by 1983	22	5	10	23	57	61	50	81	19
(h) Percentage survival of plants in (a) (g/a × 100)	81.5	71.4	38.5	78.3	87.7	81.3	65.8	61.8	65.5
(i) Expected time for complete turnover (years) (4/(100 − h) × 100)	21.6	14.0	6.5	19.3	32.5	21.4	11.7	10.5	11.6
(j) Total plants recorded during study	81	10	46	62	126	109	166	278	77
(k) Percentage annual mortality of all individuals (f/j × 100)	25.9	20.0	45.7	24.2	32.5	30.3	36.1	29.9	23.4

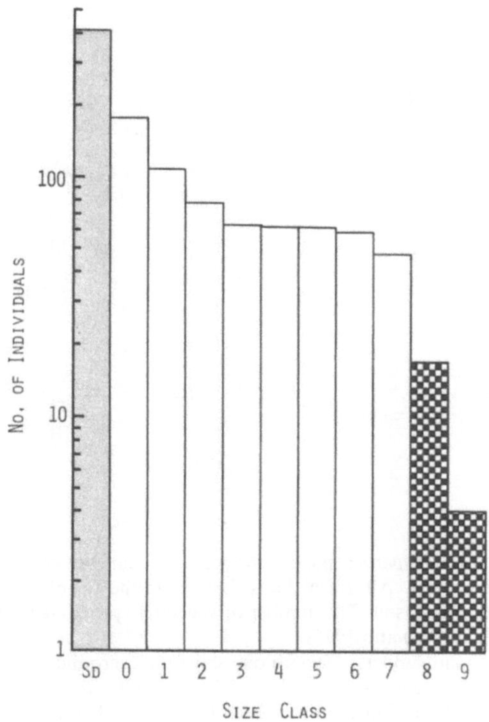

Fig. 8. Size-class structure of a population of *Allium victorialis* ssp. *platyphyllum* at a site (1 × 1 m²) near Saruyama-misaki, Noto Peninsula, Ishikawa Prefecture. Individuals belonging to size-class 8 or 9 are sexually mature. The mean seed output of the plants per m² was calculated from the mean seed number borne by an individual plant and the mean number of individual mature plants per m² (Kawano and Nagai 1975).

asexual propagule formation from the beginning to the end of the season is illustrated in Fig. 11.

The same techniques used in the studies of *Erythronium* have likewise been applied to *Disporum*. All sampled individuals of *Disporum* present per unit area were sorted into different size-classes according to the number of leaves per plant. It was confirmed that there is a conspicuous correlation between the number of leaves per plant and individual biomass, reflecting their growth stages (unpublished data).

The population structures discovered for this species are shown in Fig. 12. Without exception, intermediate size-classes were always predominant in all the wild populations surveyed. Flowering individuals usually appeared above size-class 5. A similar population structure was also found in *Disporum sessile*, which shows exactly the same manner of production of asexual propagules. In order to clarify the origin of such population structures, all the different size-classes of *Disporum smilacinum* were cultivated separately in an experimental garden, and the derivation of each size-class was carefully traced. Figure 13 summarizes the results

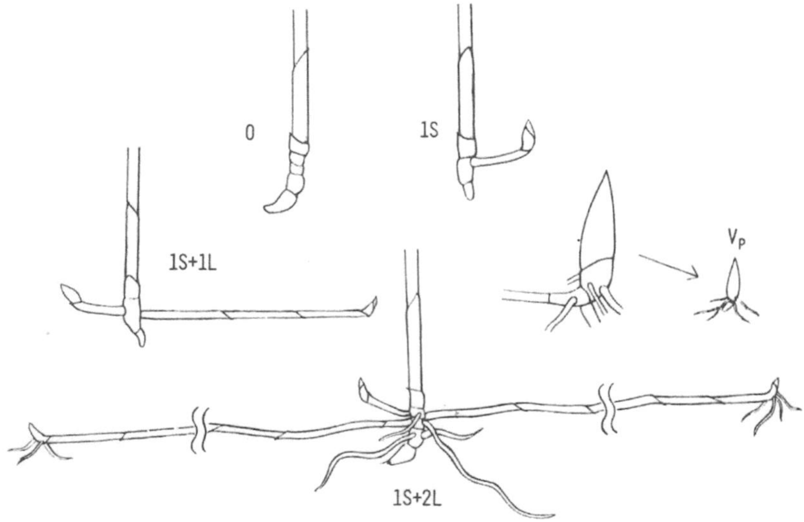

Fig. 9. Root systems of *Disporum sessile*, showing the number of underground shoots produced at different stages. All apical buds formed at the tip of subterranean runners separate at the end of the season. The number of vegetative propagules produced per plant varies from one to four (Kawano 1975).

0: without underground runners; 1S: with one short underground runner; 1S + 1L: with one short and one long runners; 1S + 2L: with one short and two long runners; Vp: vegetative propagule produced at the tip of each underground runner.

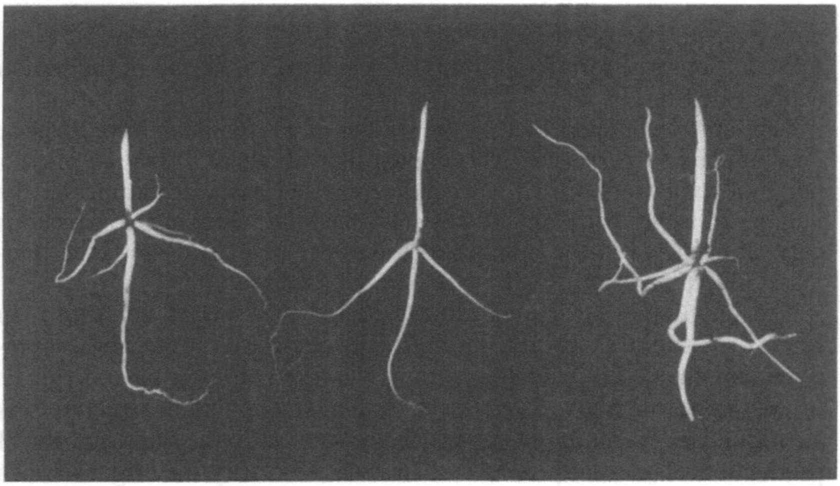

Fig. 10. Vegetative propagules of *Disporum smilacinum*, collected in December.

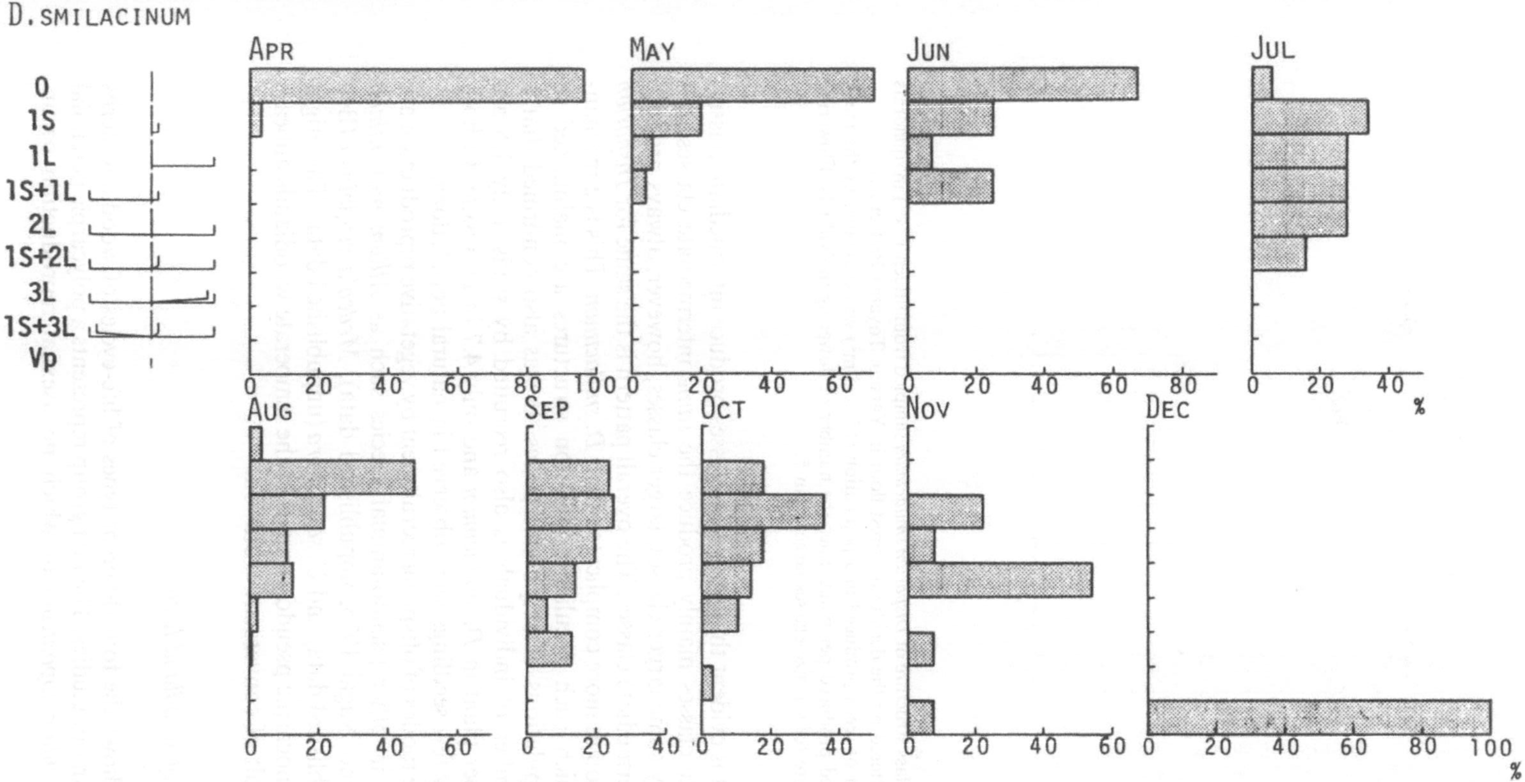

Fig. 11. Seasonal changes in frequency of plants with different numbers of underground ramets in *Disporum smilacinum*. At the end of the season all apical buds formed at the tip of underground runners became separated and completely independent. As a result each asexual propagule migrates and is relocated somewhat distantly from its parent plant.

0: plants without underground runners; 1S: those with one short runner; 1L: with one long runner; 1S + 1L, with one short and one long runners; 2L: with two long runners; 3L: with three long runners; 1S + 3L: with one short and three long runners; Vp: finally separated (independent) apical bud (vegetative propagule) (see also Fig. 10). Each month at least two samples (each 1 × 1 m^2) were taken from the habitats; the plants were sorted into each type and their proportions calculated.

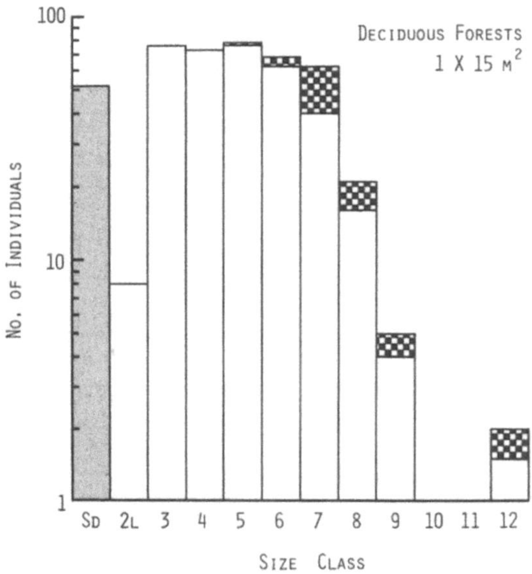

Fig. 12. Size-class structure of *Disporum smilacinum* sampled from fifteen 1 × 1 m^2 quadrats randomly scattered on the deciduous forest floor in Yatsuo, Toyama Prefecture.

The number of seeds produced in a population (15 × 1 m^2) was estimated by the mean number of seed produced per plant times the number of flowering individuals. Flowering individuals were seen in size-classes larger than 5.

obtained. It is evident that small size-classes produce intermediate classes; intermediate classes mainly produce the *same* intermediate classes, but occasionally some larger classes; larger classes, however, always produce smaller intermediate classes. The overall pattern is the same for *Disporum sessile*, although more complicated than *D. smilacinum*. This is a mechanism by which such peculiar population structures are maintained in natural populations of *Disporum* species. It was also confirmed that a limited number of individuals is also recruited by seeds: only 1.5 are produced per plant in *D. smilacinum* and only 4.7 in *D. sessile*. In both species very few seedlings were observed in natural populations.

The same modes of offspring recruitment by vegetative reproduction are also found in other pseudo-annual species such as *Allium monanthum* (Kawano and Nagai 1975, unpublished data), *Medeola virginiana* (Bell 1974, unpublished data) and *U. sessiliflora* (unpublished data). The origin of such monocarpic pseudo-annuals in the temperate woodlands appears to be a result of convergent evolution.

Life-cycles of woodland herbs

Figure 14 shows the four different types of life-cycles of woodland herbs discovered in our studies. The first group represents a polycarpic perennial like *Erythronium japonicum* in which no asexual reproduction occurs

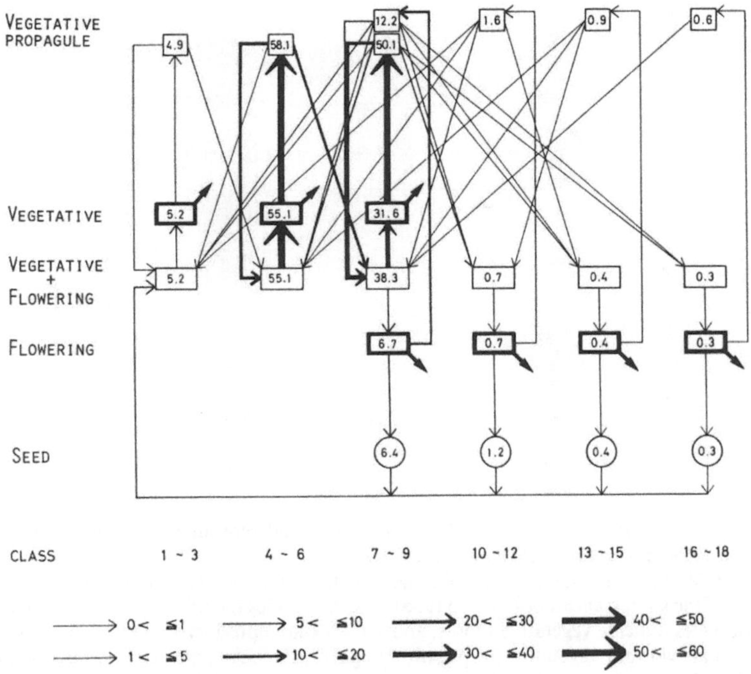

Fig. 13. The rates of origin of each size-class from its parent individual in *Disporum smilacinum*.

The horizontal axis specifies size-classes discriminated on the number of leaves per plant. Arrows indicate the directions of sexual or asexual propagule formation. The figures and thickness of arrows show the birth rates of certain size-classes in percentage. All the parental plants of vegetative and flowering stages decay at the end of the season (shown in thick boxes). The figures (in percentage) given in the circles and boxes indicate the number of seeds and/or vegetative propagules produced (expressed here in relative value). These often exceed those of parental plants: this is due to the numerous offspring produced per parental plant.

except for very abnormal cases (Kawano 1982) and offspring for new generations are normally recruited by seeds only (Kawano et al. 1982a). This type was also found in *Fritillaria koidzumiana*, *Trillium apetalon* and *T. tschonoskii* (unpublished data). Also referred to this type are evergreen herbaceous species such as *Ophiopogon japonicus* and *Liriope spicata*, which have extremely long underground stolons or rhizomes often extending over several meters. But this is doubtlessly a kind of underground branching and I do not regard it as vegetative reproduction *sensu stricto*, even though occasionally rhizome segments are cut by accident and become independent units.

The second group shows a polycarpic perennial with remarkable vegetative reproduction, as was found in *Erythronium americanum* (unpublished data). No cormlet formation occurs in small size-classes of juveniles, but one to four cormlets are produced in intermediate size-classes. Vegetative reproduction very rarely takes place in mature flowering individuals.

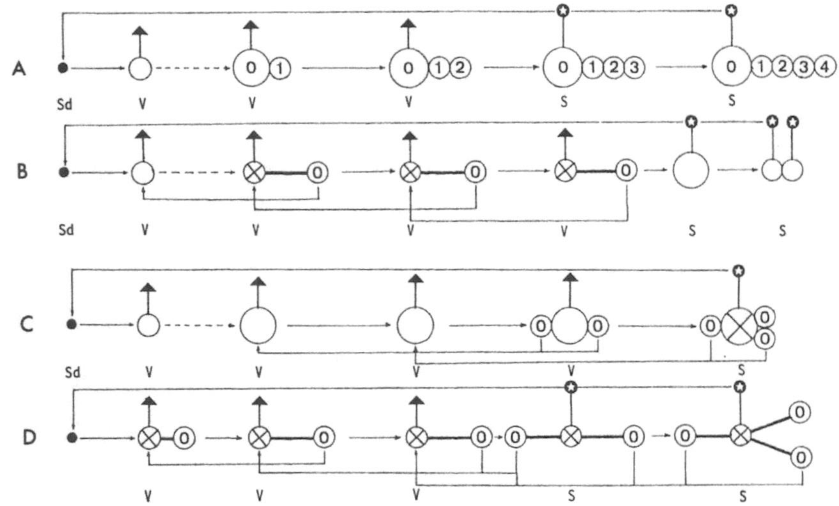

Fig. 14. Four representative types (A–D) of life cycles and reproductive systems found in temperate woodland perennials (Kawano 1975).

Sd: seed; V: vegetative stage; S: sexual stage. The figures in the diagram denote the age of the parts of the subterranean organs produced, and X specifies the decay of the parent plant. Filled triangles indicate vegetative shoots, and stars sexual reproductive organs. A: polycarpic perennials without vegetative propagation (e.g., *Erythronium japonicum*, *Smilacina japonica*); B: polycarpic perennials with vegetative propagation, especially in plants of the young vegetative stage (however, large mature bulbs also (rarely) split into two, e.g., *Erythronium americanum*); C: monocarpic perennials with vegetative propagation (e.g., *Cardiocrinum cordatum*); D: monocarpic pseudo-annuals (e.g., *Disporum smilacinum*, *D. sessile*, *Allium monanthum*, *Medeola virginiana*, species of *Uvularia*).

The third group shows a monocarpic perennial, represented by *Cardiocrinum cordatum*, in which vegetative reproduction through bulblet formation occurs in sexually mature individuals after flowering.

The fourth group is a typical pattern for monocarpic pseudo-annuals, exemplified by various species such as *Disporum smilacinum*, *D. sessile*, *Uvularia sessilifolia*, *U. perfoliata*, *Medeola virginiana* and *Allium monanthum* (Kawano and Nagai 1975, Kawano 1975, and unpublished data).

LIFE CYCLE PROCESSES AND OFFSPRING RECRUITMENT IN PLANTS

The flow-diagram given in Fig. 15 illustrates several significant stages in the life cycles of plants (including woodland plants) and all possible pathways of offspring recruitment by means of sexual and asexual reproduction. According to our recent studies, at least seven different types of life cycle processes in plants are recognized. Seven different types of offspring recruitment (summarized in Table 3; see also Fig. 15) have so far been identified. Only cases in which separation of the portions of plants

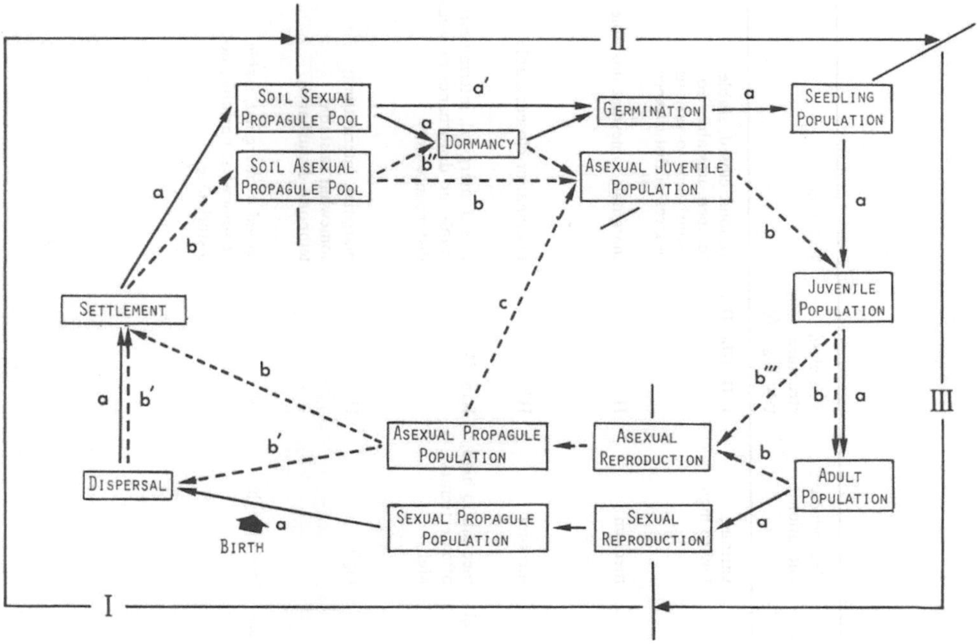

Fig. 15. Flow-diagram illustrating the basic life cycle processes in plants. Phase I: sexual or asexual propagules borne from the parental plant to the establishment in the soil layer; phase II: established propagules in the soil layer to germination or sprouting; phase III: seedlings or asexual juveniles to adult individuals. Complex combinations in the modes of offspring recruitment are recognized. For further details see the text.

or asexual propagules from the mother plant takes place regularly within every single season are cited here as examples of "asexual reproduction" *sensu stricto*. Typical size-class distributions of the various populations so far examined are shown in Fig. 16.

(i-A) Plants with Ia–IIa–IIIa pathway of offspring recruitment are typical amphimictic species and include various annuals, biennials and monocarpic or polycarpic perennials; examples of this type are numerous.

(i-B) Species with both chasmogamous and cleistogamous flowers may be referred to this type.

(ii) Plants with two pathways of recruitment (Ia–IIa–IIIa and Ib–IIb,b″–IIIb,b‴) are polycarpic or monocarpic perennials which propagate by producing seeds as well as small bulblets or cormlets attached to the parent bulb, but such asexual propagules often remain dormant in the soil over one season. Such triploid (and thus sterile) species as *Iris japonica* and *Lycoris radiata* could also be referred to this type: these propagate themselves only by daughter bulblet or runner formation, however.

(iii) Monocarpic pseudo-annuals with two pathways of recruitment (Ia–IIa–IIIa and Ib,b′–IIb–IIIb,b‴) propagate by producing seeds as well as vegetative propagules at the tip of underground stolons; but all

Table 3. Life history characteristics of herbaceous plants

Modes of the offspring recruitment	Examples	Life cycle	Reproductive life span	Reproductive output	Reproductive allocation	Size-class structure (see Fig. 16)	Habitats
(i-A): Ia–IIa–IIIa	various	annual biennial perennial	monocarpic polycarpic	sexual propagules only	various; low to high	I, II, IIIa, IIIb	various; closed, stable to open, unstable; no disturbance to various degrees of disturbance
(i-B): Ia–IIa–IIIa (chasmogamy + cleistogamy)	*Impatiens biflora*[9] *Impatiens pallida*[9] *Polygonum hydropiper*[10] *Polygonum thunbergii*[5]	annual	monocarpic	ch > cl	medium	II	marginal; somewhat unstable
	Viola nipponica[1]	perennial	polycarpic	ch < cl	medium	II?	marginal or closed; stable
(ii): Ia–IIa–IIIa Ib–IIb,b″–IIIb,b‴	*Allium victorialis* ssp. *platyphyllum*[2] *Arisaema urashima*[4] *Asarum caulescens*[4]	perennial	polycarpic	sexual > asexual	medium to high; often extremely high	IIIb, IV	closed, marginal, intermittent; stable; no disturbance or weak disturbance only
	Cardiocrinum cordatum[1]		monocarpic				
Ia → X Ib–IIb–IIIb,b‴	*Lycoris radiata* (3X)[1] *Iris japonica* (3X)[1]	perennial	(polycarpic)	asexual only	high	IV	marginal, intermittent; somewhat unstable, temporal disturbance
(iii): Ia–IIa–IIIa Ib,b′–IIb–IIIb,b‴	*Adoxa moschatellina*[1] *Allium monanthum*[2] *Disporum sessile*[5] *Disporum smilacinum*[5] *Medeola virginiana*[1,6] *Uvularia perfoliata*[1,7] *Uvularia sessilifolia*[1]	pseudo-annual	monocarpic	sexual ≷ asexual	low to high	IV	closed or marginal, stable or somewhat unstable; no disturbance

(iv):							
Ia–IIa–IIIa Ic–IIb–IIIb,b‴	*Heloniopsis orientalis*[3]	perennial	polycarpic	sexual > asexual	medium	IIIb	closed, stable; no disturbance
(v):							
Ia–IIa–IIIa Ib′–IIb–IIIb,b‴	*Dioscorea bulbifera*[8] *Laportea bulbifera*[1] *Lilium lancifolium*[1] *Sedum bulbiferum*[1]	perennial	polycarpic	sexual > asexual	medium?	IIIb, IV?	various; closed, stable to intermittent, unstable; no strong disturbance or somewhat regular disturbance
(vi):							
Ia–IIa–IIIa Ib–IIb–IIIb,b‴ Ib′–IIb–IIIb	*Allium grayi*[2] *Allium vineale*[11] *Pinellia ternata*[1]	perennial	polycarpic	sexual < asexual	high	IV?	closed, but intermittent disturbance
(vii):							
Ib′–IIb–IIIb	*Deschampsia alpina*[14] *Festuca vivipara*[12,13]	perennial	(polycarpic)	asexual only	low to medium?	IV?	open but stable; no disturbance

[1]Kawano et al. unpublished; [2]Kawano and Nagai 1975; [3]Kawano and Masuda 1980; [4]Takasu unpublished; [5]Kawano and Hiratsuka unpublished; [6]Bell 1974; [7] Whigham 1974; [8]Hori unpublished; [9]Schemske 1978; [10]Hiratsuka unpublished; [11]Iltis 1949; [12]Hulten 1962; [13]Hulten 1968; [14]Kawano 1966.

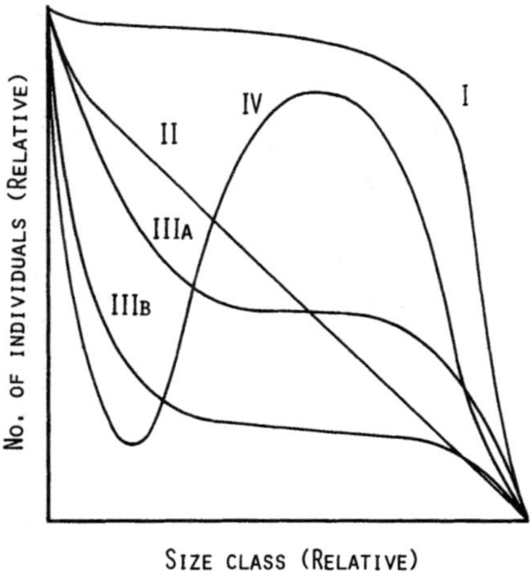

Fig. 16. The four different types of stage-class or size-class structures found in the woodland plants examined. For details see the text and Table 3.

vegetative propagules regularly become separate and independent at the end of every season.

(iv) Plants of this type are similar to (ii), but possesses two pathways of recruitment (Ia–IIa–IIIa and Ic–IIb–IIIb,b‴). Examples are perennials which propagate by producing seeds as well as plantlets at the tip of old decaying leaves (e.g., *Heloniopsis orientalis*). These vegative propagules often begin to assimilate themselves even while attached to the parent plant, and the plantlets always settle nearby it.

(v) Polycarpic perennials which propagate themselves by amphimixis and by apomixis (such as bulbil formation in the leaf axils of stems or at the tips of leaves) are referred to this type. The bulbils are dispersed around the parent plant, but loss by animal predation seems to be high since they contain ample nourishment.

(vi) This is a group of ruderal species with three different types of reproductive systems, represented by *Allium grayi* (Kawano and Nagai 1975) and *A. vineale* (Iltis 1949) for example. These species produce seeds, bulbils, and also bulblets. The bulbils produced in the flower heads often exceed 100 in number and are dispersed around the parent plant; they possess an analogous function to seeds.

(vii) Agamospecies such as *Deschampsia alpina* and *Festuca vivipara* are referred to this type, most of them occur in the alpine-arctic zone where severe environmental conditions prevail (Kawano 1966, Lee and Harmer 1980, Hulten 1962, 1968).

CONCLUDING REMARKS

The differentiations of complex life history patterns in the plants mentioned above may possibly be related to the diversity of environment factors in their respective habitats and also to changes in these factors which have occurred in their past evolutionary histories. Judging from the various life history characteristics of the woodland plants revealed by the present studies, the species have diverse phytogeographical origins. Species such as *Pacysandra terminalis*, *Heloniopsis orientalis*, *Mitchella undulata*, *Oxalis griffithii* represent, for example, the members of warm-temperate vegetation of Arcto-Tertiary origin (Axelrod 1966, unpublished data), although they often occur in montane deciduous hardwood and/or coniferous forests (or even rarely in alpine meadows above 2,500 m in elevation). The distribution of *Heloniopsis orientalis* in montane coniferous forests and also in alpine grasslands on the snowy Japan Sea side of Honshu, appears however, to be a consequence of ecotypic differentiation (unpublished data). No significant morphological differences were noted between the lowland and alpine populations, however (Kawano and Masuda 1980). A very similar example has been noticed also in an 'evergreen' herb species, *Shortia soldanelloides* (Diapensiaceae), which likewise extends over a broad altitudinal gradient from the lowlands to the alpine zone on the Japan Sea side of Honshu (unpublished data). Other evergreen herbs such as *Liriope*, *Ophiopogon* and *Epimedium* are typical warm-temperate elements, all the species being strictly confined to lowland deciduous hardwood or (occasionally) evergreen forests.

Among other herbaceous species with a 'deciduous' growth habit, at least two different groups can be recognized: typical cool-temperate and boreal elements. *Disporum sessile*, *D. smilacinum*, *Smilacina japonica*, *Tovara filiformis* and numerous 'shade-intolerant' and 'semi-shade tolerant' species may be referred to the former group; *Streptopus streptopoides*, *Clintonia udensis* and *Maianthemum bifolium* may be referred to the latter boreal groups (Kawano et al. 1968, Utech and Kawano 1975, Utech, 1973).

Nowadays the above-mentioned species occasionally grow side by side in the same plant communities, but the coexistence of such different phytogeographic elements is possibly a consequence of the reshuffling of vegetation by repeated late-glacial and/or post-glacial migrations caused by climatic changes during the past 15,000 years in the Japanese Archipelago (Watanabe 1977, Tsukada 1981a, b, 1982a, b). The structures of the deciduous hardwood and montane coniferous forests (e.g., *Cryptomeria japonica*) are complex, with diverse floristic elements of different life-forms and growth-habits on the forest floor.

In any case, what has been found in our studies (Kawano 1970, 1975, Kawano et al. 1982a b, Kawano et al. 1983) is the occurrence of diverse types of 'temperateness', in Axelrod's sense (Axelrod 1966), developed and differentiated among the temperate woodland herbs in their various life history traits. The mechanisms responsible for the evolution of such life history patterns of temperate woodland plants must, however, await future studies.

ACKNOWLEDGEMENTS

I would like to thank my colleagues and students who have worked with me on this project for the past 20 years, especially Hideki Takasu, Yokio Nagai, Junzo Masuda and Akira Hiratsuka, without whose unfailing assistance and stimulating discussions at the various stages of studies this work would not have been accomplished. I also wish to thank James White for his critical reading of the manuscript and valuable comments.

REFERENCES

Abrahamson, W.G. (1975a) Reproductive strategies in dewberries. Ecology 56: 721–726.

Abrahamson, W.G. (1975b) Reproduction of *Rubus hispidus* L. in different habitats. Amer. Midl. Nat. 93: 471–478.

Abrahamson, W.G. (1979) Patterns of response allocation in wildflower populations of field and woods. Amer. J. Bot. 66: 71–79.

Abrahamson, W.F. (1980) Demography and vegetative reproduction. In Solbrig, O.T. (Ed.), Demography and Evolution in Plant Populations, pp. 89–106. Blackwell, Oxford.

Axelrod, D.I. (1966) Origin of deciduous and evergreen habits in temperate forests. Evolution 20: 1–15.

Baskin, J.M. and Baskin, C.C. (1979a) Studies on the autecology and population biology of the monocarpic perennial *Grindelia lanceolata*. Amer. Midl. Nat. 41: 290–299.

Baskin, J.M. and Baskin, C.C. (1979b) Studies on the autecology and population biology of the weedy monocarpic perennial, *Pastinaca sativa*. J. Ecol. 67: 601–610.

Beattie, A.J., Breedlove, D.E. and Ehrlich, P.R. (1973) The ecology of the pollinators and predators of *Frasera speciosa*. Ecology 54: 81–91.

Bell, A.D. (1974) Rhizome organization in relation to vegetative spread in *Medeola virginiana*. Jour. Arnold Arb. 55: 458–468.

Bernhardt, P. (1977) The pollination ecology of a population of *Erythronium americanum* Ker. (Liliaceae). Rhodora 79: 278–282.

Bierzychudek, P. (1981) Pollinator limitation of plant reproductive effort. Amer. Nat. 117: 838–840.

Bierzychudek, P. (1982) Life histories and demography of shade-tolerant temperate forest herbs: A review. New Phytol. 90: 757–776.

Cody, M.L. (1966) A general theory of clutch size. Evolution 20: 174–184.

Evenson, W.E. (1981) Experimental studies of reproductive energy allocation in plants. In C.E. Jones and R.J. Little (Eds.), Handbook of Experimental Pollination Biology, pp. 249–274. Van Nostrand Reinhold Co., New York.

Gadgil, M. and Bossert, W.H. (1970) Life historical consequences of natural selection. Amer. Nat. 104: 1–24.

Gaines, M.S., Vogt, K.J., Hamrick, J.L. and Caldwell, J. (1974) Reproductive strategies and growth patterns in sunflowers (*Helianthus*). Amer. Nat. 106: 14–31.

Grime, J.P. (1979) Plant Strategies and Vegetation Processes. Wiley, Chichester, UK and New York.

Gross, K.L. (1981) Predictions of fate from rosette size in four "biennial" plant species: *Verbascum thapsus*, *Oenothera biennis*, *Daucas carota*, and *Tragopogon dubius*. Oecologia (Berl.) 48: 209–213.

Grubb, P.J. (1976) A theoretical background to the conservation of ecologically distinct groups of annuals and biennials in the chalk grassland ecosystem. Biol. Cons. 10: 53–76.

Harper, J.L. (1967) A Darwinian approach to plant ecology. J. Ecol. 55: 247–270.

Harper, J.L. (1977) Population Biology of Plants. Academic Press, New York.

Harper, J.L. and Ogden, J. (1970) The reproductive strategy of higher plants. I. The concept of strategy with special reference to *Senecio vulgaris* L. J. Ecol. 58: 681–698.

Harper, J.L., Lovell, P.H. and Moore, K.G. (1970) The shapes and sizes of seeds. Ann. Rev. Ecol. Syst. 1: 327–356.

Hickman, J.C. (1975) Environmental unpredictability and plastic energy allocation strategies in annual *Polygonum cascadense* (Polygonaceae). J. Ecol. 63: 689–701.

Hickman, J.C. (1977) Energy allocation and niche differentiation of four co-existing annual species of *Polygonum* in western North America. J. Ecol. 65: 317–326.

Hultén, E. (1962) The Circumpolar Plants. I. Almquist & Wiksel, Uppsala.

Hultén, E. (1968) Flora of Alaska and Neighboring Territories. Stanford Univ. Press, Stanford.

Inouye, D.W. and Taylor, O.R. (1980) Variation in generation time in *Frasera speciosa* (Gentianaceae), a long-lived perennial monocarp. Oecologia (Berl.) 47: 171–174.

Iltis, H. (1949) An immigrant conquers a continent: the story of the wild garlic. Sci. Monthly 118: 122–128.

Jaksic, F.M. and Montenegro, G. (1979) Resource allocation of Chilean herbs in response to climatic and microclimatic factors. Oecologia (Berl.) 40: 81–89.

Janzen, D.H., de Vries, P., Gladstone, D.E., Higgins, M.L. and Lewisohn, T.N. (1980) Self- and cross-pollination of *Encyclia cordigera* (Orchidaceae) in Santa Rosa National Park, Costa Rica. Biotropica 12: 72–74.

Kawano, S. (1966) Biosystematic studies of the *Deschampsia caespitosa* complex with special reference to the karyology of Icelandic populations. Bot. Mag. Tokyo 79: 293–307.

Kawano, S. (1970) Species problems viewed from productive and reproductive biology I. Ecological life histories of some representative members associated with temperate deciduous forests in Japan. J. Coll. Lib. Arts, Toyama Univ. 3: 181–213.

Kawano, S. (1975) The productive and reproductive biology of flowering plants. II. The concept of life history strategy in plants. J. Coll. Lib. Arts, Toyama Univ. Nat. Sci. 8: 51–86.

Kawano, S. (1981) Trade-off relationships between some reproductive characteristics in plants with special reference to life history strategy. Bot. Mag. Tokyo 94: 285–294.

Kawano, S. (1982) On the abnormal vegetative growth and reproduction in *Erythronium japonicum* Decne. (Liliaceae). J. Phytogeogr. & Taxon. 30: 53–55.

Kawano, S. and Hayashi, S. (1977) Plasticity in growth and reproductive energy allocation of *Coix ma-yuen* Roman. cultivated at varying densities and nitrogen levels. J. Coll. Lib. Arts, Toyama Univ. Japan. 10: 61–92.

Kawano, S., Hiratsuka, A. and Hayashi, K. (1982a) The productive and reproductive biology of flowering plants. V. Life history characteristics and survivorship of *Erythronium japonicum*. Oikos 38: 129–149.

Kawano, S., Ihara, M. and Suzuki, M. (1968) Biosystematic studies on *Maianthemum* (Liliaceae-Polygonatae) II. Geography and ecological life history. Jap. J. Bot. 20: 35–65.

Kawano, S. and Masuda, J. (1979) The productive and reproductive biology of flowering plants. VI. Assimilation behavior and reproductive allocation of *Coptis japonica* (Thunb.) Makino (Ranunculaceae). J. Coll. Lib. Arts. Toyama Univ. Nat. Sci. 12: 49–63.

Kawano, S. and Masuda, J. (1980) The productive and reproductive biology of flowering plants. VII. Resource allocation and reproductive capacity in wild populations of *Heloniopsis orientalis* (Thunb.) C. Tanaka (Liliaceae). Oecologia (Berl.) 45: 307–317.

Kawano, S., Masuda, J. and Takasu, H. (1982b) The productive and reproductive biology of flowering plants. IX. Further studies on the assimilation behavior of temperate woodland herbs. J. Coll. Lib. Arts. Toyama Univ. 15: 101–160.

Kawano, S., Masuda, J., Taksau, H. and Yoshie, F. (1983) The productive and reproductive biology of flowering plants. XI. Assimilation behavior of several evergreen temperate woodland plants and its evolutionary-ecological implications. J. Coll. Lib. Arts, Toyama Univ. 16: 31–65.

Kawano, S. and Matsuo, K. (1983) Studies on the life history of the genus *Plantago*. I. Reproductive energy allocation and propagule output in wild populations of a ruderal species, *Plantago asiatica* L., extending over a broad altitudinal gradient. J. Coll. Lib. Arts, Toyama Univ. 16: 85–112.

Kawano, S. and Miyake, S. (1983) The productive and reproductive biology of flowering plants. X. Reproductive energy allocation and propagule output of five congeners of the genus *Setaria* (Gramineae). Oecologia (Berl.) 57: 6–13.

Kawano, S. and Nagai, Y. (1975) The productive and reproductive biology of flowering plants. I. Life history strategies of three *Allium* species in Japan. Bot. Mag. Tokyo 88: 281–318.

Kawano, S. and Nagai, Y. (1982) Further observations on the reproductive biology of *Erythronium japonicum* Decne. (Liliaceae). J. Phytogeogr. & Taxon. 32: 90–97.

Kawano, S., Okumura, H. and Masuda, J. (1980) The productive and reproductive biology of flowering plants. VIII. Assimilation behavior and dry matter allocation of *Hepatica nobilis* v. *japonica* f. *variegata* (Makino) Kitam. (Ranunculaceae). J. Coll. Lib. Arts. Toyama Univ. Nat. Sci. 13: 33–46.

Kitamura, S., Murata, G. and Koyama, T. (1967) Coloured Illustrations of Herbaceous Plants of Japan. Monocotyledoneae. Hoikusha, Osaka (in Japanese).

Lack, D. (1954) The Natural Regulation of Animal Numbers. Oxford Univ. Press, London.

Law, R. (1975) Colonisation and the evolution of life histories in *Poa annua*. Ph.D. thesis, Univ. of Liverpool.

Lee, J.A. and Harmer, R. (1980) Vivipary, a reproductive strategy in response to environmental stress. Oikos 35: 254–265.

Leverich, W.J. and Levin, D.A. (1979) Age-specific survivorship and reproduction in *Phlox drummondii*. Amer. Nat. 113: 881–903.

Li, H.L. (1952) Floristic relationships between eastern Asia and eastern North America. Trans. Amer. Phil. Soc. 42: 371–424.

Lundegårdh, H. (1921) Ecological studies in the assimilation of certain forest-plants and shore plants. Svensk. Bot. Tidskr. 15: 45–95.

Lundegårdh, H. (1924) Der Kreislauf der Kohlensäure in der Natur. G. Fischer, Jena.

MacArthur, R.H. and Wilson, E.O. (1967) The Theory of Island Biogeography. Princeton Univ. Press, Princeton.

Ogden, J. (1974) The reproductive strategy of higher plants. II. The reproductive strategy of *Tussilago farfara* L. J. Ecol. 62: 291–324.

Ohwi, J. (1965) Flora of Japan. 2nd Ed. Shibundo, Tokyo (in Japanese).

Pianka, E.R. (1970) On *r*- and *K*-selection. Amer. Nat. 104: 592–597.

Pitelka, L.F. (1977) Energy allocation in annual and perennial lupines (*Lupinus*: Leguminosae). Ecology 58: 1055–1065.

Rabotnov, T.A. (1964) The biology of monocarpic perennial meadow plants. Bull. Moscow Soc. Nat. 69: 57–66. (in Russian).

Rabotnov, T.A. (1978a) Structure and method of studying coenotic populations of perennial herbaceous plants. Sov. J. Ecol. 9: 99–105.

Rabotnov, T.A. (1978b) On coenopopulations of plants reproducing by seeds. In Freysen, A.H.J. and Woldendorp, J. (Eds.), Structure and Functioning of Plant Populations, pp. 1–26, North Holland Publ. Co., Amsterdam.

Roos, F.H. and Quinn, J.A. (1977) Phenology and reproductive allocation in *Andropogon scoparius* (Gramineae) populations in communities of different successional stages. Am. J. Bot. 64: 535–540.

Salisbury, E.J. (1942) The Reproductive Capacity of Plants. Bell, London.

Schaffer, W.M. (1974) Optimal reproductive effort in fluctuating environments. Amer. Nat. 108: 783–790.

Schaffer, W.M. and Gadgil, M.D. (1975) Selection for optimal life histories in plants. In Cody, M.L. and Diamond, J.M. (Eds.), Ecology and Evolution of Communities, pp. 142–156. Belknap Press, Cambridge, Mass.

Schaffer, W.M. and Schaffer, M.V. (1977) The adaptive significance of variations in reproductive habit in the Agavaceae. In Stonehouse, B. and Perrins, C.M. (Eds.), Evolutionary Ecology, pp. 26–276. Macmillan, London.

Sharitz, R.R. and McCormick, J.F. (1973) Population dynamics of two competing annual plant species. Ecology 54: 723–740.

Schemske, D.E. (1980) Evolution of floral display in the orchid, *Brassavola nodosa*. Evolution 34: 489–493.

Schemske, D.E., Willson, M.F., Melampy, M.N., Miller, L.J., Verner, L., Schemske, K.M. and Best, L.B. (1978) Flowering ecology of some spring woodland herbs. Ecology 59: 351–366.

Seybold, A. and Eagle, K. (1937) Lichtfeld and Blatt farbstoffe. I. Planta 36: 491–515.

Silvertown, J.W. (1982) Introduction to Plant Population Ecology. Longman, New York.

Solbrig, O.T. (Ed.) (1980) Demography and Evolution in Plant Populations. Blackwell, London.

Solbrig, O.T., Jain, S., Johnson, G.B. and Raven, P.H. (1979) Topics in Plant Population Biology. Columbia Univ. Press, New York.

Soule, J.D. and Werner, P.A. (1981) Patterns of resource allocation in plants, with special reference to *Potentilla recta* L. Bull. Torrey Bot. Club 108: 311–319.
Sparling, J.H. (1967) Assimilation rates of some woodland herbs in Ontario. Bot. Gaz. 128: 160–168.
Thompson, D.A. and Beattie, A.J. (1981) Density mediated seed and stolon production in *Viola* (Violaceae). Amer. J. Bot. 68: 383–388.
Threadgill, P.F., Baskin, J.M. and Baskin, C.C. (1981) The ecological life cycle of *Frasera caroliniensis*, a long-lived monocarpic perennial. Amer. Midl. Nat. 105: 277–289.
Tsukada, M. (1981a) The last 12,000 years – The vegetation history of Japan II. New pollen zones. Jap. J. Ecol. 31: 201–215 (in Japanese).
Tsukada, M. (1981b) *Cryptomeria japonica* D. Don I. Pollen dispersal and logistic forest expansion. Jap. J. Ecol. 31: 371–383.
Tsukada, M. (1982a) Lae-Quaternary development of the *Fagus* forest in the Japanese Archipelago. Jap. J. Ecol. 32: 113–118.
Tsukada, M. (1982b) Late-Quarternary shift of *Fagus* distribution. Bot. Mag. Tokyo 95: 203–217.
Utech, F.H. (1973) A biosystematic study of the genus *Clintonia* Raf. (Liliaceae: Polygonatae). Ph.D. Thesis, Washington Univ., St. Louis.
Utech, F.H. and Kawano, S. (1975) Biosystematic studies in *Streptopus* (Liliaceae: Polygonatae) I. Morphological variation of the Asian *S. streptopoides* (Ledeb.) Frye & Rigg. Act. Phytotax. & Geobot. 27: 1–20.
Watanabe, N. (Ed.) (1977) The Quaternary Period: Recent Studies in Japan. Univ. Tokyo Press (in Japanese).
Weller, S.G. (1980) Pollen flow and fecundity in populations of *Lithospermum caroliniense*. Amer. J. Bot. 67: 1334–1341.
Werner, P.A. (1975) Prediction of fate from rosette size in teasel (*Dipsacus fullonum* L.) Oecologia (Berl.) 20: 197–201.
Werner, P.A. and Caswell, H. (1977) Populations growth rates and age vs stage-distribution models for teasel (*Dipsacus sylvestris* Huds.). Ecology 58: 1103–1111.
Whigham, D. (1974) An ecological life history study of *Uvularia perfoliata* L. Amer. Midl. Nat. 91: 343–359.
Willson, M.F., Miller, L.J. and Rathcke, B.J. (1979) Flower-visitors, nectar production and inflorescence size of *Asclepias syriaca*. Canad. J. Bot. 57: 1380–1388.
Zeide, B. (1978) Reproductive behaviour of plants in time. Amer. Nat. 112: 636–639.

BURIED VIABLE SEEDS IN TEMPERATE FORESTS

NOBUKAZU NAKAGOSHI

ABSTRACT

The buried seed population has great significance for plant demography, since it is regarded as the potential or initial coenopopulation. Buried seeds in the diverse communities in the temperate forest in Japan were analyzed. The density of buried seeds increases in the early stage of secondary succession, comes to a maximum in the secondary forest and decreases later in the climax forest. This fluctuation of densities is caused by a high seed production of the species in the felling site and secondary forest, and a low seed production in the climax forest. The seed banks of the main species consist of three types: no reserve of seeds in the growth season (Type A), reduced seed bank (Type B) and permanent seed bank (Type C). The main species are classified into nine groups based principally on their seed bank type and life form. Each of these nine functional groups has its own reproductive pattern. Each functional group is present at a particular stage during the process of vegetation change. It is suggested that these diverse patterns were evolved in forest vegetation with a mosaic structure having cyclic vegetation changes: forest vegetation is a complex association consisting of species which have diverse life history patterns.

INTRODUCTION

Ecological studies on the component species of plant communities have recently begun from the viewpoint of reproductive and demographic biology (Harper 1977), but ecological studies on the seeds of component species are rather few. The present study is intended to elucidate the dynamics of seeds, especially of buried seeds, and to place them in an ecological and phytosociological perspective. Buried seeds are defined in this paper as the seeds buried alive in the litter and soil. This buried viable seed population, in contrast to the aboveground growing vegetation, is regarded as the potential or initial (Rabotnov 1969) plant community.

The study of buried seeds linked with vegetation phytosociology originates with Nakagoshi and Suzuki (1977). However, the history of the

White, J. (ed.), The Population Structure of Vegetation.
 ISBN 978-94-010-8927-2

study of buried seeds can be traced back to Darwin (1859) who stated that numerous seedlings emerged from ground cleared of its vegetation cover. Brenchley (1918) made the first reliable study of the ecological connection between buried viable seeds and plant communities in arable fields and pastures at Rothamsted, England. Buried seeds have since been studied in various types of plant communities, such as arable lands, pastures, meadows, moors and heath, but mostly by agriculturalists concerned with herbaceous plant communities. Buried seeds in forests have not been studied so intensively. The study of buried seeds among different aged stands gives some clues to the mechanism of forest succession, as Oosting and Humphreys (1940) and Livingston and Allessio (1968) demonstrated.

The present study deals with the occurrence of buried seeds in diverse communities of temperate forest regions and with the various functional groups classified by the ecology of their seeds. Since each functional group has its own reproductive biology, an analysis of the community dynamics of the temperate forests is attempted, based on the ecological behaviour of the functional groups identified.

AREA STUDIED

The Hiba Mountains are situated about the center of the Chugoku district, in southwestern Honshu, Japan. Topographically the mountains are hilly peneplains and the highest peak ascends to 1,299 m above the sea. An area above 900 m altitude was selected as the site for the present study. The climate of this area is the Japan Sea type with deep snow in winter from early December to March. In the period 1974–1978 the annual precipitation was 2,523 mm and the annual mean temperature was 9.6°C at the nearest weather bureau (810 m above the sea). The whole study area belongs to the cool temperature zone, its climatic climax being beech (*Fagus crenata*) forest. The actual vegetation of the study area is represented by grasslands, deciduous scrub, conifer plantation and deciduous forests. The woody plant communities cover 93.1% of the study area, grow continuously, and exist in a mosaic or patchy pattern. The herbaceous plant communities cover the rest of the area (6.9%) and have been described phytosociologically by Nakagoshi (1981b). The soil is brown forest soil over almost all the area; black ando soil is restricted to the peaks.

METHODS

The vegetation was analyzed according to the vegetation concepts and methodology of the Zürich–Montpellier school of phytosociology (Braun-Blanquet 1964). Full details of stand location and phytosociological classification are given in Nakagoshi (1984b). Thirteen stands belonging to all the eight phytosociological units recognized in the mesic forest were chosen for the investigation of buried seeds. In August 1977 the age of each stand of undisturbed climax forest was estimated by counting the annual rings of the largest trees. The ages of other stands were known from dates of cutting or planting. All the component species of the 13 stands were ordinated using the summed dominance ratio (SDR_4) of Numata and Yoda (1957): this is the sum of relative cover, relative density, relative frequency and relative height, divided by 4.

First year seedlings were counted on two 1 m^2 quadrats in each stand from May to June 1978. The densities of these seedlings was observed until November 1979. The root lengths of seedlings were measured at the mineral soil on the forest floor, below the litter layer.

Soil samples for seed analysis were collected in late November 1977, after seed dispersal (Nakagoshi 1980a) and before the onset of heavy snow. The soil was sampled again in late May 1978, before the next seed dispersal. The depth distribution of buried viable seeds in soil was preliminarily investigated and large numbers of seed of various species were observed in the 0–10 cm soil layer in the beech forest (Nakogoshi 1981c). This layer consists of the litter and the organic soil.

It was noticed that the root length of emerged seedling is less than 10 cm in all cases (see Table 3 later). It was decided that a soil sampling depth of 0–10 cm would be suitable to census the buried seed bank adequately. A soil volume of 80 litres (0.8 m^2 surface area $\times$ 10 cm deep) was collected from a beech stand. A curve relating the number of species retrieved to soil volume was constructed from increasing aliquots of this 80 litre sample. A comparison of the numbers of species of buried seed found and the number of species growing in the aboveground vegetation indicated that a volume of 12 litres of soil contained the same *number* of species as those fruiting in the vegetation. This species number was irrespective of species identity. From these results, the volume of soil sampled was fixed at 40 litres (0.4 m^2 $\times$ 10 cm) in each stand (Nakagoshi 1981c). This soil was collected in 10 random samples of equal size. It was considered that this volume of soil was sufficient to compare the buried seeds and the vegetation phytosociologically. Two sets of samples were collected in five of the communities and a single sample in each of the other three.

The buried seeds were separated from the soil in order of size by using four different (4, 2, 1.5, 1 mm) mesh sieves. Woody species were difficult to identify by germination tests because of their low germination rate; their seeds were identified by external morphology, and their viability by the existence of an embryo or an endosperm. Tetrazolium chloride staining was also used with seeds whose viability was difficult to check by germination. In the glasshouse, germination tests were made with those seeds which went through the 1 mm mesh sieve: most of these were herbaceous species. Another germination test was made with fresh seeds collected from plants growing in the stands. The germination tests were carried on for a period of about 100 days in each case. (Fruits such as achenes, samaras and nuts are referred to as seeds in this paper).

Figure 1 shows the survival process of the plants treated in the present study. The buried seeds in winter correspond to the buried seeds in late November 1977, the dormant seeds in the growth season correspond to the

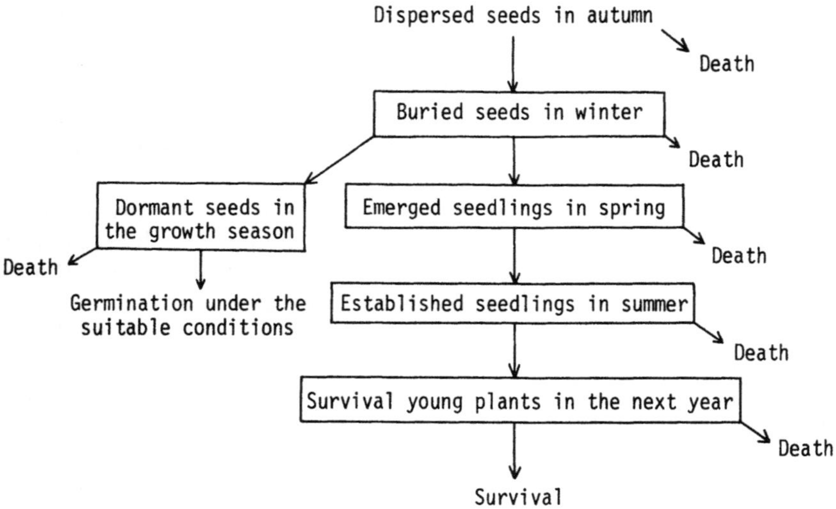

Fig. 1. Initial stages of the survival process of seed plants in the temperate forests recognized in the present study. Each stage enclosed by the solid line was observed in detail.

buried seeds in late May 1978 and seedlings in spring and summer correspond to those in May and August in 1978 and 1979.

THE PLANT COMMUNITIES AND THEIR SUCCESSIONAL SERIES

Seven communities of mesic woody vegetation were recognized in the study area by the tabular method of Braun-Blanquet (1964). Their successional interrelationships are shown in Fig. 2. Full phytosociological tables are shown in Nakagoshi (1984b). The following notes characterize them briefly.

The *Rubus crataegifolius–Aralia elata* community is a scrub which appears just after the clear cutting of forests and also in natural gaps of forests. Two initial stages are recognizable, corresponding to the developmental stages of the vegetation. The *Macleaya cordata* unit (community initial: **RAM**) is a sparse scrub at the earliest stage of a secondary succession in which *Rubus crataegifolius* and herbs mainly grow. A sampling stand 3 months old in August 1977 was set up at a felled conifer plantation. Another unit of *Actinidia arguta* (**RAA**) is a scrub in the next stage of the succession after clear cutting. Dominant species include *Aralia elata*, *Rhus javanica*, *Rubus palmatus* and *R. crataegifolius*. Two sampling stands, 6 and 11 years old, were set up.

The *Corylus sieboldiana–Sasa palmata* community (**CS**) develops at logging sites and the margins of forests. The shrub layer is dominated by *Corylus sieboldiana* mixed with some other juvenile trees, and a small bamboo *Sasa palmata* is dominant under them. It is a Mantel (edge) community of the forest. A sampling stand, 15 years old, was set up at the

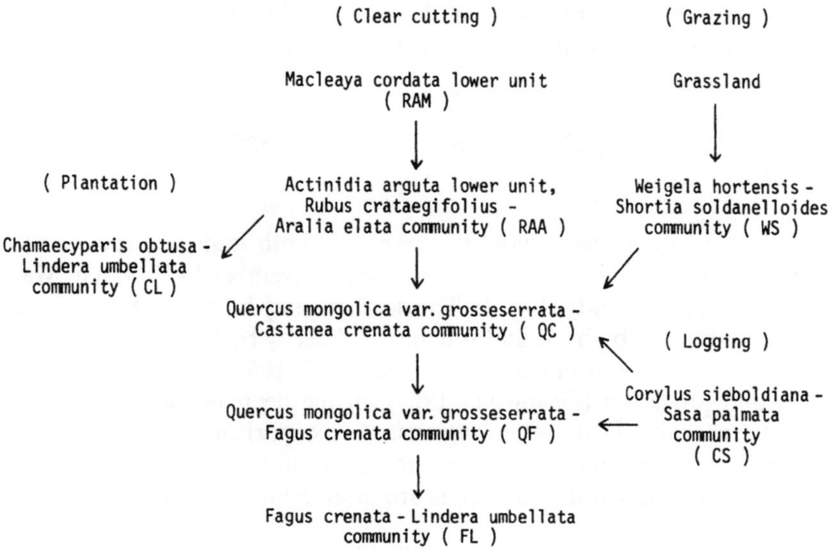

Fig. 2. Successional series in the phytosociological units of woody vegetation promoted by four kinds of disturbances (in parentheses).

side of a beech forest. The *Weigela hortensis–Shortia soldanelloides* community (**WS**), which is locally distributed near the peaks, is a substantial scrub. Topographical factors and grazing (Itow 1963) may retard its succession to forest. Two sampling stands, 36 and 41 years old, were set up.

The *Quercus mongolica* var. *grosseserrata* (*Q. crispula*)–*Castanea crenata* community (**QC**; Castaneo-Quercetum crispulae Horikawa and Sasaki 1959) is a secondary forest, and is the most commonly distributed vegetation in the study area. The tree layer is dominated by *Acer* spp., *Carpinus* spp., *Castanea crenata* and *Quercus mongolica* var. *grosseserrata.* The primeval forests might have been destroyed by cutting for charcoal for the iron industries in the past, and thus this forest has come to be widely distributed in this area. Two sampling stands, 46 and 48 years old, were set up. The *Quercus mongolica* var. *grosseserrata–Fagus crenata* community (**QF**) is a seral secondary forest; oak is dominant in the tree layer, and beech grows under it. One sampling stand, 58 years old, was set up. The *Fagus crenata–Lindera umbellata* community (**FL**; Lindero umbellatae-Fagetum crenatae Sasaki 1970) is a climatic climax forest. Two sampling stands, older than 240 and 270 years, were set up. The *Chamaecyparis obtusa–Lindera umbellata* community (**CL**) is an evergreen conifer plantation for forestry. Two sampling stands, 42 years old, were set up.

The hierarchical synsystematic classification of the above mentioned communities in the Chugoku district is given by Miyawaki (1983).

SEED BANKS OF FOREST COMMUNITIES

'Seed bank' means the buried viable seeds which supply the seedlings when conditions are suitable for germination. This is the initial coenopopulation

(Rabotnov 1969). In this section the seed banks of temperate woody vegetation in Japan and of the boreal and temperate forests of other regions are discussed.

Buried viable seed populations in the forest communities of Chugoku district.

A distinct seasonal decrease of the numbers of both seeds and species was found in the buried seeds: both were richer in November 1977 than in May 1978 in each community (Fig. 3). The decrease may be ascribable to death in winter, damage by birds and rodents, and decay by fungi and bacteria. The effects of microtopography (Harper et al. 1965), chilling (Johnson 1975) and predation (Gashwiller 1967) on the decrease of buried seeds have been reported in other regions. A comparison of the densities between these disappeared seeds and the population of the growing vegetation has shown that the seeds are more numerous than the vegetation.

The species composition is not parallel between the vegetation and the buried seeds, even in the same community. Søorensen (1948)'s coefficient of the species composition between the vegetation and the buried seeds was 44.7%, averaged over all 13 stands, with a range 32.7–66.7%.

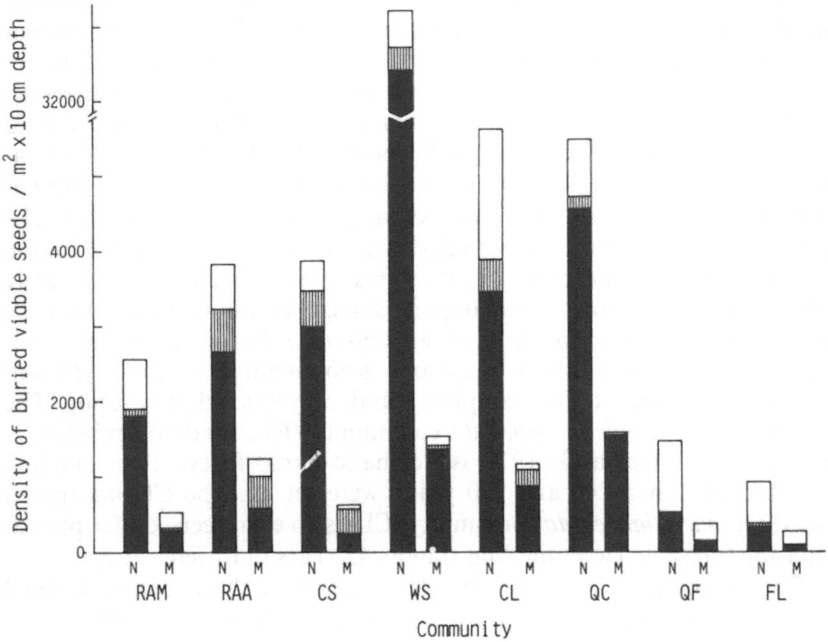

Fig. 3. Densities of the buried viable seeds in the eight plant communities in two different seasons, late November 1977 (N) and late May 1978 (M). Seeds were classified into three categories: the seeds of the fruiting species in the community (■), the seeds of the vegetative species in the community (▥) and the seeds of species not presently growing in the community (□). The initials of communities correspond to those of the text.

This difference is mainly caused by seed invasion of species which are not components of the vegetation; their disseminule forms are mostly anemochore and endozochore.

The number of buried seeds of the fruiting species (Nakagoshi 1980b) in the community increases in the early stage of secondary succession after clear cutting, reaches a peak in the *Quercus–Castanea* community, and decreases in the *Quercus–Fagus* community and especially in the climax *Fagus–Lindera* community (Fig. 3). This may be caused by a high seed production of the herbs in the *Rubus–Aralia* community, and particularly by the trees of the *Quercus–Castanea* community. These phenomena were also observed in the secondary succession to oak-hickory forest in North Carolina (Oosting and Humpherys 1940) and pine forest in Massachusetts (Livingston and Allessio 1968). The greatest number of seed grains, 33,220 per m^2 was recorded in the *Weigela–Shortia* community in November 1977, caused by the seeds of *Weigela hortensis*. In the *Chamaecyparis–Lindera* community only 6% of the buried seed population excluding *Chamaecyparis obtusa* was of local origin; this indicates the low seed productivity of the component species under the extremely shaded condition of the dense plantation.

Buried seeds in several forests of the northern hemisphere

In the warm temperate forests of Japan, the density of buried seeds in the climax *Castanopsis* forest is lower than that of secondary pine forests (Hayashi 1977, Nakagoshi 1981a). Among the pine forests, the forest of Miyajima Island has less buried seeds than the mainland forest (Nakagoshi 1984a). This may have been caused by the poor floristic composition of vegetation and the lack of grassland, pasture and arable land in the island. Generally, weeds in arable land or grassland species always produce a large number of seeds (Salisbury 1942), and their airbone seeds are dispersed and buried in the neighboring forests.

In temperate North America the climax oak-hickory forest has 1,181 buried seeds per m^2 (Oosting and Humphreys 1940) and the maple forest has 378 (Marquis 1975). In the Pacific conifer forests in Oregon and British Columbia, the number of buried seeds varies from 206 to 3,447 per m^2 (Kellman 1970, 1974, Strickler and Edgerton 1976). The boreal spruce forest in Russia contained from 1,300 to 5,000 buried seeds; in this case the dominant seeds were those of the felling site (Karpov 1960). The boreal conifer forest in Sweden has 700 buried seeds per m^2 (Granström 1981). The subalpine conifer forest in Colorado has 28 on average (Whipple 1978). In the subarctic woodlands of Canada buried seeds were absent (Johnson 1975): Johnson concluded that the lack of viable seeds was caused by the short growing period and its effect on germination. These data suggest that buried seeds decrease in number in proportion to the degree of latitude and elevation. But the tropical rain forest is noted by Thompson (1978) as another forest of low buried seed density. If his assumption is correct, the temperate forests have the richest seed banks of native species among the forests of the Northern Hemisphere.

Among the buried seeds in the forests of Japan, Europe and North America (excluding subarctic Canada) the invaded seeds of weeds or of felling site species were commonly found.

REPRODUCTIVE PATTERNS OF THE MAIN COMPONENT SPECIES

Seeds are not only disseminules for colonization but also the means of maintaining populations in the vegetation. Each component species has its own life history, and the significance of the buried seeds for population maintenance differs among them. The reproductive features of the component species were mainly examined in terms of the pattern of occurrence of buried seeds, and species were classified into 10 types based on the characteristics of their seeds and life forms.

Occurrence of the main component species as buried seeds

In the temperate forest of Japan, the seed dispersal season of forest species is usually the period from late summer to autumn (Nakagoshi 1980a), although a few spring herbs disperse their seeds in late spring. The seed banks of almost all species are enlarged with fresh seeds from October to November, and the buried seeds decrease in number after this highest density period. Three types of seasonal aspect of buried seeds were recognized in these communities in the period from 1974 to 1978, shown diagrammatically in Fig. 4; full details are given elsewhere (Nakagoshi 1984b). In late May the seed banks of species can be classified into three types: Type A (no reserve of seed), Type B (reduced seed bank) and Type C (permanent seed bank). Forest species, most of which are woody, have less varied seasonal aspects of seed banks than grassland species (Thompson and Grime 1979). Typical species classified according to these types and to their life form are listed in Table 1. The life forms are closely connected with the pre-reproductive life and total life span (Harper and White 1971, 1974).

Ordinations of species were carried out on the vegetation of the stands; 15 and 10 high-ranking species were selected as 'main component species' in the two *Quercus–Castanea* stands and in the other eleven stands respectively. (The difference in the number of species reflects the fact that the *Quercus–Castanea* stands had a larger number of component species than the others.) A subgroup of 53 species was then selected as the 'main component seed reproducing' plants. Among these 53 species the buried seeds of only 47 species were detected in the present study during 1977–1978, as shown in Table 2, where each species is ecologically arranged into one of nine groups with the 'typical' species (listed in Table 1) in the lead. The herbaceous species of Type B, e.g., *Eupatorium chinense* var. *simplicifolium* and *Miscanthus sinensis*, produce a large number of anemochoric seeds, which are widely distributed. These species, however, are absent in the vegetation of the *Chamaecyparis–Lindera* and the *Fagus–*

Table 1. Typical species classified by both the seasonal aspects of their buried seeds and their life forms among the main component species of the temperate forests of the Chugoku district. For the explanation of Types A, B and C see Fig. 4.

Seasonal aspect of buried seeds	Life form		
	Herb	Shrub	Tree
Type A	*Erigeron sumatrensis*	*Corylus seiboldiana*	*Fagus crenata*
Type B	*Eupatorium chinense* var. *simplicifolium*	*Weigela hortensis*	*Acer rufinerve*
Type C	*Disporum smilacinum*	*Lindera umbellata* *Rubus crataegifolius*	–
Usually no seed production	–	*Sasa palmata*	–

Lindera communities, which indicates that their dispersed seeds are unsuccessful in establishing in these communities (see also Table 4 later). On the other hand *Disporum smilacinum* is able to germinate in the *Fagus–Lindera* community. Among the shrubs *Weigela hortensis* and *Rubus crataegifolius* have contrasting reproductive patterns: the former survives by producing numerous seed whose mortality is relatively high, while the latter maintains its 'potential' aboveground population by persistent buried seeds at low density. The group including *Lindera umbellata, Rhus ambigua* and *Cornus kousa* is the commonest group both in the shrub layer of forests

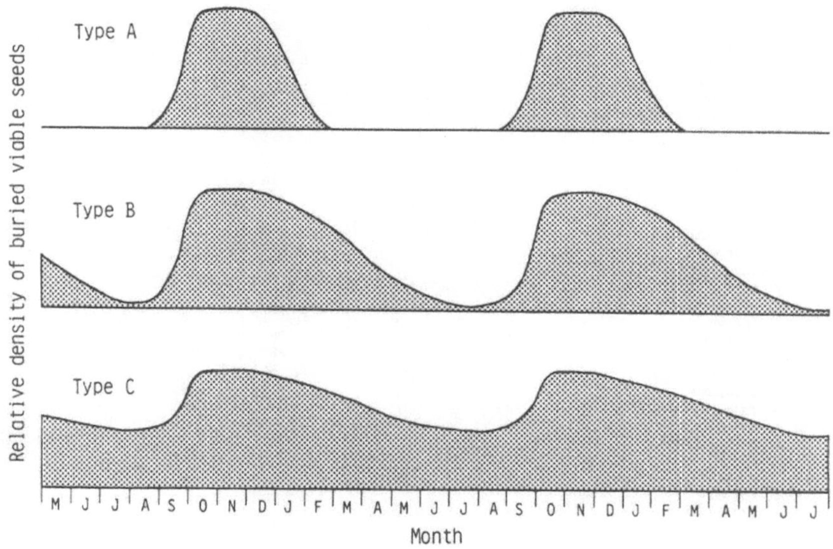

Fig. 4. Three types of seasonal aspects of the buried viable seeds of the component species of the temperate forests. Type A: occurrence of buried seeds in the seed-dispersal and inactive seasons only; Type B: occurrence mainly in the seed-dispersal and inactive seasons; Type C: permanent seed bank throughout the year.

Table 2. Buried viable seed densities of the main component species per $0.8\,m^2 \times 10\,cm$ depth in eight plant communities in two different seasons, late November 1977 and late May 1978. The component species are classified into nine groups according to the seasonal patterns of their buried seeds, their life forms and their ecological features. The initials of communities correspond to those of the text. During 1977–78 the buried seeds of only 47 of the 53 main component were detected. Data for three communities in which soil samples of $0.4\,m^2$ were taken are scaled-up for comparison with those in which $0.8\,m^2$ samples were taken.

Community	RAM		RAA		CS		WS		QC		QF		FL		CL	
Month	Nov.	May	Nov.	May	Nov.	May	Nov.	May	Nov.	May	Nov.	May	Nov.	May	Nov.	May
Year	'77	'78	'77	'78	'77	'78	'77	'78	'77	'78	'77	'78	'77	'78	'77	'78
Species																
Erigeron sumatrensis		2														
Digitaria adscendens	202															
Eupatorium chinense var. *simplicifolium*		2	662	99	1430	42	10		20		386		7		22	
Miscanthus sinensis			397	8	50		3748	409	339		10		2		344	
Solidago virga-aurea var. *asiatica*			41		130		15		40	2	30		10			
Lysimachia clethroides	10		195	140			5						1			
Polygonum cuspidatum	1120	86													640	
Macleaya cordata	10				28											
Carex dolichostachya var. *glaberrima*			60								10					
Disporum smilacinum			2		24	10	12	2	14	3	10	2	2			
Viola grypoceras	128	158	325	158	120	96	126	62	145	26	22	12	10	2	126	75
Viola vaginata									4					1	20	5
Smilacina japonica											2	2				
Weigela hortensis			326	169	316	50	20235	445	20				1			
Leucothoe grayana var. *oblongifolia*							1454	188								
Rubus crataegifolius	50	2	135	40	16	2	20		15	4	10	2	5	3	137	10
Rubus palmatus	2	14	25	13	60	46	5		20	2	200	48	64	23	359	174
Aralia elata	20	4	5	4	10	2	1		10	2	6	2	5	2	50	1
Rhus javanica			2	1			5	1	7	1	2		6	2	21	1
Actinidia arguta	10	2					1		1	2	10		4	1	6	1
Lindera umbellata			118	62	12	2	6	1	34	9	12	4	9	6	11	1
Rhus ambigua	12	6	83	60	16	4	6	2	29	3	36	24	126	65	30	14
Rubus pectinellus	10	2	5	2	24	22			10	3	60	42	30	14	108	65
Cornus kousa			2	2	28	24	1		25	12	22	6	9	1	2	2
Rhus trichocarpa		10	3	1	16			1	13	2	2	2	4	2	8	6
Symplocos coreana			5	6	10		10		6	2	10	6	7	1		1
Skimmia japonica			5	2	4		13	1	70	2	24	4	2			
Acanthopanax sciadophylloides			5	1	2				5	2	58	16	85	2	2	
Sorbus commixta			15	12			10		9	3	2	2				
Lygustrum tschonoskii				2	4			1	2	1	2	2		1		
Ilex crenata						2	29	2	2	1	10	2				
Viburnum erosum				1		2			18	2	2				1	
Prunus grayana			25	2						5				3		
Corylus sieboldiana					92		14		2				1			
Viburnum furcatum					6				3		2		1			
Cephalotaxus harringtonia var. *nana*			1						8							
Acer rufinerve			1		2	2	20	5	1	1	8		2		1	
Clethra barbinervis			415	27	290	196			2982	931	82		7			
Carpinus tschonoskii			16	14	2		2		289	211	20	4	7	1	2	
Carpinus japonica			2		4		1	1	54	21			2		6	
Acer sieboldiana				1					28	2	4		5		3	
Schizophragma hydrangeoides			160						78	8						10
Acer mono					8				3				1			
Chamaecyparis obtusa	474	44													2490	550
Fagus crenata											4		27			
Quercus mongolica var. *grosseserrata*									1		26					
Castanea crenata									12							
Others	10	80	72	167	396	2	827	111	67	10	16	120	293	97	112	
Total	2058	412	3068	994	3100	504	26576	1232	4386	1273	1100	302	735	227	4501	932

and among the buried seeds (in both two seasons) in the study area. The seeds of the next group *Viburnum furcatum*, *Corylus sieboldiana* and *Cephalotaxus harringtonia* var. *nana* are heavier than the other endozoochoric seeds and have less dormancy. Among the trees, *Acer rufinerve* and *Fagus crenata* have different reproductive patterns from each other. The former disperses its samaras broadly and some of them remain dormant during the growth season. Species similar to *Acer rufinerve* are *Carpinus* spp. and the other *Acer* species which compose the tree layer of the *Quercus–Castanea* community. The nuts of *Fagus crenata* disperse in a narrow range from a parent tree and germinate beneath the snow without resting in winter. Nuts of Fagaceae in temperate forests have no dormancy, and seed banks of species of this family are absent in summer. Emerging seedlings of these level of have a high frost hardiness.

Functional groups according to the ecological characteristics of seeds

The ecological characteristics of seeds treated in the present study are the seasonal aspects of buried seeds (Fig. 4, Thompson and Grime 1979), the germination ratio (Grime et al. 1981) in the growth season (i.e., the 'negative dormancy ratio'), the seed weight (Salisbury 1942, Harper et al. 1970, Baker 1972, Nakagoshi and Soga 1981), the length of primary root of emerged seedlings and the disseminule form (Ridley 1930, Luftensteiner 1979, Nakagoshi and Soga 1981). The germination tests on the four species of Fagaceae were made in winter, immediately after the material had been obtained on the forest floors. These characteristics of seeds are interrelated with one another. Table 3 shows the ecological characteristics of seeds of 51 of the main component seed-reproducing species. These species are classified into nine groups based on differences in the behaviour of their seeds and on their life form (Rabotnov 1981). These nine and the *Sasa palmata* functional groups are summarized below.

***Erigeron* type**: An annual weed, *Erigeron sumatrensis* has no buried seed bank in the closed forests. The seeds of this weed germinate easily and make rapid growth after invasion into a vegetation gap or disturbed site which has a sunny and bare ground surface. At these sites the weed produces many seeds in a short period, becomes dispersed widely, and once again invades fresh gaps. Another weed, *Digitaria adscendens* belongs also to this type. Some buried seeds of these annual weeds are contained in forests (Oosting and Humphreys 1940, Livingston and Allessio 1968, Kellman 1970, 1974, Nakagoshi 1981a, 1984a) but their densities are fairly low.

***Eupatorium* type**: *Eupatorium chinense* var. *simplicifolium* colonizes disturbed open habitats. This perennial herb produces numerous seeds until its vitality is reduced by shrubs or juvenile trees. Table 3 shows the other six herbs belong to this type in the study area. Seeds of species included in this type (e.g., *Eupatorium* spp., *Solidago* spp., *Anaphalis margaritacea, Digitalis purpurea, Polypremum procumbens, Chamaenerion augustifolium, Epilobium* spp., *Oenothera* spp., *Juncus* spp., *Carex* spp., *Agrostis* spp., *Andropogon* spp., *Deschampsia caespitosa, Festuca* spp., *Panicum* spp. and

Table 3. Ecological characteristics of seeds and seedlings of 51 main component species (excluding two bamboo species) belonging to the nine types of functional groups defined in Table 1. The life forms refer mainly to Raunkiaer (1934)'s system.

Species	Seasonal aspect of buried seeds*	Germination ratio (%)	Air dried seed weight (mg)	Length of primary root of seedling (mm)	Disseminule form#	Life form**
Erigeron sumatrensis	A	100	0.02	1.8	AN	Th
Digitaria adscendens	A	100	0.57	4.5	AN	Th
Eupatorium chinense var. *simplicifolium*	B	100	0.40	7.6	AN	H
Solidago virga-aurea var. *asiatica*	B	100	0.24	4.9	AN	H
Macleaya cordata	B	85.0	0.15	4.5	AN	H
Lysimachia clethroides	B	100	0.26	8.5	AN	H
Polygonum cuspidatum	B	100	0.97	8.8	AN	H
Carex dolichostachya var. *glaberrima*	B	100	0.96	7.4	AN-HY	H
Miscanthus sinensis	B	100	0.87	8.4	AN	H
Viola vaginata	C	55.0	1.37	20.0	AU-MY	G
Viola grypoceras	C	28.3	0.83	7.9	AU-MY	H
Oxalis japonica	C	84.7	0.34	9.5	AU-CL	G
Disporum smilacinum	C	63.0	15.83	19.0	EN	G
Smilacina japonica	C	77.8	32.46	26.4	EN	G
Maianthemum dilatatum	C	76.0	17.26	25.3	EN	G
Weigela hortensis	B	100	0.11	4.5	AN	N
Leucothoe grayana var. *oblongifolia*	B	100	0.04	2.5	AN	N
Salix daisenensis	A	100	0.16	4.2	AN	N
Aralia elata	C	17.9	0.99	7.6	EN	Mc
Actinidia arguta	C	23.5	1.43	8.6	EN	Mcl
Rhus javanica	C	10.0	1.06	8.5	EN	Mc
Rubus crataegifolius	C	7.5	0.62	6.5	EN	N
Rubus palmatus	C	12.5	1.29	9.9	EN	N
Viburnum erosum	C	31.5	7.03	19.9	EN	N
Lygustrum tschonoskii	C	45.9	49.19	22.3	EN	N
Symplocos coreana	C	48.9	37.38	21.0	EN	N
Cornus kousa	C	52.1	39.77	27.0	EN	Mc
Acanthopanax sciadophylloides	C	60.4	10.28	12.1	EN	Mc
Ilex crenata	C	43.8	11.85	13.0	EN	N
Rhus trichocarpa	C	25.0	14.40	15.5	EN	Mc
Rhus ambigua	C	31.0	31.84	18.4	EN	Mcl
Skimmia japonica	C	64.2	21.77	25.5	EN	N
Sorbus commixta	C	40.6	2.22	10.1	EN	Mc
Prunus grayana	C	35.8	38.90	29.5	EN	Mc
Rubus pectinellus	C	30.5	2.94	19.0	EN	Ch
Lindera umbellata	C	42.0	76.59	25.0	EN	N
Viburnum furcatum	A	94.0	37.40	30.3	EN	Mc
Corylus sieboldiana	A	98.0	522.26	49.5	EN	N
Cephalotaxus harringtonia var. *nana*	A	90.0	704.95	65.0	EN	N
Clethra barbinervis	B	100	0.16	7.5	AN	Mc
Acer rufinerve	B	86.6	21.88	21.0	AN-HY	M
Acer sieboldianum	B	85.0	14.40	17.4	AN-HY	Mc
Acer mono	B	90.2	52.11	26.0	AN	M
Schizophragma hydrangeoides	B	100	0.43	4.2	AN	Mcl
Carpinus tschonoskii	B	88.3	8.47	26.5	AN	M
Carpinus japonica	B	79.0	3.86	22.5	AN	M
Chamaecyparis obtusa	B	95.0	7.50	14.5	AN	M
Fagus crenata	A	100	275.00	> 52.0	CL	M
Castanea crenata	A	100	840.67	> 60.0	CL	M
Quercus dentata	A	100	1256.45	> 63.0	CL	Mc
Quercus mongolica var. *grosseserrata*	A	100	2155.99	> 83.0	CL	M

* See Fig. 4.

\# AN: Anemochore, AN-HY: Anemo-hydrochore, AU-MY: Auto-myrmecochore, AU-CL: Auto-clitochore, EN: Endo-zoochore, CL: Clitochore.

** Raunkiaer (1934)'s life form; Th: Therophyte, H: Hemicryptophyte, G: Geophyte, Ch: Chamaephyte, N: Nano-phanerophyte, Mc: Micro-phanerophyte, Mcl: Vine, Micro-phanerophyte, M: Meso- or Mega-phanerophyte.

Poa spp.) are dominant among the buried seed populations in the other temperate and boreal forests (Oosting and Humphreys 1940, Karpov 1960, Livingston and Allessio 1968, Kellman 1970, 1974, Strickler and Edgerton 1976, Hayashi 1977, Hill and Stevens 1981). These herbs of the *Eupatorium*-type are the component species of herbaceous communities of grasslands, abandoned arable lands, forest felling sites and river banks.

***Disporum* type**: *Disporum smilacinum* produces less seeds per individual each year than the herbs of the former two types make, but its seeds are persistent and dormant in the growth season. In the temperate forests of Japan, this geophyte has a large proportion of the plant body as subterranean organs (Kawano 1975). This herb maintains its population with persistent seeds, rhizomes and shade-tolerant seedlings under *Sasa palmata* and tall trees. The long root of its seedling is effective in establishing it on the thick litter of the deciduous forests. *Viola grypoceras*, *Smilacina japonica* and *Maianthemum dilatatum* belong to this type. Two spring plants, *Viola vaginata* and *Oxalis japonica* which flower in early spring are included in this type due to the presence of buried seeds in the growth season.

***Weigela* type**: *Weigela hortensis* is similar to the *Eupatorium*-type in that it produces a great number of anemochoric seeds and disperses them widely. Following competition between it and newly invaded trees (e.g., *Acer* spp. and *Carpinus* spp.) the shrub is usually defeated. It seeds do not establish in the closed forests at all (Tables 2 and 4). The other shrubs belonging to this type are *Leucothoe grayana* var. *oblongifolia* and *Salix daisenensis*.

***Rubus* type**: The buried seeds of *Rubus crataegifolius*, most of which are dispersed by birds, are detected in various communities (Table 2), but they do not germinate in later seral communities. This bramble is one of the typical species (Miyawaki 1983) which regenerate rapidly in forest gaps and felling sites. At the earliest stage of revegetation, its dormant buried seeds germinate and establish after the former forest has been destroyed. *Aralia elata*, *Actinidia arguta*, *Rhus javanica* and *Rubus palmatus* belong to this type. Seeds similar to this type are also dominant together with those of the *Eupatorium*-type in the buried seed populations of the boreal and temperate forests: *Sambucus cerulea*, *Aralia hispida*, *Zanthoxylum* spp., *Mallotus japonicus*, *Rhus* spp., *Ulex gallii*, *Rubus* spp., *Rosa* spp. and *Ribes* spp. (Oosting and Humphreys 1940, Karpov 1960, Livingston and Allessio 1968, Kellman 1970, 1974; Strickler and Edgerton 1976, Moore and Wein 1977, Hayashi 1977, Hill and Stevens 1981, Nakagoshi 1981a, 1984a). *Prunus pensylvanica* (Marks 1974) is also included in this type.

***Lindera* type**: *Lindera umbellata* is commonly observed among a wide range of vegetation from the *Rubus–Aralia* community to the *Fagus–Lindera* community, as the member of both the vegetation and the buried seed population. It has a persistent bank of heavy seeds and shade-tolerant seedlings, whose long roots penetrate thick litters. It has wide and random seed dispersal by birds living in the forests. The woody species in the study area belonging to this type are listed in Table 3. Their seed productivity is higher in the open habitats than in the closed forests.

Table 4. Seedling densities per 4 m² of the main component species which emerged in eight plant communities in spring 1978. † indicates seedlings which survived to reproduce in 1978, ▼ indicates the presence of established seedlings in 1978, and ★ indicates the presence of young plants which still survived in 1979. All the emerged seedlings without a mark died before they became established. In five communities seedlings were counted in two separate stands, giving a total area of 4 m²; the data from the three remaining communities, in each of which one stand was examined, are scaled-up for comparison.

Symbol of species	Species	Community							
		RAM	RAA	CS	WS	QC	QF	FL	CL
Es	*Erigeron sumatrensis*	4†							
Da	*Digitaria adscendens*	190†							
Mc	*Macleaya cordata*	2▼							
Pc	*Polygonum cuspidatum*	28★							
Rt	*Rhus trichocarpa*	2★							
Lc	*Lysimachia clethroides*		27▼						
Cd	*Carex dolichostachya* var. *glaberrima*		1▼						
Rj	*Rhus javanica*		1★						
Rc	*Rubus crataegifolius*	8★	1▼						
Ae	*Aralia elata*	4★	1						
Rp	*Rubus palmatus*	2★	6★		1				
Ec	*Eupatorium chinense* var. *simplicifolium*	6★	12▼		1	1			
Sv	*Solidago virga-aurea* var. *asiatica*	2★	1▼	2	2	1	2		
Ms	*Miscanthus sinensis*		7		65★				
Vg	*Viola grypoceras*	16★	2▼	2	2	8▼	4		6
Ra	*Rhus ambigua*	2▼	1					1▼	1
Ic	*Ilex crenata*	2	14★	10★	14★	6▼	10▼		
Ck	*Cornus kousa*		1		1	2	2	1	
Sy	*Symplocos coreana*		1	2▼		1★		1	
Lu	*Lindera umbellata*		24★	4▼		9★	2	4	1
So	*Sorbus commixta*		3★		3▼	1			
Wh	*Weigela hortensis*		12★	4	28		2		
Cb	*Clethra barbinervis*		10▼	16		16	6		
As	*Acer sieboldianum*		2	2▼				1	
Ar	*Acer rufinerve*		11★		2		2	1	
Am	*Acer mono*		1★					1	
Sh	*Schizophragma hydrangeoides*		12★		1▼	18★			2
-	*Quercus mongolica* var. *grosseserrata*		1★			1	2		
Ds	*Disporum smilacinum*		19★	2	9★	11★	8▼	1	
Ve	*Viburnum erosum*			2		5			
Vf	*Viburnum furcatum*			4★			2★		
Sk	*Skimmia japonica*				6★	3	5▼		
Ru	*Rubus pectinellus*					1★	2	1	6
Lg	*Leucothoe grayana* var. *oblongifolia*				39▼				
Sd	*Salix daisenensis*				21▼				
Ct	*Carpinus tschonoskii*					2	2		1
Cj	*Carpinus japonica*					3			
Lt	*Lygustrum tschonoskii*					2★			
Cc	*Castanea crenata*					2▼			
Fc	*Fagus crenata*						2★	2★	
Md	*Maianthemum dilatatum*							2▼	
Sm	*Smilacina japonica*							1★	
Pg	*Prunus grayana*							1	
Ac	*Acanthopanax sciadophylloides*							1	

***Corylus* type**: Fruiting plants of *Corylus sieboldiana* occur at the margins and fresh gaps of forests. The nuts of this shrub have little dormancy and are often eaten by rodents. Survival of its seedling depends on their shade tolerance. Two woody species, *Viburnum furcatum* and *Cephalotaxus harringtonia* var. *nana*, are loosely included in this type.

***Acer* type**: In the *Quercus–Castanea* community, *Acer rufinerve* produces its samaras and disperses them broadly. Woody species bearing similar winged disseminules, such as *Clethra barbinervis*, *Acer sieboldianum*, *A. mono*, *Carpinus tschonoskii* and *C. japonica*, invade the scrub, germinate quickly and grow to dominance. The seedlings of this type have low shade tolerances and do not establish even in the *Quercus–Castanea* community (Table 4). *Schizophragma hydrangeoides* (vine) and the tree *Chamaecyparis obtusa* belong to this type in a wide sense. Trees of *Pinus densiflora* (Nakagoshi 1981a), *Betula papyrifera* (Livingston and Allessio 1968, Kellman 1970, 1974) and *B. pubescens* (Hill and Stevens 1981) are typical species of this type.

***Fagus* type**: *Fagus crenata* produces heavy non-dormant clitochoric nuts, gradually colonizes the neighboring communities and the gap of the beech forest itself, and finally becomes dominant in the tree layers. This beech takes many years to extend its territory because of its slow growth and inefficient dispersal mechanism. Most of its nuts and seedlings are food of the herbivores living in the forests. However, in mast years of large nut production a few of them survive and grow to become members of the vegetation. Three species of Fagaceae, *Quercus dentata*, *Q. mongolica* var. *grosseserrata* and *Castanea crenata* also belong to this type.

***Sasa* type**: In the mesic forests of temperate Japan, shrubs of Bambusaceae, *Sasa* spp. and *Sasamorpha borealis* always exist in the lower shrub layer and characterize the forest communities. The density of their culms is fairly high in any community, and they grow even in the climax beech forests (Sasaki 1970). *Sasa palmata* in the study area flowers only after an indefinite time span of years and usually propagates itself by rhizomes. The small number of spring plants in the deciduous forests is considered to be due to the occurrence of this bamboo. Another species of this type is *Pleioblastus chino* var. *viridis* which occurs from grassland (Nakagoshi 1981b) to the *Weigela–Shortia* community. The tolerance of this species to browsing is higher than that of the trees and shrubs in the forests.

SURVIVAL IN THE INITIAL PHASE OF LIFE

A small number of species have emerged seedlings disproportionate to their numbers as buried seeds. This suggests selection for emergence of seedlings of each species in the community. Germinating seeds which have overcome one environmental barrier may experience a failure to emerge at the next barrier. The causes of the failure may be shade or drying and mineral deficiency caused by lack of sufficient root systems to penetrate the litter.

In the *Rubus–Aralia* community which has a sunny floor and thin litter many seedlings were observed. Densities of the seedlings decrease with secondary succession to forests. The emerged seedlings of the main component species are listed in Table 4. Seedling survival of each species was also observed and is shown by the symbols in Table 4. Two *Erigeron*-type weeds emerged in a high ratio and finally some of them reproduced. Most of the seedlings of the *Eupatorium*-type established in the *Rubus–Aralia* community, but died in the later seral communities. A few seedlings of *Miscanthus sinensis* survived until the next year. Seedlings of the *Rubus*-type established and survived only in the *Rubus–Aralia* community. A large portion of the emerged seedling populations of the *Weigela*-type died because of drought in the *Weigela–Shortia* community. Almost all species of the *Disporum*- and *Lindera*-types emerged in the seedling population of the communities and some seedlings survived in the forest communities (QC, QF and FL) and especially in the *Rubus–Aralia* community. The forest floor environments of the *Quercus–Castanea* community are not suitable for seedlings of the *Acer*-type. No seedlings established in the *Chamaecyparis–Lindera* community, mainly because of the shortage of sunlight.

The survival of seedlings from emergence to the establishment stage was analyzed in relation to light. Relative light intensities were measured over established seedlings, both inside and outside the stand at 1 m height above the ground. Generally, the large and tall seedlings which originated from the heavy seeds tolerate the shade, as shown in Fig. 5. Seedlings of herbaceous species of the early stage of succession have a low shade tolerance in the initial phase of life, although herbs of the *Disporum*-type are shade plants from the seedling stages. Among the woody species, seed size is clearly related to the degree of shade tolerance, a fact already

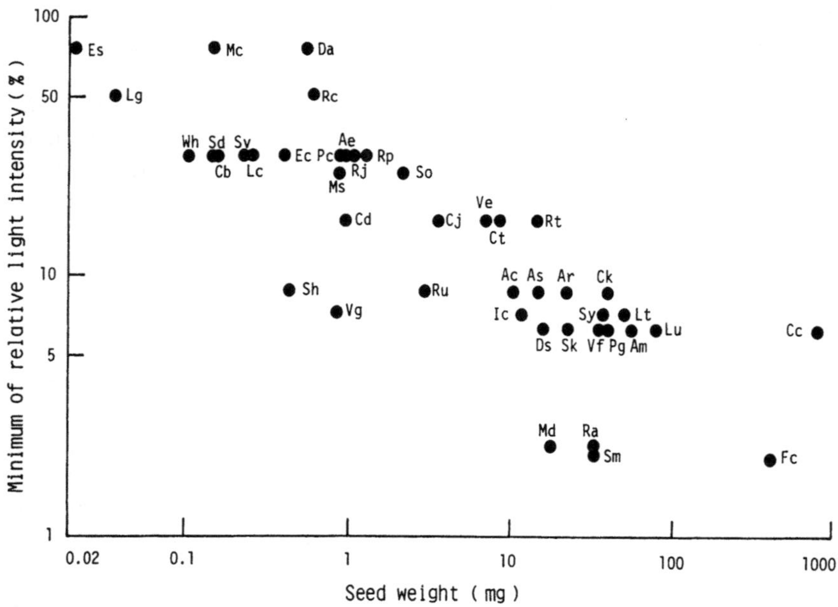

Fig. 5. Seed weight and seedling establishment in the temperate forests. The minimum of the relative light intensity measured in each community in August 1978 is regarded as the critical sunlight level for each seedling. The symbols of species correspond to those of Table 4.

confirmed by experiments with some woody species by Grime and Jeffrey (1965).

The second and sometimes critical factor in the decrease of the number of seedlings is the browsing of herbivores: seedlings of species of the *Fagus*-type are damaged fatally by them.

COMMUNITY DYNAMICS AND THE FUNCTIONAL GROUPS

There are several vegetation changes in the study area: secondary succession, regeneration of the climax forest through gap formation, logging of big trees, retrogressive succession near the pasture and man-made changes in the conifer plantation. In these community dynamics, the 10 types of functional groups have their own patterns.

Schematic diagrams of three selected kinds of community dynamics are shown in Fig. 6. A secondary succession after clear cutting occurs by the following process. Firstly this succession is promoted by the *Rubus*-type with its dormant buried seeds, the *Erigeron*-type with its newly invaded seeds and the *Eupatorium*-type with both buried and invaded seeds: as a

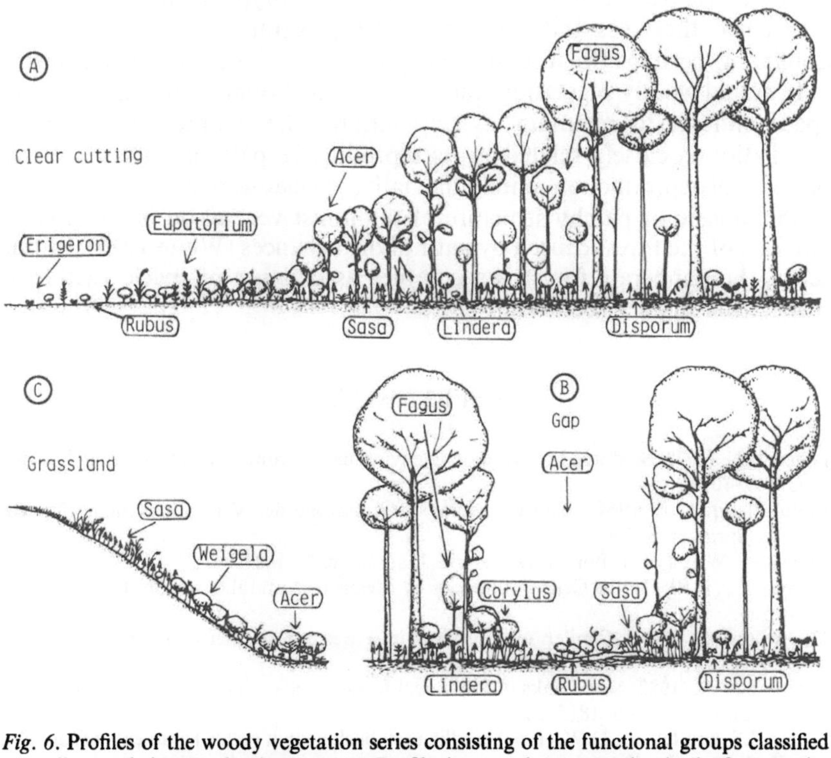

Fig. 6. Profiles of the woody vegetation series consisting of the functional groups classified according to their reproductive patterns. Profile A: secondary succession in the forest series after clear-cutting; profile B: regeneration of the climax beech forest which has gaps caused by the death of large beech trees; profile C: the zonation of the vegetation near the peak where cattle graze.

result the *Rubus–Aralia* community develops. In the next stage, the *Acer-*, *Lindera-* and *Sasa*-types invade by their different mechanisms. At the closed-forest stages, the dominant species gradually change from the *Acer*-type to the *Fagus*-type in the tree layers. The *Disporum*-type grows continually during the final stage of succession (Fig. 6A). These alternations of the types are similarly observed when large forest gaps occur, as in the *Fagus–Lindera* community, when there is a gap caused by the death of a big (older) beech (Fig. 6B). The regeneration cycle of beech forest with gap formation already been reported in England (Watt 1947). In the present study, the vegetation recovery of a fresh gap is succeeded in the order: *Rubus*-type with buried seeds, *Sasa-* and *Corylus*-types with vegetation recovery from shade-tolerant shoots, *Acer*-type with airborne seeds and finally the *Fagus*-type with clitochoric nuts, shade-tolerant seedlings and saplings. The recovery process of a site with selected logging of the tall trees (i.e., artificial gap formation) is similar to the natural one. The zonation from grassland to the small *Quercus–Castanea* community emerges according to the degree of grazing pressure. The *Wiegela*-type maintains its community by its high seed production, quick germination and rapid growth; otherwise the *Sasa*-type survives by its active and non-grazed rhizomes (Fig. 6C).

In the evergreen conifer plantation, the *Rubus*-type having a persistent seed bank, the vines and the species with prostrate forms are relatively dominant, perhaps because of the disturbances of forestry management.

In actual woody vegetation, each type of functional group has time and space enough to maintain its population. The maintenance of each population is closely related to its reproductive patterns. These diverse patterns of reproduction among the native species were probably evolved in the mosaic or patchy structure of the forest vegetation in the diverse habitats of the forest caused by natural disturbances (White 1979). In this sense, the temperate forest is a complex association of species with distinctive life history features.

REFERENCES

Baker, H.G. (1972) Seed weight in relation to environmental condition in California. Ecology 53: 997–1010.

Braun-Blanquet, J. (1964) Pflanzensoziologie, Grundzüge der Vegetationskunde. 3rd ed. Wien: Springer. 865pp.

Brenchley, W.E. (1918) Buried weed seeds. J. agric. Sci. 9: 1–31.

Darwin, C. (1859) On the Origin of Species by Means of Natural Selection. London: John Murray.

Gashwiler, J.S. (1967) Conifer seed survival in a western Oregon clearcut. Ecology 48: 431–438.

Granström, A. (1982) Seed banks in five boreal forest stands originating between 1810 and 1963. Can. J. Bot. 60: 1815–1821.

Grime, J.P. and Jeffrey, D.W. (1965) Seedling establishment in vertical gradients of sunlight. J. Ecol. 53: 621–642.

Grime, J.P., Mason, G., Curtis, A.V., Rodman, J., Band, S.R., Mowforth, M.A.G., Neal, A.M. and Shaw, S. (1981) A comparative study of germination characteristics in a local flora. J. Ecol. 69: 1017–1059.

Harper, J.L. (1977) Population Biology of Plants. London: Academic Press 892pp.
Harper, J.L., Lovell, P.H. and Moore, G. (1970) The shapes and sizes of seeds. Ann. Rev. Ecol. Syst. 1: 327–356.
Harper, J.L. and White, J. (1971) The dynamics of plant populations. Proc. Advan. Stud. Inst. Dyn. Numbers Pop., Oosterbeek 1970. pp. 41–63.
Harper, J.L. and White, J. (1974) The demography of plants. Ann. Rev. Ecol. Syst. 5: 419–463.
Harper, J.L., Williams, J.T. and Sagar, G.R. (1965) The behaviour of seeds in soil I. The heterogeneity of soil surfaces and its role in determining the establishment of plants from seed. J. Ecol. 53: 273–286.
Hayashi, I. (1977) Buried viable seed population (Japanese). In Numata, M. (Ed.), Succession of Plant Community and Its Mechanism, pp. 193–204. Tokyo: Asakura.
Hill, M.O. and Stevens, P.A. (1981) The density of viable seed in soils of forest populations in upland Britain. J. Ecol. 69: 693–709.
Horikawa, Y. and Sasaki, Y. (1959) Phytosociological studies on the vegetation of Geihoku district, Hiroshima Prefecture (Jap. with Engl. summ.). Sci. Res. the Sandankyo Gorge and the Yawata Highland. Hiroshima. pp. 85–107.
Itow, S. (1963) Grassland vegetation in uplands of western Honshu, Japan II. Succession and grazing indicators. Jap. J. Bot. 18: 133–167.
Johnson, E.A. (1975) Buried seed populations in the subarctic forest east of Great Slave Lake, Northwest Territories. Can. J. Bot. 53: 2933–2941.
Karpov, V.G. (1960) On the quantity and the species composition the viable seeds in the soil of spruce forests of Piceetum myrtillosum type (Russian with Engl. summ.). Trudy Moskow Obshch. Isp. Prirody 3: 131–140.
Kawano, S. (1975) The productive and reproductive biology of flowering plants II. The concept of life history strategy in plants. J. Coll. Lib. Arts, Toyama Univ. 8: 51–86.
Kellman, M.C. (1970) The viable seed content of some forest soil in coastal British Columbia. Can. J. Bot. 48: 1383–1385.
Kellman, M. (1974) Preliminary seed budgets for two plant communities in coastal British Columbia. J. Biogeogr. 1: 123–133.
Livingston, R.B. and Allessio, M.L. (1968) Buried viable seed in successional field and forest stands, Harvard Forest, Massachusetts. Bull. Torrey Bot. Club 95: 58–69.
Luftensteiner, H.W. (1979) The eco-sociological value of dispersal spectra of two plant communities. Vegetatio 41: 61–67.
Marks, P.L. (1974) The role of pin cherry (*Prunus pensylvanica* L.) in the maintenance of stability in northern hardwood ecosystems. Ecol. Monogr. 44: 73–88.
Marquis, D.A. (1975) Seed storage and germination under northern hardwood forests. Can. J. For. Res. 5: 478–484.
Miyawaki, A. ed. (1983) Vegetation of Japan. Vol. 4. Chugoku (Jap. with Germ. summ.). Tokyo: Shibundo 540pp.
Moore, J.M. and Wein, R.W. (1977) Viable seed populations by soil depth and potential site recolonization after disturbance. Can. J. Bot. 55: 2408–2412.
Nakagoshi, N. (1980a) Phenology of forest plants on the Hiba Mountains in southwestern Honshu, Japan (Jap. with Engl. summ.). Hikobia 8: 399–415.
Nakagoshi, N. (1980b) Phenology of forest communities on the Hiba Mountains in southwestern Honshu, Japan. Bull. Biol. Soc. Hiroshima Univ. 46: 12–18.
Nakagoshi, N. (1981a) Notes on the buried viable seeds in soils of forest communities on Mt. Futatabi, Kobe (Jap. with Engl. summ.). Studies on Vegetation and Soil of the Permanent Nature Reserve Area in Mt. Futatabi, Kobe. Rep. 2nd Investig. Kobe. pp. 69–94.
Nakagoshi, N. (1981b) A study of the herbaceous plant communities on the Hiba Mountains in southwestern Honshu, Japan (Jap. with Engl. summ.). Bull. Biol. Soc. Hiroshima Univ. 47: 5–14.
Nakagoshi, N. (1981c) Buried seeds in forest soils (Japanese). In Numata, M. (Ed.), Science of Seeds, Ecological Aspects, pp. 101–124. Tokyo: Kenseisha.
Nakagoshi, N. (1984a) Ecological studies on the buried viable seed population in soil of the forest communities in Miyajima Island, southwestern Japan II. Hikobia 9: 109–122.
Nakagoshi, N. (1984b) Buried viable seed populations in forest communities on the Hiba mountains, southwestern Japan. J. Sci. Hiroshima Univ., Ser. B, Div. 2, 19: 1–56.
Nakagoshi, N. and Soga, S. (1981) A study of the alpine vegetation on the Shirouma

Mountains in central Japan, with special reference to seed ecology (Jap. with Engl. summary). Hikobia Suppl. 1: 341–358.

Nakagoshi, N. and Suzuki, H. (1977) Ecological studies on the buried viable seed population in soil of the forest communities in Miyajima Island, southwestern Japan (Jap. with Engl. summ.). Hikobia 8: 180–192.

Numata, M. and Yoda, K. (1957) The community structure and succession of artificial grasslands I (Jap. with Engl. summ.). J. Jap. Soc. Herb. Crops and Grassl. Farm. 3: 4–11.

Oosting, H.J. and Humphreys, M.E. (1940) Buried viable seeds in a successional series of old field and forest soils. Bull. Torrey Bot. Club 67: 253–273.

Rabotnov, T.A. (1969) On coenopopulations of perennial herbaceous plants in natural coenoses. Vegetatio 19: 87–95.

Rabotnov, T.A. (1981) Viable seeds in composition of coenotic populations as indicator of plant species strategy of life (Russian with Engl. summ.). Bjul. mosk. Obshck. Ispyt. Prir. Otd. biol. 86: 68–78.

Raunkiaer, C. (1934) The Life Forms of Plants and Statistical Geography. Oxford: Clarendon 632pp.

Ridley, H.N. (1930) The Dispersal of Plants throughout the World. Ashford: L. Reeve. 744pp.

Salisbury, E.J. (1942) The Reproductive Capacity of Plants. London: Bell 244pp.

Sasaki, Y. (1970) Versuch zur systematischen und geographischen Gliederung der japanischen Buchenwaldgesellschaften. Vegetatio 20: 214–249.

Sørensen, T.A. (1948) A method of establishing groups of equal amplitude in plant sociology based on similarity of species content and its application to analyses of the vegetation on Danish commons. Biol. Skr. K. danske Vidensk. Selsk. 5(4): 1–34.

Strickler, G.S. and Edgerton, P.J. (1976) Emergent seedlings from coniferous litter and soil in eastern Oregon. Ecology 57: 801–807.

Thompson, K. (1978) The occurrence of buried viable seeds in relation to environmental gradients. J. Biogeogr. 5: 425–430.

Thompson, K. and Grime, J.P. (1979) Seasonal variation in the seed banks of herbaceous species in ten contrasting habitats. J. Ecol. 67: 893–921.

Watt, A.S. (1947) Pattern and process in the plant community. J. Ecol. 35: 1–22.

Whipple, S.A. (1978) The relationship of buried, germinating seeds to vegetation in an old-growth Colorado subalpine forest. Can. J. Bot. 56: 1505–1509.

White, P.S. (1979) Pattern, process, and natural disturbance in vegetation. Bot. Rev. 45: 229–299.

PATTERNS OF FLOWERING PHENOLOGY IN COMMUNITIES, POPULATIONS, INDIVIDUALS, AND SINGLE FLOWERS

RICHARD B. PRIMACK

ABSTRACT

The flowering phenology of vegetation is examined at four successively finer levels of organization. First, communities are contrasted for their gross differences in flowering patterns. The difficulties of singling out the main environmental factor controlling these patterns is pointed out. Second, the theories which attempt to explain the flowering sequence of each species within a community (competition for pollinators, correlations between flowering and fruiting times, and physiological ecology) are reviewed in relation to methods for testing the validity of these theories. Third, the extent of individual variation in flowering time within species is considered, in order to determine the potential for natural selection to adjust the flowering time of a population. And fourth, the longevity of individual flowers is reviewed, since the individual flower is the basic unit of all flowering patterns. The review emphasizes the need to develop theories which can be statistically tested using data from either tagged individual plants in natural conditions or experimentally manipulated individuals.

INTRODUCTION

One of the most striking aspects of natural history is the pattern of flowering, leafing out, and fruiting which occurs on annual cycles in most parts of the world. These changes in the appearance of vegetation are obviously closely associated with annual changes in the weather. Even in wet tropical areas, annual fluctuations in rainfall serve to trigger bursts of flowering activity. The cycle of change in vegetation is a good indicator of the growth of agricultural food crops. For thousands of years, man has been making observations on these changes, in order to know when to plant seeds, and when to do other agricultural jobs in which timing is critical. Presently, a major effort is being made in evolutionary ecology to describe these phenological patterns in a precise way and to determine the causes of these patterns in relation to animal pollinators and fruit

White, J. (ed.), The Population Structure of Vegetation.
 ISBN 978-94-010-8927-2

dispersers. Such an effort represents the fulfilment of a human effort lasting thousands of years, for man has always wondered at the cycles of change, of birth and death, which are associated with the seasons. This fascination with the patterns of change has been one of the major underlying themes of religion and art.

The purpose of this paper will be to examine the flowering phenology of vegetation at successively finer levels of organization. First, communities will be contrasted for gross differences in their patterns of flowering. Second, the sequence of flowering of species within communities will be examined. Third, the extent of variation in flowering time within species will be considered, in order to determine the potential for natural selection to adjust the flowering time of a population. And fourth, the longevity of individual flowers will be reviewed, since the individual flower is the basic unit of all flowering patterns.

This review will be selective, since there is now available a large volume of indirectly related literature on phenology, in particular on plant physiological ecology and detailed records of plant flowering behavior. In addition, Bawa (1983) has recently reviewed the subject of flowering phenology in tropical plants, with a particular emphasis on breeding systems and frequency, timing, and duration of flowering; Rathcke (1983, 1984) and Rathcke and Lacey (in press) have reviewed models of species patterns in flowering phenology. For recent reviews of pollination ecology, readers are directed to the works of Kevan and Baker (1983) and Jones and Little (1983). This review will place emphasis on standardizing statistics, outlining methodology, and developing rigorous experiments. Such an emphasis is needed since much of the current discussion in flowering phenology suffers from accepting casual observations or uncritical observations as tests of hypotheses.

COMMUNITY DIFFERENCES

Communities differ in the patterns of flowering phenology of their constituent species. The relative role of individual selection factors, such as weather, light for photosynthesis, water availability, pollinator availability, the abundance of herbivorous insects and fruit disperser availability, cannot usually be easily distinguished. Since these selective factors often vary together, it is difficult to distinguish single selective factors which are controlling differences between the flowering phenology patterns of communities. Community patterns will only be treated briefly here, since this subject has been extensively discussed elsewhere (Robertson 1924, Koelmeyer 1959, Janzen 1967, 1974, Daubenmire 1972, Medway 1972, Lieth 1974, Frankie and Baker 1974, Opler et al. 1976, 1980, Heithaus 1974, Stiles 1977, Heinrich 1976, Putz 1979, Parrish and Bazzaz 1979, Arroyo et al. 1981, and others). Many of these studies have used tagged individual plants rather than quadrats or transects. This procedural rigor represents an important criterion on which the results of each study should be evaluated.

Examples of generalizations coming out of this type of research are the following:

(1) Eastern North-American woodland herbs bloom in the spring before the forest canopy is closed in by leaves, in contrast with disturbed habitats in which flowering is distributed over the growing season (Robertson 1895, 1924, Heinrich 1976).

(2) The same spring-flowering species flower earlier in the Southeastern United States than in the Northeastern United States (Robertson 1924).

(3) Major studies using tagged trees have compared the flowering phenology of trees, shrubs, and treelets in tropical wet and dry forests in the Costa Rican lowlands (Frankie 1974, Opler et al. 1976, 1980). Seasonality of flowering in the wet forest was much greater for trees than shrubs. A pronounced flowering season was apparent in the dry forest but much less obvious in the wet forest. In the dry forest, trees tend to have one annual period of flowering in the dry season, in contrast with shrubs which often have two episodes of flowering per year and flower during the wet season.

(4) Phenological studies were undertaken in the Andes of central Chile along a transect from 2000 m in the subandean scrub through a cushion plant zone to the limits of the subalpine zone at 3600 m (Arroyo et al. 1981). Flowering in the subandean scrub reaches a peak in late December during the time of maximum temperature, but peak flowering in the subalpine does not come until late February, after temperatures are declining. The average length of flowering per species increases with altitude. The changes in phenological patterns with altitude are associated with a decrease in the number of specialized bees with altitude and a decrease in the visitation rate to the flowers by insects (Arroyo et al. 1982, in press).

SPECIES PATTERNS WITHIN COMMUNITIES

The phenological behavior of each population is most effectively described by observations of marked individual plants constituting the population (as in Opler et al. 1980). The population description consists of the number of flowers open per day over the flowering season. The main consideration when combining the data from individual plants is deciding whether to simply combine the raw data, giving the largest individuals in the population the greatest importance in the description of the population, or to normalize the data in some way to treat all individuals as contributing equally to the population description. These data can be used to contrast species within the community for the date of onset of flowering, date of finishing of flowering, the dates of maximum and modal flowering, the duration of flowering, and the standard deviation, the skewness, and the kurtosis of flowering patterns. Using such detailed analyses, Thomson (1980) has calculated skewness for wildflowers in the Rocky Mountains, and has found flowering phenologies to be frequently skewed to the right.

Detailed studies of flowering at the community, family, and genus level frequently show a sequence of flowering of successive species (Heithaus

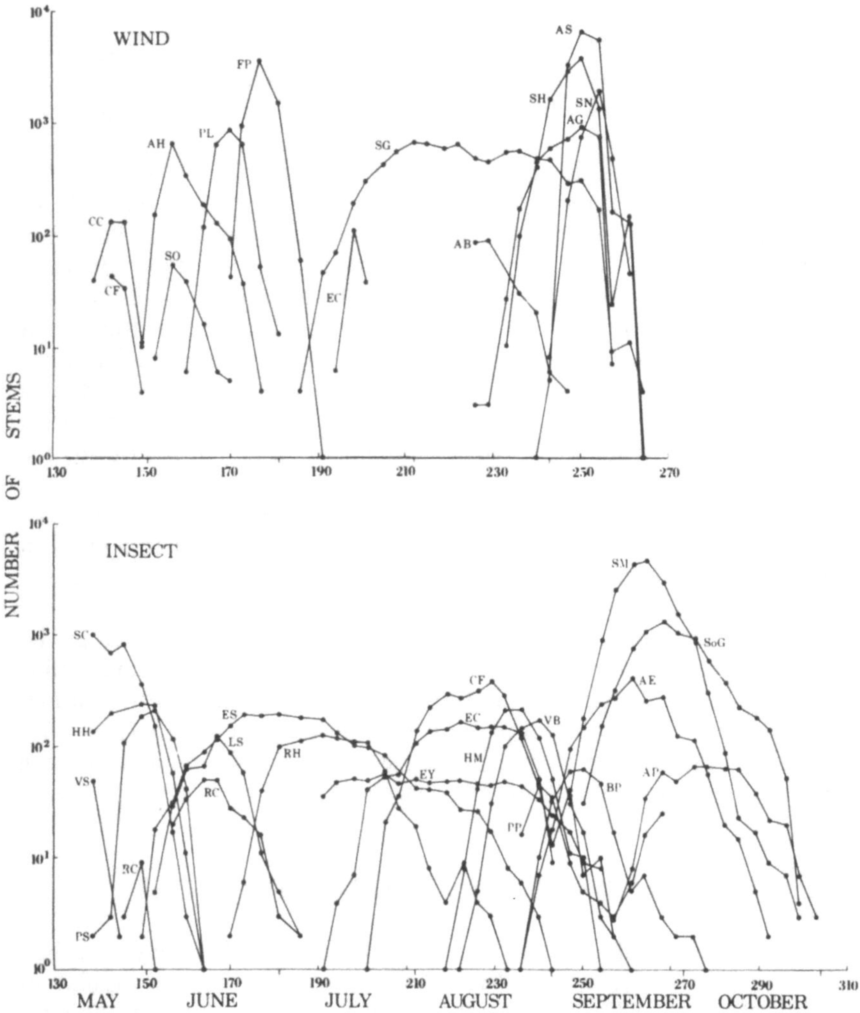

Fig. 1. Phenological curves for wind- and insect-pollinated prairie plants (from Rabinowitz et al. 1981). The number of flowering stems in the 800-m transect is plotted. For clarity, only species with a maximum of >40 stems observed on a single day are shown. Logarithmic transformation makes the generally bell-shaped curves appear more abrupt and flat topped. Species abbreviations are the initial letters of the generic and specific names, which are given in the original article.

1974, Pojar 1974, Frankie 1975, Reader 1975a, b, Stiles 1975, 1977, 1978, Heinrich 1976). These patterns often appear quite dramatic graphically (Fig. 1). This order of flowering seems to require an explanation of the pattern. Reasons are sought to explain why a community has such a regular sequence of flowering in its individual species. The following arguments have been presented to explain the sequential flowering of species within communities:

(a) Competition for pollinators
(b) Correlations of flowering and fruiting phenologies
(c) Physiological explanations.

Competition for pollinators

The flowering of plants in succession over the growing season has been explained by Robertson (1895, 1924), Mosquin (1971), Pojar (1974), Reader (1975a, b), Heinrich (1975, 1976), Zimmerman (1980), and others on the basis of competition for pollinators. By staggering the flowering times, it is assumed that competition for pollinators is reduced. Species avoiding competition for pollinators by flowering alone will have increased seed production. As an example of such a study, Parrish and Bazzaz (1979) compared the flowering phenologies of four plant communities, a prairie, winter annual weeds, spring annual weeds, and successional perennial weeds using niche widths and niche overlaps calculated from the equations of species packing theory. While the concept of this study is quite original, the small numbers of species in the communities (5, 3, 3, and 21) make the conclusions of this study not convincing. Also, the use of species packing theory has been seriously questioned due to its many unrealistic assumptions (Heck 1976). In particular, the impact of bimodal distributions of flowering, as is found in many communities, skewed flowering of individual species, and the flush of flowering in the spring, make species packing theory as currently constituted of little value in analyzing flowering phenology.

Attempts have been made to provide valid statistical tests to determine whether a sequence of species is flowering in a statistically regular manner that can be distinguished from a random pattern (Poole and Rathcke 1979, Cole 1980). The analysis by Poole and Rathcke (1979) showed that the well-known example of sequential flowering in Costa Rican hummingbird-pollinated plants (Stiles 1975, 1977, 1978) could not be distinguished from a random distribution. In his reply, Stiles (1979) pointed out that the Poole–Rathcke model assumes a continuous growing and flowering season, which is not valid in Costa Rica, nor probably any temperate or tropical region. Stiles (1979) felt that his example was still valid, based on his extensive knowledge of the natural history, and that the statistical test as presently constituted was not flexible enough to test his example. While his criticism seems valid, one can only wonder how to develop a model that is complex enough to handle any natural plant community. Cole (1981) follows Stiles' (1979) suggestions by providing a statistical test of the data by breaking the year into a wet and dry season, which gives quite different results than Poole and Rathcke (1979). Cole (1981) points out that there are many difficulties constructing analyses of this type of data.

An excellent recent paper presents a comparison of the phenological properties of wind and insect pollinated plants occurring on the same prairie (Rabinowitz et al. 1981). The assumption of this study is that if the animal-pollinated species are competing for pollinators, the flowering phenologies of these species should show less overlap than the phenologies of wind-pollinated species. Wind-pollinated species are expected to flower at random, without regard to the behaviour of other wind-pollinated species. The results show that both wind- and animal-pollinated species are randomly distributed over the flowering season (Figure 1). Also, both groups of species have similar amounts of overlap between groups of randomly selected species. So there is no evidence that plant species are competing for pollinators on a community basis. Rabinowitz et al. (1981) point out that the analysis is not totally satisfactory for three reasons. First, wind-pollinated species may not flower randomly: there may be selection to flower when pollen of other species will not contaminate the stigmas, which would cause regular patterns of flowering. Second, community-wide studies such as this consider plants pollinated by all types of pollinators. The theories of plant competition for pollinators have been based on groups of plants utilizing the same large pollinators, such as bumblebees or hummingbirds. Consequently, by studying a community pollinated by large bees, small bees, all types of flies, butterflies, and other insects, there is no reason to expect any distinct phenological patterns. Third, the wind-pollinated and insect-pollinated species are not truly comparable since the former group is mainly monocots and the latter group is mainly dicots. Additional fourth and fifth arguments could be added to their list. Fourth, the distribution of species is considered along the untransformed dimension of time. However, as a simple inspection of their flowering phenology data shows, species flowering is concentrated in two distinct periods: late May to early June and late August to early September. A fifth problem of searching for community-wide phenological patterns is that these present North American communities are probably not co-evolved groups of species, but are a recent assemblage of species that have colonized the area from diverse source areas following the retreat of the glaciers (Davis 1976). Further, the activity of man has drastically altered the ecology of the prairies, changing the relationships of pollinators and plants. These criticisms of the fine project by Rabinowitz et al. (1981), which also could be made for other studies of heterogeneous communities, illustrate the limited usefulness of community-based phenological studies.

A profitable avenue of research is a careful study of pairs or groups of species which share the same pollinators and flower sequentially, as determined by statistical tests (Estabrook et al. 1982), to distinguish that two phenological patterns are distinct. As an example, in a study of hummingbird-pollinated plants in tropical montane vegetation, there was a sequence of species flowering over the year (Feinsinger 1978). However, the flowers of rare plant species were visited actively by hummingbirds, even when a dominant plant species with abundant nectar was flowering. Species of hummingbirds shifted between plant species over the year, depending on nectar availability, but the flowers of all plant species were visited. Consequently competition for pollinators does not seem to be the explanation for the observed phenological patterns.

A well-documented case of competition for pollinators is the study of two sequentially flowering Colorado wild-flowers, *Delphinium nelsoni* and

Impomopsis aggregata, which compete for pollination by hummingbirds (Waser 1978a, b). Flowers that open during the period of overlap have reduced levels of seed set in comparison with flowers that open during the period of non-overlap. Reduced seed set is apparently the result of inter-specific pollen transfer by the hummingbirds. Seed set was reduced experimentally by performing inter-specific hand-pollinations on caged plants. As an interesting addition, it was also found that while the species are reducing each other's seed set during the period of overlap, they are also acting mutualistically to attract hummingbird pollinators (Waser and Real 1979). The weakness of these studies is that the data were all gathered at the population level using census plots, rather than using tagged individuals, so that the results cannot be considered in terms of differential fitness of individual plants within the population.

Mutualistic interactions by which species pairs receive more visits when they flower together than when they flower alone have been documented in such species as hawkweeds (Thomson 1978), understory herbs in the genus *Costus* (Schemske 1981), and Rocky Mountain meadow wild-flowers (Thomson 1980). These studies suggest that models of the disadvantage of rare species and rare morphs in pollination systems may have to be re-evaluated.

The studies reviewed here advance their arguments primarily on a species or community basis, but natural selection acts on variation among individuals within a population. To be a truly evolutionary argument, this explanation should be re-stated and tested in the field on an individual basis. The hypothesis would be that reduced seed production occurs because pollinators are not making sufficient visits to individual plants flowering during the period of overlap. During the period of overlap, the pollinators will be making visits between the species, giving plants of neither species adequate pollination. Selection may be particularly against the earliest individuals in the population due to pollinators not yet having discovered the rare plants in flower of this newly flowering species. Theoretical models of the fitness of rare plants have shown how selection can operate against these plants (Levin and Anderson 1970, Levin 1972, Bobisud and Neuhaus 1975). However the field data showing that pollinators may be a limiting factor in seed production are still limited (Bierzychudek 1981).

Correlations of flowering phenology and fruiting phenology

Timing of fruit maturation is a character which is presumably under strong selection pressure. Distinct patterns of fruit maturation have been determined for bird-dispersed fruit of the eastern North-American deciduous forest (Thompson and Willson 1978, 1979, Stiles 1980). These patterns may be determined by such factors as avoidance of mammalian seed predators and attack by micro-organisms, the abundance of summer and winter resident bird populations, and the timing of the fall bird migration. Both time of fruiting and the duration of fruiting may be part of this 'strategy' (Howe and Estabrook 1977). It seems certain that for many species, selection for timing of fruiting is developmentally correlated with the timing of flowering, since a fixed period of time may be required for the fruit to mature. Selection for a shift in the time of fruit maturation would be expected to have an immediate corresponding effect on flowering time. Consequently, flowering times may actually be explained by selection for timing of fruit maturation rather than as a separate character (Snow 1966, Janzen 1967, Paltridge and Denholm 1974, McKey 1975).

A test was made of this theory using the flora of Britain (Ross-Craig 1948–1973, Clapham et al. 1952) and the flora of South Florida (Tomlinson 1980). For 115 insect-pollinated species chosen by a regular sampling method from the flora of Britain, it was found that there was a highly significant negative non-parametric correlation ($r = -0.41$, $p < 0.01$) between the volume of the fruit (as calculated from length and width measurements) and the date of flowering (Primack, unpublished); flowering occurs earlier in the year for species with large fruits, and flowering occurs later in the year for species with small fruit (Fig. 2). Since the size of the fruit is certainly related to the time required for the fruit to mature, it seems certain that the type and volume of fruit a plant produces represents a major controlling factor in the timing of flowering.

For the tree flora of South Florida, data were extracted for 40 species on fruit volume and the number of months the fruit takes to mature (Primack, unpublished). A strong non-parametric correlation ($r = 0.47$, $p < 0.001$) exists between these two variables, supporting the view that the type and size of the fruit ties together the phenology of flowering with the phenology of fruiting.

An interesting and peculiar situation exists for the red mangrove, *Rhizophora mangle* (Rhizophoraceae). Flowers are produced and open throughout the year, but fruits are initiated only during a period of a few months

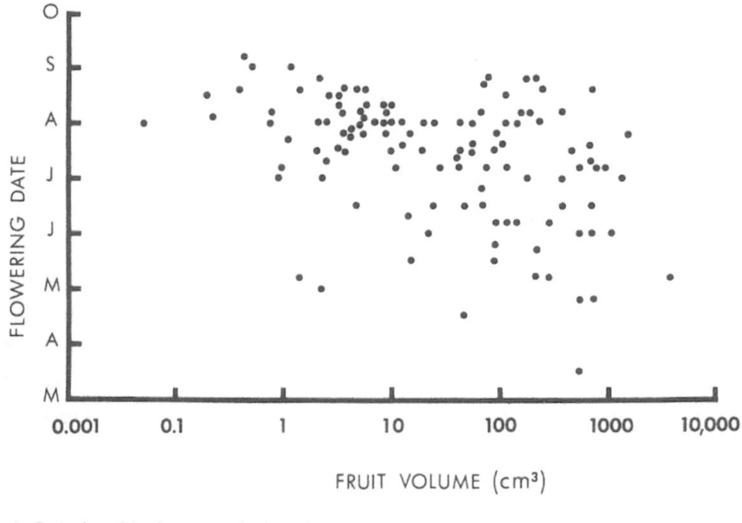

Fig. 2. Relationship between fruit volume and the flowering date for species in the British flora (Primack, unpublished). See text for an explanation of the methods.

(Gill and Thomlinson 1971). The function of the flowers produced at other times of the year is not known.

Physiological explanations

Limitations of light, water, and temperature may restrict flowering of each species to particular times of the year (see Lieth 1974, Heinrich 1976, Kawano et al. 1982 for reviews). The period at which each species flowers may be determined by peculiarities of its physiology, morphology, and natural history. The period of these limitations may extend far into the growing season, if flower buds can only be formed once the conditions for growing begin. Many plant species have dormant flower buds, which allow the species to respond rapidly to physical conditions suitable for flowering and fruit formation. In many temperate species, flower buds are dormant in the winter and begin to develop only upon the return of warm weather in the spring when the occurrence of frosts is less frequent (Lindsey and Newman 1956, Risser and Cottam 1967, Reader 1975b). The effects of temperature on flowering time can often be described precisely (Lindsey and Newman 1956).

Rainfall after a dry period is the triggering mechanism in the development of dormant flower buds and new flowers of many tropical species and desert species in areas with more or less pronounced dry seasons (Holttum 1973, Beatley 1974, Opler et al. 1976). This initiation of flowering following rainfall is an important factor for ensuring synchronization of flowering within populations (Opler et al. 1976). The control mechanism has been cleverly taken advantage of by Augspurger (1981) to induce selected shrubs of *Hybanthus prunifolius* (Violaceae) to flower out of synchrony with the natural population. This was accomplished by watering plants during the dry season. These artificially flowering plants had lower seed set and higher levels of fruit predation than plants in the control population.

Erythrina poeppigiana (Leguminosae), a tropical tree, provides an example of how increased annual variation in weather can increase flowering synchrony within the population. This tree was studied on a transect from humid to seasonally dry climatic conditions (Borchert 1980). The tree varied from being evergreen and asynchronously flowering in humid conditions, to being deciduous and synchronously flowering in the seasonally dry conditions.

The availability of sufficient light to maintain active rates of photosynthesis may restrict the flowering and growing season of wildflowers of deciduous woodlands. This closure of the forest leaf canopy is a regular annual phenomenon (Schemske et al. 1978, Kawano et al. 1982) and is linked to abortion of fertilized ovules in late-opening flowers (Schemske 1977). Species differ in their ability to maintain photosynthetic rates necessary for growth once the canopy has closed; species may be divided into shade-tolerant and shade-intolerant species.

Adverse weather conditions may affect the ability of pollinators to fly and visit flowers over the course of the growing season. Conditions may often be too cold for the pollinators of many spring wild-flowers (Heinrich 1975, Schemske 1977, Schemske et al. 1978). Cold weather conditions early and late in the growing season and also in the middle of the summer at higher elevations cause reduced visitation rates to flowers in the Andes (Fig. 3) (Arroyo et al., in press).

Many species show patterns of reproduction on a several-year basis. This phenomenon is found in both temperate regions in many perennial herbs (Tamm 1956, 1972) and in families such as the Fagaceae in temperate woodlands, the Dipterocarpaceae in tropical rain forests, and the Gramineae in montane grasslands. Explanations for this phenomenon of mass fruiting at intervals have included the need to build up energetic reserves necessary for fruit formation which are released by environmental triggers for flowering, unusual weather conditions causing the abortion of reproductive structures, the requirement of populations to flower at high density in order to insure successful pollination, and selection for mass fruiting to satiate density-dependent seed predators (Salisbury 1942, Janzen 1967, 1970, 1974).

Certain tropical species may show several episodes of flowering and fruiting over one season (Opler et al. 1980, Bullock and Bawa 1981). The selective value of episodic flowering has been interpreted by Gentry (1974) and Bawa and Webb (1984) as a bet-hedging strategy to ensure at least occasional small reproductive success. Models of community flowering phenology should consider also these tropical species which flower either several times per year or only once in several years.

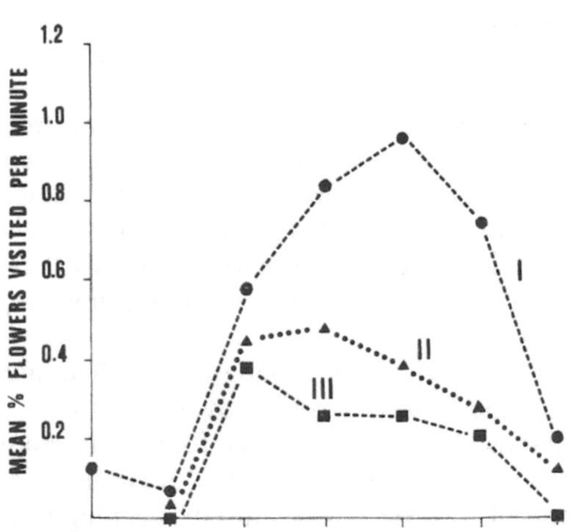

Fig. 3. Mean monthly visitation rates (mean % of flowers visited per minute) at three altitudinal levels in the Andean zone in Central Chile. I = 2200–2600 m; II = 2700–3100 m; III = 3200–3600 m (Arroyo et al., in press).

VARIATION AMONG INDIVIDUAL PLANTS

Timing of flowering

An individual plant produces flowers which open on successive days. A complete census of this plant involves counting every flower open on the entire plant on every day that the plant is in flower (Fig. 4). Such a census yields data on the date of first flowering, the date of last flowering, and the date on which the most flowers are open (date of maximum flowering). It is not readily apparent which of these statistics is the most appropriate for describing the flowering date of an individual, since each of these three statistics describes slightly different properties of the individual. The date of first flowering represents the date on which the plant first enters the reproductive population, while the date of maximum flowering represents the date on which the plant makes its greatest contribution to the phenology of the population. Of course, variation in these three dates are probably highly positively correlated among themselves, suggesting perhaps that using any one of the three statistics would be acceptable. However, until additional field studies are complete, this is probably premature. The censusing of flowers also allows the modal date of flowering to be calculated; this is the date on which exactly half of the flowers have been produced (assuming that flowers are short-lived and have a similar longevity). For a plant which produces a few flowers per day over many weeks or months, the day of modal flowering may be more significant in its natural history than the day of maximum flowering.

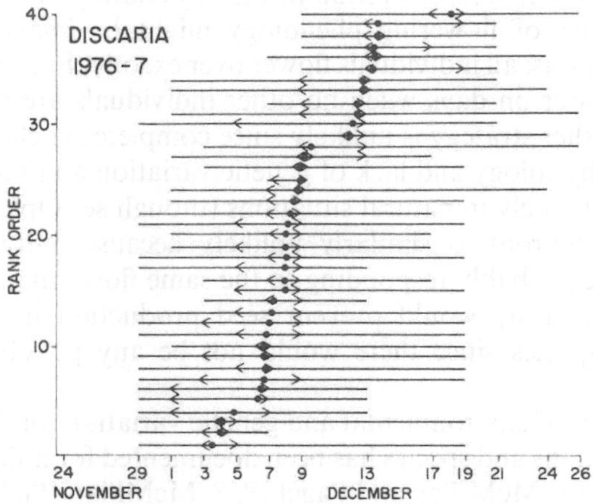

Fig. 4. Flowering dates for 40 *Discaria toumatou* (Rhamnaceae) shrubs from New Zealand in 1976–1977 (from Primack 1980). Dates on the axis are days on which the population was examined. Each horizontal line represents the total duration of flowering for a plant. The heavy dot is the day of maximum flowering. The brackets enclose the dates on which the number of flowers open on the plant was at least 50% of the number of flowers open on the maximum flowering date.

The duration of flowering may be an appropriate statistic for comparative purposes for species which begin and stop flowering abruptly, and have about the same number of flowers open per day; these are species with a platykurtic distribution of flower production. However, with species which have a slightly or strongly leptokurtic production of flowers, other statistics must be used for comparing individuals in a population; in these species, there is a relatively brief period of intense flowering which is preceded and (or) followed by long periods in which only a few flowers are open. Such a phenomenon was found in New Zealand shrubs (Primack 1980). Describing the entire flowering period of such a plant as its flowering duration is misleading since the great majority of flowers are only open during this brief phase of strong flowering. In a complete flower census of an individual plant, it is possible to calculate the mean, standard deviation, skewness, and kurtosis of flowering phenology, for that individual, using the methods described in statistics books, such as by Sokal and Rohlf (1969). However, this has not yet been done for an individual plant.

The statistics on duration of flowering of individuals can be used to describe the extent of overlap or synchrony in flowering time within populations (Primack 1980). The degree of overlap in flowering among individual plants in a population gives some measure of the degree of gene exchange possible within the population. Also, the degree of overlap in flowering within the population gives some measure of the ability of the plants in a population to interact with density-dependent pollinators and density-dependent flower and ovule predators. Using a simple model of overlap in flowering time between pairs of individual flowering shrubs, differences in flowering time were shown to have the potential of reducing gene exchange by up to 70% (Primack 1980). More complicated statistical models could consider the actual number of flowers two plants have open on the same days as well as overlap in the days both plants are in flower.

Two extremes of flowering phenology might be imagined to occur within populations: all individuals flower over exactly the same days vs. all individuals flower on days when no other individuals are in flower. The existence of either strategy is unlikely since complete synchrony implies a precision of physiology and lack of genetic variation and microsite variation which is unlikely in natural situations (though see Opler et al. 1976). Complete asynchrony is similarly unlikely because individuals in the population are probably responding to the same flowering cues. Further, complete asynchrony would prevent seed production in an obligately out-crossing species since there would not be any possibility for gene exchange.

The existence of environmental and genetic variation for flowering time within populations and species has been documented for numerous species (for examples, see McMillan and Pagel 1958, McMillan 1967, Harris 1970, McIntyre and Best 1978). Many species show changes in flowering patterns over geographical areas and even small microsites (Jackson 1966, Harris 1970, Hodgkinson and Quinn 1978). Fruit trees also show genetic variation in flowering times (Janick and Moore 1975). Recently, attention has focused on the flowering phenology of tropical trees. Evidence for stabilizing selection on flowering time has been provided for the tropical shrub *Hybanthus prunifolius*, as a result of density-dependent pollinators and seed predators (Augspurger 1979, 1981). Variation in phenology between sexes has been documented in such dioecious species as *Ficus* (Moraceae) (Valdeyron and Lloyd 1979), *Juniperus* and *Salix* (Falinski 1980), the tropical tree *Jacartia dolichaula* (Caricaceae) (Bullock and Bawa 1981), and the New Zealand herb *Gingidia decipiens* (Umbelliferae) (Webb 1976). In *Jacartia*, male trees have an extended and asynchronous flowering period while females bloom somewhat synchronously over a shorter period of time. These differences in the flowering phenology of male and female trees were interpreted in terms of sexual selection theory (Bullock and Bawa 1981).

Several possible hypotheses can be considered to explain the existence of genetic variation for flowering time within natural populations:

1. Yearly variation in weather results in selection for early flowering in some years and late flowering in other years.
2. Variation in flowering time might result from the higher fitness of early and late flowering individuals (Schemske 1977). This would occur if the plants which flower early and late in the season experienced greater gene flow of pollen due to a lower density of flowering plants in comparison with the time of peak flowering for the population.
3. Genetic variation in flowering time might exist because of annual fluctuations in the population size and time of peak density of pollen, ovule and fruit predators and fruit dispersal agents. These annual fluctuations in animal populations may favor early flowering plants in some years and late flowering plants in other years, leading to disruptive selection. The evidence to support this hypothesis does not exist. In fact, the experiment by Augspurger (1981) indicates that predators may select against asynchronous individuals.

An interesting consequence of variation in flowering phenology within populations is that the possibilities of gene exchange between all individuals within the population are greatly reduced (McMillan and Pagel 1958, Primack 1980, Somers and Grant 1981). Assortative mating occurs among those individuals that flower at the same time. Since variation in flowering time is often consistent in successive years, these patterns of assortative mating will also be repeated in successive years. Variation in flowering time will have to be considered, along with the tendency of nearest neighbors to mate (Levin and Kerster 1974), as a factor reducing panmixis in models of population genetics.

In order for selection to cause changes in the flowering time of a species the following conditions must exist:

1. There must be variation in flowering time among individuals within a population.
2. Some of this individual variation must be genetically based.
3. A selective agent, whether physical or biological in nature, must act on this variation in flowering time such that individual variation in seed production is related to variation in flowering time.

Within populations, individuals will also show variation in their tendancy to flower. The percentage of individuals flowering in the population will affect the density and total number of flowers in the population. Within populations small individuals have a lower probability of flowering than larger individuals (Hillman 1962, Primack 1979). In dioecious species, male plants may flower almost every year while female plants may flower only once every few years (Lloyd and Webb 1977, Bullock and Bawa 1981, Bawa et al. 1982).

Duration of flowering

Individual plants within a population can differ in the duration of flowering (Gentry 1974, Primack 1980, Bullock and Bawa 1981, Augspurger 1981). The duration of flowering is positively correlated among plants in successive years; plants flowering over a long period in one year tend to do so in subsequent years (Primack 1980). The duration of flowering of an individual plant is almost certainly related to the number of flowers per plant. For plants producing the same number of flowers, genetic variation may also exist in the degree to which flowers open simultaneously or sequentially. Selection on such genetic variation would serve to shift individual plants and ultimately the population between the two extremes of mass-flowering and sequential flowering. The subject of the variation of duration of flowering among plants is largely unexplored and may turn out to be a profitable area of inquiry.

The duration of flowering may have significant consequences for the relative importance of the breeding success via pollen and ovules for dichogamous plants. Plants which undergo changes in sexual expression during the course of one flowering season must flower over a long enough period of time for these changes to occur. For example, many Umbelliferae species show simultaneous protandry or protogyny of all individual flowers on a plant (Ponomarev 1960, Cruden and Herman-Parker 1977). In many monoecious species, plants may undergo dramatic changes in sexual expression during the flowering season as flowers of first one morph open, followed by flowers of the other morph (DeJong 1976, Bawa 1977, Primack and Lloyd 1980). Perhaps individuals of these species have a longer flowering duration than related non-dichogamous species. Further, timing and position effects of flowers within an inflorescence have been shown to be very important in determining which flowers develop into fruits (Stephenson 1979, 1980, 1981, Wyatt 1982). Again, individuals must have a sufficiently long flowering time for such effects to be important.

LONGEVITY OF INDIVIDUAL FLOWERS

The flowering phenology of an individual plant is a function of the schedule of production of flowers and the longevity of individual flowers. Similarly, the phenology of the population can be described as a function of the schedule of flower production for all plants and the longevity of individual flowers. Considerable variation exists in the longevity of individual flowers, from several minutes to months, and this is important in understanding pollination ecology as a dynamic process. The longevity of the flowers is a factor in determining how many visits each flower receives by pollinators. Also, flower longevity can influence the level of out-crossing and the overall floral display of the plant by influencing the number of flowers open at any one time on the plant. Despite the significance of flower longevity, this topic has only been considered briefly in the botanical literature, by Kerner von Marilaun (1902). In this classical work he makes the observation that species with long-lived flowers, such as orchids, produce relatively few flowers which require insects for pollination. These species have flowers which remain open until they are pollinated.

Variation in flower longevity can have a significant impact on phenological patterns and the floral display. Habitats can differ markedly in the longevity of flowers (Table 1, Primack 1985). Species with 1-day

Table 1. Flower longevity in days by habitat, calculated as the mean of individual species means (Primack, 1985). Separate values are presented for flowers and Compositae inflorescences.

	Flower longevity in days				
Habitat	Mean	S.D.	N	Range	Source
Tropical forests					
Dipterocarp forest (Malaysia)	1.0	0.0	6	1–1	Chan and Appanah (1980)
Tropical dry forest (Costa Rica)	1.1	0.3	38	1–2	Bawa (unpublished)
Tropical rain forest (Costa Rica)	1.3	0.6	37	1–3	Bawa (unpublished)
Mangrove (Australia)	3.3	2.2	8	2–8	Primack et al. (1981)
Montane					
Chilean subandean					
Flowers	4.2	3.0	24	1–15	Primack (unpublished)
Compositae	6.9	2.7	5	3–10	
Chilean subalpine					
Flowers	6.9	2.9	9	4–12	Arroyo et al. (1981)
Compositae	13.6	5.9	3	8–20	
New Zealand grasslands					
Flowers	5.9	4.2	24	1–19	Primack (1983)
Compositae	9.2	4.3	3	5–13	
New Zealand subalpine					
Flowers	7.8	4.0	16	3–15	Primack (1983)
Compositae	12.9	2.3	11	10–15	
Temperate forest					
Spring (Illinois)	6.9	3.9	7	2–12	Schemske et al. (1978)
Early summer (Virginia)	5.7	3.0	32	1–14	Primack (unpublished)
Late summer (Virginia)	2.5	1.4	34	1–6	Primack (unpublished)

flowers predominate in Lowland Costa Rican Tropical Rain Forest (27 of 37 species), the Lowland Malayan Dipterocarp Forest (6 of 6 species), and the Lowland Tropical Costa Rican Dry Forest (32 of 38 species). The species with the longest lasting flowers in these habitats had flowers which lasted 3 days only. This short flower longevity is not simply a characteristic of tropical trees, because none of the eight Australian Mangrove species had flowers which lasted only 1 day. The average longevity for mangrove species is 3.3 days. In contrast, in the North American Temperate Forest environment, species flowering in the spring and early summer have flowers which last on the average 6.9 and 5.7 days respectively. This average is more than twice as long as the flower longevity of 2.5 days for species flowering in the late summer. In the spring and early summer, five species have flower longevities greater than 10 days: *Dicentra canadensis* (12), *D. cucullaria* (11), *Habenaria orbiculatus* (12.7 days), *Monotropa uniflora* (14), and *Lysimachia quadrifolia* (10.1). In contrast, none of the species flowering in the late summer have flower longevities greater than 9 days. Two altitudinal comparisons can be made with the data. Species of the high altitude Chilean Subalpine site have longer lasting flowers (6.9 days) than the adjacent lower Subandean Scrub site which has an average flower longevity of 4.2 days. All of the species at the Subalpine site have flowers which last at least 4 days, while at the Subandean Scrub site 13 of the 24 species have flower longevities less than 4 days. Further, the higher altitude New Zealand Subalpine site has greater values than the lower Montane Grasslands site for the longevity of individual flowers (7.8 vs. 5.9). The point should be clear that flower longevity is a character which is highly variable within habitats and is dramatically variable among habitats. Flower longevity is a variable which is rarely included in flowering phenology studies despite its basic importance.

An impression of many botanists is that flower senescence follows pollination in the field, in the same way that flower senescence follows hand-pollination in the greenhouse. This seems reasonable since the senescence of the flower would end the transpirational water loss and metabolic costs of maintaining the fresh flower. Nectar production is a further additional drain on the reserves of the plant. In addition, when the flower undergoes the post-pollination changes, pollinators will be less likely to visit the flower and may be directed to other open flowers on the same plant. Flowers might be adapted to wait until sufficient pollen has been transferred to the stigma, after which the flower will senesce. By this argument, when pollinators are abundant, flower longevity should be shorter than when pollinators are scarce. However, despite this impression that flowers *should* die following pollination, there is in fact little evidence to suggest that this occurs under field conditions. The influence of pollinator activity on flower longevity and consequent flowering phenology needs to be investigated.

In a study of the whorled loosestrife, *Lysimachia quadrifolia* (Primulaceae) (Primack, unpublished), a self-incompatible perennial herb, the most distinct pattern evident is the continuous decline in flower longevity from an average of 12 days early in the season to 7 days late in the season

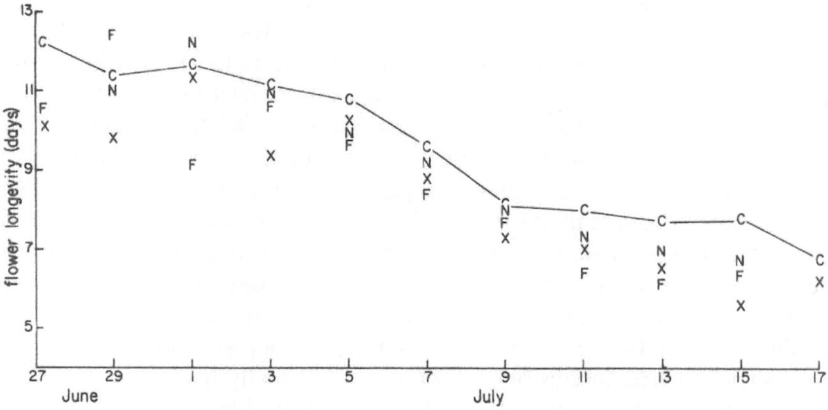

Fig. 5. Mean longevity of individual flowers of *Lysimachia quadrifolia* (Primulaceae) opening on successive days in the vicinity of Mt. Lake Biological Station in Pembroke, Virginia (Primack, unpublished). All flowers were on wild plants and exposed to normal insect visitors. Newly opened flowers were assigned at random to be pollinated by hand using pollen from the nearest neighboring plants (N) and pollen from plants more than 10 m away (F or X). Flowers pollinated by hand using pollen from more than 10 m away were separated into those that went on to become fruits (F) and those that did not form fruits (X). Flowers exposed to normal visits only are also shown (C). Each value represents a mean of between 5 and 25 flowers.

(Fig. 5). Flowering longevity is not a fixed characteristic of this species, but varies strongly over the course of the season. It is not known at present what factors cause this variation from day to day, but increases in the daily temperatures, changes in pollinator abundance, and decrease in daily humidity may be important factors. Whorled loosestrife flowers receiving hand cross-pollinations plus normal insect visits had average flower longevities of 0.5 days less than flowers receiving only normal insect visits (Fig. 5). The large sample size of the experiment allows these differences to be detected, though these differences are far less than what was expected. However, even though hand cross-pollination decreases flower longevity in this species, there is no difference in flower longevity between those hand-pollinated flowers which went on to become fruit and those hand-pollinated flowers which did not develop into fruit. Pollination has this slight effect on flower longevity, while other factors determine whether a pollinated flower will develop into a fruit.

CONCLUSIONS

Large differences in communities exist in the pattern of flowering phenology. Such differences primarily occur due to cyclical yearly changes in the weather which determine the time plants can grow and animals can be active. Within communities, the sequence of flowering of species will be determined by a complex interaction of factors such as time of flower-bud formation, weather suitable for pollinator flight, the seasonal energy budget of the plant, the build-up of populations of seed and ovule predators, competition for pollinators with other species, and the correlation between flowering time and fruiting time. Attempts to analyze flowering phenology at the community level in terms of niche partitioning of pollinator resource will probably not be successful due to the complexity of factors which affect flowering phenology. Further, plants in a diverse community pollinated by different types of pollinators are not competing for the same resource in any case. All of the factors mentioned above, and probably others as well, combine with pollinator availability to exert a selective force on variation in the flowering phenology of individual plants within a population and populations within a community.

The mathematical pecularities of phenology, such as bimodal flowering peaks and skewed species phenologies, can not really be dealt with in the present models of communities. The most profitable approach towards analyzing phenological patterns will probably come from detailed studies of populations of several sequentially-flowering plant species sharing the same pollinators. Such studies should determine if individuals flowering during the period of species overlap have reduced seed production. Further studies of the mathematical characteristics of phenology curves need to focus on the individual plant, in addition to the population, because selection operates on variation among individual plants in characters correlated with seed production and fitness.

A profitable means to investigate the selective forces acting on flowering phenology is to manipulate flowering phenology experimentally, as was done through artificial watering by Augspurger (1981). Various methods can be tried to induce plants to flower earlier or later than normal. For example, plants can be shaded with screens or given extra light with electric lights to alter the photoperiodic response. Plants can be heated with portable heaters or cooled with ice packs. With all these treatments, the investigator must try to insure that the treatment is affecting only flowering phenology and not other aspects of the biology of the plant, such as energy budgets. The reproductive success of individuals involved in these experiments should be measured in terms of pollination success (number of visits per flower per unit time; number of pollen grains deposited per stigma; number of pollen tubes per style), as well as seed production. Such partitioning of reproductive components is needed, since any plants which have been altered so much as to change their flowering time, would probably be expected to have a low seed production due to physiological imbalances. Experimental designs in which the manipulated plants are expected to have the highest reproductive success might be the most convincing.

The flowering phenology of the individual plant is determined by the production schedule of flowers and the longevity of individual flowers. Physiological mechanisms which control these two components of individual flowering phenology need to be investigated. In the field, we need to know how pollinator activity and the amount of pollen transferred affect the longevity of flowers, and the flower production, seed set, and subsequent survival of the plant. By this combination of investigating individual variation within population and physiological mechanisms controlling individuals, a comprehensive picture can be developed of the flowering phenology of a population and ultimately of an entire community.

ACKNOWLEDGEMENTS

The author wishes to thank Kamal Bawa, James White, David Inouye, David Lloyd, Eric Godley, Sally Zielinski, and Beverly Rathcke for valuable discussion and comments on the manuscript and Kamal Bawa for the use of unpublished data on flower longevity from Costa Rica.

REFERENCES

Arroyo, M.T.K., Armesto, J. and Primack, R. (in press) Community studies in pollination ecology in the high Andes of central Chile. II. Effects of temperature on visitation rates and pollination possibilities. Plant Systematics and Evolution.

Arroyo, M.T.K., Armesto, J.J. and Villagran, C. (1981) Plant phenological patterns in the high Andean Cordillera of central Chile. Journal of Ecology. 69: 205–223.

Arroyo, M.T.K., Primack, R.B. and Armesto, J.J. (1982) Community studies in pollination ecology in the high temperate Andes of central Chile. I. Pollination mechanisms and altitudinal variation. American Journal of Botany. 69: 82–97.

Augspurger, C.K. (1979) Mass-flowering of a tropical shrub (*Hybanthus prunifolius*): Influence on pollinator attraction and movement. Evolution. 34: 475–488.

Augspurger, C.K. (1981) Reproductive synchrony of a tropical shrub: Experimental studies on effects of pollinators and seed predators on *Hybanthus prunifolius* (Violaceae). Ecology. 62: 775–788.

Bawa, K.S. (1977) Reproductive biology of *Cupania guatemalensis* Radlk. (Sapindaceae). Evolution 31: 52–63.

Bawa, K.S. (1983) Patterns of flowering in tropical plants. In C.E. Jones and R.J. Little (Eds.). Handbook of Experimental Pollination Biology, pp. 394–410. N.Y.: Von Nostrand Reinhold Co.

Bawa, K.S., Keegan, C.R. and Voss, R.H. (1982) Sexual dimorphism in *Aralia nudicaulis* L. (Araliaceae). Evolution 36: 371–378.

Bawa, K.S. and Webb, C.J. (1984) Flower, fruit, and seed abortion in tropical forest trees: Implications for the evolution of paternal and maternal reproductive patterns. American Journal of Botany, 71: 736–751.

Beatley, J. (1974) Phenological events and their environmental triggers in Mojave Desert ecosystems. Ecology 55: 856–863.

Bierzychudek, P. (1981) Pollinator limitation of plant reproductive effort. American Naturalist 117: 838–846.

Bobisud, L.B. and Neuhaus, R.J. (1975) Pollinator constancy and survival of rare species. Oecologia. 21: 263–272.

Borchert, R. (1980) Phenology and ecophysiology of tropical trees: *Erythrina poeppigiana* O.F. Cook. Ecology 61: 1065–1074.

Bullock, S.H. and Bawa, K.S. (1981) Sexual dimorphism and the annual flowering pattern in *Jacaratia dolichaula* (D. Smith) Woodson (Caricaceae) in a Costa Rican rain forest. Ecology 62: 1494–1504.

Chan, H.T. and Appanah, S. (1980) Reproductive biology of some Malayan dipterocarps. I. Flowering biology. Malaysian Forester 43: 132–143.

Clapham, A.R., Tutin, T.G. and Warburg, E.F. (1952) Flora of the British Isles. University Press, Cambridge.

Cole, B.J. (1981) Overlap, regularity, and flowering phenologies. American Naturalist 117: 993–997.

Cruden, R.W. and Herman-Parker, S.M. (1977) Temporal dioecism: An alternative to dioecism. Evolution 31: 863–866.

Daubenmire, R. (1972) Phenology and other characteristics of tropical semi-deciduous forest in north-western Costa Rica. Journal of Ecology. 60: 147–170.
Davis, M.B. (1976) Pleistocene biogeography of temperate deciduous forest. Geoscience and Man. 8: 13–26.
DeJong, P.C. (1976) Flowering and sex expression in *Acer*. A biosystematic study. Meded. Landbouwhogesch. Wageningen. Ned. 76-2: 1–201.
Estabrook, G.F., Winsor, J.A., Stephenson, A.G. and Howe, H.F. (1982) When are two phenological patterns different? Botanical Gazette 143: 374–378.
Falinsk, J.B. (1980) Changes in the sex- and age-ratio in populations of pioneer dioecious woody species (*Juniperus, Populus, Salix*) in connection with the course of vegetation succession in abandoned farmlands. Ekologia Polska 28: 327–365.
Feinsinger, P. (1978) Ecological interactions between plants and hummingbirds in a successional tropical community. Ecological Monographs. 48: 269–287.
Frankie, G.W. (1975) Tropical forest phenology and pollinator-plant coevolution. In L.E. Gilbert and P.H. Raven (Eds.) Coevolution of Animals and Plants, pp. 192–209. Austin: Univ. of Texas Press.
Frankie, G.W. and Baker, H.G. (1974) The importance of pollinator behavior in the reproductive biology of tropical trees. Am. Inst. Biol. Univ. Nal. Auton. Mexico 45. Ser. Botanica. 1: 1–10.
Gentry, A.H. (1974) Flowering phenology and diversity in tropical Bignoniaceae. Biotropica 6: 64–68.
Gill, A.M. and Tomlinson, P.B. (1971) Studies on the growth of red mangrove (*Rhizophora mangle* L.). 3. Phenology of the shoot. Biotropica 3: 109–124.
Harris, W. (1970) Genecological aspects of flowering and vegetative reproduction in *Rumex acetosella* L. New Zealand Journal of Botany. 8: 99–113.
Heck, K.L. (1976) Some critical considerations of the theory of species packing. Evol. Theory 1: 247–258.
Heinrich, B. (1975) Energetics of pollination. Annual Review of Ecology and Systematics. 6: 139–170.
Heinrich, B. (1976) Flowering phenologies: bog, woodland, and disturbed habitats. Ecology 57: 890–899.
Heithaus, E.R. (1974) The role of plant-pollinator interactions in determining community structure. Annals of the Missouri Botanical Garden. 61: 675–691.
Hillman, W.S. (1962) The Physiology of Flowering. Holt, Rinehart & Winston, New York.
Hodgkinson, K.C. and Quinn, J.A. (1978) Environmental and genetic control of reproduction in *Danthonia caespitosa* populations. Australian Journal of Botany. 26: 351–364.
Holttum, R.E. (1973) Plant Life in Malaya. London: Longman. 254pp.
Howe, H.F. and Estabrook, G.F. (1977) On intraspecific competition for avian dispersers in tropical trees. American Naturalist. 111: 817–832.
Jackson, M.T. (1966) Effects of microclimate on spring flowering phenology. Ecology. 47: 407–415.
Janick, J. and Moore, J.N. (Eds.) (1975) Advances in Fruit Breeding. West Lafayette, Indiana: Purdue University Press.
Janzen, D.H. (1967) Synchronization of sexual reproduction of trees within the dry season in central America. Evolution 21: 620–637.
Janzen, D.H. (1970) Herbivores and the number of tree species in tropical forests. American Naturalist 104: 501–528.
Janzen, D.H. (1974) Tropical black-water rivers, animals and mast fruiting by the Dipterocarpaceae. Biotropica. 6: 69–103.
Jones, C.E. and Little, J. (1983) Handbook of Experimental Pollination Ecology. N.Y.: Von Nostrand Reinhold Co.
Kawano, S., Hiratsuka, A. and Hayashi, K. (1982) The productive and reproductive biology of flowering plants. V. Life history characteristics and survivorship of *Erythronium japonicum*. Oikos 38: 129–149.
Kevan, P.G. and Baker, H.G. (1983) Insects as flower visitors and pollinators. Annual Review of Entomology 28: 407–453.
Kerner von Marilaun, A. (1902) The Natural History of Plants. Blackie, London.
Koelmeyer, K.O. (1959) The periodicity of leaf change and flowering in the principal forest communities of Ceylon. Ceylon Forester. 4: 157–189.

Levin, D.A. (1972) Low frequency disadvantage in the exploitation of pollinators by corolla variants in *Phlox*. American Naturalist 106: 453–460.
Levin, D.A. and Anderson, W.W. (1970) Competition for pollinators between simultaneously flowering species. American Naturalist. 104: 455–467.
Levin, D. and Kerster, H. (1974) Gene flow in seed plants. Evolutionary Biology. 7: 139–220.
Lieth, H. (Ed.) (1974) Phenology and Seasonality Modelling. New York: Springer-Verlag.
Lindsey, A.A. and Newman, J.E. (1956) Use of official weather data in spring time-temperature analysis of an Indiana phenological record. Ecology 37: 812–823.
Lloyd, D.G. and Webb, C.J. (1977) Secondary sex characters in plants. Botanical Review 43: 177–216.
McIntyre, G.I. and Best, K.F. (1978) Studies on the flowering of *Thlaspi arvense* L. IV. Genetic and ecological differences between early and late flowering strains. Botanical Gazette. 139: 190–195.
McKey, D. (1975) The Ecology of Coevolved Seed Dispersal Systems. In L.E. Gilbert and P.H. Raven (Eds.), Coevolution of Animals and Plants. pp. 159–191. Austin: University of Texas Press.
McMillan, C. (1967) Phenological variation within seven transplanted grassland community fractions from Texas and New Mexico. Ecology 48: 807–813.
McMillan, C. and Pagel, B.F. (1958) Phenological variation within a population of *Symphoricarpos occidentalis*. Ecology 39: 766–770.
Medway, Lord (1972) Phenology of a tropical rainforest in Malaya. Biol. J. Linn. Soc. 4: 117–146.
Mosquin, T. (1971) Competition for pollinators as a stimulus for the evolution of flowering time. Oikos 22: 398–402.
Opler, P.A., Frankie, G.W. and Baker, H.G. (1976) Rainfall as a factor in the synchronization, release and timing of anthesis by tropical trees and shrubs. Jour. Biogeog. 3: 231–236.
Opler, P.A., Frankie, G.W. and Baker, H.G. (1980) Comparative phenological studies of treelet and shrub species in tropical wet and dry forests in the lowlands of Costa Rica. J. Ecol. 68: 167–168.
Paltridge, G.W. and Denholm, J.V. (1974) Plant yield and the switch from vegetative to reproductive growth. J. Theor. Biol. 44: 23–34.
Parrish, J.A.D. and Bazzaz, F.A. (1979) Difference in pollination niche relationships in early and late successional plant communities. Ecology 60: 597–610.
Pojar, J. (1974) Reproductive dynamics of four plant communities of southwestern British Columbia. Canadian Journal of Botany 52: 1819–1834.
Ponomarev, A.N. (1960) Concerning proterandry in Umbelliferae. Dokl. Akad. Nauk. SSSR. (Bot.) 135: 750–752.
Poole, R.W. and Rathcke, B.J. (1979) Regularity, randomness, and aggregation in flowering phenologies. Science. New York. 203: 470–471.
Primack, R.B. (1979) Reproductive effort in annual and perennial species of *Plantago* (Plantaginaceae). American Naturalist 114: 51–62.
Primack, R.B. (1980) Variation in the phenology of natural populations of montane shrubs in New Zealand. J. Ecol. 68: 849–862.
Primack, R.B. (1983) Insect pollination in the New Zealand mountain flora. New Zealand Journal of Botany 21: 317–333.
Primack, R.B. (1985) Longevity of individual flowers. Annual Review of Ecology and Systematics 16 (in press).
Primack, R. and Lloyd, D. (1980) Andromonoecy in the New Zealand montane shrub manuka, *Leptospermum scoparium* (Myrtaceae). American Journal of Botany 67: 361–368.
Primack, R.B., Tomlinson, P.B. and Duke, N. (1981) Floral morphology in relation to pollination ecology in five Queensland coastal species. Austrobaileya 1: 346–355.
Putz, F.E. (1979) Aseasonality in Malayan tree phenology. Malayasian Forester 42: 1–28.
Rabinowitz, D., Rapp, J.K., Sork, V.L., Rathcke, B.J., Reese, G.A. and Weaver, J.C. (1981) Phenological properties of wind and insect-pollinated prairie plants. Ecology 62: 49–56.
Rathcke, B.J. (1983) Competition and facilitation among plants for pollination. In L. Read (Ed.), Pollination Biology, pp. 305–329. Academic Press, Orlando.
Rathcke, B.J. (1984) Patterns of flowering phenologies: Testability and causal inference

using a random model. In D. Strong, D. Simberloff and L. Abele (Eds.), Ecological Communities: Conceptual Issues and the Evidence pp. 383–393. Princeton University Press, Princeton.

Rathcke, B.J. and Lacey, E.P. (1985) Phenological patterns of terrestrial plants. Annual Reviews of Ecology and Systematics 16 (in press).

Reader, R.J. (1975a) Competitive relationships of some bog ericads for major insect pollinators. Canadian Journal of Botany. 53: 1300–1305.

Reader, R.J. (1975b) Effect of air temperature on the flowering date of dogwood (*Cornus florida*). Canadian Journal of Botany. 53: 1523–1534.

Risser, P. and Cottam, G. (1967) Influence of temperature on the dormancy of some spring ephemerals. Ecology. 48: 500–503.

Robertson, C. (1895) The philosophy of flower seasons, and the phaenological relations of the entomophilous flora and the anthophilous insect fauna. American Naturalist 29: 97–117.

Robertson, C. (1924) Phenology of entomophilous flowers. Ecology 5: 393–407.

Ross-Craig, S. (1948–1973) Drawings of British Plants. G. Bell, London.

Salisbury, E.J. (1942) The Reproductive Capacity of Plants. Bell, London.

Schemske, D.W. (1977) Flowering phenology and seed set in *Claytonia virginica* (Portulacaceae). Bulletin of the Torrey Botanical Club 104: 254–263.

Schemske, D.W. (1981) Floral convergence and pollinator sharing in two bee-pollinated tropical herbs. Ecology 62: 946–954.

Schemske, D.W., Willson, M.F., Melampy, M.N., Miller, L.J., Verner, L., Schemske, L.M. and Best, L.B. (1978) Flowering ecology of some spring woodland herbs. Ecology 59: 351–366.

Sokal, R. and Rohlf, F.J. (1969) Biometry. W.H. Freeman, San Francisco.

Somers, G.F. and Grant, D. (1981) Influence of seed source upon phenology of flowering of *Spartina alterniflora* Loisel and the likelihood of cross pollination. Am. J. Botany 68: 6–9.

Snow, D.W. (1966) A possible selective factor in the evolution of fruiting seasons in tropical forest. Oikos. 15: 274–281.

Stephenson, A.G. (1979) An evolutionary examination of the floral display of *Catalpa speciosa* (Bignoniaceae). Evolution. 33: 1200–1209.

Stephenson, A.g. (1980) Fruit set herbivory, fruit reduction, and the fruiting strategy of *Catalpa speciosa* (Bignoniaceae). Ecology. 61: 57–64.

Stephenson, A.G. (1981) Flower and fruit abortion, proximate causes and ultimate functions. Annu. Rev. Ecol. Syst. 12: 253–279.

Stiles, F.G. (1975) Ecology, flowering phenology and hummingbird pollination of some Costa Rican *Heliconia* species. Ecology. 56: 285–301.

Stiles, F.G. (1977) Coadapted competitors: The flowering seasons of hummingbird-pollinated plants in a tropical forest. Science 198: 1177–1178.

Stiles, F.G. (1978) Temporal organization of flowering among the hummingbird food plants of a tropical wet forest. Biotropica 10: 194–210.

Stiles, F.G. (1979) Regularity, randomness, and aggregation in flowering phenologies: A reply. Science. 203: 471.

Stiles, E.W. (1980) Patterns of fruit presentation and seed dispersal in bird-disseminated woody plants in the eastern deciduous forest. American Naturalist. 116: 670–688.

Tamm, C.O. (1956) Further observations on the survival and flowering of some perennial herbs, I. Oikos 7: 273–292.

Tamm, C.O. (1972) Survival and flowering of some perennial herbs. II. The behavior of some orchids on permanent plots. Oikos 23: 23–28.

Thomson, J.D. (1978) Effects of stand composition on insect visitation in two-species mixtures of *Hieracium*. American Midland Naturalist 100: 431–440.

Thomson, J.D. (1980) Skewed flowering distribution and pollinator attraction. Ecology 61: 572–579.

Thompson, J.N. and Willson, M.F. (1978) Disturbance and the dispersal of fleshy fruits. Science. 200: 1161–1163.

Thompson, J.N. and Willson, M.F. (1979) Evolution of temperate fruit/bird interactions: phenological strategies. Evolution. 33: 973–982.

Tomlinson, P.B. (1980) The Biology of Trees Native to Tropical Florida. Petersham, Mass: P.B. Tomlinson (privately published). 480pp.

Valdeyron, G. and Lloyd, D.G. (1979) Sex differences and flowering phenology in the common fig, *Ficus carica* L. Evolution. 33: 673–685.
Waser, N.M. (1978a) Competition for hummingbird pollination and sequential flowering in two Colorado wildflowers. Ecology. 59: 934–944.
Waser, N.M. (1978b) Interspecific pollen transfer and competition between co-occurring plant species. Oecologia. 36: 223–236.
Waser, N.M. and Real, L.A. (1979) Effective mutualism betweeen sequentially flowering plant species. Nature. 281: 670–672.
Webb, C.J. (1976) Flowering periods in the gynodioecious species *Gingidia decipiens* (Umbelliferae). New Zealand J. of Botany 14: 207–210.
Wyatt, R. (1982) Inflorescence architecture: How flower number, arrangement, and phenology affect pollination and fruit set. American Journal of Botany 69: 585–594.
Zimmerman, M. (1980) Reproduction in *Polemonium*: Competition for pollinators. Ecology 61: 497–501.

PLANT POPULATIONS AND VEGETATION IN RELATION TO HABITAT, DISTURBANCE AND COMPETITION: PROBLEMS OF GENERALIZATION

P.J. GRUBB

ABSTRACT

Plants belonging to different species respond to particular environmental conditions, and to particular forms of disturbance and competition, so differently that generalization is difficult. Moreover, populations of different biotypes within a single species may also respond differently. It is more profitable to summarize the ways in which plant populations may differ in their response to habitat, disturbance and competition than to recognize a very small number of major plant-types or 'strategies'. Emphasis in this chapter is placed on the great variety of 'key characters' apparently needed for survival in the face of different environmental conditions, and particular forms of disturbance.

INTRODUCTION

To think of vegetation as made up of *populations of individuals* rather than species was a natural approach for all those involved in the management of forest trees from very early times, and this approach has certainly been explicit in the writings of foresters since the middle of the 18th Century (cf. Duhamel du Monceau 1760). Academic ecology, on the other hand, took root in studies of vegetation that was largely herbaceous, and in which individuals were hard to identify. Despite the early advocacy of Clements for studies of plant demography (White 1985), only sporadic efforts were made within the mainstream of the subject to study populations. The paper by A.S. Watt (1947) on 'pattern and process in the plant community' was of crucial importance in encouraging ecologists concerned with all kinds of vegetation to adopt the foresters' perspective in the sense of studying the process of regeneration, based on the life-cycles of individual plants. However, it took another thirty years for sufficient studies in demography to accumulate so that Harper (1977) could produce a connected account of 'plant population biology' on a broad basis.

Much of our thinking as ecologists is still centred on the species as the self-evident unit within a vegetation type, and a review of past ideas on

White, J. (ed.), The Population Structure of Vegetation.
 ISBN 978-94-010-8927-2

plants in relation to habitat, disturbance and competition is bound to reflect this bias. Plant ecologists, over the years, have formed a spectrum of individuals from those who have found every species interestingly different in terms of function and requirements (and who have liked to emphasize the differences) to those who have not been stimulated by nature's infinite diversity, and who have sought to interpret the world of plants by recognizing a small number of major types. The value of the first approach has perhaps been greatest in studies on mechanisms of co-existence (cf. Grubb 1977) but its potential danger has been failure to suggest a sufficient series of generalizations to form a coherent body of ecological theory. The value of the opposite extreme has been in building up just such a body of theory, but its potential danger has been to overlook so much information that only a very partial understanding of nature has been achieved.

The early plant geographers and plant ecologists did seek to generalize but tended to recognize large numbers of basic plant types, e.g., Humboldt's 16 'physiognomic types' (1805), Grisebach's 60 (1872) and Warming's 12 'growth forms' (1890), all documented by Warming (1909). Such schemes may be constrasted with the three basic 'strategies' (explerents, violents and patients) proposed by Ramenskii (1938), the two basic types of selection (*r* and *K*) put forward by MacArthur and Wilson (1967), the three basic species-types (colonizing, equilibrium and resistant) recognized by van Valen (1971), the three basic 'strategies' (ruderal, competitor and stress-tolerator) suggested by Grime (1974) and the three 'demographic strategies' (exploitation-selected, saturation-selected and adversity-selected) proposed by Whittaker and Goodman (1979).

The purpose of this paper is to review the basic issues involved in past attempts at describing plants in relation to habitat, disturbance and competition, and to try to shift the emphasis from the species to the population of individuals. The review is restricted to vascular plants.

PLANTS IN RELATION TO HABITAT

Tolerance and requirement in respect of single factors

When classification of plant species by habitat was first formalized by ecologists several terms implying affinity for particular conditions were introduced, e.g, xerophilous and halophilous, but these were gradually dropped when it was found experimentally that assocation with particular conditions in the field did not mean that plants had a physiological requirement for those conditions. When freed from competitors, xerophytes grew well without drought, halophytes without salt and so on. There gradually came into common usage terms based on tolerance such as drought-tolerant, salt-tolerant and so on, and there is no doubt that the concept of tolerance is extremely valuable in this context.

However, tolerance is an inadequate concept for two reasons. First, many plants do positively require for maximum growth rate the conditions with which they are associated in the field and do not merely tolerate them. Thus plants from areas with a very wide range of temperature, e.g., western North America, turn out to have very different temperature optima for growth (Osmond, Bjorkman and Anderson 1980). At least some plants associated with shade do not merely tolerate it, but yield poorly in full sunlight unless provided with extra mineral nutrients (Murray and Nichols 1966). Of the plants found on calcareous soils (pH 7–8) most that have been tested have grown larger on soil of pH 5–6, but some have certainly grown larger on their native soils (Paul 1969). Although most plants cannot withstand prolonged submergence in water, many hydrophytes are dependent on it for support. There is thus an element of *requirement* for certain conditions for survival as well as tolerance.

Secondly, it is found that when two plants have the same tolerance-limits they are not necessarily most abundant in the same fraction of their tolerance-range. This point can be conveniently illustrated in respect of soil pH in Britain in the results of Grime and Hodgson (1969). Among the species they found over the whole pH range *Agrostis tenuis* had a clear maximum at 4–6, while *Festuca ovina* had maxima at 3–4 and 6.5–7.5.

These limitations to the value of the concept of tolerance apply not only to treatment of species but also discussions on populations of particular genotypes. Two examples must suffice. The temperature regimes required by Norwegian and Portugese races of *Dactylis glomerata* are quite different (Eagles and Williams 1969); a Portuguese race was found to require lower temperatures (in the field it grows in winter) while a Norwegian race was found to require higher temperatures (in the field it grows in summer), and each can oust the other under appropriate conditions. The cyanogenic and acyanogenic races of *Trifolium repens* have essentially the same tolerance-ranges, but different patterns of abundance-distribution, with the cyanogenic more prevalent at warmer and drier sites (Burdon 1983).

Because 'tolerance' is an adequate concept, and because 'adaptation' is such a controversial concept (Harper 1982), I shall refer to plant species or genotypes as being *suited* to particular conditions. This is not to imply that a plant population has been in any sense optimized, merely that its constituent individuals have certain characters which enable them to survive under those conditions (in the presence of competitors), while other plants fail to do so. These characters I refer to not as 'adaptations' but as *key characters*. When these are listed for plants suited to various conditions, it is not implied that all plants suited to a particular set of conditions need all the 'key characters', only that they must have one or more of them. The easiest sets of key characters to deal with are the physiological or morphological characters that must be shared by all plants in a given set of conditions whatever their demographic characteristics, e.g., plants on salt-rich soils which must have some means of coping with salt whether they are annuals, biennials or perennials. Where there is little variation in the life history and demographic characters listed, e.g., for plants floating on still water (Table 1), those characteristics are likely to be as essential as the physiological and morphological characters listed.

Table 1. The key morphological and physiological characters, and common life history and demographic characters of plants suited to various environmental conditions, based chiefly on Osmond et al. (1980) and Walter (1968, 1973).

Key physiological and morphological characters	Common life history and demographic characters
Slender supply of mineral nutrients	
Accumulation of large amount of C per unit N during its residence-time in a leaf Long leaf-life; leaves tough and of low nutritive value Much redistribution of nutrients from ageing tissues High root weight ratio; mycorrhiza and/or nitrogen-fixing symbionts In some case extra nutrients from especially large nutrient-rich seeds, or from insects trapped or from ants provided with shelter	Mostly perennials with extensive vegetative reproduction and rare establishment from seed. Annuals rare when perennials form matrix competing for nutrients even after burning opens stand temporarily above ground, e.g., pine barrens and savannas of eastern U.S.A. and many heaths and heathy forests in Australia, but can be abundant where perennials sparse or absent e.g., on overgrazed sand dunes and shallow soil-pockets on rocks
High salt concentration	
Various combinations of the following: Salt-sequestration in vacuoles of large cells in succulent organs Salt-exclusion by roots Salt-secretion via glands or via evanescent, inflated leaf hairs	In moist tropics all perennials (mangrove trees, epiphytes, chenopods on seasonally dry land behind mangroves). In temperate coastal areas and semi-deserts range from annuals and biennials to long-lived shrubs
Seasonal drought	
(a) Therophytes Plants of nutrient-poor sites must produce seed when tiny (1–5 cm) but those on other soils can be large to very large	(a) Extent of dispersal in time and space very variable
(b) Other life-forms (most have rather high cytoplasmic tolerance of desiccation) 1. Ephemerals (pass drought as bulbs or corms) 2. Poikilohydric (mostly herbs) 3. Deciduous (mostly woody, often spiny on stems, protecting soft leaves) 4. Sclerophyllous (trees, shrubs and grasses; leaves hard, long-lived; roots often deep) 5. Evergreen malacophylls (mostly shrubs and herbs; leaves soft, last one year or less, root systems wide but often shallow) 6. Succulents have CAM; (mostly very slow-growing; often protected by spines or camouflage) 7. Phreatophytes (roots very deep, subsoil water throughout drought). 8. Thallus-like plants on rocks in seasonal rivers (Podostemonaceae)	(b) Life lengths form continuum from facultative annuals to perennials with wide clonal spread lasting 10^2–10^3 years. Inverse relation between life length and frequency of conditions required for establishment from seed; latter may be very rare (< 1 occasion in 50 yr?)

Table 1. Continued

Key physiological and morphological characters	Common life history and demographic characters
Waterlogging (often with partial submergence)	
Effective internal supply of air to submerged roots from leaves or pneumatophores via aerenchyma Tolerance of anaerobiosis in root cytoplasm, especially perhaps the accumulation of ethanol	Vegetation of lake and river edges composed mostly of perennials, with clonal spread and dispersal of fragments in currents; establishment from seed apparently rare. Mires and flooded forests similar, possibly with more species of short-lived plants
On surface of still water	
Bouyant on basis of large air spaces and unwettability of upper parts. Life-form either reduced frond at water surface (e.g., *Azolla*, *Lemna*) or rosette above water with balancing roots below (e.g., *Eichhornia*, *Pistia*)	Reproduction almost entirely vegetative; dispersal by animals and currents; conditions for establishment from seed (or spore) apparently rare and not understood
Low temperatures	
Cold-stability of membranes and other enzyme systems In evergreen species reduced rates of transpiration once stomata shut, and lesser reduction in permeability of roots to water at low temperatures; often leaves small, perhaps reducing chance of desiccation in parts far from midrib Retention of frost-excluding leaf-bases and folding of older leaves over stem apices at night Shoots condensed into cushions which lose less heat by convection	Mostly perennials; few annuals (hemi-parasites) or biennials (Gentianaceae, special mycorrhiza). Seedlings can be abundant at sites well provided with nutrients, rare at 'infertile' sites, longevity and frequency of establishment of new adults unknown. Reduced dependence on animals for pollination and dispersal
High temperatures	
Heat-stability of membranes and other enzyme systems In some gaping of stomata and increased evaporative cooling of leaf In some leaves near-vertical and/or highly reflective and so absorbing less radiation, especially at midday In some C_4 photosynthesis eliminating wasteful competition of O_2 on RuBP carboxylase	Whole range of life histories involved from annuals dependent on summer rain in prairies, forests of monsoon regions and some semi-deserts to long-lived shrubs and trees of semi-deserts and desert wadis and oases
Deep shade	
Steep rise in photosynthetic rate at low irradiance values Low respiration rate and compensation point Low root weight ratio In some high specific leaf area	Long-lived perennials mostly, often with widely spreading clones and no obvious specialized means of seed-dispersal. Annuals confined to moist sites well provided with nutrients (e.g., *Impatiens* spp.)

Under other conditions where there is a wide range of demographic characters, e.g. with seasonal drought, the physiological and morphological key characters depend on the demographic characters. In Table 1 (and two later tables) I have used the word 'common' rather than 'key' before the 'life history and demographic characters' in the heading, in order to emphasize their variable status.

To provide critical evidence that a certain character is a 'key' one in the sense defined above is indeed a demanding task, and Osmond, Bjorkman and Anderson (1980, pp. 44–47) have given a clear example of what is needed from a physiological point of view, based on the work of J.R. Ehleringer on the dense hair-covering on the leaves of the semi-desert form of the shrub *Encelia.* Ideally this physiological approach should be linked with a comparative demographic study of the fates of individuals from suited and unsuited populations sown into an appropriate area.

It is important to realize that plant populations vary greatly in the breadth of not only their ecological tolerance or 'realized habitat niche' but also in the breadth of their physiological tolerance or 'potential habitat niche'. It should not be thought that a plant population which is suited to one extreme cannot also be suited to the opposite. Although critical investigation of wide-ranging species has usually shown them to be composed of several to many distinct populations (biotypes), it is also true that single populations (biotypes) can be very widely tolerant, e.g., in *Epilobium angustifolium* one biotype was found to grow effectively on soil of pH 3.5–9.0 (van Andel, Bos and Ernst 1978). The ability of plants to be suited to opposites is most convincingly shown where individuals tolerate opposite extremes in one life-time. Thus individuals of *Hedera helix* form extensive carpets in extreme shade, but grow vigorously and flower in full sunlight. Many perennials survive alternating drought and waterlogging in certain types of seasonal tropical vegetation (Beard 1944) and at periodically flooded, low-lying sites in semi-desert (Specht 1972). Most impressive are the cushion-plants of the alpine belt in the tropics and sub-tropics which suffer 'summer by day and winter by night' with leaf temperatures reaching almost 50°C around noon and well below 0°C in the hours of darkness (Ruthsatz 1978). We should also remember the dune annuals like *Myosotis ramosissima* which normally complete their life-cycles on soils very low in available nitrogen and phosphorus, but which have the potential to develop high relative growth rates if fertilized (Boorman, 1982). The concept of the 'trade off', which implies that being suited to one condition necessarily involves not being suited to the opposite, is widely diffused in current literature but is not universally applicable.

It is important also in delimiting the tolerances of plant populations in respect of single factors that these factors do not act independently. Interaction is common, for example between shade and drought, and between shade and nutrient supply (Ellenberg 1982).

Distribution in respect of factor-complexes

It is inevitable that we should emphasize single environment factors when working out key characters of plants growing in various conditions, but as ecologists we are often confronted with factor-complexes rather than single factors, e.g., high latitude or high altitude rather than cold, or ridge-top versus gully rather than water-supply or pH. The logical approach to ordering plants in terms of these factor-complexes is 'direct gradient analysis', which was reviewed at length by Whittaker (1967), and which is used in a later section of the present paper.

Concepts of 'optimum' and 'stress'

As ecologists began to sort out the effects of factor-complexes, there arose the concept of 'optimal conditions' for plant life and the concept of 'adverse factors'. An explicit example of such thinking is seen in Beard's (1944) classification of vegetation-types in the Neotropics. His 'rain forest' occupied the optimum conditions (favourable temperatures, no drought, no waterlogging), while his formation-series reflected the response of the vegetation to various 'adverse factors': lack of rainfall, flooding, low temperatures, excessively high humidity, wind and salt.

By the 1960s the term 'stress' was widely used for any of the adverse factors which would kill or reduce the growth of a plant not specifically 'adapted'. In the *Annual Reviews of Plant Physiology* in 1961–69 a wide range of factors was considered in the 'stress physiology' section, including freezing, high altitude, drought, ionizing radiation and air pollutants. In 1972 Levitt published an explicit comparison between this use of 'stress' by biologists, and the meaning of 'stress' to physicists.

The most awkward problem for those who wish to classify plant populations in respect of 'stress' arises from the fact that conditions which amount to 'stress' (i.e., reduce growth) for one population or individual do not necessarily amount to 'stress' for another. This follows from the cases of requirement rather than tolerance listed on p. 597. The generalized concept of 'stress' adopted by Levitt (1972) and many others cannot be sustained. It is perhaps significant that since 1971 the relevant articles in the *Annual Review of Plant Physiology* have appeared under the heading of 'Environmental Physiology' or 'Population and Environment'.

Grime (1979) has given a new meaning to 'stress'. He defines it as 'the external constraints which limit the rate of dry matter production of all or part of the vegetation'. He then lists characteristics of plants which are 'stress-tolerators', e.g., low maximum potential relative growth rate (RGR), long-lived plant parts and limited phenotypic plasticity. The trouble with this approach is that it isolates a subset of the 'stresses' of Levitt (1972) and other earlier workers, i.e., those associated with a reduction in growth rate in the field. It ignores the key characters of plants suited to a wide range of conditions that are stresses (*sensu* Levitt) for many plants, e.g., high or low temperatures, waterlogging, submergence and salt. There is no evidence that perennial light-demanding herbs of temperate regions have lower maximum RGR or longer leaf-life, or less phenotypic plasticity than those of tropical regions, but they are certainly 'stress tolerant' *sensu* Levitt because they are tolerant of winter-cold. Similarly plants which grow submerged, or at least on waterlogged soil, do not have the characteristics of 'stress tolerators' listed by Grime, but are certainly tolerant of conditions which reduce the growth of other plants.

Another long-standing difficulty with classification of plants in respect of 'stress' is the ability of some plants to avoid the stress, e.g., the ability of phreatophytes in a desert or semi-desert to reach and use subsoil water, or the ability of many plants on soils with little available nitrogen to harness nitrogen-fixing bacteria, actinomycetes or cyanophytes as symbionts. Such plants inevitably tend to be faster-growing than the plants around them which do not have access to subsoil water or nitrogen fixed by symbionts. In the extreme case where the stand is composed almost entirely of the 'stress-avoider' (e.g., in the case of *Alnus crispa* at Glacier Bay in Alaska; Cooper, 1939) growth can be so fast that it will not appear by Grime's definition of 'stress' that any potentially adverse factor is involved at the site. It would be better if the 'stress' axis in Grime's scheme was renamed 'productivity' axis, since that is what it really is.

For the reasons indicated the concepts of 'optimum' and 'stress' are rejected in the following analysis.

Edge-plants

One major kind of differentiation between plants is whether they occur (in the natural landscape) in the vegetation-types that cover most of the landscape or in narrow ecotones, e.g., at the edges of forests along cliffs or rivers, or at junctions with grassland or savanna. Much attention has been given to edge-plants in central Europe since Tüxen (1952) formalized the structure of the forest/cliff or forest/grassland ecotone by recognizing the 'Mantel' (mainly of shrubs) and the 'Saum' (mainly of herbs). Dierschke (1974) has developed this approach with precise measurements of conditions. In fact there can be a subtle gradient of growth form within the Mantel element as shown by Sloet and Adriani (1971, p. 204).

Examples of edge-plants are well known at forest/savanna boundaries in the lowland tropics (White 1983) and at forest-grassland boundaries on tropical mountains (Grubb and Stevens 1985). Although many species seem to be primarily edge-species with only occasional occurrences within the more extensive communities, some are known to occur commonly in the extensive vegetation-types of one zone while remaining edge-species in another. In New Guinea some species are edge-plants in one altitudinal zone but not in the next higher up, e.g., *Coprosma discoloris* convincingly so in the Lower Montane, and usually so in the Upper Montane, but diffused through the forest in the Subalpine (Grubb and Stevens 1985).

The whole issue of edge-plants, and the characteristics which separate them from plants of gaps in the nearby extensive vegetation types, is worthy of much closer attention from ecologists.

PLANTS IN RELATION TO VEGETATIONAL DYNAMICS

Types of vegetational dynamics

Three basic types of change in vegetational composition are recognized here: fluctuation, succession and regeneration. *Fluctuation* is shown by vegetation undergoing irregular changes in relation to weather conditions or the impact of animals (including man). *Succession* is defined as directional change, and may be primary (starting from newly available, naturally bare substrata) or man-induced (starting on surfaces denuded by man). 'Succession' does not imply that necessarily species are gained and lost through time in a sample of moderate size, merely that there is at least a directional change in the relative abundance of species. *Regeneration* is the sum of the processes which lead to a community having an approximately constant composition in the long term and over a large area despite the death of individuals from old age and the periodic destruction of part of the biomass by outside influences such as wind, fire or flood: regeneration may involve 'internal successions' (Curtis 1959, p. 292) of species on sites where the earlier cover has been destroyed, but equally may involve plant-for-plant replacement without any successional tendency. The processes of regeneration and fluctuation merge in the case of annuals, e.g., therophytes in a semi-desert, and the processes of regeneration and primary succession merge in those parts of the world with frequent severe seismic activity and renewal of vegetation on the bare subsoil of landslips, e.g., in many forested parts of Chile (Veblen and Ashton 1978). In the past internal successions and man-induced successions have commonly been lumped as 'secondary successions'.

Categorization of plant populations in relation to fluctuation is best subsumed in the definition of their tolerance or requirements for adult growth and regeneration. Categorization in relation to succession and regeneration is best considered after reviewing the concept of disturbance.

Concept of disturbance

'Disturbance' has been defined by Grime (1979) as 'the mechanisms which limit the plant biomass by causing its partial or total destruction', and this definition is accepted in the present review, with the caveat that senescence of individuals is excluded. Disturbance must be initiated from outside the plant community; it may involve killing of plant parts *in situ* (e.g., extreme temperatures, lightning strikes, effects of pathogens), severance of plant parts followed by decomposition *in situ* (some short-lived high winds, hail-storms), loss by combustion, severance and removal from the site (e.g., cutting of forest, mowing or grazing) or burial followed by death and decay (if any) *in situ* (e.g., heaping of soil by animals, deposition of blown sand and deposition of slipping or falling rock debris).

The review of White (1979) illustrates the fact that the regeneration of vegetation is very often triggered off by disturbance rather than by the senescence of individuals, which was emphasized by Watt (1947) in his seminal paper on pattern and process. It is also true that disturbance triggers off many types of primary succession, e.g., where a slope is denuded of its former cover of vegetation by a landslip or the former cover is buried under a mudflow, lavaflow or new banks of silt and gravel in a river. Similarly the coming of an ice sheet and the smoothing-off of a rock face, or deposition of a moraine, constitutes a disturbance.

The all-pervading role of disturbance has led Ramenskii (1938), MacArthur and Wilson (1967) and Grime (1979) to adopt it as one of the primary axes of differentiation among plant populations in particular types of vegetation. We must now review some of the difficulties in this approach. They stem from the fact that disturbance can be of very different types, differing in frequency, extent and intensity. A basic division is into continual and periodic disturbance. Continual disturbance produces distinctive, long-persistent types of vegetation, whereas periodic disturbance produces transient gaps in which regeneration occurs.

Several different kinds of continual disturbance are found: grazing and trampling, destruction by wave action or flood, burial by sand or mud, and burial plus downhill slippage in a scree. As shown in Table 2, the key characters of plants suited to these various kinds of disturbance are very different.

Within the category of periodic disturbance there is also a great deal of variation in type, matched by variation in plant response. The simplest kind of variation is variation in the size of the area disturbed. Foresters have long appreciated that tree species can be arranged along a spectrum from 'tolerant' species that can grow up through a minimal gap (perhaps where there has been no disturbance but simply death of an older tree 'on its feet') to 'intolerant' species which can take no shade at all. Ellenberg (1982, p. 82) has categorized central European trees on a six-point scale of this type. 'Tolerance' in this sense was originally thought to be concerned wholly with tolerance of shade, but it has increasingly been appreciated that tolerance of below-ground interference by other plants can be equally important. There is some evidence that species in grassland can be differentiated along a similar axis of 'size-of-disturbance', and that in this case too tolerance of below-ground interference is as important as tolerance of shade (Fenner 1978). Grazed or mown grasslands often suffer not only continual disturbance all over but also periodic disturbance locally, e.g., through the formation of soil heaps by gophers or moles, or the deposition of faeces by cattle. The composition of the vegetation is distinctively changed on mole-heaps (Knapp 1959), and a cyclic series of changes is now known to be associated with dung patches (Burdon 1983).

In some kinds of vegetation, particularly those which are prone to periodic destruction by fire, the operative variable differentiating between plant species or populations is not size of disturbance but frequency of disturbance. The characters which determine whether species benefit from a little or a lot of disturbance are quite different in this case from those that are important where size of disturbance matters (Table 3).

The final variable in disturbance, its intensity, is a particularly important variable in the case of fire; yet a further set of vital characters is found

Table 2. The key morphological and physiological characters, and common life history and demographic characters, of plants suited to various types of continual disturbance (based chiefly on Ellenberg 1982).

Key morphological and physiological characters	Common life history and demographic characters
Grazing	
Possession of meristems close to ground; leaves in rosette or tussock arrangement; in extreme cases unpalatable or spiny; ability to withstand continual reduction in shoot/root ratio	Mostly perennial with annuals, biennials and pauciennials in interstices and fluctuating in relative abundance. Dispersal very limited; seeds often not specialized to animals or wind
Trampling	
Roots able to penetrate compact soil and survive at low oxygen concentrations in soil; shoots able to resist bruising	Life-length and extent of clonal spread very varied
Burial by sand	
Shoots able to penetrate deposited sand and produce adventitious roots in newly deposited material; in extreme cases young shoots or leaves very tough and streamlined	Perennials, mostly relatively long-lived. Seed dispersal by wind or animals
Burial and dragging downhill by scree	
(a) Rhizomes or roots able to resist breaking with downhill movement, and put up new shoots between stones	(a) Long-lived perennials, often with wide clonal spread
(b) Quick maturation in opportunists of temporarily stable microsites (species often shared with cliffs)	(b) Annuals and biennials, widely dispersed
Frequent rearrangement of humus-poor river gravels by floods	
Relatively rapid growth and seed set	Effective seed dispersal, often by wind
Yearly scour of drift-line by winter floods or tides	
Very rapid growth, using abundant nutrient supply and seed-set in first season	Annuals, dispersal often in drift material or water

in plants suited to very intense (i.e., very hot) fire (Table 3). The intensity variable can also be important in the case of continual disturbance, especially with grazing (Table 2).

We are bound to conclude that arrangement of species or populations along a single axis of more or less disturbance is unlikely to be adequate in any scheme intended to have world-wide application. However, we shall certainly find it useful to adopt 'disturbance' as a major kind of differentiation among plants at any one point along an environmental gradient, provided we consider separately the effects of continual and periodic disturbance.

Table 3. The key morphological and physiological characters and common life history and demographic characters of plants suited to high levels of disturbance: (a) plants of large gaps rather than small (mostly from Wells 1976, also Marks 1975, Bazzaz 1979 and Coley 1983), (b) plants suited to burning at short intervals rather than long (from Noble and Slatyer 1980), (c) plants suited to intense burns rather than mild burns (from Daubenmire 1968a and Gill 1981).

Key morphological and physiological characters	Common life history and demographic characters
(a) *Plants of large gaps*	
High relative growth rate (high photosynthetic rates per unit leaf area, low root weight ratio in juvenile; little 'investment' in protection against herbivores)	Life length and time to fruiting very variable; cf. some *Cecropia* spp. in Venezula lasting 3–5 years (Uhl et al. 1981) and various timber trees in W. Africa lasting 150 years (Jones 1955).
Short mean leaf-life and continual extension growth (rather than flushing)	Shorter lived, smaller plants often have long-lived seed in soil.
Greater desiccation tolerance	Dispersal by animals or wind.
Often space-filling crown (long internodes or petioles, open habit with radial rather than bilateral symmetry of shoots)	In temperate zone shorter lived plants often herbaceous and sometimes 'biennial' (truly pauciennial). Normally no annuals (*Senecio sylvaticus* is an exception).
Often light wood with high conductivity	
(b) *Plants of frequently burnt sites*	
Ability to recover vegetatively or invade quickly by seed	Usually long-lived grasses most abundant (as in much savanna) with some short-lived woody plants and dicot herbs
If regenerating vegetatively, an ability to maintain plant size despite frequent loss of 'capital' by combustion (explains loss of *Pteridium* from many areas)	Where soil is too nutrient-poor for grasses (e.g. pine barrens of eastern U.S.A.) annual burns may give patches of perennials (*Carex*, legumes, mosses, lichens) and numerous individuals of annual dicots, e.g., species of *Gerardia* and *Sabatia*
If regenerating by seed, an ability to become established and produce mature seed before next fire	
(c) *Plants of intensely burnt sites*	
Thick bark or vascular bundles buried deeply in trunk as in tree-ferns, palms or *Xanthorrhoea*	Life lengths very variable, often long
Possession of buds at or below ground level (seen in many woody plants as well as herbs); exact height of buds becomes critical with the most intense burns	

Facultative and obligate association with disturbance

Many species are facultatively associated with disturbance. For example, many species that arrive in the second wave of invaders on newly exposed glacial moraines at low altitudes in the Northern Hemisphere (such as *Dryas* and *Salix* spp.) have more permanent homes in shallow accumulations of soil on cliffs and rock outcrops where the later successional species of moraines cannot become established. Similarly, on a lava flow in the

Cameroons, the earliest invaders (which were ousted later) were found to include species with permanent homes as epiphytes nearby or as lithophytes at higher altitude (K. Domke in Tüxen, 1975, p. 21). In New Zealand some early successional species of landslips in the moist part of the country have more permanent homes in drier parts of the country where they are not ousted by species from the moister parts, e.g., *Leptospermum ericoides* and *L. scoparium* (Burrell 1965). Further examples emerge in later sections.

The species which certainly show obligate association with disturbance are some of those typified in Table 2, i.e., those associated with continual destruction of strandlines, continual rearrangement of gravels, continual deposition of sand or silt and continual scree activity. It is possible that many species which follow disturbance of forests or grasslands by fire, tree-fall or animal activity are obligately associated with disturbance, and are never found in permanently available sites such as forest- and river-edges, cliffs or mires, but there is little precise information available on this point.

Origin of early-successional species in man-induced successions

There has been little evidence in the literature that ecologists have appreciated the fact that the species which are most abundant in the early stages of succession on abandoned fields and forest-clearcuts are quite different from those which are most abundant in the largest natural clearings in the forests that have been replaced by the fields or clearcuts. When working in Lower Montane Rain forests in New Guinea in 1970–1, I was impressed by the fact the tree species most abundant in the newly created secondary forest in the study area were never or almost never found in the largest natural tree fall gaps in the primary forest, but were recruited from forest-grassland margins, river gravels or landslips (Grubb and Stevens 1985). The only herbaceous plant to become established very quickly on the ground in clear-felled forest (*Crassocephalum crepidioides*) was not found in any natural habitat in the study area, and was probably brought in by man. I have seen a similar phenomenon in the Rio Negro region of Venezuela, where the succession on old fields has been described by Uhl et al. (1981). The early herbaceous invaders like *Borreria* spp. have probably come in from non-forested sites in another part of the country, and some of the shrubs like *Clidemia sericea* appear to have come from the forest-savanna boundary further north (O. Huber, personal communication).

This general issue has been discussed critically by Marks (1983) in respect of old-field succession in eastern North America. There is no sharp distinction between forest-gap and old-field species but many species can be assigned decisively to one category or the other. As Marks has emphasized, the physiological and demographic characteristics that cause a species to belong to one or other of the groups have still to be worked out with certainty. In particular, it is not yet clear what characteristics enable some species or populations to 'explode' from communities on forest edges, river gravels etc. into the man-disturbed landscape, while other species in these specialized communities fail to do so.

The essential point for this review is that it is important not to make a naive equivalence between forest-gap and early successional species of secondary successions, thinking of them both as being simply associated with strong disturbance.

Effects of disturbance on pattern of distribution on an environemntal gradient

The observations on *Leptospermum* spp. in New Zealand, recorded above (p. 607) illustrate nicely the broadening of tolerance with disturbance. Werner and Platt (1976) made similar observations on the breadth of tolerance of water relations in *Solidago* spp. in North America; in old fields they had appreciably wider tolerances than in prairies. Ellenberg (1958) reported analogous European examples in respect of soil pH.

Most plant populations seem to have wider physiological tolerance than ecological tolerances, as evidenced by the growth of so many species in botanic gardens well away from their natural habitats, and the effects of disturbance on tolerance may well turn out to be widespread in natural and semi-natural communities.

Plants in relation to environmental gradients and disturbance

The chief purpose of this section is to illustrate the fact that, possibly with a few exceptions, plants dependent on disturbance can be found at all points along an environmental gradient. Two examples must suffice – one concerning periodic disturbance and regeneration, the other continual disturbance and primary succession.

Daubenmire (1966, 1968b) has described for a long environmental gradient in eastern Washington State (U.S.A.), running from higher altitude sites (cooler and wetter) to lower altitude sites (hotter and drier) not just the distribution of tree species but also, on the basis of size distributions of populations at particular sites and a knowledge of their life-history traits, an analysis of which species can maintain populations in the absence of disturbance, and which cannot. In this case the most important natural disturbances are fires. His results are summarized in Table 4. Only at the very dry end of the gradient is there no species present, dependent on disturbance. Several analogous types of vegetation, with single tree species, are known, e.g., *Juniperus virginiana* forming 'cedar glades' on blocky limestone in the eastern U.S.A., or *Betula pubescens* ssp. *tortuosa* in subarctic Scandinavia. It is very likely that in all these extreme forests there are shrubs, herbs, bryophytes or lichens dependent on disturbance for their persistence, and that is certainly so in the *Betula* forests mentioned where disturbance is caused chiefly by periodic outbreaks of voles (Callaghan and Emanuelsson 1985). In Table 4 it is particularly notable that some species are able to act as late-successional on one part of the gradient, and early-successional on another (analogues of the *Leptospermum*

Table 4. An analysis of tree species on part of a long gradient of drought and temperature in eastern Washington State (after Daubenmire 1966); species acting as early-successional on one part of the gradient and late-successional on another part are shown in bold type. Species are ordered according to decreasing drought and temperature down the table.

After recent disturbance	No recent disturbance
Pinus ponderosa	*Pinus ponderosa*
Pinus ponderosa	*Pseudotsuga menziesii*
Larix occidentalis *Pinus contorta* and *P. monticola* **Pinus ponderosa** **Pseudotsuga menziesii**	*Abies grandis*
Abies grandis *Larix occidentalis* *Pinus contorta* and *P. monticola* **Pseudotsuga menziesii**	*Thuja plicata*
Abies grandis *Larix occidentalis* *Picea engelmannii* *Pinus contorta* and *P. monticola* **Pseudotsuga menziesii** **Thuja plicata**	*Tsuga heterophylla*
Abies grandis *Larix occidentalis* *Picea engelmannii* *Pinus contorta* and *P. monticola* **Pseudotsuga menziesii**	*Abies lasiocarpa*
Abies lasiocarpa *Larix occidentalis* *Picea engelmannii* *Pinus contorta* and *P. monticola*	*Tsuga mertensiana*

spp. in New Zealand mentioned on p. 607). Although much less is known about the distribution of different populations within species in relation to disturbance, there is certainly evidence for differentiation within species, based on longevity, time to reproduction etc., as shown for *Taraxacum officinale* (Solbrig and Simpson 1977) and *Poa annua* (Law 1979). There seems to be no reason why such differentiation should not occur in species found at all points along most environmental gradients, including perhaps the single tree species of certain extreme sites. The vegetation type where the role of disturbance is still least certain is semi-desert (cf. Wells 1961, Roughton 1972, Vasek 1980).

The example concerning continual disturbance is based on sand-dune vegetation. Here the plants of wholly mobile dunes or mobile dune-tops are obligately associated with continual disturbance, i.e. burial and erosion involving wind-blown sand. Such plants can be found not only in relatively moist and cool climates but also in extremely hot and dry semi-desert climates. Examples for western North America and southern Australia are shown in Table 5. The same principle is shown by the

Table 5. An analysis of plants on unstable and stabilized dunes in two parts of the world in relation to the amount of drought experienced. (a) is based on Franklyn and Dyrness (1969) for the coast of Oregon (rainfall ca. 1800 mm yr^{-1}) and on Chadwick and Dalke (1965) for part of Idaho (rainfall ca. 370 mm yr^{-1}). (b) is based on Specht (1972) for the coast of South Australia (rainfall 400–600 mm yr^{-1}) and on Buckley (1981) for the Simpson Desert (rainfall ca. 150 mm yr^{-1}).

	Much disturbance by blown sand	Less disturbance (stabilizing)
(a) *Western North America*		
Strong drought	*Elymus flavescens*	*Artemisia tridentata*
	Psoralea lanceolata	*Prunus virginiana*
		Purshia tridentata
Little drought	*Carex macrocephala*	*Gaultheria shallon*
	Glehnia leiocarpa	*Pseudotsuga menziesii*
	Poa macrantha	
(b) *South Australia*		
Strong drought	*Calotis erinacea*	*Abutilon otocarpum*
	Zygochloa paradoxa	*Triodia basedowii*
Moderate drought	*Spinifex hirsutus*	*Casuarina stricta*
		Melaleuca lanceolata

existence of distinct pioneer species on screes at both low altitudes and very high altitudes, e.g., *Rumex scutatus* and *Viola cenisia* respectively in central Europe (Ellenberg 1982), and by the presence of distinct pioneers at damp versus dry, or nutrient-rich versus nutrient-poor sites on glacial moraines, screes and volcanic debris (Grubb 1985).

In short, it is not true that the combination of low productivity ('stress') and disturbance is non-viable for plants, as Grime (1979) has alleged. Southwood (1977) has quoted various examples of animals associated with this combination of conditions.

Further kinds of differentiation between species

In the examples given in Tables 4 and 5 just one or a few species have been plotted at approximately the same position in a two-dimensional arrangement based on distribution along an environmental gradient and response to disturbance. In fact it is usual to find several to many species at approximately the same point, especially in the herb and shrub layers in temperate communities, and equally in the tree layer in warm temperate and tropical communities. How may these species at one point in the two-dimensional plot be differentiated? Differentiation in phenology is of great importance in some types of vegetation. For example, in many prairies there are waves of species, defined by times of leafing and flowering, which are successively taller through the season (Curtis 1959). In other cases seasonal changes in canopy height are not marked but phenological separation can still be effective (Berendse 1983; Burdon 1983). Phenological separation in such functions as nutrient uptake has also been found between co-existing tree species, e.g., by Rogers and

Westman (1979) for certain *Eucalyptus* spp. in Queensland. Comparable differentiation in phenology must occur between populations within particular species in many kinds of vegetation.

Subtle differentiation in growth form is probably more important for co-existence than has been considered in the past. For example, slight differences in canopy height may be important for co-existence in closely-grazed turf; the taller-growing species can suppress the shorter-growing but never oust them as long as grazing continues, while the lower-growing species are rarely damaged by the grazer, especially if the grazer is non-selective as cattle are (Grubb 1984). As for shrubs, subtle differences in growth form between widespread northern European species (relating to height and to ability to persist as straggling shoots among the larger crowns of other species) are indicated in the work of Sloet and Adriani (1971). Recently experimental evidence has been produced for the view that there is greater complementarity in growth requirement between late-successional species than between early-successional species in tall-grass prairie (Parrish and Bazzaz 1982); the complementarity may involve both phenology and subtle differences in growth form. This idea should certainly be followed up in other types of community.

After differentiation in growth form and phenology, it is often found that many species still overlap widely in range. It follows from what I have written earlier (Grubb 1977) that differentiation in requirements for regeneration (other than degree of disturbance) is important here. Examples of factors that matter are whether the year of gap formation is wet or dry, hot or cold, whether the soil surface is rough or smooth, bare or covered by litter, and whether the gap is reached by specific herbivores or micro-organisms or not. Examples of this kind of differentiation in 'regeneration niche' are given for short-lived chalk-grassland plants in Britain by Grubb (1984). The need for very subtle characterization of the season and soil surface in differentiating species is beautifully illustrated by the work of Breman et al. (1980) on annual plants of savanna and the savanna-desert transition in West Africa. In perennial communities, especially forests, there is an important degree of differentiation between species according to the *part* of the gap which they require. Oldemann (1978) has categorized the distinctive parts of a tree-fall gap in a Tropical Rain forest, and Putz (1983) has given examples of species suited to different parts. As to 'onward growth', after establishment, Boukoungou, Raynal and Geis (1983) have shown for temperate forest the kind of differentiation that can be found between trees at one site in their responses to weather and pests. There is a potentially infinite number of axes for differentiation during regeneration, and there is no virtue in trying to sift out only one axis here.

PLANTS IN RELATION TO COMPETITION, INTERFERENCE, DOMINANCE AND RELATIVE ABUNDANCE

Competition and interference

Competition has come to have three different meanings to plant ecologists:

(i) the relationship between any two species not symbiotic with each other and capable of occupying the same landscape unit, considered over the whole life cycle,

(ii) the sum of the processes by which the yield of one plant is reduced as a result of another plant being present, and

(iii) the 'tendency of neighbouring plants to utilize the same quantum of light, ion of a mineral nutrient, molecule of water or volume of space' (Grime 1979).

The first meaning seems to be the oldest. A term is certainly needed for the relationship in question, and there is great value in thinking of plants as competing at all stages of the life cycle (cf. Salisbury 1929, Levins and Culver 1971). I therefore accept that meaning. It is important to realise that in nature the balance between two species is often determined by the presence of third parties; this point is elaborated elsewhere (Grubb 1985).

The second meaning for 'competition' has been given the name 'interference' by Harper (1961), and I shall use that term. There is much variation in the extent to which interference effects are reciprocal, as shown by Harper (1977). It is essential to appreciate that plants cannot be ordered in respect of their potential for interference without stipulating the conditions for growth; there are many experimental results published, showing that the outcome of interference between two species or two biotypes depends on temperature, water supply, pH and nutrient-regime (Clements, Weaver and Hanson 1929). It is also essential to realize that susceptibility to above-ground interference does not necessarily parallel susceptibility to below-ground interference (Newman 1983). Similarly the potential to inflict above-ground interference does not necessarily parallel the potential to inflict below-ground interference. Plants may inflict severe below-ground interference where they are not even represented by shoots, e.g., in the gaps between shrubs in semi-desert; Friedman, Orshan and Ziger-Cfir (1977) found for an area of the Negev Desert where *Artemisia herba-alba* had a crown cover of only 16% that the removal of its shoots and roots from five areas of 6.25 m^2 led to a tenfold increase in the yield of annual herbs. It is also true that low-growing plants can seriously reduce the performance of taller plants above them; the effect of grasses on orchard trees in moist climates is well-known (Atkinson and White 1980) and more recently Knoop and Walker (1985) have shown for South African savanna that removal of grasses can induce a massive increase in shoot growth in the trees in a matter of weeks. Where savanna grasses are overgrazed, impenetrable scrub develops (Walter 1973, p. 336). Reference to such effects reminds us that interference between plants is of great importance all along any environmental gradient, and not just under 'optimal' conditions *sensu* Beard (1944); the point was well made in respect of nutrient-poor sites by Newman (1973).

The third meaning of competition cited above is not usually an operational one because we simply do not have the information necessary to interpret most recorded events of interference.

Grime (1979) has suggested that competitiveness be used as a major axis for differentiating plant species or populations. I cannot agree. Both competition in the broad sense, and interference, are important in three contexts: in confining the distribution of plants along environmental gradients, in affecting their performance in relation to different levels of disturbance, and in determining their relative abundance where a particular combination of environmental conditions and degree of disturbance is found. The last case is discussed elsewhere (Grubb, Kelly and Mitchley 1982).

Competition can be important not only in limiting the distribution of a plant population on that side of its range where it has the potential to grow faster, but also on the side where it tends to grow more slowly. Two examples must suffice to make this point, one concerning an altitudinal gradient and one a gradient of water supply.

The first example is based on the '*Nothofagus* gap' on the western side of the South Island of New Zealand (Wardle 1964). North and south of the 'gap' it is possible to recognize a lowland type of forest rich in tree species, a montane belt in which *Nothofagus* spp. and *Weinmannia racemosa* are the principal trees, and a subalpine belt in which *Nothofagus menziesii* is the principal species and *Weinmannia racemosa* is normally absent. In the '*Nothofagus* gap' the climate, rock and topography are very similar to those found in the areas where *Nothofagus* is present to the north and south; the reason for the gap in the distribution of *Nothofagus* is not understood, but presumably has a historical basis. Where *Nothofagus* is absent, the lowland forest is much as elsewhere, but *Weinmannia* is the principal species in both the montane belt and the somewhat truncated subalpine forest, forming a timberline at ca. 750–760 m, whereas the normal upper limit of this species is ca. 600–630 m. There is thus a *prima facie* case that the upper limit of *Weinmannia racemosa* is determined by competition from *Nothofagus menziesii* wherever the latter species is present. Similarly in the less oceanic part of the '*Nothofagus* gap', *Podocarpus hallii* forms the overstorey of the forest at 600–900 m, whereas to the north and south this zone is dominated by *Nothofagus solandri* var. *cliffortioides*, and *Podocarpus hallii* is not found above ca. 600 m. 'Nature's experiments' of this kind make one wonder how many other species are confined by competition to the lower altitudinal belts on mountains.

The second example is based on the distribution of trees at the forest-steppe boundary in western North America. The tree species which are most resistant to summer drought and winter cold (*Pinus ponderosa*, *Juniperus monosperma*, *J. pinchotii* and various *Quercus* spp.) form rather open woodlands on soils of all textures where the rainfall is higher than that experienced in the steppe (Daubenmire 1968, Walter 1968). However, in the Great Plains region they are confined to the rocky soils of steep scarps, and there some of them occur with a yearly rainfall as low as 300 mm; before man intervened the fine-textured soils of the plains were wholly covered with treeless prairie. That the *Pinus*, at least, can survive the drought experienced by plants on the fine-textured soils of the dry prairie region, provided grasses are removed while the trees are very young, is shown by what happened in one plantation described by Wells (1965). The trees grew to a height of 7 m in 30 years, and survived the most severe droughts on record. Probably the prairie concerned is naturally treeless because tree seedlings grow so slowly in the presence of grass that they never escape from the killing fires that spread through the grass quite frequently. Indeed, Wells argued that grasses could be seen as competing with woody plants through their formation of an inflammable matrix around them. Suppression of the growth of tree seedlings by grasses may also make it virtually impossible to escape from grazing mammals.

It is also important to understand that the impact of competition on the distribution of a plant population in vegetation subject to varying degrees of disturbance can be such as to limit its performance on the side of greater disturbance as well as the side of lesser disturbance. It is generally appreciated that early-successional annuals and biennials cannot become established under later-successional perennials, and that the same is true of 'intolerant' trees under 'tolerant' trees, but it is often not appreciated that at an early stage in succession annuals can suppress the performance of biennials and perennials. This much was shown by experiments in which annuals were removed from old-field sites by Raynal and Bazzaz (1975). In a long-term outdoor experiment at Cambridge *Dactylis glomerata*, a fairly small-seeded species seen to act as a pioneer at field sites, was able to suppress the growth of the larger-seeded late-successional *Arrhenatherum elatius* in the first year or two after planting but gave way to *Arrhenatherum* by about the fifth year (Grubb 1982). It seems likely that, when'tolerant' trees are mixed in forest gaps with 'intolerant' trees, their growth can be suppressed but I know of no evidence on this point. However, it is well established that in man-induced succession on old fields and clearcuts 'mid-successional' shrubs can delay the establishment and expansion of late-successional trees for very long periods (Franklyn and Dyrness 1969); some decades may be needed before the shrub cannopy breaks up and lets in the trees.

All of the points made above about the nature of competition between species are likely to apply with equal force to the relations between different populations within species. As yet there have been rather few relevant studies on wild plants, e.g., Eagle and Williams (1968) on *Dactylis glomerata*, Snaydon (1971) on *Trifolium repens*, Solbrig and Simpson (1977) on *Taraxacum officinale*, and many more are needed, preferably carried out under field conditions with comprehensive monitoring of the environment.

In so far as both competition in the broad sense and the more restricted phenomenon of interference are important at all points on an environmental gradient, and at sites suffering all degrees of disturbance, and in so far as the outcome of mutual interference depends so much on environmental conditions, I think it is quite wrong to use 'competitiveness' as an axis for differentiating species, or to recognize a 'strategy' of 'competitor'.

Dominance

According to Clements et al. (1929, p. 4) the terms dominant and suppressed were first used of individual plants, in fact by foresters in Denmark between 1780 and 1796. Such a usage has persisted with foresters, and has been adopted recently by a wider range of ecologists in studies on the effects of crowding of plants, e.g., by Mohler et al. (1978).

Despite the historical precedent, the concept of dominance has been used by ecologists chiefly for relations between species, and with two extreme meanings, both of which were established fifty years ago. Braun-Blanquet (1932) used it to refer to the proportion of the total biomass present, irrespective of life-form; in recent years this meaning has been adopted particularly by those interested in formal studies on 'dominance-diversity' relations (cf. Whittaker 1965). The other meaning can be traced back to early descriptions of certain life-forms and physiognomic types dominating the appearance of a community (cf. Warming 1909), and implications about certain life-forms determining what others may do. Weaver and Clements (1929, p. 126) explicitly recognized dominance as the relationship between life-forms. The two meanings of dominance diverge most clearly when the plants which determine the future development of a site represent comparatively little of the total biomass, e.g., the invading *Sphagnum* individuals in a forest about to be turned into a blanket bog or raised bog.

Dominance is not a necessary term in 'dominance-diversity' studies because the two essential elements of diversity have their own well-established names 'richness' and 'equitability'. In contrast it is useful to have a term for the relationship between life-forms, and I therefore propose to perpetuate the term dominance in this context (Grubb 1984). The adjective 'dominant', so often applied to the most abundant species in a life-form, is conveniently replaced with 'principal'.

Probably the original usage of dominant, in the context of individuals, can go on side by side with the usage in the context of life-forms without any serious harm.

Control of relative abundance

As indicated on pp. 609–610 it is usual to find several species at any one point on an environmental gradient and at a site with a particular degree of disturbance. These may be conveniently separated according to their relative abundance, i.e., their abundance relative to one another, as done by Grubb et al. (1982) for various communities, e.g., in their Fig. 2 for calcareous grassland. 'Abundance' is here taken in the broadest sense to include measures of cover, above-ground biomass or basal area as well as number of individuals.

Ordering by relative abundance will depend on the time of sampling (reflecting the 'phenological niche'), the growth form, the potentials for interference of the relevant species at the point on the gradient concerned, and various axes of the 'regeneration niche' other than the response to degree of disturbance. The potential for interference cannot be defined without reference to the weather experienced for some time past, and the length of significant lag effects is likely to differ between plant types at one site, as shown by Noble (1977) for semi-desert in South Australia. In this way the 'habitat niche' (including response to fluctuating weather) is bound to merge with the 'regeneration niche'.

It is clear to all that most environmental gradients are complex in nature. I have tried to show that the phenomenon of disturbance is also very complex. It has to be emphasized that the control of relative abundance is even more complex. To understand it under a given set of conditions provides an outstanding challenge.

CONCLUSIONS

1. Plant populations are distributed according to requirements as well as tolerances. Conditions which suppress the growth of some plants are actually required by others. There can be no satisfying, generally applicable concept of 'stress'. The key characters of plants suited to various 'stress' conditions are very different (Table 1). It is not possible to characterize usefully a single 'stress-tolerator' type of plant when 'stress' is defined as it has been by most ecophysiologists in the past. The most useful way of differentiating species in relation to environmental conditions is to order them along gradients of factor-complexes. 'Edge-plants' which are naturally confined to narrow ecotones form an interesting special case.

2. Disturbance-events affecting vegetation are of two main types – continual and periodic. Continual disturbance can be of several very different kinds (Table 2). Periodic disturbance can vary significantly in extent, frequency and intensity. The key characters of plants suited to various disturbance-regimes are very different (Table 3). It is not possible to characterize a single super-type of plant that is suited to strong disturbance. Many plants are facultatively associated with disturbance (whether in primary or man-induced successions) and many have wider tolerance-ranges under more disturbed conditions. Man-induced successions do not simply repeat natural internal successions or primary successions of near-by sites. In most vegetation types it is possible to arrange species in a two-dimensional framework of reference based on environmental gradients and response to either disturbance during natural regeneration processes, or response to continual disturbance, e.g., on sand dunes.

3. Competition is best defined as the relationship (considered over the whole life cycle) between two plants that have the potential to occupy a given landscape unit. The ability of one plant to reduce the performance of another plant growing with it is best called interference. Both competition and interference, so defined, are important in three contexts: (i) determination of the distributions of populations along enviromental

gradients (including limitation on the side of the range where the individual plants grow more slowly), (ii) determination of performance along disturbance gradients (including limitation on the side of more disturbance), and (iii) determination of relative abundance under a given set of conditions. The outcome of both interference and the whole of competition between two species depends on the conditions. It is misleading to describe a single type of plant as a 'competitor'.

4. The term 'dominance' is best reserved to describe the relationship between life-forms, or between individuals within a population of one species. Control of relative abundance at a given point on an environmental gradient with a given degree of disturbance depends on the time of sampling (and thus phenology), the growth forms of the plants involved and aspects of the regeneration niche other than response to extent, frequency or intensity of disturbance.

5. Generalizations are essential to the development of any science. The question at issue is what kinds of generalizations are best for plant ecology. I suggest that it will be hard to sustain the recognition of two or three basic types or 'strategies' as a satisfying approach in the context of the whole world's vegetation because such types can represent adequately only a fraction of the major functional types, and must be defined either too vaguely or in a way subject to endless exceptions. I suggest that in contrast generalizations about kinds of ecological differentiation, such as are set about above, can be sustained and do have very general applicability.

6. Further work should be aimed at disentangling the major kinds of differences in life-form, phenology and requirements for regeneration (including life history traits and demographic properties) that enable plants in various kinds of community to coexist indefinitely, and determine their relative abundance under particular conditions. Special attention might be paid to plants of edges as opposed to natural gaps, and to plants able to increase their range after disturbance where others cannot.

ACKNOWLEDGEMENTS

I am indebted to many friends for comments on early drafts, especially Dr. J.J.N. Campbell, Dr. H.J. During, Professor J.P. Grime, Dr. E.I. Newman and Professor C.B. Osmond.

REFERENCES

Andel, J. van, Bos, W. and Ernst, W. (1978) An experimental study on two populations of *Chamaenerion angustifolium* (L.) Scop. (= *Epilobium angustifolium* L.) occurring on contrasting soils, with particular reference to the response to bicarbonate. New Phytologist 81: 763–772.

Atkinson, D. and White, G.C. (1980) Some effects of orchard soil management on the mineral nutrition of apple trees. In D. Atkinson, J.E. Jackson, R.O. Sharples and W.M. Waller (Eds.), Mineral Nutrition of Fruit Trees, pp. 241–254. Butterworths, London.

Bazzaz, F.A. (1979) The physiological ecology of plant succession. Annual Review of Ecology and Systematics, 10: 351–371.

Beard, J.S. (1944) Climax vegetation in tropical America. Ecology 25: 127–158.

Berendse, F. (1983) Interspecific competition and niche differentiation between *Plantago lanceolata* and *Anthoxanthum odoratum* in a natural hay field. Journal of Ecology, 71: 379–390.

Boorman, L.A. (1982) Some growth patterns in relation to the sand dune habitat. Journal of Ecology, 70: 607–614.

Boukoungou, G.J.E., Raynal, D.J. and Geis, J.W. (1983) Tree population dynamics in relation to climate and forest history in the Oswegatchie Plains, northern New York. Vegetatio, 54: 37–59.

Braun-Blanquet, J. (1932) Plant Sociology (English translation by G.D. Fuller and H.S. Conard). McGraw Hill, New York.

Breman, H., Cissé, A.M., Djiteye, M.A. and Elberse, W.T. (1980) Pasture dynamics and forage availability in the Sahel. Israel Journal of Botany, 28: 227–251.

Buckley, R. (1981) Soils and vegetation of central Australian sandridges. III. Sandridge vegetation of the Simpson Desert. Australian Journal of Ecology, 6: 405–422.

Burdon, J.J. (1983) Biological flora of the British Isles: *Trifolium repens* L. *Journal of Ecology* 71: 307–330.

Burrell, J. (1965) Ecology of *Leptospermum* in Otago. New Zealand Journal of Botany, 3: 3–16.

Callaghan, T.V. and Emanuelsson, U. (1985) Population structure and processes of tundra plants and vegetation. In J. White (Ed.), The Population Structure of Vegetation, pp. 339–439. Dordrecht: Dr. W. Junk.

Chadwick, H.W. and Dalke, P.D. (1965) Plant succession of dune sands in Fremont County, Idaho. Ecology, 46: 765–780.

Clements, F.E., Weaver, J.E. and Hanson, H.C. (1929) Plant Competition. Carnegie Institute of Washington.

Coley, P.D. (1983) Herbivory and defensive characteristics of tree species in a lowland tropical forest. Ecological Monographs, 53: 209–233.

Cooper, W.S. (1939) A fourth expedition to Glacier Bay, Alaska. Ecology, 20: 130–155.

Curtis, J.T. (1959) The Vegetation of Wisconsin. University of Wisconsin Press, Madison.

Daubenmire, R. (1966) Vegetation: identification of typal communities. Science, New York 151: 291–298.

Daubenmire, R. (1968a) Ecology of fire in grasslands. Advances in Ecological Research, 5: 209–266.

Daubenmire, R. (1968b) Forest vegetation of eastern Washington and northern Idaho. Technical Bulletin of the Washington Agricultural Experiment Station 60: 1–104.

Dierschke, H. (1974) Saumgesellschaften im Vegetations – und Standortsgefälle an Waldrändern. Scripta Geobotanica, 6: 1–246.

Duhamel de Monceau (1760) Des semis et plantations des Arbres et de leur Culture.

Eagles, C.F. and Williams, D.H. (1969) The effect of environment on the competitive ability of two natural populations of *Dactylis glomerata* L. Journal of Agricultural Science, Cambridge, 72: 271–272.

Ellenberg, H. (1958) Bodenreaktion (einschliesslich Kalkfrage). Handbuch der Pflanzenphysiologie, 4: 638–708.

Ellenberg, H. (1982) Die Vegetation Mitteleuropas mit der Alpen in ökologischer Sicht, 3rd edn. Ulmer, Stuttgart.

Fenner, M. (1978) A comparison of the abilities of colonizers and closed-turf species to establish from seed in artificial swards. Journal of Ecology, 66: 953–963.

Franklyn, J.F. and Dyrness, C.T. (1969) Vegetation of Oregon and Washington. United States Department of Agriculture Forest Service Research Paper (PWN-80), i–vi, 1–216.

Friedman, J., Orshan, G. and Ziger-Cfir, Y. (1977) Suppression of annuals by *Artemisia herba-alba* in the Negev Desert of Israel. Journal of Ecology, 65: 413–426.

Gill, A.M. (1981) Coping with fire. In J.S. Pate and A.J. McComb (Eds.), The Biology of Australian Plants, pp. 65–87. University of Western Australian Press, Nedlands.

Grime, J.P. (1974) Vegetation classification by reference to strategies. Nature, London, 250: 26–31.

Grime, J.P. (1979) Plant Strategies and Vegetation Processes. Wiley, Chichester.

Grime, J.P. and Hodgson, J.G. (1969) An investigation of the ecological significance of lime-chlorosis by means of large-scale comparative experiments. In I.H. Rorison (Ed.), Ecological Aspects of the Mineral Nutrition of Plants, pp. 67–99. Blackwell, Oxford.

Grubb, P.J. (1977) The maintenance of species-richness in plant communities: the importantance of the regeneration niche. Biological Reviews, 52: 107–145.

Grubb, P.J. (1982) Control of relative abundance in roadside *Arrhenatheretum*: results of a long-term garden experiment. Journal of Ecology, 70: 845–861.

Grubb, P.J. (1984) Some growth points in investigative plant ecology. In J.H. Cooley and F.B. Golley (Eds.), Research Trends in Ecology in the 1980s, pp. 51–74. Plenum, New York.

Grubb, P.J. (1985) The ecology of establishment. In A.D. Bradshaw, D. Goode and E. Thorpe (Eds.), Ecology and Landscape Planning, in press. Blackwell, Oxford.

Grubb, P.J., Kelly, D. and Mitchley, J. (1982) The control of relative abundance in communities of herbaceous plants. In E.I. Newman (Ed.), The Plant Community as a Working Mechanism, pp. 79–97. Blackwell, Oxford.

Grubb, P.J. and Stevens, P.F. (1985) The forests of the Fatima basin and Mt Kerigomna, Papua New Guinea, and a Review of Montane and Subalpine Forests elsewhere in Papuasia. Department of Biogeography and Geomorphology Publication (BG/5). Australian National University, Canberra (in press).

Harper, J.L. (1961) Approaches to the study of plant competition. In F.L. Milthorpe (Ed.), Mechanisms in Biological Competition. Symposia of the Society for Experimental Biology, 15: 1–39.

Harper, J.L. (1977) Population Biology of Plants. Academic Press, London.

Harper, J.L. (1982) After description. In E.I. Newman (Ed.), The Plant Community as a Working Mechanism, pp. 11–25. Blackwell, Oxford.

Jones, E.W. (1955–6) Ecological studies on the Rain Forest of southern Nigeria. IV. The plateau forest of the Okomu Forest Reserve. Journal of Ecology 43: 564–94, 44: 83–117.

Knapp, R. (1959) Untersuchungen über den Einfluss von Tieren auf die Vegetation. I. Rasengesellschaften und *Talpa europaea*. Angewandte Botanik, 33: 177–189.

Knoop, W.T. and Walker, B.H. (1985) Interactions of woody and herbaceous vegetation in a southern African savanna. Journal of Ecology, 73: 235–254.

Law, R. (1979) Ecological determinants in the evolution of life histories. In R.M. Anderson, B.D. Turner and L.R. Taylor (Eds.), Population Dynamics, pp. 81–103. Blackwell, Oxford.

Levins, R. and Culver, D.C. (1971) Regional co-existence of species and competition between rare species. Proceedings of the National Academy of Science, U.S.A. 68: 1246–1248.

Levitt, J. (1972) Responses of Plants to Environmental Stresses. Academic Press, New York.

MacArthur, R.H. and Wilson, E.O. (1967) The Theory of Island Biogeography. Princeton University Press. Princeton, N.J.

Marks, P.L. (1975) On the relation between extension growth and successional status of deciduous trees of the northeastern United States. Bulletin of the Torrey Botanical Club, 102: 172–177.

Marks, P.L. (1983) On the origin of the field plants of the northeastern United States. American Naturalist, 122: 210–228.

Mohler, C.L., Marks, P.L. and Sprugel, D.G. (1978) Stand structure and allometry of trees during self-thinning of pure stands. Journal of Ecology, 66: 599–614.

Murray, D.B. and Nichols, R. (1966) Light, shade and growth of some tropical plants. In R. Bainbridge, G.C. Evans and O. Rackham (Eds.), Light as an Ecological Factor, pp. 249–263. Blackwell, Oxford.

Newman, E.I. (1973) Competition and diversity in herbaceous vegetation. Nature, London, 244: 310.

Newman, E.I. (1983) Interactions between plants. Encyclopaedia of Plant Physiology, New Series 12C: 679–710.

Noble, I.R. (1977) Long-term biomass dynamics in an arid chenopod shrub community at Koonamore, South Australia. Australian Journal of Botany, 25: 639–653.

Noble, I.R. and Slatyer, R.O. (1980) The use of vital attributes to predict successional changes in plant communities subject to recurrent disturbances. Vegetatio, 43: 5–21.

Oldemann, R.A.A. (1978) Architecture and energy exchange of dicotyledonous trees in the forest. In P.B. Tomlinson and M.H. Zimmermann (Eds.), Tropical Trees as Living Systems, pp. 535–560. Cambridge University Press.

Osmond, C.B., Bjorkman, O. and Anderson, D.J. (1980) Physiological Processes in Plant Ecology. Springer, Berlin.

Parrish, J.A.D. and Bazzaz, F.A. (1982) Competitive interactions in plant communities of different successional ages. Ecology, 63: 314–320.

Paul, P. (1969) Études expérimentales sur le déterminisme de la composition floristique des pelouses xérophiles. I. L'action autecologique comparée du sol en conditions naturelles et controlées. Oecologia Plantarum, 4: 47–70.

Putz, F.E. (1983) Treefall pits and mounds, buried seeds, and the importance of soil disturbance to pioneer trees on Barro Colorado, Panama. Ecology, 64: 1069–1074.

Ramenskii, L.G. (1938) Introduction to the Geobotanical Study of Complex Vegetations. Selkhozgis, Moscow. (in Russian).

Raynal, D.J. and Bazzaz, F.A. (1975) Interference of winter annuals with *Ambrosia artemisiifolia* in early successional fields. Ecology, 56: 35–49.

Rogers, R.W. and Westman, W.E. (1979) Niche differentiation and maintenance of genetic identity in cohabiting *Eucalyptus* species. Australian Journal of Ecology, 4: 429–439.

Roughton, R.D. (1972) Shrub age structures on a mule deer winter range in Colorado. Ecology, 53: 615–625.

Ruthsatz, B. (1978) Las plantas en cojin de los semi-desiertos andinos del Noreste Argentino. Darwiniana, 21: 491–539.

Salisbury, E.J. (1929) The biological equipment of species in relation to competition. Journal of Ecology 17: 197–222.

Sloet van Oldruitenborgh, C.J.M. and Adriani, M.T. (1971) On the relation between vegetation and soil-development in dune-shrub vegetations. Acta botanica neerlandica, 20: 198–204.

Snaydon, R.W. (1971) An analysis of competition between plants of *Trifolium repens* L. populations collected from contrasting soils. Journal of Applied Ecology, 8: 687–697.

Solbrig, O.T. and Simpson, B. (1977) A garden experiment on competition between two types of the common dandelion (*Taraxacum officinale*). Journal of Ecology, 65: 427–430.

Southwood, T.R.E. (1977) Habitat, the templet for ecological strategies? Journal of Animal Ecology, 46: 337–365.

Specht, R.L. (1972) The Vegetation of South Australia, 2nd edition. Government Printer, Adelaide.

Tüxen, R. (1952) Hecken und Gebüsche. Mitteilungen des Geographische Gesellschaft, Hamburg, 50: 85–117.

Tüxen, R. (1975) Dauer-Pioniergesellschaften als Grenzfall der Initialgesellschaften. In W. Schmidt (Ed.), Sukzessionsforschung, pp. 13–30. Cramer, Vaduz.

Uhl, C., Clark, K., Clark, H. and Murphy, P. (1981) Early plant succession after cutting and burning in the upper Rio Negro region of the Amazon basin. Journal of Ecology, 69: 631–649.

Van Valen, L. (1971) Group selection and the evolution of dispersal. Evolution, 25: 591–598.

Vasek, F.C. (1980) Early successional stages in Mojave Desert scrub vegetation. Israel Journal of Botany, 28: 133–148.

Veblen, T.T. and Ashton, D.H. (1978) Catastrophic influences on the vegetation of the Valdivian Andes, Chile. Vegetatio, 36: 149–167.

Walter, H. (1968) Die Vegetation der Erde im Ökophysiologischer Betrachtung. II. Die gemässigten und arktischen Zonen. Fischer, Jena.

Walter, H. (1973) Ibidem. I. Die tropischen und subtropischen Zonen, 3rd edn. Fischer, Jena.

Wardle, P. (1964) Facets of the distribution of forest vegetation in New Zealand. New Zealand Journal of Botany, 2: 352–366.

Warming, E. (1909) Oecology of Plants. Clarendon, Oxford.

Watt, A.S. (1947) Pattern and process in the plant community. Journal of Ecology, 35: 1–22.

Weaver, J.E. and Clements, F.E. (1929) Plant Ecology. McGraw-Hill, New York.

Wells, P.V. (1961) Succession in desert vegetation on streets of a Nevada ghost town. Science, New York, 134: 670–671.

Wells, P.V. (1965) Scarp woodlands, transported soils and concept of grassland climate in the Great Plains region. Science, New York, 148: 246–249.

Wells, P.V. (1976) A climax index for broadleaf forest: an n-dimensional ecomorphological model of succession. In J.S. Fralish, G.J. Weaver and R.C. Schlesinger (Eds.), Central Hardwood Forest Conference, pp. 131–176. Department of Forestry, Southern Illinois University, Carbondale, Illinois.

Werner, P.A. and Platt, W.J. (1976) Ecological relationships of co-occurring goldenrods (Solidago; Compositae). American Naturalist, 110: 959–971.
White, F. (1983) The Vegetation of Africa. UNESCO, Paris.
White, J. (1985) The census of plants in vegetation. In: J. White (Ed.), The Population Structure of Vegetation, pp. 33–88. Dordrecht: Dr. W. Junk.
White, P.S. (1979) Pattern, process and natural disturbance in vegetation. Botanical Reviews, 45: 229–299.
Whittaker, R.H. (1965) Dominance and diversity in land plant communities. Science, New York, 147: 250–260.
Whittaker, R.H. (1967) Gradient analysis of vegetation. Biological Reviews, 42: 207–264.
Whittaker, R.H. and Goodman, D. (1979) Classifying species according to their demographic strategy. I. Population fluctuations and environmental heterogeneity. American Naturalist, 113: 185–200.

POPULATION DYNAMICS OF AUSTRALIAN PLANT COMMUNITIES, WITH SPECIAL REFERENCE TO THE INVASION OF NEOPHYTES

O.B. WILLIAMS

ABSTRACT

The colonization of Australia by the British, starting in 1788, led to the imposition of agriculture and other European activities on the landscapes, the permanent displacement of the Aboriginal people, prior residents for more than 30,000 years, and a massive never-ending flow of neophytes. Rarely have these neophytes invaded unmodified plant communities; some indigenous species assumed neophytic status. Differences between human impacts before and after 1788 are described, particularly in relation to fire and disturbance regimes. Demographic studies in representative Australian vegetation have explained the importance for neophytes of these differences in human impacts. Research and management procedures designed to block the ingress of unwanted neophytes into pastures of sown neophytes might be applied successfully to natural and semi-natural vegetation under threat of invasion.

INTRODUCTION

Australia was settled by the British in 1788 and was appropriated in its entirety as a dependency of the United Kingdom in 1829; settlement was rapid between 1830 and 1880, with little effective resistance from the Aboriginal people who had been in residence for more than 30,000 years previously. There are reasonable descriptions of Australian landscapes in the early days of European settlement and considerable documentation of the development of these landscapes for European farming, the major land use today (Davidson 1981).

Since 1788 each activity of European peoples has spread rapidly to all parts of the continent because the topography is subdued, there are few impenetrable forests, deserts can be avoided, the few perennial rivers are easily crossed, there are no hostile natives, there is a common language and political system, and the land mass with the island state of Tasmania is not shared with other nations. From 1788 until federation of the Autralian colonial states, the flow of new plant propagules into Australia

White, J. (ed.), The Population Structure of Vegetation.
ISBN 978-94-010-8927-2

was almost uncontrolled. The first trickle of plants was represented mainly by species from Britain, few of which grew without massive and continuous disturbance of the native vegetation and soils.

By the mid-1800s this trickle had become a torrent and by the 1880s it was a flood (Rolls 1969). Plant propagules from Europe, North, Central and South America, Africa and the Middle East in particular were introduced by individuals and so-called 'acclimatisation societies'. Many introduced plant species failed to establish, some established and conferred considerable economic benefits on society; others caused, and continue to cause, substantial economic losses. Plant propagules also entered as impurities in crop and pasture seeds, contaminants in drought fodders (such as hay imported from Chile), in jute and other packaging materials, in hair of horses, fleeces of sheep and goats and in saddles of camels. Plant quarantine restrictions came into force progressively from 1st July, 1901 but the quarantine inspection system for plants did not function fully until 1908.

In discussion of plant invasions a number of terms are used and their definition follows: Aliens as defined by Michael (1981) can be exotic species (plants introduced from overseas) as well as indigenous species (Australian native species) that have successfully taken up residence in plant communities in which there is good reason to believe that the species are newcomers. Sometimes there can be difficulties in distinguishing with certainty whether or not the plant community so invaded by an indigene is man-modified or natural. Indigenous species need not be exclusive to Australia. The term neophyte is the same as alien (Michael 1981) and therefore covers successful exotics and native (or indigenous) invaders. Neophytes are now recognised as being naturalised species.

Successful invasion of a native plant community by an exotic neophyte can be recognised, and is generally beyond dispute, but there are occasional difficulties (Everist 1960, Kloot 1984). Michael (1981) noted that present populations of both *Oxalis corniculata* and *Cynodon dactylon* include alien and native forms (equivalent to exotic and indigenous forms). *Tribulus terrestris* is another example (Squires 1979). Obviously the alien form is now naturalised, and the native form, should it successfully invade a plant community in which it has not been resident, would also be naturalised and would become a neophyte. Other native (indigenous) species which invade and become neophytes include *Acacia baileyana*, *Eremophila mitchelli* and *Sclerolaena birchii* (Auld 1981).

The apparent ease with which so many neophytes have invaded and become part of Australian landscapes and by implication the apparent weakness of the pre-1788 plant communities to repel these invaders belies the intensity of post-settlement disturbances. In this chapter the demographic behaviour of a range of species is described in their community context, both before and after European occupation, this 'apocalyptic event for Australian ecosystems' (Adamson and Fox 1982). The contribution that plant demography can make in studies on neophytes is also assessed.

MAJOR AUSTRALIAN VEGETATION TYPES ABOUT 1788

The commentary and map sheet 'Natural Vegetation' prepared by Carnahan (1976) show the distribution and character of the vegetation before it was changed by British settlers and their Australian descendants; this scheme has been modified from Specht (1970). A generalised version of this map together with abbreviated descriptions of the major groupings of natural vegetation are provided by Carnahan (1977): this map and its major groupings provide the framework for *Australian Vegetation* (Groves 1981). In contrast, the map accompanying *The Vegetation of Australia* (Beadle 1981), shows the major vegetation types for Australia about 1788.

Plant communities ca. 1788 were assemblages of species that were survivors or the descendants of an ancient flora that had experienced far-reaching and drastic climatic and anthopogenic influences over millennia (Barlow 1981, Smith 1982, Lange 1982, Singh 1982, H.A. Martin 1982, Barker and Greenslade 1982). Major influences in the development of the vegetation as it was seen ca. 1788 were the fire regimes of Aboriginal populations for possibly 10,000 years in a climate not too dissimilar from the post-1788 period, with short periods of wet years and long stretches of dry years. A cold arid period 20,000–15,000 years ago may have been responsible for the extinction of many species that had survived since the late Tertiary (Walker 1982). The Australian landscapes ca. 1788 were well-stocked with perennial vegetation and were accompanied by a unique fauna. Australian isolation and the absence of migratory herds of non-endemic hooved ungulates adapted to grazing and browsing steppe and grassland made entry of external herbaceous elements difficult or, at best, slow (A.R.H. Martin 1982). Although Martin was referring to the largely autochthonous desert biota, the same argument could be made for mesic environments. There was no tradition of Aboriginal villages and agriculture with systems of shifting cultivation to aid entry of ruderals.

INVASIVE CHANGES IN SOME INDIGENOUS SPECIES BEFORE EUROPEAN INTERVENTION

The population biology of the species in Australian plant communities ca. 1788 reflected the influence of fire, climate, Aboriginal people, and indigenous vertebrates and invertebrates. Fire appears to have been a major ecological and evolutionary factor in Australian vegetation (Gill et al. 1981) with extreme susceptibility to fire exhibited by such species as *Eucalyptus regnans*, *Callitris columellaris* and *Atriplex vesicaria*; species such as mallee *Eucalyptus*, some *Eremophilia* spp. and *Acacia* spp. regenerated from lignotubers or rootstocks. Yet other species such as *Atriplex nummularia* were fire-retardant. Paradoxically there are fire-susceptible species that are fire-dependent. The population structure of plant communities differed, depending on age since last fire, and the fire mosaic enabled this spread of population structure to be expressed across a landscape. The fire mosaic was as important to the continued existence of the plant (and animal) species as fire *per se* and the fire regime determined the success of the species. Climate, a major long-term factor, operates through fuel production and provides conditions under which fires can occur and plant recruitment or regeneration take place. Aboriginal people, in using fire, were both ecological and evolutionary agents: "in Australia every day for millions of days countless fires had been lit or enlarged for countless purposes, and many of these fires had unintended effects" (Blainey 1975).

Grazing by vertebrate herbivores, such as the larger marsupials, is presumed to have a substantial effect at the species level and to interact strongly with fire. Chiswell (1982) invokes such grazing in the pre-European period for the maintenance of open forest in what is now the Pilliga Scrub. A markedly different effect of such grazing is the actual example of *Eucalyptus regnans* seedlings of small or scattered cohorts 'being remorselessly grazed by wallabies (*Wallabia bicolor*), which eliminated all but the most protected plants' (Ashton and Willis 1982); grazing pressure of this kind on small cohorts would accentuate the tendency for one-age stands produced by occasional catastrophic fires to predominate.

The southern closed-forests characterised by *Nothofagus* (Antarctic Beech) are dominated by *N. cunninghamii* in the southern region and *N. moorei* in the northern region. Howard (1981) has reviewed their distribution, structure, floristics, regeneration and succession. These forests are self-regenerating, and *Nothofagus* is capable of invading the understorey of adjacent *Eucalyptus* tall open-forest. The contraints on extension of *Nothofagus* are primarily climatic, but there are influential interactions with fire, competition from other genera, and soil fertility; colonizers within the *Eucalyptus*-dominant communities are susceptible to fire. Although the mature closed-forest of *Nothofagus* itself is fire-resistant, severe fires may destroy parts of it at the margins. The holocaust fires such as those of 1939 and 1983 are particularly destructive but, depending on locality, are thought to be more characteristic of post-European settlement than earlier, when fires are thought to have been more frequent, of low intensity and more patchy. Successional observations following the 1939 fires in Victoria suggest that upwards of 40 years might elapse before local canopy closure of *N. cunninghamii* occurred under tall open-forest. The northern *N. moorei* can spread to disturbed areas where competition is lacking, but is out-competed by species of the northern closed-forests (rainforest).

The interplay between *Eucalyptus* tall open-forest and *Nothofagus* is described by Ashton (1981): the successional timescale for invasion by species such as *Nothofagus* is 100 years, leading to complete replacement by 200 years and disappearance of the *Eucalyptus* trees at 400 years as they reach the end of their life-span. Fire frequency (a function of rainfall) can affect this successional sequence. Ashton suggests that *Eucalyptus* forests were rare in the distant past, with rainforests occupying the mesic zone and *Eucalyptus* woodlands the xeric zone. *Eucalyptus* dominance was favoured by Aboriginal fire regimes over millennia and took place at the expense of rainforest. The fire-regeneration capacity of *Eucalyptus* species such as *E. regnans* is based on seed retention (in capsules) on the tree: without fire there is sporadic release, but after fire millions of seeds are quickly released per hectare. (This process occurs also in shrub species of *Banksia*, *Casuarina* and *Hakea* (Gill 1981)). Death of all *E. regnans* and sterilisation of the soil surface is followed by massive seed germination in the ash bed (Ashton and Willis 1982). Rapid juvenile growth of seedlings, aided by the temporary high-fertility status of the soils then occurs. A fire frequency of once per 200 to 400 years is postulated for *E. regnans*. In addition to this rare event there were in the past occasional ground fires which seldom consumed more than the ground layer of grasses, shrubs and small trees. Increased fire frequency over the past century is considered responsible for the formation of a *Eucalyptus* scrub instead of regenerated forest.

Woodlands dominated by *Eucalyptus* species exhibit the full range of response to fire already described. Species of *Acacia*, *Casuarina* and *Callitris* can also occur as dominants or co-dominants; these species and associated understorey species have a wide range of responses to fire (Hodgkinson and Griffin 1982). Some *Acacia* species regenerate from rootstocks and seed, others are killed and regenerate from massive seedling populations *via* hard seed in the soil. *Callitris* is susceptible to fire at all stages from seedling to mature tree, but maintains its often dominant position in landscapes where fire is endemic by prodigious seed production. Lacey (1972) reports estimates of 5.4 million seeds from a mature tree in a good year, 24,000 seeds in a poor year, and juvenile stands of 120,000 stems per hectare (in excess of 2.5 million stems per hectare have been recorded). The dense stands of *Callitris* effectively prevent growth of grasses and limit fuel loads; fires sweeping towards these thickets are halted by them, the fire front killing only the trees at the margins, much as has been described for *Nothofagus*. Self-regenerating stands of hard-seeded *Acacia* species only need occasional inputs of recruits, and, unlike the *Eucalyptus* species, destruction of individuals or complete stands need not be the end of *Acacia* occupancy. As already described, the pre-European fire mosaics could have given well-stocked stands of adult trees with short-lived cohorts, and thickets or dense adult stands which fires had bypassed; seedling replacement of dead plants or stands could be expected on the basis of reported population studies (Burrows 1973, Cunningham and Walker 1973). The vegetative reproduction from the root system (root suckering) which occurs following mechanical disturbance in species such as *A. harpophylla* is an introduced, European phenomenon, unlike suckering after fire in *A. dealbata*, which, together with its concomitant seedling populations reveals a complex demographic and genecological situation (A.O. Nichols personal communication).

Shrublands of the mallee *Eucalyptus* and *Acacia aneura* type differ substantially in their response to fire, even though the challenge of fire was common to them. Responses of mallee *Eucalyptus* are similar to those of heathland species, with lignotubers (= rootstock) providing the regenerative mode for above-ground biomass. Studies by Parsons (1968), Wellington (in Parsons 1981), and Noble (1982) show that the seedlings observed after fire are short-lived in the presence of regenerating adults. Presumably, destruction of adults is a pre-condition to successful recruitment. Multiple-burn regimes, applied experimentally by Noble, give some idea of the measures required to prevent regeneration of adult plants; no regeneration followed four successive autumn (April) fires using straw for fuel, but some 50% of plants regenerated following four successive spring (October) fires. The occurrence of the appropriate meteorological events to provide four fuel loads for four successive autumn fires in a semi-arid region is unlikely. Repeated cutting can also cause death, but this was an unlikely pre-European condition.

Herblands of the hummock grassland type are distinguished by highly inflammable species such as *Triodia* and *Plectrachne*, which, although well spaced, can carry fire in seasons when dry annual plants provide fuel between the hummocks. The life-span of *Triodia* is up to 25 years. The plant shows progressive death of the interior part and an expanding ring-shaped margin; a rate of expansion of 1 m in 3 years has been recorded. Eventually the ring breaks up and another (plant) starts. Spectacular death of whole stands has been observed after drought. In contrast, post-fire populations usually originate from seedlings. Hummock-grasslands appear to be the most frequently burnt of Australian arid plant communities. Fire mosaics are now being re-established in Aboriginal lands. The tussock grasslands of the *Astrebla* and *Dichanthium* type burn when dry, but the role of fire in the pre-European period is not clear; certainly fires swept across the formation, but fire frequency is unknown, as is the influence of Aboriginal burning in producing a fire mosaic which controlled wild fire. The demography of one *Astrebla* grassland is well documented (Williams and Roe 1975), and a series of demographic studies, including the seed bank component, on this grassland and two more-northerly grasslands on the north-south cline, is nearing completion (D.M. Orr personal communication). Because of a long history of grazing by domestic livestock, these grasslands should be considered as modified.

DEVELOPMENT OF AUSTRALIAN VEGETATION AFTER 1788: THE INFLUENCE OF NEOPHYTES

The fact that fires were widespread and frequent in the pre-European period is well-known, but the emphasis now given to fire and Aboriginal 'fire-stick farming' as major influences over millennia in the development of Australian plant communities and their population structure is comparatively recent. So, too, is the attention now given to the devasting effects of European occupation post-1788. Adamson and Fox (1982) have described the nature and intensity of this impact on a continental scale; 'The European invasion was a watershed beyond which Australian ecosystems are permanently changed. A revolution has occurred and no new steady state is in sight. Completely new ecosystems are being created from the interaction between human activities, the original organisms, the many successful immigrant organisms, and the changed environmental factors.

Virtually the whole continent is in flux because the new impact is open ended and because not all changes instituted earlier have worked throughout the system. It is salutary to realise that the life-span of individuals of many Australian plants is greater than total time since the European arrival, so that plant communities contain abnormal numbers of ageing individuals inherited from before European settlement. In such vegetation, hidden changes have yet to become fully manifest'. *Acacia sowdenii* (Lange and Purdie 1976) and *A. burkittii* (Crisp and Lange 1976) are two such species in semi-arid South Australia.

It can be argued that this concern with species and individuals in their community context has heightened interest in, and examination of, the major formative environmental factors affecting populations.

Adamson and Fox also refer to the infancy of 'ecological theory suitable for handling revolutionary transitions such as the impact of European settlement in Australia'. Indeed, the prevalence of disturbed vegetation and the absence of an ordered sequence of species (i.e., succession) culminating in a community dominated by a suite of long-lived species which could then restrain the subordinate species can explain, in part, the reluctance of Australian plant ecologists to use 'successional' concepts. The reasons for the belated recognition and analysis of the *revolutionary* state are many, and include pre-occupation with pastoralism and forestry in a research and political environment concerned more with increase in production than with conservation. The consequent lack of demographic studies in non-pastoral environments and the absence of ecological theory based on the performance of Australian plant communities in the revolutionary state are important deficiencies now being recognised.

Fire and disturbance regimes

Fire practices in the post-European settlement era have been described for the major Australian plant communities (Gill et al. 1981). Summer fire was suppressed in the predominantly winter rainfall region of southern Australia in the interests of human safety, prevention of loss of fencing, buildings, livestock, forests and forage. Deliberate autumn burning of alpine grazing lands continued until recent times. Cool-season fires are not possible in southern grazing lands because the vegetation is green and succulent, temperatures are low and rain is frequent. Summer is the period of high fire danger and every few years there are major bush fires with the energy output of atomic bombs. Systems of preventative and prescribed burning have been developed in wooded lands to head off these holocausts. The fire history of a contemporary landscape can be seen as an outcome of past Aboriginal fires which produced mosaics of the type already described and modern fire suppression for forestry, pastoral and agricultural activities but with occasional bush fires or grass fires; currently in wooded landscapes a programme of cool-season fires is practised to reduce fuel loads in the following summer and prevent uncontrollable bush fires.

In a cautionary pyric tale from a calibration of the grass trees

Xanthorrhoea preissii and *Kingia australis* growing in forest near Perth, Western Australia, Lamont and Downes (1979) noted that the frequency of flowering responses induced in each of them by fire over the 150 years since European settlement in 1829 was 22 and 12 times, respectively. This compares with only three times and once, respectively, in the 120 to 150 years before settlement.

In the summer-rainfall mesic region of northern Australia burning is conducted on an annual or near-annual basis which has little or no relationship with Aboriginal fire regimes; time of burning varies from late summer through the dry winter to early summer. In the semi-arid summer rainfall region the fire regimes are more opportunistic with the procedures in recent years tending towards those used by the Aboriginal people. In northern grazing lands the purpose in burning in late summer or early winter is to remove dead leaf material of poor nutritive value to livestock and promote leaf production using stored soil moisture. In southern grazing lands, in normal years, the domestic herbivores are effective in reducing fuel loads to such a level that fire, either lit by human or by lightning strike, does not reach disaster level. Many examples of the use of fire in research on the management of rangeland communities in semi-arid Australia are given by Harrington et al. (1984).

The additional major disturbances are grazing and trampling by alien herbivores such as sheep, cattle, rabbits, goats, camels and donkeys, the ever-spreading systems of arable agriculture, pasture improvement, forestry, extractive industries and constructional activities (roads, railways, pipelines). In general terms these types of disturbance occur world-wide and there are few peculiarly Australian practices that need description. Examples linking this disturbance with European settlement are given by Gillespie (1981) and Harrington et al. (1979).

Where all the indigenous vegetation is removed by such activities, the sites are made highly receptive to invasion by neophytes (Everist 1960).

Neophytes in natural or semi-natural plant communities

Few neophytes have invaded apparently undisturbed native vegetation in Australia, and examination of those sites where aliens are found generally produces evidence of various types of disturbance (Amor and Piggin 1977). Three examples follow; two involve acknowledged exotic neophytes and one involves an indigenous neophyte.

Pinus radiata, the Monterey Pine, is unquestionably an exotic, and its ca. 1954 invasion from a 1934 plantation into an adjacent woodland dominated by *Eucalyptus rossii*, and *E. macrorhyncha* is well documented (Burdon and Chilvers 1977, Chilvers and Burdon 1983). Demographic information for *P. radiata* shows that its incursions have been episodic, commencing some 10 years after abundant *P. radiata* seed became available on the plantation trees. The age structure was dominated by cohorts from the early 1970s, with cohorts before and after this short period being represented by small numbers of individuals. The 1981 cohort, although more numerous than the total number of individuals that had established during the previous 10 years, had almost been eliminated some 8 months later by drought. When the *P. radiata* plantation was clear-felled in 1977, the colonists themselves contributed progeny to the invading population. Chilvers and Burdon (1983) postulate that the mixed *Eucalyptus-Pinus radiata* woodland will become *P. radiata* dominant unless pests or diseases not yet resident in Australia intervene, or unless the mixed woodland experiences a wildfire of the type that recently destroyed *P. radiata* plantations in the region without destroying the fire-adapted *Eucalyptus* spp. Two major causes of *P. radiata* invasion have been identified, both of which have altered the pre-existing disturbance regimes in the *Eucalyptus* woodland. First, the change in the natural fire frequency in the *Eucalyptus* woodland by fire prevention measures permitted the adjoining *P. radiata* plantations to grow fire-free to the seed production stage. Second, in the period 1934 to 1954 substantial populations of the European rabbit, *Oryctolagus cuniculus* in the *Eucalyptus* woodland eliminated seedling *P. radiata*; the rabbits were themselved eliminated by the introduction of the myxoma virus. Later, depredations of the feral pig eliminated some cohorts of *P. radiata.*

The fire-adapted South African exotic neophyte *Ehrharta calycina* (veld grass) has invaded *Eucalyptus–Banksia–Casuarina* woodland, a fire-adapted community, leading Baird (1977) to conclude that the type of fire regime is much more critical for the continuing dominance of the various indigenous species than for *Ehrharta.* Exactly what sort of disturbance permits entry of *Ehrharta* in the first place is not clear, although ploughed fire breaks and fire frequency combined with openness of the tree stratum may be influential. Once established, *Ehrharta* offers severe competition through a root system, which unlike that of the resident indigenous species, thoroughly explores the soil profile. Luxuriant foliage effectively smothers seedlings of the native species and prevents their recruitment.

The indigenous neophyte *Pittosporum undulatum* is a fire-susceptible species which has recently invaded natural bushland and woodland because of successful fire preventation programmes (Adamson and Fox 1982). This invasion and the subsequent challenge to *P. undulatum* by the alien *Ligustrum lucidum* and *L. sinense*, both fire-susceptible, are further cases of contemporary invasions, long-delayed by management procedures that have been changed. A further example involves the agency of the European blackbird *Turdus merula* and fire prevention measures in converting *Pittosporum undulatum* into a neophyte which threatens the long-term future of *Eucalyptus* forests (Gleadow and Ashton 1981, Gleadow 1982).

Neophytes in man-modified plant communities

Most neophytes have invaded and are still invading plant communities where the impact of European activities has been substantial. The invasion of cultivated land, over-grazed pasture and roadsides by weedy neophytes is appreciated and understood (Parsons 1976, Tothill et al. 1982) and will not be discussed here. There is, however a paucity of records describing the circumstances surrounding the initial invasion of weedy neophytes. This information could be valuable in the design of methods for their control. For example, the biological control of *Opuntia* spp. may have been expedited by knowing their point of departure from the Americas (Everist 1960). Furthermore, the taxonomy of weedy neophytes needs investigation: Groves and Cullen (1981) note that *Chondrilla juncea*, a weed of agricultural lands and roadsides in south-eastern Australia, is represented by 3 forms and that the rust fungus *Puccinia chondrillica* which was introduced in a biological control programme is a natural enemy only of the commonest form. The cultivation and fertiliser practices which appear to control *C. juncea* in its southern European homeland cannot be used, but Australian population studies show that effective control can be attained with a combination of competing plant species of Mediterranean origin such as *Trifolium subterraneum*, and natural enemies from Mediterranean Europe (Groves and Cullen 1981).

Neophytes and their derivatives have been used as the principal plants of Australia's sown pastures. Endeavours to construct stable plant communities from these species have approached, in part, through plant population studies. This can be seen, for example, in the series of reviews on *Medicago sativa* (lucerne) (Leach 1978), *Trifolium subterraneum* (subterranean clover) (Rossiter 1978), *Stylosanthes humilis* (Townsville stylo) (Gillard and Fisher 1978) and *Macroptilium atropurpureum* (siratro) (Jones and Jones 1978), as well as in the experiments on the control of nitrophilous weedy neophytes by perennial neophytes such as *Phalaris aquatica* (Michael 1968) and the grazing management experiments comparing various shown neophytes (Jones and Evans 1984). Jones and Mott (1980) and Hodgkinson and Williams (1983) have examined the demographic approach as it has been used in research on management problems in pastures composed of recent neophytes.

CONCLUSIONS

An appreciation of the substantial impact of Aboriginal man on Australian vegetation as seen ca. 1788 is a necessary prerequisite to an understanding of the immense success of deliberate and accidental neophyte invasions into Australian plant communities after 1788. These successes are attributable to the widespread and severe impact of European colonists who have, in less than 200 years, all but eliminated unmodified plant communities from alpine, mesic and semi-arid Australia. Altered fire regimes (to the point of their prevention altogether), and grazing by alien herbivores have been influential factors leading to substantial floristic changes and neophyte invasion even in the least-modified of vegetation assemblages.

Demography has become a significant strand in Australian science: contemporary studies on the least-modified of plant communities with few or no neophytes and on sown pastures constructed entirely of neophytes generally utilise demographic parameters. With few exceptions such investigations are aimed directly at management of the vegetational and faunal complex. In the least-modified of plant communities, whose conservation is generally the major interest of vegetation scientists today, the invasion of neophytes (including indigenous species), is associated with fire prevention or fire regimes which appear to differ from those which gave rise to the particular community in the first place. There are immense difficulties in assessing the specific role of fire in the Aboriginal era. Intelligent experimentation should, however, enable a rational anti-neophytic strategy to be devised for plant communities whose conservation value is high. Research and subsequent management designed to use plant species to block neophyte invasions into plant communities deserve to be accorded as high a priority in the so-called 'natural' plant communities as is accorded to it in sown pastures.

REFERENCES

Adamson, D.A. and Fox, M.D. (1982) Change in Australasian vegetation since European settlement. In J.M.B. Smith (Ed.), A History of Australasian Vegetation. McGraw-Hill Book Co., Sydney. 5: 109–146.

Amor, R.L. and Piggin, C.M. (1977) Factors influencing the establishment and success of exotic plants in Australia. In D.T. Anderson (Ed.), Exotic Species in Australia – Their Establishment and Success. Proc. Ecol. Soc. Aust. 10: 15–26.

Ashton, D.H. (1981 – Tall open-forests. In R.H. Groves (Ed.), Australian Vegetation. Cambridge Univ. Press, Cambridge. 6: 121–151.

Ashton, D.H. and Willis, E.J. (1982) Antagonisms in the regeneration of *Eucalyptus regnans* in the mature forest. In E.I. Newman (Ed.), The Plant Community as a Working Mechanism. Brit. Ecol. Soc. Special Publ. Ser. 1: 113–128.

Auld, B.A. (1981) Aspects of the population ecology of galvanised burr (*Sclerolaena birchii*). Aust. Rangel. J. 3: 142–148.

Baird, A.M. (1977) Regeneration after fire in King's Park, Perth, Western Australia. J. Roy. Soc. W. Aust. 60: 1–22.

Barker, W.R. and Greenslade, P.J.M. (Ed.) (1982) Evolution of the Flora and Fauna of Arid Australia. Peacock Publications, Frewville. 392pp.

Barlow, B.A. (1981) The Australian flora; its origin and evolution. In Flora of Australia. Aust. Govt. Publ. Serv., Canberra. 1: 25–75.

Beadle, N.C.W. (1981) The Vegetation of Australia. Cambridge Univ. Press, Cambridge. 690pp.

Blainey, G. (1975) Triumph of the Nomads: a history of ancient Australia. MacMillan, London. 285pp.

Burdon, J.J. and Chilvers, G.A. (1977) Preliminary studies on a native eucalypt forest invaded by exotic pines. Oecologia (Berlin) 31: 1–12.

Burrows, W.H. (1973) Regeneration and spatial patterns of *Acacia aneura* in south west Queensland. Trop. Grassl. 7: 57–68.

Carnahan, J.A. (1976) Natural vegetation. In Atlas of Australian Resources, Second Series. Dept. of National Resources, Canberra.

Carnahan, J.A. (1977) Vegetation. In D.N. Jeans (Ed.), Australia: A Geography. Sydney Univ. Press, Sydney. 9: 175–195.

Chilvers, G.A. and Burdon, J.J. (1983) Further studies on a native Australian eucalypt forest invaded by exotic pines. Oecologia (Berlin) 59: 239–245.

Chiswell, E.L. (1982) The Pilliga story. Forest & Timber 18: 23–24.

Crisp, M.D. and Lange, R.T. (1976) Age structure, distribution and survival under grazing of the arid-zone shrub *Acacia burkittii*. Oikos 27: 86–92.

Cunningham, G.M. and Walker, P.J. (1973) Growth and survival of mulga (*Acacia aneura* F. Muell. ex Benth.) in western New South Wales. Trop. Grassl. 7: 69–77.

Davidson, B.R. (1981) European Farming in Australia: an economic history of Australian farming. Elsevier, Amsterdam. 437pp.

Everist, S.L. (1960) Strangers within the gates: plants naturalized in Queensland. Qld. Nt. 16: 49–60.

Gill, A.M. (1981) Patterns and processes in open-forests of *Eualyptus* in southern Australia. In R.H. Groves (Ed.), Australian Vegetation. Cambridge Univ. Press, Cambridge. 7: 152–176.
Gill, A.M., Groves, R.H. and Noble, I.R. (Eds.) (1981) Fire and the Australian Biota. Aust. Acad. Sci., Canberra. 582pp.
Gillard, P. and Fisher, M.J. (1978) The ecology of Townsville stylo-based pastures in Northern Australia. In J.R. Wilson (Ed.), Plant Relations in Pastures. CSIRO. AUST., E. Melbourne 22: 340–352.
Gillespie, P.D. (1981) Development of gully erosion at the head of Bango Creek near Yass, New South Wales. J. Soil Cons. N.S.W. 37: 6–12.
Gleadow, R.M. and Ashton, D.H. (1981) Invasion by *Pittosporum undulatum* of the forests of central Victoria. I. Invasion pattern and plant morphology. Aust. J. Bot. 29: 705–720.
Gleadow, R.M. (1982) Invasion by *Pittosporum undulatum* of the forests of central Victoria. II. Dispersal, germination and establishment. Aust. J. Bot. 30: 185–198.
Groves, R.H. (Ed.) (1981) Australian Vegetation. Cambridge Univ. Press, Cambridge. 449pp.
Groves, R.H. and Cullen, J.M. (1981) *Chondrilla juncea*: the ecological control of a weed. In R.L. Kitching and R.E. Jones (Eds.), The Ecology of Pests: Some Australian Case Histories. CSIRO AUST. E. Melbourne 2: 7–17.
Harrington, G.N., Oxley, R.E. and Tongway, D.J. (1979) The effect of European settlement and domestic livestock on the biological systems in poplar box (*Eucalyptus populnea*) lands. Aust. Rangel. J. 1: 271–279.
Harrington, G.N., Wilson, A.D. and Young, M.D. (Eds.) (1984) Management of Australia's Rangelands. CSIRO, AUST., E. Melbourne. 354pp.
Hodgkinson, K.C. and Griffin, G.F. (1982) Adaptation of shrub species to fire in the arid zone. In W.R. Barker and P.J.M. Greenslade (Eds.), Evolution of the Flora and Fauna of Arid Australia. Peacock Publications, Frewville. 15: 145–152.
Hodgkinson, K.C. and Williams, O.B. (1983) Adaptation to grazing in forage plants. In J.G. McIvor and R.A. Bray (Eds.), Genetic Resources of Forage Plants. CSIRO. AUST., E. Melbourne. 6: 85–100.
Howard, T.M. (1981) Southern closed-forests. In R.H. Groves (Ed), Australian Vegetation. Cambridge Univ. Press, Cambridge. 5: 102–120.
Jones, R.M. and Evans, T.R. (1984) Problems associated with grazing experiments involving the integration of different pastures – a case study in south-east Queensland. Trop. Grassl. 18: 39–45.
Jones, R.J. and Jones, R.M. (1978) The ecology of Siratro-based pastures. In J.R. Wilson (Ed.), Plant Relations in Pastures. CSIRO. AUST., E. Melbourne. 23: 353–367.
Jones, R.M. and Mott, J.J. (1980) Population dynamics in grazed pastures. Trop. Grassl. 14: 218–224.
Kloot, P.M. (1984) The introduced elements of the flora of southern Australia. J. Biogeog. 11: 63–78.
Lacey, C.J. (1972) Factors influencing occurrence of Cypress Pine regeneration in New South Wales. Forestry Commission of New South Wales Tech. Paper No. 21., 17pp.
Lamont, B.B. and Downes, S. (1979) The longevity, flowering and fire history of the grass-trees *Xanthorrhoea preissii* and *Kingia australis*. J. Appl. Ecol. 16: 893–899.
Lange, R.T. (1982) Australian Tertiary vegetation. In J.M.B. Smith (Ed.), A History of Australian Vegetation. McGraw-Hill Book Co., Sydney 3: 44–89.
Lange, R.T. and Purdie, R. (1976) Western myall (*Acacia sowdenii*), its survival prospects and management needs. Aust. Rangel. J. 1: 64–69.
Leach, G.J. (1978) The ecology of lucerne pastures. In J.R. Wilson (Ed.), Plant Relations in Pastures. CSIRO AUST., E. Melbourne. 19: 290–308.
Martin, A.R.H. (1982) Proteaceae and the early differentiation of the Central Australian flora. In W.R. Barker and P.J.M. Greenslade (Eds.), Evolution of the Flora and Fauna of Arid Australia. Peacock Publications, Frewville. 7: 77–83.
Martin, H.A. (1982) Changing Cenozoic barriers and the Australian paleobotanical record. Ann. Missouri Bot. Gard. 69: 625–667.
Michael, P.W. (1968) Control of the biennial thistle *Onopordon* by amitrole and five perennial grasses. Aust. J. Exp. Agric. Anim. Husb. 8: 332–339.
Michael, P.W. (1981) Alien plants. In R.H. Groves (Ed.), Australian Vegetation. Cambridge Univ. Press, Cambridge. 3: 44–64.

Noble, J.C. (1982) The significance of fire in the biology and evolutionary ecology of mallee *Eucalyptus* populations. In W.R. Barker and P.J.M. Greenslade (Eds.), Evolution of the Flora and Fauna of Arid Australia. Peacock Publications, Frewville. 16: 153–159.

Parsons, R.F. (1968) An introduction to the regeneration of mallee eucalypts. Proc. Roy. Soç. Vic. 81: 59–68.

Parsons, R.F. (1981) *Eucalyptus* scrubs and scrublands. In R.H. Groves (Ed.), Australian Vegetation. Cambridge Univ. Press, Cambridge. 10: 227–252.

Parsons, W.T. (1976) Noxious Weeds of Victoria. Inkata Press, Melbourne. 300pp (reprinted 1976).

Rolls, E.C. (1969) They All Ran Wild: The Story of Pests and the Land in Australia. Angus and Robertson, Sydney. 444pp.

Rossiter, R.C. (1978) The ecology of subterranean clover-based pastures. In J.R. Wilson (Ed.), Plant Relations in Pastures. CSIRO AUST., E. Melbourne. 21: 325–339.

Singh, G. (1982) Environmental upheaval. In J.M.B. Smith (Ed.), A History of Australasian Vegetation. McGraw-Hill Book Co., Sydney. 4: 90–108.

Smith, J.M.B. (1982) An introduction to the history of Australasian vegetation. In J.M.B. Smith (Ed.), A History of Australasian Vegetation. McGraw-Hill Book Co., Sydney. 1: 1–31.

Specht, R.L. (1970) Vegetation. In G.W. Leeper (Ed.), The Australian Environment. CSIRO, AUST. Melbourne Univ. Press, Melbourne. 44–67.

Squires, V.R. (1979) The Biology of Australian weeds: 1. *Tribulus terrestris*. J. Aust. Inst. Agric. Sci. 45: 75–82.

Tothill, J.C., Mott, J.J. and Gillard, P. (1982) Pasture weeds of the tropics and subtropics with special reference to Australia. In W. Holzner and N. Numata (Eds.), Biology and Ecology of Weeds. W. Junk, The Hague. 35: 403–427.

Walker, D. (1982) The development of resilience in burned vegetation. In E.I. Newman (Ed.), The Plant Community as a Working Mechanism. Brit. Ecol. Soc. Special Publ. Ser. 1: 27–43.

Williams, O.B. and Roe, R. (1975) Management of arid grasslands for sheep: plant demography of six grasses in relation to climate and grazing. Proc. Ecol. Soc. Aust. 9: 142–156.

PATTERNS IN VEGETATION SUCCESSION, AN ECOMORPHOLOGICAL STUDY

P. HOGEWEG, B. HESPER, C.P. VAN SCHAIK and W.G. BEEFTINK

ABSTRACT

Traditionally there are two largely independent sets of concepts for the description of assemblages of plants: those pertaining to 'populations' and those pertaining to 'vegetation'. To bridge the gap between them, we cannot confine ourselves to the original sets of concepts because each represents a different and incompatible global description, obtained by different ways of averaging. Nevertheless they describe the same 'object'. Relating population and vegetation concepts can be done via a microscopic description sufficient to generate this object, which in turn can be described (amongst others) in terms of populations and/or vegetation by appropriate methods of averaging. The relations between descriptions in terms of populations and in terms of vegetation are shown to depend on variation in the microscopic structure of plant assemblages, which should be defined in terms of local interactions between local entities such as individual plants or small areas. The concepts 'scale' and 'detail' are important in our discussion.

INTRODUCTION

The purpose of this book is to bridge the gap between two different, largely independently developed approaches to the study of assemblages of plants: analyses in terms of 'populations' and in terms of 'vegetation'. In such an effort a crucial question is whether the two sets of concepts are structured in such a way that the description of an object in terms of one of them is sufficient to *generate* a description in terms of the other set of concepts, or if, on the other hand a third set of concepts is needed by which the descriptions in terms of the two other sets can be generated. In this paper we argue that the latter is the case. Our arguments are in the first place theoretical: by analysing the concept 'pattern' and exploiting the methodological similarity of ecomorphological research with other morphological research, we conclude that both concepts 'population' and 'vegetation' are applicable to the 'macroscopic' structure of plant assemblages, but that the units in which to describe the interaction leading to this macroscopic description should be in terms of individual plants or small areas and cannot be in terms of populations or vegetation themselves.

'Ecomorphology' is the study of spatial patterns in assemblies of organisms in two dimensional space. As in all morphological research (e.g.,

White, J. (ed.), The Population Structure of Vegetation.
© 1985, Dr W. Junk Publishers, Dordrecht. ISBN 90–6193–184–3.

phytomorphology, zoomorphology, geomorphology) the basic steps in such a study are:

1. *Pattern analysis*: the representation of the spatial relations between recognisable substructures and of the ontogeny of these spatial relations.
2. *Pattern generation*: the construction of a set of units and interactions between these units such that, starting from simple initial conditions, these interactions lead to the formation of spatial structures which can be compared with the spatial relations of the object studied. In other words, to form a model of the ontogeny of the spatial relations such that the output of this model can be analysed by the same pattern analysis and representation methods as were used in analysing the original data.

This paper consists of two parts. In the first we explain the concepts underlying ecomorphological research and give a theoretical justification for the methods to be used in the second part. There we apply these concepts and methods for the pattern analysis (i.e., step 1 above) of vegetation development on initially barren sand flats and we discuss how the results obtained can be used to shape the second step of the full research program. Scale and detail play an important role in our treatment. The ecomorphological analysis is mainly in terms of the detection and representation of structural features of the dataset. An analysis in terms of autecological properties of the various species is not included because the analysis of the dataset is meant to illustrate the concepts developed in the theoretical part of the paper rather than as a complete analysis of one particular succession series.

CONCEPTUAL BASIS

The concept of pattern

There are two entirely different concepts called 'pattern'. The fact that some similar techniques are used in both contexts should not obscure the differences between these approaches (Kendall 1972).

Pattern as a statistical concept is defined as divergence from randomness. This concept is used in the context of vegetation studies by, for example, Greig-Smith (1964, 1979), Goodall (1974), Goodall and West (1979); random is mostly defined as Poisson distributed. The data are assumed to be a sample of an infinite set of similar data and the purpose is to obtain a characterisation of this set independently of the particular sample under consideration. The results are of the form: "there is/is-not a pattern in these data" (i.e., 1 bit of information). By stating how the data diverge from randomness (e.g., clumped, overdispersed) one or a few more bits of information are obtained. Moreover, by analysing the data at different scales, this amount of information is obtained for each of these scales. Along the same lines, considering the data as a sample, one can also search for periodicities in the data. In all cases pattern is a global property, measured at a particular scale, of a population of which the data are a sample.

Pattern as a heuristic concept is defined as a recognisable structure in the data, i.e., as a model of the data. It contrasts with the statistical concept in that the data are not considered to be a sample but are treated as the entire universe (Kendall 1972). Thus, a sample-independent characterisation is not a meaningful concept in this context; on the contrary pattern is what is generated by applying a transformation to the particular datapoints available. Pattern is here not a property of the data alone, but explicitly a property of the interfacing of the data with some transformation. The transformation used is generally a many-to-one transformation so that the resulting structure is simpler than the original data were, but the information content remains much larger than in the case of a statistical analysis. The transformations can be local or global. In the local case the transformation of a datapoint depends on nearby datapoints, whereas in the global case the transformation depends only on the datapoint to be transformed itself. Transformations also differ in that they are either chosen *a priori* or can be derived from the data (by an *a priori* transformation of the data). This is the meaning of pattern as used in picture processing (see, e.g., Rosenfeld 1969). Cluster analysis and principal component analysis also belong to the class of methods using this concept of pattern. Cluster analysis applies an *a priori* chosen transformation to the data in a local way; the resulting pattern depends on the data and the chosen transformation. Principal component analysis applies a global, data-dependent transformation to the data (projection); the axes for projection are obtained from the data as axes of maximum variation. The usefulness of the results depends largely on the recognisability of the obtained pattern. This is obvious in the context of picture processing but applies as well to cluster analysis, etc. In the latter case this amounts to the possibility of labelling the obtained clusters in a simple way (Hogeweg and Hesper 1981b), using external (i.e., not included in the pattern analysis), possibly partial, information about the data or using internal information simplifying the characterisation of the clusters such that it can be easily used. External labelling of the clusters can be based on the location of the datapoints in some other space, e.g., 'space–space' (revealing a spatial pattern) or in a state–space other than the one used in the pattern detection (e.g., abiotic environment–space if the pattern is detected in species–abundance space).

The results of a statistical analysis and a pattern analysis are often contradictory. For example, Dubes and Jain (1979) complained that data without 'clustering tendency' (i.e., not diverging from a random distribution in state–space) nevertheless yielded clusters which could be labelled with external information (Hogeweg and Hesper 1981b). Likewise, Williams et al. (1971) argued that minimal distortion is not a sensible optimisation criterion in pattern analysis of, for example, vegetation data because the pattern should be exaggerated to be heuristically useful. The infiltration of statistical notions in cluster analysis has often prohibited the optimal use of these techniques. For example, Ward's (1963) clustering criterion was originally discarded because it would generate a clear pattern even in regularly distributed data and because it deforms the state–space in a way dependent on the availability of datapoints. It turns out that these properties are the ones which make this criterion superior to many other criteria: it extracts heuristically useful patterns in which detail is shown where warranted and no distinctions in areas where few data are available (Hogeweg 1976a, Hogeweg and Hesper 1981b).

Pattern analysis in vegetation

In this paper we use the word pattern exclusively in the heuristic sense, that is, to mean 'recognisable structure in the data'. In our view it is this concept of pattern which makes sense in vegetation studies as well as in ecosystems: members of an ecosystem 'recognise' a type of vegetation by applying a transformation (their perception) and react on the results. We should keep in mind however that the various members will use different transformations.

Spatial pattern in vegetation

The study of spatial pattern in vegetation takes the form of relating point patterns in two spaces, 'space–space' (the space in which the vegetation dwells) and 'state–space' (the space describing the vegetation features of parts of the space–space). The two spaces contain corresponding sets of labelled points; the spatial pattern in vegetation can be expressed in terms of comparisons of the distance relations between points in both spaces.

In statistical pattern analysis of vegetation data neighbouring points in space–space (or points at some fixed distance) are mapped onto state–space; their distance in state–space is a measure of the amount of pattern in the data. The results are averaged over an arbitrary set of pairs of points at a certain distance (Greig-Smith 1964, Goodall 1974) as the results are to be global and sample-independent and there are indeed no criteria for selecting the pairs in space–space.

Heuristic methods for spatial pattern representation, on the contrary, start by detecting a pattern in state–space. Because of the uneven and meaningful distribution of points in state–space (as opposed to the *a priori* chosen random or grid pattern in space–space) a pattern generating transformation (for example, cluster analysis) can generate a subdivision of this space in a local, data-dependent way. The pattern obtained is subsequently mapped in space–space to yield a representation of the spatial pattern in the vegetation. This representation can itself be interpretated locally: the relationship between distances in state–space and space–space can differ for different regions in either one of these spaces. The pattern features which can be detected by this method include different types of ecotones and patchiness of the pattern. The characterisation of a pattern represented by this method depends crucially on the scale of observation in each of the two spaces. We will call scale of observation in state–space 'detail', and scale in space–space 'scale'.

During succession we may expect changes in the vegetation which affect the scale and the detail in which certain pattern features occur in the vegetation. For studying vegetation succession it is therefore useful to use data and methods which allow for analysis at different scale or detail. We use grid data and use as pattern-detection method a hierarchical clustering strategy: scale and detail can then be chosen at will.

Representing vegetation succession

Vegetation succession in permanent quadrats (PQs) characterised by their species composition can be thought of as a bundle of trajectories in *n*-dimensional state–space; in studying succession we study features of this bundle of trajectories. A way to represent the bundle of trajectories is as a directed graph (Harary et al. 1965) of which the nodes represent subregions of state–space and the arcs represent the transitions of PQs between these regions. In this graph representation the concepts parallel, convergent, divergent and reticulate succession have meaning, since these are graph theoretic terms. A given successional dataset can, however, give rise to quite different graphs depending on the subdivision of the state–space. When the state–space is subdivided into very small subregions (i.e., so small that just one space/time sample fits into it) the graph will be a disconnected one and the succession would be called a parallel succession as no divergence or convergence occur in this representation. By choosing the appropriate subdivision of the state-space (possibly of different sizes and shapes) any type of graph can be obtained from a given dataset. It is therefore crucial to define how the subdivisions are formed so that at least different datasets can be compared with respect to their successional (graph) properties relative to a comparable subdivision of the state–space. We propose to use cluster analysis for this purpose; the resulting subdivision is dependent on the pattern in the data, but is independent of their labelling (in this case their localisation in space/time). The clusters will therefore contain a mixture of trajectories and times, dependent only on their similarity; the resulting graphs therefore interrelate the space/time pattern of the data and will depend on the set of PQs as well as on the time period included in the analysis, e.g., the observed pattern in the data depends on whether or not knowledge of the future pattern is included in the analysis.

Pattern generation

The purpose of this step in morphological research is the generation of the observed spatial pattern through the interaction of autonomous elements. The elements should be defined as autonomous information processing systems (henceforward called entities) which can interact with each other in a local way. They should not be defined as parts of a 'whole', nor should they in any way observe the whole system or interact with it since this would trivialise the research effort (Hogeweg and Hesper 1979, Hogeweg 1983).

Maximal vs. minimal information models

The choice of the definition of these entities can be inspired by two quite different optimising criteria: (1) they should represent as closely as possible some subsystems of the system studied, for example, they should represent the individual plants as closely as possible; (2) they should behave as simply as is compatible with respect to generating the observed pattern. These two criteria can lead to quite different results: after all, the observed overall pattern is only a side effect of the behaviour of the plants and there is no reason whatsoever (evolutionary or otherwise) to expect that the interactions should be minimised with respect to the pattern. The purpose of studies using the first optimising criterion is to find gaps in our knowledge about the represented entities (Hogeweg and Richter 1982), whereas the second optimising criterion provides us with an idea about the complexity of the pattern and whether such patterns may be expected in many situations (Hogeweg and Hesper 1979).

Individual oriented models

In the context of both optimising criteria the entities defined in a pattern generation study of vegetational patterns can be comparable to individual plants. This choice is attractive because these are the most tightly connected informational elements in a vegetation (McMahon et al. 1981, Hogeweg and Hesper 1979, Hogeweg and Richter 1982). Models using individual plants can be best formulated in terms of the heterarchical systems simulation modelling formalism (Hogeweg 1983), using its self-structuring properties ('MICMAC modelling', Hogeweg and Hesper 1981a). The individuals cannot be lumped into populations because these are not spatially connected and cannot form the type of spatial pattern sought. The entities (individual plants) should be defined in terms of their spatial behaviour, i.e., in terms of the generation of new such entities in time/space relative to their own position in time/space. This behaviour can be expressed in terms of their morphological growth pattern (Bell et al. 1979), their seed dispersal, etc. The behaviour of the entities may, or may not, depend on the local circumstances in which the entities find themselves. In fact, one of the central questions in ecomorphology is to what extent local interactions between individuals and their environment take place and whether such interactions involve (pre-patterned or not) abiotic variation or plant-to-plant interactions. Whatever the answer to this question may be, the generation of vegetation patterns should be based on the behaviour of individual plants, not of populations, as the former and not the latter are localised in space and can therefore be elements in a spatial pattern. If no local interactions are necessary for the pattern formation, the properties of the individuals can be derived from the population properties by random drawings from the functions describing the populations. If local interactions are needed these functions should be modified according to the local circumstances. The population properties (in the sense of the average values of certain properties of all individuals) are in that case the result of the model rather than its input. We will, moreover, find a mutual dependence of vegetation pattern and properties of the populations, as they are both generated by the same properties of the individuals. Therefore, studies on the population properties of vegetations run the risk of running in closed circles: *a priori* defined vegetation types described in terms of their (average) species composition can be trivially derived from the population properties of its species as they are observed in this vegetation type. Only by fixing the properties of the individuals over a larger area in space or time than the vegetation from which they are derived can a nontrivial relation be demonstrated.

Cellular space models

Another possible choice of entities in a pattern generation study in vegetations are small areas of space ('cells') as the elements generating the pattern. The model can in that case be formulated in terms of the cellular space (cellular automaton) formalism (Burks 1970, Zeigler 1976, 1982, Hogeweg 1983). Each cell changes its state (here plant occupation) dependent on its own state and possibly the state of its neighbourhood (however defined, possibly empty). The successional model of Horn (1975) is an example of the simplest case of this model structure: every cell is occupied by exactly one plant, and the neighbourhood is empty, so that a cell transition depends only on the state of the cell itself. The results of this model are well known: starting from any initial condition the species composition converges to a 'climax' vegetation of which the species composition is independent of the initial conditions. The representation of the Horn model in the cellular space framework shows clearly its assumptions: a very localised influence of plants on their environment, a very global availability of seeds and no locally extending growth pattern. By changing these assumptions so that a non-empty neighbourhood results, this can lead to a stable spatial differentiation. However, when the transition functions of the cells are derived by averaging observed transitions over the entire area in space/time studied, and when the final vegetation is represented in terms of its species composition (irrespective of spatial patterning), any succession can be moulded into a model resembling the original Horn model, because the differences are averaged out by the dataprocessing (Hogeweg, unpublished). Such traps can be circumvented by studying both the observed patterns and the generated patterns by nonsupervised pattern analysis methods and by comparing the results; in that case comparisons are done on the basis of categories not previously recognised so that the pattern generation study cannot be trivially moulded into the expected results.

Summary of concepts

Pattern analysis is a heuristic enterprise which seeks independently recognisable patterns in the data (i.e., in a state–space); these patterns are not defined on the statistical properties of the data.

The independent recognition of the pattern can be based on the location of the data in some other space (e.g., space–space); therefore pattern analysis is a useful method for studying spatial patterns in vegetation.

Spatial patterns in vegetation are therefore the relations between state–space and space–space; they are dependent on scale of observation in both (called respectively 'detail' and 'scale').

Vegetation succession can be viewed in terms of trajectories in state–space; spatial patterns in succession are the mapping of these trajectories in space–space.

Classical features of succession are graph theoretic properties of the directed graph formed by lumping regions of state–space into nodes; these nodes are connected when a region in space–space goes from one of these areas (represented by a node) to another; the nodes can be formed by cluster analysis.

The properties of the graphs are dependent on scale and detail and on the spatial/temporal extent of the data.

The fact that various methods and various selections of the data generate different results should be used explicitly in the analysis, as these different behaviours together characterise the data.

Pattern generation studies should use autonomous entities, chosen to be as simple as possible or to represent information processing in the entities they represent as closely as possible.

Obvious choices of elements are those representing individual plants; a computationally simpler possibility is the choice of small areas as information processing units.

The resulting pattern should be studied by the same (nonsupervised) pattern analysis techniques as the original pattern, so that new features of both patterns are considered in the comparison.

THE ANALYSIS OF A VEGETATION SUCCESSION

In this section we present the results of ongoing research on the vegetation succession on low-lying sand flats in the southwest of the Netherlands (Fig. 1). These sand flats emerged above the water level after enclosure of an estuary by a dam in 1961. The tidal influence disappeared abruptly; in a few years the salinity of the surrounding water dropped to brackish values; desiccation, desalination, increased aeration of the soil and increased mineralisation of organic matter were the early factors forcing the vegetation succession. Since the time of emergence the vegetation development on these flats have been sampled by Beeftink and collaborators. Previous reports of this research include Beeftink et al. (1971, 1978), Hogeweg (1976a) and van Noordwijk-Puijk et al. (1978). Figure 2 and

Table 1 give an impression of the vegetation succession taking place on these flats (Beeftink et al. 1971; updated here to 1983).

Methods

The data consist of:

1. Yearly surveys of 33 permanent quadrats distributed over the entire area, which consists of two islands and one peninsula connected to the mainland (Fig. 1). The size of these quadrats varies from 7 to 120 m^2. As the flats were barren at the time of the initiation of the quadrats, the vegetation could not influence the choice of the locations.
2. Two grids of plots of 2 × 2 m^2; Grid 1 was sampled in 1973, 1976, 1978, 1980 and contains 150 plots (the total size is 20 × 30 m^2) and Grid 2 was sampled in 1976, 1978, 1979, 1980 and contains 48 plots (size 12 × 16 m^2). The total size of the grids is thus of the same order of magnitude as the 33 permanent quadrats.
3. Three subgrids of plots of 25 × 25 cm^2, two located in plots of Grid 1 and one located in Grid 2. They were sampled in 1976, 1978 and 1980.

The more detailed grid data cover only a very small portion of the area and they are all located on the least disturbed island (S). Nevertheless the data allow us to study the spatial/temporal development of the vegetation at several levels of resolution in both the spatial and temporal dimension. The data were cover estimates (in %) which were transformed prior to the analysis in two different ways: logarithmised (covers of $< 1\%$ were not logarithmised) or binarised.

Agglomerative (hierarchical) cluster analysis was used in order to study the pattern at different *levels of detail*. The dissimilarity measure between samples used was city-block distance and the clustering criterion was Ward's minimal increase of mean sum of squares (Ward 1963, Hogeweg 1976a). Optimal splitting levels of the dendrogram into clusters were determined by the criterion of Hogeweg (1976b), which maximises the ratio of (local) between-cluster distances and within-cluster distances, thus selecting levels of detail at which clusters are best separable.

The clusters were characterised *monothetically* and *oligothetically*. For the monothetic characterisation Kruskal–Wallis one-way analysis of variance was used (i.e., a statistical measure of rank-ordered data). In this way the importance of characters in generating cluster structure can be determined. The oligothetic characterisation was done by the minimal classification error method of Hogeweg and Hesper (1981b). By this method a small number of characters (species) are selected which are sufficient for the 'correct' classification (correct being defined by the initial, polythetic cluster analysis) of each of the objects in the recognised clusters, using a quadratic discriminant function. The results of this method are, apart from the list of selected species, a list of 'mis'-classified items, indicating the exceptional members of the (original) clusters. ('mis' refers to the original clusters, in fact the oligothetic characterisation often generates an intuitively better classification of these items (see later)).

A filtering technique for pattern enhancement, called iterative character

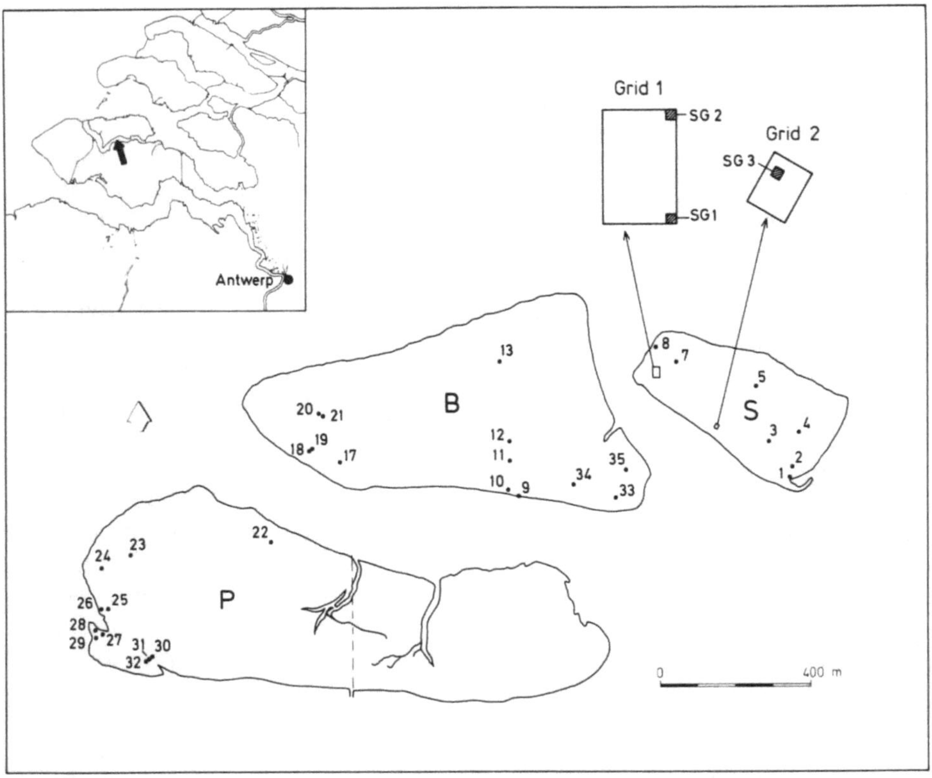

Fig. 1. Map of the Middelplaten (in the southwest of the Netherlands). P Peninsula, B Big Island, S Small Island. Numbers of permanent quadrats correspond with those of Tables 2a and 2b. The positions of the grids and subgrids are shown.

weighing (Hogeweg 1976a, b), was sometimes used to increase the interpretability of the generated patterns. In this method, characters which are 'important' (as indicated by the monothetic characterisation of clusters) for the initial clustering are weighted preferentially to generate subsequent clustering(s): in this way consistent patterns are enhanced (see below).

All analyses were performed by BIOPAT, a program system for bioinformatic pattern analysis (Hogeweg and Hesper 1972).

The pattern analysis of these data, presented here, leads to two types of results:

A representation of the spatio-temporal patterns of the vegetation independently of the species (populations) of which they consist. Such a representation makes feasible a comparison of the spatio-temporal pattern of vegetations composed of different species; this is necessary if we are to relate population characteristics with (independently derived) vegetation characteristics.

Hypotheses about how the species interact to produce the observed pattern.

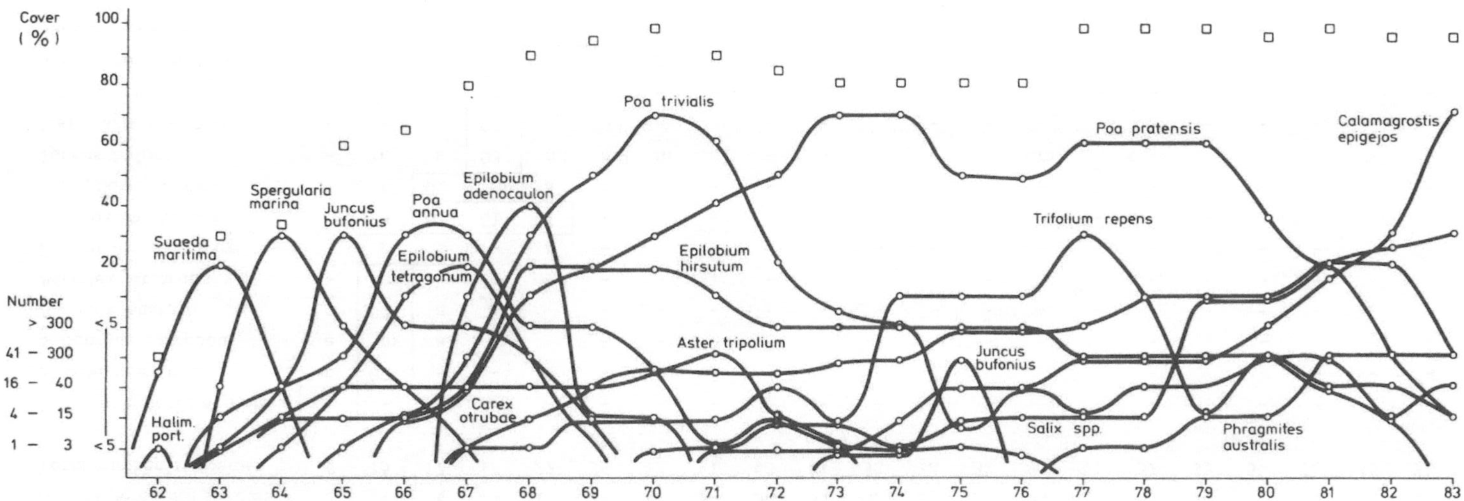

Fig. 2. Temporal pattern of species succession, showing the changing dominance of the most important species on permanent quadrat 2 on Small Island. □ indicates total % cover of all species. Compare also Table 1.

Table 1. Sequential occurrence of species on permanent quadrat 2 on Small Island. (Compare also Figure 2). Species composition is expressed as cover and number of individuals, according to the 14-point scale of Doing Kraft (1954). Number of plants when cover is less than 5%: r, 1–3 individuals; p, 4–15; a, 16–40; m, 41–300; m!, > 300. Cover estimates range from 01 to 09 in 10% increments from 5%, and 10 for cover > 95%.

Year	62	63	64	65	66	67	68	69	70	71	72	73	74	75	76	77	78	79	80	81	82	83
Cover (%)	3	30	30	60	65	80	90	95	98	90	85	80	80	80	80	98	98	98	95	98	95	95
Nr. of annual species	3	11	9	8	9	12	11	11	7	6	6	6	6	5	3	2	1	1	3	3	4	1
Nr. of perennial species	2	4	8	9	13	24	31	30	36	35	32	32	31	38	38	35	34	33	33	32	29	29
Total number of species	5	15	17	17	22	36	42	41	43	41	38	38	37	43	41	37	35	34	36	35	33	30
Halimione portulacoides	r	-	-	-	-	-	-	-	-	-	-	-	-	-	-	-	-	-	-	-	-	-
Suaeda maritima	a	02	a	-	-	-	-	-	-	-	-	-	-	-	-	-	-	-	-	-	-	-
Salicornia europaea	a	m	a	-	-	-	-	-	-	-	-	-	-	-	-	-	-	-	-	-	-	-
Atriplex hastata	r	02	a	a	-	-	-	-	-	-	-	-	-	-	-	-	-	-	-	-	-	-
Atriplex littoralis	-	m	-	-	-	-	-	-	-	-	-	-	-	-	-	-	-	-	-	-	-	-
Polygonum aviculare	-	p	p	-	r	-	-	-	-	-	-	-	-	-	-	-	-	-	-	-	-	-
Puccinellia distans	r	a	01	01	01	-	-	-	-	-	-	-	-	-	-	-	-	-	-	-	-	-
Spergularia marina	-	a	03	m!	a	r	-	-	-	-	-	-	-	-	-	-	-	-	-	-	-	-
Juncus bufonius	-	r	a	03	m!	m!!	m	p	-	-	-	r	r	m	-	-	-	-	-	-	-	-
Matricaria recutita	-	-	m	02	a	m!	a	m	-	-	-	-	-	-	-	-	-	-	-	-	-	-
Sonchus oleraceus	-	-	r	p	-	-	-	-	-	-	-	-	-	-	-	-	-	-	-	-	-	-
Matricaria maritima	-	p	a	a	02	m	p	-	p	p	r	-	-	-	-	-	-	-	-	-	-	-
Senecio vulgaris	-	p	-	r	a	a	r	r	-	-	-	-	-	-	-	-	-	-	-	-	-	-
Poa annua	-	p	a	m	03	03	m!	m!	a	r	p	r	r	-	-	-	-	-	-	-	-	-
Erigeron canadensis	-	-	-	-	r	r	-	-	-	-	-	-	-	-	-	-	-	-	-	-	-	-
Epilobium tetragonum	-	-	r	a	01	02	m	r	-	-	-	-	-	-	-	-	-	-	-	-	-	-
Cerastium holosteoides	-	-	-	-	r	02	01	01	m	a	a	r	p	-	p	p	p	-	r	-	-	-
Sagina procumbens	-	-	-	-	r	m	p	p	r	-	p	a	a	-	-	-	-	-	-	-	-	-
Epilobium parviflorum	-	-	-	r	01	02	02	01	m	p	m	a	-	r	p	r	-	-	p	-	p	r

Plantago coronopus	-	-	-	-	a	a	m	p	a	p	p	r	-	-	-	-	-	-	-	-	-	-
Sonchus asper	-	-	-	-	p	p	p	r	r	-	-	-	-	-	-	-	-	-	-	-	-	-
Epilobium adenocaulon	-	-	-	-	-	01	03	p	p	-	-	p	-	-	-	-	-	-	-	r	m	r
Epilobium seedlings	-	-	-	-	-	-	04	01	m!	-	-	-	-	-	-	-	-	-	-	-	-	-
Plantago major	-	-	a	a	m	m	m!	m!	a	a	p	p	p	r	p	a	p	r	-	r	r	p
Epilobium hirsutum	-	-	-	-	p	a	01	02	02	01	m!	m!	m!	m!	m	m	m	m	m	a	p	-
Poa trivialis	-	-	-	r	p	m	03	05	07	06	02	01	m	p	a	p	a	a	m	m	p	a
Aster tripolium	-	r	p	a	a	a	a	a	m	m	p	r	r	r	r	-	-	-	-	-	-	-
Taraxacum spp.	-	-	-	-	-	p	p	a	m	m	m	m	m	m	m	m	a	a	p	p	r	r
Poa pratensis	-	-	-	-	-	a	02	02	03	04	05	07	07	05	05	06	06	06	03	02	m	p
Bellis perennis	-	-	-	-	-	-	r	p	p	a	a	m	a	m	m	m	a	p	p	-	-	-
Cirsium arvense	-	-	r	-	-	r	r	r	p	p	p	a	m	m	m	m	m	m	m	m	m	a
Centaurium pulchellum	-	-	-	-	-	r	m	a	p	-	-	-	m	m!	-	-	-	-	r	p	p	-
Scirpus maritimus	-	-	-	-	-	-	r	r	-	p	p	p	a	m	m	m	m	p	p	m	p	r
Trifolium repens	-	-	-	-	-	r	r	p	p	p	a	p	01	01	01	03	01	p	m	a	a	p
Sonchus arvensis	-	-	-	-	-	r	p	p	p	a	a	p	p	a	a	m	m	a	a	m	m	p
Holcus lanatus	-	-	-	-	-	-	-	-	-	-	-	-	r	p	a	m	m	m	m	a	p	p
Ranunculus repens	-	-	-	-	-	r	r	p	a	m	m	m	m	m!	m!	m	m!!	m	m	m	p	p
Lycopus europaeus	-	-	-	-	-	-	r	-	r	p	p	p	p	a	a	p	m	m	m	m	m	a
Epipactis palustris	-	-	-	-	-	-	-	-	-	-	r	p	p	a	a	a	m	m	m	m	m	m
Eleocharis uniglumis	-	-	-	-	-	-	-	-	p	m	p	m!	m	m!	m	m!	m!!	01	m!!	01	m	m
Salix repens	-	-	-	-	-	-	r	r	r	p	p	p	p	p	p	p	p	01	01	01	01	01
Carex otrubae	-	-	-	-	-	r	p	a	m	a	a	m	m	m!	m!	m!	01	01	01	02	02	m
Phragmites australis	-	-	r	-	-	-	-	-	-	-	-	-	-	-	-	r	r	p	p	m	m	m
Hippophaë rhamnoides	-	-	-	-	-	-	-	-	-	-	-	-	-	r	p	p	p	p	p	p	01	m
Carex distans	-	-	-	-	-	-	-	-	-	-	-	p	r	r	p	p	p	a	a	a	m	p
Calamagrostis epigejos	-	-	-	-	-	-	-	-	-	r	r	r	p	a	a	m	m	m	01	02	03	07
Salix spp.	-	-	-	-	-	-	-	-	r	r	p	p	r	p	p	p	p	01	01	02	02	03

a-1

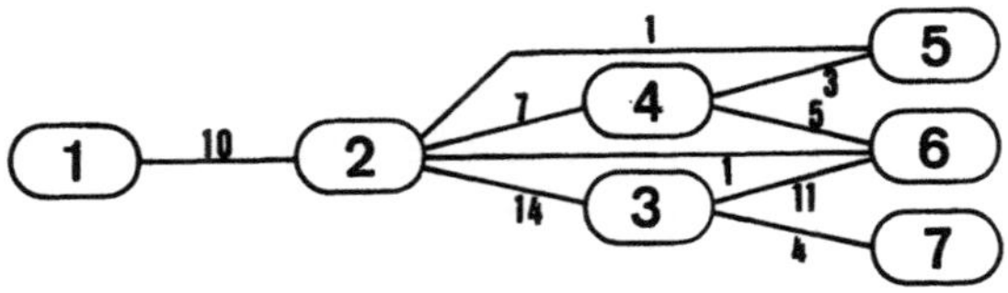

a-2

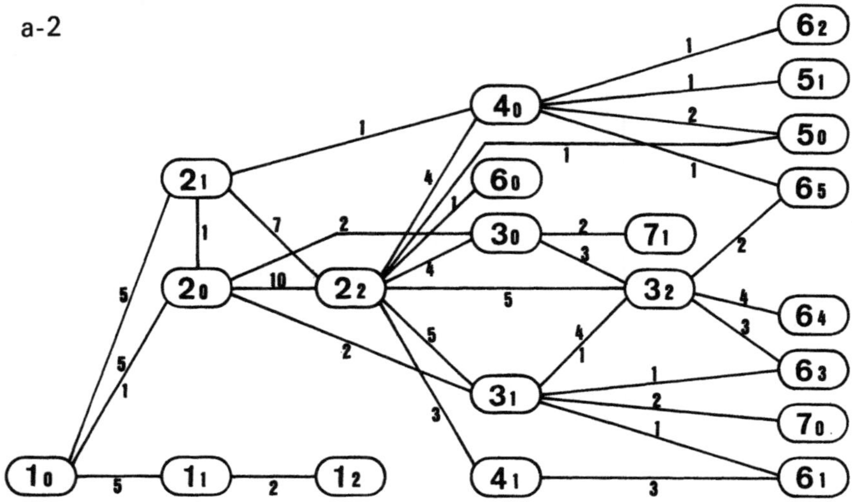

Results

Clusters of relevés were generated solely on the basis of their species composition: their localisation in space–time (including the identity of quadrats) was not used in the cluster analysis. In this section we discuss the pattern in space–time of the clusters so generated, that is, we establish a relationship between the state–space (of species composition) and space–space with the time axis.

Large scale spatio-temporal patterns using the 33 permanent quadrats

A preliminary analysis of the PQ dataset was published by van Noordwijk-Puijk et al. (1979). It was thoroughly analysed by Hogeweg (1976a) using data up to 1973 and was later updated by van Schaik (unpublished) to 1978, using identical pattern analysis methods; we shall discuss both these analyses here.

Tables 2a, b and Figs. 3a, b show the transition tables and transition graphs for the dataset up to 1973 and up to 1978 respectively. Two levels of detail are shown, corresponding to the first two optimal splitting levels of the dendrogram of the logarithmised species covers in each of the PQs in each of these periods. The differences between the two analyses are striking. The partial dataset (up to 1973, Fig. 3a) shows a network-like b-1

b-2

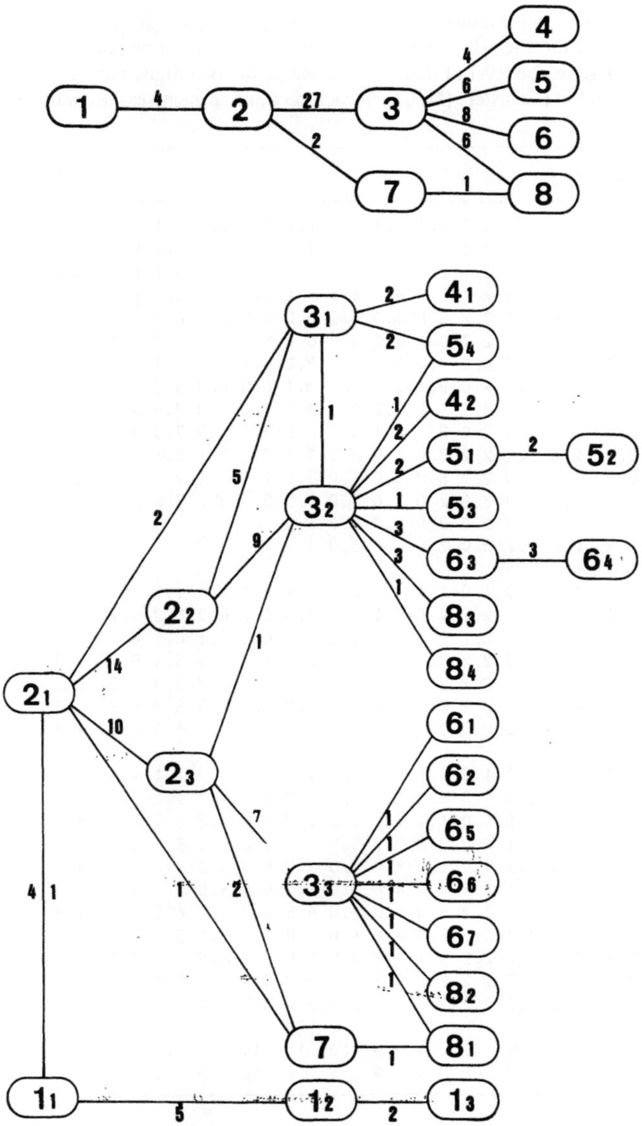

Fig. 3. Transition graphs of the permanent quadrats. (a) Dataset 1963–1973, logarithmised data (compare Table 2a). (a-1) Transition graph for 7 clusters. (a-2) Transition graph for 21 clusters. (b) Dataset 1963–1978, logarithmised data (compare Table 2b). (b-1) Transition graph for 8 clusters. (b-2) Transition graph for 27 clusters. The numerals in each cluster denote its level of detail (see Table 2).

transition pattern at both levels of detail, whereas the complete dataset (up to 1978, Fig. 3b) shows a much more tree-like transition pattern, although at each level of detail a convergence does occur. Moreover the sites united into clusters differ clearly, also for those samples included in both analyses. Both patterns are similar in showing an overall progressive divergence.

Table 2(a). Transition table indicating the cluster membership of each plot (PQ) in each year. Transition table of the 1963–1973 dataset (logarithmised data) (compare Fig. 3a). The first digit alone represents the level of detail of 7 clusters, the two digits the level of detail of 21 clusters. The number of clusters (N. Cl.) in each plot and in each year are indicated for each level of detail.

PQ	'63										'73	N.CL	
B 9	1.0	1.0	1.0	1.1	1.1	1.1	1.1	1.1	1.1	1.1	1.1	2	1
B10	2.0	1.0	2.1	2.2	4.1	4.1	4.1	4.1	6.1	6.1	6.1	6	4
B34		1.0	2.1	2.2	4.1	4.1	4.1	4.1	6.1	6.1	6.1	5	4
B33		1.0	2.1	2.2	4.1	4.1	4.1	4.1	6.1	6.1	6.1	5	4
B35	1.0	2.1	2.2	6.0	6.0	6.0	6.0	6.0	6.0	6.0	6.0	4	3
B11	2.0	2.2	2.2	3.1	3.2	3.2	6.4	6.4	6.4	6.4	6.4	5	3
B12	2.0	2.2	3.2	3.1	3.2	3.2	3.2	6.4	6.4	6.4	6.4	5	3
B17	2.0	2.0	2.0	3.1	3.1	3.1	3.1	6.3	6.3	6.3	6.3	3	3
B18	2.1	2.2	2.2	3.2	3.2	3.2	7.0	7.0	7.0	7.0	7.0	4	3
B19	2.1	2.0	2.0	3.1	3.1	3.1	7.0	7.0	7.0	7.0	7.0	4	3
B20	2.0	2.2	2.2	3.0	3.0	3.0	7.1	7.1	7.1	7.1	7.1	4	3
B21	2.0	2.0	2.0	3.0	3.0	3.0	7.1	7.1	7.1	7.1	7.1	3	3
B13	2.0	2.0	2.0	2.0	3.0	3.0	3.2	3.2	3.2	3.2	3.2	3	2
B14	2.0	3.0	3.0	3.0	3.2	3.2	3.2						
B15	2.0	2.2	3.0	3.0	3.0	3.0	3.0						
S 1	1.0	1.0	1.0	1.1	1.1	1.1	1.1	1.1	1.1	1.1	1.1	2	1
S 8	2.1	2.1	2.2	4.0	4.0	4.0	4.0	6.2	6.2	6.2	6.2	4	3
S 2	2.1	2.1	3.1	3.1	3.1	6.1	6.1	6.1	6.1	6.1	6.1	3	3
S 3	2.0	2.2	2.2	3.1	3.1	3.2	6.4	6.4	6.4	6.4	6.4	5	3
S 4	2.0	2.2	2.2	3.1	3.1	3.2	3.2	3.2	6.3	6.3	6.3	5	3
S 7	2.0	2.0	2.2	2.2	3.0	3.2	3.2	3.2	6.3	6.3	6.3	5	3
S 5	2.0	2.0	2.0	2.2	3.2	3.2	3.2	3.2	6.3	6.3	6.3	4	3
P27	1.0	1.0	1.0	1.0	1.1	1.1	1.1	1.1	1.1	1.1	1.1	2	1
P26	1.0	1.0	1.0	1.1	1.1	1.1	1.2	1.2	1.2	1.2	1.2	3	1
P25	1.0	1.0	1.0	1.1	1.2	1.2	1.2	1.2	1.2	1.2	1.2	3	1
P30	2.1	2.1	2.2	2.2	2.2	5.0	5.0	5.0	5.0	5.0	5.0	3	2
P31	2.1	2.2	2.2	4.0	4.0	4.0	4.0	5.0	5.0	5.0	5.0	4	3
P32	2.1	2.1	2.1	4.0	4.0	5.0	5.0	5.0	5.0	5.0	5.0	3	3
P29	2.1	2.1	2.2	4.0	4.0	4.0	6.5	6.5	6.5	6.5	6.5	4	3
P28	2.1	2.1	2.2	2.2	4.0	4.0	4.0	5.1	5.1	5.1	5.1	4	3
P24	2.0	2.2	2.2	3.2	3.2	3.2	6.5	6.5	6.5	6.5	6.5	4	3
P23	2.0	2.2	2.2	2.2	3.2	3.2	3.2	6.5	6.5	6.5	6.5	4	3
P22	2.0	2.2	2.2	3.2	3.2	3.2	3.2	3.2	3.2	3.2	3.2	3	2
N.CL	3	5	7	9	9	10	14	14	13	13	13		
	2	2	3	4	5	5	6	6	6	6	6		

Examining the pattern of the dataset up to 1973 (Table 2a, Fig. 3a) we can draw the following conclusions (Hogeweg 1976a):

The pattern is mainly temporal in the earlier years and mainly spatial in the later years.

Much of the spatial pattern in each of the years is due to local differences in speed of development of the vegetation (quadrats 'enter' and 'leave' clusters at different times; see Table 2a). This is most conspicuous in the highest and lowest areas where vegetation resembling the one predominant in earlier years remains for a long time; at smaller detail, differences in development rate also show up as spatial differentiation. Such differences in rate of development are strongly related to differences in height and height-associated soil processes (van Noordwijk-Puijk et al. 1979).

Table 2(b) **Transition table indicating the cluster membership of each plot (PQ) in each year. Transition table of the 1963–1978 dataset (logarithmised data) (compare Fig. 3b). The first digit alone represents the level of detail of 8 clusters, the two digits the level of detail of 27 clusters. The number of clusters (N. Cl.) in each plot and in each year are indicated for each level of detail.**

PQ	'63															'78	n.cl.	
B 9	1.1	1.1	1.1	1.2	1.2	1.2	1.2	1.2	1.2	1.2	1.2	1.2	1.2	1.2	1.2	1.2	2	1
B10	2.1	1.1	2.1	2.3	3.3	3.3	3.3	3.3	6.1	6.1	6.1	6.1	6.1	6.1	6.1	6.1	5	4
B34		1.1	2.1	2.3	3.3	3.3	3.3	3.3	6.2	6.2	6.2	6.2	6.2	6.2	6.2	6.2	5	4
B33		1.1	2.1	1.3	3.3	3.3	3.3	3.3	6.7	6.7	6.7	6.7	6.7	6.7	6.7	6.7	5	4
B35	1.1	2.1	2.3	3.3	3.3	3.3	3.3	3.3	6.6	6.6	6.6	6.6	6.6	6.6	6.6	6.6	5	4
B11	2.1	2.2	2.2	3.2	3.2	3.2	5.1	5.1	5.1	5.1	5.1	5.1	5.2	5.2	5.2	5.2	5	3
B12	2.1	2.2	2.2	3.2	3.2	3.2	3.2	5.1	5.1	5.1	5.1	5.1	5.2	5.2	5.2	5.2	5	3
B17	2.1	2.1	2.1	3.2	3.2	3.2	3.2	5.3	5.3	5.3	5.3	5.3	5.3	5.3	5.3	5.3	3	3
B18	2.1	2.2	2.2	3.2	3.2	3.2	4.2	4.2	4.2	4.2	4.2	4.2	4.2	4.2	4.2	4.2	4	3
B19	2.1	2.1	2.1	3.2	3.2	3.2	4.2	4.2	4.2	4.2	4.2	4.2	4.2	4.2	4.2	4.2	3	3
B20	2.1	2.2	2.2	3.1	3.1	4.1	4.1	4.1	4.1	4.1	4.1	4.1	4.1	4.1	4.1	4.1	4	3
B21	2.1	2.2	2.2	3.1	3.1	4.1	4.1	4.1	4.1	4.1	4.1	4.1	4.1	4.1	4.1	4.1	4	3
B13	2.1	2.1	2.1	2.1	3.1	3.1	3.1	3.1	3.1	3.1	3.1	3.1	5.4	5.4	5.4	5.4	3	3
B14	2.1	3.1	3.1	3.1	3.1	3.1	3.1											
B15	2.1	2.2	3.1	3.1	3.1	3.1	3.1											
S 1	1.1	1.1	1.1	1.2	1.2	1.2	1.2	1.2	1.2	1.2	1.2	1.2	1.2	1.2	1.2	1.2	2	1
S 8	2.1	2.1	2.3	3.3	3.3	3.3	3.3	6.5	6.5	6.5	6.5	6.5	6.5	6.5	6.5	6.5	4	3
S 2	2.1	2.1	2.3	2.3	3.2	6.3	6.3	6.3	6.3	6.3	6.3	6.4	6.4	6.4	6.4	6.4	5	3
S 3	2.1	2.2	2.2	3.2	3.2	3.2	6.3	6.3	6.3	6.4	6.4	6.4	6.4	6.4	6.4	6.4	5	3
S 4	2.1	2.2	2.2	3.2	3.2	3.2	6.3	6.3	6.3	6.3	6.4	6.4	6.4	6.4	6.4	6.4	5	3
S 7	2.1	2.1	2.2	2.2	3.1	3.2	3.2	3.2	5.4	5.4	5.4	5.4	5.4	5.4	5.4	5.4	5	3
S 5	2.1	2.1	2.1	2,2	3.1	3.1	5.4	5.4	5.4	5.4	5.4	5.4	5.4	5.4	5.4	5.4	4	3
P27	1.1	1.1	1.1	1.1	1.2	1.2	1.2	1.2	1.2	1.2	1.2	1.2	1.2	1.2	1.2	1.2	2	1
P26	1.1	1.1	1.1	1.2	1.2	1.2	1.3	1.3	1.3	1.3	1.3	1.3	1.3	1.3	1.3	1.3	3	1
P25	1.1	1.1	1.1	1.2	1.3	1.3	1.3	1.3	1.3	1.3	1.3	1.3	1.3	1.3	1.3	1,3	3	1
P30	2.1	2.1	2.3	2.3	2.3	7	7	7	7	7	7	7	7	7	7	7	3	2
P31	2.1	2.3	2.3	7	7	7	7	7	7	7	7	7	7	7	7	7	3	2
P32	2.1	2.1	2.1	7	7	7	7	7	8.1	8.1	8.1	8.1	8.1	8.1	8.1	8.1	3	3
P29	2.1	2.1	2.3	3.3	3.3	3.3	8.1	8.1	8.1	8.1	8.1	8.1	8.1	8.1	8.1	8.1	4	3
P28	2.1	2.1	2.3	2.3	3.3	3.3	3.3	8.2	8.2	8.2	8.2	8.2	8.2	8.2	8.2	8.2	4	3
P24	2.1	2.2	2.2	3.2	3.2	3.2	3.2	8.3	8.3	8.3	8.3	8.3	8.3	8.3	8.3	8.3	4	3
P23	2.1	2.2	2.2	2.2	3.2	3.2	3.2	8.3	8.3	8.3	8.3	8.3	8.3	8.3	8.3	8.3	4	3
P22	2.1	2.2	2.2	3.2	3.2	3.2	3.2	8.4	8.4	8.4	8.4	8.4	8.4	8.4	8.4	8.4	4	3
N.CL.	2	5	4	10	7	8	12	17	19	20	20	19	18	18	18	18		
	2	3	3	4	4	5	7	7	7	7	7	7	6	6	6	6		

In the period 1966–1969 the pattern coincides entirely with height gradient; height varies between 0 m and 0.8 m above NAP (Dutch Ordnance Level) and is identical in each of the land bodies. Around 1969 some island-specific vegetations are established, which differ strongly from the remaining vegetation and form separate main clusters (e.g., clusters 5 and 7 in Table 2a and Fig. 3a).

The island-specific pattern can be enhanced by iterative character weighing (Hogeweg 1976a, b): all peninsula quadrats from 1967 onward can be separated from other quadrats on the basis of their vegetation.

The network structure of the graphs is caused mainly by the establishment of the island-specific clusters mentioned above (which cut through previously established height gradient patterns) and by 'jumps', that is, by a quadrat which skips a stage in the vegetation development (e.g., B35). However, the boundary lines between height-specific clusters may shift also during the succession (e.g., B20 belongs to the 'lower' cluster (2.2) in 1964 and to the 'higher' cluster (3.0) in 1966).

Examining the pattern of the full dataset up to 1978 (Table 2b, Fig. 3b) we see that a finer spatial differentiation is made than in the previous analysis: in 1973 the quadrats are distributed over 20 clusters in this analysis, whereas there were only 13 clusters at that time in the previous analysis (in 1978 the number drops to 18). This spatial differentiation includes both island-specific differences (clusters 7 and 8 occur only on the mainland and cluster 6 only on the islands; see Table 2b) and gradient differences (for example, the subdivision of cluster 6 on the islands). The spatial differentiation shows up in this analysis already in the years which were also included in the previous analysis, that is, later-formed differentiation is projected back into the past by the cluster analysis. This could be caused by the mere availability of data: the vegetation of the later years is represented by many samples and is therefore analysed in more detail by the method. Thus when data of every year are included, the slow-down of succession could by itself be sufficient to generate a divergent succession pattern when looked upon by a method which adjusts detail on the basis of availability of data. However, in the present case the divergence is not caused by slowing down of succession: the dissimilarity of the PQs in successive years, and the turnover of species, remains on average the same during the period of observation (van Schaik, unpublished; Fig. 4a, b). Similarly if the sampling density in the later years is reduced, the resulting dendrogram still joins samples of one site before joining samples of any other site. Both these observations force us to conclude that the increase in spatial differentiation causes the seeming slowdown of succession and not the other way around. The increase in spatial differentiation after 1973 is mainly due to the differentiation of the various land bodies. This differentiation is caused by differences in management: on the peninsula cattle and sheep have been grazing since 1965, the big island is mown once a year since 1971, and the small island is left alone.

It is interesting to note that the iterative character weighing on the data up to 1973 enhanced the pattern so as to make it resemble more closely the future (not included) pattern. Contrary to the analysis of the dataset up to 1973, the analysis of the full dataset does not represent the temporally induced spatial patterns around 1970, nor the differences in succession speed which shows up as the skipping of stages. This shows clearly that an optimal representation of the data needs several subsets of the data. The temporal differences are apparently the cause of the maximum in species number occurring around 1970.

The analysis of the years separately is more difficult to interpret because we lack the historical context. In the cluster analysis the lack of historical information (i.e. the presence of samples of previous years) leads to a less consistent pattern over the years. From the cluster analysis as well as from the principal component analysis it is clear, however, that the pattern has

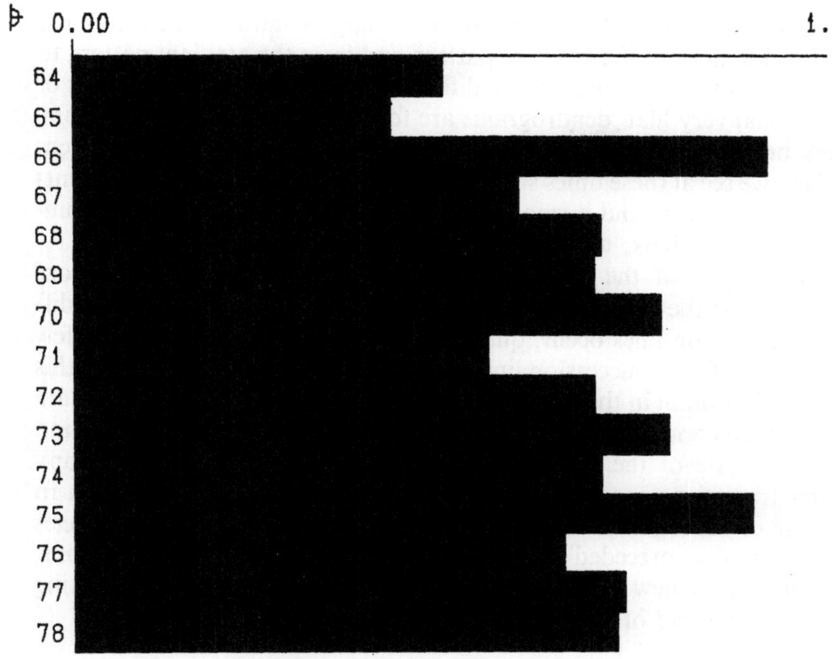

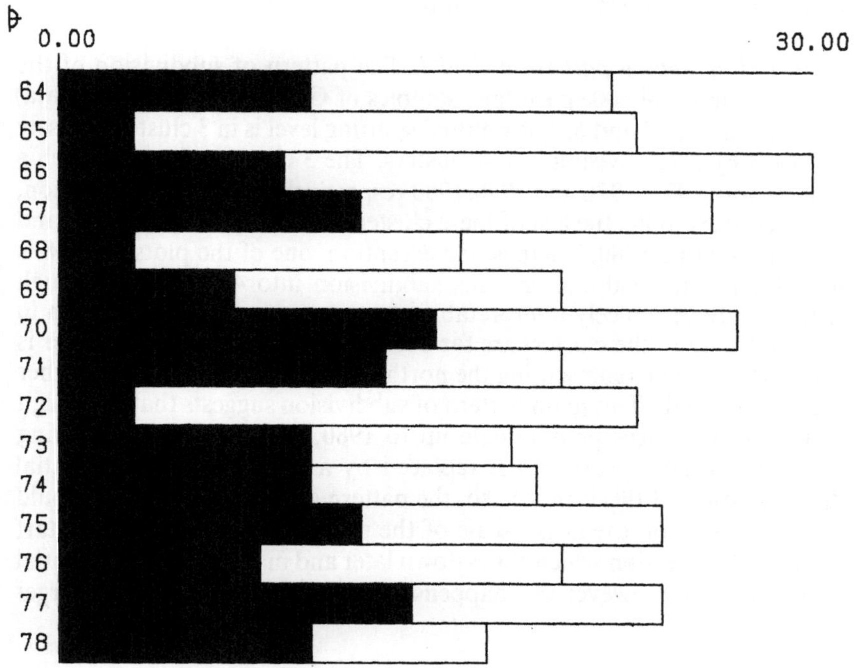

Fig. 4. Pattern of change in the permanent quadrats. (a) Average difference between subsequent years in all permanent quadrats, expressed as mean character difference. (b) Number of immigrant species (species which were not present in the previous year) (white). Number of extinct species in the subsequent year (species which were present in the previous year but not in the current year) (black). The sum of these gives the species turnover from year to year.

shifted over the years. In the periods of the largest shifts, that is, from the temporal pattern to the gradient pattern and from the gradient pattern to the island differentiation (first the differentiation of the peninsula, later of the islands) very 'flat' dendrograms are formed without optimal splitting levels because of the conflicting patterns. In the principal component analysis we see at these times smaller correlations between the components of successive years and a smaller percentage covered by the first 4 components (van Schaik, unpublished).

The analysis of the binarised data shows patterns similar to those described for the logarithmised data, but more transitions between different succession lines occur; quadrats are classified during one or a few years in a different succession line from the remaining years. Because this happens less often in the analyses in which the (logarithmised) cover data are used we conclude that abundance of species adds some information about the fate of the permanent quadrats. It is interesting that many cluster transitions are somewhat later in the binarised analysis than in the case of the logarithmised data. This indicates that changes in species composition is preceded by changes in dominance of species. Thus, immigration of new species seems to depend on changes in the existing vegetation instead of vice versa.

Medium scale pattern using the Grids

Temporal vs. spatial pattern in Grid 1. The pattern of subdivision of the dendrogram of all 600 space/time samples of Grid 1 (logarithmised data) is shown in Figs. 5 and 6. The optimal splitting level is in 3 clusters, closely followed by a subdivision into 4 clusters. The 3 clusters represent each a time period: 1973, 1976 and 1978/1980 respectively. The next subdivision, into 4 clusters, splits the last of these clusters in one representing 1978 and one representing 1980. There is one exception: one of the plots of 1978 is included in the 1980 cluster. This subdivision into 4 clusters, although semantically very nicely interpretable, does not represent an optimum in the dendrogram: the clusters are far better separated if the 1980 cluster is split into two, one representing the northern half of the grid and the other the southern half. This main pattern of subdivision suggests that, although temporal differences predominate up to 1980, the succession is slowing down in the later years and is replaced by a more pronounced spatial differentiation. If this is indeed so, the pattern changes at this scale would resemble those on the larger scale of the permanent quadrats: they start with a fast succession which slows down later and makes place for a spatial differentiation. However this happens much later here than at the larger scale.

Temporal patterns of scale and detail in Grid 1. The analysis of van Schaik (unpublished) revealed a pattern in the successive subdivision of the clusters (until 1978): the later clusters were subdivided at a higher dissimilarity level than the earlier ones whereas the overall dissimilarity was less in the later ones. Thus, if we visualise the variation in the

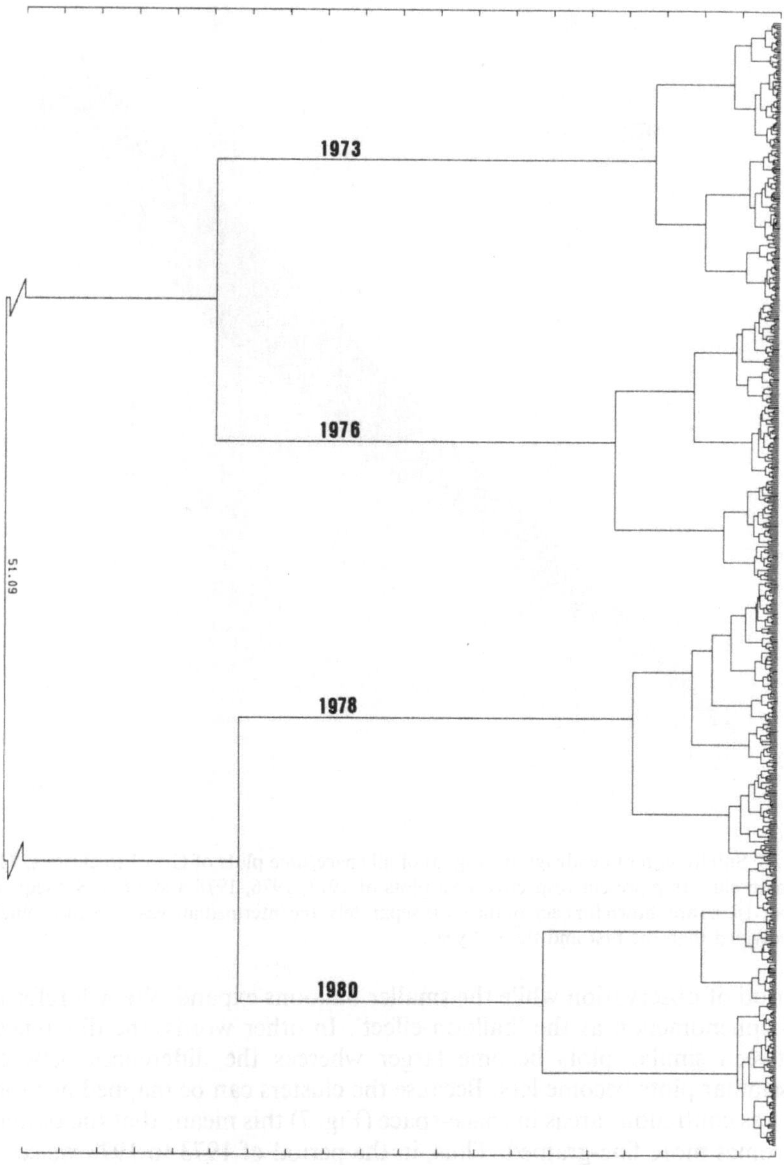

Fig. 5. Dendrogram of all 600 space–time plots in Grid 1, logarithmised data. The 4 main clusters represent the 4 different years as indicated. For the spatial pattern of the first 12 clusters see Fig. 7a.

vegetation in the form of balloons of various sizes, depicting subclusters as smaller balloons in the larger balloon which represents the total variation, the changes in the variation of the vegetation during the observed years of succession are such that the outer balloon shrinks during the

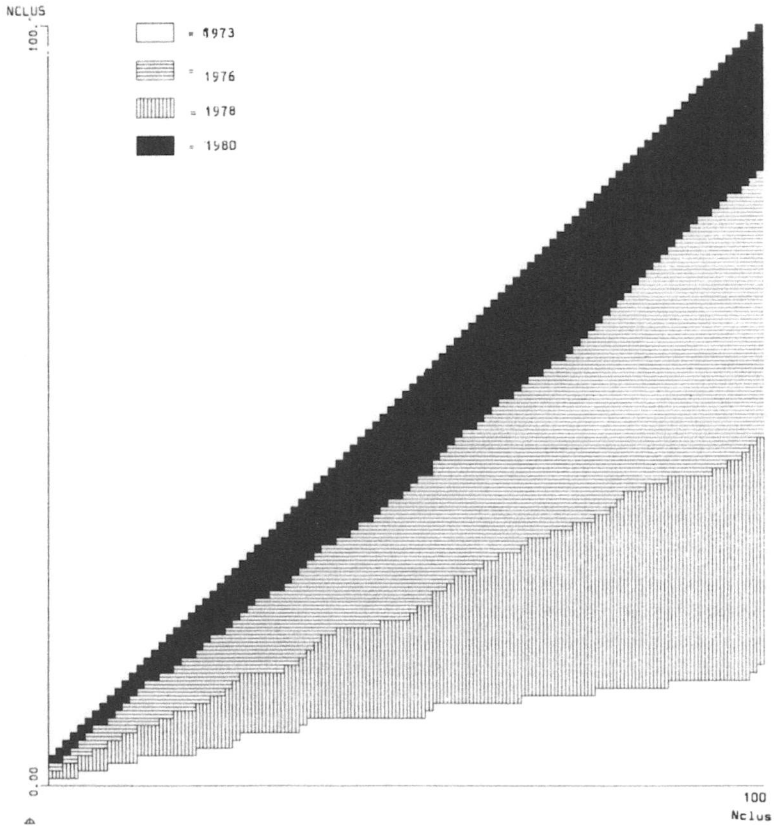

Fig. 6. Subdivision of dendrogram (Fig. 5) of all space/time plots of Grid 1 in clusters. The 4 main clusters represent respectively all plots of 1973, 1976, 1978 and 1980. Subsequent subdivisions are shown for each of the years separately: the intermediate years are more finely subdivided than the first and the last year.

period of observation while the smaller balloons expand. We will refer to this phenomenon as the 'balloon effect'. In other words, the differences between similar plots become larger whereas the differences between dissimilar plots become less. Because the clusters can be mapped as more or less contiguous areas in space-space (Fig. 7) this means that the pattern becomes more fine-grained. Thus, in the period of 1973 to 1978 we see a convergent succession at the scale of the entire grid and a divergent succession at the scale of neighbouring plots. This 'balloon effect' seemed to be an interesting interpretation of succession of a vegetation in terms of detail in state-space and scale in space-space: the overall area becomes more homogeneous but differences remain at a smaller scale (= diversity in some sense of the word). However, as the data presented here (Figs. 5, 6) show, this trend was abruptly reversed in 1980, when there was a very sharp subdivision into the northern and southern half of the grid and the differences between adjacent plots become less pronounced (Figs. 7, 8).

Thus the 'balloon effect' should be interpreted as a transition and such an interpretation seems indeed possible when the vegetational changes are analysed in terms of species composition.

Temporal patterns in species composition in Grid 1. There were only a very few species present in reasonably invariant amounts during the entire period of observation: *Juncus gerardii*, *Bellis perennis* and *Carex obtrubae*. All other species (which have been present in more than a few plots) were either rapidly decreasing or rapidly increasing during the period of observation, or reached their highest numbers in the intermediate years. The overwhelming impression which both observations of the area and casual observation of the data give is that *Calamagrostis epigejos*, and later on *Hippophae rhamnoides*, 'take over' the vegetation (Fig. 8). As both are sturdy plants this observation is easily interpreted in terms of their outcrowding the declining species (in particular *Plantago coronopus*, *P. major*,

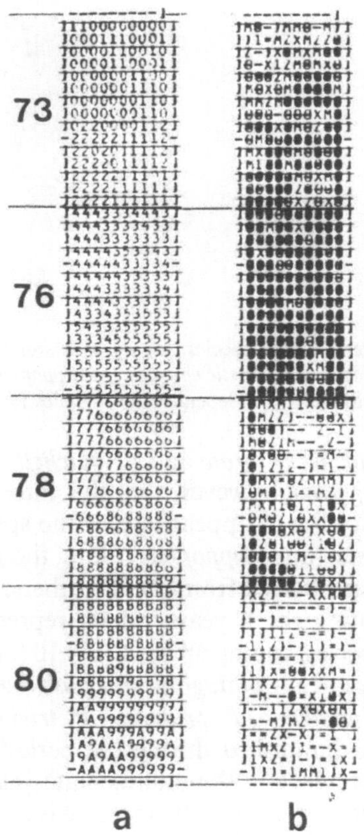

Fig. 7. Grid 1, maps of clusters and species diversity. (a) Spatial/temporal distribution of the 12 clusters of Grid 1 (numbered 0–9, A, B). (b) Spatial/temporal distribution of the number of species per plot in Grid 1 (minimum 7, maximum 23, indicated by relative degrees of shading).

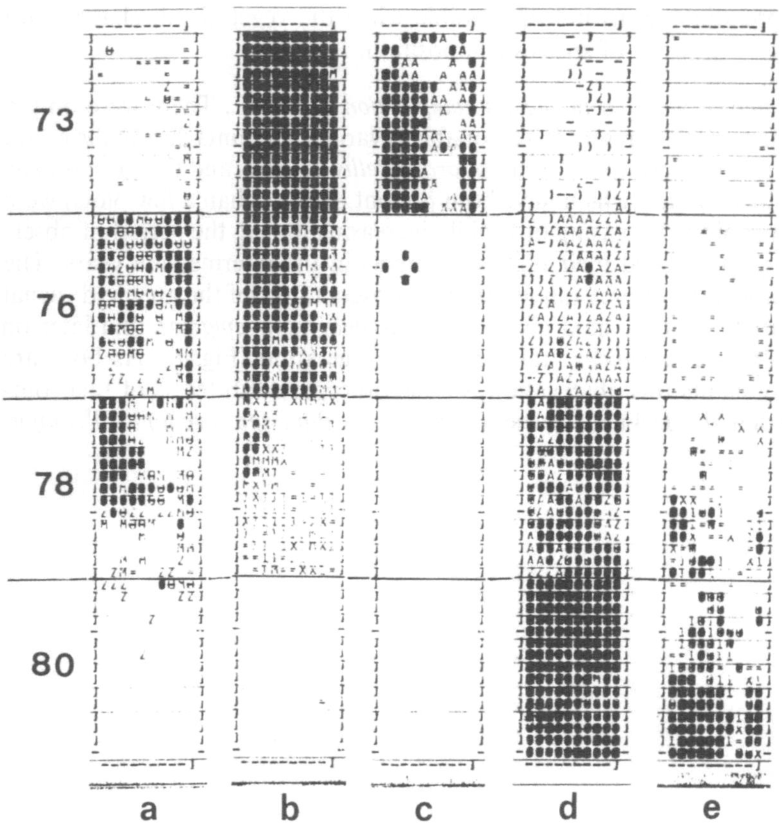

Fig. 8. Grid 1, maps of cover of some species. (a) *Agrostis stolonifera*. (b) *Plantago coronopus*. (c) *Juncus bufonius*. (d) *Calamagrostic epigejos*. (e) *Hippophae rhamnoides*. The cover of the species is scaled logarithmically; independent gray scales for the various species.

Juncus bufonius, *Centaurium littorale* and *C. pulchellum*). A more careful analysis of the data suggests, however, that this interpretation is wrong, and that on the contrary the disappearance of the species is the primary event, with *Calamagrostis* (and *Hippophae*) filling the gaps (Hogeweg and Hesper 1981b). This was inferred from the oligothetic characterisation of the 4 clusters representing the 4 years, which represented the ongoing changes in terms of 7 species, ending up with only 10 (out of 600) misclassified plots. These species are: *Plantago coronopus*, *Juncus bufonius*, *Gnaphalium luteo-album*, *Poa annua*, *P. pratensis*, *P. trivialis* and *Centaurium littorale*, species which diminished during the period of observation or reached their highest presence in the intermediate years. Although there are some strongly increasing species, the increase is a less consistent event than was the decrease of the earlier species. This observation is further corroborated by the fact that the mean cover is smaller during the intermediate years (subsequently: 78%, 63%, 76% and 78%). There was more open space in the intermediate years, which would not be expected if the new species were crowding out the older ones. The number of species also varies: it is highest in the intermediate years (see also Fig. 7b which shows how the number of species varies over the grid). Together these observations suggest that the vegetation predominated by *Plantago coronopus* starts disappearing due to some external factor. The empty space is used by several species, some already present in low numbers and others newly immigrating into the area. *Poa* species and *Agrostis stolonifera* increase faster and *Calamogrostis epigejos* lags behind, but finally seems to crowd out the other species: whether it is indeed competition which causes the decrease of those species cannot be unequivocally concluded from our data.

Spatial patterns in Grid 1. Looking at the dendrogram of the 600 space/time plots of Grid 1 (Fig. 5) in more detail, we see a local optimum for a subdivision into 12 clusters. The location of the members of these clusters in time and space is represented in Fig. 7a. In each of the years the grid shows the most conspicuous spatial differences in the north/south direction, although in 1973 the southern cluster extends somewhat into the northern area. Next the grid is subdivided in each year into 3 clusters. The third cluster splits off in all years an area in the western part of the grid, sometimes located in the northern half (1976 and 1978, clusters 4 and 7 resp.), sometimes in the southern half (1973 and 1980, clusters 2 and A resp.; see Fig. 7a). These spatial differences are due to different species in the various years, as is to be expected from the fact that only very few species remain relatively constant in occurrence during the entire period. The subdivision into 3 clusters in each of the years is characterised in terms of the species composition in two different ways. Table 3a indicates the 3 species which differ strongest statistically between each of the subsequent subdivisions, as measured by Kruskal–Wallis one-way analysis of variance, i.e., on the basis of averaged, rank-ordered data. Table 3b shows the species selected for the oligothetic characterisation of these clusters (3 in each year). Moreover the space/time patterns for a number of species are given in Fig. 8. These are interesting species because they are selected by the oligothetic characterisation, that is, they go a long way in characterising the clusters. As almost all species change strongly during the period of observation it is to be expected that the observed spatial pattern in the various years is due to species which are either increasing or decreasing. The relative invariance of the spatial pattern may seem surprising in view of these species' fluctuations.

Analysis of soil parameters as collected in the winter of 1978 indicates that this pattern follows a clear differentiation in the environmental conditions. The soil samples were taken along the edge of the grid in order to minimise disturbance of the succession due to sampling. Clustering the soil samples we find, contrary to the vegetation pattern, the most conspicuous variation in the east-west direction. The western part contains more nutrients (P, N, POC) and is wetter than the eastern part; it seems more suitable for plant growth. There are, however, clear differences also between the northern and southern part: the northern part is richer in

Table 3(a). Species characterising the spatial differentiation (3 clusters) in Grid 1 in the various years. Monothetic characterisation of the first and second subdivision in each year (Kruskal–Wallis one-way analysis of variance).

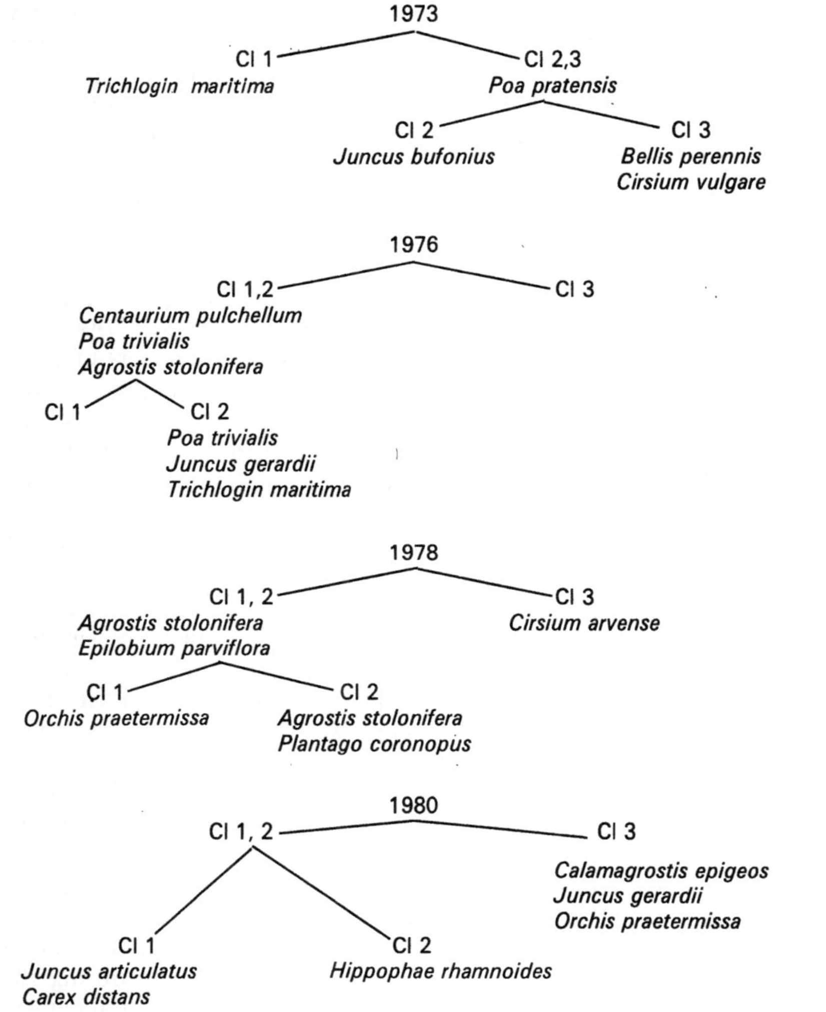

$CaCO_3$ than the southern part, presumably due to shell deposits. Apparently many species limit their range of occurrence in relation to this factor because the north–south differences predominate over the entire period. The nutrient differences sometimes become expressed in the major vegetation pattern either in the southern or in the northern area, although traces of this differentiation can be discerned in the more detailed pattern all over the grid in all years.

The spatial pattern can also be interpreted in terms of temporal differences. In particular the nutrient-related differentiation of the northern half coincides with a temporal pattern: the northwestern area is the last foothold for the declining species, in particular *Plantago coronopus* and

Table 3(b). Species characterising the spatial differentiation (3 clusters) in Grid 1 in the various years. Oligothetic characterisation of the three clusters of each year. Mean logarithmised cover of the selected species is given for each of the clusters.

Species selected by oligothetic characterisation			
1973			
Trichlogin maritima	1.3	0.5	0.5
Juncus bufonius	1.3	0.6	2.0
Plantago major	1.08	1.37	1.68
Calamagrostis epigejos	0.17	0.91	0.33
Juncus articulatus	3.00	2.00	2.25
1976			
Poa trivialis	1.05	2.30	0.1
Agrostis stolonifera	2.74	3.11	1.27
Centaurium pulchellum	0.9	2.19	0.6
Crepis capillaris	0.02	0.17	0.7
Salix repens	0.67	0.15	1.3
Plantago coronopus	3.5	4.5	3.5
1978			
Agrostis stolonifera	1.9	4.0	0.6
Bellis perennis	0.7	0.3	1.6
Juncus gerardii	1.55	2.5	2.2
Cirsium arvense	0.16	0.04	1.0
Carex serotina	0.32	0.04	0.96
Potentilla anserina	0.47	0.02	0.04
Sagina procumbens	0.15	0.25	0.6
1980			
Juncus articulatus	2.2	0.1	2.3
Epilobium palustris	1.8	1.7	0.2
Orchis praetermissa	0.6	0.3	1.6

Juncus bufonius, while the species of the intermediate period reach their highest coverage there (the intermediate period is here longest). Moreover, the area occupied by cluster 2 in 1973, which is exceptional in extending from the south into the northern area, seems to correspond with the area where *Juncus bufonius* in particular is disappearing and in which *Agrostis stolonifera* and *Calamagrostis epigejos* settle. It is located in the eastern, nutrient deficient part of the grid. This settlement pattern does not seem to be due to preferences of the settling species as these reach their highest cover in quite different areas (Fig. 8). Finally the nutrient related east-west differentiation in 1980 in the southern area is due to a preference pattern between the relatively new species *Hippophae* and *Salix repens*.

Grid 2. The pattern analysis of this (smaller) grid reinforces the conclusions of the other grid. The cluster analysis of all space/time plots splits into the 4 years for which data are available but in addition there is a fifth cluster in which the same plots of all years are collected: the *Festuca rubra* vegetation present in these quadrats remains relatively unchanged. Moreover, there are about 20 'misclassifications' among the years; these are plots which are joined to themselves in previous or future years. As in the other grid, the increase of *Calamagrostis epigejos* is the most conspicuous change over the period of observation, but is not selected to characterise the temporal changes by the oligothetic characterisation of the clusters. Instead *Trifolium repens* is selected which, like *Plantago coronopus* in the other grid, decreases strongly in the period of observation. Thus, in this grid the decrease of some species (although other ones) seems also to be the primary event and the increase of *Calamagrostis* the secondary event.

It is interestng to note that the oligothetic characterisation 'corrects' the misclassification by restoring a purely temporal pattern. Plots which were originally clustered on the basis of very local characteristics (causing them to join the same quadrat in a different year) are transferred to the year cluster, on the basis of the species found to be characteristic for that year.

Small-scale pattern using the subgrids

The pattern analysis of the subgrids shows an increase in the small-scale pattern in 1978, as compared to 1976 and 1980. This can be concluded from the optimality of the splitting levels of the dendrogram as well as on the basis of the increase in significant correlations (phi coefficients for these binary data) between plant species. The differences between small plots of different subgrids (far apart) are less in 1978: clustering all $25 \times 25\,cm^2$ plots together for each of the years shows an almost complete separation of the subgrids for 1976 and 1980, whereas in 1978 the plots of the subgrids of Grid 1 are intermingled in several clusters. These observations agree well with the pattern found in the larger grids: a more localised mosaic pattern in the transition period of which the parts of the mosaic occur all over the area.

CONCLUSIONS AND DISCUSSION

All these observations seem to fit into the following scenario: vegetation changes are initiated by the spontaneous decline of some species. This decline may be due to external changes or to exhaustion of the environment by these species themselves. The open areas are colonised by a relatively large number of species. The species which develop fastest are those which seem to have little preference for the differences occurring in the area. Those that do show such preferences develop slower but can finally outcompete the other settlers, and possibly other species. When they do, a strong, spatial differentiation becomes visible, coinciding with pre-existing spatial differences, such as soil differences; it is therefore similar in different periods, although different plants then dominate the vegetation. The transition phase shows a less clear large-scale spatial differentiation because of the conflicting patterns of settlement and dominance and because of the relatively large number of species without particular preferences. Groups of 'older species' or groups of species which can occur together form mosaics in that period. This gives rise to the 'balloon effect' described. If such a mosaic pattern should become stable, the vegetation would show a combination of convergence at a larger scale and divergence at a smaller scale. In principle this may happen at one scale at some level of detail while at other scales or levels of detail other patterns occur. However the present data analysis shows that such patterns can occur at a rather wide range of scales, almost simultaneously.

The oligothetic characterisation of the clusters focussed attention on a few species which by themselves are sufficient to characterise the clusters based on all species. Pattern generation experiments using physiological and autecological (demographic, morphological) data can best start with these species. In particular, it would be interesting to see whether physiological characterisation of *Calamagrostis epigejos* and the species it replaces (*Plantago coronopus* and *Juncus bufonius* in Grid 1 and *Trifolium repens* and *Poa pratensis* in Grid 2) supports our hypotheses that the decline of the latter species is the primary event, with *Calamagrostis* filling the gaps. In such a model, storage of nutrients will play an important role.

In the case of minimal information models we think in the first place of models based on the hypothesis of the alternation of dominance of faster growing species which do not differentiate between small-scale differences in the environment, and slower growing species with very specific (environmental) preferences. To this end we would define a set of species randomly, taking into account a trade-off between growth rate, carrying capacity and niche width. Although these concepts do not have an operational definition (i.e., cannot be measured in the field) they can be used in such a simulation: the appropriateness of the concepts can then be tested by the comparison of patterns generated in the simulation and the role of the various 'species' therein, with the patterns and species observed in vegetation. These species would each have their own optimum in a spatially and temporally varying environment. We would test whether these conditions are sufficient to generate the patterns described above.

An obvious problem in such simulations is the fact that we want to observe the effect of small-scale interactions on larger-scale patterns and thus need a very large model. MICMAC modelling can in principle resolve this problem, by generating simplified representations of larger patches of the environment so that larger-scale pattern can be simulated (Hogeweg and Hesper 1981a).

REFERENCES

Beeftink, W.G., Daane, M.C. and de Munck, W. (1971) Tien jaar botanisch-oecologische verkenningen langs haet Veerse Meer. Natuur en Landschap 25: 50–65.

Beeftink, W.G., Daane, M.C., de Munck, W. and Nieuwenhuize, J. (1978) Aspects of population dynamics in *Halimione portulacoides* communities Vegetatio 36: 31–43.

Bell, A.D., Roberts, D. and Smith, A. (1979) Branching patterns: the simulation of plant architecture. J. Theor. Biol. 81: 351–375.

Burks, A.W. (1970) Assays on Cellular Automata. University of Illinois Press.

Doing Kraft H. (1954) L'analyse des carrés permanents. Acta Bot. Neerl. 3: 421-424

Dubes, R. and Jain, A.K. (1979) Validity studies in cluster methodologies. Pattern Recognition 11: 235–254.

Greig-Smith, P. (1964) Quantitative Plant Ecology, 2nd ed. Butterworths, London.

Greig-Smith, P. (1979) Pattern in vegetation, J. Ecol. 67: 755–779.

Goodall, D.W. (1974) A new method for the analysis of spatial pattern by random pairing of quadrats, Vegetatio 29: 135–146.

Goodall, D.W. and West, N.E. (1979) A comparison of techniques for assessing dispersion patterns, Vegetatio 40: 15–27.
Harary, F., Norman, R.Z. and Cartwright, D. (1965) Structural Models, an Introduction to the Theory of Directed Graphs. Wiley, New York.
Hogeweg, P. (1976a) Topics in biological pattern analysis. Thesis, University of Utrecht.
Hogeweg, P. (1976b) Iterative character weighing in numerical taxonomy. Comput. Biol. Med. 6: 199–211.
Hogeweg, P. (1983) Heterarchical systems as simulation modelling formalism, Pergamon Encyclopedia of Systems and Control. (In press).
Hogeweg, P. and Hesper, B. (1972) BIOPAT, program system for bioinformatic pattern analysis. Bioinformatica, Utrecht.
Hogeweg, P. and Hesper, B. (1979) Heterarchical, self-structuring simulation systems: Concepts and applications in biology. In Zeigler, B.P. et al. (Eds.), Methodologies in Systems Modelling and Simulation, pp. 221–231. North-Holland Publ. Co., Amsterdam.
Hogeweg, P. and Hesper, B. (1981a) Two predators and one prey in a patchy environment: An application of MICMAC modelling. J. Theor. Biol. 93: 411–432.
Hogeweg, P. and Hesper, B. (1981b) Oligothetic characterisation of clusters. Pattern Recog-Recognition 14: 131–136.
Hogeweg, P. and Richter, B. (1982) INSTAR, a discrete event model for simulating zooplankton population dynamics. Hydrobiologia 95: 275–285.
Horn, H.S. (1975) Markovian properties of forest succession, In M.L. Cody and J.M. Diamond (Eds.), Ecology and Evolution of Communities, pp. 196–213. Princeton University Press.
Kendall, M.G. (1972) Cluster analysis. In Watanabe, S. (Ed.), Frontiers in Pattern Recognition, pp. 291–310. Academic Press, New York.
McMahon, J.A., Schimpf, D.J., Smith, K.G. and Bayn, R.L. (1981) An organism centered approach to some community and ecosystem concepts. J. Theoret. Biol. 88: 287–308.
van Noordwijk-Puijk, K., Beeftink, W.G. and Hogeweg, P. (1979) Vegetation development of salt-marsh flats after disappearance of the tidal factor. Vegetatio 39: 1–13.
Rosenfeld, A. (1969) Picture Processing by Computer. Academic Press.
Ward, H.H. (1963) Hierarchical grouping to optimise an objective function. J. Amer. Stat. Ass. 58: 236–244.
Williams, W.T., Lance, G.N., Dale, M.B. and Clifford, H.T. (1971) Controversy concerning the criteria for taxonomic strategies. The Computer J. 14: 162–165.
Zeigler, B.P. (1976) Theory of Modelling and Simulation. Wiley, New York.
Zeigler, B.P. (1982) Discrete event models for cell space simulation. Int. J. Theor. Physics 21: 21: 573–588.

INDEX

No attempt has been made to index every occurrence of all terms. Page numbers shown in boldface type indicate the first page of a chapter or of a section in a paper in which a topic is dealt with repeatedly.

Age-class distribution 26, 59, **413**, **531**
Age determination 413
Age-state 36, 64, 69, 118, **121**, 145, **180**, **208**, **241**, **347**, 355
Age-state ontogeny **130**, **151**, **167**, **180**, **209**, **229**, **247**, 354
Age-state spectra **146**, **180**, **210**, **230**, 354
Age structure **26**, 36, 45, 54, 247, 270, 276, 279, 283, 630
Allocation of dry matter, energy or nutrients
18, 424, **517**
(see also Reproductive allocation)
Allozyme variation 455
Annual grasses **317**
Animal activity/grazing 27, **49**, 62, 67, 137, 161, 188, 298, 300, 317, 327, 342, 353, 357, 362, 406, 433, 475, 490, 509, **626**
(see also Seed dispersal and predation)
Biological control 632
Biomass **100**, **182**, 275, 300, 306, 333, 345, 510
Breeding system 384, 499, 528, 541, 583
Bud bank 20
Cedar glades 373
Census of plants 2, 16, 17, **33**, 115
characteristic features 34
history **37**
Chalk grassland 65, 72, **343**
Clone formation **23**, 103, 147, 213, **233**, **241**, **272**, 299, 306, **415**
(see also Particules)
Cluster analysis **639**
Coenopopulations 10, 63, 71, 114, **121**, **143**, **179**, **207**, **229**, **241**
age state index **158**
balance lability **159**
calculation units **146**, 120
large life cycle 126, 188, **199**
ontogenetic change **158**
specific age level 199
types 126, **182**
Coexistence of species 8, 16, 103, 289, 342, 366, 401, 512, 575
"laws of coexistence" 15
Colonization 59, **266**, 406, 425, 433
Commensalism 404
Competition 6, 38, 44, 70, 73, 102, 104, 217, 273, 277, 300, **314**, 330, 336, 344, 366, 387, 401, 404, 459, 510, 575, **612**
Control of relative abundance 615
Critical size for flowering 357, 533
Cushion plants 408, 600
Demographic stochasticity 6
Desert Botanical Laboratory **45**, 62
Desert and semidesert vegetation **45**
Die-back model of trees 27
Dispersal **17**
(see also Seed dispersal)
Disturbance 8, 129, 136, 317, 322, 378, 459, 491, 510, **603**, **629**
Dominance 615
Ecological and physiological amplitude 7, 249, 442, 453, 457, 469, 473, **595**
Ecomorphology **637**
Ecophysiology of species 160, 277, 279, 387, 393, 403, 406, 410, **423**, **441**, **457**, 483, **491**, **503**, 579, **596**, 665
Ecotpyic variation 7, 221, 236, 355, 385, **441**, **453**, 468, **472**, 545, 597
Edificator species 5, 16, 103, 162, 196
Endemic species 375
Environmental gradients **442**, **608**
Ephemeroids 127, **225**
Exclosures 43, **50**, 57, 61, 72, 322
Factor complexes **601**
Fecundity (see Seed production)
Fire in vegetation 128, 498, 500, 606, **625**
Flowering 25, 153, 210, 357, 383, 391, 451, 468, **572**
Fluctuations 72, **135**, 202, 260, 294, **307**, 603
Forest herbs 140, **225**, **241**, **516**
Functional description of vegetation 503, **517**, **553**, **595**
Gaps in vegetation 274, **341**, 362, 568, 606, 611
Genetic neighbourhood 456
Growth forms **90**, **226**, **407**, 503, 559, **561**, 611
(see also Plant morphology)
Hemiparasites **463**
Herbicides 21, 323
Inland dunes **265**
Interactions between populations 431
Invasion 24, 96, 102, 279, 314, **323**, **490**, **624**

Jornada Range 43, **52**
Key characters **597**
Kranz syndrome 393
Life cycles 16, 22, 50, 99, 258, **407**, **507**, **516**, **538**
Life-history characteristics of species 7, **17**, 42, 50, 53, **126**, **145**, 175, **296**, **302**, **334**, **505**, **515**, **542**, **598**
Life tables 53, 67, 74
diagrammatic **348**, 376, 385, 392, **465**, 495, 497
Longevity 18, 26, 47, 53, 55, 151, 173, **189**, **211**, 301, 329, 364, **415**, 626, 628
of flowers **584**
Long-rhizome grasses and herbs **174**, **180**, 271, 407
Maturation 294, **305**
Microclimate 225, 346, 359, 361, 374, 450, 483, 519, 532, 566, 579, 599
Mortality (see Survivorship)
Mutualism 404, 577
Natural selection 41, 73, 221, 401, 403, 452, 500, 506, 511, 583
Nature conservation 21, **26**, 632
Nature of vegetation **2**
individualistic concept 3, 112
Neophytes 490, **623**
Ordination of vegetation 442, 558, 601
Pantograph 44, 51, 53, 55, 68, 74
Particules **145**, **164**, **174**, **195**, **208**, **233**
Patchiness in vegetation **270**, **317**
(see also Gaps in vegetation)
Pattern in vegetation 75, **638**
(see also Spatial pattern of vegetation)
Permanent quadrat **1**, **33**, **90**, **111**, **244**, **315**, **645**
Phenology 26, 260, 317, 428, 508, **516**, **572**, 610
Photography of vegetation **48**, 59, 62
Photosynthesis 161, 346, 403, 424, 450, 509, 517, 566, 579
Phytosociology **1**, **15**, **95**, **122**, **182**, **214**, 252, 257, **265**, 298, 343, **442**, **465**, 481, **554**
Pioneer species 22, **266**
Plant associations 3, **21**, **95**, 252, 257, **266**, **443**, **465**
Plant morphology **23**, 35, 50, **64**, 71, **122**, **145**, **213**, 229, **241**, 357, 403, 414, **450**, **467**, 503, **508**, 533
modular growth 411
Plant plasticity 35, 39, 237, 401, 455, 459, 499, 506, 601
Plant population
delimitation 10, 35
dynamics 116, 135, **199**, **213**, 260, 273, 282, 449, **465**, 472, 494, 499, 534
projection of future numbers 48, 251, 331, 499
Pollination 384, 406, **575**
Polycormones **241**
Prairie glacial marshes **293**
Productivity 24, 100, **182**, 300, **306**, 333, 345, 347
Principal components analysis 639
Psammophyte vegetation **265**
Quadrat
chart 1, **41**, **60**, 116, 208, 229, **245**, 266
denuded 42, 92
list 1, 41, 55
(permanent – see Permanent quadrat)
Rangelands **42**, 68, 490
Range reference areas **57**
Regeneration **508**, 603
Regeneration niche 363, 611, 616
Reproduction (see Seed production)
Reproductive allocation 278, 386, **426**, **521**, **525**, **542**
Root depth 285
Ruderal vegetation 96, 99, 407, **507**
Santa Rita Range 43, 48, **50**
Saum Species **22**, 602
Seedling
demography 61, 69, **267**, **352**
survivorship 46, 63, 65, **115**, 267, 386, 451, **564**
Seed
dispersal **18**, 149, 266, 302, **319**, **349**, **359**, 532
dormancy 21, 99, 128, 150, 301, 335, **379**, **492**, **558**
germination 19, 20, 50, 128, 150, 162, 266, 277, 285, 298, 302, 351, **379**, **388**, 451, 453, **465**, **492**, 562, 627
in the soil **20**, 128, 150, 188, **266**, 271, 298, **302**, **319**, **341**, **350**, **382**, **390**, **465**, **492**, **551**
predation 149, 162, 358, 467, 475, 496, 532
production 136, **147**, 162, 171, 236, 274, 277, 334, **348**, 358, **425**, 450, **492**, **521**, 580, 627
seed bank types **558**
Shifting and shuttle species 341, 479
Shoot density 128, 162, **189**, **214**, **247**, **272**, 308
Shoot-root ratio 100, 104, 457, 517
Short-lived perennials **345**
Short-rhizome herbs **162**
Shrub demography **51**, 61, 70

Size classes of plants 250, 256, 278, 286, 385, **530**, **541**
Sociological groups **95**, **225**, **298**, **559**
Soils
fertility 442, **446**
moisture 160, 268, **285**, **318**, 346, 361, 374, **378**, **491**, **598**, 662
nutrients 104, 269, 343, 402, 446, 492, **598**, 662
Spatial pattern of vegetation 102, 106, 112, 217, 237, **241**, **270**, **313**, 365, 404, 430, 511, **640**
Spatio-temporal pattern **649**
Species diversity 105, **298**, 345, 375, **511**
Square-foot density 56
Steppe vegetation 112, **143**
Strategy types 17, 114, **126**, 143, **302**, 342, **363**, 403, 405, **507**, **596**
Stress **507**, **601**
Succession in vegetation 2, 3, 17, 27, **41**, **57**, **89**, 112, 137, 139, **251**, **266**, **293**, 313, 375, 554, **603**, **626**, **641**
Summer annuals 93, **345**, **388**, **467**, 498
Survivorship 43, **50**, 69, 72, 115, **153**, 167, 172, 175, **277**, 322, **348**, **419**, 449, 482, **529**
Syndynamics of vegetation 17, 28, **95**
Synusia **225**
Tap-rooted perennials **166**, 233, 284
Tolerance 507, **596**, 614
Tundra **399**
Tussock grasses 44, **53**, **147**, **180**, **208**, **266**, **317**, 408
Vegetation
cycles **298**
dynamics **269**, **293**
structure **265**, **313**, **341**, **359**, 442, 445
Vegetative propagation 20, **23**, 66, 103, **147**, **166**, 171, **194**, **233**, **241**, **271**, 401, 405, 409, **419**, **536**, 627
Vivipary 20, 233, 401, 407, 431
Winter annuals 73, 93, **277**, **317**, 363, **377**, 496, 600

MIX
Papier aus verantwortungsvollen Quellen
Paper from responsible sources
FSC® C105338

If you have any concerns about our products,
you can contact us on
ProductSafety@springernature.com

In case Publisher is established outside the EU,
the EU authorized representative is:
Springer Nature Customer Service Center GmbH
Europaplatz 3, 69115 Heidelberg, Germany

Printed by Libri Plureos GmbH
in Hamburg, Germany